Millennial Biology: The National Science Foundation and American Biology, 1975-2005

Donald J. McGraw

Millennial Biology: The National Science Foundation and American Biology, 1975-2005

Springer

Donald J. McGraw
Ephraim, UT, USA

ISBN 978-3-030-56366-0 ISBN 978-3-030-56367-7 (eBook)
https://doi.org/10.1007/978-3-030-56367-7

This Springer imprint is published by the registered company Springer Nature Switzerland AG
The registered company address is: Gewerbestrasse 11, 6330 Cham, Switzerland

This book is dedicated to my wife Laura;
our daughters Adrienne, Holly, and Rachel;
grandsons Sage, Hawthorn, and Patrick;
granddaughter Hazel

Foreword

The global dominance and unprecedented share of the world's wealth with which the United States emerged from World War II set the stage for large-scale experimentation in many areas. In those heady years, we began to consider ourselves the "world's policeman," played a key role in establishing the United Nations, and thought hard about how the nation's efforts in many fields, including science and technology, could be strengthened for the public benefit. It even seemed possible to leaders like President Lyndon Johnson that we could begin to rectify the social inequities of the past, given our background of growth and prosperity.

Science and technology had played such an obvious role in making possible our wartime success that the idea of strengthening these fields to enable future progress was widely accepted. Unlike many other countries, which conducted their research primarily in government-funded centers, we considered our universities to be the place where most discoveries would be made. The relationship between the newly established National Science Foundation (1950) and the various science and engineering communities that it serves is what Donald McGraw, the author of this outstanding study, has aptly termed the *intricate dance*. This endless dance has led to spectacular improvements in most fields of science, but especially in the life sciences, and the present work provides a deep and accurate analysis of how these improvements came about.

As participants in this endless dance, we often failed to recognize the major trends that continued to take place while the dance itself continued. One of the most important features of this book is to show how these trends affected the organization of science and provided the background for various outcomes that became possible over time. From the start, it was inevitable that universities would play a major role in shaping the Foundation and its activities; at the same time, their own programs were progressively shaped by decisions made at the Foundation. At an institution like Stanford University, for example, Fred Terman, a long-time university officer who served as Provost from 1955 to 1965, eagerly sought grants from the Department of Defense and then from the NSF as it grew, building the University into a science and technology powerhouse by the end of his tenure. Terman, along with William Shockley, was largely responsible for the formation of the Silicon Valley complex nearby, another critical factor that facilitated the University's rapid growth in technical fields. Along the way, some worried about the blurring of the traditional lines between the supposedly free inquiry characteristic of universities and the goal-driven research in corporations: Terman's model worked. And Stanford certainly was not alone in taking advantage of the opportunities afforded by that period.

At the same time, the heavy concentration of government funding on science and technology and the increasing dependency of universities on such funding worried many observers of American higher education. Particularly, they were concerned by the fate of the social sciences, arts, humanities and other areas that had been nurtured by universities for many centuries. To the extent that the primary driver for many universities was to secure funds, there arose a strong tendency to concentrate on the fields for which they could obtain external funding. In doing so, they ran the danger of creating unbalanced program in which many traditional fields of fundamental importance could be ignored. This has proved a real and lasting concern, one that has become acute even in public universities, where the proportion of state funding has rapidly decreased from the 1950s onward.

Any thoughtful university student would hope that these institutions will continue to pursue goals much broader than simply obtaining funds from the available external sources for their programs. However it be defined, "pure" science remains deeply important for the advancement of society as a whole. We clearly need to protect the ability for independent investigators to pursue the kind of theoretical research that led to the development of the transistor, the discovery of the role of nucleic acids in organisms, or to understand and measure the flow of energy through ecosystems. As the brilliant and societally aware corporate leader Roy Anderson once pointed out to me, one important reason that corporations continue to fund universities is to enable their scientists and engineers to continue to produce important results that might not be forthcoming from the corporations own, necessarily highly focused, research programs.

Over the course of its 70-year history, the National Science Foundation, operating within in a context that has at times become intensely political, has had to be an adept partner in the *intricate dance* that was shaping it and its partners, adjusting to particular times and the new influences they brought with them. Among the major problems with which the Foundation had to deal were the tension between funding large vs. small science; who would fund medical science; how and what kinds of education would be funded; how to provide large, expensive facilities for those who needed them; to what extent it should help to create and support centers in particular fields; whether to support applied research; and to determine an appropriate role for it in funding the social and behavioral sciences.

As I am a first-hand observer of these complex developments, I am personally grateful that Donald McGraw has given us this detailed and masterful history of the National Science Foundation during the key three decades between 1975 and 2005. Within the Foundation, this important period began with bringing together the social and behavioral sciences along with the life sciences into a single directorate, the Directorate of Biological, Behavioral, and Social Sciences (BBS). At the same time, funding for the medical sciences was largely moved to the National Institutes of

Health (NIH). Those who conceived the merger considered that the combined division would provide opportunities for synergistic growth in many of its constituent fields. Eventually, many of these hopes were met in a way helpful to the development of science overall.

By charting the history of this merger, McGraw ably makes it clear how much discussion and indeed infighting was behind what seemed a fairly simple and logical step to many of us at the time. Adding to this impression was the fact that Guy Stever, the then NSF director, often spoke of it as a primarily administrative matter, claiming that there had been little dissent about the merger. In fact, there was a great deal of dissent! The Foundation's primary clients – the universities and, in this instance, particularly the universities' social science faculties – were deeply concerned with what the new combination might mean to them.

Their concerns were exacerbated by the fact that scientists in other fields tended to regard the social sciences as "soft" and marginal to the scientific enterprise as a whole. In fact, the wide vision of the world held by the behavioral and social sciences is and has always been fundamentally important for the future development of the life sciences. Social sciences were often considered more "applied" than others were and thus often expected to provide results of immediate societal interest and importance. For the NSF, their theoretical underpinnings were more important, and the Foundation has through the years played a fundamental role in developing them. Most of the funding for these fields from other sources has been to obtain practical results that could be applied in the "real world." Over the years, the social sciences have become ever more useful in their applications, supported by an ever-greater understanding of their theoretical bases.

Clearly, the emphases placed by some of the social sciences on longitudinal studies and long-term trends has a salubrious effect on the development of the environmental sciences as well as the life sciences generally. As an example of the dangers of short-term thinking, it is significant that the observations of carbon dioxide levels in the atmosphere at Mauna Loa in Hawaii, arguably providing the most portentous scientific data of our time, were almost halted after a few years as people tried to move their funding to other purposes. In international science circles, America has gained a reputation for short-term funding and a perceived inability to make long-term commitments to international programs, tied to endless reexamination in succeeding budget cycles.

From the 1970s onward, the NSF played a dominant role in the development of the life sciences and how they were organized. The Foundation continued to do so when the new combined division was established and does so today as well. From the molecular, cellular, and organismal levels of organization to the population level (and, in a sense, on to the social sciences) the new possibilities and developments were nurtured and made accessible by innovative Foundation programs. None of the new areas could have reached full development – molecular biology, cell biology, development, neurobiology, integrative biology, or population biology – without NSFs active participation in the development of these fields and the provision of funding for them. In addition, population biology, systems biology, integrative biology – all of the new way that we now look at biological organization – all depended heavily on generations of intelligent and insightful NSF staff members for major parts of their conception and their subsequent development.

From the 1970s onward, the development of molecular biology provided a major impetus for this growth. Applied to many fields of the life sciences, the new insights gave us understanding to a degree that could not have been imagined twenty years earlier. Although the field of molecular biology can justifiably be considered to have begun at Cambridge University in the 1950s, it is one in which the United States has played a dominant role for the following half century. Also important in the expansion of the life sciences were the assembly and use of large data sets, often longitudinal, and the vertical integration of biology as a whole. Key also was the subsequent organization of the NSF Computer Science Division, in which Biology Division head Mary Clutter, encouraged by NSF Director Erich Bloch, played a key role. The same talented pair also took the fundamental steps that led to the development of the Internet itself. They had the foresight to see that its basic technology had a much wider role to play than that represented by the military purposes for which it was invented.

The environmental sciences, which grew rapidly in size and scope from the 1970s onward, were intricately connected with, and partly based on, the social sciences. As they grew, the importance of bioinformatics, large databases, and the application of computers to their problem areas played increasingly important roles. The results they achieved were soon viewed as critical to understanding human existence on earth and securing its future.

Within the life sciences generally, the continuing rise of interdisciplinary studies, together with developments such as increasing computerization, the use of model systems, the increased use of databases, the rise in longitudinal studies, and the establishment of effective Science and Technology centers may all logically be regarded to some degree as having been influenced by the temporary "marriage" between the life and social sciences during the period they were grouped into a common Directorate, 1975–1992.

Strong demands for practical results have been brought to bear on the NSF from the time it was established, and they will doubtless continue far into the future. Notably, the Research Applied to National Needs (RANN) program, established in response to these pressures in 1972, was an attempt to channel as much NSF funds as possible into "useful" projects. Its formation was one element in the endless struggle between the practical and the theoretical that often colors our views of any scientific effort. From the start, the RANN program had a troubled history politically, with Director Richard Atkinson shutting it down in 1981, after nine tumultuous years. It seems that we will never learn that the distinction between completely open investigations ("pure" science), said to be conducted in an "ivory tower," and applied science is by no means black and white. Apparently, these terms are too useful for political warfare in the sciences to be abandoned.

The NSF has from the start played a strong role in funding education, primarily at graduate and postdoctoral levels, thus greatly strengthening our national effort in many fields by encouraging the best qualified to pursue them or remain within them. The graduate and postdoctoral fellowships that the Foundation has provided have been of great importance to many, including myself, and remain so today; they comprise a significant fraction of the Foundation's budget. For the twenty-first century, however, the Foundation's development and introduction of the Science, Technology, Engineering and Mathematics (STEM) concept of primary and secondary education has been deeply influential, helping the United States to overcome the quality of science and technology education everywhere.

Overall, during its 17-year history, the BBS has clearly played an important role in stimulating interdisciplinary efforts, even though its primary "dance partners," the universities, have often exerted a strong pull in the opposite direction. They have for the most part maintained traditional, narrow, discipline-based departments that serve as the primary source of salary advances, tenure, and other advancements for their faculty members. The close connection between NSF staff, often on short-term leave from academia, with their colleagues in the universities has clearly been an important factor in broadening thinking both at the Foundation and in the universities.

In general, the Foundation has tried to balance the continued support of disciplinary research (often small grants) with the development of new kinds of interdisciplinary research (often supported by large grants). Through their support for broadly conceived research projects, programs, and centers, they have assisted greatly in bringing about the kind of broad thinking that almost everyone considers important, despite the intellectual and institutional inertia that often tends to slow progress. Universities continue to struggle with the same challenges, but they too are innovating, continuing to alter and enlarge the scope of their programs to meet the demands of a rapidly changing world. In 1989, the National Research Council study I chaired concluded importantly that increasing the amount and scope of interdisciplinary work was one of the most important improvements that could be made within the life sciences. Notwithstanding the general agreement with this proposal, academic rewards continue to be meted out in traditional ways that make the framework itself resistant to change.

A serious challenge for the new NSF Directorate of Biological, Behavioral, and Social Sciences, as it was for many aspects of American life, arose when Ronald Reagan became our 40th president in 1981. The social sciences, already mistrusted by many Republicans, became a target for budget reduction or even elimination (were they a covert form of socialism? Did we need them?). Within the NSF budget, as elsewhere, budget "hawks" targeted these fields particularly. The runaway inflation that the Reagan administration confronted was clearly an important issue for the country, and areas seen as superfluous, together with any programs that had been launched recently, became easy targets whatever their intrinsic importance. The shared delirium of the 1950s and 1960s, in which it seemed for a time that there was plenty of money for anything we wanted to get done, was starting to come to an end. Thus the realization that we were only one of a community of nations was bolstered by the horrors of the unwinnable Vietnam War. Regardless of how badly we wanted to continue to be regarded as the world's dominant superpower, we came to the initially unconscious realization that we were only one of a community of nations, and not to be the world's dominant superpower forever, regardless of how badly we wanted to retain that status. We had only so much money to use for any purpose, whether domestic or international.

The various controversies about what the NSF ought to be funding gained strength in 1984, when President Reagan appointed Erich Bloch – the first NSF director whose background was in industry. Bloch, who served a full 6-year term, presiding over an agency that experienced a significant degree of tumult during his tenure. More than almost any other director of the Foundation, he was decisive, making numerous, often beneficial, changes in the organization of the agency. Many of us regretted Director Bloch's elimination of the NSF International Division, which was rationalized by the pious hope that the other divisions would take up international programs suited to them. In a world where international cooperation in science has never been more important, they failed to do so effectively. The NSF has continued to impose severe limits on international funding of any kind, which seems to me to be unfortunate given the universal nature of science.

At about the same time that Erich Bloch was appointed Director of the NSF, I was appointed Chair of the Advisory Committee for the Division of Biological, Behavioral, and Social Sciences. Since it was a time when the Reagan administration was attacking the social sciences vigorously, we were thrust into the position of defending our existing programs. I soon learned that there were many excellent reasons for doing so, and we used them as diligently as possible in maintaining funding for all areas.

From the 1970s onward, the environment *per se* emerged as an important area of focus for both theoretical and practical reasons. Correspondingly, the NSF began to fund studies, programs, and centers intended to help us understand and deal with the problems that we were facing. Clearly, the rapidly growing human population coupled with our demands for ever-increasing consumption were endangering our life-support systems. During the 40 years following the establishment of the NSF (1950–1990), the global population had doubled, growing from 2.5 billion to 5 billion, and that of the United States increased over the same period from 150 million to 250 million, and the results of this growth were evident everywhere one cared to look.

As funding for environmental studies and focused facilities grew, some corporations and individuals began to worry that they might lead to an increase in regulation and thus limit their short-term profitability. Unfortunately, this problem has reached runaway proportions in the half century since the Nixon administration, when it first began to limit certain kinds of government funding and studies. Given the nature of science, its politicization is extremely unfortunate, and yet it is a fact with which we must continue to live. Indeed, in the only country in the world where over a third of the people reject evolution as an explanation for the diversity of life on earth, grant titles have had increasingly to be "sanitized" to protect the integrity of the Foundation, and indeed of science, as a whole.

The kinds of environmental problems that were gaining attention could now be attacked with systems thinking at all levels, and the NSF played a dominant role, employing several different strategies, in helping to make just that happen. Given the steady ascendency of molecular biology, the new pull at the upper end of biological integration led to disagreements over which parts of biology deserved the most funding, with the molecular biologists jealous of the major biome studies that have so clearly increased our understanding of the planet that is our only home. In any case, the Foundation nurtured both ends of the levels of organization in the life sciences and helped the overall discipline grow in a way analogous to the growth of the physical sciences, largely in Europe, in the first half of the twentieth century.

As the costs for conducting science increased, so did the overheads charged by universities and other recipients, a strong tension arose between funding "big science" *vs.* "small science." As the NSF has grown to its present budget of about $8 billion, about a fifth of the NIH's $40 billion, it has definitely been able to fund larger projects and bring people together more effectively than it could earlier. In this way, it has realized the objectives of the original merger of the life sciences with the social sciences that was attempted nearly a half century ago, and, by continuing to be an experienced practitioner of the necessary *intricate dance*, it continues to facilitate enormous benefits for us and for all the world. At the same time, the NSF has retained to the extent possible its funding for "small science" projects carried out by individual investigators, although even for these the amounts of money involved have grown ever larger over the years.

In 1992, as described in meticulous detail in this book, the life sciences and social sciences were separated into different directorates once again, reorganization that led to months of controversy both within and outside of the Foundation. Most of the social sciences were moved to a new Directorate of Social, Behavioral, and Economic Sciences, where it was hoped that they would be able to gain more prestige and additional funding for the scholarly areas gathered together in the new directorate. Although it has remained relatively small, NSF funding for the social sciences was of key importance. It represents a disproportionately large portion of the funding for basic research in many areas of the social sciences, even though the amounts of money involved are small. For this reason, NSF funding in the behavioral and social sciences is critically important for the progress of these fields, and playing political games with that funding was a disaster.

When the Foundation removed most of the social sciences in a directorate of their own in 1992, behavioral sciences, with their close connection to neurobiology, were retained with the other life sciences in the new Division of Biological Sciences (BIO). Whatever the configuration of the programs at different times, the availability of substantial grants tended to facilitate interdisciplinary studies, smaller grants mostly continuing to address specific, short-term efforts. Importantly, the relatively small funding the NSF has provided for the social sciences themselves has often been virtually the only funding available for basic investigations in these fields.

For the life sciences, Mary Clutter effectively guided the formation of a newly delimited Directorate, establishing the basis for the rapid expansion of effective funding over the decades to come. Considering biology, medicine, and agriculture – the subjects of the NRC committee I chaired in the late 1980s – there is and has been large-scale Federal funding for decades. Indeed, our committee concluded those areas that were most interesting intellectually and growing most rapidly, such as neuroscience and developmental biology, were already swimming in money:

additional funding could not have helped them much, at least proportionately. Overall, some fields that were both interesting and significant for people, such as the environmental and biodiversity sciences, were not and are not being funded according to the need for what they were capable of producing.

Within the new Biology Directorate, intelligently conceived programs continued to expedite the application of findings from molecular biology on a massive scale to solve problems in a way that we had considered impossible as recently as the 1950s. As the United States played the key role in developing and applying this information, the NSF fostered the new applications and experiments that led to the new level of understanding that was both well-conceived and intelligently applied. All parts of the life sciences became better understood as a result of these advances, with the role of the NSF indispensable at every level.

In this Foreword, I have been able to touch only on some highlights derived from or related to the rich feast that Don McGraw has provided for us: his concept of *Millennial Biology*. Anyone interested in the history of science or its application to the problems of the present will clearly find as much satisfaction as I have in deepening their understanding of the Foundation at a critical period of its development. These pages provide a strong basis for dealing with the future, one that will no doubt offer challenges even greater than those discussed so ably in this useful book.

President Emeritus, Missouri Botanical Garden Peter H. Raven
St. Louis, MO, USA

Preface

This history is of the biological sciences directorate at the National Science Foundation (NSF) from the time it was the Directorate of Biological, Behavioral, and Social Sciences (BBS), 1975–1992, to that of its successor, the Directorate for Biological Sciences (BIO), 1992–2005. This study was done during the years 2008–2012 under a contract, which had been created by James P. Collins, then assistant director for BIO, and Marc Rothenberg, then historian of the NSF, now retired. Three questions were asked of the author through that agreement. The first of those made reference to the biology directorate prior to and during a reorganization in 1975 (Chap. 1) before which it had long been known as the Directorate for Biological and Medical Sciences (BMS). BMS had also included the social and behavioral sciences.

This book was written to answer the three questions of the contract. The first of these:

> The claim made by the NSF administration was that bringing together [the] diverse fields [of biological, behavioral, and social sciences – and excluding medical sciences] under one umbrella would lead to the interaction, branching, and convergence of disciplines to produce new, hybrid disciplines. Did this actually occur? What was the impact on biology of the later removal [1992] of the social and behavioral sciences from the directorate?

The answers to these questions are given in Chaps. 1 and 2, primarily, but particular aspects of them are scattered in many other locations, as well. The effects of what became a split in 1992 of the behavioral sciences, specifically, over two new directorates, one of which focused on the social sciences and the other on the biological, had repercussions that echoed well into the twenty-first century. Numerous references are made to the behavioral sciences, part of which came into BIO, and of the social sciences, all of which went to the Directorate of Social, Behavioral, and Economic Sciences (SBE; Chap. 2). Their effects on biology are considered throughout this study. After Chap. 2, the focus is exclusively on the biological sciences.

The second question was closely related to the first, as this work is fundamentally an organizational history, but which considers much in the history of biology in the late twentieth/early twenty-first century:

> In the 1975 reorganization, biomedical research was removed from the NSF. How did this action impact the support of biological research at the NSF and the growth of General Medical Sciences at the National Institutes of Health [NIH]?

The answer to this question (Chap. 1) has shown that even after the reorganization, biomedical research was still sponsored, in a much less overt way, by not only BBS and BIO, but by other directorates, such as engineering, but always adhering to the notion that fundamental science – basic and, later applied, research – was never to engage directly the area of human *medicine*. Once again, however, over the three decades covered, a relationship between NSF and NIH was always present and the applicability of basic research knowledge might easily have bearing on human biomedicine.

The third question asked:

> How did the NSF respond to, assist and promote the changing nature of biology during these years? When the NSF chose to intervene in the field, how successful was it?

The answers to this two-part, and very crucial, question are spread throughout the entire substance of this study. There had been, since the founding of NSF in 1950, a relationship between the NSF and the members of its various science and engineering *communities,* especially the academy, that the Foundation serves. I have termed this the *intricate dance*. The dance floor of both basic and, later applied, research was the landscape upon which the two partners, NSF and its individual communities, carried on the relationship that is central to the growth of the biological sciences, as the one of interest here, but applies to all the disciplinary directorates.

The *intricate dance* is an inextricable feature of what was essentially an unnumbered fourth question:

> In analyzing these issues, as well as other issues the contractor deems important for gaining an understanding of the history of biology and the NSF, the contractor should consider the larger contexts of biological research being supported and conducted under auspices other than the NSF, federal support for science more generally, and the administrative history of the NSF.

These matters have been given extensive consideration in this study as the administrative history is integrated with the nature of the biological sciences. This is so because the Foundation is the *prime supporter for basic research for the biological community in America*. That group is mostly composed of the academic sector, but includes others such as other federal agencies, nongovernmental organizations, independent institutions, etc.

As the research, analysis, and subsequent writing developed, it occurred to me that what I was discovering was both an ever-changing organizational history *and* the history of a group of disciplines that we term biology, and those from 1975 into the new millennium. The two dance partners are so intertwined that it is impossible to extricate one from the other; and that close relationship has continued to develop over the life of the National Science Foundation and the biological community. It is what each says to the other, what each asks of the other, that makes for the dynamic dance. Funding research and the growth of biology as we have experienced it then is what issues from the interaction.

This is both an organizational and a disciplinary history, but almost exclusively a view through the eyes of but one of the two partners: the federal. As I conclude in the Epilog, the view from the community is yet to be told, so the concept of *Millennial Biology* developed here is partial and calls for yet more study.

Ephraim, Utah
January 2020

Donald J. McGraw

Acknowledgments

It took the help and thoughtful suggestions of a great many individuals to bring this work to fruition, and over several years, and I am deeply grateful for all they have done. One always prepares to write this section with considerable trepidation for fear of missing some person who did so much to aid the effort. However, I will attempt to list all those who did give me support, information, and encouragement. Should I miss someone, please know that it is only through error and by no means intent.

First, I wish to thank the two individuals most closely tied to the creation of the contract under which this work was performed. The genesis of the idea for a study of this directorate and during the years covered was James P. Collins, who was Assistant Director of the Directorate of Biological Sciences (BIO) at NSF when the contract began (2008), but who completed his time at BIO and returned to his home institution, Arizona State University, the following year. Working with Jim was Marc Rothenberg, then Historian of the NSF, who held that position as of the early manuscript writing period and who was my contact, colleague, and guiding editorial hand over the first several years of a demanding project. I am obliged to both and keenly appreciative of their aid and oversight.

I also thank most heartily the 29 scientists I interviewed, who are present or former figures in one or another of the three incarnations of the biological sciences directorate at NSF for more than three decades. Some have retired, some have moved on to other venues, and yet others are still present. Within all three groups many have held more than one position at the Foundation, so I will not try to place them with any given administrative unit other than the biology directorate alone. Alphabetically, they include: Eve Barak, Eloise "Betsy" Clark, Jim Collins, former NSF Director Rita Colwell, Mark Courtney, Todd Crowl, Machi Dilworth, Mary Clutter, Jim Edwards, Penny Firth, Henry Gholz, Phil Harriman, Maryanna Henkart, Carter Kimsey, David Kingsbury, Sonya Mallinoff, Christopher Platt, Jim Rodman, Joann Roskoski, David Schindel, Judy Skog, Bob Sterner, Fred Stolnitz, Joanne Tornow, Bruce Umminger, Judy Verbeke, Quentin Wheeler, John Wooley, and Bill Zamer. As the text will demonstrate, the interview transcripts were among the most useful of all the materials I studied for this project. Many of these individuals before and after the formal interviews provided more information as they thought of it or as I asked for elaboration. They were invaluable all.

Beside then Director Jim Collins, others in the Office of the Assistant Director for Biological Sciences include(d): Sarita Shephard, Joanne Rodewald, and Joann Roskoski, to all of whom I offer my warmest thanks.

Without the detailed aid of the NSF Library staff, this research would have been sorely lacking in some of the most important documents present there or ordered on Interlibrary Loan by those dedicated professionals. They include: Tina Gheen, Sue Olmstead (most especially!), Sarah Miller, and others whose names remain unknown to me. I am indebted to each and every one of them and thank them most graciously.

Others located throughout the Foundation were also helpful and include: Angel Lane, Jannele Gosey, Dorothy "Dottie" Holmgren, Christina Bartlett-Whitcom, Ellen Weir, Greg Smith, Linda Boutchyard, Wanda Ward, Curtis Klema, Mary Lou Tillotson, Jeffrey Lawson, Dana Wilson, Dan Wiegmann (BIO), Judy Hayden, Susan Mason, Greg Warr (BIO), John Perhonis (retired), Shayna Daniel, and Lyn Orrell. I thank them one and all.

Persons from other federal agencies to whom I am grateful included: John Rees, Richard Mandel, Barbara Harkins, Stephen Greenberg, and David Porter all at the National Library of Medicine; David Cantor at the National Institutes of Health History Office; and many unnamed but very helpful individuals at the National Archives and Records Administration II (College Park, Maryland). As well, and as the accommodations were better than any hotel, I thank manager Dick Sage and a great many unnamed others at the Federal Deposit Insurance Corporation Student Training Center housing unit in Arlington where I stayed on three occasions while doing my research.

Helpful faculty and staff from a number of universities include: Robert F. Murphy of Carnegie Mellon University, Dave Colman and Alan Covich of the University of Georgia, Luis Sayavedra-Soto of Oregon State University, Diana Buchwald of the California Institute of Technology, Norma Brinkley of the University of Maryland (College Park), Denise Karns of the University of Wisconsin (Madison), James Pritchard of Iowa State University, and Tyna Chu of Arizona State University.

Yet more to whom I offer my to thanks are: Jennifer Williams of the American Institute of Biological Sciences; Emilie Benoit, Chris Spencer, and Tanisha Howard of Ubiquis Inc.; and Bob Waide and George Garcia of the Long Term Ecological Research Network (LTER) Office.

I am especially thankful to Peter Raven, Director Emeritus of the Missouri Botanical Garden in St. Louis, for his most informative and thoughtful – and, yes, kind – Foreword to this study. Dr. Raven will be met many times in this history given his many roles at NSF, and well beyond in service to American and world science.

Finally, moral and intellectual support came from my family (named on the Dedication page) and friends including: Jerry Singleton (deceased), André Doyon, Barbara Noble, Susie Hancock, and Peter Degen. I thank them all most sincerely.

Note on Sources

The matter of discovery of the source materials used in this study and, more notably, assigning permanent loci for long-term provenance presented and presents, respectively, considerable difficulty. This is so because the National Science Foundation (NSF) and its major units, Directorates, have been very irregular with regard to saving, curating, and archiving documents of historical importance. Additionally, the Foundation has not been consistent in keeping a historian of science on its permanent staff. Yet further, the history of science office at the time this research was done had only the most meager storage space for records. The Library at NSF headquarters in Arlington, Virginia, however, has preserved a number of documents of use to this history, but by no means a large number.

Having said that, source materials for the earlier periods considered here are, in fact, more easily found. The National Archives and Records Administration in College Park, Maryland, NARA II, is where archived documents are to be found for the NSF. However, those records present for the period beginning in 1975 are far fewer in number than those preceding that year. Nonetheless, I was able to derive value from those documents present. Beginning after that year, source materials are found in many locations. Some were then at NARA II. Some uncurated boxes held by the Historian of the NSF were of very considerable utility, but at the time of this onsite research (2008–2009), the files were mostly recent donations from various retired or retiring persons in the biological sciences directorates. As such, endnote references to the chapters present here will often refer to a given person's name and her or his "files." These typically come from the uncurated boxed materials. Their future curation and archiving remain unknown after 2009.

In many other cases, ephemera, the origins of which are entirely unknown, have been useful and, at times, crucial. These, along with copies of all other documents used in this study are held in this author's care and tied to the title of this study. They can be consulted by other researchers through a request to me at: donaldjmcgraw@icloud.com. The challenge of ephemera and their provenances is to be seen in many of the figures where location is simply unknown.

Yet other major resources, primary through tertiary, are to be found in the Library of the NSF in Arlington; this is true, for instance, for the Minutes of the National Science Board and many others, such as the *Annual Reports* of the NSF.

One well-curated collection is that termed, in endnotes here, the BIO Lateral Files. These are very modern and reside in lateral filing cabinets within the main office of the Directorate for Biological Sciences (BIO). Of especial note are the Minutes of the Advisory Committee for BIO, which cover most of the years from 1992 to 2005 (and beyond).

The excellent collection at the National Library of Medicine (of the National Institutes of Health) held but little of use to the present history, but discussions with individuals there did prove of value for matters discussed in Chap. 1.

Liberal use of the World Wide Web/Internet has been made as this is a modern history, and much of what resides there, as a primary source or higher, is easily found *if* the electronic pages are still active. The research for this work was done in 2008–2009, with some minor additional work thereafter. Writing was done over the next several years, but publication did not come about until 2020. Website addresses were not tested for activity in mid-2019, but where they are no longer active, the reader is commended to use of the waybackmachine.org site, which archives billions of web pages. The real possibility of the transient nature of websites has been considered and so all, ideally, materials used from the Web/Internet have been printed and added to my files for this book. It is likely that some of these documents have not been printed and saved in paper form inadvertently, but at the same time many of the sites are intended by the federal government, universities, scientific societies, and others to be permanent. These are less ephemeral sites and they are listed in the endnotes among all other cited references. In this changing era of how documents are stored and retrieved, the historiographical community and others recognize that problems of this nature are still in flux of solution.

Some resources, such as hardbound and paperback books, are to be found in my library, but many are also in scholarly libraries nationwide and readily retrieved.

Of great value to this study have been the 28 formal, audio-recorded interviews. Each was done in both analog and digital form. These were all transcribed and, in some cases, edited by the interviewees for clarity prior to my use of the finished transcriptions; though I still maintain the rough forms. The names of all of those interviewed are given in the Acknowledgments section here as well as in the main text and many endnotes. Other transcribed interviews were available from NSF's then Historian and they are credited where used. In a very few cases, formal interviews were not done, but the notes taken of the conversations are referred to by the individuals' names, unless confidentiality was requested (extremely rare circumstances). Wherever possible, statements made in interviews, which may be of potential contention, I have sought to prove by reference to the formal written record. In some cases, I have used the broad consensus in the oral record. However, given the poorly curated resources I often had to work with, I have had to resort to presenting many statements as opinion when they may well have been fact that I could not otherwise substantiate. The tapes and digital form of the interviews, as well as the transcripts, reside with the other files for this work.

Small historical nuggets arose during the research, which have highly varied provenances; they are explained in the appropriate endnotes. Also, in the Acknowledgments, I thank many persons, some authors of well-regarded works, but I want to again here thank Toby Appel for her *Shaping Biology*, an absolutely crucial work that helped lay many foundational aspects of the present history.

Finally, taking from endnote 41 in Chap 5, I need to make clear why my sources present so much difficulty in writing this history.

> It was a time, as we have seen, in which many files "stored under the stair case" (literally) at 18th and G [streets] were lost, or disposed of (?), as the move to Arlington progressed. But the activity of disposing of files, and thus history, was not limited to moves alone. Bruce Umminger, long a senior leader within the biology directorate, recounts the story of one program director, a rotator (name withheld here), from a major northeastern university who said there was going to be a "catharsis" in the then PCM division [Physiological, Cellular and Molecular Biology] of past award files including all the comments on the assessment documents, temporary and permanent personnel names, etc. – the "jackets" – because "we're worried about what we're funding today [not] what we funded ten years ago." To which Umminger said, many years later: "my God, what a purge of information!" Further, Umminger said that "[PCM] was probably the biggest division in all of NSF at that time, that's why it split up…[it] just got too big. In that division, *everything got chucked*." Umminger interview; emphasis added.

Contents

About the Author

Donald J. McGraw is a biologist and historian of science with broad training in the sciences, extensive administrative experience, and service in varying capacities with federal agencies, colleges and universities, and NGOs. He holds a BS in biology from California State Polytechnic University (Pomona), and an MS in Bacteriology and Public Health from Utah State University (Logan). His PhD is in biology and the history of science from Oregon State University (Corvallis). He has taught at such colleges and universities as Franklin (IN), Bard (NY), St. Thomas (MN), and the University of San Diego (CA) where he was Associate Provost and Professor of Biology for 16 years. His broad biological training and expertise in general biology, botany, microbiology, and history of biology have given him a solid foundation in the life sciences.

His more than fourteen seasons as a Ranger/Naturalist with the National Park Service in three national parks and monuments provide him with extensive natural history knowledge. Dr. McGraw served his active duty in the past and is currently a Captain (Colonel) with the US Public Health Service Commissioned Officer Corps Ready Reserve. He has served on the Board of Directors of Sigma Xi-The Scientific Research Society, and as an officer in the Pacific Division of the American Association for the Advancement of Science. He has also served as both a reviewer and a contractor for the National Science Foundation and as a reviewer for the National Endowment for the Humanities. He has been a grantee of Eli Lilly and Company, the National Science Foundation, a contractor with the US Fish and Wildlife Service and is continuously listed in various of the *Who's Who in America* publications since 1984. Dr. McGraw has also been recognized with the Meritorious Public Service Award by the US Navy – the Navy's highest civilian honor – for his work in Naval Reserve Officer Training Corps (NROTC) education nationwide.

His more than 100 publications and presentations include two previous books in the history of dendrochronology, several book chapters, and refereed articles (*Isis*, *American Scientist*, etc.), proceedings, reviews, and others.

Abbreviations[1]

AAAS	American Association for the Advancement of Science
AAM	American Academy of Microbiology
AB	Animal Behavior Program
AC	Advisory Committee
AD	Assistant Director
AEC	US Atomic Energy Commission
AIBS	American Institute of Biological Sciences
AID or USAID	US Agency for International Development
AMA	American Medical Association
ARS	USDA Agricultural Research Service
ASM	American Society for Microbiology
ATOL or AToL (see also ToL)	Assembling the Tree of Life
BBS	Division of Biological, Behavioral, and Social Sciences
BE	Biocomplexity in the Environment Initiative
BIO	Directorate for Biological Sciences
BIR	Division of Biological Instrumentation and Resources
BMS	Division of Biological and Medical Sciences
BNS	Division of Behavioral and Neural Sciences
BOB or BoB	Bureau of the Budget
BPEC	Biotechnology Process Engineering Center
BRI	Division of Biological Resources and Instrumentation

[1] The federal government uses an extensive number of abbreviations and acronyms. Because of that, this history also employs them to reduce the number of words overall in use here. However, this can present a challenge to the reader. Therefore, it is strongly suggested that the pages of this section be copied and kept alongside this work as the reader progresses through it as a ready reference to these abbreviations.

BRR	Biological Research Resources
BSCC	Biotechnology Science Coordinating Council
BS&I	Biotic Surveys and Inventories Program
BSR	Division of Biotic Systems and Resources
CAREER	Faculty Early Career Development Grant Program
CENR	Committee on Environment and Natural Resources
CEPRAP	Center for Engineering Plants for Resistance Against Pathogens
CISE	Directorate of Computer and Information Sciences
CIBIO	Cyberinfrastructure for the Biological Sciences Program
CLMIB	Center for Light Microscope Imaging and Biotechnology
CLS	Committee on Life Sciences
CLSH	Committee on Life Sciences and Health
CMB	Center for Molecular Biotechnology
CN	Cellular Neurosciences
CNH	Dynamics of Coupled Natural and Human Ecosystems Program
COSEPUP	Committee on Science, Engineering and Public Policy
COSSA	Consortium of Social Science Associations
COV	Committee of Visitors
CPRB	DoE/NSF/USDA Joint Program on Collaborative Research in Plant Biology
CRB	Conservation and Restoration Biology Program
CSA	House Committee on Science and Aeronautics
CST	House Committee on Science and Technology
DBI	Division of Biological Instrumentation and Resources
DCB	Division of Cellular Biology
DD	Division Director
DEB	Division of Environmental Biology
DDIG or DIG	Doctoral Dissertation Improvement Grant Program
DMB	Division of Molecular Biology
DNA	Deoxyribonucleic Acid
DoD or DOD	US Department of Defense
DoE or DOE	US Department of Energy
EC or E.C.	European Commission
EF	(Virtual) Division of Emerging Frontiers
EHR	Directorate for Education and Human Resources
ENSO	El Nino/Southern Oscillation
EoL	Encyclopedia of Life
EPA	US Environmental Protection Agency

ER	Environmental Research
ERC	Engineering Research Center
ERE	Environmental Research and Education Initiative
EROC	Ecological Rates of Change Program
ES	Ecological Studies Cluster
EST	Expressed Sequence Tag(s)
Evo-Devo	Evolution of Development
FAQ	Frequently Asked Questions
FCCSET	Federal Coordinating Council for Science, Engineering, and Technology
FY	Fiscal Year
GAO	General Services Administration
GBIF	Global Biodiversity Information Facility
GCC	Global Climate Change
GCRP	Global Change Research Program
GEO	Directorate for Geosciences
GenEn	Genome-enabled Environmental Science and Engineering
GLOBE	Global Learning and Observations to Benefit the Earth
GLOBEC	Global Ocean Ecosystems Dynamics Program
GPRA	Government Performance Results Act
GWU	George Washington University
HBP	Human Brain Project
IAB	Integrative Animal Biology Program
IBN	Division of Integrative Biology and Neurosciences
IGB	IGB Products, Limited
IID	Advisory Panel for Instrumentation and Instrument Development
IOB	Division of Integrative Organismal Biology
IPB	Integrative Plant Biology Program
IRRPOS	Interdisciplinary Research Relevant to Problems of Our Society
IWG-MG	Interagency Working Group on Microbial Genomics
IWG-PG	Interagency Working Group on Plant Genomics
K-12 or K12	Kindergarten through 12th grade
KDI	Knowledge and Distributed Intelligence Program
LExEn	Life in Extreme Environments Program
LMER	Land-Margin Ecological Research Program
LOME	Land/Ocean Margin Ecosystems Program
LTER	Long Term Ecological Research Program
LTPEB	Long Term Projects in Environmental Biology Cluster

LTREB	Long Term Research in Environmental Biology Program
MACOS	Man – A Course of Study
MB or Mb	Megabase(s) in DNA
MCB	Division of Molecular and Cellular Biology
MGP	Microbial Genome Program
MIP	Microbial Interactions and Processes Program
MIT	Massachusetts Institute of Technology
MO	Microbial Observatories Program
MOLE	Molecular Evolution Program
MOU	Memorandum/a of Understanding
MPIWG	Microbe Project Interagency Working Group
MPS	Directorate of Mathematics and Physical Sciences
MRE	Major Research Equipment Program or Account
MREFC	Major Research Equipment and Facilities Construction Fund
NAPA	National Academy of Public Administration
NASA	National Air and Space Administration
NAS	National Academy of Sciences
NBII	National Biological Information Infrastructure
NCEAS	National Center for Ecological Analysis and Synthesis
NCGA	National Corn Growers Association
NEON	National Ecological Observatory Network
NGO	Nongovernmental Organization
NIDA	National Institute of Drug Abuse
NIGMS	National Institute for General Medical Science
NIH	National Institutes of Health
NIMH	National Institute of Mental Health
NIST	National Institute of Standards and Technology
NLM	National Library of Medicine
NMNH	National Museum of Natural History (Smithsonian)
NOAA	National Oceanic and Atmospheric Administration
NPGI	National Plant Genome Initiative
NRC	National Research Council
NRL	Naval Research Laboratory
NSB	National Science Board
NSTC	National Science and Technology Council
OECD	Organization for Economic Cooperation and Development
ONR	Office of Naval Research
OPUS	Opportunities for Promoting Understanding Through Synthesis Grant Program
OSTP	Office of Science and Technology Policy
OST	Office of Science and Technology

OSHA	Occupational Safety and Health Administration
OTA	Office of Technology Assessment
P and B	Physiology and Behavior
P and E	Physiology and Ethology
PBPE	Population Biology and Physiological Ecology Program
PCAST	President's Council of Advisors on Science and Technology
PCM	Division of Physiological, Cellular, and Molecular Biology
PD	Program Director
PEET	Partnerships for Enhancing Expertise in Taxonomy
PGR	Plant Genome Research Program
PHS or USPHS	US Public Health Service
PI	Principal Investigator
PO	Program Officer
PSAC	President's Science Advisory Council
QTL	Quantitative Trait Locus
R&D	Research and Development
RANN	Research Applied to National Needs
REU	Research Experiences for Undergraduates Grant Program
RCS&E or RCSE	Research Collections in Systematics and Ecology Program
RNA	Ribonucleic Acid
RUI	Research in Undergraduate Institutions Grant Program
SBE	Directorate for Social, Behavioral, and Economic Sciences
SBIR	Small Business Innovation Research Program
SES	Division of Social and Economic Sciences
SEP(s)	Special Emphasis Panel
SGER	Small Grants for Exploratory Research Program
SIG	Shared Instrument Grant Program
SPB	Systematic and Population Biology Cluster
STC	Science and Technology Center(s)
STIS	Science and Technology Information System
TF	Task Force
TIGR	The Institute for Genomic Research
ToL or TOL or ATOL	Tree of Life Program or Assembling the Tree of Life
U.C.	University of California
USDA	US Department of Agriculture
WG	Working Group

List of Figures[1]

[1] All figures in the text, except where otherwise indicated, are taken from the National Science Foundation and are in the public record. The term 'specific provenance unknown' indicates that the figure was found as an ephemeron.

Chapter 1
The Year 1975

> *We shall not cease from exploration,*
> *And the end of all our exploring*
> *Will be to arrive where we started,*
> *And know the place for the first time.*
> T.S. Eliot

1.1 Introduction

If Eliot was correct, to "know the place for the first time" still lies in the future for the history of the US National Science Foundation (NSF). Unique in the American experience as a federal agency of the US Government that supports scientific research financially but does not engage in scientific work itself, the NSF's history is known only in part. That history is constantly being created and only slowly being chronicled. That it is a vibrant, expanding, and living entity makes the final telling of its story impossible. From much that has been written about its beginnings, and for just some of its component parts, much can be learned. However, such a great deal is yet to be told that the telling of but one significant feature of the whole story demands following just a single thread through only a part of the Foundation's broad fabric. That is the purpose of this history: to come to "know the place for the first time," specifically for the NSF's role in the history of the biological sciences from 1975 to 2005.

For nearly half of the time covered in this study, the social and behavioral sciences were housed administratively with the biological, but only a select portion of their history will be discussed. They are necessarily considered to an extent here, though, for the reason that they had certain *clear influences on the directions biology took* over the early part of these decades. The biological sciences will occupy essentially all of this study as, by the latter quarter of the last century, they came to dominate the center of *all of science itself.* If the founding of the NSF in 1950 was couched in an era of physics, especially atomic physics, certainly by the *fin de siècle* of the twentieth century and the beginning of the twenty-first, biology was, and remains, the queen of sciences for the predictable future.

D. J. McGraw, *Millennial Biology: The National Science Foundation and American Biology, 1975-2005*, https://doi.org/10.1007/978-3-030-56367-7_1

Details of NSF history prior to 1975 more broadly can be found in a number of accounts, several historians having undertaken to chronicle the history of the Foundation. To them the interested reader is commended.[1] The *official* early historians of the NSF itself penned crucial studies: J. Merton England and George T. Mazuzan. Several other book-length works by Milton Lomask and Dian Olson Belanger tell also the broad early story (Lomask) or focus on certain threads and bring the reader up to the 1990s for engineering, in particular (Belanger). There is also a collection of other historical studies that run from short journal articles to larger works. Since the social and behavioral sciences will be only slightly covered here, reading the work of Larsen, told by one who lived those sciences at NSF for a number of years, is recommended. He discussed this polyglot collection of specialties from prior to the establishment of the Foundation until 1991–1992. That is the year in which they and the biological sciences split from one another entirely in the management structure of the Foundation. The present history makes particular note in this and, especially, the following chapter of the *interface between the biological and the social sciences* from 1975 to 1991–1992. Also recommended is a book by Mark Solovey, *Social Science for What? Battles over the Public Funding for the "Other Sciences" at the National Science Foundation* (Cambridge, MA: MIT Press, 2020).

[1] J. Merton England, *A Patron for Pure Science: The National Science Foundation's Formative Years, 1945–1957* (Washington, D.C.: National Science Foundation, 1982); George T. Mazuzan, *A Brief History [of the National Science Foundation]* (Arlington, VA: National Science Foundation, 1988): the history covers the period from WWII to about 1985; H. Guyford Stever, *In War and Peace: My Life in Science and Technology* (Washington, D.C.: Joseph Henry Press, 2002) (see especially Chapter 7ff.); Milton Lomask, *A Minor Miracle: An Informal History of the National Science Foundation* (Washington, D.C.: National Science Foundation, 1976); Dian Olson Belanger, *Enabling American Innovation: Engineering and the National Science Foundation* (West Lafayette, IN: Purdue University Press, 1998); Otto N. Larsen, *Milestones and Millstones: Social Science at the National Science Foundation, 1945–1991* (New Brunswick, NJ: Transaction Publishers, 1992); Detlev W. Bronk, "The National Science Foundation: Origins, Hopes, and Aspirations," *Science* (1975):409–441; Daniel J. Kevles, "The National Science Foundation and the Debate over Postwar Research Policy, 1942–1945: A Political Interpretation of *Science—The Endless Frontier*," *Isis* (1977):5–26; Daniel Lee Kleinman, *Politics on the Endless Frontier: Postwar Research Policy in the United States* (Durham, NC: Duke University Press, 1995); Robert Franklin Maddox, "The Politics of World War II Science: Senator Harley M. Kilgore and the Legislative Origins of the National Science Foundation," *West Virginia History* (1979):20–39; George T. Mazuzan, "Good Science Gets Funded…: The Historical Evolution of Grant Making at the National Science Foundation," *Knowledge: Creation, Diffusion, Utilization* (1992):63–90; Toby A. Appel, *Shaping Biology: The National Science Foundation and American Biological Research, 1945–1975* (Baltimore, MD: The Johns Hopkins University Press, 2000); see also the NSF website "Selective Bibliography" at http://www.nsf.gov/about/history/bibliography.jsp and the chapter footnotes sections of this text for additional materials. Much of the main text of this introduction is based upon a synthesis of the works listed in this footnote. When England's history was published, the then National Science Board chair, Lewis M. Branscomb, made copies available to all Members of the Board. He not only called the work "an excellent piece of scholarship," but made an important point: one must read it because "it shows the extent to which many issues facing the Board and the Foundation are of an ongoing nature." Such was the case, as will be evidenced throughout the present history. See the origin of the quotation in NSB, "Minutes," NSB-83-92, March, 22, 1983, pg. 2-83:3.

It is the most modern source on the history of the social sciences at NSF and is something of a companion study to the present work.

The most important historical study pertinent to the present work is that by Toby A. Appel, *Shaping Biology: The National Science Foundation and American Biological Research, 1945–1975*. The present history is, in many ways, a continuation of Appel's work and brings the status of America's biological sciences into the twenty-first century as they were funded, affected, and, to a very real degree, interactively guided by NSF.

The central purpose of this widely constituted present chapter is to help set the stage for the reader for the beginning of this 30-year history. It examines important events in the earlier history of the Foundation because they play strongly upon the development of the various biology directorates. Such issues as education, applied research, medical science, the National Institutes of Health, the beginnings of biotechnology, and other matters must be considered. In Chap. 2, the social and behavioral sciences will be considered based upon their *influence upon biology* at NSF.

1.2 Arenas of Contention

The NSF was founded in 1950, but no history of it would be complete without reviewing some crucial parts of it. And even though this study takes up only the more limited history of the biological sciences at the Foundation, and that starting at a much later date, it still requires some prefatory commentary for readers unfamiliar with it.

The organic act creating the Foundation allowed it to organize its internal structure as it saw fit.[2] It could create *divisions* (administrative units) that best answered its fundamental mandate to support financially the basic scientific research enterprise in America. It was clear, though, that the legislation called for the support of certain areas: *medical research*; mathematics; physical, engineering, and *biological sciences*; and, *other sciences*. A division of scientific personnel and education (scholarships and graduate fellowships) was also among the early units.[3]

Physicist Alan Tower Waterman was chosen in 1951 by then President Harry F. Truman to head NSF, but Waterman chose not to create one particular mandated division, *medical research*. That would have profound effects on the history of the NSF, especially before and to a limited extent at and just after a major structural reorganization that took place in 1975. His argument was simple enough: the National Institute*s* (plural by 1950) of Health was already extant and deeply involved in supporting medical research, not surprisingly. The small amount of support for medical research (as *medical science*, not *medicine*) that did emanate from NSF was

[2] The National Science Foundation is governed by the National Science Foundation Act of 1950, as amended (42 USC 1861–75), as well as other statutes.

[3] Mazuzan, "Brief History," pg. 6.

via a division devoted primarily to biology.[4] Waterman did not intend that NSF does *no* research in basic medical sciences but just that it does not try to take on the rapidly growing NIH in so doing. Certainly, NSF had to avoid any research dealing with the *practice of medicine*; strictly a mandate of the NIH. While much of the history of those institutes does not concern us here, by 1951 and Waterman's decision *not* to establish a division of medical research, the NIH had already become something of an "empire."[5]

Waterman also was not excited by the notion of an education program at NSF (the fellowship program, in particular), and the tension in that area arose in connection with the matter of medical research. In the first budget requested under Waterman's leadership, $1.2 M had been asked for medical research (and also $2.9 M for biological and $4 M for mathematical, physical, and engineering sciences—one division—and $0.5 M for operations) as part of a total request of $8.6 M.[6] The maximum request at that time could not exceed $15 M annually. But legislation amended the organic act in 1953 removing that stricture.[7] The Bureau of the Budget (BoB), later to become the Office of Management and Budget (OMB), proposed eliminating the medical item altogether, given all that was ongoing at the rapidly growing NIH. BoB had planned to cut $2 M altogether (more than half of that was earmarked for medical research) from the total $8.6 M, but after much wrangling, Waterman won the day when BoB backed down from its intended cuts. The final budget was done in a ratio of 1:2:3, that is, medical, biological, mathematics/physical science/engineering. Waterman had insisted that "'a comprehensive program of support for basic research,' [was] not simply one of 'filling gaps'."[8] That is, the NSF was to take the leadership role planned for it in the organic act and not merely plug holes in what basic research funding was needed in order to support an NIH or a USDA or any other agency.

If the matter of basic medical research was one of concern to this study, and more of that further below, there were others in the long-held "arenas of contention" both prior to 1950 and after: that is to say, what should make up the NSF and what should be left out. These decisions would ring down through the years concerning the two major reorganizations noted above.

The Executive Branch of the US Government typically makes policy, but in how the Legislative Branch—Congress—writes laws, policy can certainly inhere in them or come out of them, or both. That was the case with the organic act for NSF, for instance, and it would be even more obvious when the House Committee on Science and Aeronautics (CSA) was created in July of 1958. When chaired by

---

[4] Ibid.

[5] See especially Donald C. Swain, "The Rise of a Research Empire: NIH, 1930 to 1950," *Science* (1962):1233–1237.

[6] England, "A Patron," pg. 147.

[7] Mazuzan, "Brief History," pg. 7.

[8] Ibid., pg. 148.

Representative George P. Miller (D-CA) in the 1960s, the Committee "provided a focal point for assessing the entire federal system of research support."[9]

An early difficulty with regard to providing that research support swirled around the matter of the federal government entering into the arena of education, long a domain held to be a matter of states' rights as well as those, more especially, of local school boards for kindergarten to 12th grade (K–12). This issue would come up numerous times in the history of NSF and reached a boiling point by the time of President Ronald Reagan's administrations in the 1980s. But even earlier, this issue became one of those Pandorean dilemmas that, if only now past history, haunts the NSF even today and ripples across much of the history presented in this study.

Since the later 1950s and after the shock of the world's first man-made satellite, the Soviet's *Sputnik*, and the subsequent gearing up of science and engineering education, the Foundation had supported some 53 different precollege curricular projects that spanned all of the sciences.[10] For instance, physics, mathematics, and chemistry education were expanded, and there was the establishment of summer institutes for K–12 teachers.[11] Among the programs counted in that list was one well-known example. It was one that was only slightly controversial due to its human evolution content. It was titled the Biological Sciences Curriculum Study, BSCS. But another one in particular, "Man: A Course of Study," better known as MACOS, led to repercussions of truly immense proportions that had lasting effects on the very structure and operations of NSF—and the biological sciences directorate—in later years.

MACOS was one of the science education programs that can be traced back to the rapid expansion of a series of more successful such programs that were created in the wake of *Sputnik*. The first incarnation of the President's Science Advisory Committee (PSAC) was formed within a month after the launch of the satellite during the Dwight D. Eisenhower presidency and went on to good service in both the John F. Kennedy and Lyndon Baines Johnson administrations that followed. As a former Director of NSF, Richard C. Atkinson, said in later years, "[i]n particular, science education blossomed." Indeed, by 1968, the organic act was strongly changed (by the Daddario-Kennedy amendment discussed in detail below) to specifically focus more strongly on applied research, as well as science education, as charges to the NSF.[12] The second Director of the Foundation, Leland John Haworth, stated that:

> [i]t is important to realize that the revised act not only makes changes in the NSF charter and operations, but that it represents clear congressional endorsement of the importance of

---

[9] Appel, *Shaping Biology* pg. 155.

[10] Stever, *In War and Peace*, pg. 229.

[11] Richard C. Atkinson, from a paper read at the Colloquium Series on the History of Science and Technology, University of California, Berkeley, November 10, 1997, and published in the *Proceedings of the American Philosophical Society* (1999): "The Golden Fleece, Science Education, and U.S. Science Policy," at http://www.ucop.edu/pres/comments/gfleece.html for the complete paper. Quotation from pg. 1 (online version).

[12] Ibid.

the central functions of the Foundation, namely, the advancement of science *and science education.*[13]

That reemphasis on science education was directly tied to the *Sputnik* scare and to the fact that there had been only "a trickle of science education activities in the early years."[14] But "science education is a lightning rod all across Washington. People [in education] are accused of having very little data and very strong opinions."[15] MACOS became such a lightning rod.

Peter B. Dow has recounted the history of MACOS in a number of publications and noted that, in September of 1964, Jerome Bruner of Harvard took a leave of absence to "launch a project at [then] Educational Services Incorporated" which eventuated into MACOS.[16] Bruner had already laid out much of the program from before 1959 as part of his self-chosen response to *Sputnik* and his perception of needs in this country's science education efforts. He had also published a book that contained the "intellectual underpinnings" of the subject the same year (1963) that NSF, he, and others began work on MACOS.[17] Dow felt that the ideas incorporated into the course of study were "exhilarating," but it did not turn out that way in the end.

MACOS, premiered in 1970, was challenged that same year in a school in Florida and went totally defunct in a shower of implosions by the mid-1970s. The reasons, on a national educational level, were two: primarily, the content of the material itself and also the fact that the "mood of the country was changing [due to d]istrust of federally funded materials" such as those in mathematics where the "curricula were not developing basic skills, and the materials were too esoteric."[18] Within other circles, however, there was a more feared issue: the political threat to peer review at NSF.

Still, the issue of "local values" to school boards, towns, school districts, etc. was of great importance to many of the citizens of the country whose taxes paid for the work of and support from NSF. The Florida instance was but the first of many and is also a good example of the notion of threats to local values. Reverend Don Glenn, a Baptist minister, visited at the school in Lake City, Florida, to see his sixth grade

[13] National Science Foundation, *Annual Report for 1968*, "Director's Statement" (Washington, D.C.: US Government Printing Office, 1969), pg. xii. Emphasis added.

[14] Atkinson, "Golden Fleece," pg. 1.

[15] Platt interview.

[16] See http://www.nationalacademies.org/sputnik/lappan3.htm which describes a symposium hosted by the National Academy of Sciences on the occasion of the 40th anniversary of the launch of *Sputnik I*: "Reflecting on Sputnik: Linking the Past, Present, and Future of Educational Reform," 1997. Peter Dow presented a paper there entitled "The MACOS Materials: How Success Can Go Awry." See above cited website. His earlier work on the subject is *Schoolhouse Politics: Lessons from the Sputnik Era* (Cambridge: Harvard University Press, 1991) and covered the MACOS matter in considerable detail. It has been reprinted by Replica Books, Bridgewater, New Jersey, in 2000.

[17] Jerome Bruner, *The Process of Education* (Cambridge, MA: Harvard University Press, 1963). The quotation comes from Dow's address at the 40th anniversary symposium. See footnote above.

[18] Dow, symposium paper, pg. 2 (online version).

daughter in her class. It was the fall semester of 1970 when MACOS made its first appearance. Upon seeing the MACOS curriculum, he formed a local community committee that accused the program of advocating "sex education, evolution, a 'hippie-yippee philosophy,' pornography, gun control, and Communism," warning that MACOS was a "threat to democracy"![19] A series of other negative responses to MACOS led to a "full scale Congressional debate…in both houses in 1975."

What was so "threatening" about the program? MACOS was designed as a 1-year course that would aid children in "think[ing] about the nature of human beings and what is unique about being human."[20] The course made comparisons to the behavior and basic biology of a variety of organisms, including salmon, herring gulls, and baboons. It also included a film on the life of the Netsilik Eskimos—the nexus of the coming troubles for NSF.[21] MACOS showed the traditional life of the Eskimos before the acculturation brought on by the appearance of European peoples, said then Director of the NSF, H. Guyford Stever, by way of lament in his memoirs. The Netsiliks were very insular and so could be compared at a most basic human level to other early societies no matter the harsh environmental conditions under which they lived. When MACOS was introduced in 1970, some 1700 schools took it up (many more added it soon thereafter), typically with positive outcome. But rumblings, like that in Lake City, Florida, soon began, so Stever reviewed the program closely. As he said:

> I then learned that one feature of family life it showed was that anyone who was too old and infirm for the vigorous life of these people was put on an ice floe and sent out to sea. That was not very pleasant, especially teaching it to fifth and sixth graders.

"But there was worse," Stever went on: "[f]or example, if the wife of a hunter was ill and couldn't go on a hunt with her husband, he borrowed someone else's wife."[22]

As disturbing as Stever found this, he found worse the "vivid hunting scenes" with the killing of seals and the subsequent popping out of the eyeballs as a delicacy: those were shown being consumed on the film that accompanied the MACOS package. It did not surprise Stever when he learned that some children had fainted in the classroom at the screening of the documentary. He concluded that "while the science was right, the politics was not."

But that was not the whole story. An internal investigation of the approving of the grants for the MACOS project "turned up serious irregularities" which "lethally… subverted the [F]oundation's peer review system."[23] Stever was even more shocked when he questioned the project's NSF program officer, who responded in regard to

---

[19] Ibid.

[20] Ibid.

[21] The film was a work funded by the Ford Foundation Project that had been ongoing at Educational Services Incorporated (EDI) to which Bruner had come in 1964. EDI is today the Education Development Center in Newton, MA. See www.edc.org.

[22] Stever, *In War and Peace*, pg. 230.

[23] Ibid., pg. 231.

certain irregular grant approvals that "in effect, [she said] 'oh, yes, because my supervisors told me that we had to have a perfect case or the grant would not be approved'." Stever wrote to Congress indicating his intent to form an NSF committee to consider all the issues tied to the MACOS debacle and its implications for peer review at the Foundation.

A presidential appointee, Bob Hughes, was then serving as an assistant director at NSF and chaired the study, though he was infrequently present for the group's work. Since he was the chair, the committee's published findings then became known as the Hughes Report. As it happened, it was the first thing that the new deputy director, Richard Atkinson, read upon his arrival at the Foundation in mid-1975. He recounted that:

> [t]he report did not deal with the philosophical criticisms of NSF [over the MACOS incident and other matters]. Instead, it discussed NSF's business dealings and the appropriateness of its peer-review procedures as they applied to NSF curriculum projects.

Atkinson went on to say that he was called to Capitol Hill a few weeks later to testify about peer review as it was practiced across NSF as a whole. He was representing still then Director Guyford Stever who was away on a trip at the time. Several senators were "sympathetic," but long-time nemeses of NSF, Representatives Robert Bauman (R-MD) and John B. Conlan (R-AZ), called the Hughes Report "a pack of lies" and, as such, that NSF was "deliberately misleading Congress."[24]

In addition to all of this, an extremely conservative employee of the NSF had told Conlan about MACOS and the investigation: the "awful things" that NSF was doing. That led to Stever being called on the carpet a few weeks later at a second meeting before the same subcommittee of the House Committee on Science and Technology (CST), which approved (and still as recently as 2009 approves) funding for NSF and which had excoriated Atkinson just days before as he had stood in for the traveling Stever.

Every member of the subcommittee was there for the second convening and the audience area was packed. The matter of peer review was most in question, but so were the nature of MACOS and the role of the social sciences at NSF. Stever's honest and forthright answers satisfied the subcommittee for the moment, but he, as with Atkinson before him, was attacked further by both Conlan and fellow conservative Bauman. At the Fiscal Year 1976 (FY76) appropriations discussion, Conlan described MACOS as:

> a course for 10-year olds mainly about the Netsilik Eskimo subculture…[including] stories about…cannibalism, adultery, bestiality, female infanticide, incest, wife-swapping, killing old people, and other shocking condoned practices…It is absolutely unacceptable for NSF to continue using taxpayer's money for aggressive promotion and marketing activities for their own preferred social studies courses, undercutting competition from regular textbook publishing houses.[25]

---

[24] Atkinson, "Golden Fleece," pg. 7 (online version). The first quotation given here is that made by the senators and the second by Atkinson.

[25] Stever, *In War and Peace*, pg. 233.

One might fairly ask whether Congressman Conlan was more concerned about the content of MACOS or the loss of profits to the business sector in the form of the publishing community! At any rate, a House vote taken after "acrimonious debate" about MACOS favored Stever and NSF by 215 to 196 concerning whether or not Congress should review *all NSF curriculum projects* (the Conlan amendment). Instead, it approved Bauman's amendment that Congress review *all NSF research proposals in all fields*! That amendment, as Stever said, "thankfully" died in the conference committee.[26]

(There remains the question of whether the whole MACOS flare-up would have occurred quite as it did without the firestorm that Conlan might be said to have touched off. In Ken Heckler's history of the CST, he noted that Conlan only rose late in the 4-day series of meetings of the Committee as it was considering the FY75 budget. Conlan began "rather mildly" but soon began to speak of the Netsilik film as portraying "shocking film segments displaying gore and immoral acts." The MACOS program was becoming moribund by that time at any rate, and, for the CST, they deemed it to have been "too innovative." The "mark-up" activities, a late stage in preparing a bill for congressional action, did increase the science education's proposed budget from $70 M to $90 M, but, as Committee Chair Representative James W. Symington [D-MO] stressed, that should be for "*general* science education."[27])

As a result of the second hearing and due to the way Stever was treated by various congressmen, the NSF director suggested that he would request a General Accounting Office (GAO) investigation of the issues at NSF. GAO accepted that request and worked on it during the summer of 1975. But even by the fall of that year, little was heard of it.

Finally, on one day in January 1976, Atkinson was called into the office of sympathetic Symington. There, the congressman had Atkinson read the "bad news" of the GAO report. Later that same evening, Atkinson assembled trusted colleagues at NSF to review the report, and the upshot was that not only was the "bad news" of the GAO report correct but that it had "merely scratched the surface."[28] The "bad news" was that NSF was doing a poor job in its *business practices* and was failing to do appropriate audits, and, GAO noted inappropriate expenditures were found, as well.

While nothing was considered criminal, the NSF's governing body, the National Science Board, had to approve (as a matter of routine in those days) *all* NSF grant expenditures. In the case of MACOS, it was found that the peer reviews sent to the Board concerning the funding of the program were "redacted by program officers so that they were highly selective, emphasizing positive assessments and deleting negative ones." None of this had been presented in the Hughes Report as some members

---

[26] Ibid.

[27] Kenneth Heckler, *Toward the Endless Frontier: History of the Committee on Science and Technology, 1959–1979*, US House of Representatives (Washington, D.C.: Government Printing Office, 1980), ppg. 515–516; emphasis added.

[28] Atkinson, "Golden Fleece," pg. 8.

of that internal NSF investigatory committee "conspired among themselves to cover up the problems." Hughes himself, often away on other business, was unaware of these shenanigans. But Conlan and Bauman knew of these activities "within hours" after they had occurred at NSF due to their own inside informants.[29]

Atkinson said that NSF's response to the Congress regarding the very unflattering GAO report was done with such "candor," that it "took the wind out of our critic's sails." Two individuals at NSF were eventually put on administrative leave, and one of them was later terminated. The science education programs underwent revision within their policy guidelines, and new leadership was brought in.[30]

Peer review is a matter of the highest concern at NSF and totally integral to how it operates. The MACOS affair was not the only time that a collision of politics, programs, policies, and peer review would come to the fore. Along with the changes made in the curricula projects' review matter, changes were also made in the peer review process across NSF as a whole. Atkinson said years later that:

> [t]he program officers had, and still have, great flexibility. They solicit peer views [*sic*] for a given proposal, and then use the information—as they judge appropriate—to decide whether or not to fund the project…[and though p]rogram officers should have that kind of authority…there is also a need for oversight. Accordingly, we established an audit office that did random samples of peer reviews to ensure that they were being used appropriately.[31]

Even further, NSF changed the procedure for soliciting of peer reviews. Whereas applicants for a grant could request a copy of the reviewer's comments with her/his identifying information redacted, the redactions led to too many problems. This was solved by having reviewers write their reviews in such a way that their identities were protected but that the applicants could still see a copy of the reviewer's comments unedited.

Science education and peer review were not the only issues to have repercussions in and after 1975; so also was a matter at the very heart of the National Science Foundation's *raison d'être*: what kind of research it would fund.

The MACOS affair brought out certain issues that would dog NSF, and especially education, for years. A history of science and engineering *education* as a component of the Foundation has yet to be written. Only little more will be said of it in this history at appropriate places, but there were other matters that came up in the earlier years that profoundly changed the structure of NSF during the 1975 reorganization. These affected the biological, social, and behavioral sciences—and every other science, mathematics, engineering, and operating division. The birth of a program titled RANN, Research Applied to National Needs, came closely tied to events with regard to education. Over the years of 1969 to 1977, as Mazuzan has so artfully stated it, there "blossomed…one of the biggest controversies in the history

[29] Ibid., pg. 9.

[30] Ibid.

[31] Ibid.

of the agency."[32] The question of including *applied research* support to the NSF's charge was an old question that dated back to Senator Harley Kilgore's (D-WV) intent in the 1940s that an NSF should include *both* basic and applied research. He so stated it in his first bill of 1942 toward the creation of such an agency. Contrary to that thinking was the vision of Vannevar Bush that the best science to be supported would be basic research only. (Bush was known for his report that led to the establishment of NSF, "Science—The Endless Frontier."[33] Kilgore's significant intellectual input to its eventual creation is often minimized, however.[34])

The big controversy that "blossomed" was the establishment of RANN: note that it did not use the phrase *applied research*, though. RANN was not the first of its ilk to be promulgated by NSF as a predecessor program, IRRPOS, or Interdisciplinary Research Relevant to Problems of Our Society, had been created a few years earlier. RANN came after IRRPOS as Congress funded IRRPOS at $6 million for its last year (FY70) when NSF redirected that to the newly forming RANN. But both of those applied research programs could have, indeed, used that once unutterable term due to one major change that had come about just prior to the creation of IRRPOS. (Interestingly, even at the time of this writing, the circumlocution of "use-inspired basic research" can still be heard at NSF.[35])

It was in 1963 that an arm of the CST, the Subcommittee on Science, Research, and Development, came into existence and was chaired by Representative Emilio Q. Daddario (D-CT). It began a review of the NSF in 1965 and ended it by amending the NSF charter in July of 1968. It was then that President Johnson signed the amendment legislation known as the Daddario-Kennedy Act. *Johnson, with the stroke of his pen, redirected NSF into applied research officially for the first time.* Also, as a consequence of how monies would be spent in coming years, he left little funds for the doing of science and engineering *education* at the Foundation.[36] The amendment both authorized and encouraged applied research support as an element of the complex roles of the NSF in American science policy and funding practice.

IRRPOS was the first attempt by NSF to take on the very different area of applied research, as contrasted to the more familiar territory of funding basic research. The Daddario-Kennedy Act was signed in July, and, by the fall of 1968, IRRPOS was established. It had been the brainchild of the Engineering Advisory Committee of NSF and would have been "IRPOS" had not outgoing Director of the Foundation, Haworth, inserted the word "Relevant" into IRPOS to name it IRRPOS. Haworth served a single term (6 years by law) as director from July of 1963 to June of 1969. His integrity was highly valued, as was his objectivity and attention to details,

---

[32] Mazuzan, "Brief History," pg. 21.

[33] Vannevar Bush, *Science, the Endless Frontier: a Report to the President* (Washington, D.C.: Government Printing Office); see OCLC 1594001.

[34] Donald J. McGraw, "Vannevar Bush, Senator Harley M. Kilgore and the Birth of the National Science Foundation: Second Thoughts," manuscript in preparation (2012); not published.

[35] The term was used in a presentation at the Advisory Committee meeting of the Directorate for Biological Sciences (BIO AC) on April 29, 2009; the author was present.

[36] Appel, *Shaping Biology*, pg. 235.

among other traits.[37] He was, as were so many directors in the early history of NSF, a physicist and had worked at MIT on radar during WWII. Haworth also, like so many of his cohort (including Waterman earlier), served on the Atomic Energy Commission, having been appointed to it by President Kennedy in 1961. As NSF director, he followed through well as a man after the heart of Waterman with regard to the Foundation's basic research funding orientation, but "Daddario" thrust him into a new sphere where Waterman would not have wanted to tread.

Prudently or not, applied research became a part of the Foundation. NSF was, from July 1968 onward, in the business of supporting applied as well as basic research, a very major shift for the agency. IRRPOS and RANN both took advantage of this new opening but with questionable results.

As Mazuzan said in summing up the IRRPOS story:

> [IRRPOS] reflected the traditional approach of the agency by responding to proposals from the scientific community rather than the agency stimulating specific research proposals. The emphasis of the awarded grants was in the areas of environmental quality and urban growth and management.[38]

IRRPOS ran for two fiscal years and then expanded into RANN. And it was the third Director of NSF that ran with it wholeheartedly.[39] Texan William D. McElroy was a biochemist who had degrees in biology from Stanford University, Palo Alto, California, and Reed College, Portland, Oregon, and a doctorate from Princeton University, New Jersey. He served about half of his term as NSF director (from July of 1969, only weeks after Haworth left, to 1972) when he accepted the Chancellorship of the University of California, San Diego. He stayed there from 1972 to 1980 leading that major research university, which had long-term ties to NSF as grants' recipient and in other roles, as shall be recounted. He again, like so many others of the era, came out of the Office of Naval Research but, unusually, had experience at the National Institutes of Health, as well.[40]

Otto Larsen was forthright about RANN when, in his history of the social sciences at NSF, he summed up his section on that story by entitling it "RANN: Born 1 February 1971—Died 15 September 1977." But, of course, things did happen under the RANN program as later Director McElroy found "semantic triumph" in the phrase "research applied" to national needs, as opposed to "applied research."[41]

What was wrought by RANN? The question is important to this history because, even though it "died" in 1977, the very notion that the combination of the Daddario-Kennedy amendment and NSF's answer to that via IRRPOS and RANN meant that

---

[37] Maurice Goldhaber and Gerald F. Tape, *Leland John Haworth, 1904–1979: A Biographical Memoir* (Washington, D.C.: National Academy of Sciences Press, 1985).

[38] Mazuzan, "Brief History," pg. 21.

[39] Syl McNinch, Jr., Ed. by David E. Gould and Louise McIntire, "The Rise and Fall of RANN" unpublished, in-house NSF document; July 31, 1984.

[40] See Donald R. Helinski, "Biographical Memoirs: William D. McElroy," *Proceedings of the American Philosophical Society* 149 (2005):104–106, and J. Woodland Hastings, "William David McElroy," *Biographical Memoirs*," National Academy of Sciences (2004):164–183.

[41] Larsen *Milestones and Millstones*, pg. 93.

the types of grant applications that could be made to the Foundation and the types of competitions that it would itself offer up to the scientific and engineering communities for consideration could include applied research proposals as outright acceptable.

The goals of RANN were to:

> [i]dentify national needs not being addressed by existing research agencies; provide early warning of potential national problems; and initiate assessments and research that address these needs and problems.

Further, it was to:

> [s]horten the lead time between basic scientific discoveries and relevant practical applications, and serve as a bridge between the Foundation's basic research programs and the development, demonstration, and operational programs of Federal mission agencies, State and local governments, and industry.[42]

Present in this mandate are several terms that are of particular interest here. First, applied research was now an acceptable phrase at NSF (though it still grates some traditionalists to this very day[43]) and, second, the mention of *mission agencies*. While it is true that NSF certainly has a "mission" to support research, the NIH is a classic example of what is typically termed a "mission agency": recognize, define, and attack, intramurally and/or extramurally, a problem (a given disease, say, in the NIH example), and eradicate it or find a cure for it, if at all possible. The matter of mission agencies was one of the things that made the NSF truly unique in the American experience: NSF supports research, but does not itself *do* research, certainly not in the intramural sense of field or laboratory research seen at, say, NIH. On the other hand, the mission agencies do, NIH, USDA, Naval Observatory, the National Laboratories, and the list goes on.

After the demise of RANN, the basic philosophy of the acceptability of applied research at NSF continued and remains a strong and fundamental aspect of the structure and organization of the Foundation today. By the death date of RANN, both IRRPOS and RANN together provided a number of years of, what one might term, experiential education for NSF in learning how best to support applied research and how best to design programs inviting research proposals that may be basic, applied, or both.

For such areas as engineering, RANN brought about highly positive results, as Belanger has noted in numerous instances. For but one example, the successes in earthquake engineering, as stated by her:

> By 1976 the National Science Foundation could boast that RANN's earthquake engineering program had developed computer programs on soil response that were used by 'virtually all' engineering consulting firms that did construction-related earthquake analysis.[44]

---

[42] Ibid., ppg. 93–94.

[43] Various discussions between a number of NSF scientists and the author.

[44] Belanger, *Enabling American Innovation*, pg. 158.

The earthquake work represented about half of all RANN's funds at the time. RANN also included excavation technology programs, pollution eradication programs, robotics, solar energy efforts, urban problems of a variety of sorts, and much else.[45]

RANN was in a moribund state by the time of the major NSF reorganization of 1975: its role with regard to the biological sciences, specifically, will be considered in chapters to come. By 1975 and 1976, RANN was being deconstructed and parts of it "went to related in-house disciplines." But those dealing with energy research specifically, and it was a large component by that time, were "transferred to the Energy Research and Development Administration."[46] RANN was being phased out as, by then director, Richard Atkinson had chosen a new direction for the NSF,[47] as Larsen has opined.[48] Further, Belanger has considered the matter in detail and averred that "RANN represented a clash of cultures," not only in the issue of applied versus basic research, but with regard to Atkinson's views of what NSF should be and how it should be effecting its mandates: Atkinson "questioned both RANN program content and management."[49]

The most valuable and heavily committed programs under RANN were divided up over a number of locations both within and without NSF's purview. Some of these have been mentioned above, but others went in disparate directions. In an NSF internal report, it was stated that the program in fire research went to the then National Bureau of Standards, while Chesapeake Bay environmental research went to the Environmental Protection Agency. At least four programs remained at NSF, with the most important being the Earthquake Hazards Mitigation Program, it representing a commitment of about $13 million, or about 20% of the last RANN budget. There is no mention of any programs going to NIH.[50]

Another report spoke of the nation's "economic competitiveness," and it was in this connection that:

> RANN…was a harbinger of the Foundation's effort of a decade later to link academic basic research with industry to help stem the crisis the nation faced in international competitiveness.[51]

While many features of international competition can be tied directly to basic scientific research and its oftentimes follow-on applied form, this history will look at one example in particular, biotechnology, in this and a later chapter.

MACOS and RANN are given considerable shrift in this chapter as their effects over time at the Foundation were profound. They played upon the very psyche and character of the NSF, and their influence will be seen in continuing roles for years

---

[45] Ibid., Chapters 3 and 4, *passim.*

[46] Mazuzan, "Brief History," pg. 22.

[47] Larsen, *Milestones and Millstones*, ppg. 91–92.

[48] Ibid., Chapter 5 *passim.*

[49] Belanger, *Enabling American Innovation*, ppg. 118–119ff.

[50] McNinch, "Rise and Fall".

[51] Mazuzan, "Brief History," pg. 22.

in the biological, as well as social, sciences. The decisions made on all sorts of issues were often swayed by the way that MACOS and RANN had affected the agency.

1.3 Life Begins at 1975

The NSF was built, in the spring of 1975, upon the framework of having a director, aided by a deputy director and advised by a governing committee, the National Science Board. Subsidiary to the Office of the Director, there were other organizational units, too (see Fig. 1.1). Under that senior administrative layer came a series of *divisions* that would, after late 1975, morph considerably into what became known as *directorates*, each led by an assistant director (AD) of a given directorate. The pre-1975 divisions were based upon the disciplines supported by the Foundation, but those various sciences and engineering divisions were all grouped under a directorate led by the Assistant Director for Research (Fig. 1.1, second column from the right). In other words, the heart of the NSF, the sciences, was felt to be best placed *subordinately* under the title of research instead of that of each discipline (or a group of closely related disciplines, e.g., the biological sciences) in its own higher-level category. The concept of research, then, was such that a solitary assistant director was responsible for research in *all* realms of science, engineering, and technology. That was a rather notable challenge for a single individual. Such a person was likely to have been a disciplinary-based scientist or engineer and could hardly have known the content, as might be hoped, for each and every one of the disparate fields that constituted the NSF.

As can be seen in this first figure, there was a Division of Biological and *Medical* Sciences but another for the Social Sciences: the latter had been somewhat elevated by then from an earlier, lower standing. Note, too, that such areas as environmental science were separated from the biological sciences, computer research from engineering and from mathematics and physical sciences, and so forth. While this arrangement had been in effect for some time, the realization that it might not be the best way to organize the NSF became clear by 1975 and even earlier for biology, as will be recounted.

The concept of a division then was as a subordinate creature to the more exalted concept of research as one of the then (spring 1975) five areas mentioned earlier and which were headed by assistant directors. The NSF would be a very different beast by the autumn of that same year when the business of *administering research* fell to *each* new *directorate*. In so doing, such research administration would be done discipline by discipline (or more correctly for biology functional area by functional area) with specialists in their own fields dealing directly with their own research award and support activities. The significance of the restructuring of the NSF in 1975 can hardly be overstated.

The Division of Biological and Medical Sciences (BMS) had long been in existence up to early 1975: it would be radically restructured and renamed by the close

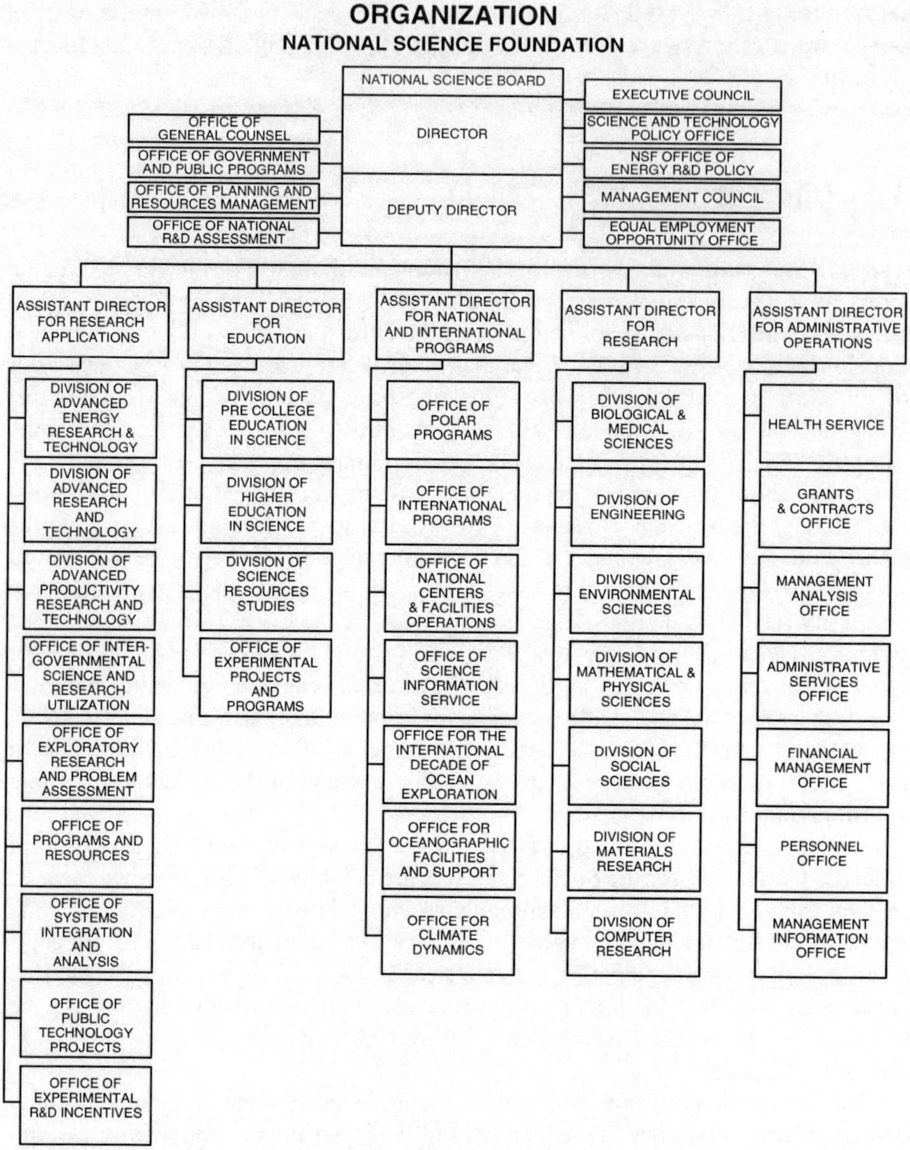

Fig. 1.1 Organization chart for NSF. In-house telephone directory, April 1975

of that year. It became the newly established Directorate of Biological, Behavioral, and Social Sciences (BBS), a combination that lasted until the social sciences attained their own directorate in 1992 at the next major reorganization.

At the time, Physicist H. Guyford Stever was the fourth Director of the National Science Foundation. His term in office was from February of 1972 until August of 1976. President Richard M. Nixon announced Stever's appointment at a news conference in November of 1971, and the Senate, without dissent, later confirmed him. He thus would remain in office long enough to see the fruits of the reorganization in which he played a significant part. In a press release announcing the changes of 1975, Stever claimed that the reorganization was "under active consideration for more than a year" and that it would strengthen a busier NSF and provide "an effective management structure for dealing with current and future operations."[52]

However, in a later November 1975 address to the National Academy of Engineering, he was more candid, acknowledging that:

> the force of politics on internal administration—'closer scrutiny by Congress' and 'tighter priority-setting by the administration'—with budget cutbacks, questioning of research programs, and criticism of NSF management [were all partly responsible for the reorganization].

The questioning of management was due to the MACOS peer review affair. Further, Senator Edward William Proxmire (D-WI), who went by William during his life, was only one of many "force[s] of politics" that plagued Guy Stever in his time as director. More will be said about Proxmire later, as he had distinct effects on the biology directorates. Stever joked at the meeting of the engineering academy that he decided against titling his talk "Science—In Search of the Golden Fleece" (named for Proxmire's award to those agencies which, he held, wasted government funds) because Stever was well aware that there was a "new climate" being felt across the land and accountability would be one of its characteristics, hence, in part, a reason for the reorganization of 1975, noted in Stever's quotation above.[53]

Although reorganization of the entire NSF had been in contemplation "for more than a year," that for the BMS had been discussed within the division since at least 1971 and very seriously underway by 1973.[54] In 1971, then Betsy Clark's Deputy Director of BMS, John W. Mehl, wrote a proposal for reorganization for that division that began:

> The objectives to be served by an organizational structure can be viewed as primarily managerial or primarily scientific. In the first case, the intent can simply be to establish a clear chain of command within the organization that allows top management to fix responsibility for particular functions and decisions.

---

[52] Belanger, *op. cit.*, *Enabling American Innovation*, pg. 125.

[53] Ibid.

[54] There are a number of ephemeral pages from an unknown source that provide hints as to the directions a new BMS may take. However, the earliest of these is from 1971 and which provenance is known: John W. Mehl, Deputy Director, Division of BMS, "Proposed Reorganization of Biological and Medical Sciences," April 15, 1971 (NARA, Box 11, File "BMS Reorganization 1971–1972"). NARA is the National Archives and Research Administration. This in-house white paper was apparently prepared for use within the Division or possibly for the director of NSF; there is no letter of transmittal. At any rate, no earlier documentation has been found concerning the concept of the reorganization for the biology directorate. On the contrary, numerous memos, some with date and author and others with ephemeral pages, on the subject are seen in greater numbers starting in 1973 when it seemed to become clear that a reorganization was imminent. These will be cited, as appropriate, in other footnotes.

> Considerations of management should, however, be secondary to or should serve the purposes for which the research divisions exist, which are scientific. Within BMS, our interest should be to identify that research in biology which can most effectively improve the description of biological systems in quantitative terms and predict the course of changes in such systems under different conditions; to identify the people who can most effectively carry out this research; and to provide support for the research in such a way that its progress is facilitated to the maximum extent with available funds.[55]

Mehl, a very thoughtful individual, laid out the basic philosophy from which higher authority could begin to make moves toward a *new* BMS. Indeed, his thinking could well have applied to the entire NSF. In his 16-page white paper, Mehl considered a number of defining points for BMS, such as problem areas with regard to then and future structural issues; the content of science to be placed in any new structure suggested; the need for special programs (the International Biological Program, e.g., discussed more fully in later chapters) and ways to manage these; how to assign disciplinary problems to aspects of structure; and so forth.

By 1973, the "defining points" process was well underway and would be present in a form that Stever could work with once he became director. When Stever arrived in Washington to take the directorship, he averred that his first year (1972–1973) was "uneventful," at least on the political front. He may well have expected it to be otherwise given the level of political tenseness in US society at the time. Politics, both internal and external to the NSF, affected Stever's term in office greatly. On the matter of science, however, it was a highly productive year as "we learned what moves mountains, probed the stunning violence of the universe, and explored the coldest, driest, windiest, and highest continent [Antarctica, long a province of the NSF's research interests]."[56]

The budget of the NSF in 1951 was $225,000, only enough to pay for the administrative start-up costs. In FY52, the first year the support of a part of the nation's science activity was undertaken, it was $3 M, but, had Vannevar Bush had his way, the very first year's budget would have been over $33 M. *Sputnik* changed everything in 1957, though, such that by FY73, the first budget year under Stever's guidance, it had grown to $610 M: science had gained great importance in American society since 1957.

A much larger budget was one thing, but, by Stever's time, many political issues had come to the fore concerning the structure of the NSF and its management, to say nothing of budget tightening. Of great importance was the fact that Stever's appearance at the Foundation in 1972 fell right in the middle of a nation caught up in a plethora of paroxysms and much high hysteria (demonstrators lying in front of troop trains, for instance), given the:

> protests revolving around the war in Vietnam, civil rights, environmental pollution, and the development of nuclear power plants…[These issues] sparked a reexamination of conventional policy structures for science and created a demand for expanded scientific advice. There was a general increase in the public scrutiny of science, and scientists were increas-

[55] Mehl, "Proposed Reorganization," pg. 1.

[56] Stever, *op. cit.*, *In War and Peace*, pg. 179.

> ingly viewed as an interest group susceptible to the same politics of open debate as everyone else. Skepticism about the claims of science grew, and Congress and the general public began questioning whether Government expenditures for research were benefiting the scientists and their institutions more than the public. And the scientific community began to worry that such questioning might well erode their support structure within the American political system. Many leaders within the science policy community thus came to view this as a period of 'crisis' [1965-1975] for science in the United States.[57]

So, while Stever may have had a placid year politically in 1972–1973, it could not last. Many of then former President Johnson's long-mandated "Great Society" programs were already draining coffers, some moneys of which might have gone to the NSF. Those programs also established many entitlement programs that would last well into the future. Additionally, something new due in significant part to the Great Society milieu was the call for "broad public support…for research to be *socially relevant*."[58] This would later have a direct impact on the nature of the 1975 reorganization.

The year 1973 would bring another early "crisis" situation for Stever and the NSF as President Nixon sacked the President's Science Advisory Committee. The Committee was a child of the Eisenhower years of 1953 to 1961 when Nixon himself was vice president. Not only was the Committee eliminated, but Nixon also abolished the Office of Science and Technology, both of which had been in existence for some years. This was a pair of policy voices muted at a time when they were both much needed to aid in explaining to the public the value of the new research into recombinant DNA (deoxyribonucleic acid, the stuff of genes), human subjects' research, fetal research (NIH), radioactive fallout from nuclear testing, air pollution, and environmental contamination.[59]

The assistant director of biology, Betsy Clark, held that NSF was, historically, "basically apolitical and the program directors were very well insulated from The Hill or from any political pressures…That held sway until Nixon arrived…and then it began to crumble."[60] Clark went on to aver that:

> [Biochemist William D.] McElroy [who preceded Stever] was a good politician. He really did an amazing job working The Hill and everybody else in town. And he always said that he had never run into such a tough situation as he had dealing with Haldeman and Ehrlichman [assistants to Nixon and central players in the Watergate debacle] because they really controlled everything.

It was in this "time of tumult," as Stever dubbed it, in which he and all of the NSF and many another agency had to function in those years. Vice President Gerald Ford

---

[57] Stine, *op. cit.*, "History of Science Policy," pg. 57. The use of the term "crisis" can be found in Don K. Price, *op. cit.*, "Science at a Policy Crossroads." Stever, in his memoirs, also chose to cite much of Stine's "History" comments at pg. 182. Stine had read the transcript of Stever's memoir (personal communication between the author and NSF Historian Marc Rothenberg, January 19, 2010).

[58] Ibid., Stine "History," pg. 57. Emphasis added.

[59] Ibid., pg. 58.

[60] Betsy Clark interview with the author, May 26, 2009, by telephone.

took on the presidency when Nixon resigned in mid-1974, but it was nearly 2 more years, until May of 1976, until Ford signed into law the new Office of Science and Technology Policy (OSTP). Thus he returned a science presence in the White House that Nixon had eliminated.[61] At the same time Ford established the OSTP, he also returned to existence the science advisor to the president position. For the purposes of the present history, Stever can be left to tell much of his own story for those years of tumult through the pages of his memoir, *In War and Peace.* But by March of 1975, President Ford had appointed Atkinson as Deputy Director of the NSF to serve under Stever.[62] The Senate confirmed him in May.[63] (An acting deputy director, the mathematician Lowell Page, had preceded Atkinson.)

Atkinson was a psychologist, one of the first social scientists to hold the deputy's position and, later, the first such to hold the director's post. That was a propitious sign for the social sciences just starting to make some headway of note in the Foundation. A very long history of leadership at NSF had been in the hands of physical scientists: echoes of the naval research office and atomic energy agencies' respective make-up. How the fortunes for the biological and social sciences accelerated to new heights in the final quarter of the twentieth century in the existence of the NSF may be shown to have ties to the appearance of Atkinson as director in the post-1975 structure of the Foundation.

The budgetary news was as grim in 1975 as Stever had described it in the previous decade, the 1965–1975 "time of tumult." All national research and development expenditures, both basic and applied, and for all agencies whether independent (NSF) or under a given federal department, when taken together in 1975 were expected to reach just over $34B. However, in 1967 dollars, to which the 1975 budget was compared, the hoped-for 7% increase over the budget of 1974 was, in fact, 3% *under*—inflation having consumed the rest. Inflationary losses in budgetary figures would plague the NSF, as well as the entire government, for a very long time to come. Planning under such conditions, plus the uncertainty of any year's budget until passed by the Congress and signed by the President, made doing science very difficult. For many grantees, they could not be sure whether they would be funded over multiple years (many had continuing grants) and, if so, at the levels they needed, a truism for many different periods in NSF history. As it happened, the NSF (and all of government) was operating under a *continuing resolution* in the latter months of that very year of 1975 until the appropriations bill could be passed. The "continuing resolution" way of life for the federal government's operations was not an uncommon occurrence at many different times in American history right up to the present.[64]

---

[61] Public Law 94–282.

[62] NSB, "Minutes," March 20–21, 1975, 171:6.

[63] NSB, "Minutes," May 15–16, 1975, 173:20.

[64] NSB, "Minutes," September 18–19, 1975, 175:19. For a notion of how common such continuing budget resolutions (CRs) were, consider the following: "Continuing resolutions date from at least the late 1870s, and have been a regular part of the annual appropriations process for over 50 years. In fact, with the exception of three fiscal years (FY1989, FY1995, and FY1997), at least one con-

Basic research, *per se*, was imbedded with *all* federal R&D (the full $34B), but it only totaled that year just over $4B, all the rest being applied, and, as always, yet another large portion was defense spending. Further, that basic research budget line was a *decrease* of 8% from FY74: a very precarious environment in which to work and one of Stever's many problems. Still, he stated "that the reorganization plan announced by him on July 10 had been well received."[65]

And, for present purposes, what the plans demonstrated was a wholly new structure as compared to that seen in Fig. 1.1. Figures 1.2 and 1.3 both convey three very important points. First, the notion that an entity termed research standing alone as a directorate was superior to the disciplinary areas was eliminated. *This was done by incorporating research management and decision activities, as well as funding lines, into each of the new disciplinary directorates*. Those directorates had replaced the old *divisions* and were more specific and less lumped. (The term *division* can still be seen in the new organization chart (Fig. 1.3), but such became subordinate to directorates.) Second, the former Division of Biological and Medical Sciences (BMS) was gone and was replaced by the new Directorate of Biological, Behavioral and Social Sciences (BBS). That implied, third, that the *social sciences* had a much higher visibility in the post-reorganization NSF while also acknowledging that *medical science* was not to be considered of such significance as had the organic act (and Vannevar Bush) given it: the NIH "empire" owned so much of the area, and logically so, already. It did not entirely disappear from NSF/BBS, however, as will be seen.

---

tinuing resolution has been enacted for each fiscal year since FY1954... From the early 1970s through 1987, CRs gradually expanded from interim funding measures of comparatively brief duration and length to measures providing funding in lieu of one or more regular appropriations bills *through the end of the fiscal year* (referred to as *full-year continuing resolutions*). The full-year measures included, in some cases, the full text of one or more regular appropriations bills and contained substantive legislation (i.e., provisions under the jurisdiction of committees other than the House and Senate Appropriations Committees). Since 1987, continuing resolutions have generally been interim funding measures with little substantive legislation...Until the early 1970s, continuing resolutions principally were limited in scope and duration, and rarely exceeded a page or two in length. They were used almost exclusively to provide interim funding at a minimum, formulaic level, and contained few provisions unrelated to the interim funding...Beginning in the early 1970s, conflict between the President and Congress over major budget priorities, triggered in part by rapidly increasing deficits, greatly increased the difficulty of reaching final agreement on regular appropriations acts. This conflict led to protracted delay in their enactment. Continuing resolutions, because they historically have been viewed as 'must-pass' measures in view of the constitutional and statutory imperatives, became a major battleground for the resolution of budgetary and other conflicts...These trends culminated in FY1987 and FY1988, following a period of persistently high deficits and sustained conflict over how to deal with them. For those 2 years, CRs effectively became omnibus appropriations measures for the federal government, incorporating all of the regular appropriations acts for the entire fiscal year as well as a host of substantive legislation covering a broad range of policy areas." The source for the foregoing comment is Sandy Streeter, *Continuing Resolutions: FY2008 Action and Brief Overview of Recent Practices: Updated January 9, 2008*, at http://www.rules.house.gov/Archives/RL30343.pdf. Emphases in original.

[65] Ibid., the "Minutes".

Federal obligations for basic research, by detailed field of science & engineering, and selected agency: fiscal years 1970-2001

[Dollars in thousands]

National Science Foundation

Field of science & engineering	1970	1971	1972	1973	1974	1975	1976	1977
Total, all fields	244,977	272,566	367,691	392,442	415,217	485,989	523,634	624,900
Life sciences, total	52,662	54,795	67,600	72,610	76,011	91,576	97,593	108,237
Biol & agricultural, total	47,176	51,876	59,747	63,497	75,796	91,576	97,460	107,11
Biol (excl environmental)	NA	NA	NA	NA	NA	NA	NA	NA
Environmental biology	NA	NA	NA	NA	NA	NA	NA	NA
Agricultural	NA	NA	NA	NA	NA	NA	NA	NA
Medical sciences, total	5,120	0	7	25	215	0	0	1,122
Clinical medical	4,120	0	7	25	187	0	0	0
Other medical	1,000	0	0	0	28	0	0	1,122
Life sciences, n.e.c.	366	2,919	7,846	9,088	0	0	133	0
Psychology, total	7,600	5,249	6,171	5,531	5,595	8,474	8,249	11,381
Biological aspects	4,210	2,973	3,541	3,307	3,628	6,149	5,697	8,368
Social aspects	1,430	2,276	2,456	2,224	1,967	2,325	2,552	3,013
Psychological sciences, n.e.c.	1,960	0	174	0	0	0	0	0
Physical sciences, total	74,649	77,550	108,900	115,239	118,600	137,581	143,471	179,180
Astronomy	19,922	23,236	25,608	27,080	30,208	28,733	31,754	41,585
Chemistry	21,741	21,665	30,171	29,966	33,370	43,439	40,570	54,387
Physics	32,892	32,649	48,985	52,660	55,022	65,409	71,147	83,175
Physical sciences, n.e.c.	94	0	4,136	5,533	0	0	0	33
Environmental sciences, total	53,031	71,502	91,026	86,344	112,280	123,218	120,789	177,627
Atmospheric sciences	22,973	29,000	33,118	38,090	43,480	44,832	45,410	61,270
Geological sciences	10,288	9,516	11,422	19,174	32,524	36,286	34,463	52,684
Oceanography	19,680	30,986	43,613	23,699	36,276	42,088	40,755	63,662
Environmental sciences, n.e.c.	90	2,000	2,873	5,381	0	12	161	11
Math & computer sciences, total	17,061	17,236	26,813	22,115	23,239	27,341	32,845	38,985
Mathematics	NA	NA	NA	NA	NA	NA	18,391	21,624
Computer sciences	NA	NA	NA	NA	NA	NA	14,444	17,361
Math & computer sciences, n.e.c.	NA	NA	NA	NA	NA	NA	10	0
Engineering, total	21,669	24,037	39,818	40,485	44,436	64,445	61,912	70,424
Aeronautical	520	1,164	1,981	209	684	22	0	588
Astronautical	0	0	24	0	56	0	0	0
Chemical	3,810	4,191	4,330	4,704	5,817	10,846	10,962	9,475
Civil	2,736	3,392	2,956	6,107	5,031	4,223	4,427	6,596
Electrical	4,953	4,795	8,724	10,328	9,622	11,257	14,972	19,495
Mechanical	3,811	3,654	4,376	5,166	7,076	6,909	7,123	6,733
Metallurgy & materials	2,740	4,695	8,486	9,504	11,666	16,088	15,475	17,802
Engineering, n.e.c.	3,099	2,146	8,941	4,467	4,484	15,100	8,953	9,735
Social sciences, total	14,234	20,225	26,117	24,689	22,629	27,127	34,217	34,525
Anthropology	3,227	3,999	4,118	3,886	3,913	3,309	3,449	3,924
Economics	4,111	4,443	4,685	4,622	5,220	6,661	6,998	8,704
Political science	1,554	706	1,395	1,431	1,290	1,361	2,249	2,348
Sociology	1,311	2,581	2,969	2,383	1,857	2,444	2,733	3,600
Social sciences, n.e.c	4,031	8,496	12,950	12,367	10,349	13,352	18,788	15,949
History	924	733	876	898	920	818	NA	NA
Linguistics	1,296	137	94	1,253	1,300	1,320	NA	NA
Other social sciences	1,811	7,626	11,980	10,216	8,129	11,214	NA	NA
Other sciences, n.e.c	4,071	1,972	1,246	25,429	12,427	6,227	24,558	4,541

Fig. 1.2 Federal obligations for basic research, by detailed field of science and engineering, and selected agency: fiscal years 1970–2001 (dollars in thousands). NSF. Division of Science Resources Studies, NSF01-333, July 2001

But how did all this shuffling and revisioning come about? A look at the BMS, with special focus on the "M" for medical, is one issue to be examined in order to appreciate the significant changes that occurred at NSF when BMS was eliminated.

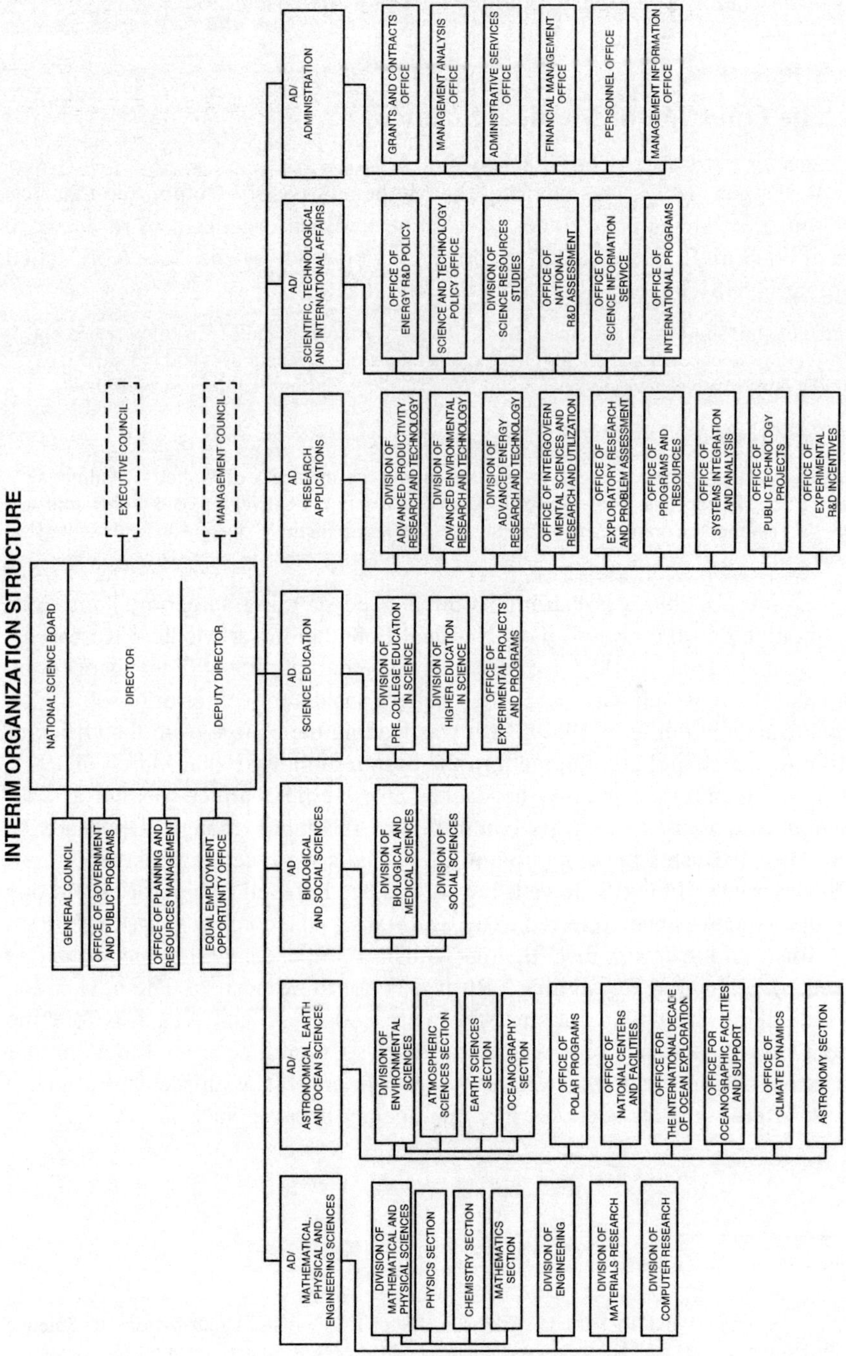

Fig. 1.3 Interim organization chart for July 10, 1975. NSF. Specific provenance unknown

Also, the social sciences underwent a change at the same time and will be mentioned further below. We dip back in time again for this discussion.

1.4 The Question of Medical Science

Walter W. Palmer, M.D., was chairing the Medical Advisory Committee that was one of four specialized committees that Vannevar Bush organized in response to President Franklin Roosevelt's call for advancing post-war science (see works cited in footnote 1). NSF historian, England, has said that a:

> board chosen from [a National Academy of Sciences] panel of names [was to] be relied on to foster pure science…[so it could] guard against the kinds of social uses and federal direction that conservatives feared would result from [Senator Harley] Kilgore's proposal.

And, averred England, the:

> fear of government control was even greater among the members of Walter W. Palmer's [Committee]. Palmer was professor of medicine at Columbia University, and all his committee's members except Linus Pauling, a CalTech chemist, were medical school professors.[66]

This group, possibly less Pauling, wanted "free standing autonomy" and even had a "fixation" on the notion of a "National Foundation for Medical Research," stated England. This would have been independent of the NSF envisioned by Kilgore and Bush, which nature was then being considered well prior to settling on the organic act's wording of 1950. Bush was leading the charge from the Office of Scientific Research and Development in the design of an NSF that March of 1945, and Palmer's Committee "caused headaches" for Bush.[67] Palmer was not as adamant about autonomy as was his committee membership, though. Upon advice from another of Bush's planning committees' chairs who were against the autonomy, Bush wrote to Frank B. Jewett. Jewett was head of Bell Laboratories and one of a group, which has been referred to by historians as the country's wartime "brain trust."[68] Bush said to Jewett that "the mechanism I propose is preferable to the one proposed by [Palmer's] Committee." Bush was not to be pushed! The next week, Bush shared his draft plan for an NSF with Roosevelt, just weeks before the President's death. The outcome was that Palmer's Committee agreed that another "Foundation" for medical research, *in addition to* an NSF with medical sciences included, offered greater sense as just one agency because such an arrangement

---

[66] England, *op. cit.* "A Patron for Pure Science," pg. 16.

[67] Ibid., pg. 16–17.

[68] William A. Blanpied and Richard C. Atkinson, "Social Scientists' Contributions to Science Policy during the New Deal," at http://www.rca.ucsd.edu/speeches/Social_Science_and_Science_Policy.pdf.

would provide "practical desirability."[69] Bush's plan was then the one accepted—and with medical sciences as part of the embryo Foundation.

When the Division of Biological and Medical Sciences (BMS) was formed at the infant NSF, the National Science Board created a fiction of sorts in order to placate the budget bureau:

> For the [FY53 budget] presentation to Congress in early 1952, the board had agreed to submit a program for biology and medicine [*sic*] as a single line item, with the understanding that the [two] programs would remain 'administratively separate' and funded at the approximate ratio of two to one [biology to medical sciences—not *medicine*].[70]

This fiction was dropped soon thereafter, however. In 1954, Waterman sought the support of the Board to combine the biological and medical sciences as a "group" (which became BMS) and to defend their budget "as such." BMS became a solid reality due to Waterman's strategy. As has been seen, the legislation that created the NSF indicated *what* sciences (plus engineering) had to be included but left the leadership of the NSF to decide *how* to organize them. Waterman's approach assured that *basic medical research* was safely ensconced in the actual organizational structure, and so those areas of medical science properly closely affiliated with biology would get the attention and financial support they needed. They included such areas as bacteriology (not the medically applied areas, though) and pharmacology (but not pharmacy), as but two early examples.[71]

In 1951, the NSF was being organized by a variety of committees, one being the Temporary Committee on Medical Research. They determined to include a variety of basic medical sciences within the structure that would house both biological and medical sciences, such disciplines as anatomy, physiology, bacteriology, pharmacology, and others. But, strangely enough, they also included medicine and surgery![72] While the other areas noted as examples had clear application to both biology *and* medicine, the areas of medicine, *per se*, and certainly surgery, were clearly the province of NIH. It should come as no surprise then that the Bureau of the Budget would have objections to funding the medical sciences portion of the requested FY52 NSF allocation. And it would not be surprising that Waterman would establish the fiction of burying the funding for the medically oriented aspects of the biological sciences within budget lines that did not leave medical sciences standing out alone and obvious to the scrutiny of BoB. BoB was, however, excellent at scrutinizing.

In May of 1951, in working on the budget for FY52, Frederick Schuldt of BoB sought to eliminate about $1 M from the NSF's first funding request that would have been used toward basic medical sciences. It was that very situation that led Waterman, as has been stated above in regard to policy-making at NSF, to object strenuously to BoB's attempted action. While BoB held that the monies would be duplicative of the work done at NIH, Waterman argued that:

---

[69] Ibid., pg. 18.

[70] Appel, *op. cit.*, *Shaping Biology*, pg. 53.

[71] Ibid.

[72] Ibid., pg. 47.

> it was the opinion of most research men familiar with both the medical and biological side [of such research] that really fundamental [=basic] research was inadequate in the field of medical research and that for the good of clinical medicine and the medical schools, this support should be given.[73]

The question of hampering NSF just as it was beginning as a new agency by allowing BoB to make policy (not its place) would "create a precedent" unacceptable to NSF, argued the Temporary Committee on Medical Research. The Committee prevailed.

However, there were members of Congress that felt as did Schuldt at BoB. But instead of supporting basic medical sciences' research, they introduced H.R. 3371 that would direct NSF medical funding to *medical education*! The argument made was that the Korean War, then ongoing, showed that the nation was short of physicians. They felt NSF could award scholarships, grants, and even facilities' construction monies for medical education. The American Medical Association, the AMA, was fearful, however, that the federal government would have too much control in the arena of medicine if allowed to take this tack in medical education, Waterman discovered.[74]

Medical school deans were not happy with H.R. 3371 either because they desired medical education funding be directed through the US Public Health Service and, so to say, its subordinate the NIH. The alternative Senate bill, "The Emergency [i.e., Korean War] Professional Health Training Act of 1951," S. 337, was put forth, but it failed. Both the NSB and Waterman were displeased with the House version as the total cost to NSF could be as much as $100 M for medical education alone and would dwarf *all* the rest of the NSF's first real working budget. And, of course, the AMA opposed anything in the medical *education* arena that smacked of federal intrusion.[75]

Finally, Waterman had a hard time finding a head for the division of medical sciences. One individual after another turned down the job offer including, most significantly, James A. Shannon, who by 1955 became the Director of NIH and "the man perhaps more responsible than any other for limiting NSF's role in medical research."[76]

Appel has devoted a chapter to the transition of the BMS to the BBS and has cited a number of other issues that contributed to the reorganization of 1975, especially with regard to biology; most of these will not be repeated here. What she did

---

[73] Ibid., Appel quoting Waterman.

[74] Ibid., pg. 48, 49.

[75] Ibid., pg. 49.

[76] Ibid. Appel details many other aspects of this story in the above cited and neighboring pages, but repetition here of all of these early activities is unnecessary to the points to be made concerning the time at and near the year 1975, this chapter's primary concern. Regarding Shannon, Appel cites "Comments on candidates for Assistant Director for Medical Research," a commentary taken from the Waterman files at NARA and dated July 17, 1951, as well as a telephone call from Dr. Detlev Bronk of the following day and mentioned in the Waterman Diary Notes in the historian of the NSF's files at NARA. The present author has not examined these.

not consider to any great degree, however, was the aspects of the "M" in BMS: the matter of what finally did become of the medical sciences.

By the 1970s, a change was looming in regard to the place of medical sciences at NSF given the existence of the NIH "empire," located only miles away, and the financial challenges created by the many crises of 1965 to 1975, among other issues. Appel noted that the:

> BMS experienced mounting difficulty clarifying its mission in the face of NIH domination of biology. [NIH's appropriations were rising from the stagnant years and, f]rom an outsider's perspective, it appeared that the biological sciences were well supported [via NIH budgets] compared to the physical sciences [supported almost entirely by NSF].[77]

The Department of Health, Education, and Welfare, which oversaw the PHS and so NIH, was in 1971 supporting "a rather surprising" 46–47% of *all academic research*. Then Acting Director of BMS, Mehl, stated (as reported by Appel) that by 1973:

> '[t]he life sciences, including *clinical medicine*, represented over 50% of federal funds to universities.' [More importantly, however, t]he problem…was that 'Congress, the Office of Management and Budget [formerly BoB], or scientists other than biologists are likely to contrast fields of science using this total base of the life sciences.'[78]

That is to say, those communities would not recognize the fundamental differences between research in clinical medicine and that in the medical sciences, nor even fully between basic and applied research; applied research was commonplace at NIH. The federal government was combining the last two into one grouping then, at any rate, for reporting purposes. Even excluding clinical medicine, NIH was still receiving 63% of 1971 federal research funds compared to NSF's meager 12% that same year! Mehl argued that biology was being biased toward medical schools and the basic biomedical sciences at such schools—vertebrate models versus plant biology and ecology, as just two examples. The very fair question was asked whether molecular and cellular research areas should be supported by NSF when other areas of biology suffered (plant science, for instance) as NIH's "empire" continued its rapid upward growth spiral. Indeed, for the first time, "NSF biologists perceived NIH as a serious problem."[79] Not surprisingly, Senator Proxmire found "duplication" of funds going to those agencies, and NSF found itself on the defensive. The duplication matter would return in a few years in a very different guise.

Taking the "M" out of BMS to create, in part, BBS was not an absolute thing in the sense that funding of basic medical research would cease once BMS expired. It would not, and has not to this day even with another major reorganization in 1991–1992. Certain basic science research at medical schools can still be funded via NSF.

[77] Appel, *Shaping Biology*, pg. 273.

[78] Ibid. Emphasis added.

[79] Ibid., pg. 274.

It may have been, as Appel has said, that the "chief impediment to NSF support of medical research was financial."[80] While that was true in that time period, over the years other factors would be added, such as all the "tensions" that arose in the "crisis" years of 1965 to 1975 and, possibly most importantly, the growth of NIH. NSF may have reasoned: why take on the "empire"? The actual funds going to basic medical research via the NSF always remained a small fraction. Indeed, Betsy Clark has opined that BMS "*never* supported the medical sciences, *per se*."[81] The NIH had become, by 1975 and after, "a big actor," and so NSF's role would be "minimal" in regard to support of basic biomedical research, even recognizing the "fuzzy" aspect of funding such research done at a medical school, she stated. Still, other issues must be taken into account. Clark has also argued that the famous Mansfield Amendment had indirect but very palpable effects on how NSF spent its funds, as Appel has also noted, in part.[82] The Amendment was passed in 1973 and modified the Military Procurement Authorization Act of 1970. The Mansfield Amendment stated that:

> none of the funds authorized to be appropriated by the act may be used to carryout any research project or study unless such project or study has a direct and apparent relationship to a specific military function or operation.[83]

The Amendment was in force for only a single year prior to modification of its language, but in that time the various mission agencies began to "define more strictly the kinds of research they considered relevant," it has been noted.[84] The NSB issued an appeal in 1974 for NSF and other agencies to increase their support for basic research in connection with their missions, and certain political figures saw the Amendment as one method of eliminating university campus disturbances, so common in the "time of tumult." It was the indirect effects of the Amendment, especially in 1974, that indicated that it was just the "tip of an apparently large iceberg" of concern about how federal agencies determined who and what to support with funds authorized to them.[85] Funding overlap between and among federal agencies became an issue to be closely monitored, and the loss of the "M" did not, as Clark has argued, "have a major impact on the type of research supported" by BMS.

Closely tied intellectually to the elimination of medical sciences, qua the term, was another feature of the BMS. As Appel pointed out in her history of the division, "one of the hallmarks of BMS was its organization." The first efforts proposed were disciplinary for both biological and medical sciences (as exemplars, genetics and anatomy, respectively), but since BMS combined both biological and medical sci-

---

[80] Ibid.

[81] Clark interview. Emphasis added.

[82] Appel, *Shaping Biology*, pg. 237, and elsewhere.

[83] As cited in Bruce L. R. Smith, *American Science Policy Since World War II* (Washington, D.C.: The Brookings Institution, 1990), pg. 81.

[84] Ibid., pg. 82.

[85] Ibid. The quotation was originally from Rodney W. Nichols, "Mission-Oriented R&D," *Science* (1971): 29, and cited in Smith. The year of Nichols' article appears to be in error in Smith, however.

ences, the notion of *functional* organization was recognized as superior and able to handle both pure biology and pure medical research in a fashion that likely would have satisfied Waterman's concerns in earlier times. Organization by "functional scheme" rather than along traditional disciplinary lines was a *significant early innovation* at NSF.[86]

The Embryologist Paul Alfred Weiss was, in 1951, Chair of the Division of Biology and Agriculture of the National Research Council.[87] It was he who suggested the plan for functional organization in NSF's BMS division, a good example, by the way, of informal external advisors to NSF, much seen over the years. This came about as Weiss:

> hoped to position the division to advise federal agencies in the postwar era by creating a set of committees to cover all of biology…[and thus] saw his ordering as a means of reintegrating the field.[88]

The brilliance of this schema cannot be overstated in the history of the biological sciences at NSF and, in later years, in the history of biology in the academy. *Integrative biology* and related matters constitute much of the discussion in later chapters in this study. While academic departments concerning the areas of the life sciences all across America remained divided along disciplinary lines (a course in zoology, a course in botany, a professor of bacteriology, etc.), and would remain so for many decades to come, NSF, and BMS in particular, blazed a new trail that, to this day, takes the life sciences at the functional level and less explicitly at the disciplinary. The academy did not follow suit until, as Appel has claimed, "the late sixties." However, that is not entirely correct, as most major (and still today only some smaller) colleges and universities have taken the step of reorganizing along functional lines. Indeed, Appel even cites works noting that the trend became much more obvious from the 1990s, not her own posited 1960s.[89]

Such an arrangement as that of Weiss' for the NSF/BMS allowed for all sorts of disciplinary integration with regard to program creation, funding of attractively different sorts of individual grant proposals, and so forth. This will be seen often throughout the present history. Given the historical period in which Weiss saw the wisdom of this approach—now seven decades ago—it can only be called visionary in the first degree.[90]

---

[86] Appel, *Shaping Biology*, pg. 63.

[87] See http://www.nap.edu/readingroom.php?book=biomems&page=pweiss.html for details on Weiss. In Chap. 5 of the present history, there is a mention of Weiss again in a footnote with regard to GST, general system theory, and the use of the name Biotic Systems and Resources Division of what was then BBS. See also the text of that chapter.

[88] Appel, *Shaping Biology*, pg. 63.

[89] Ibid., for her quotation. But the notion of the functional restructuring trend is discussed—and whom she lists—by Wade R. Roush, "Biology Departments Restructure," *Science* (1997):1556–1558. More changes by far occur well after the 1960s, as Appel has stated on her pg. 275.

[90] Ibid., with regard to the central text consulted in Appel. The comments about Weiss' insufficiently heralded thinking are my own. See, for instance, his obituary in the *New York Times*, pg. B10, September 12, 1989. See also another of his obituaries by Jane Overton, "Paul Alfred Weiss," *Biographical Memoirs*, National Academy of Sciences, (1997):372–386.

The upshot of the matter was that Weiss recommended that the BMS division be arranged into areas of molecular biology; cellular biology; genetic biology; developmental biology; regulatory biology; and group (i.e., taxonomic groups) and environmental biology (the latter two as a single entity). In July of 1952, the early leadership in BMS decided upon a slight modification of Weiss' groupings. They divided biology into seven areas: molecular, regulatory, developmental, genetic, environmental, systematic, and psychobiology. A slight but important change from Weiss was made with regard to the more "macro" end of the life sciences (here systematics and environmental), and in the recognition of fields of behavior and related areas in the functional category of what was termed psychobiology; much of which would be behavioral biology. This last, too, would have major impacts in both the 1975 and again in the 1992 reorganizations.

Also, by the fall of 1974, Betsy Clark was fairly pleading to have the word "Medical" removed from the BMS title.

> In the committee reorganizations recently undertaken by [and for] Congress, jurisdiction for research in our Division will pass to the Commerce and Health Committee—the Committee that is also responsible for the NIH budget. I do not know in detail how this will affect our budget presentations and hearings. Because of Congress' traditional concern with 'health,' the change, conceivably, could be advantageous for our research effort.
>
> In view of the potential problems we might confront in distinguishing our research from that of NIH, and in view of the long-standing desire on the part of the Staff of the Division to drop 'Medical' from the Division name, we respectfully request that we become the Division of Biological Sciences.[91]

While Clark's suggested divisional name did not come to pass, of course, just 1 year later, the "M" was, to the satisfaction of many, gone. There was nothing that existed after the 1975 restructuring that could be distinctly alone termed medical, medical research, or medical science: they, with biology, had been much integrated, and, among many other advantages, NSF carefully avoided NIH's territory.

It was evident that for the fiscal years of 1968 through 1978 the NIH "empire" was the clear-cut leader in funding *health R&D*, hardly a surprise. The crucial data were in FY68 NSF contributed over $21 M to health-related R&D, and that rose slowly over the next several years (but with inflation not accounted for). At the same time, the FY68 figure for NIH was $864 M—evidence for the term "empire"! And in subsequent years, the figure for NIH rose at a rapid rate to hit just under $2.6 *b*illion estimated for FY78. At the same time, NSF was only contributing $65 million.

Indeed, the federal R&D support for health had literally doubled within the years from 1969 to 1976. It had gone from $1.1B to $2.2B—an annual average rate of 10%—and after adjustment for inflation, it was steady at a real rate of 3%, inflation being so very high at the time. Of particular note, "*biomedical research* has grown at the fastest rate of any broad health R&D program areas." By 1976, it made up 90% of federal R&D support for health.[92]

---

[91] Division Director, BMS [Clark], to Assistant Director for Research, "Subject: Name Change for BMS," October 18, 1974. All single quotation marks and underlined emphases are in the original.

[92] NSF, note in *Mosaic* (1976): back cover.

An intriguing problem arises, however, with the loss of the "M" and the funding pattern before and after 1975 and the appearance of BBS. Even though the BBS did not mention basic medical research in any overt way, such as did the prominent "M" in BMS, there was only a slight drop in funds provided to the total health-related research enterprise in federal funding the year BBS came into existence, and then it rose again after that year. That is, FY74 ($47 M) had been the next year in a continuing, but slow, increase in funding since 1968. In FY75 it dropped to about $45 M, for the first time in at least the 8 years. However, by FY77 it had rebounded to just over $55 M. (Data are unavailable for FY76.) A funding drop in connection with the loss of the "M" is probably not surprising, but how is the continuing rise *after* FY77 and the estimated FY78 contribution of even more money explained?

That has remained something of a conundrum. Since its inception to the present day, the NSF and the NIH have maintained a relationship, sometimes notable, at other times very minimal. In the historical files of the National Library of Medicine (NLM), a unit of NIH, there exists a set of files that covers the years 1948 to 1975.[93] This file for the "National Science Foundation Special Committee on Medical Research" ends in the very year of interest, 1975. That is significant in itself as BMS had become BBS and, it is suggested, the committee seemed no longer to offer the value it may once have had to the two agencies concerned. The committee was long chaired by Joseph M. Pisani of NSF and was most active during the mid-1950s accomplishing various studies on federal programs in health activities managed by agencies of that period. The NLM archives suggest nothing about the nature of the funding for medical research provided by NSF itself to the overall health-related R&D national budgets, how it was used, for what sciences, nor any other enlightening information, and especially not in or near the year of 1975.[94]

Guy Stever was thinking a lot about the 1975 reorganization prior to that calendar year though as, by late January, he had the need to respond to a request from Ronald W. Lamont-Havers, then Acting Director of NIH. Lamont-Havers had made a request to Stever that led the latter to say:

> In response to your request of January 3 that I name a National Science Foundation representative to the Diabetes Mellitus Coordinating Committee, I question the need for formal representation from the Foundation. Although some research supported within the [BMS] may have a bearing on basic mechanisms in endocrinology, genetics or metabolism [all of which could be termed basic medical sciences' research] which relate to diabetes mellitus, it is neither strongly oriented toward diabetes mellitus as a health program nor extensive…

---

[93] See http://oculus.nlm.nih.gov/cgi/f/findaid/findaid-idx?c=nlmfindaid;id=navbarbrowselink;cginame=findaid-idx;cc=nlmfindaid;view=reslist;subview=standard;didno=nsf.

[94] The author had extensive email discussions with historians of medicine at the NLM and during several visits and examination of the Committee records through 1975 and could discover no evidence that some agreement had been made between NSF and NIH concerning the elimination of BMS (especially the "M") and its rebirth as BBS. Paper copies of emails and location information (see footnote just above) are on file with the historian of NSF in connection with this present contracted history, as well as copies in the author's files. He can be contacted at donaldjmcgraw@icloud.com.

[and so] it is doubtful that such representation would contribute significantly to the specific objectives of the [NIH] Coordinating Committee.[95]

It would appear that while, as Appel has said, financial considerations were possibly paramount in considering how much NSF would contribute annually to health-related research from the earliest period of NSF's life, especially when comparing it to what Congress awarded the NIH to expend, the "crisis" components of 1965–1975, and other issues noted above, may be joined by what appears to be Guy Stever's desire to downplay the "M" on the way to his soon-to-be-proposed BBS. Certainly, he was aware of Betsy Clark's feelings and that of her division staff. It appears that he was seeking a lesser and lesser role in biomedical research for NSF compared to NIH. Disease research (diabetes mellitus, here) was to be left to NIH, as was mandated by law anyway.

Stever went on to say in the letter to Lamont-Havers that Clark would check into the matter of what kinds of grants and programs might exist concerning medical sciences' research applicable to diabetes mellitus. Indeed, she did just that, and in her letter 10 days later to Stever, she provided a listing of all possible projects then operating that *might* pertain to diabetes mellitus. She was cautious in saying that "we think [these projects] have *some relevance* to diabetes mellitus."[96] In her appended listing, she cited a total of some 24 projects running from as little as $15,000 over 2 years to as high as $72,000 for the same time frame, though the latter was atypical as most were in the $30–40,000 range. All but one was for 2 years; the single exception was an 18-month award. None of the projects' titles contained the words diabetes mellitus, but that was in keeping with the notion that the researchers were doing "basic medical sciences' research" and, simply enough, it was good politics to avoid mentioning a specific disease—the pretty much exclusive province of NIH. Furthermore, if the researchers hoped to win these awards from NSF, they would have chosen to exclude a disease name from their proposed titles or would have been coached by NSF (BMS/BBS) to have done so. If they wished a disease name in the title, they could submit their proposals to NIH vice NSF. Some of the titles to be found in Clark's listing were of projects (though maybe not all 24) that were obviously of value to anyone working in diabetes mellitus research, for instance, "Mechanism of Insulin Secretion" (University of Connecticut; $52,500) and "Regulation of Gluconeogenesis in Crustacean Skeletal Muscle" (University of Alaska; $40,000) as just 2 of the 24 projects listed. But these projects also had importance in other areas. The Alaska study, for example, had applications to general biology, arthropod biology, fisheries, biological oceanography, and probably several other basic sciences totally devoid of a medical connection. Still, such connection may well have come out of the research efforts and so was fundable under

[95] H. Guyford Stever to Ronald W. Lamont-Havers, January 21, 1975. O/D Central File, Comm. 2-8-B. Copy used by author now on deposit with the NSF historian and author; see fn above.

[96] Betsy Clark to Ronald Lamont-Havers (cc to Stever), January 31, 1975. O/D Central File, Comm. 2-8-B. Emphasis added.

the rubric of "basic medical sciences' research" if, indeed, that particular grant was so handled.

The loss of the "M" then was a complex issue, financial considerations being one part, but requirements in the organic act in another (although an actual level of funding for medical science was never promulgated precisely in the Act), as were the many crises of 1965–1975, the continuing and massive growth of the NIH "empire," and the structural problems in BMS to which Stever had alluded and that contributed to the birth of the BBS.

Yet another matter was how one counted funding for basic medical research. That is to say, as NSF went on to support projects in the areas of interest to both science and medicine, the oft-mentioned bacteriology, anatomy, pharmacology, etc., could other areas have counted in the total annual contributions by agency to medical research? Could, for instance, some other part of NSF's activities be included along with what biological sciences offered? It is believed that the answer is yes as circumstantial evidence has arisen to suggest so, though it remains inconclusive. This is so because how categorization was done in the 1970s is unknown and so the following represents an informed estimation of how such categorization was likely done.

Without further sure knowledge, the continuation of funding of monies in basic biological sciences and applied medical research in engineering (and other areas?) to the overall annual total health-related R&D effort in the country likely included a highly mixed bag when NSF was listed among contributing agencies (along with such examples as the Department of Defense contributions). And, crucially, the NSF was charged with putting together the annual report for Congress on health-related funding provided by *all* the federal agencies (NIH included!), so it would have been quite simple to determine just which grants and programs the NSF would choose to list in the contributions that it felt were basic medical research monies.

For instance, the engineering groups at NSF funded studies on instrumentation that could or did have direct medical applications though, once again, it is unlikely that the grant title language would have included anything overt, vis-à-vis medicine. In correspondence with the author, Mark Courtney suggested that engineering might indeed have been supplying monies, via its funded grants, that might have application to medicine. In proof of that, Gil Devey, then at NSF in engineering, told the author that, for instance, "support for biomedical activities in Engineering at NSF began circa early 1964 and continues until now [mid-2008]." He went on to say that:

> [a] most significant single project was [a]...survey of R&D and commercial production of diagnostic sonographic instruments. This resulted in a 42-page report published in November 1973 that included a target specification for such a device.[97]

[97] In a long series of email correspondence between the author and individuals in the then (mid-1970s) BMS/BBS and the then engineering division, one email and one live conversation are most important to this conclusion: Mark Courtney's oral conversation with the author on February 10, 2009, and an email to the author from Gil Devey (engineering division) of May 8, 2009. See G.W. Stroke, et al., *Ultrasonic Imaging and Holography: Medical, Sonar, and Optical Applications*

These instruments are used routinely today to study an embryo/fetus non-invasively in a gravid woman—easily considered as human medical diagnostics, if not research.

It should be considered that Betsy Clark did not think that the engineering argument made here *is* the likely answer to the continued level of funding to medical research seen after 1975–1976, though. She has stated, "NSF supported a whole bunch of scientists who were in medical school[s]."[98] She gave as an example the work done by Arthur Kornberg (later below) at the Medical School of Washington University in St. Louis, Missouri, which was very basic research in DNA replication. If, as she has averred, NIH were to have reported NSF funding for Kornberg as a faculty member at a medical school, which would have been likely, then it would seem that NSF was supporting medical research, *per se.*[99]

In Guy Stever's long and literate "Director's Statement" in the NSF's *Annual Report 1975*, he mentioned contentious areas such as RANN; he mentioned the social sciences and their continuing challenges; he mentioned science education; and he mentioned the physical sciences and the biological ones, as well. He did not pen the word medical even once.

In the crucial year of FY74, that is just before the "M" disappeared from the division-cum-directorate for the biologically related sciences, the total monies spent on *health-related* research and development for the then Department of Health, Education, and Welfare, was some $2.75B. (Total R&D for the full government in *all* matters was just under $17.5B.) The share spent by NIH was just under $1.75 *b*illion, while that spent by NSF was only $47 *m*illion of all health R&D.[100] NSF's contribution to total R&D that year, health included, was a little over $556 M, so the $47 M was on the order of less than 9% of NSF's budget. And it must be remembered that that was *basic* medical research on, as an example, bacteriology that might be useful to medicine as much as to general microbiology. The same may be said for anatomy, physiology, pharmacology, and a number of other disciplines. How much of the research was, in the very purest sense of the term "basic medical research," may not be knowable. And, further, the clarity of the budget language was such that the NSF directors only very loosely tied any one pot of money for medical

(New York: Plenum Press, 1974); this is the book form of the "Proceedings of the United States-Japan Science Cooperation Seminar on Pattern Information Processing in Ultrasonic Imaging," 1973. There is also some very interesting commentary on the Proceedings at http://www.ieeeghn.org/wiki/index.php/Oral-History:George_Wilhelm_Stroke#NSF_Committee_on_Ultrasonic_Imaging.

[98] Clark interview.

[99] Ibid.

[100] Data are derived from a table prepared by NSF as part of its organic act obligation to provide information annually with regard to federal R&D programs. See the letter from Charles A. Falk, director of the Division of Science Resource Studies of NSF, to Dewitt Stetten, Jr., Deputy Director for Science at NIH dated July 29, 1977, and its appended tables. A somewhat similar chart can be seen in NIH, "Basic Data, NIH 1975," a small annual handout booklet made for general use by NIH and given to interested parties. Other similar tables from a variety of sources at NSF and NIH were also checked for accuracy of all these figures.

research to clearly identifiable goals, for example, research on blood substitutes for war or other emergency use, basic knowledge in nutrition and biochemistry, etc.[101]

The research community learned, over the years, just what kind of research project proposals to aim at NSF as compared to the types more likely to be funded by the NIH. Also, to the present time, both NIH and NSF will suggest to possible grantees the wisdom of seeking funds from the other of these two agencies if the project appears to fit one better than the other.[102] And, so while support for basic medical sciences research never actually went away from the days of the BMS to those of the BBS, it was downplayed in the arena of titles for programs, offices, divisions, etc. Nonetheless, grants could be sought that had a strong applied connection (in any area, not just the biological sciences), and could even state in the body of their respective texts the possibility of the research findings having application to human medicine, but the titles were always such that they were written to best fit the expectations of NSF or NIH, whichever was appropriate. Grants funded by NSF for FY75 in the biological sciences totaled some 1760 summing to nearly $78 M; medical aspects were not mentioned in the official documents of the time.[103]

The loss of the "M" then helped determine the nature of the newly formed BBS, but another area would still plague many in the biological sciences that might have wished that it had gone the way of the "M"—but they would have to wait more than another 15 years.

1.4.1 The Social Sciences: First Thoughts

In contrast to the disappearance of basic medical sciences as part of what was moving from BMS to BBS, the long-lived social sciences' "Office" survived to become a division of the new BBS Directorate. This was because of one man in particular, sociologist Harry Alpert.[104]

Alpert, who had been working on public opinion and statistics for several federal agencies (including the Bureau of the Budget) during the war years, came to NSF in the beginning to work with its Program Analysis Office. He was set to the task of studying the role of NSF in the social sciences and the then current status of those disciplines. Vannevar Bush had not done anything along these lines as he worked to create the NSF, but did favor the kind of work Alpert undertook and told Congress as much.

---

[101] Appel, *Shaping Biology*, pg. 51.

[102] Conversations between the author and numerous NSF employees from October 2008 and May of 2009.

[103] NSF, "Annual Report 1975," Table 1, pg. 4.

[104] Alpert has been discussed in a recent article. See Mark Solovey and Jefferson D. Pooley, "The Price of Success: Sociologist Harry Alpert, the NSF's First Social Science Policy Architect," *Annals of Science* (2011):229–260. I thank Mark Solovey for providing a copy of his and Pooley's paper.

Alpert, politically astute, often used Bush's words to Congress in support of the social sciences when "defensive action" of his discipline was called for.[105] For example, he borrowed Bush's words that the social sciences required an "effective integration and partnership" at NSF. He also borrowed language from then Board chair, Chester Barnard, that only the "hard science core" of the social sciences be dealt with at NSF. Conservative congressmen might believe that the social sciences were not "pure" sciences, and so Alpert calculated that use of the hard sciences phrase would fend off any untoward attacks.[106] While the strategy was excellent, the results varied markedly over the years.

Alpert spent a year preparing his report on how NSF should integrate the social sciences and what kind of program they should undertake to meet the expectation of the organic act, they having to be hidden within what became the infamous "other sciences" category of the organic act. Wisely, Alpert suggested that programs and budgets for the social sciences be "pursued cautiously and experimentally" for 3 years. He also suggested that whatever was undertaken it be "methodically [methodologically?] rigorous" and characterized by "objectivity, verifiability, and generality," as England cited Alpert's recommendations.

England made clear the objections of scientists outside the social science fields by stating that:

> [Alpert's rationale] was not unlike the one natural scientists used in justifying their own *basic* research programs. Instead of talking about the cultural value of knowledge, they often spoke of its practical social utility—usually unanticipated, sometimes long deferred, but nonetheless certain.[107]

This made it easier for the natural science community, as well as members of Congress, to appreciate that *basic* research could be and was done by social scientists as much as by natural scientists and that social science was not, *per se*, only an applied collection of activities. Few scientists of any stripe failed to recognize that a great deal of basic research may one day acquire applied utility.[108] And few would have been opposed to such an eventuality, then or at any other time.

---

[105] England, "A Patron of Pure Science," ppg. 266–267.

[106] Ibid., pg. 267.

[107] Ibid. Emphasis added.

[108] Generally held views in the broad community of science would be found to disagree with this conclusion; however commentary does exist in this area. For example, "There has been a noticeable shift in *philosophy* regarding the types of research receiving federal funding in recent years. Universities get much of their money from the [NSF]. Research at the [Lawrence] Berkeley National Laboratory [LBNL] is funded primarily by the Department of Energy and the [NIH]…[In the case of the Human Genome Project] in order to gain a better understanding of the nature of chromosomes, simpler forms of life (e.g., fruit flies, nematode worms, and yeast cells) have been extensively studied as part of the Human Genome Project. One ultimate *goal* of this ambitious program is to be able to cure genetically-caused illnesses such as Cystic Fibrosis and sickle-cell anemia through new medical techniques such as gene therapy…*Industry* does little basic research today. Due to the competitive nature of the business world, commercial research tends to emphasize projects requiring *less than 10 years* to develop a new product or process. Businesses simply cannot afford to engage in long-term research projects. As a result, *universities* and *government*

Alpert's proposal that the initial program and budget be modest represented clear thinking about the nature of social sciences at the time. The Ford Foundation was one of the primary philanthropies supporting social science studies, unlike the federal system, but asked that they be *problem-oriented* as opposed to *research-oriented*, a circumlocution, it would appear, to avoid the term applied. Alpert's view was to avoid supporting research on subjects that would attract what could become negative public attention—such as research on sex or politics.

Appel found, for instance, that there had been much stability in the structure of the NSF during its first two decades, especially for the biological sciences. Minor infrastructure changes were not uncommon, but the need for a major reorganization did not become evident until the early 1970s. It became more and more compelling by 1975, given the many crises during the decade of "tumult." The "tensions," as Stever very properly called them, were mounting and putting considerable pressure on the entire NSF.

Once the Daddario-Kennedy amendment was passed in 1968, it was fully legitimate for the NSF to fund applied research. To understand how the Foundation supported any sort of science meant looking at how NSF was organized to carry out the new mission, as well as how it responded to the many other public and federal pressures.

> NSF had begun as an agency solely concerned with undirected basic research, which it claimed to be the foundation upon which applied science and technology rested…[After "Daddario" t]he research divisions, while continuing to fund only basic research, were forced to justify new funds by focusing on 'emphases' that were based upon specific *practical* needs. *The history of biology at NSF after 1975 would therefore incorporate an entirely new element, namely NSF's relation to the growth of new* industrial biotechnology.[109]

This argument by Appel is confirmed in many ways. Not least of these is by Stine in his history of federal science policy written for the Task Force on Science Policy.

> *Nothing* heightened the debate over Government regulation of basic research more than the advent of recombinant DNA (rDNA) research during the early 1970s.[110]

This statement was the last in the series of such statements tied to the 1965–1975 "crisis" period that Stine put forward as having played a role in the 1975 reorganiza-

---

laboratories (such as LBNL) are left with the responsibility to carry out basic research and long-term applied research." Emphases in original. See http://www.lbl.gov/Education/ELSI/research-main.html. The view in the scientific community is not restricted to the United States, either. As another example, "'What I do, when I don't know what I'm doing.' That's basic research, according to the German Physicist Wernher von Braun. And applied research? Is that what one does, when one knows who one's doing it for? Many people think Germans love to organize more than anything - to put everything into tiny little boxes, figuratively of course. It's a cliché, certainly, but if this tendency to pigeonhole really did apply to one area, it would be around the question of 'basic' versus 'applied' research. The borders between the two are really more fluid than ever, if there ever were firm borders at all." Quotation taken from http://www.study-in-germany.de/english/print/2.60.261.html.

[109] Appel, *Shaping Biology*, pg. 269. Emphasis added.

[110] Stine, "History of Science Policy," pg. 69. Emphasis added.

tion. He had already provided a list of difficulties from the period which included Vietnam and the anti-war movement; the Great Society and "socially relevant research"; the environmental movement; the rise of the Sierra Club, the Environmental Defense Fund, the Natural Resources Defense Council, and Friends of the Earth; debate over the supersonic transport airplane; two studies (Hindsight and TRACES, not discussed in the present history) on the role of basic research in technological innovation; the Mansfield Amendment of 1969, which defined the military's relationship to basic research; Nixon's anti-science actions; the energy crisis (including the oil embargo of October 1973, which stressed America severely); Nixon's "war on cancer"; and RANN.[111]

It is no surprise that Stine chose to end with what he felt was a final precipitating factor in the changes NSF knew it would have to make to withstand both federal and public demands in regard to doing science in America: biotechnology. Much more will be said about the social sciences in Chap. 2.

1.5 Biotechnology's Birth and the 1975 Reorganization

In January of 1973, the first Asilomar Conference took place. A small town on the Monterey Peninsula of California, Pacific Grove, has a conference facility named Asilomar, which hosted a large group of biological scientists that had determined they would *self-regulate* experimental research in recombinant DNA, then burgeoning rapidly. Biologists Paul Berg, Maxine Singer, and Richard Roblin organized the conference.

> Berg was a biochemist who discovered the mechanisms by which DNA and RNA (ribonucleic acid) direct the synthesis of proteins in living systems [and in] 1972 he and his colleagues at Stanford University synthesized the first recombinant DNA (rDNA).[112]

Recognizing the "potential physical and ethical hazards" posed by this new technology, Berg, with his like-minded allies Singer and Roblin, organized the first Asilomar Conference to discuss this danger and to push for self-regulation within the biological sciences community. Berg went on to win the Nobel Prize for his work in protein synthesis and rDNA in 1980. (More will be said in several following chapters about biotechnology and its industry.)

Already by the later 1960s, major discoveries had been made with regard to understanding DNA, RNA, the genetic code, and the many enzymes that catalyzed the reactions of these genetically important molecules. But, "if viruses could be created in a test tube, what would happen if the synthetic creations turned out to be pathogenic [disease-causing] for humans or animals?"[113] The public had already been introduced to these imaginable potentials through the pages of popular writer

[111] Ibid., ppg. 57–70

[112] http://profiles.nlm.nih.gov/CD/Views/Exhibit/narrative/biographical.html.

[113] http://profiles.nlm.nih.gov/CD/Views/Exhibit/narrative/dna.html.

and physician Michael Crichton's novel, *The Andromeda Strain*, of 1969—and many more people would become aware of it when the Hollywood movie dramatized it further in 1971—merely a year before Berg's first success with creating only too real recombinant DNA, which had been the theme of both the book and the film. "Public fears were fed by science fiction scenarios such as…*The Andromeda Strain*."[114] Once more, this sort of "tension" in the greater society (which neither Stine nor Stever considered) had had a compelling effect in this country. The power of popular culture on science policy can often go unnoticed.[115]

A group of 150 scientists turned out for the first Asilomar Conference early in 1973, and some 16 journalists covered the 4-day event. There had already been in effect a "voluntary moratorium" on rDNA research, and the self-regulation decision was based on the notion that the research could continue under "strict guidelines." The recommendations from the Conference attendees were sent to then NIH director, Donald Fredrickson, whom they charged with "formulating those guidelines." NIH responded and, in June 1976, released them (below).[116] The NIH was the primary financial supporter of both basic and (what little then existed, applied) rDNA research at the time. The guidelines dealt with several classes of experiments, most notably the so-called shotgun ones in which genes were inserted into the intestinal bacterium *Escherichia coli* (*E. coli*) randomly from other organisms, such as viruses. No one could know in advance what kind of organism might materialize from a shotgun experiment: it might turn out to be a new pathogen.

For this reason, the potential danger levels of the experiments were defined along with the level of physical containment required: P1, being of little worry and which could be done in a typical microbiology laboratory with the routine precautions always taken by practicing microbiologists or students, up to P4, the highest level and most contained in special physical facilities built to purpose. This general set of practices remains in operation to this time but has undergone a number of revisions as dictated by advances in our understanding and in technology.[117]

The NIH also decided to establish the Recombinant DNA Molecule Program Advisory Committee in October of 1974—the "RAC," as it became known when the name was shortened to Recombinant DNA Advisory Committee.

A second Asilomar Conference took place in February of 1975 and was financially supported by NSF and by NIH's National Cancer Institute. The 155 invited

---

[114] Ibid.

[115] Donald J. McGraw, "Hunting Live Dinosaurs in the Grand Canyon in 1937: On the Role of the Popular Press and the Advancement of Science." In *Proceedings of the Second Conference on Scientific Research in the National Parks*, Vol. III: 235–252 (USNPS-NTIS: Washington, D.C., 1980; 12 vol., microfiche). There are many such instances in the literature. For example, see also the origin of the newer term "cyberspace" which derived from William Gibson's 1984 novel *Neuromancer* (http://seedmagazine.com/content/print/whe_science_asks_what_if).

[116] Ibid., website reference pg. 2.

[117] See especially Maxine F. Singer, "A Summary of the National Institutes of Health (USA) Guidelines for Recombinant DNA Research," *Gene* (1977):123–139. Singer's article is also available online at http://profiles.nlm.nih.gov/DJ/B/B/S/G/_/djbbsg.ocr. The current, i.e., 2010, guidelines can be found at http://oba.od.nih.gov/oba/rac/guidelines_02/NIH_Guidelines_Apr_02.htm.

guests included not only scientists but also leading individuals from industry, the press, government officials, university administrators, lawyers, and others. It was that meeting which led to the release, in 1976, of the first incarnation of the guidelines.

In providing background for the Task Force on Federal Science Policy, Stine, citing others, called the rDNA controversy "an unprecedented chapter in the annals of basic scientific research and the governance of modern science as a social institution." And he saw the move from science led by physicists to one where a "new group of players—for the most part biologists"—were important in formulating science policy.[118]

There were yet other precipitating factors for the 1975 reorganization. Appel noted that:

> [t]he distinction between basic and applied research, which NSF had long depended upon had become increasingly murky. Retrospective studies of the genesis of technological breakthroughs, such as the military's Project Hindsight reported in 1966, showed that practical achievements did not always depend on prior basic research but often rested on prior technology.[119]

Indeed, "science and technology were in a complex dialectical relationship. Technology could redirect science as well as the other way around," Appel continued.

A crisis had been building in the BMS division that was peculiar to it specifically, along with the now long list of precipitating factors driving toward reorganization of all of NSF in 1975. The BMS had a long-standing advisory committee from the broader science sector that was terminated in 1972 as a casualty over efforts "to make the NSF operations more fair and open."

> For two decades, BMS had functioned as an informal network in which program officers—almost entirely white men—cultivated friendly relations with grantees and [graduate student] advisors. With little constraint, they had selected acquaintances as panel members[,] white males from major universities…Reviews of proposals remained strictly confidential… By the early 1970s…women, African Americans, and Hispanics, protested…[the] old-boy network…and the NSF hierarchy found itself on the defensive.[120]

This casualty problem's proximate cause was, ironically, the appearance on the scene of Nixon's view that some sort of "sunshine legislation" was needed in government that would make its activities more transparent to the public. While the act did not come out until September 1976,[121] the threat was there from at least February 1972, when the Advisory Committee had met in a tense atmosphere. Appel described a growing "hostile tenor" between the BMS and its Advisory Committee over issues such as those just cited, and the notion of opening the meetings to let the sun shine

[118] Stine, "History of Science Policy," ppg. 69–70. Stine was citing Judith P. Swazey et al., "Risks and Benefits, Rights and Responsibilities: A History of the Recombinant DNA Research Controversy," *Southern California Law Review* 51 (1978):1019–1078; see pg. 1019 therein.

[119] Appel, *Shaping Biology*, pg. 269.

[120] Ibid., pg. 270.

[121] P.L. 84–409, 94th Congress, S. 5.

in was not well received by the NSF staff. So, "in order to protect the sanctity of panels to hold closed meetings," the NSF eliminated committees at the division level and supplanted them with ones at the level of the Assistant Director for Research, "whose meetings could safely be made public." The problem for the BMS was that, after some 20 years of having an advising structure, no "NSF advisory group represent[ed] all of biology," as Appel reported.[122]

And, even more, BMS had become a "jumble of disciplinary programs…and functional programs…with overlapping jurisdictions…[leading] BMS staff [to agree] that this subdividing of biology needed a *thorough rethinking*...as [the] BMS [structure] did not reflect contemporary biology."[123]

To add to all the commotion of the times, Senator William Proxmire, created what he termed the "Golden Fleece Award." From March 1975 until December of 1988, Proxmire chose a number of government-sponsored programs, research projects, grants, etc. that he felt "fleeced" the government (i.e., the people) out of the "gold" (taxpayers' money) that should not have been spent in a manner he felt was wasteful. Most regrettably, he chose the National Science Foundation to receive the very first of those awards that March in the fateful year of 1975. That particular fleecing, he announced, was for the $84,000 study by Elaine Hatfield, and colleagues, of the University of Wisconsin, Madison, about why people fall in love. For the next 13 years, Proxmire was a public critic of the NSF's biological and social sciences programs, especially. He hoped to "galvanize public opinion against wasteful spending"; he did, but unfairly so as often as not. And, among all things, Proxmire was chair of the appropriations subcommittee that included the NSF's budget and finances from late 1972 onward for a number of years![124]

The Golden Fleece "Awards" were wildly popular with the disgruntled populace who had been suffering the problems of Vietnam, and all the other crises of 1965 onward in so many areas of daily life.[125] While the awards were given to any govern-

---

[122] Appel, *Shaping Biology*, pg. 272.

[123] Ibid., pg. 271. Emphasis added.

[124] http://www.taxpayer.net/proxmire; the site is no longer in existence. There is a photocopy of the two-page website in the National Archives and Records Administration II (NARA II) collection on the history of NSF. This NARA collection is spotty, at best, and covers the years from the founding of NSF to about 1995. Those of note to this study are to be found in Stack 130, Row 37, Compartments 15–17. Compartment 15 contains Shelves 1–7 with Boxes 1–21; Compartment 16, Shelves 1–7 with Boxes 22–42; and Compartment 17, Shelves 1–4 with Boxes 43–53. These 53 boxes make up the whole collection of NSF documents, which are held by NARA II in College Park, MD. The NSF has a policy that when records reach 10 years of age that are held in their headquarters in Arlington, VA, they are to be transferred to NARA. This is not a perfect system, however. Boxes of particular interest to this history are numbers 11, 13, 24, 30, 31, 49, and 52. Early history of the "Golden Fleece" is in Box 49, 1982 Target Files and 1983 Target Files. Hereinafter, citations concerning NARA II will be given by the formula, NARA Box [number], File [name], as Stacks and Rows do not change for NSF materials, and finding the proper Compartments and Shelves can be done by consulting the information earlier in this footnote.

[125] Elaine Hatfield, "The Golden Fleece Award: Love's Labours Almost Lost," *APS Observer* (Association for Psychological Science) (2006); online version at http://www.psychologicalscience.org/observer/getArticle.cfm?id=2001. Hatfield recounts a contest run by the *Chicago*

ment agency that came under Proxmire's scrutiny, they were mostly aimed at those that Proxmire thought the public would agree with concerning truly wasteful decisions. High cost, in actual dollar terms, was not a requirement. He gave one to the US Navy that awarded $792 for a "designer doormat" and had, as his personal favorite, the "study to find out whether sunfish that drink tequila are more aggressive than sunfish who drink gin." As shall be seen on a number of occasions, these decisions could very often be defended quite easily: Proxmire did not inform the public about the *logic* behind many of these funding decisions. At the end of his career in Congress (1989), the Golden Fleece Awards passed into history.[126] The general apprehension concerning funding decisions at NSF—and elsewhere in government—did not pass away quite so soon, though.

1.6 The Biological Sciences After the Reorganization of 1975

The Assistant Director for Research, the ADR position (Fig. 1.1), was eliminated, and the funds, management of them, and all related decisions went into a series of new directorates, each headed by an assistant director (AD; universally pronounced "A-dee") appointed by the President of the United States. Each AD had direct access and reporting responsibilities to the Director of the NSF.[127] Each directorate had personnel in a sort of business, or "front office" capacity to handle the actual funding transactions peculiar to that directorate. It was the same in every directorate, whatever the subject area.

Many reasons for this reorganization have been elucidated above, but one more factor that Appel had turned up in her research was the fact that the AD for Research "had acquired too much power."[128] And given the long history of the NSF having been led by physical scientists and having long and strongly supported those disciplines, it is not entirely surprising that the argument was made that those fields had been "favor[ed]."

The BBS came into existence, and the BMS died on July 10, 1975, with the release of Stever's memo to all concerned that the NSF had undergone a major reorganization. The word "Medical" disappeared and the term "Social Sciences" appeared. That memo, too, was the only NSF-wide document to appear from the reorganizational efforts, unlike the long and complex build-up with a great cast of players and events that would characterize the 1991–1992 reorganization (Chap. 5).

Tribune in which call-in votes were requested to see which of the two, Proxmire (and his belief that the public was being fleeced) and Hatfield (and her work on the subject of the grant), was "right." The number of votes called in was "a massive number," and the upshot of it was that Proxmire won by 87.5% to Hatfield's 12.5%!

[126] From the two-page photocopy referenced in footnote 92 above.

[127] The term AD remains in use to the present day, though they are no longer appointed by the President and confirmed by the Senate.

[128] Appel, *Shaping Biology* pg. 275.

It is for this reason that Guy Stever was said by some to have been "the only player in the action" in 1975.[129]

But the view of "only player" for Stever's role did not fit the biological sciences for, as earlier noted, Deputy Director of BMS, John Mehl, had written the first white paper on how that division might be reorganized as early as April 1971. In September 1973, Clark wrote a memo to the full BMS staff in which she averred that:

> AD/R [Assistant Director for Research—the division soon to disappear] has suggested that the BMS [budget request] write-up for Congress is likely to be more effective if we adopt the format used by MPS [Division of Mathematical and Physical Sciences] in which sections [an organizational category] do the presentation. This raises immediate, intermediate, and more long-range problems: 1. By October 15 we must present new section titles…*I favor three sections.*[130]

Clark went on in her memo to describe what she wanted from the personnel in BMS. What transpired was a series of responses to Clark and to Mehl in late 1973. The latter seemed to have been coordinating the effort that eventuated into the three biological divisions within the directorate, and a fourth, which covered the social sciences.[131] Within 2 days of Clark's memo, the Associate Program Director for the Metabolic Biology Program, Cecilia W. Spearing, was on top of the matter by offering an organizational structure that had, in one variant or another, been on the minds of many since 1971 and definitely since the memo series of 1973:

[129] Otto Larsen in his *Milestones and Millstones* hints at this by stating "Out of the anxiety stirred by basic-applied adjustments, a second *shock* came to social science in the form of a *drastic reorganization* in July 1975," pg. 101; emphases added. Larsen suggested that the reorganization efforts did *not* include persons from the social sciences' areas, at least with regard to creating and putting forth the Stever memo announcing the reorganization. In other words, they had been in the dark until all of NSF read the Stever memo. At first blush, this seems difficult to believe, as Betsy Clark and her assigns had been looking at the potential structure of a new BMS from as early as 1971 and in earnest by 1973. Given where the social and behavioral sciences were lodged, vis-à-vis BMS, however, the social sciences may well have been kept in the dark by Clark et al. See main text above and relevant footnotes in this group.

[130] Division Director, BMS [Clark], to BMS staff, "Congressional Budget Presentation," Sept. 25, 1973. Underline in original; italics emphasis added.

[131] There are a total of some 14 memos in a National Archives' file from the various program directors in BMS to either Clark or Mehl, all dated in either September or October 1973, that appear to be in response to Clark's call for suggested features of a new structure for BMS. (This collection of memos may not be complete, as many of these files do not appear to be so.) All responses came from strictly biological areas, save one: that from the psychobiology and neurobiology group (Jacob Beck, James H. Brown, and Fred Stollnitz, of whom more later; see also footnote 104 above). Dated October 5, 1973, the memo was not unlike those from other groups in which a number (usually three to four) of new divisions were suggested. However, interestingly, there were no memos whatsoever from the social or behavioral sciences groups in the Archive file! Was it possible that these areas were, indeed, totally left out of the deliberations as suggested by Larsen in his history (footnote 96 above)? NARA Box 11, File "BMS Reorganization – start 9/73."

I. Biology of the Cell

a. Energy Transduction (Bioenergetics)
b. Molecular Genetics
c. Development
d. Human Cell
e. Macromolecular Biology

II. Organismal Biology

a. Nutrition and Metabolism
b. Physiology
c. Neurobiology
d. Psychobiology

III. Population Biology

a. Ecology
b. Evolutionary Biology
c. Ecosystems Analysis
d. Popwulation Genetics[132]

While Spearing's BMS colleagues responded with variations in the titles and content of the divisions, central to the discussion was the idea of Weiss' much earlier "functional" aspects of biology versus disciplinary areas. The following day, for instance, a memo from the staff in Ecology was sent to Director Clark with another interpretation of how to reorganize BMS and was not terribly unlike that of Spearing. The memo offered three sections also (as per Clark's desire) for the biological sciences: "Cellular and Subcellular Biology…Physiological Processes and Behavior… [and] Ecology and Systematic Biology." There were different program areas, too, but still within what one could expect.[133] A dozen more memos followed with yet other interpretations, all grist for the mill for both Clark and Mehl. All memos contained suggestions, often lists as in Spearing's but for a few which were a bit more philosophical. For instance, the Program Director of Metabolic Biology, Elijah Romanoff, had to say that:

> [t]itles of sections are important because they tend to identify the programs to be listed under a given title even though the program may have a scientific content that is anything but specific. The continuum that is biology emphasizes connotative description rather than denotative values.[134]

That was an excellent observation by Romanoff. But there was some dissension. Three signatories among that group in the much-contested area of "psychobiology and neurobiology programs" (as the titles then stood) took exception to the general movement for three sections and constituent parts: they did not follow in Weiss'

[132] Associate Program Director, Metabolic Biology Program [Cecilia W. Spearing], to Division Director, BMS [Clark], "Reorganization," Sept. 27, 1973.

[133] Staff, General Ecology to Division Director [Clark], "Reorganization of BMS," Sept. 28, 1973.

[134] Program Director, Metabolic Biology [Romanoff] to Division Director [Clark], "Reorganization of Division into Sections," Oct. 2, 1973.

earlier functional approach, saying that, for instance, the proposed section "'Community Processes' sounds like sociology rather than biology."[135]

Mehl responded to one memo in the series and made his feelings clear, not unlike that of the psychobiology and neurobiology group, that "functional" grouping had its problems. Seeking some other titling approach, he was, apparently, in a minority as the Weissian vision prevailed and the historical record is mute to any follow-on to Mehl's objection.[136]

Mehl may have been alert to difficulties, even prescient, in the area of titles as just a year later, in October of 1974, Edward P. Todd, the Acting Assistant Director for Research (still then in existence) noted that there had been a change. The:

> name of the Ecology and Systematic Biology [Section would become the] Ecology and Evolution Section…[but o]bjections have been raised by several in the Foundation that in the minds of many people the word 'evolution' might not accurately indicate the nature of the research being supported [the comment itself being a circumlocution]. A review indicates that these objectives [objections?] may be valid and, therefore, to better reflect to a larger audience the nature of the work being supported by the Section, the name is being changed to Ecology and Population Biology Section.[137]

The whole matter of a federal agency supporting research in evolution would, in years to come, become a very contentious issue (especially, but by no means exclusively, during the Reagan administrations) and the necessity for alertness in naming structural units at NSF and in other federal agencies, programs, and even the titles of grant applications by individual scientists across the nation, was paramount. An anonymous wag has said that "everyone is subject to the laws of Darwinism whether or not they believe in them, agree with them, or accept them. There is no trial, no jury, no argument, and no appeal."

The resulting organization (Fig. 1.3) of the BBS Directorate was then, finally, composed of three biologically oriented divisions—Cellular and Molecular Biology, Environmental Biology, and Behavioral and Neural Sciences. This left biology in a somewhat precarious position as its identity was spread over, what Appel called "two-and-a-half divisions," because the third division, Behavioral and Neural Sciences, included anthropology and social psychology along with more traditional biological emphases (here avoiding both the terms disciplinary *and* functional). The Social Sciences became a Division on their own within the BBS Directorate but

---

[135] Jacob Beck, James H. Brown, Fred Stollnitz [Psychobiology and Neurobiology Programs] to Division Director [Clark], "Reorganization of the Division," Oct. 5, 1973.

[136] Deputy Division Director [Mehl] to Division Director [Clark], "Reorganization of the Division," Oct. 23, 1973. Mehl had said that he "remained unconvinced that the 'level of organization' offers the only possible or most logical basis for divisional organization…Can we, forgetting about names, develop any view of the way biological problems will be approached during the next few years, and use this as a basis for organizing the business of the division?" The historical record is incomplete and does not offer an answer in response to Mehl's call. Clearly, the divisions fell in with the Weissian organizational mode by the fall of 1975 when all changes done by Stever were completed.

[137] Edward P. Todd, Acting Assistant Director for Research, AD(R) Bulletin No. 74–21, Revision No. 1, "Subject: Name Changes in Biological and Medical Sciences Division," Oct. 3, 1974.

with animal behavior and the forever-fought-over turf of who owned what part of the neurosciences continuing in contention (Chap. 2). The amalgam was an imperfect one at best but one that would live for the next 16 years.[138]

Such odd mixing of disciplines like economics, law, sociology, linguistics, political science, and developmental psychology with the true areas of biology left the biological sciences in a worse position than they had been in the former BMS. The social sciences' respective research methodologies and manner of graduate training and even disciplinary goals were often totally unlike biology—it being then less unified even than it had been in the BMS. "From 1976 on, there was no Foundation entity corresponding [precisely] to biology" until the second major reorganization of 1991–1992.[139] The social sciences were included in the new BBS for the purpose of continuing their need for "protective coloration" from the depredations of some in Congress wishing to see them eliminated altogether from NSF:

> The linkage of behavioral science to biology was a deliberate effort by the NSF leadership to divert attention from social science once again under strong attack from the political community. In the 1950s, social science had effectively employed convergence with natural science as *protective coloration* to gain standing in the NSF. Now, in the 1970s the strategy was to be deployed again.[140]

There was, as Edward Lurie in his biography of nineteenth-century Biologist Louis Agassiz would have it, a lack in the new BBS of "an intellectual connective tissue."[141]

Guy Stever had told the National Science Board at its meeting in late September of 1975 about how his reorganized NSF "had been well received," but his memo to the workforce of the Foundation had been public since July 10, and the structural features of NSF were being fine-tuned thence forward. There had been only one meeting of the NSB from mid-May until that in September, and there had been no general discussion, Foundation-wide, of the reorganization in any open or broadly constituted manner. That would contrast very sharply with how matters proceeded in the reorganization of 1991–1992. The view of what had to be told to the public shareholders of their democracy in action by Guy Stever's NSF was a very much more closed (or just merely more simple?) matter than would become the norm in later years: sunshine was yet around the corner. Indeed, the director was not beholden to tell even the NSB, nor consult them before the fact, about reorganizing the NSF as it was left to the director's discretion as to how to do the jobs organizationally with which it was charged. While it was true that Nixon's Sunshine Act did not come out until September of 1976, when he was no longer in office and the Foundation's reorganization was by then a fait accompli, the need of the govern-

[138] Ibid. and also NSB Min. 9/18–19/75, 175:29 (Appen. B). Note that the term "AD: Biological and Social Sciences" in Figure 1.3, text above, drawn from the NSB Appendix B is not quite the final phrase as BBS becomes the true name of the Directorate.

[139] Appel, *Shaping Biology*, pg. 275.

[140] Larsen, *Milestones and Millstones*, pg. 102. Emphasis added.

[141] Edward Lurie, *Louis Agassiz: A Life in Science* (Baltimore, MD: Johns Hopkins University Press, 1988), pg. 85.

ment to be more transparent was clearly there by 1975, indeed well before in the Nixonian era. One may have guessed that Stever would have been very open with his quite massive changes, but such does not appear to have been the case.

In fact, Stever spoke very casually about the reorganization in his memoirs moving smoothly back and forth from the time of his appointment in February 1972 up to and through the reorganization period. For instance, he recorded that NSF had "[t]hankfully…remained untouched by this mess" when in reference to the infamous Watergate Scandal that broke quietly at first on June 17, 1972, with the arrest of the Watergate office complex intruders to the Democratic National Committee headquarters. That event and all that it precipitated would finally topple Richard Nixon from office on August 8, 1974.[142] Stever noted the oil crisis of 1973 and then the creation of the BBS which he set in *1973* when he appointed its "first" director, Betsy Clark. Why, in his memoirs, did he say 1973 and not 1975 with the July 10 memo announcing the reorganization of all of NSF, biological and social sciences included?[143]

A simple error, perhaps, but the answer might have been one of either forgetfulness on the part of Stever or an attempt to simplify his memoirs for the interested reader. More likely, however, is that the reorganization was staged in pieces over several months from July 1975 through the end of that October. In actuality, it had begun for BMS well before: from 1971 through 1973, as has been seen. First, there was the announcement of the appointment of Clark in mid-1973 as the *Division* Director of BMS.[144] She did not, could not, become an AD of a directorate until those units were created some 2 years later, of course. In his news release, Stever noted that Clark replaced Harve J. Carlson, whose career at NSF is well covered in Appel's *Shaping Biology* and who retired from NSF in June of 1972. However, Clark had moved up through the ranks at NSF (below)—not impossible then, but an uncommon occurrence in later years when NSF became much larger. The newly created "front office" of the BMS in April of 1973 had a number of other employees in addition to Clark and prior to two other senior managers joining it by fall of 1975.[145]

---

[142] Stever, *In War and Peace*, ppg. 214ff; quotation on pg. 216.

[143] Stever, at his page 220, makes the statement that he "created a new directorate for biological, behavioral, and social sciences, appointing Eloise Clark as its first director."

[144] National Science Foundation, "Dr. Eloise Clark Named NSF Division Director," news release #73–144. NSF did not routinely provide month and day, only year in the document reference formula (here, 73–144).

[145] Untitled memo to Eloise Clark from Dora J. Hruz, April 24, 1973, providing a listing of the persons in the "front office" with more biographical detail for each than is given in this note. The workers included Norma Middleton, with the Division since 1963, as a secretary often promoted; Barbara Brown, since 1971, as a clerk-typist who knew short-hand and moved up the secretarial ladder to become division deputy John Mehl's senior secretary; Sukari Smith, since late 1971, secretary and the individual who managed all international travel grants; Barbara Mutz, since mid-1972, with great strengths in mathematics, the travel clerk; Mary Chezmar, since mid-1972, secretary; Pamela Swinney, new at the time and had come as secretary from the Metabolic Biology group to the front office; and Dora J. Hruz, senior person in the office (present at NSF since mid-1963) and who, not unlike so many of the early scientists at NSF, had a background with the AEC. Two rotating clerk-typists rounded out the persons in the Division (soon to be Directorate) office.

The second evidence that Stever's comments in his memoirs are confusing is that it was not until his formal announcement via a Staff Memorandum of September 26, 1975, that he had reorganized NSF and had first created and then quickly renamed the new BBS Directorate.[146] The BBS was officially born that October:

> My organization memorandum of July 10, 1975 (O/D 75-31) announced the *overall* reorganization of the National Science Foundation and the establishment, among the directorates, of the Directorate of Biological and Social Sciences. This memorandum announces the reorganization of [that] Directorate. The name for the Directorate, effective this date, is Biological, *Behavioral*, and Social Sciences…and will become effective October 31, 1975.[147]

He went on to say that the Directorate would be composed of the four divisions first fairly well defined by late 1973 through BMS staff efforts and argued that the:

> restructuring will provide the Assistant Director…with an effective management structure for dealing with current and future developments in these fields of science.

And which it generally did, for some 16 years to follow, before a new structure was called for as the fields covered within the BBS themselves required major changes at NSF by the beginnings of the 1990s.

In the OAD, Office of the Assistant Director for BBS, Betsy Clark would continue as "acting" AD, Stever stated, and she would be aided by Richard C. Atkinson as the official new deputy assistant director. Enoch L. Dillon was designated as planning and evaluation officer. In the same memo, the Division of Physiology, Cellular and Molecular Biology was established. Its division director would have direct reports that included the heads of the Cellular Biology Section and the Physiology and Biochemistry Section. (Biochemistry was in BBS, not with chemistry in the then also new physical sciences directorate where pure chemistry was couched.) "The Division of Environmental Biology is established," the memo went on to read. Reporting to its director would be program officers heading categories termed Systematic Biology Program; Ecology Program; Ecosystem Studies Program; and Biological Research Resources Program.

Two more *divisions*, not purely biological, were also announced: Behavioral and Neural Sciences (including *programs* in Neurobiology; Sensory Physiology and Perception; Psychobiology; Social Psychology; Anthropology; and Linguistics) and Social Sciences (including *sections* in Economics and Quantitative Methods and Sociological and Political Sciences); the latter division would be under the direction of Howard H. Hines, who had been in that leadership role prior to the birth of BBS. The final words in the memo are interesting: "My search for an outstanding biologist to fill the position of Assistant Director is proceeding." It was not a given

[146] Office of the Director [Stever], NSF, "Staff Memorandum: Subject: Reorganization of the Directorate for Biological, Behavioral, and Social Sciences," Sept. 26, 1975; O/D 75–44.

[147] Ibid. Emphases added. The apparent confusion of use of two dates as "effective" ones is that the name was effective in September, but until the actual restructuring of the directorates' personnel, offices, tasks, projects, programs, sections, and divisions, the date in October really represents the founding date of BBS as it would exist into future years.

on September 26, 1975, apparently, that Clark would be that person. She was, however, on the scene in the BMS since 1969, and her work and leadership qualities, since her appointment as head of that division in 1973, were clearly known. Stever shortly thereafter did choose her, in fact, to lead the new BBS.

Eloise "Betsy" Clark was born on a farm in Abingdon and graduated from Cleveland High School in Bristol, both in the Commonwealth of Virginia. Her BA (1951) was earned in biology at Mary Washington College of the University of Virginia in Fredericksburg, where she was elected to Phi Beta Kappa. In later years, she became a Distinguished Alumnus at Mary Washington. She received her doctorate in zoology (specifically, in classical developmental biology) at the University of North Carolina (UNC), Chapel Hill, in 1957, with minors in botany and biochemistry. The minor in botany will be later seen to have been an advantage as that discipline was long underrepresented in BMS and would, under a later successor trained in plant science, be given even greater attention than Clark first gave it. She taught and did research as a doctoral student from 1953 to 1957 at UNC. She went on to do a postdoctoral period with Howard Schachman of the University of California (UC), Berkeley, in physical biochemistry—a more modern area, not classical as with her bachelor's studies. However, both she and Schachman worked together in the laboratory of Arthur Kornberg at Washington University in St. Louis, Missouri, as Schachman had a sabbatical there. Kornberg, along with Severo Ochoa of New York University, shared the Nobel Prize in Medicine or Physiology for their work in the biological mechanisms of the synthesis of DNA only a year later (1959) just after Clark and Schachman returned to Berkeley where Clark spent 1 more year. Her postdoctoral work was funded by an NIH fellowship specifically in microbiology in Kornberg's laboratory and in biochemistry and virology in Schachman's. In an interview with the author, Clark has justly said that the era was a "revolutionary period in molecular biology."[148] Her first teaching position came as an instructor at Columbia University in biology in 1959 (the first woman faculty member ever in that department), and she was promoted to assistant professor (1960) there and finally to associate professor in 1966.[149] She held that post for the next 3 years.

Betsy Clark came to NSF in 1969 for 2 years as a rotating (thus temporary) program director and ended up staying on for 14 years! It was early in the period of affirmative action at the Foundation, and female program directors were still uncommon, though women were almost always either assistant or associate program directors at that time.[150] Those positions would later disappear, however, in the various incarnations of the biology directorate specifically, though not more generally across NSF.[151] Her division directorships included, in order, that in developmental biology; then biophysics; and, finally, molecular biology. By 1973, Clark was over-

[148] Betsy Clark interview with the author, May 26, 2009, by telephone.

[149] Information on Clark's career is drawn from a mix of the "News Release" (#73–144), pg. 2; from the "BGSURA Newsletter," Bowling Green State University (2003):1, 4; and the interview.

[150] For a history and timeline of affirmative action in the United States, see http://www.infoplease.com/spot/affirmative1.html.

[151] Clark interview.

seeing the administration of some 3400 research grants and controlling an operating and grant budget of over $57 M. Another nearly $10 M, which she also oversaw, was outgoing to the International Biological Program (the IBP) through BMS.[152] She then became the *functional* deputy AD for BMS for about a year even though Richard Atkinson was officially, but in name only, the deputy AD. That arrangement was because "we ran a pretty slim ship in those days." She has said that her training in both classical areas of biology, such as the developmental biology of the time, and in more modern areas—her work on DNA with Schachman and Kornberg—prepared her to deal with all areas of the directorate when she became AD of the new BBS in the fall of 1975.[153] That was an era when a goodly amount of classical biology was essentially disappearing, or had long before been displaced, across the disciplines globally, as is made apparent in much of this study. It was early in the period which I have termed *Millennial Biology* and which concept will be developed throughout the remainder of this history.

During the overall NSF reorganization leading to the BBS, Clark averred that Stever and his deputy director, Philip Smith, seemed to realize only later in the process that they had "left biology and the social sciences out of the equation [budget request]"![154] They had been busy with the physical sciences, mathematics, oceanography, and engineering: biology and the social sciences were, as she stated, "secondary." Eventually, various organizational options for them were considered, but the budget for the social sciences was only about $40 M: small at the time when compared to other areas of the Foundation. So Stever and Smith decided it was not large enough to warrant its own directorate and therefore placed the social sciences' "Office" with biology, Clark averred.[155] It was their desire also to increase the scope of psychology, Atkinson's disciplinary area: his presence did have influence in that, Clark has explained.[156] Indeed, she has stated that Atkinson did much to organize the psychology *program*, along with the directorate overall, even though he had not yet officially moved to NSF at the time. Since most of the funding for psychology in those years was coming from NIH, it is not surprising that it was strongly clinically oriented. Atkinson pushed for more *basic* research in the field.[157] Clark has explained that as a result of those efforts, neuroscience and psychobiology were

[152] The IBP dealt with "a multi-national, integrated research effort involving environmental management and human adaptability." See "BGSURA Newsletter," fn 149 above, pg. 1. More will be said about the IBP later. While not directly controlled by the NSF, the IBP would morph into an NSF-supported program of massive size and worldwide importance in the areas of ecology and environmental sciences: LTER, the Long Term Ecology Research program, about which much will be said in later chapters. Dr. David Coleman of the University of Georgia completed a history of the IBP just as this present work was being written. See citation in later chapters.

[153] Clark interview.

[154] Ibid.

[155] See Appel's endnote 4 on page 309 where she mentions the fact that in 1958 the Office of Social Sciences came into existence.

[156] Clark interview.

[157] Ibid.

transferred from the Office under biology to the aborning behavioral and neurosciences area. Additionally, social psychology and anthropology were moved from the social sciences to the new behavioral sciences. Finally, a new area of cognitive sciences, which Atkinson wanted to see created and tied organizationally to the neurosciences, was also placed into the behavioral sciences (following chapter).[158] All of these efforts by Atkinson reflected his strong mathematical orientation, Clark has opined. And given that he was keen on the rise of the new field of neuroscience, he saw that bringing together that area with social psychology and anthropology "created a natural 'behavioral sciences' grouping as a bridge with biology and social sciences."[159]

Why were the social and behavioral sciences *really* placed with biology? Clark has claimed that they were *not* placed with biology in order to shield them from assault from outside pressures, such as by certain members of Congress, as argued above (and numerous times again later in this text). "Well, I don't think that is a valid argument," she told the author.[160] She went on to say that protection was not "a major thing," but "it was mostly really a default." The concept of "protective coloration" was not the reason, if Clark is to be believed. But Appel's history has also upheld the notion that protection *was*, in fact, the case, because "the social sciences represented a small and *politically vulnerable* part of NSF, they were placed with the biological sciences *in part* to protect them."[161] Appel does say "in part," and so it seems fair to presume that the small budget and *vulnerability* were two good reasons for the Stever/Smith decision, *contra* Clark's belief. As will be seen, much evidence suggested the need for "protective coloration." Notably, Clark did point out that since the social/behavioral sciences were placed with biology in the new BBS, the Foundation had the chance to put either Clark *or* Howard Hines, director of the social science areas, in as AD. She concluded: "I think that they felt that having a social scientist in charge would not be appropriate because biology was so much bigger." Further, she stated, "the truth of the matter is that I had some impact on gaining resources for biology…[and, also, senior management] had gained some confidence in my leadership." Finally, and it cannot be discounted, "I think they also liked the idea of having a female in the AD position": Clark was the only one at the time.[162]

Within the new BBS, Clark found herself handling some 14 different disciplinary areas in 15 programs and was supervising a staff of some 158 persons. She had also "developed guidelines for establishing review panels and [additionally] soliciting reviews by mail." Inviting review panel scientists to come to NSF headquarters to adjudicate grant proposals was a concept pioneered by the BBS, which remains

---

[158] Ibid.

[159] Personal communication with Betsy Clark, spring 2012.

[160] Clark interview. She reiterated that feeling once again ibid., when she said that "I continue to believe that promotion [of the fields] was more important than protection."

[161] Appel, *Shaping Biology*, pg. 275. Emphases added.

[162] Clark interview.

central to the workings of its successor, BIO, from 1991 to the time of this writing. Some other directorates, even as of this recording, do not employ the panel concept. Also, it was she who "established programs in neurobiology and sensory physiology…and a post-doctoral program encouraging a molecular biological approach to the study of plants...and [expanded the area of] ecological sciences [by aiding in establishing the Long Term Ecological Research—LTER—program; later chapters]."[163]

The year 1975 and the birth of BBS, like other years at NSF, was complex at many levels. The late Stephen Jay Gould's argument that "[c]urrent utility may not be equated with historical origins" (he was speaking of organic evolution) seems to fit well with what had occurred in the historical origins of the biological *and* sociological sciences at NSF by 1975. There was need for the reorganization in responding to "current utility" necessitated in these, and other, disciplines such that *applied* research could fit in better than it had from 1968 to 1975. Additionally, shuffling of the programs, sections, divisions, etc. of the old BMS into the new BBS would, as Stever had proclaimed, "provide…an effective management structure for dealing with current and future developments in these fields of science."[164]

Finally, Betsy Clark has pointed out that the reorganization of BMS into BBS, let alone so much of the rest of the Foundation, allowed for better *oversight*—and that was pleasing to the Congress.[165] The BBS set new directions beyond this, however, and many of those threads can still be seen in the BIO of the present time. For instance, since its earliest days, the biology directorate has been notable within NSF and among other federal agencies in bringing in *panelists* to assess proposals. A level of vetting does occur prior to any given panel's sitting, but the duty of the panelists can be very demanding, especially in large, popular programs with extensive numbers of grant proposals worthy of the most meticulous review. Thus, the biological sciences within BBS funded 1760 projects totaling almost $78 M, while the social sciences funded 649 at just over $26 M in FY76. To compare these figures with, say, the second most supported area (biology was the highest category Foundation-wide in both numbers and funds for that year), mathematical sciences, the awards totaled 851 at, however, a level of only about $16.5 M. Mathematics does not require the funds that both laboratory instrumentation and field time necessitate in biology and similar sciences. Fiscal Year 1975 was the third year in a row, at least, that biology was the leader with regard to both the numbers of grants funded and actual monies provided. Times, indeed, had changed since the heyday of physics in the first two decades of the life of NSF. The social sciences, on the other hand, remained almost static.[166]

In matters strictly biological, the new BBS moved confidently. NSF took up the International Biological Program, IBP (1964 to 1974), from its earlier status: it

---

[163] "BGSURA Newsletter," pg. 1.

[164] Stever, "Staff Memorandum," O/D 75–44, pg. 1.

[165] Clark interview.

[166] Ibid., pg. 4, Table 1.

eventually morphed into the LTER, as will be seen in later chapters. BBS also took notice of areas that had not up to then been of such compelling interest. One of these was biological systematics collections (which can be used, for one example, as a monitoring device for environmental/ecological changes over time) and plant sciences. Indeed, with regard to those fields, Clark has opined that it was both Stever and his associate AD, Smith, that were the "power brokers" with regard to moving those disciplines forward financially. That was possible as the "brokers" aided Clark in carrying the data and her arguments to the federal budget office to raise funding for the plant sciences.[167] Grant support under those headings included rapidly expanding work in the basic nature of photosynthesis and nitrogen fixation (bacteria making crucial-to-life nitrogen gas into usable form in the root nodules of members of the bean and pea family). Among the ends sought were novel ways to trap solar energy and ways to reduce artificial fertilization of soils with industrial nitrogen compounds.

The *coming together* of molecular biology and of "big biology" including systematics (the naming and classifying into evolutionarily based categories of all living things) and ecology, as well as environmental science (that may include ecology and much else) at BBS and other directorates of NSF, was an indication of major changes in the world of science. Much more of this later.

Well before the early 1970s, there was considerable interest in the notion of inter- or multidisciplinarity in science. In some cases, two or more diverse disciplines had come together to solve complex problems, but intellectual "territoriality" is a strong force in science, and that made it a difficult matter to undertake *interdisciplinarity*. Various sciences' particular methodologies, means of teaching and training their undergraduates and graduate students, long-term standing of each discipline in the inertia-bound academy, and approaches to job placement of the different disciplines all conspired to keep matters "mono-disciplinary" much of the time. This issue is taken on in following chapters.

The effort to structure the NSF along functional rather than disciplinary lines was only partially successful. The academy was not moving along at a fast enough pace to make its science departments unified by function rather than discipline, so the NSF had to fight hard to make its programmatic offerings sensible and accessible to academics. Though it succeeded in large part, it was not easy, and it took well beyond 1975 to reach a much higher level of interdisciplinarity. (Terms such as cross-, multi-, and transdisciplinary have also been used.) This issue is a *central part* of *Millennial Biology*, as will be demonstrated.

Among the themes of this study will be the efforts to tie together "big" (ecology and environmental biology) and "little" (molecular and cellular level biology) aspects of the life sciences and, to some extent, the involvement of the social sciences. In 1975, it was still a task of considerable difficulty.

---

[167] Personal communication with Betsy Clark, spring 2012.

1.7 First Appearances of Big Biology

As with all federal agencies, developing the annual budget is a large, complex, and essential activity at NSF. The FY76 budget can be offered as an example of issues that swirl around the budget preparation process. The process may approach 18 months to 2 years in length, so by no later than mid-1974, the FY76 budget planning would be underway. Preliminary guidance to all division directors was issued in a memo of July 1974 from the then Acting Deputy Assistant Director for Research (see Figure 1.1) concerning the "budget justification statements."[168] The acting AD noted that:

> [t]wo Panels of the Advisory Committee for Research have criticized our descriptions of Significant Recent Achievements or 'nuggets'...However, nugget writing takes on added importance in view of the adverse publicity recently given to several research grants whose objectives were not readily understandable. Some Panel members noted that many nuggets were not intelligible to scientists in other disciplines let alone the lay reader.

This matter of the readability of "nuggets" (seen in, among other places, the *Annual Reports*) was important as:

> [t]he Director calls attention to the need for careful writing of Summary Program Descriptions and Objectives, Significant Recent Achievements [nuggets], and the need to explain the reasons for increases and decreases in the [FY76 program and associated budget request].

The reference to recent "adverse publicity" was in connection with Senator Proxmire's infamous Golden Fleece Awards (above and again in later chapters).

Funding of the NSF from Congress each year depended upon clear and powerful language in the budget request in order to meet the needs of the Foundation in supporting grant making for the advancement of science across the nation. This issue was always foremost in the minds of leaders, in particular the director and the assistant directors. ADs routinely spoke before the congressional budget committee responsible for NSF's appropriations in earlier years and had to have a patent understanding of the issues of interest to the President and Congress. The ADs had this assignment removed from their portfolios by the later 1980s, however.

Betsy Clark, as one AD, had problems that would require her attention when it came to structuring her budgets for the future. In May of 1975, as one example, John Mehl was assigned to respond to the then editor of the journal *BioScience*, the organ of the American Institute of Biological Sciences (AIBS), one of the most important professional organizations with which BMS/BBS (and later BIO) had to deal on a regular basis. It seems that BBS was unhappy with an editorial in the journal entitled "Growing Grant Size Bias Toward Bigness." The struggle between big and little science had a long history in all of the sciences, but not least in physics in the 1950s

---

[168] Memo to Division Directors and Head of ERG Office from Acting Deputy Assistant Director of Research, "Subject: Fiscal Year 1976 Budget Justifications," July 17, 1974. Ephemeron; only pg. 1 available.

and 1960s, and, by at least the 1970s, it was becoming an issue in biology, as well. This will be considered in greater detail later.

A leading scientist of the time, Alvin M. Weinberg, nuclear physicist at Oak Ridge National Laboratory during the war years and director of that laboratory by 1955, knew a great deal about "big science." It led him to comment on it in his 1967 book, *Reflections on Big Science*.[169] He recognized, even by that somewhat early date, that "of all the sciences now supported by our society, biomedical science ought to stand first."[170] He went on immediately to say that:

> [w]e are, or ought to be, entering an age of biomedical science and biomedical technology that could rival in magnitude and richness the present age of physical science and physical technology. Whether we shall indeed enter this age will depend upon the attitude toward Big Biology adopted by biomedical scientists and government agencies that support biology.

Weinberg was keen on the ability of research in biomedical science, at a big level, to bring about the "possibility of alleviating human disease through such activity…[human disease being] obviously one of the most compelling." Indeed, the "claim to urgency hardly can be matched by any of the other fields of natural science."[171]

Just what did Weinberg mean by big science? "Science has become big in two different senses," he stated. One sense was that of the *requirements* to do a particular science: elementary particle physics requiring "extremely *elaborate equipment* and staffs of *large teams* of professionals," not to mention finances. But also, "on the other hand [second sense], the scientific enterprise, both Little Science and Big Science, has grown explosively and has become very much more *complicated*."[172] The requirement of large, expensive *equipment* used by *teams* and also the *level of complexity* of the science were Weinberg's two versions of big science. BBS, and all of NSF, attempted to balance the demands between the two. Oftentimes, "little biology" was looked upon as grants of small amounts to single PIs to do small projects that may or may not have been tied to then extant major programs or initiatives at

---

[169] Alvin M. Weinberg, *Reflections on Big Science* (Cambridge, MA: MIT Press, 1967).

[170] Ibid., pg. 101.

[171] Ibid., pg. 102.

[172] Ibid., pg. 39. Emphases added. Bigness usually implies big budgets; big staffs; big machines; and big laboratories. Big field projects should also likely be added to this list. The human genome sequencing project of more recent years is typically cited as a big biology undertaking, as is LTER. Renaissance precursors of big science are argued to go back to the astronomical observatory, Uraniborg, of Tycho Brahe (1546–1601) in Denmark; see John Robert Christianson, *On Tycho's Island: Tycho Brahe and His Assistants, 1570–1601* (New York, Cambridge University Press: 2000), but the use of the term has been primarily restricted to the era of WWII and following (http://encyclopedia.thefreedictionary.com/p/big%20science). See also "Physics at Low Temperatures" in Helge Kragh, *Quantum Generations: A History of Physics in the Twentieth Century* (Princeton, NJ: Princeton University Press, 1999):74–86. Particularly, see Peter Gallison and Bruce Hevly, eds, *Big Science: The Growth of Large-Scale Research* (Stanford, CA: Stanford University Press, 1992) for a good recent definition, but which study, interestingly, does not address big biology at all.

BBS. Approximately 80% of every budget year's total went to those types of investigations. "Big biology" would eventually come to characterize such endeavors as various genome projects and the LTER, even if it was, in the main, a *collection* of work done by many *individual* PIs and graduate students under a number of single or smaller grants; the total programmatic activity added up to very big biology, as we shall see.

Weinberg was well informed in biology, even though it was not his specialty.

> Are the biomedical sciences that relevant to the conquest of disease? To an applied scientist like me, this question seems absurd. What strikes an observer most about modern biology is how the new viewpoints have unified the subject. The genetic code [worked out by Francis Crick, *et al.,* during the early and mid-1960s] appears to be universal. The dogma of protein synthesis—DNA, messenger RNA, transfer RNA, protein—seems to be valid in almost every life form.[173]

The three steps leading to protein synthesis Weinberg recited, and often known as the "Central Dogma" or "the Holy Trinity" of biology in those years, were beginning to take center stage.[174] It would become *big biology.*

Weinberg had already written earlier (1961), more in the line of his own fields of research, that:

> When history looks at the 20th century, she will see science and technology as its theme, she will find in the monuments of Big Science—the huge rockets, the high-energy accelerators, the high-flux research reactors—symbols of our time just as she finds in Notre Dame [cathedral] a symbol of the Middle Ages…We build our monuments in the name of scientific truth, they built theirs in the name of religious truth; we use our Big Science to add to our country's prestige, they used their churches for their cities' prestige; we build to placate what ex-President Eisenhower suggested could become a dominant scientific caste; they built to please the priests of Isis and Osiris.[175]

Mehl's response to the editor of *BioScience* had to do with the IBP, the International Biological Program, which was then just ending and would be replaced a few years later by LTER. However, the writer of the editorial made a complaint that really was broader than any specific program and widely heard in coming years with regard to biotechnology and other areas supported by the BBS. He wrote:

> [t]he current trend for granting agencies to favor larger research projects threatens our least expensive and most imaginative resources—the independent investigator.[176]

---

[173] Weinberg, "Big Science," pg. 102.

[174] The term "Central Dogma" was first enunciated by Francis Crick, codiscoverer of the structure of DNA. See F.H.C. Crick, *Symposium of the Society for Experimental Biology XII,* (1958):139–163. Others, somewhat irreverently, termed the Central Dogma the "Holy Trinity"; personal experience, as the phrase was commonly used during the author's student days.

[175] Alvin M. Weinberg, "Impact of Large-Scale Science on the United States," *Science* (1961):161–164.

[176] John W. Mehl to John H. Behnke, May 14, 1975, concerning the editorial, "Growing Grant Size Bias Toward Bigness," written by a Professor Crowell; *BioScience* (1975):235.

Mehl's answer was both extensive and highly informative, not only for the journal editor and the editorial's author, but for present historiographical purposes, as well. Mehl provided a response to the editorial's title and implications:

> To the extent that larger research projects are favored, this is in part on the advice from the academic community…an argument made by biologists that the most progress is not made by the individual, independent investigator, but rather by a group of some significant size which is able to interact on a continuing basis.[177]

The relationship between funding agencies (here NSF) and the community of scientists that they serve, I have called the *intricate dance*. It is done by the two partners, and it is what establishes that research which was/is to be funded. The *intricate dance* will be referred to often in this history as it is, I argue, another central feature of what has made *Millennial Biology*. The NSF seeks to know what interests the community has at any one time and so tailors its programmatic offerings accordingly. The community informs NSF by way of conferences, workshops, or unsolicited grants in areas that it (the community) knows need to be funded. Contact, arm in arm between the two partners, is often and is invaluable. How such a relationship exists will be considered in a number of places in this text; panel membership is one method to which allusion has already been made. Panel members bring intelligence of what is happening in the community, what is "hot" in research fields, and so the guidance for NSF as to where those disciplines are headed in the immediate and, when possible, long-term future. This *dance* continues unabated to the present time.

The *intricate dance* is not without certain real or potential drawbacks, however. I enquired of one interviewee whether or not graduate advisors in suggesting doctoral dissertation research topics to their students might not bias what they chose. Certainly, it is not only common, but almost axiomatic, that a graduate student will work on some aspect of his or her mentor's grant research. In systematic biology, for example, the program officers spoke about the matter often, and it was even pointed out that the students, prior to beginning their research, were told by their advisors that they (the students) should contact NSF and inquire about the future of funding for what they wanted to undertake. This remains a problem in that too many specialists may be trained in some given area(s) to the exclusion of others. The interviewed program officer (PO) knew that his own personal bias could also be a contributing factor that could compound such imbalances in the community. This situation is probably much more real than potential.[178]

---

[177] Ibid., the letter; pg. 1.

[178] James Rodman interview with the author, May 11, 2009. Rodman, a botanist, felt that not only were specialties out of balance in numbers of practitioners in the area of systematic biology, at minimum, but he found himself "want[ing] people working on plants across the board." That could have led to an even greater imbalance, he admitted. Nevertheless, Rodman was particularly proud of the Doctoral Dissertation Improvement Grants (DDIG) that were unique to his division and unlike those elsewhere in the biology directorate. They were designed to lead to new systematists that were not simply clones of their major professors, and that helped keep the field more reasonably in balance, he believed.

Mehl continued on to address the IBP question:

> The concepts of coherent-area grants, centers and institutes have not been forced on the academic community by the granting agencies. The International Biological Program was not an 'agency concept,' but one which was developed by academic biologists, and the specific form of the U.S. effort was largely determined by a committee of academic scientists established by the National Academy of Sciences.

Mehl further responded that the:

> large biome [roughly, a major ecological community in a region] projects under the IBP have supported the research of many faculty members, graduate students, and undergraduates in a substantial number of institutions.[179]

"The average grant size for biology at NSF," Mehl stated, "has increased over the last decade, from $33,000 in 1964 to $45,000 in 1974…but the two figures are not comparable." This was so as the "average grant period also declined since 1964."[180] And those numbers did not take into account rampant inflation ("50% or greater"), either. This topic will be considered in a number of places throughout this study.

1.8 First Thoughts on Reductionism and Millennial Biology

The whole issue of big and little science would be a theme for years to come at NSF and BBS/BIO—and it continues. However, tied to that theme is another one not at first thought closely related: the theme of *reductionism* in biology.

> Reductionism encompasses a set of ontological, epistemological, and methodological claims about the relation of different scientific domains. The basic question of reduction is whether the properties, concepts, explanations, or methods from one scientific domain (typically at higher levels of organization) can be deduced from or explained by the properties, concepts, explanations, or methods from another domain of science (typically one at lower levels of organization).[181]

Simply, though somewhat crudely put, can all of biology be reduced to physical chemistry, or can biology be understood only, to paraphrase Theodosius Dobzhansky, in the light of integrative biology? There was a period after the discovery of the structure of DNA by Watson and Crick in 1953 wherein biology was headed along a strongly reductionist track for many decades. This present study asks, along with other questions alluded to above and more yet to be developed, whether the history of biology shows a continual increase in reductionist tendencies in biology into the twenty-first century or whether some other feature, possibly big biology, asserts itself at the *expense* of reductionism or, in one of reductionism's possible definitional senses, little biology. In so doing, we must also ask about the rise of *integrative biology* during the early 1990s and over the years that have followed. Also, the

---

[179] Ibid., the Mehl/Behnke letter, pg. 2.

[180] Ibid.

[181] http://plato.stanford.edu/entries/reduction-biology.

rise of *systems biology* at about the *fin de siècle* must also be plumbed. It is argued that in order to *do* the big biology that arose in the last half, and especially the final quarter, of the last century and into the present one, a transdisciplinarity including data, theory, funding, and the aims of biologists themselves would not only have to, more or less, parallel such areas as ecology, *but molecular biology*, as well. Is this integrative biology, or is it systems biology? That question is yet another that will be examined because the NSF's biology directorate, after the reorganization of 1991–1992, created a Division of Integrative Biology and Neuroscience (the redundancy inherent in the title itself had been open to question), and so is the biology of the twenty-first century the biology of integration (or reintegration from past incarnations)? Is it *holistic biology*? The increase of our understanding of "the big picture of life" could not succeed at the level it has to date without the concomitant input of considerable molecular information and inferences, but also with broad ecological understanding, and finally (?) an integration of all aspects of biological fact and theory. Is this all a return to the *integrated* biology of the nineteenth century, the biology of Darwin and his cohorts, but with new tools, new techniques, and new theories?

The very definition of reductionism in biology must be closely reexamined in the light of the rise of integration and systems approaches. In the course of this study, I offer new thought for just where biology went in the last decades of the second millennium and where it found itself in the beginning of the third. The final aim must be to define *Millennial Biology* and its relationship to the NSF, BBS, and, in the end, BIO, but it also must seek to define the New Biology of the new era qua science.

More broadly, the foregoing wide-ranging discussion of the years encompassing the major changes at NSF and BMS/BBS is intended as an aide to understand the many threads that will be examined throughout the rest of this study and to provide a foundation upon which to appreciate both where BBS/BIO went and what it was like to pass into a new millennium.

Chapter 2
The Effects of the Social and Behavioral Sciences upon Biology at NSF

A good scientist is, in fact, first a humanist and second a scientist.
Michael DeBakey

2.1 The Effects of the Social Sciences

The brilliant heart surgeon, Michael DeBakey, made an excellent point with which it is difficult to argue. If it is correct, it becomes only one of many reasons to understand what some of the "sciences of man" provide to the more clearly biological sciences. At NSF, the former were housed with the latter and had a profound influence on how biology developed at the Foundation.

For that reason, the current chapter will focus on both the social and behavioral sciences from 1975 to about 1991. But since there was some amount of divisional level and lower reorganization and reassignment of *behavioral* sciences specifically into either the new Directorate of Social, Behavioral, and Economic Sciences (SBE) or the Directorate for Biological Sciences (BIO) after 1991, those *behavioral* fields devolving permanently to BIO will be considered further, through 2005. It is *because of their organizational and philosophical ties to biology and, particularly, their* influences *upon it* that they are being considered in this study. For readers seeking more detail on these sciences (up to 1991), the history by Otto Larsen (see footnote 1 in Chap. 1) remains the primary source. See also, however, the new study coming out in 2020 from MIT Press by author Mark Solovey mentioned at the beginning of Chap. 1. The social and behavioral sciences must be examined in this chapter because it is important for a full understanding of the history of the biological sciences at NSF, the central task here. We are introduced to a number of *historical threads* central to the story of biology that had their inceptions within the social and behavioral sciences. Such threads as *increasing computerization*, *use of model systems*, *the continuing rise of interdisciplinarity*, *increased use of databases*, *longitudinal studies*, *the rise of Science and Technology Centers*, and several others are considered throughout this and later chapters. BBS was a unified structure, and so

D. J. McGraw, *Millennial Biology: The National Science Foundation and American Biology, 1975-2005*, https://doi.org/10.1007/978-3-030-56367-7_2

person-to-person and mind-to-mind interactions were present within that body that had influences across it.

I would contend to readers who wish primarily to read this study as a history of biology at NSF that they cannot skip this chapter crucial as it is to that discipline(s).

The new organizational divisions of late 1975 did not break along the lines of disciplinary content purely. Two of those areas more in the behavioral sciences—psychobiology (including animal behavior) and the neurosciences—will conclude this chapter. Those two are *closely allied to the more pure biology* discussed in the remainder of this book. Indeed, the very fact that the neurosciences reside almost squarely upon the dividing line between the social/behavioral sciences and the biological sciences has made for many challenges at NSF over time when organizational issues and funding priorities came to the fore. There is much about the brain and nervous systems of both non-human and human animals which make it extraordinarily difficult to divide its study organizationally over two otherwise disparate directorates. Much the same can be said about manifested behaviors in animals. The very act of so dividing them in the later reorganization of 1991 brought about much "tumult" on its own, as considered in Chap. 5.

2.2 A Revolving Door for Directors

Consideration of the social sciences will begin in the midst of a period of what surely seemed to be that of a *revolving door* for Foundation directors. From July 1969 to August 1984, there were five directors of the NSF, not one of whom completed a full term of 6 years. A discussion of the directors has relevance to all of the history of the Foundation, of course, but especially to biology as shall be seen repeatedly throughout this study.

The revolving door era had started, for present purposes, in 1976 when Atkinson took on the job of *acting* director until May of 1977. He had been permanent[1] *deputy* director from June 1975 to May 1977. That May he became permanent director but only remained through 3 of his 6 years in the post until June of 1980 when he left to become Chancellor of the University of California, San Diego. He was followed at the Foundation by physicist Donald Langenberg, who had to step in as acting director from June to December 1980. Langenberg was already permanent deputy director at the time, so he was on the scene at Atkinson's departure and filled both posts for a period of about 7 months. Note that the term permanent is not used at NSF in its strictest definitional form.

During those same years, the men who were *deputy* directors were sometimes permanent, sometimes acting, but not a single one of the total of seven filled out his full term either! Directors and deputy directors came and went then, each in less than 6 years. *Views and visions for the NSF necessarily changed with the changing*

[1] "Permanent" is used to mean congressional confirmation; no position was truly permanent.

personalities and agendas, in addition to their own varied disciplinary backgrounds: and many challenges inhered in such circumstances.

When Langenberg finally left NSF in December of 1982 as permanent deputy director, the Foundation would have already seen yet two more directors come and go. John Slaughter held the position from December of 1980 to that of 1982. Still another director filled the chair when Edward Knapp came in from November 1982 and left in August of 1984. That would be the close of *that* revolving door era, but it was only the first. Another was to come in later years.

That state of affairs was hardly conducive to manning a stable ship. Why was it so difficult to hold a director for his expected 6-year term during the period 1976 to 1984? *How did it affect the social and behavioral sciences, let alone biology* and the full Foundation? Each director provided his own answers to these questions as the tumultuous times extended beyond Guy Stever, 1975.

James P. Pfiffner has considered this highly unsettled period in the broader political arena that approached the end of the millennium. Directors of the NSF are appointed by the President and ratified by the Congress, and so are typically enmeshed in politics. Pfiffner has pointed out that:

> [i]n the latter quarter of the twentieth-century, Congress was transformed from a relatively consensual institution with significant overlap between the Democratic and Republican parties to an ideological, polarized battlefield with virtually no middle ground…Congressional polarization has ebbed and flowed over the history of the United States…[but] until the early 1970s, Congress enjoyed an unusual period of voting overlap in the middle of the ideological spectrum.[2]

However:

> [while there had been] plenty of ideological confrontation [for years]…*ideology did not reinforce partisanship*, because there were significant numbers of conservative Democrats and liberal Republicans who often crossed party lines in voting on important legislation [the annual federal budget, which included funding for the sciences, certainly being one of those]…But beginning in 1973 members of Congress began to vote increasingly along party lines, leading to the *intense polarization* that marked the end of the twentieth century.[3]

And, we might note, many years after that, as well. Pfiffner explained this in connection with changing demographics in the United States, especially the rise of the Republican party as a "viable political" entity due to the "migration of whites" from the North to the South in the period from the 1950s to the 1980s. These "conservative whites began to identify with the Republican Party."[4] Thus, when attempts were made to fund new or expanding programs within the social sciences at NSF, usually conservative Republican Congressmen turned a jaundiced eye to such matters and

[2] James P. Pfiffner, "Partisan Polarization, Politics, and the Presidency," in *Rivals for Power: Presidential-Congressional Relations*, ed. by James A. Thurber, Fourth Ed. (New York: Rowman and Littlefield, 2009). Quotation on ppg. 38–39.

[3] Ibid., pg. 39. Emphases added.

[4] Ibid., ppg. 39–40.

created difficulties for these human sciences. This chapter provides a number of examples.

The 1970s' and 1980s' era congresses had only partially helped offset the Republican's typically negative, or at least benignly disinterested, attitude toward the social sciences as

> the notion of defunding the left by getting rid of government programs that were purported to support the liberal agenda had taken root among members of the "new right." Their targets included the behavioral and social science programs at the NSF, the National Endowments for the Arts and [for the] Humanities, the National Institute of Education [etc.]. By the beginning of the Reagan Administration, conservatism was so strongly rooted that OMB was able to order NIMH [National Institute of Mental Health of the NIH] not to fund social science research unless it had a direct bearing on curing mental illness.[5]

The move toward "intense polarization" further favored the incumbents, who generally have the advantage anyway, of reelection. Thus:

> [f]rom 1984 to 1990 House members seeking reelection were successful 97 percent of the time…[though] Senators were a bit more vulnerable, but still quite successful, winning 86 percent of bids for reelection from 1982 to 2003.[6]

The social sciences were felt, by conservatives, to be too far left as disciplines. Indeed, by the end (1981) of President Jimmy Carter's single term, Pfiffner (citing others) noted that "most House Democrats would be on the liberal end of the spectrum and most Republicans on the right."[7]

In some senses paralleling the NSF senior leadership, stability in the form of full-term US presidents returned only in January of 1981 when Ronald Reagan began the first of his two complete terms. Concomitantly, the broader cultural milieu of America had also begun a headlong plunge into fiscal and social conservatism. The political version, already then well ensconced in Congress, was to last for most of the rest of the twentieth century and certainly into the first decade of the twenty-first. The electorate was inclined to vote conservatively. That conservative power in Congress for so long would alone suggest troubles for the social sciences, long maligned by the far right. Yet, as we shall see, the Reagan administration was not as fiscally unkind to the social sciences as might at first be presumed: his administration soon found data produced by the social sciences to be of use in the business of governing. For instance, critical data found only in the decennial census were necessary to many decisions made by Congress.[8] It was, as AD Betsy Clark said, "one of the things that aided social sciences' resurrection." However, it appeared that the Reagan administration would *start out* strongly opposed to funding the social and

[5] Johnson, "Next to Nothingness," Part III, pg. 323.

[6] Pfiffner, "Partisan Polarization," pg. 41. These data were cited by Pfiffner and which he derived from Burdett A. Loomis and Wendy J. Shiller, *The Contemporary Congress* (Belmont, CA: Wadsworth/Thomson Learning, 2004), pg. 66.

[7] Ibid., (Pfiffner) pg. 43. Pfiffner cited Sarah Binder, *Stalemate: Causes and Consequences of Legislative Gridlock* (Washington, D.C.: The Brookings Institution, 2003), ppg. 24 and 66.

[8] Clark interview.

behavioral sciences under an aggressive new director of the Office of Management and Budget, David Stockman. That was clear in the drastic budget "hit," as Clark had it, when, that first year under Reagan, social sciences' funding was reduced by about 44%, and the History and Philosophy of Science program "almost got eliminated in that period."[9]

The tumult in the directorship of NSF was still playing out its 8 years of revolving door directors until well into Reagan's first term. Engineer John Brooks Slaughter had a distinguished career, including earlier experience at NSF as Assistant Director of the Directorate for Astronomical, Atmospheric, Earth and Ocean Sciences (AAEO) from 1977 to 1979. He had degrees from Kansas State University, UC, Los Angeles, and his doctorate from UC, San Diego. He had worked at the US Navy Electronics Laboratory at San Diego and had directed the Applied Physics Laboratory at the University of Washington. He became Provost and Academic Vice President at Washington State University from 1979 to 1980 and later moved to the position of Chancellor of the University of Maryland at College Park from 1982 to 1988, among other academic positions.[10] That 2-year interim—1980 to 1982—saw him directing the NSF.

Slaughter liked his first experience at the Foundation as the AD of AAEO, referring to it humorously as the "Heaven and Earth Directorate" as it included astronomy and geology, among other fields.[11] Because he held that high position under Atkinson, Slaughter had an easy re-entry into the NSF milieu by December 1980 as agency director. But as such, he served only 2 of his 6 years as the Foundation's leader.

According to Slaughter, he was at the Physics Laboratory when he ran into Bill Nierenberg "by accident" in 1977 and what would lead to the first of Slaughter's two periods of senior office at NSF. Well-known scientist and scientific administrator, William Aaron Nierenberg, was then Director of the Scripps Institution of Oceanography (from 1965 to 1986) at UC, San Diego. Nierenberg wanted to advance Slaughter's name to NSF as the next AD for AAEO as the position was then open.

> I [Slaughter] said, 'Fine,' and I didn't think much of it at the time until I received a phone call from Dick [Richard C.] Atkinson inviting me to come back [to Washington, DC] for an interview, and about ten minutes into that interview he said, 'Let's walk over to see Frank Press,' and he announced to Frank that I was going to be his choice to be the assistant director [for AAEO].[12]

Geophysicist Frank Press was, from 1977 to 1980, the Science Advisor to President Carter and also Head of the Office of Science and Technology Policy, the OSTP. He would be the person to suggest to the president the name of any nominee

---

[9] Ibid.

[10] John Brooks Slaughter, "Computer Scientists of the African Diaspora," http://www.math.buffalo.edu/mad/computer-science/slaughter_johnbrooks.html. See also: "The Faces of Science: African Americans in the Sciences," at https://webfiles.uci.edu/mcbrown/display/slaughter.html.

[11] John Brooks Slaughter, interview by Marc Rothenberg, August 16, 2007, no accession number; original typescript held by Rothenberg.

[12] Ibid.

for any assistant director's position at NSF; ADs then still presidentially appointed. Note that Slaughter did not seek the position on his own but was sought out. He felt himself a "reasonable fit" as Assistant Director of the AAEO. However, he saw the NSF then as living a "double standard" with regard to grants being made to African-Americans (which Slaughter was himself). In 1980, when Atkinson sought a successor as he left for UC, San Diego, he thought immediately in terms of his recent lieutenant, Slaughter, by then at Washington State University as Provost. But the latter had been there for less than a year and a half. Once again, Slaughter got a call from Atkinson. And, again, Slaughter had not sought the position. For that reason, at least in part, Slaughter was able to say that he did not come with an agenda in mind when he filled the director's chair.[13]

Nonetheless, Slaughter was aware of the apparent unbalanced grant-making that left African-American applicants at a "huge disadvantage." Still, when called by the White House Presidential Personnel Office and offered the directorship, Slaughter was "frankly not excited" about it as he had been at Washington State for such a short time. Because of that, Slaughter balked at the NSF directorship offer. Indeed, the White House staff had to call him "a number [of times]" to convince him. Finally, the President had to call Slaughter personally:

Carter: Mr. Slaughter, this is Jimmy Carter. Are you in Washington?
Slaughter: Yes, sir, I'm in Washington State.
Carter: Oh, I know. I understand my people have been talking to you about becoming director of [the] National Science Foundation.
Slaughter: Yes, sir.
Carter: I understand you're reluctant to do so.
Slaughter: Well, yes, sir.
Carter: I just wanted you to know you're my choice…
Slaughter: Well, your phone call has heightened my interest [laughs].[14]

Why was Slaughter hesitant? For one thing, he had only recently taken the job of Provost at the University, and the change would represent a rather sudden departure from Pullman, Washington, to Washington, D.C. Second, with so many professional positions in so few years, yet another move was not terribly appealing to Slaughter and his family. Worse, the federal government did not, at that time, pay the moving expenses for senior federal personnel. And, possibly worst of all, the drop in salary from Washington State University would be difficult for the Slaughter family to manage. But, it would be hard to refuse a President of the United States, as his university president told Slaughter. When the appointment time came, though, and to Slaughter's unhappiness, he was "the *last* Carter appointee to be confirmed by the Senate."[15]

---

[13] Ibid.

[14] Ibid.

[15] Ibid. Emphasis added.

After just 2 years of excellent service, Slaughter chose to leave NSF, as he was disturbed by both Reagan's administration generally and especially with the US president's FY83 education budget. But there were other issues, too. Slaughter left the NSF even though he was not really "looking elsewhere." Having "turned down a number of invitations [to take other positions]," Slaughter told of one incident, among many, that turned him away from the administration. He did make it clear that he had been very happy in the job otherwise, however. Slaughter had received a "strange request" from Edwin Meese's office. Meese was then in Reagan's White House but is better remembered as Attorney General of the United States during Reagan's second term (1985–1988). An unnamed person in Meese's office told the NSF director that Meese "wants you to provide support for this [unnamed] program that I have in Michigan." Slaughter was "convinced that this person…had strong ties to the Reagan administration." The director called OMB and told them "this is not something I'm going to support." There were, Slaughter averred, "enough of those kinds of things that were happening that just made me realize that I was in the wrong place at the wrong time."[16] Politics prevailed in one of its many forms in Slaughter's passing from the leadership at NSF, but as many persons told the author, politics seldom sifted down to the program officer level, for instance, as had been the case here.

The "new federalism" of the Reagan years could be seen as different from both Ford's and Carter's administration as Reagan's FY83 budget was a virtual repeat of that of the previous year; there would be none of Carter's planned increases. There *would* be "strong support for national defense and 'letting the marketplace decide'" about many other aspects of the economy; that is, Reaganomics.[17] But, in NSF's mathematics and physical sciences' arena, an increase of 15% for FY83 would be enjoyed as compared to that of the previous year's figures. Engineering would climb by over 23%; and the Division of Scientific, Technological and International Affairs would jump by a significant 28%. All of these had direct applications to Reagan's strong military preparedness orientation. On the other hand, certain areas of biology received only 0.6% (after inflation)! But much worse, "[s]cience education programs, gutted in a controversial series of budget-paring sessions [for FY82]…would be totally eliminated in FY83." Slaughter found himself having to explain to the nation's press that "[t]he Minority Institutions Sciences Improvement Program would also be eliminated this year [FY83]."[18]

Slaughter's initial reluctance to take on the position of Director of the NSF might well be called a unique case, as might his reasons for leaving the Foundation's top position 2 years later. However, for each of the revolving door directors, there were compelling reasons why they were less than interested in the first place, or merely stood in as they were already present on site (e.g., Langenberg), or left for one

[16] Ibid.

[17] J. Raloff, "New Federalism: Old Story for Basic Research," *Science News* (1982):100–102. Quotation from boxed commentary on pg. 100.

[18] Ibid. All quotations are Raloff's.

personal reason or another. Knapp left in August of 1984 to return to the Los Alamos National Laboratory, from whence he had come to NSF. That left open the possibility for Reagan to appoint a very different type of director to NSF but one which fits the economic views of the man already then president for nearly a full term.[19]

In August of 1984, Reagan appointed the first person to be director who came from industry: Eric Bloch. Bloch would serve a full 6-year term. *For reasons to become apparent, I will argue that Bloch's tenure may have been the most important of all of the director's terms of the later twentieth century, certainly for biology, computerization and computer science, and possibly for much else at the Foundation.*

Under each of the revolving door directors, all of NSF had to respond to the personality, interests, and agendas of those several leaders, as well as their superiors, presidents Ford, Carter, and Reagan, let alone an always chameleonic NSB and a rotating program officer system within the rank and file of NSF itself. New faces were constantly to be seen. Each director, Atkinson, Langenberg, Slaughter, and Knapp, book-ended by Stever and Bloch, brought different experiences and had different views of what NSF should be.

Richard C. Atkinson, however, represented a rather notable case for this part of the history of the BBS and so requires a detailed analysis.

2.3 Richard Atkinson: First Social Scientist Director of the NSF

Otto Larsen termed the 1970s a time of "relevance, RANN, [and] reorganization."[20] While RANN has been discussed earlier, it had a profound effect lasting long after it concluded. The social sciences,[21] in particular, were strongly affected by NSF's deliberate move toward *applied research* that had been called for in the 1968 Daddario-Kennedy amendment. Indeed, the term *applied research* and the social sciences became *intertwined* for many years, and the effect upon each, but especially the latter, was not always a desirable one.

Atkinson, as Larsen opined, "spent considerable time trying to achieve acceptance in the eyes of the physical scientists."[22] This was so as Atkinson was a very different creature in the appraisal of many employees of the NSF: members of the so-called hard science community that included, most especially, physicists, chemists, oceanographers, geologists, astronomers, mathematicians, as well as engineers. He was less suspected in the biological community and, not surprisingly, much

---

[19] Jack Renirie, "President Announces Intention to Nominate Bloch as NSF Director," news release NSF PR 84-36, June 8, 1984, pg. 1.

[20] Larsen, "Milestones and Millstones," Chap. 5, ppg. 91–127 *passim.*

[21] The term "social sciences" will be used loosely, particularly in this chapter, to include also the behavioral sciences for brevity unless the behavioral sciences need to be specifically referred to in the text.

[22] Larsen, "Milestones and Millstones," pg. 91.

welcomed by the social scientists. But for Atkinson, following a parade of physical scientists into the leadership of NSF, he apparently did feel he had to "achieve acceptance." He reminisced about that very matter years later as he told the story of visiting Columbia University just shortly after being appointed permanent director of NSF and of meeting the renowned physicist Isidor Isaac ("I.I.") Rabi. Rabi had worked on both radar and the atomic bomb during World War II and received the Nobel Prize in Physics in 1944. Atkinson recorded that Rabi told him (Atkinson) that he had heard "only the best things about [me]…and, by the way, what field of physics was I in?"[23]

During the Atkinson years (mid-1976 to mid-1980), RANN was almost totally dismantled, but the notion of applied research had by then already become embedded in the character of NSF and would stay on until the present day, though ever-changing, ever reinterpreted. It must be remembered that in Stever's later years at NSF, he was also science advisor to President Ford. *Deputy* Director Atkinson, then, had a lot on his plate and came to understand the NSF very well and that through the filter of a social scientist. Whether applied research had become embedded or not by Atkinson's time as *director*, he demonstrated considerable hubris in saying in the Director's Statement in the *Annual Report 1979* that "[p]ractical benefits will come later, *if at all.*"[24] And although RANN would expire officially, though not be fully eliminated, on September 15, 1977, Atkinson was already Director of NSF by then for over a year. He had to, as a "soft scientist,"[25] deal with the realities of the new direction dictated by Daddario, IRRPOS and RANN—1968–1977, and beyond. In 1976, Congress added 15% to the NSF budget request specifically for RANN research in areas of social concern (e.g., welfare) and national administrative interest (e.g., productivity problems).

At that time, what came to be called the Simon Report was the single most damning negative perception in print of RANN regarding the social and behavioral sciences. While Atkinson was still the deputy and had not as yet become the acting director, he initiated (1975) the effort of analyzing the status of the social and behavioral sciences at NSF. He set Herbert A. Simon, a mathematically oriented social scientist who would win the Nobel Prize in economics in 1978, to write the report along with a committee of social scientists from *outside* of NSF.[26] The work

[23] Richard C. Atkinson, *The Pursuit of Knowledge* (Berkeley: The University of California Press, 2007), pg. 28.

[24] National Science Foundation, *Twenty-Ninth Annual Report for Fiscal Year 1979* (Washington, D.C.: US Government Printing Office, 1980), pg. viii.

[25] "Soft" was a term long considered pejorative by the social science community and, as it became more quantitative over the later twentieth century, less justifiable, if it ever had been. Furthermore, the term did not fit Atkinson well as his interests were, in fact, in the *quantitative* aspects of psychology.

[26] See Simon's partial autobiography at http://nobelprize.org/nobel_prizes/economics/laureates/1978/simon-autobio.html.

was done under the aegis of the National Academy of Sciences/National Research Council.[27] In effect, it was a report card.

The members of the Simon Committee were derived from academicians' ranks across the country. The Committee met formally in each of the last 4 months of 1975 and January of 1976, precisely at the time the overall reorganization of NSF was concluding; thus their work contributed, in part, to that effort. The members examined "more than 150 published and unpublished reports and documents" of relevance to their task and finally included some 300 persons that were interviewed or in other ways contributed to the final product.[28] Simon and the members of his independent panel reviewed each one of the social and behavioral science programs then extant at the Foundation. The members of the NSB received their copies of the report in February of 1976.[29] In the following September, Herbert Simon appeared at the Board meeting to answer questions.[30]

The Simon Report examined a number of issues, not least of which was the tie between social sciences' programs and basic versus applied research at NSF; the *intertwining* noted above. Also, the Report provided analysis and criticism of every one of the programs. The Simon Committee's views on basic and applied research were highly useful for Atkinson's efforts at redirecting the NSF, at least to some degree, toward being not only more tolerant of the social sciences but to embrace the methods and resultant data they could offer.

The "Principal Findings and Recommendations" numbered some 16 in all: 10 in basic areas and 6 in applied. The Committee found that the "quality of basic research in the social and behavioral sciences supported by the Foundation [was] generally excellent " but that many fine proposals were rejected due to lack of funds (Finding 1). Large-scale projects were "discouraged, rejected or reduced" (Finding 2), and only 30% of qualified social scientists were awarded grants versus 58% in the other sciences (Finding 3)—a clear bias, it was implied, that the social sciences were underfunded and undervalued. The Committee discovered (Finding 4) that certain social sciences were in especially dire straits with regard to funding: anthropology, psychology, and political science. Political science, at least, had been (and remains) a long-time favorite target of the conservative constituency in Congress. In the Report's Finding 5, the Committee suggested the establishment of programs in sensory physiology and perception and in memory and cognitive processes (those suggestions were taken shortly thereafter). They also found that the NSF was missing opportunities in areas of survey research, *long-term research or longitudinal studies*, and *development of large databases* (Finding 6). The latter two would become major areas in not only the social and behavioral but also the biological sciences: *one of the social sciences' effects upon the biological.* While some disciplines in the

[27] Many reports for NSF-connected persons and programs, and for Congress and other science agencies, are done by the NAS/NRC as an independent authority.

[28] "Simon Report," pg. 3.

[29] NSB, "Minutes," 179:11, June 29, 1976.

[30] Ibid., 183:3, September 21, 1976.

social sciences were found by the Committee to be quite quantitative, such as economics and sociology, others were felt to need more resources to help them to move from strongly qualitative to more quantitative research, such as sociological field research and ethnographic analyses (Finding 7). Panels were needed for all programs not yet served by them (Finding 8), more community input was needed across the board (Finding 9), and more representation to the NSF director and Congress should be forthcoming from the social scientists and the appropriate NSF administrators of those disciplines (Finding 10).[31]

The six remaining recommendations emanating from the Report dealt with applied research. Finding 11 noted that the RANN work was "variable in quality… and, on the average, not impressive." The *intertwining* of social science research and RANN was not the way to satisfy Daddario-Kennedy, it appeared. The decision to eliminate RANN was, in part, a decision Atkinson would make. What's more, he could show that it would be upheld by data found by the Simon Committee. Further (Finding 12), the procedures for developing RANN programming failed to involve applied scientists sufficiently enough. And since RANN was under the direction of only a few staff members, the healthy direction of research funds granted was under question (Finding 13). The final three findings all addressed failures in RANN as it intertwined with the social sciences. RANN did not survive once the Simon Report was in Atkinson's hands. It was Finding 13 that was especially critical:

> RANN procedures now reflect excessive confidence in the ability of a few staff members to determine the proper direction for research programs. Staff pressure to tailor research proposals to narrowly specified programs, planned from the top down, plays too large a role in RANN funding decisions, and response to genuinely unsolicited proposals plays too small a role.[32]

The Simon Report was not without one NSB member issuing a formally titled "Minority Report." Dr. Anna J. Harrison, a distinguished chemist long on the faculty of Mt. Holyoke College, Massachusetts, and on the NSB from 1972 to 1978, felt that *communications' problems* within the social science areas of NSF must be solved as one aspect of the findings of the Report. Enhancing the social sciences by increasing funding and staffing in the still extant Directorate of Research Applications (Fig. 1.1) was another. She also wanted to see more large-scale endeavors, as well as to review the social science areas regularly.[33]

When Atkinson became director, he determined to eliminate RANN. As he was a protégé of Senator Edward Kennedy, and even though the Senator was co-sponsor of the Daddario-Kennedy amendment, Kennedy supported him in that plan.[34] But it was not until long after Atkinson's departure in March of 1981 that the last of the dismantling of RANN was completed and the funds that had been tied to RANN's social science programs, as well as the Program Officers who headed them, would

[31] "Simon Report," ppg. 5–7.

[32] Ibid., pg. 8.

[33] NSB, "Minutes." 184:28–29, October 22, 1976.

[34] Larsen, "Milestones and Millstones," pg. 100.

be absorbed into two of the four divisions of the BBS. The earlier Division of Social Sciences had already been absorbed, of course, and would be renamed, with the short-lived title, Division of Economic and Social Sciences, SES, very soon thereafter.[35]

The social and behavioral sciences had become then very much *intertwined* with applied research, technology, and, by extension, engineering. The Simon Report seconded that view. While intertwined they may have been, they were still lodged with the biological sciences for the value to be found in "protective coloration." Although talk had been ongoing for years about the possibility—and desirability—of a social and behavioral sciences' directorate independent of biology, not all those concerned about the organization of NSF overall, the psychologists and the anthropologists/archaeologists, wanted it. One reason that first group was in opposition was that the biopsychology (later, psychobiology) program was tied to the biology division of Behavioral and Neural Sciences and had never been in a social science division. Anything that revolved around the brain/mind was ever on that fenceline that divided the life and social sciences' organizational hierons and to which we return below. And, too, the "protective coloration" issue continued to be of value for psychology, especially. Only somewhat humorously did a social scientist describe the 1970s as a decade "of approach and avoidance for behavioral and social scientists."[36] More, too, will be said of the archaeologists.

With the 1975 reorganization, Larsen argued that the social sciences actually dropped down one level in standing as they had been a division on a par with biology when both were under the pre-1975 Director for Research structure. As only two of the four divisions within BBS, it is true that the social and the behavioral sciences were, after 1975, one step further from the ear of the director. Added to that, the AD of BBS was, from 1975 to 1991, always a biologist, never a social or behavioral scientist.[37]

The 1975 reorganization split the programs in the old unit for the social sciences. Howard H. Hines had been the director of that group, which would, only months later, become the Division of Social and Political Sciences. Hines expressed his unhappiness with the split in November of that year:

> The Division of Social Sciences regrets losing—to the extent, I hope it's a very small one, that we are 'losing'—our oldest program, Anthropology, and our newest, Linguistics as well as the one which probably has more interaction with the other Social Science Programs than any other, namely Social Psychology.[38]

In fact, Hines and his colleagues did not lose anthropology, linguistics, or social psychology exactly, as they were still nearby, though not under *his* leadership. They had been moved into the second of the two social science divisions in the also then new BBS; this is the Division of Behavioral and Neural Sciences. That second

[35] Johnson, "Next to Nothingness, Part II," pg. 265.

[36] Ibid.

[37] Larsen, "Milestones and Millstones," pg. 101.

[38] Ibid.

division was led by Richard T. Louttit, who came to the NSF from the NIH. Although anthropology, linguistics, and social psychology are mentioned slightly below, as they did affect biology, the bigger *behavior* programs (animal behavior, psychobiology, and neuroscience) came to exist at that time and had very direct relevance to biology, as shall be seen.

In 1977, Louttit, in his first program report for director Atkinson, had defined the scope of his division. Louttit said that:

> We [in the division] are concerned with understanding human and animal behavior, and the biological basis underlying that behavior…The two programs in the neurosciences focus on the structure and function of the nervous system and on how we receive, encode, and process information from the environment through our sensory systems.[39]

Speaking in terms of "*biological* determinants of behavior" and "development of behavior in humans and *animals*," every effort was being made to tie social and behavioral sciences to biology and thus "divert attention" from those disciplines: they were under attack again by the political community. There seemed a continuing need for "protective coloration."[40]

The formal birth of the behavioral sciences had been accomplished during the creation of two of the four divisions newly appearing in BBS. They would last, in one form or another and under one rubric or organizational heading or another to the time of this writing. They did not receive more funding, by percentage of request at least, than the Division of Social and Political Sciences. However, that latter division probably required more protective coloration than the behavioral sciences, as members of Congress continued to attack the social sciences, in particular. Bruce Umminger, a program director in the biological sciences, and at NSF for over a quarter century, was hardly alone in pointing out the necessity of protecting the social sciences at the annual budget creation exercise: "we [BBS] spent much of our time trying to make a case for the social sciences…nobody was really questioning biology…but they were always going after the social sciences."[41] But both social and behavioral sciences got about 95% of their requested funds from 1976 to 1980—a very good showing.[42]

Though the period from Guy Stever's departure to the appearance of Erich Bloch was that of revolving door directors, change also occurred in the leadership of the BBS. Betsy Clark had been for some years the division head of BMS and was by 1975–1976, first acting, then permanent, AD of the new BBS, as has been recounted.[43] However, two more ADs were to follow in BBS prior to the reorganization of 1991. The first of those was David Kingsbury (1984–1988). The second was Mary Clutter, who continued through the 1991–1992 reorganization to 2005. All three of

---

[39] Ibid., pg. 102. Larsen quoting Louttit.

[40] Ibid. Emphases added.

[41] Bruce Umminger Interview with the Author, May 13, 2009, by telephone.

[42] Larsen, "Milestones and Millstones." pg. 102.

[43] NSB, "Minutes" 183:5, September 21, 1976. Betsy Clark was officially nominated by the president on July 2, 1976 for the permanent position of Assistant Director of the BBS.

these individuals are considered throughout this history. Kingsbury's shorter term was especially intimately tied to the social and behavioral sciences. He has told this author that he and Larsen got along "fabulously" as Larsen helped biologist Kingsbury to understand the social and behavioral sciences.[44] Larsen "would give me papers [on those fields] to read [every night]." In the morning, the two would discuss them with Larsen playing instructor and Kingsbury playing student.[45] Mary Clutter, AD for BBS just prior to 1991–1992, would play a smaller role vis-à-vis the social sciences. She, too, was a biologist and will become a central figure in most of this historical study—but, as shall be demonstrated, she was a major contributor to the rise of *Millennial Biology* overall.

Other matters were afoot during the Atkinson era, and that of his successors, that will rise in significance as this history proceeds. An issue of considerable importance at NSF over time was the move from disciplinarity toward *interdisciplinarity*. In this connection, Atkinson cited the thoughts of Derek de Solla Price, physicist and historian and philosopher of science, who had said that if the growth rate of American science were allowed to flatten there would be a "relentless increase in narrow specializations in scientific disciplines [and] a decline in interdisciplinary cross-fertilization." Some of the most notable scientific achievements of the last third of the twentieth century came, and still comes, from efforts highly interdisciplinary in nature. *NSF organized itself both to encourage this type of research and also responded to it as the scientific community itself tended in this direction.* The dual themes, among others, of the rise of interdisciplinarity and also of the reciprocal conversation between NSF and its constituencies, or communities—the *intricate dance*—will resurface repeatedly below.[46]

2.4 The Social and Behavioral Sciences at NSF from 1975 to 1991: Effects on Biology

In an old Arabic adage, it is said that "some days are honey, some days onions." It had been that way for decades for the social and behavioral sciences at NSF, if not more broadly felt as well by certain politicians and others in a *complex society that needed social data to be understood*. The "time of tumult" clearly proved society's need for the various disciplines of the social sciences and what they could provide for the American populace, its govern*ment*, and its gover*nance*. But it took years for them to be included fully into the realm of the sciences: that came finally in BBS, of

[44] Kingsbury interview, *op. cit.*

[45] Ibid.

[46] The quotation is to be found on pg. x. The thoughts are Derek de Solla Price's, as Atkinson acknowledged, but the words quoted here are those of Atkinson. See Price's *Science Since Babylon* (New Haven, CT: Yale University Press, 1977) for context.

course, where they were on a titular par, at least, with the "harder" biological sciences.

The social and behavioral sciences are very useful, indeed, but must be discussed here *qua* disciplines in a limited context: what role(s) did they play overall that were manifested with regard to the *biological sciences* then and later? As the primary focus of this history is upon the several biological sciences and their organizational units over three decades, one might ask if the social sciences affected the biological sciences in their maturation both at NSF and in the community. In fact, they did. We have already seen some of that evidence in regard to the *rise of applied research* at the Foundation, which had been so *intertwined* with the social sciences, as well as the rise of *longitudinal research*. To more of those effects we now turn.

2.4.1 Long-Term Studies

Sociology, *qua* discipline, was never a particularly large program in the earlier years of BBS, but that would change by the mid-1980s. The budget in FY84 was just under $3 M and so the Program was in the top two or three in BBS for funding at that time.[47] In earlier days, some of sociology's programmatic work was quite significant in producing both basic knowledge and easily applied data and data analyses. Those data/analyses could be directed to specific federal agencies or to the government as a whole when interacting with foreign powers, as just one example. Indeed, as the authors of the Simon Report noted in 1976:

> [h]istorically, the sociology program [at NSF] has paid a good deal of attention to *increasing the rigor of research methods and of theorizing, finding new methodologies for analyzing data*, and improving the quality of data.[48]

After all, as has been said anonymously, "[r]aw data, like raw sewage, requires some processing before it can be spread around."

Studying social institutions included "discern[ing] basic patterns of human social organizations." The interest lay in "the search for general patterns and principles… the dependence on reliable and reproducible data describing social institutions [and] the systematic test of general principles against reliable data."[49] To the extent that a portion of those data were more quantitative and thus amenable to placement in *databases*, sociology and related disciplines were moving in a coordinate fashion with biology and may have been ahead of it.[50]

---

[47] David Kingsbury to Members of the BBS AC, January 25, 1985, "Reading Materials for Meeting on February 4–5, 1985." A series of unnumbered pages, "profiles," for the 18 programs (equally divided between social sciences and behavioral sciences), followed.

[48] Simon Report, pg. 41. Emphasis added.

[49] NSF, "Annual Report FY76," pg. 66.

[50] Edward O. Wilson's review of the book edited by Richard C. Lewontin, *Population Biology and Evolution: Proceedings of the International Symposium, June 7–9, 1967* (Syracuse, New York:

The Simon Report pointed out that a special project had been funded within the Sociology Program termed the "Assessment of Survey Practices and Data Quality in Scientific Surveys of Human Populations." The Report averred that such a project would go a long way in the area of *longitudinal studies* as the Simon Committee felt strongly about such research approaches. Indeed, NSF became noted for these long-running studies in many other areas: the Man in the Arctic Program was such a project of the period. Significantly among those is the Long-Term Ecological Research Project (LTER), which is discussed at length later in this history and which, as it matured, added a sociological component, especially in the more recently included urban ecological sites. LTER remains a vigorous program to the time of this writing and, as in sociology and other social sciences, employs *long-term studies*. Furthermore, it brought into a mostly non-human ecology project a component very much in the tradition of human ecology. Finally, LTER shares with the social sciences the use, the necessity, of *large longitudinal databases*.

The Simon Committee was quick to point out that they, by no means, were suggesting that longitudinal and large-scale studies should become the bread and butter of the division but rather continue to do as NSF had done for its first quarter century—and continues to do to the present: support "the best-conceived, best-justified projects that come to it as unsolicited proposals."[51]

Still, by 1978, Atkinson and the editors of the NSF *Annual Report* emphasized other large, longitudinal studies that continued to provide valuable data for the nation's governance:

> the accumulation of data over a sufficient period of time...permit[s] the exploration of trends and the examination of relationships before and after critical changes in societal conditions.[52]

Edward O. Wilson, entomologist and founder of sociobiology, would surely have concurred: *interdisciplinary* research brought together sociology, economics, and biological data for his own interests and those of many in the communities.[53]

Within the governing of the NSF by the NSB, another sociological issue had been playing out in the 1970s and 1980s. As comparative psychology began to cross the line from non-human animals to humans, the NSB thought it well to establish a

---

Syracuse University Press, 1970). Review in *Science* (1970):1184. It was Wilson who said that, "[a] science can be said to enter maturity when its theories become predictive." Much truth could be seen in that with regard to many aspects of the growth of biology from the later twentieth century into the first years of the following millennium.

[51] Ibid., pg. 44 under the section on "New Opportunities and Directions".

[52] NSF, *Twenty-Seventh Annual Report for Fiscal Year 1977* (Washington, D. C.: Government Printing Office, 1978), pg. 75.

[53] The notion of the great importance of interdisciplinary research in the social sciences was explored in Karl W. Deutsch et al., "Conditions Favoring Major Advances in Social Science," *Science* (1971):450–459. They noted that "Interdisciplinary work has been a major intellectual source of contributions throughout the period [1900–1965]; responsible for nearly one-half of all advances from 1900 to 1929, it produced nearly two-thirds of the total thereafter." Quotation from pg. 458.

statement on human subjects' research for the first time. They did so in April of 1978 with a resolution stating that research on human subjects "shall be conducted in a manner that adequately protects the rights and welfare of [those] subjects" and should conform closely to what the National Institutes of Health was doing at that same time.[54] The NIH had been developing proposed regulations for the protection of human subjects in research situations; an issue strongly tied to a number of areas of biology and the more social endeavors (e.g., educational research). Today a complex area of regulations and paper trails, the then new matter, was easily handled by the National Science Board: it took the *thinking of NIH* and agreed to follow that by "conform[ing] as closely as possible with [their policies] and [NSF] shall impose on recipients of [Foundation] awards a minimum of different or additional requirements and paperwork."[55]

That decision was particularly interesting as it evidenced, at least for a second time in as many years, NSB's tendency to subjugate its decision-making in some matters to that of NIH. Where matters of much of human science was concerned—the old issue of the late 1940s and the beginning days of the NSF in the early 1950s—still held great sway, NIH was the organization to mimic when it came to issues touching upon medicine or closely related human science. The second case of the period, rulings by NIH in the use of recombinant DNA (genetic manipulation), will be considered in greater detail in later chapters and will expand discussion upon this conforming tendency.

The matter of surveys remained important in the division. By the time of the *Annual Report* for 1981, there was a growing realization that "missing data" was a distinct problem when seeking the most information possible from a data set. At that time, a National Academy of Sciences panel had looked into the matter and recognized that "[t]he loss of data has the effect of decreasing sample size." Therefore, better statistical tools to deal with this were sought, as an NAS recommendation. The National Academies pursued the issue in the following years and NSF provided grants for sociologists researching the topic.[56]

---

[54] NSB Resolution 78–45, April 20–21, 1978; 197th Meeting of the NSB. See also Edward A. Knapp, Director NSF to Members of NSB, "Protection of Human Subjects," NSB-83-210, July 21, 1983. The memo was a follow-on to H.R. 2788, 98th Congress, 1st Session, in which Senator Albert Gore introduced the bill: "To Establish the President's Commission on the Human Application of Genetic Engineering," April 27, 1983. By mid-1983, a presidential commission had taken on the matter government-wide and established a Model Federal Policy, to which NSB/NSF would adhere. The Board adjusted its wording in the April, 1978 resolution to conform to the Model, but little else changed: NIH would still be the agency to whom NSF would look to for detailed guidance.

[55] NSB, "Minutes," 197:9, April 19, 1978; NSB Resolution, Res-78-45. The area of the protection of human subjects is best covered, for the purposes of this history, as part of the discussion of recombinant DNA research in the chapters on biology (main text below) under BBS and under the later BIO directorate.

[56] National Science Foundation, *Thirty-First Annual Report for Fiscal Year 1981* (Washington, D.C.: Government Printing Office, 1982), ppg. 42–44.

There were three Sociology Program Officers by the early and mid-1980s making it one of the larger SES divisional programs by not only funding, but by number of POs, as well. By 1987, Phyllis Moen, then at Cornell University, had taken over, first as assistant director and then as director of the program. Conditions were difficult for the social and behavioral scientists at NSF, it was known, as she had been given advice by a colleague there to: "[k]eep your back to the wall and keep looking in both directions."[57] To that, she expressed in her autobiography, "[h]e was right!" While she found it challenging to work with "[t]he mostly male, mostly natural scien[tists]," she did find that BBS AD, and biologist, Mary Clutter, "broadened my understanding not only of sociology, but also of social research more generally and of science policy 'writ large.'"[58] Much more of Mary Clutter later.

2.4.2 *Interdisciplinarity*

Interdisciplinarity had long been a way of life in the social sciences. It was not uncommon then that such things as shared projects among disciplines would be both expected and desired. However, a decade later archaeologists would be one of *only two* social science disciplines to vote a *strong nay* to the plan to separate social and behavioral sciences from biology in order to create a new directorate for each. (Psychologists were the other, as has been alluded to above; see more on this matter in Chap. 5). Why that should have been *may* be due to certain peculiarities of the discipline. There has long been a separation between physical and cultural anthropologists. Also, both anthropology and archaeology have long maintained a somewhat curious vision of their own existence such that some aspects of what they do in their research *they themselves do not consider science*. As recently as 2010, that had become an issue, dividing members of these fields as the American Anthropological Association (AAA) decided to "strip the word 'science' from a statement of its long-range plan."[59] The logic was that the Association would no longer "advance anthropology as a science but rather…focus on 'public understanding.'" That was reversed some days later due to the outcry from their community. The split within the AAA had begun with the new millennium, at least, but just how such earlier views, if present in the BBS, could have played into the 1991–1992 reorganization is unclear. What is known is that some degree of separation between the "science-based" (physical) anthropologists and those doing research in *cultural* areas (not held to be science), specifically, had been in existence for some time.

Withal, by FY84, the anthropology/archaeology program was one of the most heavily funded of all those in SES. It had attained a level of well over \$6 M and had made 544 awards that year alone. One of the oldest of all the social science areas at

[57] Phyllis Moen, *Constructing a Life Course* (London: Haworth Press, 2001), pg. 103.

[58] Ibid.

[59] See http://www.nytimes.com/2010/12/10/science/10anthropology.html?_r=1.

NSF, having been established in 1954, the program had six staff officers by the mid-1980s, again among the largest for all of SES. It should be noted that the anthropology program was difficult to separate from either the behavioral sciences *per se* or even from biology, in some instances. Those fields included, particularly, human biology broadly, human adaptation to the environment, comparative primatology, and, especially, human origins (all NSF programmatic areas).

A human origins project was in keeping with the field discoveries of the era and had been started in FY80. One issue of *Time* magazine in 1977 was the largest seller of that year. On its cover was a picture of Richard and Mary Leakey and their research in the archaeology of human ancestors in the now famous Olduvai Gorge of northeastern Africa. Betsy Clark penned a transmittal memo to the NSB with an attached report on the subject in October of 1979. The *Comprehensive Approach to the National Science Foundation Support for Human Origins Research* made five specific recommendations: increase funds for the subject by $0.5 M per year for 5 years (to $3.5 M); research should be *interdisciplinary*; expand certain areas of research within the subject; assist those not then getting funded; and award fellowships.[60]

One of the values of having SES (and BNS, in particular) in the BBS of the period was the possibility for *greater communication among the various parts of the directorate*, including all of biology, as everyone attended the directorate-level meetings. Sharing of information allowed appropriate parceling with regard to which program would fund which successful grant application. Once the directorates split in the beginning of the next decade, a greater sort of intellectual distance was to be found (which was also a continuing theme in Larsen's history).

2.4.3 Big Science

There was another and very important developing trend toward what has been called *big science* ongoing in the social sciences during the decade of the 1970s. It may well have been that the types of expenditures changed the landscape slightly away from smaller, unsolicited project proposals that had been and would continue to remain the "bread and butter" of NSF. So-called *big science* will be examined again in biology in later chapters but requires defining, as understood at the National Science Foundation at this point when the term was first used extensively in the economics area. Alvin Weinberg's 1967 book defining big science broadly (cited earlier) found in the NSB early advocates for such science. Leland Haworth, an earlier director of NSF and thus an ex officio member of the Board (1963–1969), had been praised later (1979) by the NSB as a significant early supporter of big science. Haworth's inclinations were contemporaneous with that of Weinberg's in the

[60] National Science Foundation, *A Comprehensive Approach to the National Science Foundation Support for Human Origins Research* (Washington, D.C.: NSF, October, 1979); see also Eloise Clark to NSB, October 16, 1979, NSB-79-399, NSB Files.

same matter.[61] And, too, Director Stever had seen the first flexing of the Daddario-Kennedy amendment that emphasized applied, and therefore often necessarily *big* science, research.

At that time then, the NSB was concerning itself with this matter also, by having created an "Ad Hoc Committee on Big and Little Science." Its first two meetings had occurred in the fall of 1978. This Committee was formed in "preparation for a response" to the House Committee on Science and Technology's request for the FY79 budget authorization bill for NSF.[62] The CST wished to see no proposals for what might be termed big science *without* an agency first establishing procedures for setting the appropriate funding parameters: NSF had not yet done so. The Ad Hoc Committee already had in hand a study paper from its staff on the matter, so NSB felt they had a good case to take to the Congress and the OMB. A separate NSB Working Group report on big and little science had noted that an increasingly larger amount of the NSF's budget was being committed to "fixed items" (*long term* and often *big*) and that the number of such items was on the increase.[63] They suggested "clear, tough-minded policies [and] tougher decisions" for the future in this arena. Also, the Working Group wanted NSF to be more "assertive" with OMB and the CST with regard to increasing funding such that "very large-scale projects" (read big science) could be undertaken. The Group, along with its Ad Hoc Committee, *defined* big science by stating that it was to be viewed *at NSF* as having *all* of the following properties:

a. Large-scale commitment of financial resources
b. Investment of capital in facilities and major equipment
c. A duration of several years or more
d. Continuing expenditures for maintenance, replacement, operating costs, and research budgets[64]

The NSB went on to say that big science programs could be one of two types: those that could be accommodated within budget levels (known as "add-ons") and those that were so exceptional that a special proposal to the president and Congress would be made independently of the main annual budget process. An extensive and detailed policy framework and procedural guideline for administering big science followed the definition.

The main budget for FY79 for all of BBS was $157 M. The *total* NSF budget request for the next year was $1.006B—the first billion-dollar budget request in NSF's history. They would not make that milestone that year if the House had the final say, but it does not. The CST authorized nearly what NSF requested, $999.5 M,

---

[61] NSB, "Minutes," 205:2, May 22, 1979.

[62] NSB, "Minutes," 202:16, February 5, 1979.

[63] The final report for Working Group (WG) 3 had been announced in the NSB "Minutes" for February 19, 1979 at 203:11. There were two other groups at the time assigned duties on long-range planning,;WG 1 would include BBS, among other directorates. More below.

[64] NSB, "Minutes," 203:31–34, February 19, 1979, as Appendix D and as document NSB-79-65 dated January 19, 1979 (the actual Board meeting date, as opposed to the published date of the Minutes: here, February 19).

with BBS having lost some $14 M on the House floor in debate. However, the final say comes from the Senate, and they added several new programs, including Ethics and Values in Science and Technology and Memory and Cognitive Processes. This brought the authorization to NSF of $1.0095B—above the Foundation's original request.[65] The Conference Committee report recommended $1.0077B, and on August 2, President Carter signed the bill making it finally the first billion-dollar budget for NSF.[66] That was Carter. Ronald Reagan had not yet arrived on the scene, and, when he did, he would reject Carter's *entire* national budget!

Larsen noted the trend toward big *social* science when he listed a number of very large grants, not much seen in the social sciences prior to the 1970s. In 1975, something over 9% of the budget went to grants over a quarter-million dollars each. In 1976, it was 10%. By 1978, it had grown to nearly 12% in a budget approaching twice that of 1975. *Longitudinal database support would be an example of big social science*. However, a quick glance at the list brought up a difficult point which had been in the wind at NSF since the early 1950s: an apparent undue concentration of grants made to geographically limited locales (the Northeast) or to institutions (such as Yale and Stanford) which had already fared well over time due to capturing many Foundation awards. For the time being, the issue was set aside, but it would have to be put back on the table in the not too distant future.

Two areas that prospered were not surprising given the Simon Report suggestions: social indicators (not discussed here) and national databases. Also, econometrics was a "principal beneficiary," too. Indeed, there was protective coloration and the riding of coattails' value in this because the largesse for those three fields "blunt[ed] political criticism" in later years and thus "salvaged [other contentious, or smaller] disciplinary programs in the 1980s."[67] But the funding was dwindling: for instance, the number of unsolicited proposals, in all areas covered by BBS, increased by 48%, from 4993 to 7400, between 1976 and 1978. However, and not unexpectedly, the success ratio for grants awarded moved from 40% to 29% in the same period.[68]

How the NSF addressed the CST at the annual budget authorization event was also in a state of flux. For many years, ADs were charged with presenting their individual directorate budget requests to the CST but, about 1980, that ceased to be done. Thereafter, only the director of NSF presented the budget for the whole of the Foundation. The now-defunct meetings between the ADs and the CST had been important, Betsy Clark said:

> because that really meant that each program would basically write a mini program review and put forward its best projects out of which we would then build testimony to present to the [CST. When that changed] the director was also changing [Langenberg was acting] and so the director's Program Reviews were discontinued. *The whole formality of requiring*

[65] NSB, "Minutes," 206:11, June 25, 1979.

[66] H.R. 2729 became P.L. 96–44, once signed by the president.

[67] Ibid., pg. 108.

[68] National Science Foundation, *Twenty-Eighth Annual Report for Fiscal Year 1978*, (Washington, D.C.: Government Printing Office, 1979), pg. 61.

reviews for both the [National] Science Board and the Congress were essentially thrown by the wayside.[69]

That situation seems odd, in retrospect, as it was very much a question of *accountability* and America was heading toward what has been termed the "conservative revolution." It would thoroughly intensify within a few years. "The political climate for social science at NSF had changed dramatically," Larsen stated. His predecessor added that it would be "difficult to make a holding action…[let alone] get an increase that exceeds inflation."[70] They were prescient, indeed.

2.5 Increasing Quantitation

During that same several year period, Clark was able to hire, as Deputy Assistant Director for BBS, Robert Rabin, whom she appointed on November 21, 1976. Rabin had come from the Department of Energy (DoE) at that time but had been involved with the RANN program at NSF in earlier years and left the Foundation for DoE when RANN was eliminated. Phil Smith, Deputy Director of the NSF in later 1976, had suggested Rabin to Clark.[71] The year of FY79 was also one in which the Division of Social Sciences underwent a name change to the Division of Social and Economic Sciences (SES) that took effect on July 1, 1979. This had been presaged by changes that were set in motion in October of 1977 when the BBS had its two divisions of social and behavioral sciences. At that time, in the social science division, there was the section that included economics, geography and regional science, and political science. The second section included sociology, social indicators and special projects, law and social science, and History and Philosophy of Science.[72] But the then long-lasting name of the special projects and social indicators programs changed. It had become the Measurement Methods and Data Resources Program. The name was apt, because it was more representative of where support by NSF to the social science and economics' communities was moving, solidifying, or expanding: *quantitative directions*.[73] Even Galileo would have agreed: "measure what is measurable, and make what is not so." It may be, too, that a major reorganization for all of NSF was being considered by Atkinson, even though it was late in his foreshortened tenure at the Foundation (though he did not yet know that). However, the reorganization idea was dropped at that time.[74]

---

[69] Clark interview. Emphasis added.

[70] Larsen, "Milestones and Millstones," pg. 117.

[71] Clark interview.

[72] NSF, "National Science Foundation Organizational Development, October 1, 1977—September 30, 1978." 'Organizational Changes,' 3.c.

[73] NSB, "Minutes," 203:8, February 19, 1979.

[74] Ibid., 203:10.

By the time SES had been established in 1979, NSF funding concentrated on "continued improvement of statistical methods for the analysis of data." Also the efforts to expand large, longitudinal databases would continue. Studies in *methodologies* sought to "create partial substitutes for laboratory controls where possible, and to identify specific threats to valid inference where laboratory control is impossible."[75] There were values to many life sciences in that work.

Within the next several years, when the *Annual Report for Fiscal Year 1981* emerged, economics, like its relative sociology, was shown to be looking for patterns in societal behavior. *Increasing use of the computer*, as a tool for social science was part of the new directions seen in economics, sociology, political science, and in the History and Philosophy of Science, the authors of the report stated.[76] It became increasingly more difficult as interdisciplinarity grew in all the sciences to categorize research into one or the other of *continually more blurred disciplinary boundaries*, one of the great trends in the history of NSF.

A 5-year continuing grant to the Social Science Research Council would support the Center for Coordination of Research on Social Indicators approved by the Board in May of 1978. At NSB's November, 1979, meeting, the Center's grant was approved for 5 more years.[77] One of the many tasks of the Center was to discover whether or not a *model* using data *from the biological and physical sciences* could be developed *for the social sciences.* The relationship between the two areas was clearly reciprocal.

Another SES program, human geography and regional science, was doing well enough that its budget increased by nearly 30% in one 2-year period. In FY82 it had been some $700,000, but by FY84, it had risen to just over $1 M, and competitive grants were to be seen in numbers exceeding 150 in that year.[78] The funding was spread fairly evenly over a number of areas, but one of which was the program in cartography and *geographic information systems* (GIS), which would play a very large role, especially by the 1990s, in the biological sciences and ecology.[79] In fact, so important had GIS already become more broadly by 1988 that "NSF ha[d] established a new university-based center for the analysis of maps and other geographic data."[80] The GIS center, for example, was established as a consortium and was to have a life of 5 years. It was located at UC, Santa Barbara (its center), the University

[75] Ibid., ppg. 63–64.

[76] National Science Foundation, *Thirty-First Annual Report for Fiscal Year 1981* (Washington, D.C.: Government Printing Office, 1982), pg. 41.

[77] NSB, "Minutes," 211:14–16, January 21, 1980. The SSRC was requested to put two non-social scientists on its advisory board for the SI work: Richard Bolt, physicist of MIT, and Stephen H. Schneider, climatologist and Deputy Director of the National Center for Atmospheric Research. The latter was deeply involved in statistics, measurement, and estimation, all "concerns of importance to the field of social indicators," averred the Board; see pg. 211:15.

[78] Kingsbury, "Reading Materials".

[79] Personal knowledge.

[80] *National Science Foundation Annual Report 1988* (Washington, D.C.: Government Printing Office, 1989), pg. 33.

of Maine at Orono, and the State University of New York at Buffalo. Termed the National Center for Geographic Information and Analysis, it established computerized databases for GIS information. The databases' applications were in a variety of sciences: artificial intelligence, ecology, cartography, meteorology, political science, and resource management. GIS was such a new area at the end of the ninth decade of the last century that the *Annual Report* averred that "[m]any researchers predict that the impact of GIS on geographical analysis will be as important as is the telescope's role in astronomy."[81] A superb prediction!

2.6 The Intricate Dance

The program in History and Philosophy of Science was not like others at NSF because federal support for most all other areas in history was the province of the National Endowment for the Humanities, of which history, broadly defined, is part. The history of science, however, has especially close ties to the sciences themselves and was believed to fit with the social sciences at NSF particularly well. *As it eventuated, the program turned out to have many connections to both the sciences in general, and to biology in particular, in the sense of historical* drivers *at the Foundation.*

HPS was small in 1976. The Simon Report noted that the program supported work in three fields in the area: history of science, philosophy of science, and sociology of science. Only later did history of technology become one of the main four in more recent times. However, in the mid-1970s, virtually nothing was being done even in the latter of the three fields then supported: sociology of science. This the Simon authors lamented. The sociology of science would not be a growing field until some years into the future when it became an area of much greater interest to the HPS community beyond the walls of NSF. Later support for sociology of science represents *an excellent example of the NSF listening to the desires of its constituents* in the community from across the nation: the *intricate dance* between the Foundation and the communities it serves. Still, some sociology of science was being done via a joint program with the science policy group and others within the sociology, though not HPS, program.[82]

Research in HPS would suffer funding difficulties at the beginning of the first Reagan administration (1981) as its support was severely cut by OMB's Stockman. Betsy Clark had to find ways to keep the program from outright elimination, and so, in connection with historian of science and program officer Ron Overman, they created the Summer Research Program (SRP) for NSF, as a whole, and so created a source of funds such that HPS would not be exhausted by the many demands that

---

[81] Ibid.

[82] Simon Report, pg. 44.

small program could not afford to support.[83] *SRP would be used by many others in BBS, not just HPS*. Interestingly, Clark has held that "once Stockman had made his major cut, he sort of lost interest in trying to go for the elimination." She felt that was so, "because I guess he knew he would generate more bad blood than he'd expected."[84]

Several new programs were authorized for the FY80 budget, including that on Ethics and Values in Science and Technology, an offshoot in part, from the Asilomar Conferences on rDNA. By the late fall of 1983, Robert Rabin was functioning as acting assistant director with the departure of Betsy Clark; NSF Director Knapp had appointed Rabin to stand in, while the search for Clark's replacement (David Kingsbury) was found. Rabin felt that he did not have as much of a grasp of the HPS program and its relationship to other aspects of BBS as he might wish. Furthermore, Clark had not revised the Charter of the BBS' early version of an Advisory Committee for one final time before leaving for her next position. Clark left that duty, appropriately, to her permanent replacement. However, the effort to find that person was moving at a "glacial pace," noted Rabin. Thus, he and his colleagues in the BBS' higher leadership felt the need for a "broadly constituted" advisory group to help them bring together aspects of what HPS *could provide for appreciating the work of all of BBS*.[85] As he wrote to philosopher of science Thomas Kuhn, enormously influential in the HPS community at the time, and well beyond as author of the signal work, *The Structure of Scientific Revolutions*, Rabin made a statement that must have ranked as one of the most unambiguous possible in the matter of *interdisciplinarity* when he said that:

> [w]e seek to understand how the issues or questions within the domains of the social, behavioral, cognitive and information sciences are linked to life's individual and collective structures and functions, and the influence of their environs. We think, then, that conceptual breadth at this level of counseling is important, particularly since we are called upon to represent BBS as a whole in the competitive dynamics for status and resources involving all the other research directorates at NSF.[86]

Kuhn had given "past help [which had been] greatly appreciated," Rabin concluded. His invitation was for Kuhn to join a temporary advisory committee to meet twice yearly. Kuhn aided NSF by joining the committee along with Kenneth Prewitt,

---

[83] Clark interview.

[84] Ibid.

[85] In this section of the main text, the term advisory committee has been placed in all lower case letters suggesting, as Rabin claimed, that a properly constituted and chartered AC (capitals; formal committee) would not come to be until Clark's replacement had been named. However, this cannot be correct as the Minutes of the NSB's 250th meeting (NSB-84-71, pg. 2–84:5, section g.) state that the first BBS AC meeting had been held on January 26–27 and that botanist and director of the Missouri Botanical Garden, Peter Raven, was chosen Chair. Indeed, there is a cover memo from Raven to Rabin of July 13, 1984, that precedes the materials mentioned in the footnote immediately below. The exact date of formalizing the BBS AC for that period remains elusive.

[86] Robert Rabin to Thomas Kuhn, November 14, 1983. An ephemeron found in the Mark Courtney collection in the Office of the Historian of NSF.

nationwide President of the Social Science Research Council, and others.[87] In a memo to the committee, Prewitt indicated that Kuhn had "brought to our attention the consequences of the budget cuts for the number and size of grants to the [HPS] program."[88] However, Kuhn's association with the committee was not entirely a happy one, and he resigned in late May of 1985—as had Prewitt just 2 weeks before.[89] These events may have been harbingers of things to come during the Kingsbury years and are discussed further in later chapters in connection with Kingsbury's precipitous exit as AD, himself.

The first 2 years of the Reagan administration were difficult for programs in SES. For the HPS Program, in particular here, the Consortium of Social Science Associations (COSSA),[90] a Washington, D.C. "watchdog" group, pointed out that there was "a reduction in the success rate of submitted proposals in 1981."[91] But, they averred, the "most telling evidence of the impact of the budget cuts [show that in] a number of programs, the value of the average grant declined by as much as

[87] Kenneth Prewitt was not an employee of the NSF, but as president of the SSRC, he played significant advisory roles over time for the social and behavioral sciences at the Foundation. He was a faculty member at the University of Chicago from 1965–1982 and had taught also at Stanford, Washington University, and two universities in Uganda. He was a federal employee as director of the US Census Bureau from 1998 to 2001. He became dean of the graduate faculty at the New School University in 2001–2002. He held numerous awards and other positions throughout his career, which, as of the time of this writing, finds him on the faculty of Columbia University.

[88] Kenneth Prewitt to the Advisory Committee for BBS, May 29, 1984.

[89] Thomas Kuhn to David Kingsbury, May 23, 1985. Kuhn said to Kingsbury that: "I am still uncertain, as I was when you asked me to experiment for a year [being a member of the committee], whether I could be of use to [it], even under the best of circumstances. But the circumstances have proved far from the best and I am unwilling to continue trying." Kuhn felt that he could not fit himself into the "scheduling of the BBS meetings and the communication with Committee members [that] continues to be managed as it has been…If I had been given any reason to suppose that my anguished protest about the management of arrangements for the last meeting had done any good, I might have held on for a bit longer." Apparently, meeting schedules and cancelled meetings were poorly announced, and Kuhn became irritated. He copied the then chair of the AC, Peter Raven, long associated with NSF, and to Kenneth Prewitt. Kingsbury's response of June 11 was gracious and understanding, another letter of the same date went to Kenneth Prewitt who, himself, had resigned from the Committee on May 6 giving as his reason the fact that he had recently taken a Vice Presidency at the Rockefeller Foundation and was pressed for time. Whether or not that was the whole story from Prewitt, there is a question as to whether Kingsbury was handling his job as BBS AD well or not. Kingsbury's forced departure from NSF is discussed in later chapters, but this incident may, or may not, be representative of issues that arose in greater numbers later.

[90] COSSA was established in May of 1981, so was new on the scene as a purposeful ally to the social and behavioral sciences at NSF. It was initially established to function as a lobbying force to defeat an amendment sponsored by Representative Larry Winn (R-KS). His efforts to defund the SBS at NSF were in keeping with the directions of the new Reagan administration, especially in the form of the "order" given by OMB Director David Stockman for such defunding. Herbert Simon (main text) was in the forefront against the defunding effort, and Cora Marrett (see beginning of this chapter) was the first director of COSSA. The relationship between the rise of COSSA and the roles of Simon and Marrett are all closely tied to the history of the social and behavioral sciences at NSF both as SES within BBS and, after 1992, with the SBE directorate.

[91] *COSSA Washington Update*, May 18, 1984, pg. 3.

50% or more"! Overall, in SES, the "average grant was 40% smaller in 1983 than it had been in 1980."[92] For HPS, specifically, however, the figures were more dismal. The high during the period 1978–1983 for average annual awards peaked at $12,500 but dropped almost continuously to a near low of merely $5400 in 1983. This appeared to have a chilling effect in the academy as competitive proposals received dropped from a total of 209 in 1978 to only 139 in 1983: the community understood full well the implications of Reaganomics.[93]

In 1987, HPS, the oldest program in the SES division, moved out of the social sciences arena altogether and into the newly developing area in BBS termed the Studies in Science, Technology, and Society program (STS). The STS, in turn, became part of a new Division of Instrumentation and Resources (DIR), about which more will be said in later chapters.[94]

The Directorate appeared at the time as shown in Fig. 2.1.

Fig. 2.1 Organization chart for Directorate of Biological, Behavioral, and Social Sciences. (NSF. Specific provenance unknown)

Directorate for Biological, Behavioral, and Social Sciences

a. The Directorate for Biological, Behavioral and Social Sciences was established replacing the Directorate for Biological Sciences on October 31, 1975. The Directorate was reorganized primarily on a disciplinary basis into the following organizational elements:

Division of Physiology, Cellular and Molecular Biology
- Cellular Biology Section
- Biochemistry Physiology Section

Division of Behavioral and Neural Sciences
- Neurobiology Program
- Sensory Physiology and Perception Program
- Psychobiology Program
- Social Psychology Program
- Anthropology Program
- Linguistics Program

Division of Social Sciences
- Economics and Quantitative Methods Section
- Sociological and Political Sciences Section

Division of Environmental Biology
- Ecosystem Studies Program
- Ecology Program
- Systematic Biology Program
- Biological Research Resources Program

b. The Memory and Cognitive Processes Program was established in the Division of Behavioral and Neural Sciences on June 2, 1976.

[92] Ibid.

[93] Ibid., Tables 1 (pg. 4) and 3 (pg. 5).

[94] Larsen, "Milestones and Millstones," pg. 177. The Division was broadly announced in the *National Science Foundation Annual Report 1988* (Washington, D.C.: Government Printing Office, 1989), pg. 45.

2.7 Entre Acte

To this point, a few remarks on some of the programs in the social sciences have been made with consideration to their relationship with biology, in particular. More will be discussed in following pages. But with the dawn of the 1980s, many aspects of the by then Division of Social and Economic Sciences would have to make the necessary adjustments to a new world, or at least a new national, view. The years of FY80 through FY82 were particularly notable as a sort of *entre acte* with the coming "Reagan revolution" as Ronald Reagan assumed the presidency on January 20, 1981. *All* of the social sciences would, at first at least, be shaken to their very foundations that January (see the "Slaughter shock" below) but would rebound when their usefulness to the 8-year long Reagan era became apparent to both the president and to his administration.

But other matters of one kind or another were superimposed upon the change in the White House at that same time. One event that affected SES was a change of command when Otto Larsen took over the directorship of that division from his predecessor, Herbert L. Costner, indirectly effective July 31, 1980. Larsen would remain until July of 1984 when he left to return to his professorship in sociology at the University of Washington.[95] The change was "indirect" as there had been a short interim with a stand-in in the person of Bertha W. ("Bel") Rubenstein, an anthropologist, as acting director. Upon Larsen's arrival, she returned to the BBS AD's front office and her permanent position there.[96] Larsen, a sociologist with a doctorate in the subject from the University of Washington, had some previous federal experience having been appointed by President Lyndon Johnson to the 1960's era Commission on Obscenity and Pornography and also disciplinary scientific society experience having served in the American Sociological Association as its president, from 1972 to 1975.[97] Larsen praised Costner's leadership when he noted that "[m]any of Costner's imperatives became key points for collective action by social scientists in the 1980s": the *intricate dance*. But the conservative turn already underway before Reagan's appearance solidified the social scientists as they circled the wagons for self-preservation in their "depressed status."[98]

Henry W. Riecken had been assistant director for the former social sciences group under the BMS in the 1960s and, by the time of Larsen's arrival, had been gone from NSF for several years. Still, he communicated with Larsen upon the latter's assumption of the SES leadership, telling him that:

---

[95] Larsen had been made Senior Staff Associate for BBS, an external advisory position, on January 4, 1983, after leaving the director's post. NSB, "Minutes," NSB-83-58, February 22, 1983, 1–83:5. Note that the pagination system for the "Minutes" changed at about that time period.

[96] See other footnotes to the present chapter; also see Larsen's, "Milestones and Millstones," pg. 119, as well as his Chap. 6, *passim*.

[97] Larsen, pg. 119.

[98] Ibid.

> I gather that the current leader of the *know-nothing contingent* is Mr. [Representative] Ashbrook…Bel Rubenstein tells me that he is currently preparing a slightly revised version of his *chronic amendment* to reduce the social science budget.[99]

As if NSF's social sciences had not already suffered enough under Senator Proxmire and Representative Ashbrook in the previous decade, it appeared both would continue to be "chronic" irritants on into the 1980s. Larsen would be no less plagued by them than was Costner. And there were other significant changes that preceded the beginning days of the Reagan administration: Larsen would speak in terms of there being "more millstones than milestones…in the offing"—the reverse of the phrase which was the subtitle of his book on the social and behavioral sciences at NSF.[100] Some of those "millstones" require mention.

Larsen had been interviewed twice by Atkinson, who sought to hire the former because the director believed that the social and behavioral sciences could be split away from biology and that Larsen, Atkinson believed, was the man to help in doing so. However, Atkinson told Larsen that such a matter would be "discussed later" as there was "no use upsetting Betsy now," referring to Eloise Clark. But "later never happened, at least with Atkinson," Larsen said.[101]

In another personnel action in June of 1980, Atkinson himself left his position as director of NSF. It was not until 2 months later that Larsen, hired by Atkinson, arrived to fill the SES division directorship. It was necessary to seek an interim or temporary Foundation director, then, and that would be in the person of Donald Langenberg, who had just been appointed permanent *deputy* director of NSF and immediately went into the position of *acting director*. A North Dakotan, Donald N. Langenberg, earned his bachelor's degree from Iowa State University (1953), a Master's from UC, Los Angeles, and his doctorate from UC, Berkeley (1959), all in physics. In 1960, he began a 22-year career as professor of physics at the University of Pennsylvania, Philadelphia, and, in 1976, would also become a professor of engineering there. He had visiting professorships at both the University of Oxford in England and the Ecole Normale Superieure in France. His formal beginnings at NSF came with his appointment by then President Carter to the deputy director position in July 1980. Langenberg would stay in the position of deputy director until leaving (before completing 6 years) in December of 1982 to take up the Chancellorship of the University of Illinois at Chicago.[102] But his first duty as acting director would last for some 6 months until the arrival of the new permanent director, John Slaughter, in December of 1980.

Langenberg's 6 months' stint as acting director then, insofar as BBS and the social sciences were concerned, was carried out in what seemed an *air of change*

---

[99] Ibid.

[100] Ibid., pg. 120.

[101] Ibid.

[102] See http://www.uic.edu/depts/lib/specialcoll/services/lhsc/ead/003-20-03f.html, http://en.wikipedia.org/wiki/Donald_N._Langenberg, and http://www.presidency.ucsb.edu/ws/index.php?pid=45045.

about to affect those disciplines. He convened a meeting in mid-September of 1980, of all the advisory committees (ACs) of the various directorates. His agenda was composed of just three, but a very big three, items: long-range planning, the FY81 budget, and *organizational structure*. The BBS was still then composed, of course, of the four divisions (Fig. 2.1). One of those was the SES, another was for the behavioral sciences (BNS), and the other two were in pure biology.[103] However, it must be said that Langenberg partnered with chemist M. Kent Wilson, who then headed the NSF's Office of Planning and Resources Management and who held an "acerbic view of social science."[104] Wilson worked on the matter of long-range planning by continuing to "draw up maps" for NSF's future.[105] Larsen's job interview had included heavy hints about a reorganization that would separate social and behavioral sciences into a new directorate independent of biology. By September 5, 1980, Langenberg issued a draft document to the NSB that called for several major changes, and among those was *a new social sciences directorate*. He proposed one for engineering, too: Slaughter would later bring the engineering one to fruition (March 8, 1981). But not so the new social science directorate as Langenberg, with plan in mind, would just then be replaced by Slaughter. Slaughter began his directorship in an organization evidencing a long list of the persons who had just left it for other venues, Atkinson, of course; George Pimentel, the chemist who had been Atkinson's deputy director since 1977; and Norman Hackerman, who had been on the NSB for 12 years and had been its chairperson for his final six.

The new chair of the NSB had been chosen: Lewis M Branscomb, Chief Scientist at the IBM Corporation. Not surprisingly, he was very much committed to applied science—as was Slaughter.[106] Slaughter titled his first Director's Statement in the latest *Annual Report* "The Continuing Responsibility." In keeping with that sentiment, he spoke of increasing programmatic activity in science education and concerns over the health of science and engineering in the long term, there being a shortage then of such professionals, a change from just a few years before. He also averred that much of the 1970s had been taken up in "evolving a mechanism for supporting *applied* research" but argued that the "differences [between basic and applied research] are not nearly as great as the names suggest."[107] It seemed that applied research had finally been fully integrated into the NSF culture (at least according to Slaughter). That was especially so when, by August of 1981, four criteria for selection of research projects had been approved by the NSB with the simple phrase: "[projects] may have both long-term and short-term value."[108] The *similarity* of basic and applied research was so much the case for Slaughter that he

---

[103] Donald N. Langenberg memo to members of the NSF Executive Council, August 25, 1980.

[104] Larsen, "Milestones and Milestones," pg. 132.

[105] Ibid., pg. 134.

[106] See http://en.wikipedia.org/wiki/Lewis_M._Branscomb and many of the other references mentioned in this list of footnotes.

[107] National Science Foundation, *Thirtieth Annual Report for Fiscal Year 1980* (Washington, D.C.: Government Printing Office, 1981), ppg. vii–viii. Emphasis added.

[108] NSB-81-384.

said that "[r]ecognizing this, we intend to merge our applied research programs into the various research directorates"—a *major organizational change* at the Foundation. He also noted that there was a well-established precedent for such a merger: that of engineering and materials science earlier at NSF.[109] Indeed, the other big change he planned (taking his cue from Langenberg) was to establish a separate directorate for engineering; and that, of course, came to pass.[110] He closed by emphasizing that those changes would not affect the traditional budgetary roles for basic versus applied research as the House Committee on Science and Technology had long made it clear that both aspects of research must be maintained. The world was a competitive place, and "competition…is almost invariably associated with opportunities," Slaughter reminded his readers. So important has business been as a central aspect of American life that on December 12, 1980, the "University and Small Business Patent Procedures Act," was passed by Congress. Better known as the Bayh-Dole Act, for its sponsors Senators Birch Bayh (D-IN) and Robert Dole (R-KS), the Act would be of significance to NSF in coming years, not only in biology (particularly biotechnology) but in other sciences, too.

The final time that the term "applied" was to be seen in use within the social and behavioral sciences themselves at NSF was in the early 1980s when its last incarnation, "applied psychology," was dropped.[111] Titling a program as *applied* no longer seemed necessary: research *naturally included both basic and applied*. The integration appeared to be complete, at least for BBS.

The *entre acte* was not only a period of changes in personnel and other features at the Foundation but one in which a new approach to the American presidency would emerge in the form of the "Reagan revolution."[112] On Slaughter's first day on the job, a stunning incident, alluded to earlier, occurred. Slaughter was summoned to the OMB, and so, with Langenberg (serving again in his regular position of deputy director), the two walked the short distance over to that office's building; NSF then still in downtown Washington, D.C. Hugh Loweth, a budgeteer at NSF and well known to Slaughter from the latter's AD days, greeted the two and told the new director and his deputy that David Stockman had "ordered" the NSF to "close down its science education and behavioral and social science programs [*in toto*] and gave

[109] Belanger, "Enabling American Innovation," pg. 164.

[110] NSF, "Annual Report 1980," pg. viii.

[111] Larsen "Milestones and Millstones," pg. 177.

[112] One other personnel action was of note in the nearly three-decade life of the NSF in 1980, however. Doris McCarn, the longest serving employee in the Foundation, retired on February 29, 1980. McCarn had served as secretary to every one of the first five directors of NSF. She had come, with Alan Waterman, from the Office of Naval Research as he became the inaugural director in April of 1951. The Board commended her: see NSB, "Minutes," 213:4, March 24, 1980 and Appendix A (pg. 213:22) with a poem, a paean to her. The poems for departing employees became "milestones" in the long history of the NSB "Minutes." A number of these had been penned by the pseudonymous "George David Oswald Eliakim Longfellow"; see, for example, that done for the latest "class" of new Board members in NSB, "Minutes," 215:21, (Appendix C), May 19, 1980. The NSB was not without humor.

[Loweth] the timetable"! This I have referred to as the "Slaughter shock." Slaughter said:

> Hugh, suppose instead of closing them down, I just reduce the budgets of some of the other areas at least to keep [them] alive [?] And he looked me straight in the eye, and he said, 'John, any effort on your part to do that will be unquestionably denied.' And that was the first indication, that was the first charge I had when I took my new job, and it occupied a considerable amount of the time I was there. That set the tone to a large extent for my service as director, and the hardest thing I ever did was to go out to Wisconsin Avenue, where the science education program was located, and tell the 125 or so people out there that their jobs were going to be abolished.[113]

With regard to a one-term Democratic Congressman from Ohio, Robert N. Shamansky, Slaughter said that the representative "lambasted me and said that I had—his words—I had been consorting with those juvenile cretins at the OMB" in Slaughter's efforts to save science education.[114]

Stockman's reputation had its effects not only at the director's level but also at that of the AD for BBS. Betsy Clark has said that:

> we were wary when we heard Stockman was going to be budget director because he had been staff assistant for a congressman [she remembered him as John Anderson (R-IL), but it was actually John Ashbrook, noted above] who had been one of the few people on The Hill who was writing into the Congressional Record [comments about] so-called…'silly grants.'[115]

Clark further pointed out that Stockman had:

> his famous 'Black Book' in which he laid out, prior to Reagan's administration, the things that were going to get reduced, and the Black Book is where the plan [for budget reductions] was laid out: 44% reduction for the social sciences the first year, down again by that same amount the next year, and they would've been phased out in year three.[116]

The period just after Reagan took office was one felt by not only Slaughter and Clark but the NSB, too. Their ire became very much elevated! Adopted by the Board at its March 1981 meeting was a "Statement on [NSF] Budgets for Fiscal Years, 1981, 1982, and Beyond." Branscomb shared the final wording of the document with all Board members in April of 1981. The gist of the statement was that the Board recognized the "emergency nature of the economic situation" in the country and "the vigorous remedies the President seeks." However, though the Board supported the reduction of the NSF budget, it was seriously concerned about two things: the increasing obsolescence of research equipment in university laboratories and the

---

[113] Slaughter interview.

[114] Ibid.

[115] Clark interview. Stockman had served as an assistant to Representative Anderson (footnote further below) who chaired the House Republican Conference; see also following footnote.

[116] Ibid. See also William Greider, "The Education of David Stockman," *The Atlantic Online* (1981):22 ff (online pagination) at http://www.theatlantic.com/past/docs/unbound/flashbks/classics/stockman.htm.

As Clark correctly remembered, Stockman was indeed assembling "dozens of position papers on program reductions," pg. 12 (online version) of the *Atlantic* article.

potential loss of the possibility to "take advantage of emerging research opportunities," a highly significant area of the Foundation's fundamental concerns.[117] For the social and behavioral sciences, in particular, the *Board pressed the issue of the great importance of neurophysiology and cognitive sciences* and the new economics (read Reaganomics), the latter very possibly a sop to the President. The full statement turned out to be a very strong one for the NSB, brilliantly crafted in its political astuteness, yet making it clear just what NSF contributed to the nation by its support of research in all areas.[118]

In an overview of the annual budgets of the SES division of BBS for the period from Fiscal Year 1980, less than 2 years before Reagan's presidency began, through 1991, it is seen that the low point was the FY82 budget, the one Stockman wanted to see sans any presence of social and behavioral science—*and* science education. While it was a nadir at the time, the budget did begin to climb once the value of social science data was realized in the Reagan administration. Even so, the budget (in 1982 constant dollars) for SES *never* recovered its high mark (for the period) of the FY80 budget. Figure 2.2 provides greater detail for the period FY80 through FY89 with regard to the distribution of SES funds into each of its nine major program areas. The actual percentages of what both SES and BNS funds were for these years in NSF's overall budget ran from a high of 6% in FY80 to a low of slightly over 3% by FY89. It was on an almost steady downward trend.[119]

While Langenberg, as acting director, finally ended up deciding not to attempt a substantial reorganization of the BBS, at least, Slaughter was expected by the Reagan administration to abolish a significant part of the Foundation's programming, that is, social and behavioral sciences and science education. After many discussions with academics in the social sciences and with economists, some even in the White House, Slaughter felt "justified" in keeping those disciplines and did exactly that! He noted that the science education issue received more press than the other two (MACOS, see Chap. 1, was still fresh in many minds), but both Democrats and some Republicans in Congress were behind him concerning retention of the social sciences. This was because these individuals "all felt that Stockman had gone too far."[120]

[117] Lewis Branscomb memo to the NSB, NSB-81-151, April 2, 1981, with attached "Statement" (NSB-81-150).

[118] By May 21, NSB Chair Branscomb prepared an extensive statement to his colleagues on the Board in the form of a "Report" after his first 2 years as Chairman. He pointed out that the policy decisions (above) concerning the social and behavioral sciences (and science education) were of "importan[ce]" to the commitment of the NSB because it was a significant "matter for the Board to have determined [a policy] in a clear and formal way." NSB, "Minutes," 235:25 (Appendix D), June 21, 1982.

[119] Larsen, "Milestones and Millstones," Table 6.1, pg. 173.

[120] Ibid. David Stockman was a native of Fort Hood, Texas, the US Army's largest installation but was educated in the public schools of Michigan. His B.A. was from Michigan State University at East Lansing in 1968 followed by graduate education (no degrees) at Harvard in 1968–1970 and again from 1974–1975; he also studied at the Harvard Divinity School. From 1970 to 1972, he worked as a special assistant to Representative John B. Anderson, who ran in the 1980 Republican

Research Funding by Social Science Programs, 1980s
(Current $ in millions)

Fiscal Year	Econ-omics	Meas-Data	Soci-ology	Poly Sci.	Law So. Sc.	Hist. Phil.	Geog-raphy	Dec. MgtSc	Reg. Policy
80	12.2	5.0	3.9	3.6	0.9	1.5	1.6	-	2.6
81	9.4	3.9	3.0	2.9	0.9	1.1	1.2	0.4	2.7
82	6.3	2.9	2.2	2.1	1.1	0.9	0.7	0.4	0.9
83	7.2	3.3	2.5	2.4	1.2	1.1	0.8	0.9	0.9
84	9.4	3.7	3.1	2.8	1.4	1.3	1.0	1.0	1.1
85	10.1	3.4	4.6	3.3	1.9	1.7	1.3	1.4	1.1
86	10.6	3.3	3.5	3.5	1.9	1.7	1.5	1.9	-
87	13.6	3.7	4.0	3.8	1.8	-	1.7	2.5	-
88	12.7	2.4	3.7	3.9	1.9	-	2.7	2.7	-
89	12.9	2.4	3.7	3.8	1.9	-	2.9	3.3	-
$	104.4	34.0	34.2	32.1	14.9	9.3	15.4	14.5	9.3
%	39	13	13	12	6	3	6	5	3

Fig. 2.2 Research Funding by Social Science Programs, 1980s (current dollars in millions). (NSF. Budget Books for Division of Social and Economic Science, 1980–1987)

Simultaneously to the Stockman "order," there was still discussion ongoing at NSF about a new directorate for the social and behavioral sciences that finally led Betsy Clark to contact certain congresspersons to indicate that the neural psychology group in BBS was against *any* reorganization (at least at that time).[121] Her message, Larsen reported, was sent in order "to clue Congress in on the reorganization because she was opposed." But NSF had already, by early 1981, stated there would be no major reorganization as the time was "not propitious," said Langenberg.[122] The struggle over psychology's close linkage to biology and the fact that much of the funds for social science were coming via applied research projects, a legacy of RANN, made the reorganization untenable in NSF's thinking. However, and very importantly, *the academic community had become aroused* and pressure mounted for the reorganization throughout the decade of the 1980s.

With Slaughter's policy decision to merge *applied* research into the directorates and thus join that to each directorate's efforts in *basic* research, SES underwent the biggest change anywhere in the Foundation's organization chart (Fig. 2.3). That was true with regard to programs and personnel, but not with regard to funding. As

primary against Ronald Reagan. After more House experience, Stockman served two terms as a Representative (R-MI) resigning before the close of his second term to take the directorship of OMB on January 27, 1981, thus becoming one of its "most powerful and controversial" directors. He resigned from that position in August of 1985 and so did not serve the remainder of Reagan's second term. The resignation followed certain negative press and Stockman's own admission that no one really understood Reagan's "supply-side…and trickle down" economic theory. See http://en.wikipedia.org/wiki/David_Stockman. See also the article by Greider in the *Atlantic Monthly*, *op. cit.,* marking the beginning of his slide into eventual oblivion from Reagan's revolution. Also, see http://bioguide.congress.gov/scripts/biodisplay.pl?index=S000935.

121 Larsen, "Milestones and Millstones," pg. 138.

122 Ibid., pg. 139.

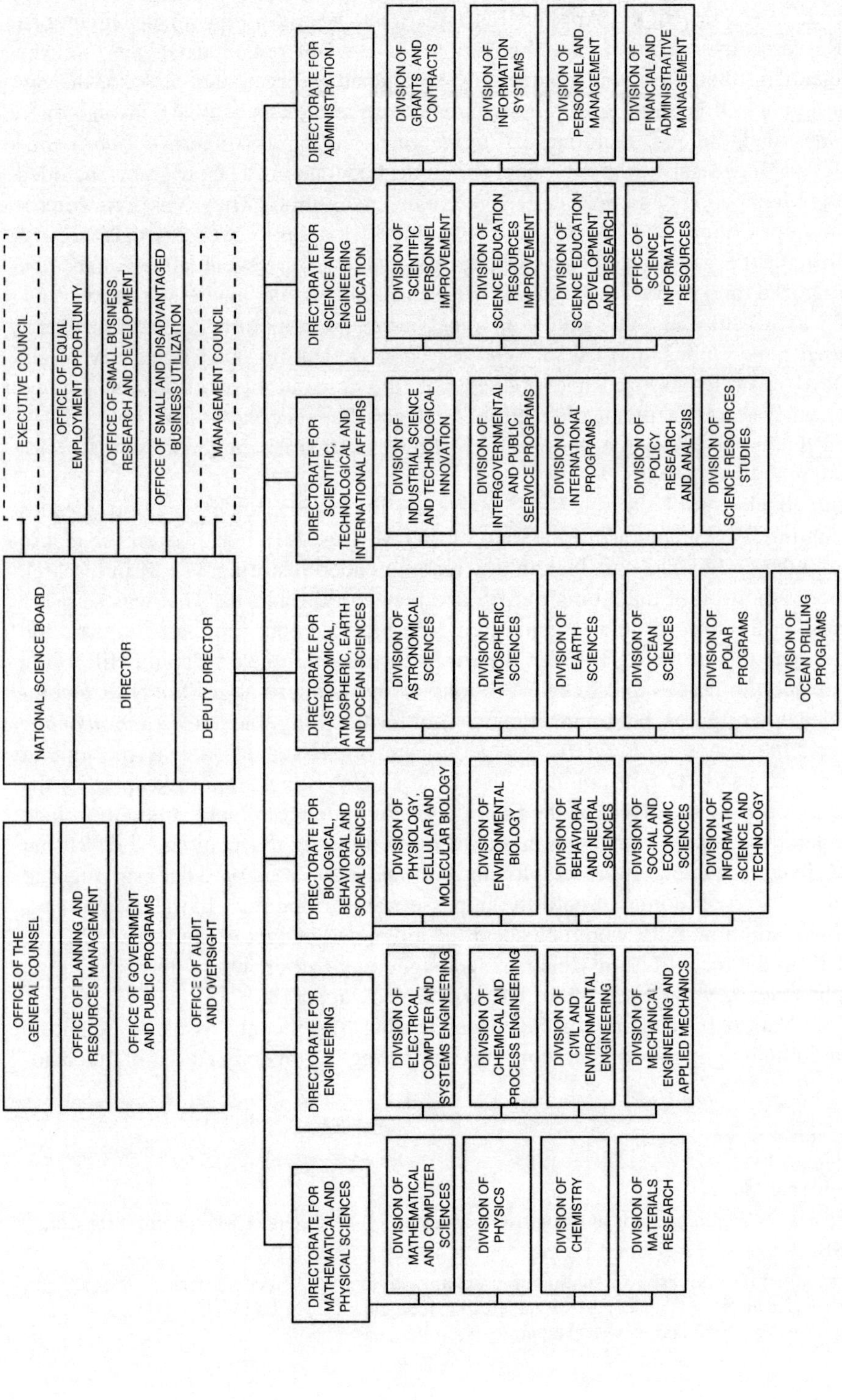

Fig. 2.3 News release on new organization chart for March 4, 1981. (NSF PR81-19)

Larsen stated it, in connection with the new policy: "[a]ll were called but few served."[123] The Division of Applied Research (not to be confused with the Directorate of Research in the pre-1975 organization chart) went out of existence with the Langenberg/Slaughter small, but highly significant, reorganization. Also, the by-then defunct Directorate of Engineering and Applied Sciences of Atkinson, which became Slaughter's Directorate of Engineering, led to SES *inheriting many programs* and, necessarily, the personnel and funds tied to them. The programs included Public Management and Service Delivery, Law and Public Policy Research, Micro-Economic Policy and Regulation, Resource Economics, and Decision and Management Science (see below). Titles and contents were eventually changed and reorganized (and so do not fit Fig. 2.3 precisely). The whole business of those transfers was a "cultural jolt" and, with time, even the sometimes-contentious term *applied research* began to fade across the NSF organization as a whole.[124] After all, as NSB chair Branscomb had pointed out, the separation of basic and applied research was an "artificial distinction" and was "unnecessary—and unhelpful—when the applied research is done by institutions [mainly universities] that don't intend to do the applying"![125]

Budget plans for FY81 and FY82 saw rescissions in SES and behavioral sciences accounting for a great part of the total of $62 M (Fig. 2.2).[126] However, the House Committee on Science and Technology recommended restoring $12 M in research support, a portion of that going to SES and behavioral sciences. That was part of a broader restoration of funds leading to a 26% increase between the administration request and that of the CST. Later, the conference committee on Capitol Hill trying to balance the House and Senate versions of the annual budget for NSF recommended other figures, but floor action was not forthcoming. Indeed, *no action of any sort on the entire national budget* was occurring! Even as late as the end of November, a *second* continuing resolution was being considered to keep all of the federal government solvent as the full budget continued to remain stymied. In fact, President Reagan did sign a continuing resolution through December 15, 1981, but many federal agencies were employing only skeleton staff by Thanksgiving and expected to go on doing so until the impasse was surmounted. In the second continuing resolution, NSF would be accorded a planned budget of some $973 M: the drop from the recent billion-dollar budget signed by Carter (but by then having no weight since he was gone from office) was already apparent.[127]

The National Academy of Sciences, fearing significant losses to scientific research funding in the country, convened a meeting in November 1981 of 100 lead-

---

[123] Ibid., pg. 144.

[124] Ibid., ppg. 144–145.

[125] Belanger, "Enabling American Innovation," *op. cit.*, pg. 147; Belanger was quoting Branscomb.

[126] NSB, "Minutes," 224:7, April 17, 1981.

[127] Continuing resolutions are a continuing problem with the federal budget because they are still being seen as recently as the congressional effort to pass the FY 2011 budget. See http://www.aaas.org/spp/rd/. The 2012 budget was even more of a roller coaster.

ing scientists and engineers and invited the major press organizations to attend.[128] The purpose was to try to come to grips with what the rescissions would mean for FY82 but as much as to understand the meaning for science and engineering in future years with Reagan as president. No one could divine then that those years would total to nearly a decade. Eventually, the final budget for FY81 would undergo rescissions in the amount of $15 M for social and behavioral sciences. By the FY82 budget, which was Carter's that was cancelled by Reagan, another $47 M was to be taken from those two areas, though that did not occur. In the end, the total cuts to social and behavioral sciences were a little over $14 M.[129] However, those figures were in current dollars. The rise in the total budget of NSF from 1967, $239 M, to 1981, $897 M, hardly compared to constant dollars (1972 value), however. In those same years, the real budget, measured in purchasing power, went from about $300 M to only about $460 M.[130] That for social sciences, specifically, went from $55 M to only $124 M during those same *14 years*![131]

The *entre acte* of 1980–1981 itself began to fade and the social sciences went on with their work for their various constituencies across the nation. Stunningly, the well-known economist and Nobelist of the time, Milton Friedman, advocated that, as Larsen said: *all of* "NSF should be eliminated." That was so as, according to Friedman, "[p]rojects undertaken by the government cost much more than the same projects in private hands"—a Reaganomics key belief.[132] And David Stockman was just beginning his more than 4 years as Director of OMB. Intriguingly, though, "[u]nsuspected allies" of the social sciences appeared during the early months of the first Reagan administration, as well documented by Larsen and unnecessary of repetition here, and this allowed for the continuing operation of the two related divisions of SES and BNS.[133] It should be reiterated, though, that when one particular anti-social sciences' amendment was in the House for discussion, "the floor erupted with more speeches *in affirmation* of social science at NSF than had ever been made

[128] NSB, 'Minutes,' 231:7 (and fn 1 on that page), January 25, 1982 (NSB meeting of November 19–20, 1981). Larsen, apparently working from a different version of the "Statement" cover memo than the April 17 one cited here, gave a date of May 1 and referred to the "Statement" as a "Mayday" call for consideration of the NSF's financial needs. See: Larsen, 'Milestones and Millstones,' pg. 156. At that same time, Philip Handler was retiring from the leadership of the NAS and "came close to accusing the [NSB] of cowardice" as the Board spoke in terms of a financial "emergency." "Emergency? What emergency?' asked Handler. 'The fact that the NSF was given but 24 hours, when the board [*sic*] was not in town, to defend its right to support social science and science education, inter alia, is preposterous[,]'" noted Larsen (pg. 158).

[129] Johnson, 'Next to Nothingness,' Part III, pg. 325–326.

[130] NSB, *Science Indicators 1982*, pg. 245, Appendix Table 2–13.

[131] Ibid., pg. 246, Table 2–14.

[132] Ibid., pg. 146. The *Science Indicators* issue for 1983 did point out that "[i]n 1980, industrial sources of R&D expenditures exceeded Federal Government sources for the first time in the last two decades." Hardly an insignificant point in the view of followers of Reaganomics.

[133] Larsen details this period very well in his book following events on an almost daily basis. See his ppg. 146 ff. See also his discussion on the Winn Amendment, named for Representative Larry Winn, Jr. (R-KS), ppg. 168ff.

in congressional history"![134] Stockman's approach had "gone too far," apparently, and a sufficient number of those in Congress were, indeed, aroused. By early May 1981, after a very tense several months, the NSB took the congressional mandate for NSF to continue supporting social and behavioral sciences to heart by going so far as to *add new programs*. NSB Chair Branscomb wrote in the pages of *Science* that NSF intended to initiate programs in neuroscience and cognitive science and expand in economics research.[135] It came to pass.[136]

The SES remained busy with a vision to the future during those years, as well. For a period of time prior to February of 1982, there had been discussions in SES about establishing a new program, Decision and Management Science (DMS), which would "support theoretical and empirical research on decision and management structures, processes, models and methods." The Board unanimously approved the DMS.[137] And it was a *highly interdisciplinary* program, too. It was a "collaborative effort by four NSF directorates: Mathematical and Physical Sciences; Engineering; Scientific, Technological and International Affairs," as well as BBS.[138] Some $500 k was allotted to begin the program; each of the directorates contributing $100 k with the remainder coming from the SES division.

There was a critical addendum to the *entre acte* coming just a short time later. By late 1982, Knapp took over the director's position at NSF, and Betsy Clark had left the BBS directorship in the hands of acting AD Richard S. Nicholson.[139] Knapp, another physicist, was appointed by Reagan and announced in a short notice that, as the *New York Times* writer Philip M. Boffey said, "attracted little attention."[140]

---

[134] Larsen, "Milestones and Millstones," pg. 171. The failed amendment was that of Representative Winn.

[135] Lewis Branscomb, "NSF Budgets: Fiscal Years of 1981, 1982, and Beyond," *Science* (1981):515.

[136] NSB, "Minutes," 221:9, January 19, 1981 Minutes of the November 20–21, 1980 meeting of the Board.

[137] NSB, "Minutes" 233:15, March 22, 1982; see also the document numbered NSB-82-18.

[138] National Science Foundation, *Thirty-Second Annual Report for Fiscal Year 1982* (Washington, D.C.: Government Printing Office, 1983), pg. 27.

[139] This cannot be correct with regard to Nicholson, mentioned by Boffey in the article cited in the immediately following footnote, as Robert Rabin was Clark's Assistant AD at the time and much other evidence cited in several locations in this current chapter indicates that it was Rabin, not Nicholson, who took over as acting AD with Clark's departure. Rabin was the signatory (as "Acting Assistant Director") for the Minutes of the January 26–27, 1984 meeting of the BBS AC, at any rate. See copies of the Minutes in the lateral files of the main office of the current AD for BIO. We also know that Nicholson held a position of "Staff Director" by the middle part of 1984, as he addressed the House Ways and Means Committee then; see the NSB, "Minutes," 8–84:5, September 24, 1984. The matter of what appears to be an error on the part of Boffey regarding the Nicholson/Rabin confusion may or may not bring into question the veracity of other apparently factual material to be seen in the Boffey article (following footnote and used in the main text above). Richard Nicholson is not to be confused with John Nicholson, who was then Director of Social Sciences at the Research Council of Canada and was visiting NSF during this period. The latter's name appears in documents of that time.

[140] Philip M. Boffey, "New Leader for the Science Foundation," *New York Times* (1982), Dec. 8: see http://www.nytimes.com/1982/12/08/us/new-leader-for-science-foundation.html. See also rele-

Knapp was already at the NSF when his appointment as director began on November 3. However, that would not be official until the Senate Committee on Labor and Human Relations approved him, as it did on April 13, 1983, but a good half-year after he assumed the directorship. It was at that meeting that Senator Ted Kennedy sought and received an assurance from Knapp that he would *not* politicize the Foundation as certain earlier "allegations and charges" had suggested he might.[141] Though Knapp said that he let certain Democrat-appointed senior persons go at NSF because "I wanted my own team," others did not feel that was the reason.[142] The announcement of Knapp's appointment published in *Science* was more informative than the White House announcement, not unexpectedly. *Science* writer, John Walsh, reviewed Knapp's personal history of 24 years at Los Alamos National Laboratory, his friendship with George Keyworth (below), and other details.[143] Boffey had said that "[o]bscurity is almost a way of life at the [F]oundation," and that "Dr. Knapp is somewhat obscure even in the scientific community." He was not a candidate when it was apparent that Slaughter was leaving. Rumors had it that all presidentially appointed persons at NSF who were "holdovers from the Carter Administration" were to be "clean[ed] out." Knapp "scoffed" at that and held that "I'm not about to have any head-hunting campaigns at the [F]oundation."[144] But within days, he reversed that seemingly adamant statement letting several Carter appointees go, not because he wanted his "own team." That became clear in a number of publicly available sources, including *Science* itself.[145] Furthermore, Betsy Clark opined that "I think [Knapp] was *charged* with getting rid of the people who'd been appoint[ed] by Democrats."[146] Boffey spoke of the NSF having suffered ridicule by Congress over the years citing time-worn negative statements about social science grants: funds to study why people fall in love was the perennial favorite, and MACOS had not been forgotten. Still, as the *Times* author pointed out, NSF had not suffered unduly in the budget purges of the previous 2 years, with two exceptions: science education and, of course, social sciences. The latter rescission was a "move widely interpreted as retaliation against liberal activists." But, as has been seen above, a fair amount of that was restored, and Knapp was quick to point that out. Even with that, though, "the [F]oundation [still] provides less than one-fifth of the Federal Government's total support for basic research," stated Boffey.

Boffey opined that Knapp had been picked as the new NSF director because the latter was a friend of George A. Keyworth II, Reagan's science advisor. Keyworth and Knapp had worked together in the past at Los Alamos where Knapp had spent

vant comments in the immediately preceding footnote.

[141] Colin Norman, "NSF Nominee Wins Committee Approval," *Science* (1983):485.

[142] Ibid.

[143] John Walsh, "Los Alamos Alumnus Touted as NSF Chief," *Science* (1982):662.

[144] Boffey article.

[145] John Walsh, "Knapp Reinterprets Excellence at NSF," *Science* (1983):990–992; see especially the first paragraph on pg. 990.

[146] Clark interview. Emphasis added.

his entire scientific career up to the NSF appointment. Still "leaders of the scientific community" saw him as a "solid scientist" and Frank Press, president of the National Academy of Sciences, felt Knapp could turn out to be a "great director."[147] Possibly so, but none of the *Science* articles cited above praised him: they were merely neutral. NSB Chair Branscomb seconded the Press comment: a circling of the wagons. We do know that Knapp found the social sciences challenging to comprehend.

> I really don't understand [the social sciences], and I'm working very hard to do that, to see where they are quantitative, where they are qualitative, where real data can be used, where they are productive in terms of scientific output. But I am learning…I'm a novice. I've never looked carefully into the social sciences to understand their successes, their failures, their criticisms.

And he further felt that *if* the data of those fields were *not* quantitative, "it could be extremely damaging."[148] He did make an attempt to learn about the social sciences, however, as he became the first director to visit (in May of 1983) an NSF-supported social science site, the Institute of Social Research at the University of Michigan. Later he joined the NSB's visit to the National Opinion Research Center in Chicago.[149]

When Betsy Clark left the Foundation, the proposals being handled by the BBS Directorate had gone up to some 8500 annually. This suggests that growth at NSF, and particularly in BBS, was not on an especially steep curve, and "tight budgets" and "political pressure" in those years (in the decade from the oil crisis to early Reaganomics) played major roles in keeping both the social sciences and the biological sciences from the spectacular upward growth the latter, at least, would show from the 1990s onward. Just one decade earlier (1973), the proposal numbers had been at 3400. So, within 10 years, the grants awarded had increased by only about 2.3 times over an entire decade. The year 1983 was also the year that NSF's budget topped $1 billion. By comparison, the budget for the "elephant in the room," as many referred to NIH, was just under $5 billion that same year.[150]

Near the end of her tenure at NSF, one of Clark's duties was to co-sign, along with Director Slaughter, the "Charter of the Advisory Committee [AC] for Biological, Behavioral and Social Sciences" (but see also the section on HPS above). It has been pointed out earlier that ACs of several types have played (and continue to do so) *a very significant role* in the operations of the National Science Foundation. AC Charters were the governing documents for how those committees were to function, how to be constituted and staffed, and how meetings were to be undertaken, etc. As required by federal law, the Charters of any ACs (from all agencies) were and are reviewed, revised, if necessary, and resubmitted to the National Archives and Records Administration by agency directors at certain intervals.[151]

---

[147] Ibid.

[148] Cited in Larsen, "Milestones and Millstones," ppg. 179–180.

[149] Ibid., pg. 181.

[150] See http://opa.faseb.org/pdf/DoublingEliminationChart.pdf.

[151] The "Federal Advisory Committee Act" (P.L. 92-463) originated in 1972 and has undergone a

NSF *advisory panels* also existed, and still do, at the level of specific programs; they, too, are subject to federal law.

The social and behavioral sciences survived the period of what I have called the *entre acte* and those disciplines affecting biology that still remain to be discussed soldiered on from 1975 to 1992 each on its own historical trajectory but always under protective coloration and the riding of biology's coattails.

2.8 Increasing Computerization, Modeling, and Artificial Intelligence

Information Science and Technology (IST) had not been covered in the Simon Report as it came into existence some years after that study had been published. In March 1981, NSF put out a news release mentioning not only the establishment of the new Directorate for Engineering (above) but also that of the transfer of the IST program to BBS.[152] The IST would become a new, *fifth division of BBS*. Later in time, IST would be reformed and reintegrated into the new BIO once BBS was phased out, but the content and orientation of IST would hold a *profound future* for BBS, especially at first, as the use of computers in the sciences grew extraordinarily rapidly. That was especially so at BBS as all areas saw expansion evermore strongly into larger databases and the kinds of research that they would allow. A new organizational structure of all of NSF that March of 1981 showed BBS with its new fifth division included.[153] It was not until 1986, however, that the new Directorate for Computer and Information Science and Engineering (CISE) was established removing IST from BBS, arguably an odd location for its placement in the first place (below). But *that* "consolidat[ion]" of computer activities at NSF would be absolutely crucial in coming years and was one of then director Erich Bloch's major accomplishments. But 1986 was yet a few years in the future.

In the newly established IST division at BBS, "inquiry about information phenomena deals with the structure of information, its storage, movement, and

number of amendments since that time. The full act and the history of its changes over time, with or without annotation, can be seen at: http://www.gsa.gov/Portal/gsa/ep/contentView.do?contentType=GSA_BASIC&contentId=11635. As an example of this rather formal procedure, I use the letter from Director Edward A. Knapp to Ronald L. Martinson, Special Assistant to the Archivist at the National Archives and Records Administration (NARA), February 22, 1984, which carried out the General Services Administration (GSA) requirement of accumulating and preserving these letters of transmittal and their appended committee and/or panel lists. This information, once approved, was/is also published in the "Notice of Establishment" section of the *Federal Register*. In the letter used here as exemplar, Knapp formed nine advisory panels in areas in the biological sciences and abolished others.

[152] Ralph Kazarian, "National Science Foundation Reorganization Announced," NSF PR81-19, March 4, 1981, pg. 1.

[153] Ibid., pg. 2.

manipulation."[154] Further, "[i]n effect, research in information science is entering a new, broadly based phase." At the time, NSF and the academic communities it served, were some years from becoming a fully computerized organism. There was not a computer on everyone's desk at the Foundation then, as there would be in the years after 1984–1985 when Erich Bloch took the helm. Accessible mainframes had been in use at NSF for many years, of course. But in FY81, the *Annual Report* continued to speak in terms of the study of information as though NSF was entering a brave new world; in many respects, it was. Still, the NSB statement on budgets noted above held that the "Nation's economy [was] continuing to shift from manufacturing to services and notes that two-thirds of the U.S. workforce is already engaged in the information sector."[155] Something like CISE was on the horizon even as IST was aborning.

The *Annual Report* for FY82 reminded the reader just how crucial information science was becoming: "[m]aterial and energy resources are the physical capital of our society, information resources its intellectual capital." Indeed, NSF was the principal supporter for research in IST, as opposed to purely mission-oriented research then supported and done by the Department of Defense (weapons' guidance, for instance) and the National Library of Medicine (improving information systems).[156] By FY83, "bioware," as well as hardware, was under study to better appreciate the comparison between information processing in humans and in machines, which explained the otherwise seemingly odd decision to place IST in BBS. Continuing work in natural languages (lodged with the social and behavioral sciences) and computer code was foremost in such work on bioware.[157] The notion of bioware was continued in research over the next several years because the relationship between natural and artificial languages required thorough understanding in the design of software for computation. Thus, by 1984, NSF was funding research "focused on information processing in such activities as categorizing and pattern recognition; [and] in learning, memory and problem solving."[158] Such work would be significant in later years as NSF and the communities' interests continued ever more rapidly toward large-scale databases.

Clearly, IST was doing well in the eyes of the budgeteers in Congress: the FY85 budget for the division was nearly 50% ($3 M) above that for the previous year.[159] And other related areas were active, too. The Decision and Management Science Program dealt with, among much else, decision modeling. Decision models were of great importance, as were other types of models and methods, including game the-

---

[154] Ibid.

[155] NSB, "Statement on Budgets," March 20, 1981, pg. 6.

[156] National Science Foundation, *Thirty-Second Annual Report for Fiscal Year 1982* (Washington, D.C.: Government Printing Office, 1983), pg. 40.

[157] National Science Foundation, *Thirty-Third Annual Report for Fiscal Year 1983* (Washington, D.C.: Government Printing Office, 1984), ppg. 41–42.

[158] National Science Foundation, *Thirty-Fourth Annual Report for Fiscal Year 1984* (Washington, D.C.: Government Printing Office, 1985), ppg. 33–34; quotation on pg. 34.

[159] Robert Rabin, "Memorandum to the BBS Advisory Committee," July 3, 1984, pg. 2, item 6.

ory, which would become increasingly important later in the biological sciences, such as in genetics and evolution.

Early grants in the Linguistics Program were of a "somewhat traditional focus" and with a "heavy emphasis on descriptive" aspects.[160] While the Simon Committee found that artificial intelligence (AI) approaches to linguistics was poorly represented in 1976, that would change in later years. Given the close association with cognitive psychology and computerization of natural languages, the Committee argued that a very broadly constituted panel for judging grant proposals would be needed representing many pertinent fields. By Erich Bloch's time, the field had matured strongly in the area of computerization to study the structure of language. Indeed, by 1985, at least one academic linguist felt compelled to argue that the program at the Foundation seemed to him to exhibit "an overwhelming emphasis on the tools of research (computers, networking systems…etc.) [to the] almost total neglect of substantive research questions."[161] The field was already moving headlong into the AI arena, and the relationship to linguistic study via use of computers was the future of the field, at least at NSF. Thus, in that same year of 1985, then BBS AD David Kingsbury and Donald Kennedy, President of Stanford University, were in communication concerning AI. Kingsbury said that:

> we have been following this whole research area for some time, and we are attempting to take a leadership role…in this newly emerging field…I am also aware that the prototype group, and the acknowledged leader, is at Stanford (together with SRI and Xerox [corporations]).[162]

Kingsbury was in answer to a letter[163] from Kennedy and was then attempting to put together a workshop at the Foundation to explore the directions that AI was assuming, and to "build the case for the intellectual coherence of the total undertaking." Kingsbury, whether correct or not in his assessment, made a revealing statement that goes far in indicating how much *interdisciplinarity* had yet to go in this particular field:

> [t]his is an activity which I have been encouraging in the Division of Information Science and Technology [IST] where the division director has taken a very active role in the integration of the computer science people (who by the way *do not completely understand this approach to software and operating system development*), the linguistics program and the cognitive sciences programs.[164]

Whatever the case with understanding, the fact remained that aspects of the social sciences were crucial in the history of the development of the biological sciences at NSF. The same could certainly be said for the behavioral sciences, themselves even more closely aligned with biology. While the behavioral sciences'

[160] Simon Report, pg. 37.

[161] David Lightfoot (Linguistics Program, University of Maryland, College Park) to Eric Bloch, June 13, 1985.

[162] David Kingsbury to Donald Kennedy, March 1, 1985, pg. 1.

[163] The letter has not been found, but can be inferred from other internal evidence.

[164] Ibid. Emphasis added.

intellectual alignment had a long history with biology, organizational alignment would not happen for several more years.

2.9 The Behavioral Sciences at NSF from 1975 to 1991

Darwin once spoke very biologically of thought as a "secretion of [the] brain." Consideration of that "secretion" is an excellent way to envision at least a part of what intrigued practitioners of behavioral science at the *fin de millennium* with much else that made and makes up the several disciplines that fall under that heading. Because the behavioral sciences border so closely on a dividing line with the biological sciences, the matter of how they were dealt with in the various organizational hierarchies and something of what they accomplished over time at NSF and within the communities they served are discussed here. After the reorganization that concluded in 1992, a number of specialties in the behavioral sciences did *not* move into the new directorate that came into existence for the social sciences (SBE) but rather moved into that of the equally new Directorate for Biological Sciences, *because they were viewed as properly one of them*. Such areas of behavior are later considered in the history of BIO chapters up to the close of this study from 1992 to 2005.

This section examines two areas closely for the years prior to 1992: psychobiology (including animal behavior and ethology) and the neurosciences. If one takes the National Academy of Sciences (NAS) as an arbiter of what constituted *science* when NSF was founded in 1950, the two groups of sciences, social and behavioral, were as if one—and, yet, not science at all! Strange as it may seem, it would not be until as late as 1971 that the NAS actually "created a class of membership for behavioral and social scientists," a year in which both were finally anointed as sciences, in fact.[165] Even in the present history, as seen in the earlier sections of this chapter, the social and behavioral sciences were shown to be almost invariably treated as a single, unified entity. However, this was not always the case, even prior to 1992. For instance, Senator Proxmire's Golden Fleece Awards much more commonly went to projects that best fit under the title of some one or another discipline within the social sciences, *sensu stricto*, rather than under the behavioral sciences, but not always. Were the latter more sheltered under the dictum of "protective coloration" or less obvious as they rode the "coattails of biology" into the future at NSF? Or was it because the behavioral sciences were so much more like the biological rather than the social sciences that they escaped as much negative attention as they did?

Reorganizational efforts begun by NSF Director Guyford Stever in 1975, continued by Richard Atkinson in 1977, and completed—to a level—by John Slaughter in 1981, had many important effects on the social and behavioral sciences. Significantly,

[165] Johnson, "Next to Nothingness, Part I," *op. cit.*, pg. 146.

most social scientists, in particular, were as pleased as was possible with these changes, even though they did not get their desired separate directorate until the early 1990s. But one group in their midst, biopsychologists (later psychobiologists), had been most closely associated with the biological and medical sciences of the former BMS and were not particularly keen on the reorganization that led to a Division of Social Sciences within a then new BBS.[166] Indeed, it has been argued that in the early 1980s:

> [t]he unanimity among engineers and the dissention among behavioral and social scientists were significant influences that resulted in a directorate for engineering and *failure to gain a directorate for behavioral and social sciences*.[167]

It is believed that this was the first high-level discussion for the possibility of a major reorganization that, for the social and biological sciences, would not come for another decade. The unrealized directorate was envisioned to include all of SES, BNS, information science and technology, applied social science programs found in NSF's Division of Applied Research (soon itself to be eliminated), and decision and management science (a new program in FY81). The consultant who worked up the plan for NSB was Mary L. Good, who later became the chair of the Board. The history of the notable influence of Mary Good at NSF is yet to be written.

The period under Atkinson found the 1975–1976 proto-Division of Social Science being absorbed into the new BBS, and its name changed to the Division of Social and Economic Sciences or SES: Atkinson himself had been a critical factor in the establishment of economics as a robust discipline at NSF. The second new division in BBS formed at that time was Behavioral and Neural Sciences (BNS) under Richard T. Loutitt, with Atkinson's strong support, as has been recounted. The BNS was *coordinate* with SES as one of the four original divisions under the new BBS. This allowed BNS to be seen as either one of the two social/behavioral divisions in BBS *or as a more biological-like set of disciplines akin to the remaining two of the BBS divisions that were "pure" biology.*

Some 17 programs were spread over the two newly created divisions in BBS. In FY75, the BNS, specifically, included programs in social and developmental psychology, anthropology, psychobiology, neurobiology, and linguistics. The two new programs in 1976 included, as we have seen, sensory physiology and perception (SPP) and memory and cognitive processes.[168]

They survived, along with the various programs in the division of social sciences up to 1980 when both areas were threatened by the Stockman "order." At the time, there was nothing comparable to the social science's broad national association, COSSA, for the behavioral sciences, but those scientists eventually did create one.[169]

[166] Ibid., Part II, pg. 265.

[167] Ibid. Emphasis added. More can be seen on the discussion of reorganization for the social and behavioral sciences (SBS) in particular in NSB, "Minutes," NSB-80-412, October 20, 1980, 219:11ff.

[168] Ibid., pg. 265.

[169] Ibid.

Discussion about forming such an umbrella group had begun early in that year. The organization came into existence in December in Chicago and incorporated in the District of Columbia in March 1981. With but a week to spare prior to OMB's rescission effort, it joined COSSA in its battle with the administration. The Federation of Behavioral, Psychological and Cognitive Sciences (FBPCS) was formed *specifically* to represent eight scientific societies in the struggle to retain funding for NSF and other federal agencies for the disciplines.[170] Only societies were represented, not individual scientists, within FBPCS. As with the social sciences and COSSA, the FBPCS helped in saving the behavioral sciences, one of which was psychobiology.

2.10 Psychobiology

Prior to the formation of the BNS proper and also before the Simon Committee's review, psychobiology had already attained a high level of productivity and had one of the largest program budgets. Its director, Jacob Beck, wrote in August of 1974 to the division director of the still then BMS.[171] Beck provided a review of the program through a set of budgets, descriptions, and objectives. The psychobiology program supported research, at the level of just under \$5 M in FY75, in "experimental psychology and ethology" and hoped to expand its budget to \$6 M in FY76 with specific orientation toward such areas as:

> sensory and perceptual processes, learning, memory, thinking and problem solving, motivation, genetic and environmental effects on behavior, and laboratory and field studies of early experience, imprinting, orientation, reproductive behavior, and the social and communicative behavior of animals.[172]

The unifying theme of these diverse specialties was a "focus on the information-processing characteristics of organisms." Specifically, the program was concerned with "how an organism receives, processes and uses information in interacting with the environment." New methodology was coming upon the scene that led the psy-

---

[170] Ibid. At the time of this writing, the FBPCS had just voted to change its name to the Federation of Associations in Behavioral and Brain Sciences. See their website at http://www.thefederationonline.org/. The societies included the American Educational Research Association, American Psychological Association, National Conference on the Use of On-Line Computers in Psychology, Cognitive Science Society, Psychonomic Society, Society of Experimental Social Psychology, Society for Mathematical Psychology, and the Society for Psychophysiological Research.

[171] Program Director for Psychobiology to Division Director, BMS, August 12, 1974, "FY76 Budget and Nuggets." Carbon copy of original in the Division's Chronology Files, Budget Projection Folder, and Golden Nuggets File, all of which had come under the aegis, at an unknown date, of Fred Stollnitz, then Division Director for BNS. A number of file boxes long held by Stollnitz, partially organized though not accessioned, were transferred to the Historian of the NSF, Marc Rothenberg, and reside at the time of this writing in the Historian's file room. Hereinafter, such files will be referred to as the Stollnitz Files.

[172] Ibid., pg. 1 of memo.

chobiology program officers toward "exciting new advances and conceptions," including such things as Fourier analysis and signal processing and, especially notable here, the "computer control of experiments and computer simulation."[173] That was yet more evidence for the *rise of computerization* at the Foundation and in the community, even a decade before Bloch became director.

Beck's memo was an example of those called for by division directors from their program directors each year at the beginning of the budget preparation and submission process. Beck suggested that the $6 M he requested for FY76 be divided among three programmatic areas within psychobiology: cognition, perception, and sensory processes, learning and memory, and species-typical behavior, at the rate of $2 M for each of the three. The amounts for the previous two fiscal years were given for comparison when the totals for psychobiology had been smaller: FY74 totaled somewhat over $4 M; and FY75's was a figure of nearly $5 M.

The psychobiology program, during the 1970s and 1980s, presents well the crucial issue of the behavioral sciences lying at the dividing line between the social sciences, *sensu lato*, and the biological sciences. Exemplars of programmatic emphases included the "species-typical behavior" category embracing such matters as "biological clocks in circadian rhythms," "avian and fish migration," and "echolocation, acoustic and olfactory communication." All of those areas are highly *bio*logical, but they abutted against other aspects of the category such as "how social behavior is adapted to an animal's environment," an area lying at the elusive dividing line. Tied to that was animal communication, for example, which became of research interest to BNS some years later (1980).[174]

When the Simon Committee report came out in 1976, it had more to say about psychobiology than *any other programmatic area* of either the social *or* behavioral sciences, so big had the program grown over the preceding years. "The striking fact about this program," the authors stated, "is its enormous diversity." Even so, they were able to show that the many disciplines involved fitted primarily into comparative psychology, ethology, and experimental psychology.[175]

---

[173] Ibid.

[174] NSF, "Annual Report 1980," pg. 61. Her research was on the variability of the call of the black-capped chickadee. See also the intriguing graphic on pg. 62: "Sounds of Danger."

[175] Simon Report, *op. cit.*, pg. 33. The phrase "animal behavior/ethology" has been used in this history to this point, but without definition. David McFarland has defined the two closely related areas in the following way. "Evolutionary biologists (including sociologists) look at behavior in terms of its past evolution and do not address problems of present-day causation. Psychologists are primarily concerned with the proximate causes of behavior and rarely make use of rigorous argument based upon the theory of natural selection. Ethologists, on the other hand, have sought to combine the mechanistic and evolutionary approaches to behavior, asking not only how behavior is controlled but how the mechanisms evolved and why particular mechanisms appear in particular circumstances," pg. 357. He further provides four "aims" that the ethologist should pursue: "the questions of causation, development, survival value, and evolution of any behavior pattern under study," pg. 360. See McFarland, *Animal Behavior: Psychobiology, Ethology and Evolution* (Menlo Park, CA: Benjamin/Cummings, 1985). Not uncommonly, however, animal behavior and ethology are used loosely and interchangeably. Care has been taken in this study to follow the terms as used in regard *not* to formal definition but to historical circumstances as seen in the record.

The Committee noted that in the year previous to producing their document, about half of all grant awards fell into the areas of either ethology or comparative psychology of non-human species. Most of the remaining awards, they stated, fell into the broad category of human experimental psychology, though, of note, many used non-human animals in their research. About a third of the second category dealt with conditioning or stimulus-response learning theory. The remainder of that group was oriented toward cognition and psychophysics, but none were in the areas of perception or emotion.[176] However, a number of grant applications in those latter two areas had been made but were declined by the review panels. Thus the Simon Committee concluded that the "pattern of support does not reflect accurately the current research interests [in the behavioral science community at large] in experimental psychology."

The Simon authors did note that the National Institute for Mental Health and its parent, NIH, both compensated for this shortcoming within the division. That was so as the funding at NIH was dropping and NSF would, it was suggested, need to take up the slack in those fields. That was even more important as similar research supported by funding through the Department of Defense (Office of Naval Research, the Air Force Office of Scientific Research, and the Office of the Surgeon General of the Army) was being "terminated or greatly curtailed" at the time. For this reason, the Simon Report was blunt: "[i]n sum, the budget pressure on the psychobiology program is severe."[177] They also concluded that the panels had not reflected the growth of certain fields in recent times, including, especially, research on human cognition. The Simon Committee was aware of the reorganization with the newly minted BNS just coming upon the scene, and so they felt that this latest division would go far in addressing some of their concerns. They pointed out that the reorganization would:

> separate new programs for sensory physiology and perception and for memory and cognition from the psychobiology and neurobiology programs…[thus] correcting the difficulties in the present arrangement.

Further:

> [p]roposals in psychophysics will presumably be assigned to the sensory physiology and perception program, while the revamped psychobiology program will deal with behavior genetics, hormones and behavior, studies of non-human behavior, and other topics in physiological psychology.[178]

The Committee went on to deal with how other grant applications might best be apportioned over the various revamped programs that were very much interdisciplinary.

In the early days after the 1975 restructuring, BNS and psychobiology were off to running starts as the division's director, Louttit, wrote his first major program

[176] Simon Report, pg. 33.

[177] Ibid., pg. 34.

[178] Ibid.

report with considerable pride in March of 1977. "We are pleased to have this opportunity to present to you, for the first time, the programs of the [BNS]."[179] While it covered all aspects of the division, not just psychobiology, the introduction and overview text was highly informative as it dealt with the changes made in the program structure once the Simon Report had come out and BNS had been formed. "Since we are just gaining visibility as a new division, all of our programs [are presented in the report]." The division settled down to seven programs that included anthropology, linguistics (noted earlier), psychology, psychobiology, and the "newer interdiscipline of neuroscience." Observe the use of the interesting coinage, "interdiscipline." Louttit went on to explain the changes, which included removal from the earlier special projects area of linguistics; the expansion of social and developmental psychology and a greater emphasis on that area than had theretofore been the case; the new program in sensory physiology and perception; and another new program in memory and cognitive processes. "We are," he said, "concerned with understanding human and animal behavior."[180]

Animal behaviorist, Fred Stollnitz's, program in psychobiology was synopsized in the report. The first of his figures noted "topics" in the discipline: learning and memory, the new program in motivation and emotion, orientation and migration, ingestive and reproductive behavior, and animal social behavior and communication. The second figure asked four questions: What are the features of each of the mechanisms, development, functions, and evolution of behavior?[181] Stollnitz, in keeping with his long-range plan, wrote in Louttit's report of research in these areas that the results "have been *inherently fascinating* and of *clear benefit* to society."[182] Both basic and applied research was doing well in the psychobiology program and so the BBS determined to make resources available for the area of "primate behavioral research [and] breeding."[183] Primate origins were of interest, too, and hopes were high for taking a "major step in understanding primate evolution within the next 5 to 10 years."[184]

The review panels for BNS were processing a great many grant applications—some 900 during the first half of FY77. Louttit was planning to make a $1 M supple-

[179] Richard T. Louttit, "Program Report, Vol. 1, No. 1, March, 1977: Behavioral Sciences," Prepared by the Office of the Director, Office of Planning and Resources Management, Program Review Office, NSF. Quotation on pg. 1.

[180] Ibid.

[181] Ibid., pg. 37; Figures 32 and 33, respectively.

[182] Emphasis added.

[183] BBS, "5 Page Issue Paper," in-house typed, six-page document; January 30, 1978, pg. 1.

[184] National Science Foundation, *Twenty-Ninth Annual Report for Fiscal Year 1979* (Washington, D.C.: Government Printing Office, 1980), pg. 51.

mental budget request.[185] He argued to Atkinson that the Director's Reserve, in Louttit's opinion, should be used specifically:

1) to meet unexpected, but mandatory needs…(e.g., pay raise, loss of an aircraft in Antarctica,[186] etc.);
2) to meet changing demands and opportunities from the field not foreseen early in the budget cycle; [and,]
3) to put the Director's imprint on the ongoing operations.

The last of the three points was nicely calculated, it seems, as Atkinson was, after all, a social scientist. More importantly, though, he and Deputy Director Pimentel had been on record as having said they wanted to increase efforts in those fields.

Apparently, Louttit's memo of May did not receive an answer from Atkinson, so he wrote a second and more aggressive one at the end of June in which he asked for a meeting with the director.[187] While the details of that meeting are unknown, the budget for BNS for psychobiology in FY77 was well over $8 M, but for FY78, it took a precipitous drop to just under $6 M! Why? That drop was the case only for that 1 year because in FY79 it had returned to an amount more in keeping with that of FY77: $8.5 M. We know that Louttit was aggressive in his request for Director's Reserve funds, but the data for FY80 and FY81 were more comparable with the FY77 figure and did drop again in FY82 with only a slow climb back to the $8 M vicinity by FY85. However, that second drop and slow recovery can be attributed to Reaganomics and Stockman's vision.[188] It should be noted that funding for NSF during the years just prior to the Reagan revolution, and through them to 1988, saw the Foundation do far better financially than what may have been expected given Stockman's own views. NSF saw a 99% increase in funding from 1978–1988.[189]

Louttit was clearly committed to his division, as he continued to find the best way to operate it in a period of financial stress. He met with then Deputy Director Pimentel to discuss "reducing the incidence of duplicate proposal submissions to

[185] Ibid., pg. 1.

[186] The history of lost aircraft in Antarctica's difficult conditions is considered at http://www.vaq34.com/vxe6/r4d.htm, at least for the widely used R4D "Gooney Bird" transport aircraft. Louttit's use of this example was not hyperbole as the number of crashes and abandonments of just this one type of airplane used in NSF's very substantial Antarctic Program was significant and costly, especially during the 1960s.

[187] Director [BNS, Richard Louttit] to Director [NSF, Richard Atkinson], June 29, 1977, same title as the May 2 memo (footnote further above). Stollnitz Files.

[188] All financial data are from National Science Foundation, Division of Science Resources Studies, *Federal Funds for Research and Development, Fiscal Years 1970–2001, Federal Obligations for Research by Agency and Detailed Fields of Science and Engineering*, NSF 01–333 (Arlington VA: National Science Foundation, July, 2001), Table 2, page 45.

[189] See the table entitled "Distribution Within the HUD-Independent Agencies Appropriations Bill: Percentage Changes by Major Components 1978–1988," 371; 6/88 (RB4); a five-page ephemeron from the Stollnitz Files. The only agency to have a higher percentage than NSF was NASA at 120%. Five others were all below NSF, with four of those in the negative range and with the Veterans Administration being the only other positive figure at 56%.

the NIH."[190] As we have seen earlier, NIH was supporting work in the behavioral sciences as well as was NSF. Louttit's plan was one in which BNS would delay its review of a given grant proposal submitted to both agencies until after the NIH review had occurred. That allowed for reducing the onerous workload by review panels at NSF if NIH decided to support any given application and thus end duplication of effort. However, grants that NIH refused may well have been quite acceptable to NSF, for its own reasons. Additionally, as Louttit pointed out:

> [o]ur interaction with and decisions made by individual investigators make it clear that everyone does not prefer NIH support. [Since that is the case]…I don't believe that we could or should delay review of all duplicate submissions pending NIH action. However, partial and selective 'phasing' [his plan] seems to be helpful in managing a heavy workload effectively.[191]

Two programs, neurobiology and sensory physiology and perception, had "low percentages of proposals receiving optimal support." Those two were targeted for more attention: later, data supported the wisdom of that targeting.[192] "Duplicate submissions" to NSF and NIH will be discussed again in later chapters.

The foregoing commentary provides evidence of the fact that, even after BMS and the loss of the "M" word, Medical, the new BBS (and NSF more broadly) still competed with NIH in its offers of grants to applicants: some best fit one agency of the two, others the second. Further, various divisions and directorates sought methods suited to their disciplines, workloads, and other peculiarities that best judged winning proposals; Louttit's plan was one. Such methodological matters changed with time and directorate (and their subdivisions) at NSF, as shall be seen.

In his introduction to the "phasing" plan report, Louttit said that the BNS "was formed…to provide a *bridge* between the biological and social sciences."[193] The group envisioned it as shown in Fig. 2.4. He referred to a small reorganization that had occurred in 1980–1981 within BNS itself that stressed behavior most especially and the need for a "multidisciplinary division" that would include anthropology, linguistics, psychology, and neuroscience. To emphasize multidisciplinarity, Louttit noted that in FY79 and FY80, some "72 projects were supported jointly with 11 other divisions in all six directorates." Such cross-directorate activity, and the disciplinary areas for which the various directorates were responsible, gives further strong proof of the *interdisciplinary attitude* of a broad swath of staff at NSF.[194] The authors of the annual reports also recognized this interdisciplinarity issue:

---

[190] Director, [BNS, Louttit] to Deputy Director [NSF, Pimentel] via Assistant Director Clark, March 5, 1980. Stollnitz Files. Pimentel would leave NSF just 3 months later, as discussed in earlier main text.

[191] Ibid.

[192] Attachment to the Louttit to Pimentel memo of March 5. The "Enclosure" was a typed table showing the results of both phased and optimal support reviews for FY79 for the two programs mentioned in the main text above.

[193] Ibid., pg.1. Emphasis added.

[194] Ibid., pg. 2. In later years, neuroscientist and career leader in that area at BBS/BIO, Christopher Platt (see Chap. 9), argued that the existence of the several directorates at NSF actually tended to

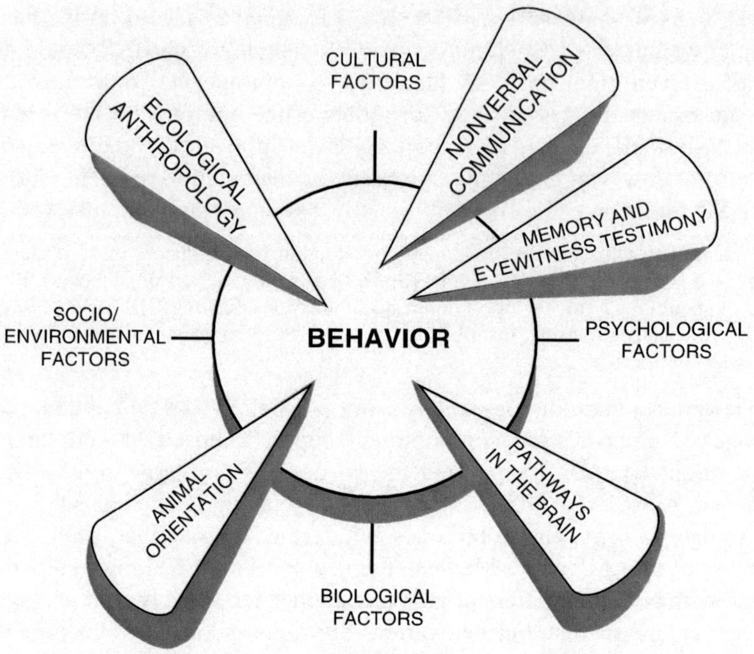

Fig. 2.4 Organizational chart for the Behavioral and Neurosciences. (NSF. Behavioral and Neurosciences Program Report 5:1, April 1981)

> Despite this apparently bewildering diversity, there is unity. The scientists who study neurons are linked to those who study social groups through their focus on living matter or its artifact…The interests represented by these relatively discrete approaches to the study of life forms have interacted, branched, and converged to produce *hybrid disciplines.*[195]

2.11 Technology Expanding

BNS moved forward with a productive set of programs, and, by 1984, Assistant Director for BBS, David Kingsbury (more on him in later chapters), was seeking to enhance the budget of SES/BNS by launching a project termed COGNET.

> There has been a spectacular growth during the past 15 years in studies of human cognition that view[s] the topic from an information processing point of view…One of the important

---

separate the various disciplines rather than unify them. He felt that what has "saved NSF, I think, has been [the] link at the Program Officer level where [they] are fluid molecules who do a lot on their own to reach across the [various] directorates." Christopher Platt Interview with the Author, February 19, 2009.

[195] National Science Foundation, *Thirty-Second Annual Report for Fiscal Year 1982* (Washington, D.C.: Government Printing Office, 1983), pg. 27.

> methods used in this research is simulation of human cognitive processes…Support [to COGNET] would be extended for a network of computers available to cognitive science researchers.[196]

Kingsbury sought $1 M to start COGNET but indicated that would be just the initial investment, though he hoped it would be a stimulus for NIH, also, to become engaged. Additionally, Kingsbury sought funding in the amount of some $2 M for "Data Resources for Research," that would include such areas as:

> new, large-scale data resources (e.g. time-use accounts [for super computer use, for instance], legal indicators, regional and state data); and augmentation of existing national data bases…and development of cooperative projects with other Government agencies… and the application of information science [to the data bases].[197]

Kingsbury, as alluded to above, was keenly oriented toward computerization of many aspects of the BBS both within NSF's walls and also with regard to the outside scientific communities it served. His memo was written in November of 1984. Erich Bloch had arrived from the IBM Corporation only 2 months before and had, of course, an even greater computer orientation.

Kingsbury sought additional funding for one other area, as well. For what he termed "Support for Priority Research" (asking for $3 M), he indicated that the monies:

> will be used to begin to support and to explore measurement and methodological issues in selected cross-disciplinary research areas identified by the [National Research Council committee currently] examining the high priority opportunities for the social and behavioral sciences over the next decade…that cross and transcend traditional disciplinary boundaries.[198]

BBS, let alone BNS and NSF more fully, was moving ever more deeply into computerization, interdisciplinary research (already well established), database-related efforts (expanding), and measurement and methodology work (long supported). No one of those areas that are trends in growth over time at BBS/BIO, and NSF in toto, and which are central threads in this history was new in the early to mid-1980s, except the expansion in numbers of desktop computers. Yet it is clear that all those trends were expanding at a rapid pace: they had profound effects throughout the Foundation. Computerization in the behavioral sciences would drive them into new and highly quantitative directions, more so than they had experienced ever before.

The NRC committee and its work, to which Kingsbury made reference, would not be completed until *calendar* 1986, but preparing budgets for *fiscal* 1986 was already facing federal agencies by November of 1984, at latest. Strikingly new cross-disciplinary areas were expected to be identified by the NRC and then supported in FY86 and beyond. A number of newly developing fields were envisioned.

---

[196] David Kingsbury to Ralph Devries, "Social and Behavioral Sciences Programs," nd (likely early November 1984). Stollnitz Files.

[197] Ibid.

[198] Ibid., pg. 2.

They included expert systems, industrial ethnography, and the psychobiology of learning, memory, cognition, and affect and motivation, among many others.[199]

Indeed, the "Program Profile" for psychobiology, which roots at NSF went back to 1952, indicated the steady gains it had been making. In FY82 its budget was over $3.5 M, in FY83 it was nearly $3.9 M, and, by FY84, it had reached just over $4.5 M. In that latter year, the number of grants funded reached 150 in that one BNS program alone.[200]

During that period, exciting new areas were emerging which BNS could support. In developmental psychobiology, the concept of the "Lifespan Approach" was one such. It involved studies in the ontogeny[201] of behavior, for many decades an interest area of the psychobiology program at NSF. Data had been forthcoming over time about how "environmental, physiological and genetic factors influence the development of species-specific behavior."[202] Even so, in at least the area of behavioral genetics, part of the Lifespan Approach, research had "diminished over the past several years [early to mid-1980s] because the behaviors being studied were not very meaningful in a biological context," Stollnitz averred in early 1986. Still, that trend had "been reversed somewhat in the last year or two."[203] And further, with the appearance of certain "genetic mutant strains" of model organisms, such strains could be used in combination with wild types of the same species "to elucidate normal mechanisms underlying behavior."

In addition to its normal routine of funding grant proposals and maintaining the two-way communication between NSF and the communities, the end of the decade saw the release of the NRC report, *The Behavioral and Social Sciences: Achievements and Opportunities*,[204] and a BNS internal report on "Research Trends and Opportunities in the Biological Basis of Behavior Program."[205] They both came out in 1988. Assessment of achievement, opportunities, and trends seemed to be of broad community interest at the time. The four authors of the BNS report seconded

---

[199] Ibid.

[200] Two-page ephemeron, "Program Profile of the Psychobiology Program," most likely written by Fred Stollnitz; nd. Stollnitz Files.

[201] Ontogeny is the collection of all aspects of development of a single organism from conception to death.

[202] Author unknown (possibly Stollnitz), "Developmental Psychobiology: Emergence of the Lifespan Approach," nd; four-page ephemeron from the Stollnitz Files.

[203] Fred Stollnitz (?), "BBS/BNS Planning for FY 1988 and Beyond," March 14, 1986; a typescript document (draft?) for budgeters in which "four questions" are posed. That on behavioral genetics is the last of the group. Stollnitz Files.

[204] *Op. cit.*, Chap. 4.

[205] Dennis L. Glanzman et al., "Research Trends and Opportunities in the Biological Basis of Behavior Program," National Science Foundation, Division of Behavioral and Neural Sciences, Fall, 1988 (termed BNS "Trends" in later footnotes). The BNS was, at that time, composed of four permanent individuals leading four programs (with a number of rotators and other permanent personnel aiding each): Dennis L. Glanzman, Neurobiology of Learning and Memory; Kathie L. Olsen, Behavioral Neuroendocrinology; Nathaniel G. Pitts, Neural Mechanisms of Behavior; and Fred Stollnitz, Animal Behavior.

the comments of Kingsbury concerning new directions in science and demonstrated the directions that BNS, specifically, was taking.

> Behavioral neuroscientists are pursuing investigations of the biological bases of behavior using increasingly sophisticated *technologies*, including those of electrophysiology, neurochemistry, immunology, neuroanatomy, pharmacology, and endocrinology. Behavioral analyses are becoming ever more *precise*, and allow for highly *quantitative* evaluations of behavior. One driving force behind this research is the incorporation of *modern technologies*, adding to the *multi-disciplinary* nature of most research efforts.[206]

If behavioral research could have ever been referred to as "soft science," that epithet was fast becoming obsolete as work was becoming ever more precise, quantitative, a user of modern technologies (the computer being only one), and multidisciplinary. Behavioral sciences were moving along with all of NSF's and the scientific research communities' directions: quantitative, computer-based, and multidisciplinary.[207] These historical threads characterize science broadly at the *fin de millennium*, both within NSF and nationwide. All of these were notable features of how NSF was influencing science in the country.

Beyond those, and peculiar to behavioral sciences (and biology and some other sciences), was that:

> [t]wo major trends are apparent in the choice of species for study. One is the utilization of *model systems* to study various phenomena, such as learning [and] often involve *simple organisms*...The second trend is to select animal preparations that are essentially *similar* to humans.[208]

The arguments and criticisms of such persons as Senator Proxmire when they railed or chortled over the study of birds, bees, and bats were also becoming rapidly obsolete. With the shift back to a "more mechanistic approach to understanding behavior," the community was focusing on "adaptive function and ultimate evolutionary explanations"; this due to the "advent of sociobiology in the mid-1970s."[209]

As the next reorganization was known to be upon the horizon, Richard Louttit provided an "Oversight Report of the BNS Division" to Bloch and Kingsbury in June of 1989. In the report, the committee chair from outside NSF noted the status

---

[206] Ibid., first page of report; all pages were unnumbered in the report. Emphases added.

[207] There are a number of studies of the role of computerization in sociological and behavioral research. See, as especially relevant: Amiram Elwork, "The Behavioral Sciences in the Computer Age," *Computers in Human Behavior* (1985):3–18. This article was the inaugural one for the then new journal. Also see Pushkala Prasad and Anshuman Prasad, "The Ideology of Professionalism and Work Computerization: An Institutionalist Study of Technological Change," *Human Relations* (1994):1433–1458; Tom Jewett, "The Dynamics of Computerization in a Social Science Research Team: A Case Study of Infrastructure, Strategies, and Skills," *Social Science Computer Review* (1991): 246–275; Rob Kling, "The Social Dimensions of Computerization," *ACM SIGCHI Bulletin* (1987): 337–339; and, Rob Kling, *Computerization and Controversy: Value Conflicts and Social Choices* (2nd ed.) (San Diego, CA: Academic Press, 1996), *passim.*

[208] BNS, "Trends," first page. Emphases added.

[209] Ibid., second page.

of each of the then four programs in BNS.[210] The 16-member committee undertook its review by examining a number of "jackets" in each field. Jackets are an NSF-wide file device set up for each proposal submitted and considered and contain all relevant documents associated with the given grant proposal.[211] They are coded and identified and available to all persons who would play a role in any decision-making with regard to the proposal. The committee was able to review some 73 jackets that had been completed and acted upon by BNS from the fall 1988 proposal-review cycle and that were spread over BNS's four main programs and a few special programs. Such reviews were (and are) typical Foundation-wide, and the results gave management personnel an outside, ideally unbiased, view as to how the program officers were doing with regard to the wisdom of choice of those proposals awarded, declined, or in other fashion acted upon. Issues of fairness at all levels (gender and race of researcher, institution, geographic spread, type of institution, requested funding amount, etc.) and alertness to directions in the disciplines, including possible new directions, as well as areas loosing relevance, could all be judged by the committee members. Again, those features considered were and remain the central methodology for decision-making for the most fundamental thing done by NSF: accept, review, and decide the fate of all grant proposals (and later manage those awarded) arriving at the Foundation each year.

2.12 Neurosciences

When the Simon Report came out in 1976, the authors had given considerable space to psychobiology. But they only allotted one paragraph to neurobiology, for the reason given in the quotation below. The use of the term neurobiology was not restricted to the Simon Committee, but the more commonly used neural science(s) or neuroscience(s) was much in force in later years. The Simon Report authors noted that:

> [t]he neurobiology program has supported research in sensory physiology, which is now in a new program; neural processing; genetic, development [*sic*] and molecular neurobiology; and behavioral neurobiology. Although it has close relations with psychology (about one-third of the research is behavioral), the neurobiology program *lies primarily in the biological sciences*. Hence it isoutside the agenda of this Committee, and the scope or emphases of its projects are not reviewed in this report.[212]

[210] Isador Gormezano, Chair, "Division of [BNS] Oversight Committee Report, June 5–6, 1989," pg. 17; attached to the cover memo from Richard T. Louttit to Director NSF via David Kingsbury, September 14, 1989. Stollnitz Files.

[211] A jacket with a red sticker on it was one typically from a congressperson asking for special attention, a great rarity according to one employee. Christopher Platt Interview with the Author, February 19, 2009.

[212] Simon Report, pg. 35. Emphasis added.

The Committee did, though, conclude the single paragraph with the remonstrance that the program was suffering from lack of funds and, given other agencies' declining support at the time, the implication was that NSF needed to fill the growing funding gap for basic research in the neurosciences.

BBS—the Directorate of Biological, Behavioral, and Social Sciences—was just that: it included the behavioral sciences *along with the biological sciences*. It is no surprise, then, that behavioral research would have such strong ties to the two divisions of more pure biology—as both science and organizational unit at NSF—and would lead the Simon Committee to necessarily leave discussion of neurosciences out of their social science-oriented report.

Steven Kornguth has considered briefly the history of the neurosciences during the last half century of the previous millennium and finds that "the understanding of neural function has advanced by quantal [*sic*] jumps unparalleled in most other areas of science."[213] Indeed, "[t]he neurosciences verge[d] on a revolution closely rivaling that in biotechnology," stated the *Annual Report 1982*.[214] Studies of the biology of the neuron, brain, and nervous system, along with several other areas (studies in molecular-level evolution, systematics, biological aspects of paleontology, genetics and the molecular features of life, DNA, genomics, proteomics, biotechnology, etc.) grew rapidly in the biological sciences, both without and within NSF, in the late twentieth century.

Kornguth argued that neurosciences advanced due to the influence, *particularly, of its growing ties to specific features of the growth of certain technologies*. The BNS internal "Trends" report of 1988 suggested much the same thing and the funding of research in the neurosciences was coupled closely with instrumentation. Research into neuromodulators, protein synthesis, long-term potentiation of nerves, computer simulations of physiology and of learning, etc. were all noted in the "Trends" report and are all highly technology-dependent areas of neuroscience research.[215] Indeed, the *Annual Report* even a decade earlier noted that:

> neuroscientists, concentrating on the nervous system/behavior link, are capitalizing upon invertebrates with simple nervous systems that are amenable to analysis in *molecular terms*.[216]

Those invertebrates included work on the leech by famed biologist Gunther Stent at UC, Berkeley, where he was attempting "to elucidate the developmental mechanisms that give raise to…neural circuits."[217] And of considerable importance at the time, too, was Eric Kandel's work on the giant axon of the sea slug, *Aplysia*, for

---

[213] Steven Kornguth, "A Journey in the Neurosciences from 1950 to 2000," *Neurochemical Research* (2000):1435–1437; quotation on pg. 1435.

[214] National Science Foundation, *Thirty-Second Annual Report for Fiscal Year 1982* (Washington, D.C.: Government Printing Office, 1983), pg. 27.

[215] BNS "Trends" Report, pg. 6, for example.

[216] National Science Foundation, *Twenty-Eighth Annual Report for Fiscal Year 1978* (Washington, D.C.: Government Printing Office, 1979), pg. 61. Emphasis added.

[217] Ibid., pg. 68.

which he would win the Nobel Prize in 2000. Shortly thereafter, the nematode, *Caenorhabditis elegans*, would become the most favored "worm" in developmental studies of all sorts.

As Kornguth has further asserted, "[n]euroscience was in its infancy at the 20th century's mid-point."[218] He was seconded, in the earlier NSF *Annual Report* for 1976, in his assessment of the rapidity with which the neurosciences were growing: "[n]euroscience is among the fastest growing of the sciences."[219] Kornguth followed up his "infancy" comment immediately with the statement that:

> [s]o was computer science, and its concurrent and equally rapid development fueled the explosion of research and understanding in the neurosciences. The insights achieved in computer sciences allowed the development of concepts such as the neural net, codes for modeling neuronal activity and tomographic instruments [such as MRI, fMRI and PET] that enabled us to accomplish [much].

The use of large and expensive instruments, including MRI (magnetic resonance imaging), fMRI (functional MRI, wherein real-time hemodynamic responses are visible as evidence of neural activity), and PET (positron emission tomography), was funded by NSF for the neurosciences, often as part of the interdisciplinary grants that covered more than one directorate: *big science*. Though it was not until the last years of the previous century that one major project began in funding neuroscience research, it still remains exemplary of NSF involvement in the area: the Center for Behavioral Neurosciences (CBN) headquartered in Atlanta, Georgia. It was established in 1998–1999.[220] The CBN is one of a series of NSF-funded Science and Technology Centers (STCs) that play a major role in BBS/BIO and other Foundation directorates. Recall also the earlier discussion of the Center for Coordination of Research on Social Indicators that had been, by FY83, funded by NSF for about 10 years.

Technologies of direct application to behavioral sciences included the growth of electron microscopy in the decade of the 1960s; that of macromolecular separations technologies and monoclonal antibodies synthesis (1970s); specialty technologies

---

218 Kornguth, "A Journey," pg. 1435.

219 NSF, "Annual Report 1976," pg. 55.

220 See http://www.cbn-atl.org/about/index.shtml. On the "About Us" page under the heading of "History of the CBN," one can read the following: "The Center for Behavioral Neuroscience was established in 1998 by a grant from the Robert W. Woodruff Foundation. In November 1999, the Center became one of the National Science Foundation's Science and Technology Centers and expanded to include seven institutions in Atlanta, Georgia (Georgia State University, Emory University, Georgia Institute of Technology, Morehouse School of Medicine, and the three schools in the Atlanta University Center: Clark Atlanta University, Morehouse College, and Spelman College) and other community partner organizations. The CBN's original scientific focus was the neuroscience of social behaviors in the areas of affiliation, aggression, fear and reproduction, and the emotional and regulatory processes that underlie them. In the intervening years, inclusion of additional behavioral neuroscientists at the participating institutions resulted in an expansion of the CBN's activities into the areas of memory, cognition, reward functions of the brain, and positive emotional states. That process continues today as the CBN responds to the changing landscape of neuroscience and the evolving needs of its member institutions."

for the neurosciences in the 1980s (the patch clamp for ion detection across cell membranes, polymerase chain reaction, cloning, as well as the MRIs and PET and desktop computers); and genome mapping in the 1990s.[221] Well before, there had been "autoradiographic and enzyme tracer techniques to delineate pathways."[222]

It was not only technology that gave such a boost to neuroscience in the last century's final decades but also the "economic and intellectual context" of the United States which "nurtured and encouraged the rapid development of the neurosciences," Kornguth has argued.[223] Also, "[t]he large increase in science funding through the [NIH] and the [NSF] during the past three decades [of the 20th century] enabled such rapid growth," he stated. The NSF realized the future potential of the field and so, in 1976, "reorganized [the programs]…to focus research on human and animal behavior and the organ-system that directly underlies behavior—the nervous system."[224]

In the milieu of the burgeoning neurosciences of the middle 1970s, it behooved the BNS leadership to look to the future with a long-range plan. In 1976, Richard Louttit sought estimates from each of his program officers, providing "their individual perspectives," of where BNS, in toto, might or should direct its efforts during the period of 1979 through 1983.[225] The assessment referred to the Simon Report, "where appropriate," but as that report did not consider all aspects of the behavioral sciences, other trend information sources were examined, as well. Those included several federal studies and also NSF internal reports.

The *intricate dance* was once again shown to be in effect, as Louttit stated that "[e]ach program's long range plan provides a valuable perspective on current and future *opportunities and needs* in the fields for which the [BNS] is responsible."[226] Exactly what the long-range plans included was, in part, consolidated into a budgetary table created by Louttit. It is given here as Fig. 2.5.[227] Two types of funding streams were suggested. The primary one ("increase") was the annual figure requested with increasing percentages over the 5-year plan period (FY79-FY83). Then already-committed budgets for the current FY77 and FY 78 were included by Louttit to show the trend line. The second stream, "add-ons," was just that extra monies sought (such as that for FY78) for "large-scale needs, which cut across all our areas of responsibilities" and which had been suggested by the NAS. It was clear that there had arisen, by that time, "[t]he need for specialized research facili-

---

221 Kornguth, "A Journey," pg. 1435.

222 NSF, "Annual Report 1976," pg. 59.

223 Kornguth, "A Journey," ppg. 1435–1436.

224 NSF, "Annual Report 1976," pg. 59.

225 Richard T. Louttit, "Division of [BNS]: Division Director's Assessment of Long Range Plans—1979–1983," a typescript document, hand-dated July 21, 1976. To it are attached the "perspectives" of each of the BNS program directors. Stollnitz Files.

226 Louttit, "Long Range Plan," unnumbered pg. 3.

227 Ibid., unnumbered pg. 2. Emphasis added.

	FY 1977	FY 1978	FY 1979	FY 1980	FY 1981	FY 1982	FY 1983
Increase	4.7	4.0	4.0	4.0	4.0	4.0	4.0
Add-on	--	3.8	3.0	2.0	1.5	.5	--
TOTAL	24.4	32.2	39.2	45.2	50.7	55.2	59.2

Fig. 2.5 Funds required for a reasonably effective research support program (dollars in millions). (NSF. Behavioral and Neurosciences Program Report 5:1, April 1981. Definition of terms "increase" and "add-on" is given in the main text)

ties in the behavioral sciences." Such things as the later Science and Technology Centers seemed to be under consideration, as the division noted that

> [the BNS] programs have never had sufficient funds to *encourage* the development of such [large-scale] facilities honestly and effectively. However, we must begin to provide such encouragement and to assist in meeting these needs now.[228]

The notion of "encouragement" is exemplary of the NSF side of the *intricate dance*: the BNS, in this case, knew what was of concern to the scientific community, and the former would, if funds should allow, *encourage* the researchers to propose large-scale projects, centers, etc., by which to carry out given studies. The Center for Behavioral Neurosciences (above) of the late 1990s would be an example from the next decade, and under BIO, of the type of thing that NSF would encourage. From the beginning of the 1980s, funds to create new centers or maintain those already extant were sometimes sought as part of the regular budget requests to Congress. Still, such funding for add-ons in the 1970s and 1980s was more typical outside the behavioral sciences than within at NSF. Astronomy, oceanography, and other large-scale sciences required, as fundamental, greater inputs of funds for basic research to create or maintain such things as observatories and research vessels, respectively. But funding for the behavioral sciences was generally insufficient to keep the annual programs going, let alone to finance add-ons. For instance:

> we are moving backward in meeting the needs of the scientific community for research proposals [as an] overall 'success rate' (i.e. what percent of applicants got some money—seldom an optimal amount) [was] 35.6% in FY 1976 versus 38.5% in FY 1972.[229]

Louttit's cover commentary for all four of the BNS-appended reports ended with him saying that:

> [t]he importance of and need for additional resources for research on animal behavior is well-presented…The unfortunate fact that a few projects in Psychobiology and in Social and Developmental Psychology have been criticized publicly by a handful of politicians

[228] Ibid. Emphasis added.

[229] Ibid.

> and journalists should in no way detract from the importance of increased support for these programs.[230]

It was at about that time (1977) that Nathaniel G. Pitts came to NSF. Pitts had a doctorate in neuroscience from UC, Davis with a specialty in pain mechanisms at the spinal cord level. He moved from the academic sector to the federal environment of research funding as he did not want to stay at the laboratory level "forever."[231]

> [NSF] wanted me to join a team of people to look at how to support this burgeoning field, neuroscience, from a basic research perspective alone. The idea of creating programs to deal with a frontier was interesting to me.[232]

Though he originally came as a 2-year rotator, he stayed on; a story not unlike many other professional employees in NSF's history—and a number of which are chronicled here. He became, at one point, a director of the Integrative [Neural] Systems program.[233] In answer to the question of whether the neurosciences were new at NSF at the time, Pitts responded: "Absolutely." Pitts recounted to his interviewer how Atkinson had recruited Louttit and started the BNS. The early days were heady, "almost like wildcatting in the oilfields," Pitts recalled.[234] Beyond the federal, the professional institution in the form of the Society for Neuroscience was new then, too, having been founded in 1969 with its first annual meeting held in 1971. It would be yet another decade before the *Journal of Neuroscience* appeared.[235] As programs at NSF and in the community were growing "by leaps and bounds," the program directors could be "choosey," within the limitations of funds, of course, as to what "opportunities" to support.

Pitts averred that accepting employment at NSF was "not considered an optimal career path" and that the period was one in which Nobel Prizes were being awarded in the *experimental* neurosciences, in particular, given just how active the field suddenly had become. Pitts himself had worked in a laboratory situation with dual Nobelist, Sir John Carew Eccles.[236] But optimal or not, they were "crazy times," said Pitts, as he had to join his colleagues to give a presentation to the NSB on the trends in neuroscience only 2 months after his arrival at the Foundation. Though at odds with Louttit's impression of what might be called sufficient funding, Pitts recalled that "our division was always getting more money than the other divisions [of BBS] because this was considered one of the frontiers, and *directors of [NSF] have always*

---

[230] Ibid., sixth unnumbered page.

[231] Nathaniel G. Pitts (retired from NSF/BNS) interviewed by Marc Rothenberg (Historian of NSF) on November 28, 2007; pg. 2.

[232] Ibid., pg. 3.

[233] Ibid., pg. 9.

[234] Ibid., pg. 4.

[235] See http://www.sfn.org/index.aspx?pagename=about_sfn, *op. cit.*, for a brief history of the Society.

[236] Pitts interview, pg. 5. For more on Australian Sir John Carew Eccles (1903–1997) and his work, see http://nobelprize.org/nobel_prizes/medicine/laureates/1963/eccles-bio.html.

pushed frontiers of science."[237] In a comparison among the four BBS divisions, that may have been true, but not when compared to divisions across NSF; and funding decisions take into account all NSF directorates once each year.

A variant of the *intricate dance* concept, at the scientist-to-scientist level within NSF, was ratified by Pitts. He called it the "old water-fountain effect." One could easily meet a scientist in another field over the water fountain or "lunch counter" such as a sociologist or an economist and thus be able to ask that individual: "What's your field doing?" and so "then you [would end up getting] a different collaboration."

> You get informal ideas about things…[and] maybe you want [the scientists you encounter at the water fountain] to sit on your panel…[you get] topflight names of good people and you make informal connections…So most certainly having all of us not just on the same floor, but in the same contiguous space [promoted these contacts and collaborations].[238]

Pitts hit upon a subject in his interview that played a distinct role, still needful of careful historical study at NSF, on how architectural realities in buildings affected interpersonal communication and allowed for or hindered collaborative, hence often interdisciplinary, interactions. Indeed, his office neighbor, a social psychologist, stated that, "there was an approach to the [physical office] space issue in government that needed to be altered for better communication and [a] more effective working environment." Pitts went on to note that "today the case is [that] those people [scientists from various disciplines] are on different floors entirely in this building [current NSF headquarters in Arlington, Virginia]." In such situations, "you certainly have different kinds of interactions…if all of biology is housed on one floor and your social scientists are [on] another floor, you're not going to have those happenstance arguments or discussions." While the issue is not a trivial one and is part and parcel of this history, its fuller explication must be left to an architectural sociologist. The notion of the *intricate dance* can be expanded with architecture in mind. It is not only crucial to how the NSF interacted/interacts with its constituent scientific communities across the nation but also how various program officers, and others, interacted with one another within the various NSF headquarters' locales themselves.

Further, there is a dictum that office space and the nature of that space in Washington, D.C., are a matter of very considerable interest to federal employees. Specifically, in the 1800 G Street NW ("18th and G") building—which NSF shared with the Secret Service—all of the assistant directors (each directorate) were on a single floor with their deputies, and the staffers were on other floors. For BBS, for instance, biology staffers were split over two floors. Pitts was located in the 18th and G headquarters during his tenure. It was an old building not in the best of condition, and the distribution of the various directorates was such that the then much smaller BBS, and BNS in particular, shared a floor with chemistry, in the physical

---

[237] Ibid., Pitts, ppg. 5–6. Emphasis added.

[238] Ibid., pg. 6. The author has also heard many references to this matter that tend to ratify that to which Pitts referred.

sciences directorate. When Pitts had a biochemical problem in neuroscience, he sought help from his colleagues in the chemistry area. This promoted more than just problem-solving: it could also engender new program ideation and creation. That was just the case, as Pitts recalled that "[i]n those days, we would have initiatives [new programs] that would…be a combination of biology and chemistry…because in the sixties and seventies [those two disciplines] were really coming together."[239] Thus, the community side of the *intricate dance* was signaling to NSF where it was going—and the Foundation was listening. While it is not the purpose of this history to engage in the matter of architectural design and communication among building occupants, it remains that Pitts was correct when he said: "As arbitrary as this seems, [the communication or lack thereof] could affect the types of science frontiers we identify, recognize, and support."[240] It is certainly an area that has been considered in the literature and is likely very critical in the history of NSF and of the *intricate dance*, in particular.[241] As Pitts pointed out, while the spatial arrangement at 18th and G was not a hindrance for the senior leadership at NSF, it was a distinct challenge for BBS. Therefore, communications were not as good as could be hoped, but that would change positively once the move to Arlington was made.

The 1977 Program Report, the one penned by Louttit, considered not only psychobiology and other areas but "neural sciences" as well. Just the year before, the President's Biomedical Research Panel opined that "[p]erhaps the ultimate challenge to biomedical research, representing the very pinnacle of our understanding of the human organism, lies in neurobiology…[t]his Panel commends neurobiology as a compelling long-range interest worthy of national attention."[242] The 1970s, then, was a growing period for the rapidly expanding neurosciences.

Pitts saw more than the results of the NSF's community dialog, as he described the smaller internal reorganizations at the programmatic level done in BNS over time. Such minor reorganizations were and are commonplace and necessary as NSF directorates move with their partners on the dance floor of fund-seeking and support of research. What Pitts also witnessed, then, was the rise of both *computerization* and the use of *model systems* and of their importance to biology. One reason that NIH, for example, did not fund heavily in certain aspects of neuroscience, which void NSF could and did fill easily, was because in Pitts' opinion NIH was *not* "interested in very theoretical concepts" in neuroscience such as "constructs of how neu-

---

239 Ibid., pg. 8.

240 Ibid.

241 See, for instance, a recent work on the question: Kyungjoon Lee et al., "Does Collocation Inform the Impact of Collaboration?" *PLoS ONE* (2010):e14279. doi:https://doi.org/10.1371/journal.pone.0014279. The totally online, free, open access, peer-reviewed journal in biology, *PLoS*, was a new concept near the turn of the millennium. See especially http://www.plos.org/oa/index.php. See also a list of papers on the general subject put together by the editors of *Science* at http://www.sciencemag.org/content/316/5827/1036.abstract?andorexacttitleabs=and&HITS=10&sortspec=date&hits=10&maxtoshow=&andorexactfulltext=and&FIRSTINDEX=0&title=team&resourcetype=HWCIT&searchid=1&RESULTFORMAT=.

242 "Report of the President's Biomedical Research Panel," DHEW Publication (OS) 76–500, April 30, 1976, pg. 34; as cited in Louttit's "Program Report," *op. cit.*, pg. 2.

rons…work." NIH was funding and doing both extra- and intramurally what was expected of it, mostly practical human medicine. For researchers at such places as "MIT [and] CalTech" who were doing "very, very theoretical" studies, NIH was not the place to which such workers would turn for support. Such investigators "easily moved into the Computer Age because they were thinking about thinking machines": they "kept getting a [funding] lifeline from NSF" as it was the only location in the federal system to provide such support.

Further, NIH did not want to support invertebrate physiologists, those who worked with "squishies and crawlies," because the types of model systems those biologists worked with and the value of that work were not understood by NIH, so Pitts opined. However, it was work with just such model systems, he pointed out, that led to Kandel's Nobel Prize. Kandel's studies in learning in the giant sea slug, *Aplysia*, made the animal famous as an earlier *model system* for neurobiology. Prior to that, Haldan Keffer Hartline had won the Nobel in 1967 for his work in the "neural circuitry of horseshoe crab vision" with support from NSF. "Crawlies" were crucial.

Both of those Nobelists were members of a small group of persons who might then have used the term neuroscientist. In Hartline's era, the term was not in use; by Kandel's it was in coinage but was not formalized. Indeed, there may have been only about "500–1000" such scientists worldwide at the time.[243] But by 1985, the Society for Neuroscience numbered some 10,000 members.[244] So, the BNS was able to say in its next long-range plan (1980) that there was:

> reasoned optimism that we will soon understand how aspects of the machinery of the brain generate coherent behavioral acts, enable learning, and act as the most efficient computer known to man.[245]

The now commonplace notion that the brain could be considered a computer is well established.[246] But the first personal computers had yet to reach the desks of NSF employees by the early 1980s.[247] With Bloch's arrival in 1984, that would change, as has been said.

As the neurosciences grew in the general community of science, so also did they grow at NSF, organizationally. If "behavior is the product" of the neurons of the brain, then "two questions suggest themselves."

---

[243] Division of Behavioral and Neurosciences, "Neurobiology Program Long Range Plan, June, 1980," unpublished, in-house, typescript document. Unnumbered pages; second page: "Until about a decade ago [i.e., 1970], there was no organized body of scientists who recognized themselves as neuroscientists (in fact the word was probably not used)." Stollnitz Files.

[244] *Op. cit.*, http://www.sfn.org/index.aspx?pagename=about_sfn.

[245] BNS, "Long Range Plan, 1980," second pg.

[246] See, for instance, P.N. Johnson-Laird, *The Computer and the Mind* (Cambridge, MA: Harvard University Press, 1988); Ned Block, Owen J. Flanagan, and Gavin Gazeldere, eds., *The Nature of Consciousness: Philosophical Debates* (Cambridge, MA: MIT Press, 1996), *passim*.

[247] Pitts interview, pg. 14.

(i) Should the neurosciences be allied with the physiological, cellular and molecular sciences at the Divisional level? [and]
(ii) Would the discipline and the NSF benefit from a subdivision of the two combined neuroscience programs into three or four smaller units?

These were the questions asked by BNS in the 1980 long-range plan.[248] Louttit and his colleagues recognized the phenomenal growth of the neurosciences and the pains that came with it. There had been a 16% increase in proposals from FY79 to FY80 (278 vs. 322), and the success rate expectedly dropped, as competition and quality/significance of proposals increased, from 20–25% in FY79 down to 16–17% in FY80. Some proposals were withdrawn by their PIs as they had already been funded by NIH; dual submissions were still then permitted. Also, the types of proposals seen at BNS at that time included a heavy preponderance in studies of neurotransmitters; it's what was hot in the community.[249] Many of these were being shown to be neuropeptides and:

> this new direction is likely to have a profound impact on research in the field. It has already spawned at least three new professional journals, over 20 national/international conferences, and altered the directions of research of literally hundreds of investigators and students.[250]

Other aspects of behavior were "hot," as well: behavioral endocrinology—hormones and reproduction, for instance. Elizabeth Adkins of Cornell University had shown that even a brief exposure in early fetal development to sex hormones "dramatically and permanently" affected adult behavior later in life. Also, just a few years prior, NSF was instrumental in helping to found the Association for Chemoreception Sciences.[251]

In the 1981 program report for BNS, there had not been any obvious restructuring that had been implied by the two questions asked the year before. NIH was still the giant in the greater National Capital Area, and so Pitts decided to arrange with Ken Surrey of NIH to do an "exchange of desks" for a year. Pitts worked with NIH proposals in the neurosciences at their locale (colloquially known as the "campus") in Bethesda, Maryland, and Surrey did the same at Pitt's desk at NSF at 18th and G. Both learned a great deal about the relative situation of his opposite: the two agencies approached matters *very* differently. As it had been since the beginning of the NSF, program officers were responsible for the review of a proposal, then for its award, and finally for all oversight, interaction with the PI, and reporting on the awarded project, *to its conclusion*. On the other hand, NIH had long separated review as one element and all other post-review issues as a second. Different sets of personnel had responsibilities for either one stage or the other at NIH, but not both, as at NSF. For that reason, the workload per program officer at the NSF was much

---

248 BNS, "Long Range Plan, 1980," third pg.

249 NSF, "Annual Report 1979," pg. 60.

250 BNS, "Long Range Plan, 1980," fourth pg.

251 National Science Foundation, *Thirty-First Annual Report for Fiscal Year 1981* (Washington, D.C.: Government Printing Office, 1982), pg. 37ff.

more onerous. For instance, in BBS in the late 1970s to early 1980s, at least, a given program officer may have been responsible for dealing with 150 proposals per year with some percentage making it to the award stage and consequent management of those till they were concluded. And each year brought on another batch. The technique was and is that still used by NSF, as of this writing.

After some 3–4 months into the exchange of desks' project, Surry had a heart attack. Though he recovered, Pitts felt that he had not prepared his NIH opposite for the "strain of the workload at NSF." Surrey, used to working under the National Institutes' approach, told Pitts that he (Pitts) would have to make work for himself because, at NIH, "there just isn't much to do." Moreover, each NIH program officer had a desktop computer by the early 1980s—but they were "using them as doorstops"! "They were bureaucrats," who, in Pitts' view, would not deign to use a computer. That is to say, the review stage at NIH was done by professional scientists, and, once an award was made, management was left to computer-savvy technocrats and secretarial workers.

Pitts did not see himself as an NIH "bureaucrat." He "realized that each division [at NIH] had a computer technician that wasn't being used by anyone," so he had the technician teach him all the major useful software—word processors and spreadsheet programs—of the period. Pitts demanded a computer of his own upon his return to NSF. At the end of the experiment of the desk exchange, Pitts and Surrey wrote their planned reports. "The bottom line was at NIH there were three and a half people that did the work that one program officer at NSF does"![252] Pitts felt that once computers did appear on the desks of NSF program officers, however, it also moved a great deal of "mundane routine secretarial" work to them. "So, I think the workload of the NSF Program officers has been exacerbated over the years with the computers."[253] Pitts was not alone in that assessment.[254]

The reason for the different approaches at NSF versus NIH can be traced back to Alan Waterman, who felt that program officers would have greater influence over the granting process if they were to do both the reviewing *and* the later follow-on management of awards. Though the workload would be heavy, the program officer would have an "impact" of the type Waterman wished to see at the new NSF. That method has survived.[255]

Computerization brought on other changes, too. Prior to their establishment at NSF, some aspects of reviewing proposals in BBS (and later, BIO) called for sending them via paper mail to outside reviewers. The return commentaries were then taken to the formal review panels, made up of outside reviewers who traveled to NSF headquarters, along with the pertinent program officer(s). The reviewing pro-

[252] Pitts interview, ppg. 14–15.

[253] Ibid., pg. 16.

[254] The author's discussions and interviews in 2009 with numerous NSF BIO employees ratified the sense that Pitts conveyed about the respective workloads and computer use features he had experienced in the early to mid-1980s.

[255] Pitts interview, pg. 16. Information in a comment made by Pitts' interviewer (Rothenberg). That the method is still extant was mentioned to the author by several interviewees.

cess was, then, a mix of live action discussions supplemented by the return mail commentaries. That was a "[v]ery beautiful scenario" but was best suited to a still small NSF that received only 20,000 to 30,000 proposals per year Foundation-wide. Once that total had gone beyond 45,000 annually by the 1990s, with little concomitant and much needed growth in total employees, the mail reviewing approach was downplayed or halted altogether. That "puts a different psychology into the mix for the program [officers]," Pitts opined.[256]

Beyond that, the number of rotators began to rise, as extra review help was clearly needed on-site. In about 1977, the percentage of "rotators"—primarily academics on professional leave—was about 30% of the scientific and engineering workforce, Foundation-wide. (By 2007, for example, the total workforce—*all* categories—was made up of 15% rotators. About 33% of scientists and engineers, specifically, were rotators, however.[257]) In 1978, the NSB adopted a resolution reaffirming the importance of rotators to the functioning of the NSF that included the statement that the number of such personnel "should not be restricted."[258] And rotators found a certain intellectual freedom in their time at NSF. "[T]here was one famous rotator here who…wrote six papers [over his two years] because he was uninterrupted by academic [demands]." He worked on NSF matters till about 8:00 p.m. and then worked on his book or paper until 11:00—his family not being present with him during the 2 rotator years.[259]

Funding for neurobiology was climbing steadily along with the numbers of rotators and the move into a more computerized environment. The program had obligations of something over $8 M in FY79, but by FY83, it had gone up to $11 M. The competitive success ratios dropped along with the rise in proposals submitted, of course. In FY79, the rate was nearly 29% but that for FY81 had already seen a drop to 20%.[260]

By mid-1982, there had been "a decade of coherent support of the neurosciences at NSF."[261] For that reason, Louttit and his deputy, Jim Brown, sought to take "actions we believe will enhance the neuroscience effort at NSF for the coming decade." Proposal numbers went up, success rates went down, but "fiscal growth has not kept pace with the increased opportunities," despite a threefold increase in the program's budget. So, Louttit and his colleagues began (in 1981) an experimen-

---

[256] Ibid., pg. 17.

[257] See http://www.nsf.gov/pubs/2008/hcsp2008/nsf_humancapitalstrategicplan_0803.pdf.

[258] NSB, "Minutes," NSB-78-360 (Revised), September 25, 1978, Appendix B, NSB/Res[olution]-78–85, August 17, 1978, 200:23.

[259] Platt interview.

[260] Jim Brown (BNS deputy division director), typescript entitled "Research Opportunities in the Neurosciences," July 1982. The document was likely a draft for a later program announcement. Data from appended Table 1, "Obligations and Competitive Success Ratios: FY 1979–1983, By Program." Stollnitz Files.

[261] Director and Deputy Director, BNS Division (Louttit and Brown, respectively) to Eloise Clark, Assistant Director, BBS, "Recommendations Emerging from the Experimental Reorganization of Neurosciences Proposal Review During FY 1982," July, 13, 1982. Quotation from pg. 1

tal five-panel review system in the neurosciences. They included panels for Developmental Neuroscience, Integrative and Motor Processes, Molecular and Cellular Neurobiology, psychobiology, and Sensory Mechanisms, Processes, and Perception.

A close reading will evidence a partial reorganization in keeping with question one of the two asked in the 1980 long-range plan (above). Even though there was no actual move to relocate the more molecular aspects of the neurosciences to one of the more biological divisions of the five divisions then extant in BBS, the title of one of the five review panels—Molecular and Cellular Neurobiology—indicated a sound arrangement for molecular neurosciences in BBS. (But matters would be quite different after the 1991–1992 restructuring; Chap. 5.) Indeed, all five were working out and the leadership recommended continuing that arrangement into FY83 with a minor "'nip and tuck'" to one or two programs. Along with two new staff requested, the neurosciences were moving into the mid-1980s splendidly, Louttit and Brown argued.[262]

Brain mapping had risen to prominence in the BNS program for 1982 with the work of Rodney Murphey of the State University of New York, Albany. He employed the common cricket as a *model system*, because it "serve[s] neurobiologists…[just as] bacteria and viruses serve molecular biologists."[263] But by 1983, brain mapping had given way to "mental imagery" in the interests of the authors of the *Annual Reports*. Albert Einstein had said that his first insight into relativity came as he imagined himself chasing a beam of light and "seeing" himself run ahead of it and thus catch up. The chemist, Friedrich August Kekule von Stradonitz (known as Kekule), dreamed of a circle of snakes biting the tails of the one ahead and so figured out the structure of the benzene ring. Those two well-known stories may have led psychologist Stephen M. Kosslyn to develop the first comprehensive theory of visual imagery. First working with funds from the NAS, Kosslyn was later supported by BNS. His book, *Ghosts in the Mind's Machine* came out in 1983.[264]

At that same time, the Board requested Betsy Clark to "begin discussions with the external community" with regard to a reorganization of BBS, though such would not come to pass for nearly another decade.[265] At any rate, NSB Chair Branscomb said that "such considerations are at a preliminary stage." But Clark would leave NSF in June of 1983 to take up her new position as Vice President for Academic Affairs at Bowling Green State University in Ohio. Still, the Board was much involved in thinking about the future of BBS in March of that year.[266] Their director-

---

[262] Ibid., pg. 2. Brown was an excellent partner to Louttit as the former had served for a time as deputy assistant director of BBS and so brought additional insights to BNS upon his return full-time to the latter position in mid-1979. See NSB, "Minutes," NSB-79-211, June 25, 1979, 206:9.

[263] Ibid., pg. 36.

[264] National Science Foundation, *Thirty-Third Annual Report for Fiscal Year 1983* (Washington, D.C.: Government Printing Office, 1984), pg. 34–35. This paragraph is a paraphrase of the text of the Annual Report.

[265] NSB, "Minutes," NSB-83-92, March 22, 1983, 2–83:6 (new pagination system).

[266] Ibid., 2–83:20–22.

ate priorities for biology included the physiology of behavior and continuing support on human origins research. That would require moving programs from BNS to one of the two biology divisions, if not a wholesale reorganization of BBS. In that latter area, research into the transformation of primate ancestors into modern humans was to be undertaken, as well as how our species moved from hunting and gathering to agriculture and thence on to large-scale urban societies.[267]

Not only was the Board interested in BBS more broadly in 1983, but so it was in BNS, in particular. A Program Review of BNS had been conducted by Louttit, Pitts, and two others.[268] Louttit indicated in his remarks to the Board that BNS would like to increase funding for instrumentation (so important had it become to psychobiology and the neurosciences), graduate student support, funding for undergraduate institutions, and human origins research (very expensive, given the field component). "The Foundation is the only organization responsible for this type of research in the U.S. and, generally in the world," Louttit averred regarding that specialty of anthropology. He also pointed out just how difficult it was to keep biologically oriented research funds separate from more clearly behavioral research: a problem for a boundary line area of science.[269]

Louttit also advocated a *change in name for the directorate*, from BBS to Directorate for Life Sciences. However, the Minutes reported that:

> [t]he Board Chairman [Branscomb] stated his enthusiasm for the name change, but noted that no action should be taken until the Directorate advisory committee (which is being formed) has a chance to consider the pros and cons.[270]

That was an example of "what if" history that tantalizes, but came to nothing. Eight more years for BBS as then constituted were still in the future. The new BBS Advisory Committee, to which Branscomb referred, was announced by Robert Rabin in the pages of the *AIBS Forum*, a publication of the American Institute of Biological Sciences, in its final issue for 1983.[271] Peter Raven, the Director of the Missouri Botanical Garden and one of America's leading biologists of the last century and into the present one, would be its chair. Raven had also just been elected president of the AIBS. The AC's first meeting was held in late January 1984. Raven will be met again numerous times in this study.

The following year saw another updating of the 5-year long-range plan, running from FY86 to FY90. In the budget for FY83, some 8% of the funds for BBS went to the behavioral sciences, specifically. It was felt by David Kingsbury in calendar 1984 that the budget for FY84 would be substantially the same. Specifically, BBS would put some $4 M into the Center for Advanced Study in the Behavioral Sciences

[267] Ibid.

[268] John Yellen remained at NSF through the 1991–1992 reorganization and became part of the new Directorate for Social, Behavioral, and Economic Sciences (SBE); he was still with SBE in 2010.

[269] Ibid., NSB-83-373, February 16, 1984, 11–83:7.

[270] Ibid.

[271] *AIBS Forum* (1983): back cover.

at Stanford University.[272] The support for the Center had been partially undertaken by the directorate to the level of $600 k annually since 1977. The FY85 budget proposal would continue that for another 5 years. An additional amount of $700 k was planned for FY86.[273]

The long-range plan for FY84-FY88 was still in process even as that next one was being written. And BBS was growing rapidly as it had become, in 1984–1985, composed of *six* divisions with a total of some 37 programs in all. The latest division came into existence when one of the two "pure" biology divisions, that of Physiology, Cellular and Molecular Biology was split into two new divisions: the Division of Molecular Biology (so large and significant had molecular biology become) and the Division of Cellular Biology.[274] (Both are discussed in the following chapter.) The commentary in the *Annual Report 1984* spoke in terms of Behavioral and Neurosciences as being composed of "*anthropology,* cognitive science, and neuroscience": psychobiology was not then a term used in many quarters, and the term anthropology was being used in a much more broad fashion than theretofore.[275] The matter of the first humans in the Western Hemisphere interested the report's authors, as did how nervous systems develop: the biology of the brain and related nervous system features was being studied by the Alan T. Waterman Award winner of 1983, Stanford's Corey Goodman. Not unlike the use of crickets by his predecessors, Goodman's work utilized the *model system* of grasshoppers. The animals have a transparent embryo, which allowed the easy use of implanted electrodes. The electrodes demonstrated the growth of neurons and the connections they made: neuronal pathways.[276]

The intensity of self-examination in the mid-1980s at NSF broadly, and in both BBS and BNS in particular, was most clearly seen in the 1984 *Status of Science Reviews*, a biennial publication for Foundation in-house purposes, and which "reflects the individual program manager's perceptions of the organization and developments in science."[277] The BNS indicated that four of its seven programs

---

[272] Kingsbury interview. It was a center that Stanford's Hoover Institute felt was "a radical leftist organization," according to Kingsbury.

[273] "BBS Long-Range Planning, FY86–90," BBS January 4, 1984. The document is located in the lateral files in the sixth floor BBS Directorate Office at NSF; hereinafter "BIO Lateral Files." For more on the Center, see http://www.casbs.org/. The CASBS has been in operation since the 1950s.

[274] Robert Rabin to Members of the BBS AC, "Items of Information," July 3, 1984, pg. 2. The two "new" divisions became official on June 25. Director Knapp had sent a memo to members of the NSB, June 7, 1984, NSB-84-181, to discuss his desire to make the split and noted that the matter had been fully vetted in the BBS with the local union, submitted to Senator Jake Garn (R-UT), Chairman of the Subcommittee on HUD-Independent agencies' Committee on Appropriations, where NSF fell along with such others as NASA, the Veteran's Administration, etc. That Subcommittee "neither questioned nor disapproved" the action. More details are given in Chap. 4 below.

[275] National Science Foundation, *Thirty-Fourth Annual Report for Fiscal Year 1984* (Washington, D.C.: Government Printing Office, 1985), pg. 28.

[276] Ibid., pg. 30.

[277] Division of Planning and Policy Analysis, Office of Planning and Resource Management, NSF,

would be discussed: anthropology, linguistics, neuroscience, and psychology. Terms no longer seen in use, at the division level, were psychobiology and, at the program level, animal behavior—which had been subsumed under the "psychology" of the 1984 *Reviews*. That, however, would change back in coming years. A greater degree of interdisciplinarity was reflected in this newer nomenclature. That was very likely due to the number of federal, scientific society, and association reports that had emerged in the early 1980s, all of which emphasized the rapid growth of the neurosciences, in the broadest application of the term. The reports further noted the great importance of the molecular aspects of behavior, the central nervous system, and features such as the growing dependence on instrumentation, among others. However, the actual dollar amount and percentage of BNS's budget provided to instrumentation went from 12% ($3.4 M) in FY78 to only 6.5% ($2.2 M) by FY83; a possible result of Reaganomics (but see below).[278] Nevertheless, BNS had a budget of nearly $32 M in FY82, a requested FY83 budget of well over $33 M and sought an FY84 budget of just under $38 M.[279] Of the three program titles still then in use in the early 1980s, neurosciences, cognitive sciences or psychology, and anthropology, the first would get the lion's share at well over $24 M, vice just over $7 M for cognitive sciences and a little more than $6 M for anthropology.

The *Annual Report 1985* saw a significant change in its published layout, with a comparatively lengthy Director's Statement by Erich Bloch and a nod toward the 35th anniversary of the Foundation. He spoke of "Key Themes and Initiatives" that included continuing support for the best science, "concern for freeing research from arbitrary limits imposed by disciplinary bounds," and cooperation in all that NSF does. In addition, Bloch emphasized engineering research centers, advanced supercomputer centers, and gave considerable space to science and engineering education—a long-term survivor of generally negative congressional pressure, not unlike that on the social and behavioral sciences.[280] All but one of the research "successes" for calendar 1984 was outside the social and behavioral sciences. The plan of the annual report was strikingly different than in, at least, the previous 10 years and was *much less informative* to its readership (having only very brief articles), from congresspersons to the general public.

The *Report*'s layout would change again in 1986. The Director's Statement ceased after that year but reappeared with Bloch's 1989 report. Other sections were new, as well, but the always-present financial statement remained at the end of the document. Most notably, the *Annual Report*s had come to resemble a corporate one so closely that it was hard to tell the new NSF slick-surfaced booklet from those seen emanating from industry. The new format may have been due to the Reagan

---

Status of Science Reviews 1984, November 1983.

[278] Ibid., Table 1, pg. 63.

[279] "National Science Foundation, FY 1984 Budget Request," an in-house administrative document, copy in Stollnitz Files.

[280] National Science Foundation, *Thirty-Fifth Annual Report for Fiscal Year 1985* (Washington, D.C.: Government Printing Office, 1986), pg. 1–2. His Statement was followed by text that was titled "Special Section," which reported on various research activities of the past year.

RESPONSIBILITIES OF BOARD, DIRECTOR, STAFF

	BOARD	DIRECTOR	STAFF
STRATEGIC POLICY	ESTABLISH/ APPROVE	SUPPORT/ GENERATE	SUPPORT
MAJOR INITIATIVES	APPROVE	GENERATE	SUPPORT
"LONG-RANGE" PLANS	REVIEW	APPROVE	GENERATE
BUDGET SUBMISSION	APPROVE	GENERATE	SUPPORT
PROGRAM IMPLEMENTATION	REVIEW OR APPROVE	APPROVE/ REVIEW	GENERATE
NSF OPERATIONS	REVIEW	APPROVE/ REVIEW	GENERATE

KEY:

ESTABLISH = ISSUE POLICY STATEMENT
GENERATE = TAKE RESPONSIBILITY FOR PRODUCING
APPROVE = CONSIDER PROPOSAL; ISSUE BINDING APPROVAL, DISAPPROVAL, OR BOTH
REVIEW = CONSIDER PROPOSAL AND PROVIDE NONBINDING RESPONSE

Fig. 2.6 Responsibilities of Board, Director, and Staff. Minutes of the Board of the NSF 1–85:12

influence and those who advised him in economics, budgets, and finance: industry can do a better job than government in many areas, and it should be allowed to do so, they held. For instance, recall economist Friedman's suggestion for the *wholesale elimination* of the NSF and the turning over of its responsibilities to the private sector! Also it may have been that NSF wished to look more "corporate" in its most visible annual publication and thus achieve some amount of "protective coloration" from the likes of Friedman and others. Finally, it may have been because the new director had come from the corporate world instead of the academy, as had most of his predecessors. Whatever the cause, it represented a fundamental change in the way the Foundation portrayed itself.

The NSB Minutes did not change, however, though in 1985 the Board did provide what may have been the best explication of what their duties were, vis-à-vis the Foundation's director and its staff. Figure 2.6 is the NSB's January 1985 table. The Board was clarifying for interested parties its role at NSF. The amendment to the organic act that would extend the range of funding that the director could approve without the Board's advance approval was then ready to be submitted to Congress. All matters financial had to live in a new environment—that of Reaganomics—and it behooved the NSF and the NSB to be very much alert to the changed milieu. The Graham-Rudman-Hollings Act had already reduced NSF's budget by 4.3%, or over \$65 M, for the FY85 budget proposal.[281]

The NSF amendment had certain limitations to it, though. The director could not make any grants or other funding available to create centers or institutes "that are

[281] NSB, "Minutes," NSB 86–32, March 21, 1986, 3–86:4.

especially sensitive for any reason."[282] If any such centers were being suggested for the social or behavioral sciences, it is a good guess that they would have been excluded under the new authorization from the director's sole purview. No such centers were known to be in consideration at the time, however. But the Board was to approve a Biological Centers Program under BBS, at least, within months (more in following chapters).[283] The first program announcement for that came out in late 1986.[284] It was aimed exclusively at biotechnology, however.

By 1987, the *Annual Report* had become strongly oriented toward the biological and other sciences and engineering, nearly disregarding the social and behavioral sciences altogether. The Board Minutes were peppered with new concerns, such as a growing emphasis on engineering and energy, though a postdoctoral fellowship program did arise for the History and Philosophy of Science program. As the Board began consideration of the long-range plans for FY88 to FY92, one of the three major themes was concerning centers. The basis for the consideration of creating centers came from the newly issued Packard-Bromley Report.

David Packard, Chairman of the Board of the Hewlett-Packard Corporation, and D. Allen Bromley, still then Professor of Physics at Yale University, were asked by President Reagan to examine the relationship among government, the corporate sector, and academia with regard to maintaining America's leadership in industry and defense. (Bromley would become the director of the OSTP in August, 1989.) The study came out as a "Report of the White House Science Council: Panel on the Health of U.S. Colleges and Universities" in February of 1986.[285] In essence, it determined that in order to bring American universities, and related entities, up to strength to meet Reagan's expectations, some $10B in support would be needed; a considerable amount of that would go to such things as *science and technology centers* (STCs). The Packard-Bromley findings ratified Representative Donald Fuqua's (D-FL) bill introduced in 1985. The "University Research Facilities Revitalization Act" (H.R. 2823) would have provided NSF, as the first of six named agencies, with funding to modernize university research laboratories. However, Packard-Bromley suggested the same amount that the Fuqua bill would provide, $10B, but had the federal government providing half that figure and indicating that the other $5B would have to come from non-governmental sources.[286] Reagan, in

[282] Ibid., January 17, 1986, 1–86:5.

[283] Ibid., November 13–14, 1986, 11–86:3. In fact, NSB Chair Schmitt, then Board Member Rita Colwell (later NSF Director), and BBS Assistant Director Kingsbury would attend the groundbreaking (November 13) for the University of Maryland/National Bureau of Standards' Center for Advanced Research in Biotechnology (CARB) in Montgomery County, Maryland. As of July 1, 1987, Colwell became director of MBI, the Maryland Biotechnology Institute, of which CARB was a part. In later years, the present history's author would serve with Colwell on the UMD President's Advisory Board for MBI, an institute supported by the NSF and the State of Maryland.

[284] National Science Foundation, *Program Solicitation: The Biological Centers Program,* NSF 86–89.

[285] James D. Savage, *Funding Science in America: Congress, Universities and the Politics of the Academic Pork Barrel* (Cambridge: University Press, 1999), pg. 90.

[286] Ibid.

his State of the Union address in January of 1987, averred that he wished to see NSF's budget doubled over the following 5 years. Further, *STCs were central to Reagan's Executive Order 12–591 that year.*[287] However jaded was the view of Stockman (by then gone from OMB) and others in the Reagan administration concerning the social and behavioral sciences (and science education), it *seemed* Reagan himself understood the importance of science and engineering to a healthy US economy and, especially, to America's military prowess, though I have no proof for this assertion. (We do know that he felt efforts to slow the effects of acid rain was wasteful of tax payers' money![288]) Indeed, Reagan's executive order became one of two authorizing documents to further the STC program's development. Thus, in FY88, the Program Solicitation for centers was open to any type of Science and Technology Centers that might be proposed: social and behavioral sciences were not excluded.[289]

Behavioral science was by no means entirely forgotten in the later 1980s, as Fred Stollnitz had prepared a set of notes in order to make an invited address to a BNS meeting with Kingsbury in March 1987 for future budget planning.[290] The specific subject was comparative analysis of behavior. At that same meeting, a colleague was to present commentary on a very new area in the behavioral sciences: analysis of cellular mechanisms of behavior. That was an example of a historical trend, to be seen most clearly in biology, that of *reductionism*. The matter of dealing with behavior at the cellular/molecular level was quite new at the time, but, as it would happen in the purely biological sciences, reductionism (already then long established in biology) would be central to the history of the broader biological sciences in the latter half of the last century. This *theme* will be examined in considerable detail in a number of chapters that follow.

The behavioral sciences were not restricted from centers. Two slick publications on the subject came out of NSF concerning science education being done within all

[287] NSB, "Minutes," NSB 87–136, June 18–19, 1987, 6–87:6.

[288] "In the early 1980s, the US government issued a report that supported the claims of environmentalists that industrial emissions caused acid rain. However, the Reagan administration resisted calls from the environmental community to strengthen the Clean Air Act. Rather, in part due to its commitment to **deregulation,** it called for further study. As a result, not until 1988 did the US government take any formal action to reduce acid rain, when it signed a United Nations-sponsored protocol that sought to freeze the rate of nitrogen oxide emissions at 1987 levels." See http://www.questia.com/library/book/encyclopedia-of-the-reagan-bush-years-by-peter-b-levy.jsp. See also for this and other science issues: http://www.issues2000.org/celeb/Ronald_Reagan_Environment.htm.

[289] National Science Foundation, "Program Solicitation: Science and Technology Centers, Fiscal Year 1988." BIO Lateral Files.

[290] An unsigned, ephemeral two-page computer printout (notable, as all his work had been in hand or typed form in prior times), that was likely written by Stollnitz, was found in his files. Dated March 23, 1987. It should be noted that Kingsbury has said to this author that, with regard to budgetary issues, "[t]he animal behavior stuff was always something that I just didn't increase. I would have liked to decrease it." Kingsbury did not explain why he felt that way, however. Kingsbury interview.

those centers then in existence, albeit they were mostly in the physical sciences.[291] However, the University of Pennsylvania, with NSF support, hosted the Center for Research in Cognitive Science. Penn also offered undergraduate education in the form of a Cognitive and Computer Science Dual Degree Program. Students could earn an undergraduate degree in computer science and a second bachelor's in linguistics, mathematics, philosophy, or psychology. A minor in cognitive science was also available. At that time, some 60 undergraduates were enrolled in the Penn program. There were 25 STCs overall, 6 of which were biological, with very strong engineering or technology associations, but only the single center for behavioral (cognitive) sciences at Penn.

By the close of the 1980s, BNS and SES made up two of the, by then, six divisions within the directorate (Fig. 2.7). The remaining four were biological or supportive in nature (for instance, DIR; see Chap. 6). However, the somewhat catchall DIR included the social area of Studies in Science, Technology, and Society (STS), primarily the old History and Philosophy of Science program.[292] Nor was there any mention of those sciences in the Minutes of the second and final BBS AC meeting of October 1989. The directorate-level split was imminent (Chap. 5).

The year of 1989 was the first time that NSF passed the $2B budget mark; it had received 37,500 proposals and made more than 16,000 awards Foundation-wide.[293] The behavioral and neurosciences budget had reached nearly $45 M, and 799 awards had been made in the area. BBS's overall budget was well over $281 M with only slightly more than $4 M having gone to the new STCs.[294] And the neurosciences had evolved, from 1976 to 1988, into a more complex grouping of programs, as seen in Fig. 2.8. And, too, David Kingsbury had left the BBS directorship under a cloud and had been replaced by Mary Clutter (following chapters).

It is likely fair to say that, by the end of the 1980s, the entire BNS was a well-oiled machine. Although minor reorganizations were always undertaken every few years, if not more frequently, the logic for that was sound: the *intricate dance* called for each partner to assess its *current status*, vis-à-vis the science being done in the community or being either responded to or being driven by NSF. The dance floor for

---

[291] Two slick, color publications were extant during that time: National Science Foundation, *Science and Technology Centers: Partners in Science, Engineering, and Mathematics Education*, and NSF, *Science and Technology Centers: Leadership in Research Technology Education*. Neither was dated, but both appear to be from the later 1980s. The former listed 25 centers, the latter, 24. Both were similar in content. Stollnitz Files.

[292] This Illustration was taken from a large group of overhead slides that were used in the May 5–6, 1989 BBS AC meeting, not one other slide of which addressed the social or behavioral sciences. Author unknown, likely Mary Clutter. BIO Lateral Files.

[293] National Science Foundation, *Annual Report 1989* (Washington, D.C.: Government Printing Office, 1990), pg. 4. Note that the style of the title had changed by then and the "Highlights" section of the preceding several years had been replaced by "Research Notes." One of those Notes addressed the first group of the STCs: 11 centers in eight states had been awarded in December of 1988. They totaled almost $25 M in NSF support. None of the Notes covered any social or behavioral science.

[294] Ibid., Table 1, pg. 75.

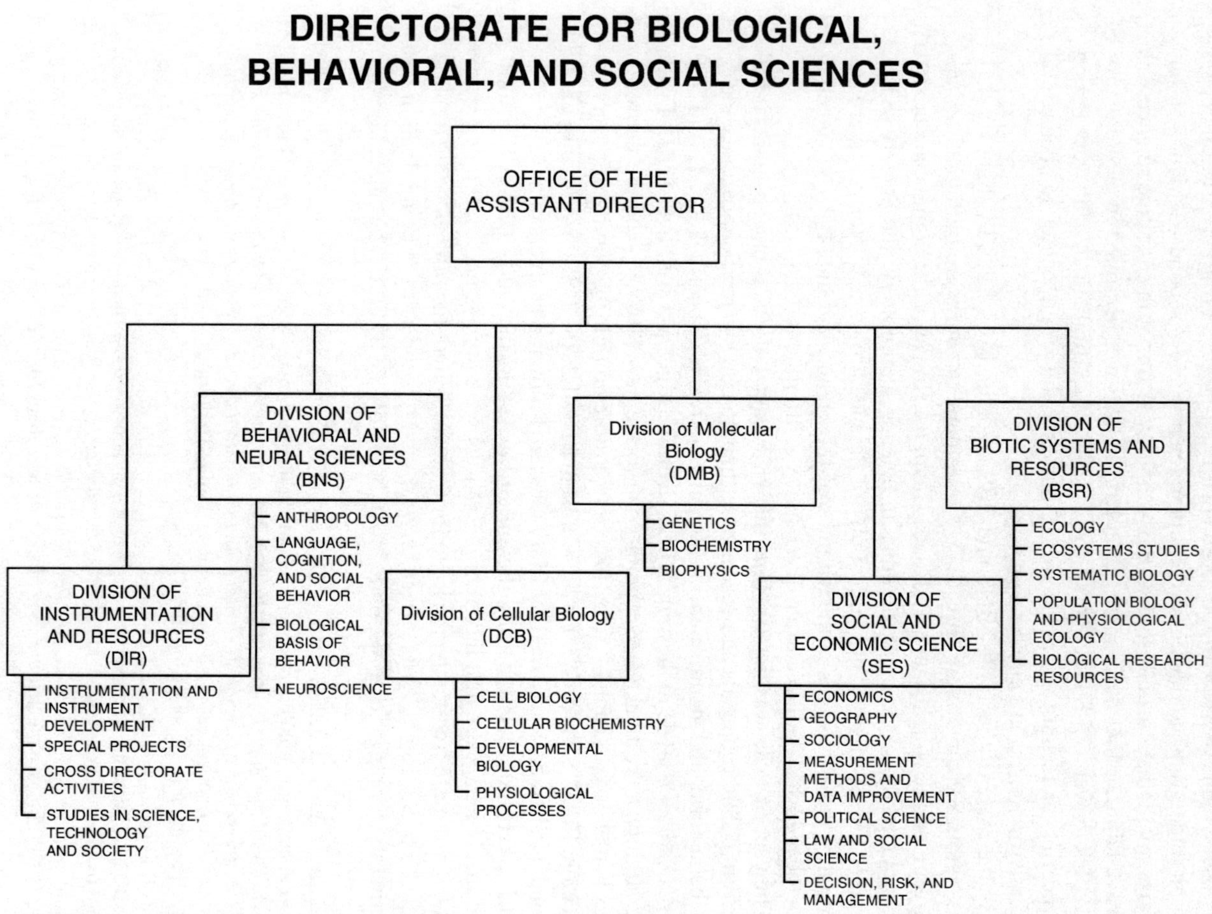

Fig. 2.7 Organization chart for Directorate of Biological, Behavioral, and Social Sciences (BBS). (NSF. Taken from a visual for the meeting of the Advisory Committee of the BBS, May 5–6, 1989)

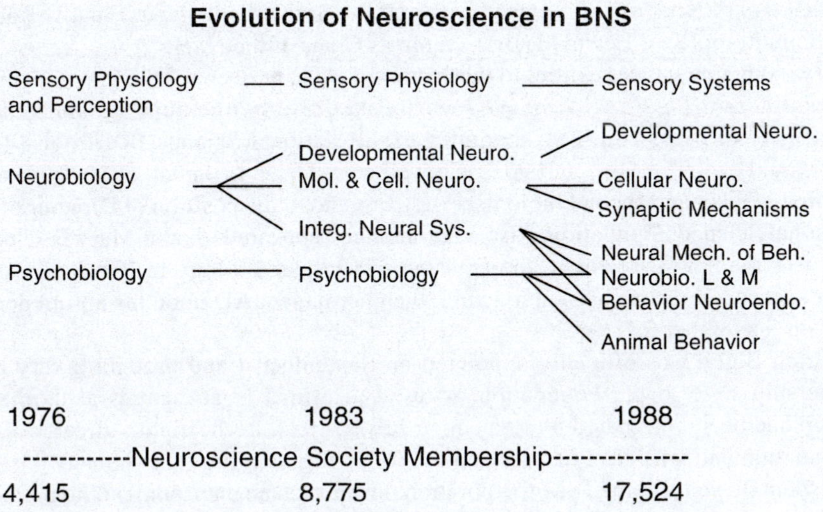

Fig. 2.8 Evolution of Neurosciences in the Behavioral and Neurosciences programs. (NSF. Taken from a visual for the meeting of the Advisory Committee of the BBS, May 5–6, 1989)

the behavioral sciences began to look more and more like biology and less like its long-time sister, the social sciences. *Behavior was being shown to be a manifestation of the genetic reality of an organism—nature—as much as, or more so, than by any social realities—nurture.*

With the passage of time, understanding behavior, just as with biology, would be through the increasing use of model systems; quantification; computerization; employment of complex technologies; database-related efforts; the rise of centers; reductionism; measurement, and methodology work; longitudinal studies; and the dawning of big science in BBS. In addition, an extra-federal burgeoning of the neurosciences and an ever-closer connection to biology put behavioral science upon the bridge that led into another realm of intellectual commitments. And, as the behavioral sciences stood upon that bridge, their "protective coloration" *did* aid them in oft-times slipping below the radar of those who opposed funding for them.

2.13 A New Director for NSF

We have seen that Betsy Clark left the BBS AD position for Ohio and Robert Rabin became acting assistant director until David Kingsbury was nominated by Reagan on March 29, 1984.[295] Kingsbury was sworn in at the June 1984 NSB meeting.[296]

[295] NSB, 'Minutes," May 10–11, 1984, 5–84:4. See section g. on that page.

[296] Robert Rabin, "Memorandum to the BBS Advisory Committee," July 3, 1984, pg. 2, item 5.

Also, Edward Knapp was replaced by Erich Bloch that September. Finally, Roland W. Schmitt replaced Lewis M. Branscomb as Chair of the NSB.[297]

Two other new faces central to this history were to be seen for the first time in the *Annual Report for Fiscal Year 1984* when Rita Colwell, microbial ecologist at the University of Maryland, was appointed to the National Science Board for a term that would expire in May of 1990.[298] Colwell would later, in the fall of 1998, become the first woman and second biologist ever to ascend to the position of Director of the National Science Foundation. Also, another name appeared: that of Mary E. Clutter. She was the new Director of the Division of Cellular Biology in BBS and would later succeed Kingsbury as, first acting, then permanent AD upon his abrupt departure on September 23, 1988.

Erich Bloch was officially appointed on September 4 and brought a very new leadership style to the Foundation, so used to primarily academics at the helm. Bloch had been nominated by Reagan on July 2, 1984, as the eighth director of the Foundation and was easily confirmed the following month by the Senate.[299] Bloch had spent 32 years at the IBM Corporation. From such an environment, no previous NSF director had ever come. While NSF had had an almost unbroken share of academic physical scientists as directors, an electrical engineer with a strong corporate background was new.

Bloch was born in Sulzburg, Germany, but went to Switzerland to take his earliest university degree, in electrical engineering, at the Federal Polytechnic Institute of Zurich. He came to the United States and took a Bachelor of Science degree in the same field at the State University of New York at Buffalo. By 1952, he was an employee in his discipline at IBM and rose through the ranks to hold several vice presidencies. He finally retired there in February of 1985 to pursue his already newly begun career at NSF. The Chair of the Board at IBM, John R. Opel, said at that time that "[h]is commitment to the enhancement of technical vitality will help ensure that IBM people remain on the leading edge of this industry."[300] That would have been a good prediction for the NSF, too. Meanwhile, the new Chair of the NSB, Schmitt, laid out what the Board expected from the next director:

> In this stage of its history, the [NSF] requires a Director who meets four important criteria: 1) a strong record of commitment to academic research and to science and engineering education; 2) a demonstrated ability to provide the dynamic leadership needed to continue

---

[297] Ibid. See Appendix A by Branscomb and Appendix D by Schmitt. Charles E. Hess was chosen as Vice-Chair of the Board at that same time.

[298] Colwell was nominated for the Board position by President Reagan on August 23, 1984, and was confirmed by the Senate on October 5. She was sworn into the Board on October 18. She was, at that time, Vice President for Academic Affairs at the University of Maryland, College Park.

[299] Jeffrey Mervis, "Erich Bloch's Campaign to Transform NSF," *American Scientist* (1988):557–561. Cited information is on pg. 557.

[300] The commentary in the main text above is a combination of ibid. and information derived from two locations: http://www-03.ibm.com/ibm/history/exhibits/builders/builders_bloch.html, from whence the direct quotation is taken, and also from Internet pages within http://www.ieee.org. See also Jack Renirie, NSF news release PR84–36, for more information concerning Bloch's biography.

> the Foundation's extensive support of these core areas; 3) an understanding of *the dependence of industry on academic research* for trained people as well as research results, and 4) extensive personal experience in managing large, *multidisciplinary* groups engaged in forefront science and engineering.[301]

Bloch brought an agenda with him for his tenure at NSF. Just how well he succeeded will be considered later, but his connection with what has at times been the world's largest computer manufacturer clearly would lead to changes in the technology infrastructure for the *management of the Foundation*. But it would also lead to the directions NSF would take in funding research that was highly dependent on *what computers could do for science*, be it social, behavioral, biological, or otherwise.

Early 1984 also saw Kenneth Prewitt, in his role on the AC of the BBS, report to that Committee the continuing "ambiguous position of the social sciences in [NSF]." For example, he noted that:

> [a]lthough the social sciences often do not generate the data upon which they depend, especially in economics, political science, and sociology, modest investments can sometimes greatly enhance the scientific value of data sets produced by such agencies…as the Census Bureau or the Department of Labor.[302]

He felt that NSF grants had made much of that possible. However, he warned against the perhaps excessive expenditure on the "generation and preservation of data sets." While he may or may not have had a point for the social sciences, biology would most definitely profit immensely due to similar investments in years to come.

With so much antipathy for the social and behavioral sciences, however, it was the budgetary matters that most worried COSSA: its director had said so in a guest editorial in *Science* fewer than 2 years previously. Roberta Balstad Miller, still then Director of COSSA but soon to be at NSF, pointed out that federal "funding for research in the social and behavioral sciences is scattered among nearly 40 agencies, programs or departments" in FY85. Further, "[w]hat is missing is a sense of the scientific and economic benefits of an integrated, collaborative national research program in [these sciences]." She argued that inefficiency was present because emphasis was being placed upon one "research enterprise" in one agency, while at the same time, similar efforts were being "dismantled in another part of the government"![303] Miller wrote a lengthy memo to Robert Rabin pointing out these problems and attached the *Science* editorial as evidence of her worries.

There were a number of reports on the status of the social and behavioral sciences that appeared in print from 1982 to 1988, and so the question arose: Were their findings being implemented? Mary Clutter, by then BBS AD, was asked by the

---

[301] Ibid. Emphases added. See also Renirie news release, pg. 2.

[302] BBS AC, "Minutes," January 26–27, 1984, pg. 6.

[303] Roberta Balstad Miller (guest editorial), "The Budget for Social Science Research," *Science* (1984). The lengthy memo preceding a photocopy of the editorial had been derived from "a somewhat longer preface to the FY 1985 analysis" of the federal budget proposal prepared by COSSA, said Miller. Miller to Rabin, May 14, 1984.

new SES Director, Miller, in March of 1989, that very thing: "What has NSF done to implement the report[s]?" That memo arrived on Clutter's desk just before the reorganization efforts of 1991–1992. Data showed that the estimated FY90 budget for the total of the social and behavioral sciences would be nearly $67 M in current dollars (but only $51 M in 1982 constant dollars). That could be compared to biology's figures of over $247 M current and $189 M constant dollars. The total, for all of NSF, was then $1.8B current dollars and nearly $1.4B in constant figures. Still, it must be noted that the financial support for the behavioral and social sciences for almost all of the eighth decade of the century remained *virtually flat* at the same time that the total for NSF continued on a nearly unbroken rise.

The continuing struggle for the behavioral and social sciences to break out of the long stagnant funding rut led to a number of letters exchanged among Miller, Clutter, Bloch, and Mary L. Good, by then chair of the NSB having replaced Schmitt. Also included was then Chair of the House CST, Representative Doug Walgren (D-PA; in office from 1977 to 1991) and with writer Joseph Berger of the *New York Times*, who had called the social sciences' budget "a decline in interest" on the part of the Foundation. The upshot of the exchanges was that the community of behavioral and social scientists set their research priorities and the Foundation responded within budget limits, according to Good.[304]

The Minutes of the BBS AC for October of 1989 do not mention the social and behavioral sciences at all. It may have been that no business was necessary of discussion at that point for SES and BNS, but it may also have indicated the drift that was beginning as the 1991–1992 reorganization discussion was already underway: SES was soon to separate from biology, but BNS would follow a somewhat different route.

But, withal, these data were just more pieces of evidence that something of momentous proportions must be undertaken to finally stabilize the social and behavioral sciences at the NSF. Indeed, by mid-1985, the NSB, along with Bloch, was considering a 5-year plan that was to begin with the FY87 planning period and extend into the future—1991—by which time the next major reorganization would have already begun, though in calendar 1985 that was not yet foreseen.[305] Roberta Miller was destined to become much more closely involved with NSF. And that destiny was to be carried through very soon after her *Science* editorial appeared: she became the new Director of SES in June 1984.[306] By 1985, the SES had been restructured under Miller and, as hinted at in the *Science* editorial, into the form shown in Fig. 2.9 here. Much of what had been familiar in the post-1975 era was still present, but those programs that were moved out of the finally dismantled RANN were now fully integrated into SES under varying titles. Also in June of

---

[304] Mary Clutter to Erich Bloch, June 9, 1989, "Re: Response to the New York Times"; Doug Walgren to Mary L. Good, June 26, 1989, concerning the third (chronologically) of the three large reports (from various organizations) then current in the social sciences; Mary L. Good to Honorable Doug Walgren, August 15, 1989, re that third report (the one discussed in the main text above).

[305] NSB, "Minutes," NSB-85-169, June 20–21, 1985, 6–85:3ff.

[306] Robert Rabin, "Memorandum to the BBS Advisory Committee," July 3, 1984, pg. 2, item 8.

Fig. 2.9 Organizational chart for the Division of Social and Economic Science (SES) as of 1983. (NSF. NSF83-72 report)

- Acting Division Director
  - Staff Associate
- Section Head, Economics & Geography
  - Program Director, Economics
  - Program Director, Applied Economics
  - Assistant Program Director, Economics
  - Program Director, Geography and Regional Science
- Section Head, Social Measurement & Analysis
  - Program Director, Measurement Methods and Data Resources
  - Staff Associate
  - Program Director, Sociology
  - Associate Program Director, Sociology
  - Professional Assistant, Sociology
  - Program Director, History and Philosophy of Science
  - Program Associate
- Section Head, Political & Policy Sciences
  - Associate Program Director, Political Science
  - Program Director, Law and Social Sciences
  - Professional Assistant, Law & Social Sciences
  - Program Director, Regulation and Policy Analysis
  - [Acting] Program Director, Decision and Management Science

1985, Bloch was joined by his new permanent deputy director, John Moore, on leave from the conservative Hoover Institute at Stanford University. Moore brought a different perspective to the senior leadership group as his background was in economics and law.[307] His was an interdisciplinary approach, and Bloch was very much alert to the importance of that issue as a crucial formal theme for NSF by its 35th anniversary; it had already been a fairly long-term reality, if not a widely pervasive one. In his first Director's Statement, Bloch stressed three issues that he termed "Key Themes and Initiatives." Of particular interest here was his "[c]oncern for freeing research from the arbitrary limits imposed by disciplinary bounds."[308]

The "Gramm-Rudman" act would, as the NSB realized, reduce the NSF budget by just over $65 M for FY86.[309] On the other hand, the Board noted that the NSF annual Authorization Act for FY86 would increase substantially the amount of funding that the director could alone approve as grants without seeking Board permission first. It had long been that the director could commit the Foundation to a single project up to $500 k for a single year and up to $2 M, total, for multi-year projects. The new levels were three times greater than they had been: $1.5 M and $6 M, respectively.[310] As it eventuated, that would allow for some very large project

307 National Science Foundation, *Thirty Fifth Annual Report for Fiscal Year 1985.* (Washington, D.C.: Government Printing Office, 1986), pg. 41.

308 Ibid. See first (unnumbered) page of the Report.

309 NSB, "Minutes," NSB-86-32, January 17, 1986, 1–86:2.

310 Ibid., 1–86:5. See also Appendix C (NSB-86-20) of those same Minutes for the formal Resolution of the Board concerning such expenditures by the director.

work in biology, in particular, in coming years. The organizational structure of the BBS by the end of the 1980s would soon change radically.

The next two chapters focus exclusively on the biological sciences from 1974 to 1991–1992. The first of these (Chap. 3) concerns the "micro" end of the biological spectrum of life with the second (Chap. 4) examining the "macro" end.

Chapter 3
Little Biology and Biology of the Little

The importance of deoxyribonucleic acid (DNA) within living cells is undisputed
J.D. Watson and F.H.C. Crick

3.1 The Importance of DNA

In Watson and Crick's words, there opened what may arguably be called the single most significant publication in the history of biology in the twentieth century. Its repercussions to this day have fundamentally changed not only biology, but also much of the everyday aspects of life for our species. When James D. Watson, a young American in England, and Francis H.C. Crick, a Briton in his homeland, announced their discovery of the structure of deoxyribonucleic acid (DNA) in the pages of the journal *Nature* on May 30, 1953, they truly began a revolution in biology and, eventually, in so much that affects human existence.[1] It is no exaggeration to say that their discovery had impacts just that profound: we continue to live our lives in an altered world, one where we have come to know a central tenet of the basic nature of nature itself.

The elucidation of the structure of DNA was only one part of a much more complex set of discoveries and new directions whose paths were just coming to be blazed. One path was of very considerable consequence. The notion of *reductionism* was very well established in the sciences long before the early 1950s—indeed, as anciently as Aristotle in *Posterior Analytics I*, for instance, or Democritus even before him. Its role in regard to "Big Biology," and a necessary corollary of "little biology," is such that we must consider the idea that *the rise of big biology*, as we

[1] J.D. Watson and F.H.C. Crick, "General Implications of the Structure of Deoxyribonucleic Acid," *Nature* (1953):964–967; see also, Watson and Crick, "The Structure of DNA," *Cold Spring Harbor Symposium on Quantitative Biology XVIII,* (1953):123–131; and, James D. Watson, *The Double Helix: A Personal Account of the Discovery of the Structure of DNA* (New York: Norton Critical Editions, 1980).

D. J. McGraw, *Millennial Biology: The National Science Foundation and American Biology, 1975-2005*, https://doi.org/10.1007/978-3-030-56367-7_3

now know it, would not have been fully possible without the *concomitant parallel increase in reductionistic thinking in biology*. The overall direction of this and the following chapter, as well as others later in this study, will begin to evidence this relationship. It is not to say that big biology, for example biological ecology, diversity of life studies, macroevolution, and related areas, would *not* have developed at all within the timeframe of interest to this book, but the evidence from within and without National Science Foundation (NSF)-funded research clearly suggests that modern big biology is intimately *intertwined and supported* by discoveries in areas of biology driven by reductionistic tendencies. Biology of the little informs biology of the big.

In regard to reductionism in biology, the following comment has been made:

> Philosophical interest about reduction in biology is pervasive throughout the history of philosophy and science. Many contemporary debates have historical analogues, reflecting long-standing controversies among biologists about the legitimacy of reductionist research strategies and modes of explanation used by different life science subdisciplines.[2]

Such debates may have *added to them* the role of reductionism in the progress of big biology; a collection of "subdisciplines" to be examined in this study. While an apparent contradiction at first blush, the evidence for the growth of areas of big biology has become *dependent upon* molecular aspects of that science, as shall be demonstrated. The research funded and promoted by NSF and its responses to the needs of the biological community, and its efforts in guiding of US (and to some degree world) biological research, provides incontrovertible evidence of this interactive fusion of molecular and macroscale biology. Macroscale biology, for instance, life diversity studies (heavily funded by NSF in recent years), today requires not just an appreciation, but a true understanding of molecular-level evolution of life forms in ecological settings. This micro–macro relationship will become obvious in coming chapters.

The terms big and little biology have multiple meanings and applications. Little biology should be thought of in two senses. First, "biology of the little," that is to say, the study of molecular biology crudely put is life reduced to physical chemistry. Second, and an entirely different use, is to say single-researcher projects *not* typically having the characteristics of "big biology" as defined by Alvin Weinberg (see earlier) in 1967 and as clarified for NSF's use by the National Science Board's (NSB's) Working Group 3 in 1979 when it considered big sociology (Chap. 2). It is worth repeating the four traits necessary to make for big science as understood by the NSB at that time, and hence by NSF. To wit:

a. Large-scale commitment of financial resources.
b. Investment of capital in facilities and major equipment.
c. A duration of several years or more.

[2] *Stanford Encyclopedia of Philosophy,* "Reductionism in Biology," 2008; see: http://plato.stanford.edu/entries/reduction-biology/.

d. Continuing expenditures for maintenance, replacement, operating costs, and research budgets.[3]

Little biology, then, is research *not* characterized by these four traits; it is, for the most part, single researchers at work on financially small-scale projects (small grants) with minimal needs for major equipment, maybe lasting several years (though commonly less), and typically without a need for continuation (sensu NSB), or, if so, that of a minimal nature.[4] In some areas, a strong argument has been made that little science—individual researchers—may require big and expensive equipment, too. Astronomy has been shown to fit such a category.[5] In later times, historians have pointed out that, in addition to the four points stated above, it would be necessary to add others. As an example, large numbers of persons working on a team or larger grouping (those at a National Laboratory, for instance), as became typical after WWII, may be part of a fuller definition.[6]

Because biology of the little is, for present purposes, research in molecular, cellular, and other aspects of sub-organismal biology such as physiology and genetics, it is the subject of this chapter. Indeed, the very title of the division of the Biological, Behavioral, and Social Sciences (BBS) in 1975, responsible for those disciplines, was reflective of them. The following chapter will consider *big biology* in its coordinate two senses of big science, that is, the four points above, *and* its second meaning of biology at the organismal level, but especially above that level (populations, communities, biomes, ecosystems, etc.). Biologists are fond of pointing out that they study one or more aspects of the science as it is laid upon a continuum from the molecular to the biospheric level with the organism, the central concern of biology, placed approximately in the middle of that continuum. We examine in the present chapter the left or micro end of that continuum. Institutional history, with its emphases on management, organization, structure, etc., remains the central thread for this and all following chapters, of course.

It would be the height of Whiggishness to suggest that there was some teleological force in operation at the NSF when biological research was being funded such that research in biology of the little would be destined to become so informative for biology of the big. However, it did happen, as shall be seen, and played a central role in defining the overall nature of biology in America by the early part of the present millennium, besides defining what the biology directorate and NSF itself would

[3] NSB, "Minutes," NSB-79-65, *op. cit.*

[4] See Toby Appel's enlightening commentary concerning NSF biologists' early views about "big biology" in her "Shaping Biology," pg. 205, *op. cit.*

[5] In Chap. 7 of Peter Galison and Bruce Hevly, eds. *Big Science: The Growth of Large-Scale Research* (Stanford: University Press, 1992), Robert W. Smith considered "The Biggest Kind of Big Science: Astronomers and the Space Telescope." He spoke in terms of the needs of single researchers, or a senior astronomer with a graduate student or two, working in optical astronomy needing time on large telescopes, among the most expensive of all scientific instruments. It was radio astronomy, not optical, that became the first example of "big science" within the broader areas of astronomy.

[6] Ibid. See especially Galison's "Introduction: The Many Faces of Big Science," ppg. 1–17.

become.[7] But, as philosopher of biology Alexander Rosenberg has pointed out, it was not a sure thing in an earlier time.

> After all, it took Frederick Sanger and a team six years [culminating in 1955] of arduous experimentation to provide the primary sequence of the first protein…characterized: insulin. At that rate of progress the molecular biologists' claims about the reduction of biological functions to chemical structure would be at best *hopes* for the next millennium.[8]

Had Rosenberg written these lines in, say, 2000, instead of 1985, he would not have made such a comment, so far had reductionism come in structuring biology as we understand it today. Explaining aspects of ecological-level biology by use of molecular-level biology is hardly new, of course, but we shall find just how significant NSF's role in making it so was the case.

> I take it that no one would deny that a major trend—probably *the* major trend—in science has been toward the explanation of the large in the terms of the small. Moreover, even before the coming of molecular biology, methodological reductionism of this kind was being practiced highly successfully in biology. Think of the importance of the classical concept of the gene.[9]

So it was seen by philosopher of biology, Michael Ruse. And so shall it be seen with the history of the National Science Foundation and the Directorate of Biological, Behavioral, and Social Sciences.

3.2 The Biological Divisions Within BBS in 1975

The previous two chapters can be viewed as introductory to the analysis of the history of biology at the NSF from 1975 to 1991. As has been alluded to, the close relationship between biology and the *social sciences* at NSF in earlier years (both before and after 1975) was such that the latter sciences had a clear influence in the operation and activities of the older BMS and the newer BBS, both organizationally

---

[7] See the concept of Whig history in Herbert Butterfield's classic, *The Whig Interpretation of History* (New York: W. W. Norton, 1965), *passim.* Butterfield dealt, in the original publication in 1931, with British history only, but later the term Whig history was applied to certain trends in the history of science, as well as other areas of study. See also the more modern examination of this issue in: Keith Sewell, *Herbert Butterfield and the Interpretation of History* (Basingstoke: Palgrave-Macmillan, 2005).

[8] Alexander Rosenberg, *The Structure of Biological Science* (Cambridge: University Press, 1985), pg. 83, in his aptly named Chap. 4, "Reductionism and the Temptation of Provincialism." Emphasis added.

[9] Michael Ruse, *Philosophy of Biology Today* (Albany, NY: State University of New York Press, 1988), pg. 25. Emphasis in original. Methodological reductionism is defined as "the idea that biological systems are most fruitfully investigated at the lowest possible level, and that experimental studies should be aimed at uncovering molecular and biochemical causes." *Stanford Encyclopedia of Philosophy*, "Reductionism in Biology," *op. cit.* See also: W. Bechtel and R. Richardson, *Discovering Complexity: Decomposition and Localization as Strategies in Scientific Research* (Princeton, NJ: Princeton University Press, 1993).

and intellectually. Evidence of this has already been seen in the growth of big science in sociology at NSF *before* its later parallel efflorescence in biology. That biology and the *behavioral sciences* were even closer cousins has already been made obvious, too. Further, most of the behavioral science programs moved to BIO after the reorganization of 1991–1992, for instance, and not to the new social, behavioral, and economic sciences directorate (SBE).

Returning to Fig. 1.3, in Chap. 1, we see the basic plan for the BBS directorate in 1975–1976 and beyond for several years. One of the then two biological divisions considered the physiological, cellular, and molecular (PCM) aspects of living creatures; the other dealt with aspects mostly at and above the organismal level, the Division of Environmental Biology, or DEB. DEB will be the subject of Chaps. 4 and 9.

3.2.1 PCM: The Division of Physiological, Cellular, and Molecular Biology

What was the state of PCM at the National Science Foundation at the time of the establishment of BBS? When Atkinson became the deputy director of NSF in June of 1975, the period was, as has been demonstrated, one of "tumultuous times." Basic research expenditures within national research and development (R&D) efforts for 1975 would drop by 8% in constant dollars from 1974.[10]

At its mid-January 1976 meeting, the National Science Board would offer an encomium to a past member of its body and a major player in the history of "biology of the little," Edward Lawrie Tatum, who had just died in November of 1975. Tatum had been a member of the NSB for two terms, from 1956 until 1968. In 1957, he had become a faculty member at the Rockefeller University studying the nutrition, biochemistry, and genetics of microorganisms. By 1958, Tatum had won the Nobel Prize in Medicine or Physiology, which he shared with George W. Beadle and Joshua Lederberg, equally giants among that group of biologists who could be said to be the leaders of reductionism in America during the period. Indeed, Tatum and Lederberg were only two among a long list of NSF grantees who would win the Nobel Prize in the 1960s, and in later decades.[11] Tatum and Beadle's work was best thought of at that time as the "one gene, one enzyme" theory of the function of genes, while Lederberg showed that bacteria could mate in a sort of sexual manner, theretofore unknown, and by such mating could exchange genetic material.

It seemed, then, manifest that the division of sub-organismal biology at NSF, even in the decades before 1975 in Tatum's day, was clearly on a path that was leading to even greater reductionism than had ever characterized biology in earlier

[10] NSB, "Minutes," NSB-75-111, March 20–21, 1975, 171:6, *op. cit.*

[11] Appel, "Shaping Biology," pg. 210. For the complete list to 2009, see: http://www.nsf.gov/news/news_summ.jsp?cntn_id=115759.

times. With the Foundation's birth only 3 years before the stunning announcement made by Watson and Crick, at least some part of the BBS was set de facto to pursue biology along reductionist lines. This has been clearly detailed by Appel. By the fiscal year 1975 (FY75) and its financial woes, however, the NSB was not only mourning the loss of early fellow Board member Tatum, but also noting in the same issue of the NSB "Minutes" that:

> [f]or the smaller grant, essentially nothing but salaries can be provided [to the researcher]. The Task Group [on "Larger but Fewer Research Grants and Equipment"—early shades of big science] felt that this trend has seriously reduced the effectiveness of the small individual grant [always the "bread and butter" of the NSF].[12]

Since the typical "little science" individual grants for a PI were then running on average below $25,000, they "provid[ed] almost no equipment or facilities." So the task group (TG) recommended that the "Foundation systematically devote more of its research funds" to those two areas.[13] It was also at that time that the Foundation had celebrated its 25th anniversary and shortly thereafter (early 1976) that Betsy Clark came to serve as acting assistant director (AD) of the new BBS. But by July 2, she became the permanent head of the directorate. By August of 1976, Stever was confirmed by the Senate as the new director of the OSTP, and Atkinson would become first acting and then permanent director of NSF (earlier chapters).

The Board was also considering the notion of a "Departmental Research Centers Program."[14] The Departmental Research Center (DRC) concept arose out of a Task Force on the future of NSF and considered that select academic departments should receive support.[15] After all, "the university sector was the largest producer of research [articles] in 1975" in a number of fields, those having accounted for some 73% of published papers.[16] While much greater than support for individual PIs, the DRCs would still not equal that of the Science and Technology Centers (STCs) that would later follow. The DRC program conceived of awarding some $40 M annually to about 150 academic departments (in all sciences and engineering) with $250 k "block grants" going to 10 departments in each of 15 disciplines, biology included. The grants would be "long-term" at 4 years with possible renewal for an additional year. NSF would set aside 5% of its annual budget for the Program.[17] Those were heady days, if not financially fat ones, for the NSF at the same time that the nation celebrated the U.S. Bicentennial.

Most significant, vis-à-vis science of the little and the power of reductionism, was the fact that the NSB had on its agenda for the August 1976 meeting, the subject

---

[12] NSB, "Minutes," NB-75-111, 178:40, Appendix E.

[13] Ibid.

[14] Ibid., November 18–19, 1976, 186:23.

[15] Task Force 74-C; June 1974.

[16] National Science Board, *Science Indicators 1976*, pg. 66. In growth after 1960, articles in the biological sciences had increased by 22% by 1962, but had increased by 141% by 1975; Table 3–21, pg. 246.

[17] NSB, "Minutes," November 18–19, 1976, 186:23.

of "Guidelines for Research Involving Recombinant DNA Molecules."[18] NSF had been immersed, if not yet deeply so, in the "age of molecular biology" for over a decade, however, as Appel has documented.[19] NSB member Joseph M. Reynolds was the chair of the Planning and Policy Committee, PPC—one of two standing committees of the NSB at the time—which was a powerful arm of the Board. It was the opinion of the PPC that the matter of recombinant DNA (rDNA) research guidelines for the Foundation should be fashioned after those of the NIH. Allowing the NIH to take leadership in this and other situations where matters such as ethical research guidelines were of concern, was typical at NSF (human subjects' research, discussed in Chap. 2, was also an example). To this approach, we shall return. Reynolds led the discussion on the agenda item and the Board unanimously adopted a resolution that stated that NSF would adhere to the recently published guidelines as promulgated by the NIH for rDNA research.[20] Reynolds also sent, via Atkinson, a letter to Donald Fredrickson, director of NIH, informing him of the Board's decision.[21] The NIH director thanked the NSB for their decision and reminded them of the "flexibility" of the guidelines, the whole new field of rDNA research being in such a state of flux.[22] The origin of the NIH's *own* guidelines had been based upon material deriving from the Second Asilomar Conference.

As has been seen, in 1973 the first of two Asilomar Conferences on biological hazards had taken place. Paul Berg's own successful work in rDNA research at Stanford in the early 1970s cued him to the need for a molecular biology community-wide discussion once humans could, as was said, "play God."[23] That first one was funded entirely by NSF.[24] The second one (1975) was of special significance here as it was directed at the new science of rDNA, specifically.[25] Recombinant DNA had become a possibility in the years between the two conferences due to the NSF-supported research of Herbert Boyer at UC, San Francisco and Stanley Cohen who, like Berg, was also at Stanford. Boyer was at a conference in Hawaii in 1972 when he met and conversed with Stanley Cohen. By 1973, a research partnership had developed between the two men, which led to their successfully inserting a piece of amphibian DNA into a plasmid (small, circular strand of DNA) of the bacterium

[18] NSB, "Minutes," NSB-76-301, September 21, 1976, 183:27. The set of Minutes for the Board for all of 1976 was among the largest for any year in that period, so active had the Board and NSF become.

[19] Appel, "Shaping Biology," pg. 209ff.

[20] *Federal Register*, July 6, 1976. See also the National Library of Medicine's Donald S. Fredrickson Papers at: http://profiles.nlm.nih.gov/FF/Views/Exhibit/narrative/rdna.html and also at: http://profiles.nlm.nih.gov/FF/Views/Exhibit/narrative/director.html.

[21] NSB, "Minutes," 183:27.

[22] Ibid., NSB-76-345, 184:6.

[23] Among many other articles on this subject, see Susan Wright, "Legitimating Genetic Engineering," *Dissent* (2001):62–69, as well as her "Recombinant DNA Technology and Its Social Transformation, 1972–1982," *Osiris* (1986):303–360, *passim.*

[24] Appel, "Shaping Biology," pg. 254.

[25] http://sciencepolicy.colorado.edu/admin/publication_files/resource-2642-2005.74.pdf.

Escherichia coli, long a *model organism/system* in bacteriology. That event, "molecular cloning," is considered the *beginning of genetic engineering*.[26] That same year, Boyer reported his findings at a June Gordon Research Conference[27] and they were published for the first time in the *Proceedings of the National Academy of Sciences* (commonly known as *PNAS* or, simply, the *Proceedings*) in November of 1973.[28] While Boyer went on to found the first rDNA research corporation, Genentech, in 1976 with venture capitalist Robert Swanson, Paul Berg and his associates were preparing for the second Asilomar Conference of February 1975. The second conference was felt necessary given the concerns that arose after the *PNAS* paper of Boyer and the general rapidity with which the whole new research arena of rDNA was growing.

In the NSF's *Annual Report* for FY76, Atkinson had a great deal to say in the "Director's Statement" portion of it. Much was said about matters in his own interest area (the social sciences) but, also, he addressed the "revolution [that] is under way today in the biological sciences."[29] One aspect was the work the Foundation had been supporting in the area of *restriction enzymes*: "an array of specialized proteins in bacteria…which cleave strands of nucleic acids [such as DNA] at highly specific points." Those were the very enzymes that allowed Boyer, Cohen, and others to begin the genetic engineering feature of the "revolution"—and for which NSF-supported investigators would share a Nobel Prize in 1978 (Chap. 7). Atkinson was able to state that such work allowed for "a chain of 20 DNA subunits [to be] sequenced in a single afternoon. Only a few years ago, it took as long as 2 years to perform this task."[30] By the new millennium, the task would be reduced to minutes, if not seconds.

He mentioned, also, *molecular cloning*: "the transfer [of] genetic material from one organism into an unrelated organism"; the type of work done by Boyer and Cohen. One aspect of that work was the possibility of "amplifying" (making many copies of) the specific gene that, say, could be transferred into cereal crops from legumes and increase the former's ability better to deal with the crucial element nitrogen.[31] Nitrogen is required in all proteins and many other organic (carbon-containing) molecules necessary to life. Plants cannot "fix" nitrogen by taking the

---

[26] Appel, "Shaping Biology," pg. 119.

[27] http://www.grc.org/meetings.aspx?year=1973.

[28] Stanley N. Cohen, Annie C. Y. Chang, Herbert W. Boyer, and Robert B. Helling, "Construction of Biologically Functional Bacterial Plasmids In Vitro," *Proceedings of the National Academy of Sciences* 1973:3240–3244.

[29] NSF, "Annual Report 1976," pg. x, *op. cit.*

[30] Ibid., pg. xi.

[31] The use of the term "amplification," prior to 1983, should not been confused with the Nobel-winning work of Kerry Mullis who created the in vitro technique known as the "polymerase chain reaction," PCR, that was used routinely to make multiple copies ("amplification," sensu Mullis) of select DNA from that time onward. Prior to 1983, the term amplification implied that DNA, placed into a *living cell*, would be amplified into multiple copies, but only a fraction as many as PCR could later provide.

Biological, Behavioral, and Social Sciences
Fiscal Years 1974, 1975, 1976
Transition Quarter (July 1-Sept. 30, 1976)
(dollars in millions)

	Fiscal Year 1974		Fiscal Year 1975		Fiscal Year 1976		Transition Quarter	
	Number	Amount	Number	Amount	Number	Amount	Number	Amount
Physiology, Cellular and Molecular Biology	994	$36.85	1,006	$41.87	1,053	$43.69	279	$12.00
Behavioral and Neural Sciences	468	15.23	547	18.50	539	19.69	140	5.47
Environmental Biology.	398	21.78	487	26.05	534	26.84	156	7.15
Social Sciences	405	18.81	359	17.75	364	18.81	100	5.11
Total	2,265	$92.67	2,399	$104.17	2,490	$109.03	675	$29.73

Fig. 3.1 Biological, Behavioral, and Social Sciences (BBS) Fiscal Years 1974, 1975, 1976, and "Transition Quarter," a 15-month year (see explanation in main text). (Source: NSF, *1976 Annual Report*)

gas that makes up about 80% of Earth's atmosphere and incorporate (fix) it into organic compounds, but certain bacteria (and grossly similar creatures called Archaea), which produce the nitrogenase enzyme, can. The work in nitrogen fixation became a notable program at BBS in years to come. Atkinson also considered in his comments the use of molecular cloning to make human insulin in bacteria—a matter that later came to fruition.[32]

Just how rapidly the biology of the little was growing in the mid-1970s can be seen in a table published in the 1976 *Annual Report.*[33] Figure 3.1, here, demonstrates the growth for the PCM division (top line) for a period of three fiscal years (and the "extra quarter" as the federal fiscal year moved to a beginning date of October 1 annually from 1976 onward). In both *numbers* of grants, nearly half of all BBS individual awards, and *total dollars* granted and projected, the PCM led the other three of the four divisions in the BBS—and by a margin of 2 or 3:1. Clearly, biology of the little, as well as little biology, was big at NSF early in the molecular revolution.

The authors of the section on PCM in the *Annual Report* stated that: "[d]isciplinary boundaries in biology have virtually disappeared," and also that those boundaries were being "approached with multidisciplinary methodologies."[34] The technologies not only included restriction enzymes and molecular cloning, but also such methods as gel electrophoresis and new uses of the electron microscope, besides the employment of many other newer tools and techniques. The interrelationship between *structure* and *function* of genes became amenable to understanding within the biology of the day. A graphic (Fig. 3.2) was incorporated into the

32 PR Newswire July 21, 1980, "Eli Lilly and company today announced that it has begun limited testing in healthy human volunteers of biosynthetic human insulin produced by recombinant DNA technology…[i]n what is believed to be the first application of [that] technology to human health problems."

33 NSF, "Annual Report 1976," pg. 55.

34 Ibid., pg. 56.

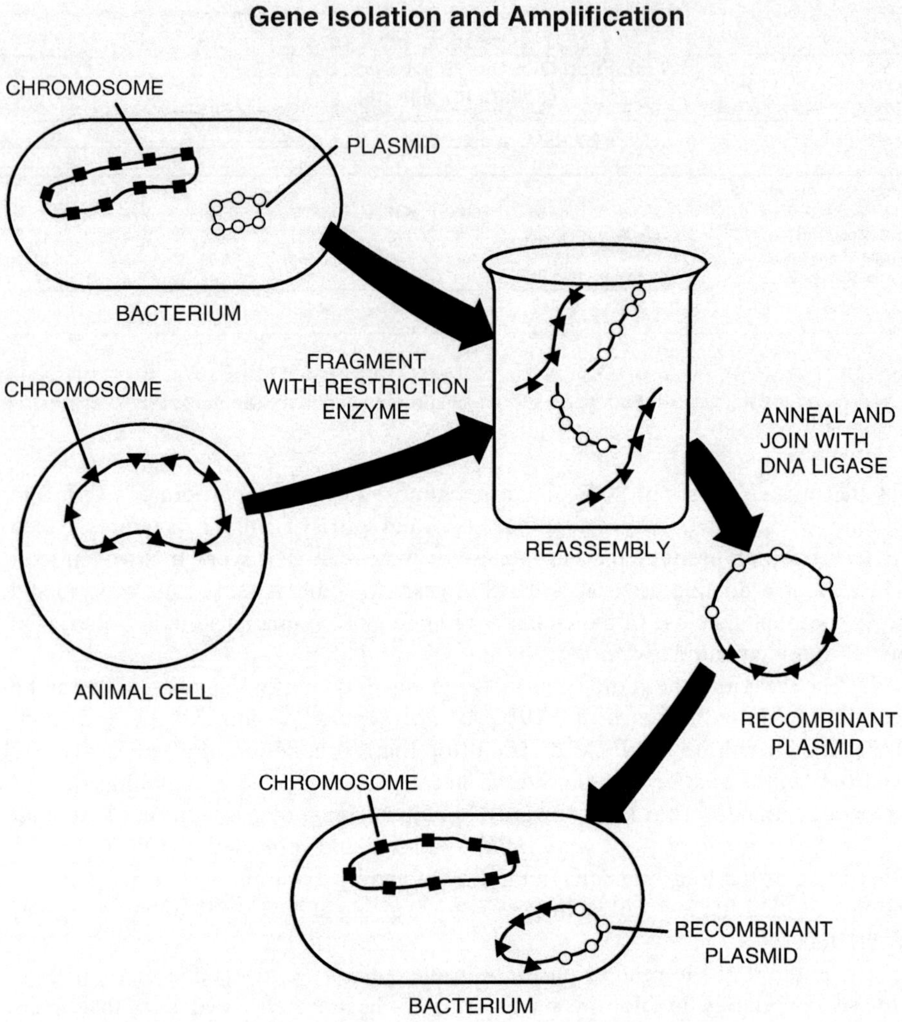

Fig. 3.2 Gene amplification and isolation. (Source: NSF, *1976 Annual Report*)

PCM division progress report, which highlighted the possibilities of probing genes in a way (restriction cleavage and amplification) never before possible. That was a spin-off, in part, of the Boyer-Cohen work. Much could be learned about the nature of genes with the new tools, and that could be extended throughout other levels of the biological hierarchy.

Research support by BBS was provided for individuals and small teams at UC, San Diego (carbon dioxide fixation in plants), at UC, San Francisco ("topological organization of multiple copies of mitochondrial genes"), at Harvard (nitrogen fixation in legumes; nature of the genes that control ribosomes; function of hemoglobin genes), at Stanford (messenger ribonucleic acid—mRNA—and how histone genes work in the most famous of the model biological systems, the fruit fly *Drosophila*), and at the University of Iowa (mechanisms seen in the mitochondrial genes), among others. At MIT, a gene was synthesized in vitro and successfully placed into a bacterium deficient in that gene and its functional duty: the function was restored. That was a highly notable achievement in what was among the first important efforts in the new area of *biotechnology*. Such placing of genes of one creature into an unrelated one would become a staple of the new *Millennial Biology*.

3.3 The Rise of Biotechnology at NSF and Its Effects

The year of 1984 was a signal one for BBS, and NSF more broadly, in regard to new programmatic initiatives, new personnel in both the directorate and the senior leadership of the Foundation, the inauguration of the BBS Advisory Committee (AC), and the first suggestions of institutions to be known as Centers. As the header above suggests, these major happenings were tightly interwoven, with *biotechnology being very much at the underpinning* of many initiatives and other organizational creations of that period. Computational biology would be both influenced by biotechnology and integrated into the weave, as well. A theretofore little-known plant, the thale cress, would appear on the scene as the model system for the genetics of flowering plants, too.

However, that notably historic year of 1984 was some time in the future as the matter of biotechnology's birth was about to take place in the middle part of the 1970s in the scientific community in general, and for NSF in particular.

The PCM Division was funding Herbert Boyer at UC, San Francisco in 1975, to make an "analysis of eukaryotic chromosome structure and function" at a monetary level approaching $100 k.[35] His basic work, which played the first role in the biotechnology revolution, was not alone at the time. Leroy Hood at the California Institute of Technology (Cal Tech), a major player in the revolution, was also being funded for his studies of genetic control in cells.[36] Yet others at Stanford were continuing their work with NSF support on a bacterial virus, or bacteriophage, known as "lambda."

Lambda had become an important tool in the early period, and beyond, of molecular biology, particularly as it could be used as a device that would transport genes from one organism into another (a process known as transduction), where those genes could be expressed, that is, their function allowed. As the American Society for Microbiology has said, lambda had a special role in "history, fame and fortune"

---

[35] National Science Foundation, *Grants and Awards Fiscal Year 1976*, pg. 2. Hereinafter, "NSF G&A [year]."

[36] Ibid., pg. 1.

with regard to the molecular, or biotechnology, revolution.[37] While many of the fundamental discoveries were already in place by the mid-1970s concerning lambda and other tools and techniques, the actual beginning of the revolution could only come to pass after the discoveries of Boyer and Cohen noted above. Some years later (1980), they would patent their method.[38]

For FY76, the PCM funded D.P. Nierlich of UC, Los Angeles at the level of $3000 to put together what became the "ICN [Pharmaceuticals Company]-UCLA Conference on Molecular Mechanisms in the Control of Gene Expression" that was held in Keystone, Colorado, later that year.[39] Many other researchers in many states across the country were receiving NSF grants that would be of significance to the revolution. Paul Tso at Johns Hopkins University in Maryland was also granted $3000 to fund a symposium: *supporting conferences in that period was a notable feature of NSF's interests.* At the Carnegie Institute of Technology in Pennsylvania, the work of Kelley and Brown on the "structural studies of DNA polymerase I of *E. coli*" was funded at the level of $77 k for 2 years.[40] These were only a few examples of the many projects then underway and funded by PCM for biology of the little.

In August 1976, as recounted above, NSF adopted (temporarily) the rDNA research guidelines established by the NIH. Both NSF director Atkinson and NIH director Fredrickson, kept in touch during early 1977 concerning the continuing efforts at formalizing a set of guidelines that would appeal to Congress and that could become law. Member of the NSB, Alexander Rich, had attended at least one of the NIH guidelines' discussions during the period and could report back to the Board about the ongoing efforts at the Institutes to refine those guidelines.[41]

---

[37] See http://www.asm.org/Division/M/fax/LamFax.html where, in a bibliographical note, the ASM points out the importance to this subject of the following articles and scientists: "J. Cairns, G.S. Stent, and J.D. Watson, *Phage and the Origins of Molecular Biology*, (Cold Spring Harbor Laboratory, 1966). Some immensely readable and entertaining chapters mentioning lambda include A. Lwoff, 'The prophage and I,' p. 88ff, E. Kellenberger, 'Electron microscopy of developing bacteriophage,' p. 116ff, A.D. Kaiser, 'On the physical basis of genetic structure in bacteriophage,' p. 150ff, and J. Weigle, 'Story and structure of the lambda transducing phage,' p. 226ff." See also: http://www.asm.org/Division/M/blurbs/Secrets.html for a list of "major discoveries made with bacteriophages."

[38] See a number of sources concerning this signal event: http://www.kauffman.org/uploadedFiles/Feldman_Maryann.pdf, http://www.ias.ac.in/currsci/mar252009/760.pdf, http://www.bizjournals.com/sanfrancisco/stories/1997/11/24/story2.html, http://bancroft.berkeley.edu/Exhibits/Biotech/25.html, among many others.

[39] NSF, "G&A," pg. 1; see also: http://agris.fao.org/agris-search/search/search.do?query=author:%22%20ICN-UCLA%20Conference%20on%20Molecular%20Mechanisms%20in%20the%20Control%20of%20Gene%20Expression,%20Keystone,%20Colo.%20(USA),%201976%20%22&from=br. The proceedings came out later as: D. P. Nierlich, W. J. Rutter and C. F. Fox, eds., *Molecular Mechanisms in the Control of Gene Expression* (ICN-UCLA Symposia on Molecular and Cellular Biology, Vol. 5, 1976).

[40] NSF, "G&A," pg. 5.

[41] NSB, "Minutes," NSB-78-62, March 23, 1978, 195:7. It is not clear whether the Board *actually dispatched* Rich to sit in on the NIH meetings or not. "Dr. Rich reported on a recent NIH meeting he attended…" is as informative as the record allows.

Fredrickson was by then serving as chairman of the federal Interagency Committee on Recombinant DNA Research and told Atkinson that a special subcommittee, including Rich, had been at work since late 1976 "reviewing present statutory and regulatory authority to govern" rDNA research.[42] The subcommittee had concluded that no single federal agency or department had "full authority" to regulate recombinant research and pointed out that a bill was being put forth by Senator Dale Bumpers (D-AR), a freshman senator, on the matter. Fredrickson considered not only the Bumpers bill[43] and its companion in the House introduced by Representative Richard Ottinger (D-NY), which was actually the un-cosponsored House Resolution (H.R.) 131,[44] but also the fact that some "12" bills to regulate rDNA research were introduced in the 95th Congress![45] Senator Gaylord Nelson (D-WI), for instance, introduced an amendment to the main bill, S. 1217, challenging it.[46] Numerous other Congressional offerings clouded the discussion.

Acrimony followed over the wording of a bill for nearly 2 more years and it did not end until mid-December 1978 when a telegram was delivered to then Secretary of Health, Education and Welfare, Joseph Califano, Jr. Califano was a Catholic of Italian descent and so the telegram humorously "purport[ed] to be from the Vatican" and was entitled "*Habemus regimen recombinatum.*" It celebrated the 3-year-long struggle to revise the NIH guidelines and thus satisfy Congress.[47] In Fredrickson's history of this multiyear incident, he has termed it both the "recombinant DNA affair," and the "rDNA controversy." The actual Congressional decision that closed the debate, for a time, was H.R. 11192, which passed the necessary committee on April 18, 1978, and provided for a "two-year interim control of all publicly and privately supported activities" in rDNA research.[48] Those guidelines (regulations to be formalized) were published in the *Federal Register* on July 28, 1978.[49] Sheldon Krimsky has characterized the debate.

---

[42] NSB, "Minutes," NSB-77-96, March 25, 1977, Appendix E, 187:31; letter from Fredrickson to Atkinson of February 18, 1977.

[43] S. 621; February 1977.

[44] See: http://thomas.loc.gov/cgi-bin/bdquery/?&Db=d095&querybd=@FIELD(FLD003+@4((@1(Rep+Ottinger++Richard+L.))+00882, item 5, which does not support Fredrickson's memory as chronicled in the text of the history cited in the following footnote. But see more comments there.

[45] See: http://profiles.nlm.nih.gov/FF/B/B/K/C/_/ffbbkc.pdf. Fredrickson's six-page typescript version of "A History of the Recombinant DNA Guidelines in the United States" is fundamental to our understanding of that subject. Online pg. 154. While Fredrickson does not itemize all 12, the NSB "Minutes" for April 25, 1977, that is, NSB-77-156, does list two others beyond the Bumpers and Ottinger offerings (at 188:11). Ottinger does seem to have introduced not only H.R. 131, but also H.R. 3191, which would, in fact, corroborate Fredrickson's recollection (footnote immediately above).

[46] NSB, "Minutes," NSB-77-386, September 21, 1977, 192:8.

[47] Fredrickson, "A History," pg. 151.

[48] NSB, "Minutes," NSB-78-242, July 13, 1978, 198:7.

[49] See Donald S. Fredrickson, *The Recombinant DNA Controversy: A Memoir,* in *Science, Politics, and the Public Interest 1974–1981* (Washington, D.C.: American Society for Microbiology Press, 2001), and the review by Sheldon Krimsky, *American Scientist* at: http://www.amErichanscientist.org/bookshelf/pub/the-rdna-debate

> The recombinant DNA wars, as they have been called, had an enormous impact on many sectors of society, including the courts, Congress, the executive branch, interagency committees, foreign governments, the National Academy of Sciences, professional societies, the new biotechnology industry, university scientists, local governments and the media.[50]

There were few institutions of society left uninvolved; certainly not among them the NSF, even if not listed by Krimsky.

However, much struggle there was to finalize a widely acceptable set of guidelines, such did come to pass, NSF using the NIH version.[51] The Foundation had to make resolutions for some time in order to use interim variants and then, eventually, the final guidelines.[52] Indeed, the NSF was taking direction from not only the scientific community, but also the NSB, when it advised the readers of the *Annual Report 1977* that "[t]he Foundation continues to support carefully safeguarded research with recombinant DNA."[53] Atkinson wondered whether these "developments are posing questions that must be dealt with ethically and politically." Once he asked the more specific question—"[c]an we embark on 'gene therapy' for disease?"—the astute observer could quickly divine an answer to his first question. Ethics and politics would be inextricably tied to the growth of the biotechnology revolution to the present day.

The PCM section of the *Annual Report 1977* had not had a great deal to say about the research projects that would make up the basis for the revolution, certainly not as compared to coming years. A single, long paragraph was devoted to the subject and continued to speak in terms of, by then already well established, restriction enzymes and in vitro techniques for doing rDNA research. More space was devoted to cell culture centers, pheromones, and biological fertilization—all those areas primarily set on the biological continuum at the cellular level, a step above the molecular.[54] On the other hand, little biology done through single-researcher PCM grants, as listed in the *Grants and Awards for Fiscal Year 1977*, occupied some 18 pages as compared to macro-level biology in the DEB where only 10 pages of state-by-state

---

[50] Krimsky, "Review," pg. 1 of website pagination.

[51] At the time that Fredrickson wrote his "A History" (footnote further above), the guidelines were just that; they had not yet become formal federal regulations. That would occur during the mid-1980s. See: http://usbiotechreg.nbii.gov/Coordinated_Framework_1986_Federal_Register.html.

The use of the term "guidelines" continued even after they became official regulations. See: Jun 26, 1986–66 "Coordinated Framework for Regulation of Biotechnology: Announcement of Policy and Notice for Public Comment," 51 *Federal Register* 23302–23393. See also following footnote. For the full online copy of the FR cited above, see: http://www.epa.gov/fedrgstr/EPA-TOX/pre1994/ost86.pdf.

[52] Resolutions were made along the way as the guidelines matured from the July 6, 1976, *Federal Register* version promulgated by NIH, through, for instance, another version in November of 1977: see NSB, "Minutes," NSB-77-474, January 5, 1978, 194:4; 194:29; and, Appendix B, 194:32 where, on November 17–18, 1977, the Board adopted the then latest version.

[53] National Science Foundation, *Twenty-Seventh Annual Report for Fiscal Year 1977*, pg. vii, *op. cit.*

[54] Ibid., ppg. 62–67.

awards' listings were to be found.[55] Furthermore, a higher number of molecularly oriented grants, as compared to those for cellular or physiological areas, were to be seen in the FY77 report in contrast to the FY76 edition.

In January of 1978, when the BBS sat down to examine where it might head by FY80, the PCM area emphasized four fields for study. Two of them were central to the issues of biotechnology: "gene splicing" and "macromolecular assemblies."[56] Indeed, the projected budget for FY79, which had already been sent to Congress, sought just over $61 M for PCM (all programs), but only a little over $34 M for DEB: but funding totals would reverse in the future as Big Biology grew. Also, later that year, the PCM leadership was about to change. Herman W. Lewis had come from Michigan State University "as a rotator for a year and stayed for twenty."[57] He headed the genetics program at BMS from 1962 to 1973, evidenced a "strong social conscience," and became a major player in the ethics of "potential risks of genetic engineering" both before and after 1973.[58] His interest in the matter even led to the establishment (in February of that year) at NSF of the Ethical and Human Values Implications of Science and Technology (EHVIST) program.[59]

The matter of rDNA was but one issue to occupy the Board's time. The NSB was highly active in the period on many other fronts, including playing its role in Congressionally inspired discussions and legislative offerings regarding equity for women in science and as NSF grantees (Title VI of the Civil Rights Act of 1964), reaffirmation of the value to NSF of rotators (no restrictions set on their numbers), setting up a new young investigators program and another for minority graduate fellowships (which drew some 4000 applicants several months later), historical celebration of the Board's 200th meeting (August 17–18, 1978), drafting of *Science Indicators 1978*, and providing Congress with the inaugural *Science and Technology Annual Report* (required as per the National Science and Technology Policy, Organization, and Priorities Act of 1976).[60]

And further, by the end of calendar 1978, the Board was addressing again the issue of big and little science as the House Committee on Science and Technology (CST) requested information from the Board in regard to big and little science for the CST's consideration of the FY79 NSF reauthorization act.[61] An NSB ad hoc committee on big and little science determined that the Board would not set financial and other guidelines for the two levels of research, but let the directorates and their leadership determine them; to that the members agreed unanimously.[62] The matter had grown crucial because "a large part of the [NSF] budget is being devoted

---

[55] NSF, "G&A FY77"; data on ppg. 103–120 versus those on ppg. 120–130.

[56] BBS, "5 Page Issue Paper," January 30, 1978, pg. 1, *op. cit.*

[57] Appel, "Shaping Biology," pg. 158. Appel interviewed Lewis on March 18, 1991.

[58] Ibid.

[59] Ibid., pg. 255.

[60] NSB, "Minutes," various for 1978–1979.

[61] Ibid., NSB-78-490, February 5, 1979 (November 16–17, 1978 meeting), 202:16.

[62] Ibid., 202:17; see also Appendix E: 202:45–47, Working Group 3.

to fixed items [e.g. astronomical observatories, and that] portion is increasing." The Board wished to be "more dynamic and assertive" in the connection of conveying to Congress and the Executive branches the significance of such a large share of the Foundation's annual budget being committed to "very large-scale projects."[63] The implication was that such large and expensive instruments, such as X-ray detectors that had proven so useful in "mapping the crystal structure of enzymes" and an "instrument combining a wide bore magnet with a Fourier transform spectrometer was ready for detailed studies of large, complex molecules," were both matters that had a role in the biotechnology revolution and could threaten "little science" if the proportion of fixed item costs was allowed to swamp single investigator awards.[64]

"Evolution is now addressed in molecular terms," and "[i]nterdisciplinary activity is accelerating rapidly."[65] Both are themes in this historical study and were matters that were apparent at the time: NSF was witnessing the continued increase in *reductionism*, *interdisciplinarity*, and the *size* of scientific research projects. Speaking of an "indirect index of the excitement" pervading biology, the NSF noted the increase in *unsolicited* grant applications even as "expectations of support" were dropping between 1976 and 1978. The increase was not a political comment made to impress upon Congress the need for more funding, as the proposal numbers spoke for themselves: proposals had gone up by 48% in that 2-year period (from 4993 to 7400)![66] Furthermore, "[c]ompetitive success ratios have dropped from 40% to 29%," even with a "substantial increase in the programs' overall budget."

In the PCM report for 1978, the authors spoke of "the 'new biology'," that is, the reductionist trend in the field devolving toward molecular levels.[67] (That same phrase of "new biology" would be used again a decade later, as shall be seen.) "Much of the current research in [PCM] grew out of the past discoveries" of the nature of DNA and "new insights" into the molecular end of biology.[68] Although it was "unfortunate" that X-ray crystallography and nuclear magnetic resonance spectroscopy had not been much used to date in molecular biology, "recent developments" were changing that situation. Work at UC, San Diego had increased the X-ray diffractometer's ability to visualize molecular structure by ten-fold and, in "just a few weeks" of work, high-quality data were derived on a given protein's structure. As well, questions of how proteins interact with other features of a cell, such as nucleic acids (DNA, RNA, etc.), were leading to answers at the University of Tennessee through the work of Ada Olins, and also that of Kensal Van Holde at Oregon State University.[69] The structure of eukaryotic chromosomes was being

[63] Ibid.

[64] National Science Foundation, *Twenty-Eighth Annual Report for Fiscal Year 1978*, pg. viii (Atkinson's "Director's Statement").

[65] Ibid., pg. 61.

[66] Ibid.

[67] But see discussion in later chapters where BIO AD Mary Clutter uses the phrase "New Biology" with initial capital letters and a highly specific meaning.

[68] Ibid., pg. 62.

[69] Ibid., ppg. 63–64.

"decoded," as was the structure of ribosomes (RNA-based assembly points for proteins). (The work of Francis Crick during the early 1960s had worked out the genetic "code," *per se*, however.) All of this work would play into the growing biotechnology revolution. And that most of it was being done on university campuses was not surprising given how the academic sector had grown. Some 36% of *all* scientific research was done in the academy in 1960. By 1978, it was at 52%! The industrial sector had dropped to only 16% of the total that year.[70] The academic sector continued to climb in its share of this productivity. By 1982, it was providing "about two-thirds of the research literature in the most influential…journals."[71]

As Director Atkinson had it, "[i]nnovation without new knowledge soon runs dry," and the new knowledge being produced through NSF support that would be utilized in the molecular revolution was quite considerable. Large, complex, and expensive equipment would be one thing much needed and, by the close of the decade, some 14 regional science centers were then in operation to allow for researcher's shared use of such notable tools.[72] Concomitant with machines were policy issues: NSF was "heavily involved" with its support of the Domestic Policy Review on Industrial Innovation—a discussion on the state of electronics—which was crucial to the instrumentation that made much of the biotechnology revolution possible.[73] Fictional detective, Charlie Chan, had long since sagely noted that "good tools shorten labor," and the significance of electronic instrumentation to the rise of biotechnology was no side issue. In fact, years before, Betsy Clark began "the practice of taxing each program a certain amount of money and forming a pool" of funds for the express purpose of having finances available to purchase expensive equipment. Kingsbury later followed that lead and pushed it further—toward a new division within BBS. The various program directors "would compete for [the] pool" and have a "reasonable chance of getting funded." It was the pool that eventually matured into the biological instrumentation program (BIP) and finally into its own division (Chap. 6).[74]

Also, the NSF undertook its first *Five-Year Outlook for Science and Technology* to be released in 1980. Gene activation, how genes are "turned on and off," was one of the issues that was addressed—an area of little biology still of great interest as of this writing. Mechanisms of gene regulation were studied by Lawrence Wang at

---

[70] National Science Board, *Science Indicators 1978* (Washington, D.C.: National Science Foundation, 1979), pg. 64.

[71] Ibid., *Science Indicators 1982* (Washington, D.C.: National Science Foundation, 1983), pg. 118.

[72] Regional science centers were merely locations for expensive equipment to be shared by many investigators and should not be confused by other types of "centers" to be discussed in the main text of this and other chapters.

[73] See the report at: http://www.iiasa.ac.at/Admin/PUB/Documents/PP-80-008.pdf.

[74] Eloise Clark Interview with the Author, May 26, 2009. Clark averred that the approach became a model for "research resources in the environmental area and for the anthropology museums and research resources, as well." The general "pool" was an ideal source for instruments, etc., which had a high price tag but were very "clearly worthy projects [on which] a program director was reluctant to spend big chunks of [line] money," and were needed for a given piece of research.

Brandeis University in Massachusetts, as he demonstrated the role of steroid hormones on protein synthesis, while George H. Jones of the University of Michigan showed that factors other than hormones can activate genes, too. For instance, Jones showed that injury to mammalian skeletal muscles induced regenerative processes: gene expression could be assessed at the level of transcription and translation—two processes central to protein production.

At the same time, the dynamics of the genome (the entirety of an organism's hereditary information) was being considered. In later years, genomics would become a central research area for the BBS and, eventually, BIO (Chap. 7)—*and form one of the most significant parts of the definition of* Millennial Biology. Indeed, it was not until 1986 that the word "genomics" was actually coined.[75] The early work was centered on transposons, small sections of DNA that could "jump" from one chromosome to another. The very stability of the chromosome had been brought into question by the discovery of those "transposable elements" by Barbara McClintock in 1948 in maize, her *model system*.[76] But it would take several decades to recognize the importance of transposons—and to honor McClintock. Not until 1983 did McClintock get the Nobel Prize. Melvin Simon's NSF-sponsored work at UC, San Diego on the bacterium *Salmonella* in the later 1970s was one of many studies that evidenced the significance of McClintock's discovery. In fact, within certain bacteria, as shown by Nancy Kleckner at MIT, "genetic engineering" occurred *in natura*, not just in vitro done by laboratory biologists, as the bacteria moved antibiotic resistance gene-carrying plasmids from one individual to another during "sexual" mating.[77] Many related works supported by PCM, once again, played notable roles in the revolution.

PCM brought out its program report in mid-1979 in which Larry Faller, the director of the biological instrumentation program (BIP), discussed the value of X-ray crystallography to the visualization of macromolecules (larger biological compounds such as proteins and nucleic acids).[78] The long-known digestive enzyme, chymotrypsin, had its active site revealed through research done with the newest X-ray devices, for instance. That was notable at the time, as fewer than 100 protein

---

[75] Satya P. Yadav, "The Wholeness in Suffix *–omics, –omes*, and the Word *Om*." *Journal of Biomolecular Technology* (2007):277. The word was first coined by Thomas H. Roderick, a geneticist at the Jackson Laboratory, Bar Harbor, Maine, in 1986, Yadav has said.

[76] See, among many other sources on McClintock: http://www.osti.gov/accomplishments/mcclintock.html, as well as http://profiles.nlm.nih.gov/LL/Views/AlphaChron/date/ and Nina Federoff's article "How Jumping Genes Were Discovered," *Nature Structural Biology* (2001):300–301, or at http://www.nature.com/nsmb/journal/v8/n4/full/nsb0401_300.html. Several books have been written on McClintock, but that by Evelyn Fox Keller is one of the best: *A Feeling for the Organism: The Life and Work of Barbara McClintock* (New York: Henry Holt and Company, 1983); but see also the more recent work by Nathaniel C. Comfort, *The Tangled Field: Barbara McClintock's Search for the Patterns of Genetic Control* (Boston: Harvard University Press Edition, 2003).

[77] National Science Foundation, *Twenty-Ninth Annual Report for Fiscal Year 1979*, ppg. 52–56.

[78] National Science Foundation, "Program Report: Physiology, Cellular and Molecular Biology," Vol. 3, No. 4, July 1979, pg. 3ff.

structures were yet known atom-by-atom. Larger-level structures in the cell, such as ribosomes, were being explored through instrumentation, too. The locations of proteins in the otherwise mostly RNA structure of ribosomes were being elucidated by NSF grantee Martin Deutch at MIT.

Program Director for Genetic Biology, Philip Harriman, described the work in protein–nucleic acid interactions in the PCM report by pointing out Olin and Van Holde's recent work (above). Their continuing support by NSF refined the nucleosome model of the chromosome (DNA wrapped around a histone protein core) to a higher degree. That was bolstered by the then more recent work of Ulrich Laemmli at the University of Geneva in Switzerland (an example of NSF grants to international scientists), who showed that removal of the histone proteins from a chromosome did not cause it to lose its relatively compact structure. Laemmli later went on to develop significant methodologies in the field and is remembered in the name of a specialized solution, the Laemmli Buffer, for dealing with cellular proteins, among other more notable discoveries.

Harriman's contribution to the PCM report reflected his own educational and interest backgrounds, he having gotten his undergraduate degree in physics from Cal Tech in 1959.[79] His doctorate was from UC, Berkeley (1964) and subsequent postdoctoral work was done at the University of Cologne in Germany, the Radium Institute of the Institut Pasteur in Paris, and the Cold Spring Harbor Laboratory on Long Island, New York. His is a good example of just how well trained and educated program officers at NSF typically were, and are. All of his positions were in the genetics of microbes, particularly viruses. His first formal job came in the biochemistry department at Duke University Medical School in 1968. He had applied for a grant simultaneously to both NSF and NIH to set up his laboratory at Duke. Both were awarded but, as the subject of the grant was the same, Harriman would have to choose the one that offered him the best situation. NIH was willing to offer more funding than was NSF, not an uncommon story, so Harriman turned down the NSF offer. After 7 years at Duke, Harriman left for a position at the University of Missouri, Kansas City, in 1975. Within a year after his arrival there, though, Harriman received an unsolicited call from Herman Lewis at NSF. Lewis had taken the division directorship position in 1976 and asked Harriman to come to the Foundation on a 1-year rotator assignment in genetic biology, Lewis' former post. As Harriman was so new at the University of Missouri, he declined the offer and Lewis said, "what about the year after?" Harriman responded, "well, get back—ask me later."[80]

As it happened, the department at Kansas City had a younger contingent of researchers (including Harriman) and an older cadre. The whole group was in "turmoil…[and so was not a] pleasant environment…there was a big battle going

[79] Philip Harriman Interview with the Author, March 13, 2009. Harriman also provided a number of smaller corrections and suggestions in his communication with the author in May of 2012. These are scattered throughout the main text where his name is mentioned.

[80] Ibid., the Interview.

on"[81]—a situation not unknown in the academy. Thus, when Lewis called Harriman once again (1977), the latter agreed. Harriman did not initially plan to stay at NSF, but he was excited by seeing the stages preceding the actual doing of research, that is, the application, reviewing, funding, and management stages. He arrived at the Foundation in fall of 1977. Mary Wolff was at the time the very experienced associate director for Harriman's program area. Since the business of becoming a rotator was "such a shock" to Harriman, given how different it was from any of his previous experiences, Wolff turned out to be a great help to him. There were ongoing review panels currently in operation, and she had created lists of potential persons to serve in new panels: "[s]o I [Harriman] just had to suggest replacements [from the lists] for people as they cycled off" of a given panel, he averred.[82] After his 1-year rotation, he was offered a second; again, not an uncommon happening at NSF. Since it was made clear to Harriman that he was "working out" well, he accepted. Wolff moved on the next year to the NIH, having told Harriman that she "trusted" leaving the Genetic Biology Program in his able hands. The trust was well founded as NSF offered him a permanent job in 1979.

Wolff would have to be replaced, of course. Though she did not have a doctorate, she was an excellent administrator, according to Harriman. DeLill Nasser, trained as a microbial geneticist, replaced Wolff. Years later, in his obituary of Nasser in the journal *Genetics* (primary publication of the Genetics Society of America), R. Scott Hawley said that "[i]f 'real genetics' ever had a patron saint, it was DeLill Nasser."[83] She had, by the time of her death, been a 22-year veteran at NSF. But in 1977, Nasser was living in San Francisco and had just divorced her husband, a professor at UC, San Francisco and desired to "get as far away from [there], as possible," Harriman recalled.[84] Having simultaneously been offered a post at NIH and NSF, Nasser found she liked what she saw of the Foundation and accepted its offer. This delighted Harriman as his background was in physics and the biology of viruses and bacteria, not genetics, *per se*. As he was striving to learn who was who in the genetics' community, he found in Nasser a colleague who already knew many of the researchers in that field. Kingsbury has said that he found both Nasser and Harriman to be "incredibly valuable to me."[85]

By 1981, Nasser was well established in the program and Harriman was offered a one-year leave to take one of the 15 Congressional Fellowships available to federal employees government-wide each year. He would be paid by NSF, but would intern in some Congressional office. It would be done under the Intergovernmental

[81] Ibid.

[82] Ibid.

[83] R. Scott Hawley, "In Memoriam: DeLill Nasser (1929–2000)," *Genetics* (2001):1389–1390.

[84] Harriman interview.

[85] Kingsbury interview. He also mentioned the fact that Nasser had been at the University of Washington, Kingsbury's doctoral *alma mater*, but they did not know each other there; rather, they met at one or more of the American Society for Microbiology meetings in later years. Kingsbury saw a great deal of Nasser after she returned to the San Francisco Bay Area and UCSF, however, he told the author.

Personnel Act, the IPA.[86] He was intrigued by the appropriations' process and also felt that taking the fellowship may boost his position at NSF, as he believed that such a post as his was partly a "grooming" process for higher leadership at the Foundation. Further, he felt that he would not have been offered the fellowship had it not seemed to NSF that he planned to "stick around for a while." On a training aspect of the Congressional Fellowship program, Harriman toured around the offices of both senators and representatives, most of whom were not, he discovered, interested in science. One who was, however, was Albert A. "Al" Gore, then chair of the House CST. That permanent Committee approved the budget of NSF, but it already had taken on two AAAS Fellows whose training schedule released them to look for Congressional offices a month before Harrison's fellowship program. For that reason, Harriman could not also be taken on.

Since Harriman wanted to be in a senatorial office, he initially leaned toward that of, most interestingly, Senator Proxmire![87] But Harriman's dream had been to serve with Senator Harrison Schmitt (R-NM), who had a BS from Cal Tech and a doctorate from Harvard in geology. Schmitt had been an astronaut and was the last human to walk on the moon. To Harriman's regret, though, his plan with Schmitt did not work out. Recognizing the potential for conflict of interest had he been in Proxmire's office, alternatively he chose instead to work in the office of the freshman Congressman from the district in Oklahoma that included the University of Oklahoma, Dave McCurdy (D). McCurdy was on the House CST under Gore's leadership. Ronald Reagan was coming into the presidency in 1981, along with many Republicans, when Harriman took up his fellowship with McCurdy, a conservative Democrat who eventually served seven terms (1981–1995).

Once ensconced in office space for McCurdy and his associates, Harriman handled all of the scientific matters, including writing position papers, dealing with lobbyists. All of the others in the office had law backgrounds. Harriman found it all "enjoyable [and] fascinating."[88] In the meantime, Nasser was handling Harriman's directorship of the genetic biology position to which the latter intended to return. As it happened, Betsy Clark's deputy assistant director of BBS left and Clark asked Harriman to take that position upon his return to NSF. Though he was not keen on taking that new position, he felt he owed NSF for having allowed him to take the fellowship the year before. In late 1981, then, Harriman became the deputy assistant director of BBS. He spent a year as deputy—it was a rotator post at that time—and enjoyed working with all of the scientists, including those in the social and behavioral areas, as well as the DEB members of the directorate. What Harriman found surprising, though, was that neither Clark nor anyone else at NSF ever asked him for a lengthy report about his time with Congress or, crucially, the CST specifically![89] The BBS front office was "very small" and Clark only sought "mechanical" assis-

[86] For more information on the IPA, see: http://www.opm.gov/programs/ipa/assignn.asp.

[87] Harriman interview.

[88] Ibid.

[89] Ibid.

tance from Harriman, even though he felt he could have been helpful in preparing budgets for the Congress' CST. Since Nasser had been running the genetics program and doing well, it became clear that Harriman would not go back to his former program director's post as her supervisor. What did come to pass, however, was that the genetics program was informally split (discussed elsewhere in this chapter) partially along eukaryotic and prokaryotic/virus lines; Nasser took the plants and animals, while Harriman took the prokaryotic bacterial and viral aspects, as well as the fungal (eukaryotic) features of the program. The split worked out "fine," he said, but it stayed formally the Genetic Biology Program: singular. Given their adjoining offices, Nasser and Harriman helped one another with panels.

In about 1990, James H. Brown took over the temporary leadership of the PCM division, of which the Genetic Biology Program was a part, recalled Harriman.[90] Jim Brown had earlier been deputy AD for BBS and had followed that by returning to his principal position as deputy director of the Behavioral and Neural Sciences program in 1979.[91] Nasser's chronicler, Scott Hawley, noted that Nasser was opposed to "bureaucrats and bureaucracy" and had told him (Hawley), when he was an NSF grant applicant in 1985, that she cared nothing about his budget that an NSF staffer had challenged, but only for the science that Hawley proposed.[92] He further emphasized her dedication to the science of genetics when she, along with Harriman, argued that genetics was an independent discipline and not just "a tool that was used by all biologists" when some "senior staff at NSF" had argued for that view. Nasser was crucial in the development of the Foundation's support of genetic studies of a weedy plant called the mouse ear cress (also known as the wall cress or, more commonly, thale cress) for genetic study. The small plant is an organism in the mustard family, Brassicaceae—*Arabidopsis thaliana*—and was destined to become the first, and still today most important, *model system* for genomic studies among the crucially important flowering plants (Chap. 7).[93] It was Mary Clutter who was the primary promoter for the use of *Arabidopsis*, however.[94] Nasser's enthusiasm for *Drosophila* genetics was as great as that for thale cress as she granted support to Seymour Benzer for his work in behavioral genetics in the fruit fly, which led to his becoming a founder of that field.[95]

Harriman later became acting director of PCM when Brown took a sabbatical at Michigan State University about 1991–1992, the period of the great reorganization (Chap. 5). Over his time at NSF, Harriman had close contact with the NIH, specifically the National Institute of General Medical Sciences (NIGMS), because it included basic and medical genetics. Indeed, Harriman had been delegated as liaison to that Institute by the NSF director. He was, too, the NSF representative on

[90] Ibid.

[91] NSB, "Minutes," NSB-79-211, June 25, 1979, 206:9.

[92] Hawley, "Memoriam," pg. 1389.

[93] Ibid.

[94] This according to the author's interview with Kingsbury.

[95] Ibid.

NIH's Recombinant DNA Advisory Committee. He sat in on their discussions of given grant applications and so became aware of who was being funded by NIH and which PIs were not; at least some of those could be funded by NSF, often as not.[96]

3.4 A New Decade: Expansion in the 1980s

As the decade of the 1970s came to a close, J. Wayne Campbell had been appointed as the new rotating director of PCM. Campbell was on leave from his position as Chairman of the Department of Biology at Rice University, Houston, Texas.[97] As well, molecular evolution and investigations into an early incarnation of what would become much later the Tree of Life program (later chapters) were glimmering on the horizon. The interdisciplinarity that led to biology of the little serving biology of the big blurs the line between reductionist features and macroscale features of biology: molecular evolution was spread over two directorates, PCM and DEB. The NSF prepared its FY80 budget and PCM remained the leader of the four BBS divisions: PCM was to be funded at the level of just over $67 M as compared to the next highest figure, that for DEB, at somewhat less than $38 M.[98]

The 1980s opened with two new situations that would strongly affect molecular biology, besides other areas. New and onerous regulations emanated from the Occupational Safety and Health Administration (OSHA), as it intended to issue revised regulations concerning toxic chemicals that would have a profound effect on teaching science and on doing research. As well, H.R. 4805, the "Research Modernization Act," was before the Subcommittee on Science, Research, and Technology of the House CST and that would curtail the use of animals in biomedical research.[99]

With regard to OSHA, Board member Alexander Rich proposed to the NSB at its February 1980 meeting that it establish an ad hoc committee to examine the OSHA plans as three issues were of significance: legal aspects; chemical aspects, that is, identifying the chemicals in question in teaching and research and their use; and, the question of carcinogenesis. Both the NAS/NRC and the NIH were looking into the matter at the same time as was the General Counsel of the NSF. As with other similar issues already discussed, it was decided that the NSB should join the NIH once again in the development of chemical use guidelines that the health agency was then drafting. The power of NIH as leader in so many areas of ethics and human subjects' research guidelines, for instance, was also to be seen with the new OSHA intentions. NSF, as the less powerful entity, once again sidled up to the "empire."[100] By

[96] Ibid.

[97] NSB, "Minutes," NSB-79-383, October 22, 1979, 209:6.

[98] NSB, "Minutes," NSB-80-58, February 25, 1980, Appendix A, 212:31.

[99] Ibid., NSB-80-119, March 24, 1980, 213:17.

[100] Ibid., 213:18–19.

the March meeting of the Board, three external experts had been invited to attend and offer advice, not in order "to seek an exemption" from the OSHA regulations, but rather to seek a:

> rational examination of those regulations as to the degree of hazard [of the pertinent chemicals] in laboratories as opposed to the general working place and that appropriate actions be taken accordingly.[101]

The NRC's draft comments from the Committee on Hazardous Substances in the Laboratory were then set to be released. Neither NSF nor the Board would get back to the OSHA matter for some time as much of the year of 1980 was taken up by several issues: the change in directors from Atkinson to Langenberg to Slaughter; a change in Board chairs from Hackerman to Branscomb; the Mount St. Helen's volcanic eruption to which much of NSF reacted quickly on the research front dispensing funding; and, the discussion of a Foundation-wide reorganization that did not materialize. However, the new directorate of engineering *did* come to pass and BBS took over several programs such as decision and management science and applied social sciences, as we have seen.

Nonetheless, John Slaughter was able to state, in 1980, that the growing importance of biology of the little for the development of biology of the big was very clear:

> The remarkable discoveries in molecular biology have *profoundly* influenced other areas of biological endeavor, such as plant science, *ecology*, and physiology, and have laid the groundwork for new insights into cells, tissues, whole organisms and their behavior, populations, communities and *larger systems*.[102]

Much of the 1950s and 1960s saw the dissection of the cell into its component parts at the subcellular and molecular levels when all the major families of macromolecules were finally defined (the nucleic acids being the last after the already well-known carbohydrates, lipids, and proteins—though the latter were still somewhat poorly known): new fields of little biology became the areas to study.

> The early 1970s marked a transition to a reconstructive phase in molecular biology. Questions were beginning to be asked how subcellular and molecular components function and interact within the heterogeneous milieu of the cell itself…These questions were concerned with the molecular architecture.[103]

Along with those technologies mentioned earlier, additional advances in visualization tools provided for video recording in real-time of cellular activities, in connection with computer graphics generation and microprocessor control of "highly sophisticated instruments as adjunct techniques"—much of which were central to PCM research areas. The number of grants in PCM versus DEB also climbed by 1981: 22 pages versus 12 in the *Grants and Awards for Fiscal Year 1981*, making the ratio higher than the comparable data for 1977 (above). The growth in biotechnol-

[101] Ibid., NSB-80-144, April 21, 1980, 214:8–9.

[102] NSF, "Annual Report 1980," *op. cit.*, pg. 49. Emphasis added.

[103] Ibid., pg. 50.

ogy, *per se*, was referred to as "explosive" by 1982.[104] And given the types of research supported, there was little question of the central significance of that line of research to biology overall. *Molecular biology was becoming big biology*, financially.

The increasing interest in those areas was being reflected in fiscal decisions, as well. Between 1963 and the then-proposed fiscal year 1981 budget, obligations for the life sciences had increased fivefold, whereas the once "queen of the sciences," physics, and the physical sciences more generally, increased only threefold. That biology would claim supremacy during the last quarter of the previous millennium and the beginning of the next was growing ever more obvious with every passing year. There was, though, a cloud over that otherwise positive statement as the NSB pointed out that:

> [s]tudies indicate that between 10 and 30% of a scientist's typical workweek may be accounted for by administrative requirements imposed due to institutional or governmental needs.[105]

Such a situation would haunt researchers well into the present century: with funding comes fiscal accountability and the close attention of the PI.

Board member Alexander Rich at MIT—a very significant figure in the history of DNA research—and David Stollar at Tufts University, Massachusetts, supported by PCM, were both looking into how cells control in which regions of the body their DNA is expressed in various tissues. They worked with what had been termed the Z form of DNA and discovered that it "may play a regulatory role in gene expression," as methylation of the Z form led to gene inhibition.[106] The methylation line of research went on to be of profound importance to twenty-first century biology. During the same period, other researchers first caused a foreign gene to be expressed in a mammal (the venerable laboratory rat as *model system*), a major leap from the earlier work done in the much less complex bacterial DNA.

In their long-range planning, the NSB identified, in early 1982, a number of "high priority initiatives." One of those was biotechnology. (Two others were noted: plant sciences, about which much more will be said; and, the neurosciences, which has been discussed in the previous chapter.) Several months prior to that, an Office of Interdisciplinary Research (OIR) had been created and placed in the new Directorate of Engineering. Along with the extant interdirectorate Interdisciplinary Research Committee, the OIR had been charged with the identification, stimulation, and coordination of work "in areas such as robotics, catalysis, resources availability, and natural hazards."[107] Biotechnology, *qua* industry, also found a home there because "as in the robotics area the [OIR] has also formed a team of biologists,

---

[104] NSF, "Thirty-Second Annual Report FY82," *op. cit.*, pg. 27.

[105] National Science Board, *Science Indicators 1980* (Washington, D.C.: National Science Foundation, 1981), pg. 72.

[106] Ibid., pg. 28.

[107] Belanger, "Enabling American Innovation," pg. 150. See her Chap. 5, endnote 55, pg. 312. The OIR was proposed by Slaughter in January 1981 and was established by late that November.

material scientists, and engineers [to] coordinate and encourage" industrial-level research and interaction into rDNA between industry and the NSF. The role for rDNA was "to increase the knowledge base," just as it was for robotics, said the OIR.[108]

At that juncture, the NSB had a series of "mini-groups" that were charged with specific duties in regard to long-range planning work. The group led by Board member John R. Hogness was responsible for the BBS. Three other Board members were on his mini-group. Hogness reported that, in working with BBS, they had "experience[d] significant difficulty" due to the "diverse and diffuse nature" of the directorate. That was a notable comment and one that may well have lingered in the collective mind of the Board over a considerable period leading up to the 1991–1992 reorganization that led to a much less "diverse and diffuse" pair of new directorates: SBE and BIO. Such was still well into the future, but having placed social and biological sciences together in the move from BMS to BBS had its challenges, even for Board members. The Hogness mini-group focused on all aspects of the science done in BBS, not just biotechnology. One thing the group discovered about BBS was that it was:

> unique not only in the limitation of the percentage of grants submitted that are funded, but also in terms of the percentage of dollars awarded as compared to those requested in the successful grant. For example, in FY 1981 the directorate received 6106 proposals of which BBS was able to fund only 34 percent; of that 34 percent, BBS was able to award only 58 percent of the dollars requested.[109]

In years to come, a 34% success rate would have been considered a high one by all involved. But in 1981, there was clearly a strong imbalance present and the mini-group saw it as a "critical problem" requiring solution. Such could come via "(1) a greater percentage of the *total* NSF effort be put in BBS and/or (2) that the BBS Directorate establish priorities more so than it has in the past."[110] Both were suggested and the second notion needed to be undertaken whether or not the first notion was also embraced, Hogness and his colleagues argued. In regard to establishing priorities, the mini-group said that BBS should:

> examine factors such as the role of NSF in supporting the area relative to other sources [e.g., NIH, USDA, etc.], the ripeness of the discipline for scientific advance, and the minimum core necessary to support a viable research effort.

It is a little more than likely that BBS had already been following such a set of considerations long before that but, if not, the Hogness guidance was clearly highly valuable. Finally, the mini-group made a "strong recommendation" that a:

> directorate-wide advisory group be established to assist in comprehensively examining priorities for BBS from a directorate vantage point, rather than separately at the division level.[111]

---

[108] NSB, "Minutes," NSB-83-92, March 22, 1983, 2–83:6 (new pagination style).

[109] Ibid., 2–83:10–11.

[110] Ibid., 2–83:11. Emphasis added.

[111] Ibid.

Such a group "should consist of 12 individuals, six of whom would examine primarily" biology and the other six the social and behavioral sciences "with the understanding that there would be a meeting of the minds between the two." Though it would not be formed in quite this manner, this was the ***first effort*** at establishing an *Advisory Committee*, *AC*, for BBS. The first, but only meeting of the Advisory Committee as then constituted, came quickly; July 28–30, 1982.[112] But that first group was hastily put together and not created in a more formal manner. That formalization would come soon, however.

The mid-1980s was a hectic period for BBS (and the NSF overall) in that Betsy Clark had just left the position of assistant director and Robert Rabin, her deputy, had been functioning as acting AD. She had been at the Foundation since 1969 and the AD since 1975. Rabin had been delegated by the director of the NSF, Edward Knapp, to put together a more permanent AC than that which had met only that single time in 1982: Rabin was seeking scientists to fill out the Committee. Further, a reorganization of BBS was in consideration, but it was dropped.[113] One of the researchers that Rabin wrote to was Lotfi Askar-Zadeh, a computer scientist at UC, Berkeley and known for his work in soft logic or fuzzy set analyses.[114] Rabin sent him a letter of invitation in December of 1983 asking him to join the planned new AC. He explained to Askar-Zadeh that he (Rabin) could do a much better job as acting AD were he to have an advisory group to aid him, but there was more to it than that.[115] Rabin noted that Betsy Clark could have put together the new AC but, knowing she would be leaving to take her new position in graduate affairs at Bowling Green State University, she preferred to let the new, permanent AD take on that duty, as has been stated earlier.

However, two reasons led to Acting AD Rabin's efforts to have a committee in place as the BBS awaited President Reagan's appointment of Clark's successor. First, the new AD appointment process was moving at a "glacial pace," said Rabin and, second, he was getting pressure from Hogness' mini-group, which made it unanimously clear that an advisory group should be "appointed without further delay."[116] Rabin pointed out to Askar-Zadeh that:

> [i]ndeed, a number of NSB members have long desired the Assistant Director to go beyond our regular panel system for peer review, and receive advice from a formally constituted group that would assist BBS in its organizational, planning and budgeting issues.

That was at the heart of the genesis for the permanent BBS (continuing into BIO) AC. Recognizing the "diverse and diffuse" nature of the polyglot BBS, "given the 25 or so disciplines in the directorate," Rabin and the five division directors agreed that, "no body of counselors can be effective or efficient if it is constituted to

---

[112] Ibid., 2–83:8.

[113] NSB, "Minutes," NSB-82-103, March 22, 1982, 233:9–10.

[114] See: http://www.cs.berkeley.edu/~zadeh/. See also: http://azer.com/aiweb/categories/magazine/24_folder/24_articles/24_zadeh.html.

[115] Robert Rabin to L. A. Zadeh, December 1, 1983, BIO Lateral Files.

[116] Ibid.

narrowly represent disciplines." Hence Askar-Zadeh, it was hoped, would not be expected to "define small angles and arcs of the budgetary pie."[117] So Rabin conveyed to him (and other invited scientists, it must be assumed as no evidence could be found) that what the AC was to be was a group that should attempt to:

> understand how the issues or questions within the domains of the social, behavioral, cognitive and information sciences *are linked* to life's individual and collective structures and functions, and the influence of their environs.[118]

It was a boldly worded goal and one that certainly tried to make interdisciplinary that group of highly "diverse and diffuse" intellectual areas. Rabin sought "conceptual breadth" for the "competitive dynamics" operating at the Foundation with regard to "status and resources involving all the other research directorates at NSF."[119] *He and his colleagues wanted, in sum, a "small group of broad thinkers."* That goal for the membership and purpose of the BBS AC obtains to this day. As Askar-Zadeh had been of aid to NSF in the past, Rabin had him on a list of persons to whom he was offering the invitation to join the new AC for its twice-annual meetings.

That first effort to form a directorate-wide AC may have come at a good time to fill gaps that seemed to be arising. As we have seen, the NSF budget had such shortfalls in the climate of the early period of Reaganomics that the program director of the Regulatory Biology Program of the PCM, for instance, was warned that its scheduled panel meeting for May 1983 might have to be canceled.[120] That warning memo came earlier in the year and Rabin was inviting several scientists, presumably, at the end of 1983 to join the AC. One must wonder whether the AC came about *only* due to the reasons given by Rabin in his letter to them, or also due to another motive altogether. Under the new stringency of funding brought on by Reagan and OMB Director Stockman, a single guiding committee would be *economically* a much more sound approach than a great many disciplinary panels that, further, *might* well be eliminated in the interest of downsizing the federal budget. While that did not happen, such a fear could have been an additional boost to establishing the *broad and singular* BBS Advisory Committee. Even though no direct historiographic evidence yet discovered appears to exist for this assumption, the Rabin letter and other documents of the period suggest it. In fact, being upset by the hint that the Regulatory Biology Program panel might not meet led two of its members (Peter K.T. Pang of Texas Tech and Colin G. Scanes of Rutgers) to send a letter to Director Knapp warning him of the damage to the peer review process at NSF that might eventuate by canceling such panel meetings. As they said,

---

[117] Ibid.

[118] Ibid. Emphasis added.

[119] Ibid., pg. 2.

[120] Bruce Umminger (Director of the Regulatory Biology Program) to Mary Clutter (Section Head for CPB), February 1, 1983, *op. cit.*

> all eight of us panel members agreed to serve on the May panel *at our own expense*. This gesture is a response to an unwelcome emergency situation and should not be interpreted as a precedent for the future.[121]

That problem was solved for the moment, at least.

Sometime later, NSB Chair Branscomb wrote a priorities' statement for the Board and chose some 14 areas as leaders in that list. It included "plant sciences and agriculture"—an early indication of growing interest in plant science. Curiously, it did not mention biotechnology/molecular biology, which had been much on the minds of both the NSB's and the NSF's leadership for several years. Also, as Slaughter had then recently said, biotechnology was having a "profound" effect on many areas, plant science included.[122] The plant sciences were on the rise at BBS, even though they would be seen as even more notable several years into the future when botanically oriented Mary Clutter became the AD after Clark's replacement, David Kingsbury, left. He had spent only 4 years in the position. It was the feeling of an unnamed BBS author, or a BBS-sanctioned writer, that little had been done in the plant sciences with regard to the *molecular* end of things.

> Until recently, relatively few plant biologists were exploiting new developments in molecular biology, and few molecular biologists had experience on plants. [There was a shortage of students in these areas and new] researchers were needed to apply techniques such as recombinant DNA to the plant sciences.[123]

While more will be said later about the plant sciences, *per se*, their connection to molecular biology would be greater in future years because NSF had begun a program of Postdoctoral Research Fellowships in Plant Biology. But, when the history of the molecular features of the plant sciences are examined, the comment about educational shortages in the quotation will be seen not to obtain. Still, the Foundation received nearly 200 applications for the plant biology fellowships in 1983 with 24 being awarded, including 14 women. Women and minority numbers in science at the time were low and both the Board and the Foundation were making all efforts to change that status quo. The point of the fellowships was to "cross-train" new doctoral graduates in plant science if they came from another background. Part of those postdocs would be in the molecular area.

> Through the development of such new technology as recombinant DNA it is now possible to identify, isolate, clone, and manipulate a wide array of genes…[which could allow for the creation of] hybrid plants…During the past 30 years, much experimental research with plant systems lacked the sophistication applied to microorganisms and animal systems. One reason for this is that tough cell walls and large vacuoles, or cavities, make plant material

---

[121] Peter Pang and Colin Scanes to Edward Knapp, January 28, 1983, cc to Clark, Clutter, and Umminger (see footnote just above) and to James H. Brown, Division Director of PCM; see also main text earlier above concerning Brown. Emphasis added.

[122] See Branscomb's "Topics for NSF/NSB Consideration" in the "Minutes," NSB-82-243, August 23, 1982, 236:17ff.

[123] National Science Foundation, *Thirty-Third Annual Report for Fiscal Year 1983* (Washington, D.C.: Government Printing Office, 1984), pg. 27.

more difficult to manipulate. These and other obstacles diminished with the use of new techniques. For example, methods to fuse and grow protoplasts.[124]

For the first time during the period of interest to this history, however, the total pagination devoted to grants and awards in BBS's annual listing was the same for both PCM and DEB. "Biology of the big" was on the rise at NSF, as will be discussed in both the following chapter and in Chap. 9. The expansion in plant biology, which had both molecular and organismal-and-above orientations, was remarked by the NSB mini-group in early 1983 prior to Clark's departure.

The BBS mini-group, previously chaired by John Hogness, now came under the leadership of Walter E. Massey, a Board member who would later become director of the NSF. Massey's close ties to BBS would be of value to the directorate when he assumed the Foundation's top position during and after the reorganization of 1991–1992. The group's comments about the fact that plant science was spread over several divisions of BBS indicate that the Board had some concerns. This led the Massey mini-group to say that they "plan to explore further possibilities for coordination of plant biology research at a future meeting."[125]

They had other concerns, too. One of those was that BBS consistently, among all the directorates, awarded funds at a lower level than their confreres (above): only about 30% of quality applicants were supported. This meant that, "BBS is faced with either increasing the success rate or adequately supporting existing grants known to be of high quality."[126] Also, BBS was concerned, the Board knew, about "how new research opportunities can be identified so that especially promising ideas are not passed over in favor of existing projects." That was one of the drawbacks of big science in that more and more funds were directed toward large projects that had lives lasting for several years: a major fiscal commitment. For the first concern, BBS was "considering targeting research areas that it believes are of special importance with the idea of initiating programs to support those areas at a few chosen institutions"—the Engineering Research Centers (ERC) and the Science and Technology Centers (STC) programs (below). For the second concern, it was hoped that the new AC to be created would assist BBS in how it could make intelligent trade-offs to best spread its funds wisely.[127] The concept of the AC would be of very considerable value, in fact, and would last until the time of this writing and likely beyond.

124 Ibid., pg. 28.

125 NSB, "Minutes," NSB-83-58, February 22, 1983, 1–83:16.

126 Ibid., 1–86:17.

127 Ibid.

3.5 Biotechnology Booms

The "Directorate Priorities, 1985–1989" for BBS were before the NSB in March of 1983 and they emphasized many areas, two of which are notable here: industry connections; and, "Genetic Approaches and Techniques…the biology of cells, the reaction of DNA structure to gene expression, and genetic interactions during the development of plant and animal systems."[128] Both of these had relevance to biotechnology.

The following year of 1984 would prove to be a major one for the early history of biotechnology at NSF, besides much else: it was one of the two areas of strongest interest to the Board that year. In fact, that interest was suggested, said Walter Massey, by Board and mini-group member, Charles E. Hess.[129] (Hess was Dean of the College of Agriculture and Environmental Sciences at UC, Davis from 1975 to 1989.) Chairman Branscomb made the Board's interest in biotechnology clear in a letter that January to George A. Keyworth II, director of OSTP, when he told the latter that each January the Board sought study projects to be done in two arenas, biotechnology being the second of the two for 1984, and which would lead to a major report forthcoming at the June meeting of the NSB.[130] (The other project was infrastructure for support of research in universities, which had been "suggested by the OSTP and other sources."[131])

Massey was chair of one of the two standing committees of the NSB, the powerful Planning and Policy Committee, as well as the BBS mini-group. That was one reason that the Board took up biotechnology as one of its major projects for the year. Interestingly, the NSB did not yet have a definition for the term biotechnology, though that may not have been surprising as the discipline had a number of definitions.[132]

---

[128] NSB, "Minutes," NSB-83-92, March 22, 1983, 2–83:20–22. Chaps. 2 and 3, herein.

[129] Ibid., NSB-84-71, February 16–17, 1984, 2–84:9.

[130] Lewis M. Branscomb to George A. Keyworth II, January 20, 1984. Reproduced in the NSB "Minutes," NSB/EC-84-2, Executive Committee, January 19, 1984, EC:84–1:12 (yet another style of pagination; used for the Executive Committee).

[131] NSB, "Minutes," NSB-84-71, 2–84:9 (see also note further above concerning "new pagination style").

[132] Robert Budd (a Briton writing in Britain) has taken up that matter, along with much else, in his "Biotechnology in the Twentieth Century," *Social Studies of Science* (1991):415–457. See also his book, *The Uses of Life* (Cambridge: University Press, 1993). The definition has "important ramifications" as the U.S. Office of Technology Assessment (OTA) pointed out that "[t]he terms used to describe biotechnology can affect research funding and the regulatory treatment of potential commercial products." Cited in Budd article, pg. 416. At the time of interest to this history, 1984, the OTA spoke in terms of both an "old" and a "new" use of the term. The old was more in keeping with the European definition that included classical fermentation technology as part of how they viewed the subject. The "new" was more directly tied to genetic engineering (sensu Boyer and Cohen). In Budd's paper, he follows the tortuous path of arguments as to the origins and proper use of the word.

Beside that rather fundamental need, Hess led a Board discussion on biotechnology matters that posed five significant questions. The first of those was which NSF directorate should take the lead in the area. While much had been done via BBS for a number of years, the engineering directorate had both a clear interest and involvement. The second question was whether centers should be developed. The third question considered the mix of basic and applied research: what percentage of each? The fourth asked about interdisciplinarity and human capital needs, with the final question concerning training. The Board discussed several of the issues and, not surprisingly, some members felt the need to focus more strongly on basic research—NSF's long-time principal charge—in this burgeoning field, as others (NSF's own engineering directorate might be in such a list) could take up the more technological and applied aspects. Much of the dialog focused upon the need for advanced chemically oriented instrumentation and, given that "infrastructure" was the second of the two major topics under discussion by the Board that year, they agreed unanimously that those two areas for consideration in 1984 would go forward.[133] Shortly thereafter, Hess was named chair of a three-person committee (including also Board members Mary Jane Osborn and Roland Schmitt, who would replace NSB Chair Branscomb in May) by Massey. Massey himself would not play a role in the work of the new Biotechnology Task Group. Four others were named to an Infrastructure Task Group that would, among other matters, look into supporting large, expensive, chemical analyses-oriented instruments that, in part, would be used in biotechnology research and education.[134]

The strong upswing in the Board's direct interest in biotechnology and the work of the task groups, along with the transition from Branscomb to Schmitt as NSB Chair, led to the Board adopting a "Statement on Research Related to Biotechnology"—the earlier projected report—when it debuted at the Board's June, 1984 meeting.[135] Occasionally, the NSB met away from NSF headquarters and that meeting had been held at the Rand Corporation in Washington, D.C.[136] Robert Rabin was at the time acting as Executive Secretary for the Board prior to returning to his regular deputy AD duties at BBS, and so was close to NSB and knew its interest in biotechnology. This knowledge he could take back to BBS with him.

The "Statement" noted that the "biotechnology revolution" had already made major contributions to agriculture, environmental quality, chemical and pharmaceutical production, and in the food and energy industries.[137] The Board authors pointed

[133] NSB, "Minutes," NSB-84-71, 2–84:10–11.

[134] Ibid., NSB-84-103, March 15–16, 1984, 3–84:2.

[135] Branscomb had come from the IBM Corporation and had left the NSB Chairmanship just before Erich Bloch became director of the Foundation. Branscomb, however, became an informal advisor to Bloch, himself a product of the IBM corporate milieu: of course, both apparently thought in similar veins. See: Erich Bloch Interview with Marc Rothenberg, Historian of the NSF, February 11, 2008.

[136] NSB, "Minutes," NSB-84-193, August 20, 1984, 6–84:2.

[137] Ibid., Appendix A, that is, NSB-84-189, 6–84:5–7. Main text comment above at 6–84:5.

out that biotechnology "promise[d] solutions to long-standing health problems." Further, they stated that the biotechnology revolution was being paralleled by the "information revolution." That discoveries in biotechnology were dependent upon advances in the biological sciences was clear; and so was the reverse. Such dual growth would lead to "new commercial applications." American leadership globally must continue and "there is widespread agreement" that both more funding and new directions in education and training would be required to maintain that leadership. For all those reasons, and more, the NSB charged the director and the staff at NSF to "develop an expanded program focusing on research and education related to biotechnology."[138] The director should be cognizant of what other agencies were doing, but NSF should focus on both basic research and education. The expanded effort should include, but not be restricted to:

- increase[d] support for areas, basic and applied, that may [otherwise] impede progress in biotechnology;
- establish[ment of] mechanisms to increase interdisciplinarity and which mechanisms may lead to multidisciplinary centers;
- establish[ment of] mechanisms to aid in education and training [e.g., postdoctoral fellowships: NIH would be the first to move on that area] of biotechnologists; and,
- establish[ment of] mechanisms to address current problems in instrumentation [equipping both new and older laboratories at universities, etc.].[139]

The last point was significant, as the NSB desired that NSF provide for "equipment items…*on standard grants*," a practice long needed by BBS and other directorates. The Statement went on to list areas of "Selected Examples of Research Opportunities," the last two words in that title being common coinage at NSF over time as it interacted with the scientific communities: NSF was ever alert to new *research opportunities*. Such opportunities included, in the biotechnology arena, bioprocessing (bioreactors, production), cellular manipulation (creation of hybrids, for instance), cloning and other techniques (gene expression, etc.), macromolecular areas (new molecule design, synthesis, enzyme modeling), and preservation of genetic diversity (germplasm collection and storage). The last of those would have implications for the big biology to come and will be considered again.

NSF was not the only agency interested in the growth of biotechnology in 1984 as several other major documents came out on the subject: they, too, would play roles in its future direction and development. One of those was a massive, 600-page report from the Office of Technology Assessment (OTA; an arm of the U.S. Congress in existence from 1972 to 1995). The report provided analyses of complex scientific information to members of Congress, and others, but that were also understandable to other non-specialists. OTA's director at the time was John H. Gibbons, an experimental physicist, and of whom more later.[140] The OTA's "Commercial Biotechnology:

[138] Ibid., 6–84:6.

[139] Ibid., 6–84:6–7. Emphasis added.

[140] Gibbons, who had in 1973 served as the first director of the Federal Office of Energy Conservation (at the time of the Middle Eastern "oil embargo" to the United States), would later serve as the director of the OSTP during a portion of the Clinton administrations. See: http://www.johnhgibbons.org/.

An International Analysis," would be read by those in BBS and many other venues where the data and conclusions were of special concern.[141] The report had been requested by the House CST and a Senate committee and had considered, primarily, the economics and international business environment for biotechnology at the time.

The year 1984 was notable in yet another connection: the organization chart for BBS saw significant "changes [which reflected] in part the growth in research on recombinant DNA and the importance of the biotechnology revolution."[142] That is, the appearance of *two new divisions* emerging from one: PCM had its ten programs spread into those new offices (Fig. 3.3; the Information Science and Technology [IST] division not shown). The first of those was the Division of Molecular Biology (DMB), and the second was the Division of Cellular Biology (DCB); both of which have been alluded to earlier. The Genetic Biology Program was split into DMB (viruses and prokaryotes, plus fungi) and DCB (eukaryotes, excluding fungi). The DMB would include programs in Alternative Biological Resources, Biological Instrumentation, Biophysics, Biochemistry, Metabolic Biology, and Prokaryotic Genetics. The DCB would be composed of programs including Cell Biology, Cellular Physiology, Developmental Biology, Eukaryotic Genetics, and Regulatory Biology.

The rationale underlying all of those changes from the PCM split was due to reasons of both *staffing and budgets*. PCM was established as a division in 1975 when BBS was formed. From that year to 1983, staff had grown from 24 to 50, the number of programs from 7 to 10, awards from 1006 to 1491, budget from nearly \$50 M to over \$87 M, and the number of proposals that the single PCM director had to deal with went from 2200 to 4230. In all those categories, the growth leap averaged 70–110% per category. The budget change from 1975 to FY84 was 250%. PCM was not only the largest division in BBS, but also among the three largest in all the NSF, in budget, and the largest in terms of proposals. That was certainly evidence that biology had overtaken physics as the "queen of sciences," at least in financing and numbers of practitioners. While the workload in PCM was 14% of that of the entire Foundation, the division had only 4% of the total staff.[143] For FY85, the estimate for the budget for DMB was well over \$63 M and that for DCB was just under \$53 M: clearly, the littlest of little biology was at the center of things.

[141] *Commercial Biotechnology: An International Analysis* (Washington, D.C.: Congressional Office of Technology Assessment, OTA-BA-218, January, 1984). We know that the report had some impact on BBS as reference to it was made in an overhead slide for a presentation by someone in the directorate in or after 1993. Ephemeral sheet titled "Biotechnology Timeline"; no other provenance data known.

[142] National Science Foundation, *Thirty-Fourth Annual Report for Fiscal Year 1984* (Washington, D.C.: Government Printing Office, 1985), pg. 19.

[143] Memorandum to Members of the [NSB] from Edward A. Knapp, Director, NSF, June 7, 1984; cover memo of one page plus five pages of background information and tables.

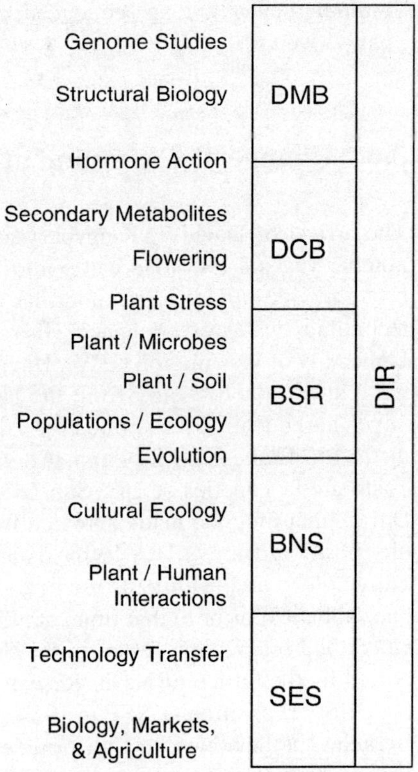

Fig. 3.3 Organization of the BBS as of 1985 (see main text for definition of terms used). (Source: NSF, *Thirty-Fourth Annual Report for Fiscal Year 1984*)

In that sort of situation, though, it can be seen that it was not just social and behavioral sciences that were given short shrift at the Foundation, but all of BBS. So long had the NSF been driven by the physical sciences and by a string of physicist directors that BBS was, in a very real sense, still looked down upon. For instance, even with all the growth in PCM, besides other programs in the directorate, funding did not represent the level of activity to which BBS had risen as a percentage of the total NSF budget. Further, BBS by that year had grown to 6 divisions and 37 programs overall. The techniques of molecular biology had become so important that they were "the driving forces behind research in *all* areas of biology" and beyond that they had "helped to *relax the once rigid boundaries* that defined the life sciences," thus bringing "new opportunities for scientists."[144] Coordinate with that was the importance of instrumentation: some \$10 M had been spent in this area in FY83, but had leapt to \$15 M by FY84—the desires of the Board to focus more on biotechnology were clearly to be seen in those remarkable numbers, but the funding share for BBS did not reflect the workload reality. Furthermore, discoveries in maize genetics with key elements having been cloned, mobile DNA (McClintock's

[144] NSF, "Annual Report 1984," pg. 20. Emphasis added.

"jumping genes") being sequenced, and the rise of *Arabidopsis*, were all seen that year, as well.[145]

3.6 Change of Command at BBS: I

The arrival of David T. Kingsbury as the new AD for BBS also occurred in that notable year of 1984. He came into the directorate at a clearly exciting time and brought with him a background that was particularly apt for that period when biotechnology was growing apace. His undergraduate work was in microbiology at the University of Washington (UW). Having just finished his masters at UW in 1964, he was called to active duty from the Navy reserves and was to be sent to Vietnam. Given his education, he applied for a direct commission, received it, and so was sent instead to Bethesda, Maryland. After his active duty in the Navy, he finished his doctorate in genetics at UC, San Diego and then did a postdoctoral period there. During that time, he made a switch from genetics to animal virology. His first faculty position came at UC, Irvine as an assistant professor in the microbiology and immunology area within the medical school. He stayed there until 1980, having had one sabbatical prior to that time, at NIH. He moved to UC, Berkeley that same year where he became a professor in the School of Public Health and did research supported by the Office of Naval Research and NIH. He remained there until he took the BBS AD position at NSF in 1984.[146] However, since he had graduate students at Berkeley, he flew back to California every other week. He kept up that travel for a year and a half and was able to use his Navy grant money, by the Navy's direction, to do so.[147] He came to the Foundation with no previous experience with it in any way, but with a good understanding of the NIH and other funding agencies.

The institute at Berkeley that Kingsbury was directing was located in an old building dating to World War II and was not in the best condition with regard to infrastructure; particularly old-fashioned was the telephone system. In October of 1983, he received a telephone call there while working in his laboratory.

> And over the central intercom, our switchboard guy, you know, comes out with 'Dr. Kingsbury, you've got a long distance phone call. It's an international…call. Please dial the switchboard.' So I, you know, get my phone and I dial the switchboard. And he said well, you've got a phone call from a doctor in Switzerland…and the person on the other end said, 'Oh, I am glad I got you. This is Ed Knapp. I'm the director of the National Science Foundation and I'm over here at CERN [European Organization for Nuclear Research] doing some experiments and I just wanted to get to you before my secretary called to set up your appointment to come and visit the Foundation.'

[145] Ibid. The genus *Arabidopsis* was so new on the scene, it must be assumed, that the authors for the PCM/BBS section repeatedly misspelled the taxon as "Arabodopsis." The error also could have been due to a typist.

[146] Kingsbury interview.

[147] Ibid.

Kingsbury was taken aback:

> My response was, 'What are you talking about?' [Knapp] said, 'Oh, I guess I'd better tell you, you know, your name was put into the White House to be the head of the biological and behavioral sciences at NSF. Are you interested in that job?' [laughter from Kingsbury] And I had never even thought about it…And he [Knapp] said, 'Oh, well I can understand that.' [Kingsbury replied,] 'I guess I ought to think about this.'[148]

Kingsbury's name had been put into consideration by then Secretary of the Navy, John Lehman. Shortly after Knapp's surprise call, his secretary did telephone Kingsbury and arranged with him to come to the Foundation for his interview in late October 1983. The interview process included one at the White House also, as ADs were presidential appointees. There, George ("Jay") Keyworth, presidential science advisor to then President Reagan, interviewed Kingsbury. Kingsbury met with many others that day, but he would also have to speak with the then head of White House personnel, John Harrington, who was unavailable. Due to that situation, Kingsbury then had to return east to Washington several weeks later to meet with Harrington for a "strict 15 minutes, no more." While Kingsbury and Knapp were having lunch together during that second trip, Harrington called Kingsbury from his office in the West Wing of the White House. They met there and part way through the interview, Harrington:

> stops…and he just looks at me [Kingsbury recalled], and he said, 'Dr. Kingsbury, I just want you to know, I am really uncomfortable with Berkeley professors who didn't vote for Ronald Reagan.' [Kingsbury laughs] I said, 'I'm sorry, what can I tell you.' And he said, 'I don't want you to tell me anything. I just want you to know.' So [the interview] ended up; 15 minutes turned into an hour and a half.[149]

During this author's interview with Kingsbury, from which the foregoing quotations were taken, I asked him how Harrington knew that Kingsbury did not vote for Reagan, and Kingsbury replied:

> He assumed I was a typical professor from a university in California. Of course I didn't vote for Ronald Reagan…I mean, it was an assumption [that I didn't vote for him, but] it was a correct one…they knew I was a [registered] Democrat [as they had done a full security check on me].

Kingsbury received a leave of absence from Berkeley for 2 years and a leave from his ONR-supported research from the Navy, though he apparently did not take that in full as his trips back and forth to California suggest. Upon his arrival at NSF, Clark had already left as BBS AD and had departed the capital for Ohio; Rabin was serving as acting AD. The two men worked together for a period of time, Rabin getting Kingsbury familiarized with the BBS. That was in February of 1984 and Kingsbury had yet to be confirmed by the Senate, so he had a flurry of visits to make with members of Congress and others. He recalled that the OMB had "devastated" the social and behavioral sciences (Chap. 2) as OMB "had gone overboard" and

[148] Ibid.

[149] Ibid.

"even OSTP knew that."[150] Indeed, "Jay [Keyworth] asked me [if I was] 'prepared to address the social sciences problem.'" The House interviewers, too, had a "keen interest" in that topic. Finally, in June 1984, Kingsbury was sworn in as the next BBS AD.

Though Rabin and Kingsbury had a "good relationship," the latter had chosen another person to fill the post of deputy AD and so Kingsbury had to look for a "good place" for Rabin. Soon thereafter, Rabin moved from the Foundation to the White House taking a position at the OSTP and remained there until he retired from federal service. The new deputy AD was to be Alan Leshner, who was then in the BBS's behavioral sciences division.[151]

Besides the focus on biotechnology at BBS, the then newly extant and now more formal AC had met in January of 1984. The chair of the AC was botanist Peter Raven. Raven was appointed chair by Kingsbury.[152] Mary Jane Osborn, a Board member, attended the new AC at that inaugural meeting.[153] The AC was to meet twice a year, according to its charter, though some years saw more meetings than that. Director Knapp was the guest speaker. He provided his expectations for the AC in saying that he hoped the group would aid in long-range planning and the ordering of priorities for BBS, in particular, among several other charges.[154] Board member Osborn noted that it was important that the directorate not be simply reactive, but proactive with the appropriate communities. It is not clear that BBS was *not* proactive; evidence suggests quite the contrary. Osborn may have merely been giving general advice to the inaugural members of the AC. She also brought up the increasing importance of plant science besides her main points of the dual revolutions ongoing in information/computerization and biotechnology. Other speakers included committee members: L. Askar-Zadeh, on information sciences; David H. Cohen, on neurobiology; Lucille Shapiro, on the genetics of bacteriophages and other viruses; Kenneth Prewitt and Otto Larsen, on the social sciences; Lawrence Bogorad, on plant biology; and, Thomas Kuhn, on history and philosophy of

---

[150] Ibid.

[151] Leshner has had an illustrious career having been a professor of psychology at Bucknell University, Lewisburg, Pennsylvania, for a decade before moving to the NSF in 1984. He later moved on to become deputy director (then acting director) of the NIH's National Institute of Mental Health (NIMH), then as director of the National Institute of Drug Abuse (NIDA) from 1994 to 2001. In December of that latter year, he became the chief executive officer of the AAAS and executive publisher of the journal *Science*, the position he holds at the time of this writing. In 2004, President George Bush, *fil*, appointed Leshner to the NSB.

[152] Personal communication, David Kingsbury to the author, email of June 9, 2010. AC Chairs were not elected by the AC membership, but appointed by the AD. Kingsbury said that "[g]iven Peter's [Raven] reputation for being a good leader and well organized[,] it seemed clear to me (with consensus from Bob [Rabin] and Otto [Larsen]) that he [Raven] would be the right Chair at this critical point in the process."

[153] She was later professor and chair of microbiology at the University of Connecticut Health Center. For more on Osborn, see: http://grad.uchc.edu/faculty/bios/osborn.html.

[154] BBS, "Minutes of the AC Meeting, January 26–27, 1984," cover memo by Robert Rabin, BIO Lateral Files. Hereinafter, these Minutes will be termed BBS AC "Minutes," [date].

science. Only Frederick Mosteller of the AC did not speak. Kuhn did not come to the second meeting of the AC in September of 1984, but resigned in early 1985 only 2 weeks after social scientist Prewitt had done the same thing. (The cause of those resignations was explained in Chap. 2 and relevant footnotes.)

After hearing from the various speakers, the full Committee discussed strategies for its future. They planned to focus on biotechnology, information science, and the interface between chemistry and biology at the fall meeting. In preparation for that meeting, Rabin sent out a memo in mid-July, 1984, to the AC members concerning matters issuing from the NSB.[155] In June, the Board had considered biotechnology and had engaged in a discussion based upon two reports it had in hand. The reports were written by staff from several directorates and reviewed by the NSB Task Group on biotechnology chaired by Charles Hess (above), Rabin explained to the AC.[156] The first of those, an executive summary with analyses and recommendations, acknowledged the existence of a "biotechnology revolution" and indicated that the Board had decided to use the Office of Technology Assessment's definition of the subject:

> Biotechnology, broadly defined, includes any technique that uses living organisms (or parts of organisms) to make or modify products to improve plants or animals, or to develop micro-organisms for specific uses.

They noted that the definition was "meant to encompass three primary groups of technologies," which included: rDNA; cell fusion (hybridoma/monoclonal antibodies) and somatic cell genetics; and, bioprocess engineering.[157] NSF would, according to its long-established practices, support biotechnology *primarily* at the basic research level and would leave to other federal agencies and non-federal institutions to support the more applied aspects. NSF's own engineering directorate would also support applied features, of course (below). In practice, NSF's financial support of later STCs dealing with biotechnology would be strongly tied to the applied end of the spectrum. The Board indicated that a "biotechnology initiative" should be developed. (Knapp had written Senator Al Gore and told him as much; Gore was deeply involved in the dual revolutions.[158]) The NSB was keen that whatever NSF did, it should be alert to the national economic growth, security, quality of life and environment, and enhancing scientific progress. The Board saw advances possible in medicine, agriculture, energy, feedstock chemicals, and microbial products such as biopolymers, adhesives, and plastics. With regard to the "initiative," NSB stated that

---

155 Memorandum to the BBS Advisory Committee from Robert Rabin, Acting AD, BBS, July 3, 1984, pg. 1.

156 Ibid. The two reports were: NSB, "Discussion Issues 1984; Research Related to Biotechnology; Vol. 1: Executive Summary/Issues Analysis, Needs, Opportunities and Recommendations," NSB-84-159, June, 1984, and the second volume of the report, which this author has been unable to locate. The NSB ended that meeting by issuing a "Statement on Research Related to Biotechnology as Adopted by the National Science Board at its 253rd Meeting on June 21–22, 1984," NSB-84-189.

157 NSB, "Discussion Issues 1984," pg. 1.

158 Ibid., pg. 2.

the areas of focus would be in increased support overall, multidisciplinarity, collaboration, postdoctoral support for cross-disciplinary approaches, university equipment acquisition, and increased support for industry/university cooperation.[159] The report covered many other areas of concern for NSF's role in supporting the biotechnology revolution. The Board, at that same June meeting, closed its business by issuing a "Statement" on biotechnology that, in essence, repeated the findings of the main report as noted in the executive summary and discussed above. The "Statement" was aimed primarily at the NSF director.[160]

Rabin's mid-July missive to the AC covered not only the newly increased Foundation/Board emphasis on biotechnology, but also mentioned infrastructure, an area of strong interest. It was to include new construction, renovation, access to supercomputers, engineering research centers,[161] and all aspects of the new biotechnology initiative. BBS had become most closely associated with the San Diego supercomputer center as computational biology was the dominant type of research done there and that by individuals who tended *not* to be computer scientists, *per se*, but rather "computational scientists [biologists with such a specialization]," recalled Kingsbury.[162]

The arrival of Erich Bloch was imminent in mid-1984 and the changes in many areas of NSF, NSB, and the BBS AC were either already happening or set to begin. Kingsbury, some months later, was confirmed by the Senate then, as has been seen. Rabin told the AC that Kingsbury would "lead the remainder of the summer's busy schedule of budget development including the biotech initiative."[163] The House and Senate appropriations' committees approved the new NSF budget for FY85 and in their conference report they *exempted BBS from budget reductions* ongoing in other areas. The other half of the "dual revolution" pair, information sciences, got a $3 M increase and the Advanced Scientific Computing Program's budget went from $20 M to $40 M, possibly in anticipation, at least in part, of Bloch's arrival, though this is an assumption on my part.

Several months later, the NSF put out a news release that indicated that the biotechnology initiative was to be made available to the interested scientific communities.[164] It noted the existence of the dual revolutions and that biotechnology was an area of "vital" interest to the country's "industrial and economic health." The announcement tied in education and infrastructural needs and the plans to meet those needs through the initiative. A few months later, the NSF released its program announcement concerning the initiative and thus its request for proposals.[165]

---

159 Ibid., pg. 7.

160 NSB, "Statement on Research Related to Biotechnology."

161 Belanger had considered these, primarily in Chap. 8 of her history of the engineering directorate.

162 Kingsbury interview.

163 Rabin memo to AC, July 3, 1984, pg. 2.

164 National Science Foundation, "News Release," NSF PR84–39, June 27, 1984.

165 National Science Foundation, "Opportunities for Support of Biotechnology Research and Related Activities from NSF," n.d.

Rabin had a good deal on his plate during the summer as Kingsbury moved into his position formally, thus slowly displacing the acting AD. Rabin was testifying in June before both the Subcommittee on Department Operations Research, and that on Foreign Agriculture of the House Committee on Agriculture. The second of those two subcommittees had conducted a series of hearings on "critical issues in agricultur[al] science and education" since 1982 and Rabin represented the NSF with regard to the roles BBS had in those matters. He conveyed to the House members present the rapid rise of biotechnology, the initiative, and related issues. As well, Kingsbury attended those hearings with Rabin and so the acting assistant director was able to introduce the new AD to the Subcommittee and, at the same time, give Kingsbury a taste of federal hearings. Much of Rabin's comments mirrored the various reports of the NSB, its "Statement," and other features of the Foundation's interest in and directions with regard to biotechnology.

The AC met again in September 1984 and Erich Bloch would start off the second meeting of the still-new Advisory Committee. He talked about "streamlining" NSF and developing new relationships with "academia, industry, and Government."[166] The AC was chaired by Kingsbury; Rabin had moved on to the OSTP. Bloch explained to the Committee that he would have to study the social and behavioral sciences' areas as he knew so little of them; he also suggested that a review of those areas be undertaken by the Committee. The interface between chemistry and biology had been examined by the AC and they had established a new initiative, Chemistry of Life Processes, along with that of biotechnology. The BBS Division of Information Science and Technology (IST), whose director was Charles N. Brownstein, had been thoroughly reviewed, too. One aspect of that was to question the appropriateness of that division being in BBS. The AC considered the matter and concluded that the division should remain in BBS as it was "both correct and important" and that it do so because "it had a direct connection to other divisions in BBS." Indeed, Peter Raven was very keen that the division remain with BBS and wrote a formal letter to Kingsbury that October urging that it stay in the life sciences directorate.[167] But IST had grown considerably and was immersed in its own revolution. Access to supercomputers had also been on the AC's agenda and that access was organized jointly between the computer and electrical engineering programs at the Foundation. That cross-directorate effort suggested the wisdom of removing IST from BBS. That would happen, but somewhat later.

While all other members were present for the fall meeting, Kuhn and Prewitt were not; they would resign formally the following March (1985), after attending the February meeting, over struggles concerning the social sciences. Bloch intended for NSF to have a broader recognition in the future and for all areas of the Foundation

[166] BBS AC, "Minutes," September 19–20, 1984, cover memo by David Kingsbury with seven pages of information and two addenda (see next footnote).

[167] Ibid. Raven's letter was one of the two addenda. The other addendum was another letter from Raven to Kingsbury indicating that the AC "unanimously agreed" that a "survey of biology," broadly understood, was "badly needed and should be conducted." Both letters were dated October 4, 1984.

to be more aggressive and proactive, but where certain programs at the agency did not stand the "test of reason," he said, they should be reduced or eliminated. He may well have been thinking of the social and behavioral sciences, my assumption, though he was circumspect enough not to have identified them by name. Still, he "stressed the fact that he need[ed] to develop his own convictions" about those fields.[168]

The February 1985 meeting of the new AC was its third and the Committee would gear up to meet its charter obligations by considering many areas that affected the directorate in the social and behavioral sciences. No biology was considered at that meeting. Kingsbury prepared the membership of the Committee by sending out a list of reading materials. The AC still numbered only eight persons (Prewitt and Kuhn still present). Four other individuals were invited guests from the scientific community; a Board member was invited as well as a scientist from the Office of Management and Budget. The meeting gave AC members and guests an opportunity to learn about the 18 programs then extant in BBS' social and behavioral sciences' divisions with each program synopsized in one to two pages. As well, the membership was provided with budget summaries for FY82–84. The following month, coincident with the two social scientists' resignations, AC Chair Raven wrote a three-page memo to Kingsbury in strong support of keeping the social and behavioral sciences in the BBS. That would last for another 5 years, but those disciplines were heading for a major change of location after the 1991–1992 reorganization.

For the NSB and their first meeting of 1985, the Board would increase its standing committees from two to three. The ever-powerful Planning and Policy Committee (PPC) became the Programs and Plans Committee. Along with it were the Education and Human Resources Committee and the Budget Committee. Below those standing committees came task groups (TG) that included the International TG, the Excellence in Science and Engineering TG, and any others that may from time to time have been needed. There was also, of course, an Executive Committee.[169] Those structural changes at the Board were necessitated by the level of activity at the Foundation in 1984: *the mid-1980s represented one of the National Science Foundation's major growth spurts*. Beyond the Board, the need for reorganization of the current PCM into two new divisions was also clear, as we have seen. And, "[p]erhaps in no other field has change come so fast or so dramatically" as in computer science, thus suggesting the need for a "home of its own" and for the likely move of BBS' IST.[170] Indeed, Bloch would spearhead a new directorate of computer science in 1986, staffed by Mary Clutter. As that year opened, and even though the BBS AC felt to the contrary, Director Bloch officially announced his plans to create the Directorate for Computer and Information Science and Engineering, CISE.[171] That move would consolidate computer science, information science, computer

[168] Ibid., pg. 3.

[169] NSB, "Minutes," NSB-85-29, January 17–18, 1985, *passim.*

[170] Belanger, "Enabling American Innovation," pg. 171.

[171] NSB, "Minutes," NSB-86-78, March 21, 1986, 3–86:4.

engineering, supercomputers, and networking into a unified directorate. It would necessarily, of course, lead to the extinction of IST in BBS, besides removing divisions from the Directorate for Mathematics and Physical Sciences, and the Office of Advanced Scientific Computing that had been in the Office of the Director. Several programs from the engineering directorate would also move to the new CISE. BBS was then out of the computer science business, at least insofar as running programs in the area.

In 1988, a commentary on Bloch's first 4 years of Foundation leadership stated that he had "overseen more changes at the agency than all the agency's other leaders combined in the past two decades"![172] But what may appear to be desirable was not a change unalloyed, as shall be seen.

3.7 Institutionalization of Biotechnology

As with other programmatic beginnings, the period of start-up for the biotechnology initiative saw much activity by the Board, by the BBS leadership and staff, by Congressional offices, and with regard to interaction with the community. After a program's start-up, less would be seen with regard to any new program until a review was scheduled or some unexpected issue came to the fore and precipitated discussion. And an unexpected issue was just over the horizon, as it happened. In the meantime, though, NSF/NSB's role with regard to biotechnology had become one of stewardship, primarily, the new discipline was by then a creature of the BBS. While the program did not fall off the radar screen by any means, much less discussion was required. Emphases at the Board level, at the Foundation level, and at the directorate level could all be reoriented to other pressing issues, program initiatives, responses to problems, and so forth.

This process of institutionalization of biotechnology can be seen in a report by Robert Rabin and Thomas Quarles, a BBS staff associate in the AD's office, produced for the FY85 in which the first grants were given under the new biotechnology program initiative.[173] In the final Report of October, Quarles noted that a database had been established to follow the awards for the program and directed interested individuals to the "central HP [Hewlett-Packard] 3000" computer that served *all* of the NSF: desktop computers were yet to come. The awards' database required a key-word guide of some six typed pages in order to enter figures into the HP 3000. Indeed, Quarles had to provide in his memo a detailed explanation of how to find and extract the data back *out* of the mainframe, so new was computer use by most biological scientists at NSF.

[172] Irwin Goodwin, "Erich Bloch: On Changing Times and Angry Scientists at NSF," *Physics Today* (1988):47–52, quotation on pg. 47.

[173] Thomas Quarles, OAD BBS, to Assistant Director, BBS, "FY 1985 Awards in Biotechnology-Related Research," October 25, 1985, two pages with addenda.

It was noted that nearly $82 M was obligated for biotechnology-related research under the new initiative that year and 1773 awards were made, though not all of those came from BBS: some came from other directorates, such as engineering, physical sciences, Slaughter's "heaven and earth directorate," and others. However, the biology divisions were the most active having awarded just under 85%, or $69 M, through 1569 grants of the total. The vast majority came through the genetics programs, now split over the two new divisions with 32% (combined) of the total funds. The relative success rate for grants, however, was less than 30% in genetics. The programs outside of BBS, but "biotechnology-related," included the new Chemistry of Life Processes and Small Business Innovation Research (the Foundation-wide, well-known, and long-lived SBIR), among others.

General management of the biotechnology initiative was felt complex enough, financially major, and of sufficient significance to Foundation activities that Bloch set up an *Office of Biotechnology Coordination*. Given that its reach was across so many divisions and directorates, the new discipline required a high level of oversight, such that the Office would "gather, collate and analyze data pertinent to grant and contract activities in biotechnology throughout the [Foundation]," stated Bloch.[174] Importantly, a definition for biotechnology was finally agreed upon and was used NSF-wide and by the new Office. It borrowed the Office of Technology Assessment's definition (above) as its first paragraph and then went on to add considerably more. Nonetheless, the single-paragraph definition was the primary one used. Of especial note, however, was the computerization of the data generated within the biotechnology initiative—not scientific data, but management data. Much of Rabin's and Quarles' first report in August of 1985 concerned the computer programming necessary to provide good outputs for analysis by BBS and others. That historical evidence suggested just how new digital data manipulation still was at BBS, even in the mid-1980s.[175]

David Kingsbury had strong interests in not only computing, but also biotechnology, of course, and an event that occurred in 1985 became an early indication of the new AD's growing ties to the biotechnology industry. The editor, Jack Melling, of the *Journal of Chemical Technology and Biotechnology* based in London, England, had decided to redirect its long-established orientation to chemical technology and the chemical industry as its strong point and to emphasize instead the increasing role of biotechnology in Europe and also in the United States.[176] The new editorial

[174] Robert Rabin and Thomas Quarles to Assistant Director, BBS, "FY 1985 Awards in Biotech-Related Research: A Progress Report," August 30, 1985. This memo was the first of the two reports made by Quarles. Rabin did not participate in the second report of October 25 (above) as he had moved on to his new position in OSTP. The memo was given as Attachment 5 to the October memo.

[175] The matter was not entirely unusual at the time. When I was hired in 1988 at the University of San Diego as the new Associate Provost, my first duty was to spearhead the computerization of student records. Continued digitization efforts then spread over all of the Student Affairs areas; later such activities moved into other functions at the University.

[176] Jack Melling, ed., "Comment," *Journal of Chemical Technology and Biotechnology* (1985):35B.

board would include increased representation from industry, and that from biotechnology, especially. The coverage would consider not only the intellectual subject area, but also a wider range with regard to geography—reaching beyond England to account for the United States' rapid development in biotechnology. For that reason, it appears, Melling wrote to Kingsbury at the end of May 1985 to invite the BBS AD to join the editorial board of the journal.[177] There were yet other features of Kingsbury's ties to industry, but they would not become apparent for nearly two more years.

Biotechnology was becoming widely visible both within and without the circle of its practitioners. The NSF *Annual Report 1986* opened with biotechnology as the first of the several stories it was to tell of Foundation-sponsored research.[178] "Decoding Genes Quickly" showcased the work of Leroy Hood, noted earlier, and Jane Sanders at Cal Tech with regard to their creation, with others, of a machine that analyzed the sequence of nucleic acid bases in DNA with great rapidity. The *gene sequencer would become one of the most important tools in the molecular revolution, and well beyond*. It was, arguably, second only to the discovery of Boyer and Cohen in significance as a tool for that revolution, at least in those earlier years. The second piece of research to be highlighted in the 1986 report was also in the molecular genetics area: the complete genome of a common cold virus was clarified by Michael Rossman at Purdue University. Using a supercomputer, he and his team were able to find the location in the viral structure where an antiviral drug attached. That locus was not in the genetic core of the virus, but rather in the protein coat it wears.

3.8 The Rise of the Centers' Concept

The mid-1980s were heady days at NSF as, in 1985, the first of the Engineering Research Centers (ERCs) was established. From 140 proposals in answer to the program solicitation, 6 had been selected.[179] The new ERCs would be *models* which BBS, and other directorates, might base various other types of *centers* upon as they emerged onto the scene. The engineering directorate would play the central role in establishing the Biotechnology Process Engineering Center (BPEC) at MIT, Belanger has noted.[180] The BPEC had education/training as a "centerpiece," and so

[177] Jack Melling to David T. Kingsbury, May 32, 1985. The letter was addressed to Kingsbury's Berkeley, California address, not to NSF, and with the return address not to the *Journal*'s London address, but to the personal one of Melling's in Porton Down, Salisbury.

[178] National Science Foundation, *Annual Report for Fiscal Year 1986* (Washington, D.C.: Government Printing Office, 1987), pg. 19. While that report would have been the 36th in the series, the title would no longer use the numbering, but merely indicate the year.

[179] National Science Foundation, *Thirty-Fifth Annual Report for Fiscal Year 1985* (Washington, D.C.: Government Printing Office, 1986), "Director's Statement," pg. 1.

[180] Belanger, "Enabling American Innovation," ppg. 228–230.

was a partial answer to the last of Board member Hess', and the whole NSB's, five significant questions of 1984 concerning biotechnology (further above). The NIH contributed training funds to BPEC for interdisciplinary and cross-disciplinary studies (the Board's "significant question" four) from the undergraduate to the post-doctoral levels. In close cooperation with internal programs at MIT, undergraduates were able to experience a type of biotechnological education not seen in the research laboratory alone. Additionally, a great many industrial and international personages visited the "highly visible [MIT] facility" and that, too, had its educational value.[181]

BPEC arose out of the ERC program and not through either the soon-to-appear initiatives of the Biological Centers (which were tied especially to biotechnology) *or* the Science and Technology Centers Programs (STC). But during the period of 1986 and 1987, a new *center's era* was arising at the National Science Foundation. The NSB found itself approving one new program after another, and most of those were *hybrid* ones that involved computerization within various fields of science such that the computer became a central element, a new scientific tool to be used in ways not seen before: the machines were crucial to the rising centers. The period of computers being relegated to their more traditional uses in organizational administration and in the historically quantitative sciences was giving way to their use in new applications. Under a program entitled "Emerging and Critical Engineering Systems," the Board approved both a new program in Computational Engineering and another at the same meeting titled the Neuroengineering Program. Further, by May, the NSB had approved the beginnings of a communication tool to be called NSFNET, a network for computer and information science.[182]

In late 1986, came the unanimous recommendation of the NSB to approve the Biological Centers Program that authorized the director of NSF to "take final actions on grants, contracts, or other arrangements" to establish biologically oriented centers.[183] The first program solicitation announcement appeared and set three submission deadlines for FY87. The goal was to "stimulate the growth of fundamental knowledge in biological research areas of importance to the continued development of biotechnology." Board member Rita Colwell (in later years to become director of NSF; as noted earlier) led the Board committee that had proposed the centers and "noted that this program is being developed as a major part of the continuing NSF initiative *on research related to biotechnology*."[184] Centers would become a major involvement of NSF, and BBS in particular, during the 1990s and later years. They would go beyond biotechnology to cover other areas of biology, as well, and those through the STC Program (following chapters). Indeed, the first would come sooner than that, though not as a direct effort from the NSF Biological Centers Program. CARB, the Center for Advanced Research in Biotechnology, was a facility sponsored

---

[181] Ibid., pg. 229.

[182] For a brief history of NSFNET, see especially: http://www.nsf.gov/about/history/nsf0050/internet/launch.htm; also see: http://www.livinginternet.com/i/ii_nsfnet.htm.

[183] NSB, "Minutes," NSB-86-229, November 13–14, 1986, 11–86:3.

[184] Ibid. Emphasis added.

jointly by the University of Maryland, the National Bureau of Standards (later to become NIST), and Montgomery County, Maryland, all contributing to the State of Maryland's Biotechnology Initiative. In later years, more centers were added to CARB to create the Maryland Biotechnology Institute, MBI (termed the University of Maryland Biotechnology Institute, UMBI, after 1993).[185] Indeed, Bloch, Kingsbury, and Colwell all attended the groundbreaking ceremony for CARB on November 13, 1985, in Baltimore and so both Bloch and Kingsbury had a good initiation into what biologically oriented centers might look like that would be founded with NSF support.

In addition to new uses of computation, Bloch, by the June 1987 Board meeting, informed the members that the 5-year planning process for FY88–FY92 would cover three major areas, one of which would be the rapidly rising "Centers and Groups."[186] More than that, the director wanted to see "cross-cutting scenarios, rather than the traditional focus on each Directorate." Later in the day, Richard Nicholson, AD for MPS, began the discussion concerning centers and groups and reminded the Board that President Reagan had announced in his January 1987 State of the Union address his intention to double the NSF's funding over 5 years. Reagan *also* spoke of his desire to see the establishment of science and technology centers. Indeed, the President's view has been considered both "radical and important."[187] *This was so much the case that he issued Executive Order 12–591, which directed not only NSF, but also other federal agencies, to develop science centers.* The NSF was to ask the NRC to do a study on the subject, and the Foundation was to develop a coordinating office for centers and to prepare a management structure to handle the new concept. At that meeting, Rita Colwell had suggested that "centers may be an important vehicle for providing new leadership" on various university campuses where most centers would come to exist. That was the case, as most awards would be made to higher educational institutions or their consortia for the new centers.[188] Moving rapidly, the Board approved the creation in August of a new unit to be called the Office of Science and Technology Centers Development (OSTCD) as the coordinating device for the coming growth of centers Foundation-wide, biology most certainly included. Colwell was congratulated by the Board at the same meeting for her new appointment to the Maryland Biotechnology Institute.[189] She was something of a centers' specialist in the midst of the Board; a value to both her *and* the NSB then—and later.

About that same time, Mary Clutter, who had been serving as Senior Science Advisor to Director Bloch, returned to her duties as director of the Division of Cellular Biology. However, there were events brewing in the background that

[185] I was a member of the MBI President's Board of Advisor's during the 1990s when Rita Colwell was the director of MBI (later, UMBI). See: http://www.umbi.umd.edu/about/images/umbi_at_20.pdf.

[186] NSB, "Minutes," NSB-87-136, June 18–19, 1987, 6–87:2.

[187] Discussion between the author and Marc Rothenberg, Historian of the NSF, Fall, 2010.

[188] NSB, "Minutes," NSB-87-136, June 18–19, 1987, 6–87:8.

[189] Ibid., August 20–21, 1987, 8–87:3.

would propel Clutter to yet a higher position as David Kingsbury came under fire during the late 1980s. But early in 1988, Kingsbury wrote to the GSA with regard to establishing an Advisory Panel for the new Plant Sciences Centers Program.[190] The panel was in anticipation of growing interest in centers, in general, and in plant science, in particular: NSF, the U.S. Department of Agriculture (USDA), and the Department of Energy (DoE) had prepared, in 1986, to establish a cooperative research program in plant science, especially in the genetics and genomics of agriculturally important plants (Chap. 9). An outline for the program had been drawn up in June of 1986.

The NSB heard a report at its March 1988 meeting on the status of the Foundation-wide STC initiative, which had been strongly spurred by biotechnology's need for a distinct research structure and, further, *that* in the new NSF world of Erich Bloch. Though President Reagan had called for centers, it was a combination of the needs of biotechnology, specifically, and Bloch's vision for NSF more generally, that built the first STCs. Bruce Umminger of BBS (more of him later in this chapter), and who was at various times assigned to either the Office of the Director or the NSB, was serving as the Executive Secretary to the Board's task force, chaired by Charles Hess, examining the balance between centers and individual grants. Umminger noted that he spent a year and a half on that task force and felt the report emanating from it was "fairly balanced." However, Umminger went on to say that:

> I think Erich Bloch had wanted a ringing endorsement of centers. And the report didn't necessarily say centers were bad, but that there should be a balance between [what they can do], certain things, [and what] individual investigators could do, [other] things…there should be a balanced portfolio…Bloch really did not like [the report].[191]

It was soon after that that Kingsbury took the opportunity to send Umminger on a one-year sabbatical to the Department of State. While Umminger wanted to do a stint at State, he was not yet ready to make that leap and thus "I got the feeling I was being sent out into purgatory"![192] Worse, Umminger was to be sent to Europe as a roving ambassador for international biotechnology policy by Kingsbury after the former was supposed to return to his acting directorship in the Division of Cellular Biology. Indeed, Umminger has said that, "I really, to this day [2009, though] I can't prove this…thought I was being punished for that report on centers"! However, after Bloch left and the report was "resurrected," according to Umminger, "people really liked it." Later, though, Kingsbury was on his way out and Mary Clutter on her way in, so she told Umminger that he was *not* going to Europe but was to return to his job, first as acting director of DCB (1988–1989), then as its director (1989–1991), as she had vacated that position to become the AD of BBS. "So [with Kingsbury's] leaving and her coming in, she more or less saved me from being, I think, farmed out forever [laughs]."[193] As well, given his strength and interest in STCs over all,

[190] David T. Kingsbury to James L. Dean, GSA, January 13, 1988.

[191] Umminger interview.

[192] Ibid.

[193] Ibid.

Umminger spent his final six and a half years (of more than 26 at NSF) in the Office of Science and Technology Centers Development, the office that had been set up to coordinate the Centers. It was lodged in the Office of Integrative Activities in the Office of the Director. Umminger liked the fact that centers could do "bigger things" than could single investigators; and that was the view of Bloch, after all.

Bloch informed the Board that 323 proposals had already been received under the new STC program announcement that had closing dates in mid-January and the first of February 1988. That number was so large that 15 panels were convened to judge the proposals. Possibly, funding for the sciences had for so long been depressed that something like an STC at one's home institution could be envisioned as a way to employ many scientists and graduate students. Clearly, an STC would be a desirable addition to a campus. Mark Twain's comment would have been most appropriate: "[f]ew things are harder to put up with than the annoyance of a good example." And among the proposals, there were numerous good examples. Some 15% of the 323 proposals were in the first cut prior to selecting a smaller group. What became the 49 semi-finalist proposals were spread over 38 institutions and were recommended for a visit by panel members. The total amount requested (though much less was finally awarded) by them came to $686M for 5 years. Additionally, 258 *more* proposals from a variety of institutions were received by NSF as *planning* grants for yet other STCs.[194]

During June and July of 1988, the review panels' members visited, as it eventuated, 48 proposed centers across the country from which a "top ten" spread over all directorates were to be chosen.[195] A second, higher-level review was to follow, but with the names of those institutional hosts kept confidential. By the end of the year, the NSF announced that, of the 48 finalists for funding to create STCs, 11 were chosen spread over 8 US states.[196] The STCs, and the program that made them possible, were intended to last for, coincidentally, 11 years under support by NSF—but a number of those founded over the years since are still in existence. Others would join them in a later competition. The first group was funded, in toto, for the amount of nearly $25 M. An article in *The New York Times* noted that centers, broadly conceived, were a cornerstone of the Reagan administration's science policy: by then 18 ERCs and five National Supercomputer Centers were already in existence. The initial 11 STCs included aspects of the sciences of superconductivity, astronomy, materials, advanced computing, storm prediction, and, of course, biotechnology. The award amounts ran from $900 k to more than $4 M per center. At Cal Tech there would be the $3 M Center for the Development of an Integrated Protein and Nucleic Acid Biotechnology, while at Michigan State University there would be another biologically oriented center for microbial ecology. Such big biology (ecology at the

---

194 NSB, "Minutes," May 10–11,1988, 5–88:3.

195 Ibid., NSB-88-154, August 17–19, 1988, 8–88:4–5.

196 Warren E. Leary, "11 Universities Chosen In Technology Initiative," *The New York Times*, December 6, 1988. The mention of eight states was not made by Leary, but is to be found in National Science Foundation, *Annual Report 1989* (Washington, D.C.: Government Printing Office, 1990), "Director's Statement," pg. 15.

organismal level and above), and similar developments, will be discussed in the following chapter and again in Chap. 9.

During 1988, the NSF also funded 20 institutions under the slightly earlier initiative, the Biological Facility Centers (BFC) program, to make "sophisticated instrumentation" available across the land. From 10 to 20 scientists from several nearby universities were expected to use the equipment in each center. Biological molecules were studied by Margaret A. Lindorfer at the University of Oregon's center, for instance. Lindorfer and others had at their disposal at the local BFC a protein sequencer to determine the amino acid make-up of a given protein, a peptide synthesizer to create "chains of amino acids made to the experimenter's specifications," and an X-ray detector to examine the crystal structure of proteins. So, while biotechnology *per se* would have its own centers, those supported under the Biological Facility Centers Program could also provide scientific knowledge of direct applicability to the molecular revolution, as well as to scientists in other areas in need of the equipment.[197]

The same reporter who stated that Erich Bloch had "overseen more changes at the agency…" ended his article's following paragraph by tempering that comment. His revelation was that:

> [w]hile committing the agency to new functions and curtailing some old ones, Bloch has earned more cheers *and calumny* than any of his eight predecessors, going back to the NSF's founding in 1950.[198]

Unlike previous directors—"backroom boys"—Bloch had a "remarkably high profile": he was "different." In fact, claimed Irwin Goodwin in his "Washington Reports" section of the newsletter *Physics Today*, he was "combative."[199] This reputation for being both "scrappy" and "self-confident" was acquired at IBM, Goodwin

---

[197] National Science Foundation *Annual Report 1988* (Washington, D.C.: Government Printing Office, 1989), pg. 32.

[198] Goodwin, "Changing Times," pg. 47. Emphasis added. Much has been said about Bloch's style, personal approach, engineering and physical sciences, and computer orientation, but under him biotechnology flourished. Other observers were assessing Bloch at the same time as was Goodwin. Jeffrey Mervis, senior editor and Washington Bureau chief for the journal *The Scientist*, noted that, "Bloch evokes strong reactions." Mervis also cited physicist Daniel Kleppner of MIT as saying, "[t]he bottom line is, Erich Bloch had a vision for NSF and it failed. I feel sorry for him." Maybe, but former director of NSF Richard Atkinson, then still chancellor of UC, San Diego, felt otherwise: "I think that Erich's been a very effective director…by getting NSF more involved in applied research he's made it a more visible player in Washington." Those figures who evoke strong reactions may evoke in their critics varying analyses. See: Mervis, "Erich Bloch's Campaign to Transform NSF," *The Scientist* (1988):557–561; quotation on pg. 557. Bloch's later interviewer, Marc Rothenberg, also asked him (February 11, 2008) how he thought he did as director—especially given Kleppner's criticism: "Well, I don't know how well I succeeded. It's not up to me to say." He felt that NSF looked different than it had after he left and that he took it in the right direction; he did not succeed in everything, he said, but that in a changing world, NSF needed to change along with it. "[NSF] had better look at itself" at any time along its trajectory as an institution. "No, you never achieve everything, or if you do, you didn't set your sights high enough, I guess." See also this author's assessment in the Epilog herein.

[199] Ibid., Goodwin, pg. 47.

argued. It showed in the regular press conferences the director called after NSB meetings: they were "no-holds-barred." Though Bloch was a registered Democrat, it was his *style* that appealed to Reagan's man, George Keyworth, and was the reason the first science advisor to Reagan recommended Bloch for the NSF directorship.[200] Running the Foundation with "an iron fist," he shook NSF from its "staid complacency," Goodwin had said that NSF personnel told him. A central unhappiness in the science community, substantially equivalent to the academy insofar as NSF's majority of funded researchers went, was the direction the Foundation was taking under Bloch. It was based on the shift in emphasis in the organic act from the 1950 founding wording to its amended form of August, 1986. The change, as we have seen, was from the familiar support of "science research and education"—primarily "little science" in the academy—toward the big, applied science of the ERCs and STCs, albeit still mostly academically hosted. It was that redirection which was irking the academician-scientists.[201] In an interview question to Bloch, Goodwin said:

> Q: Despite all your assurances, some people out there fear that NSF's support of centers will result in less funding for individual investigators. How do you allay that fear?
>
> A. [Bloch replied that] I don't know if you can allay it completely—and maybe that's not necessary.

The reason for it not being necessary, Bloch elaborated, was that less than 10% of the total award budget for NSF was going to centers. He had told that to NAS, too, as it had also inquired.[202] The NSF must support science in the "best way" and it needed a "balance among different research programs, approaches and procedures," Bloch stated. It has been argued, somewhat in this connection, that NSF itself became generally more interdisciplinary with the rise of centers.[203] It seems a logical conclusion that such would be the case given the personnel structure—different disciplinary histories of the various scientists present—at the centers and the continual *intricate dance* between the communities and the Foundation. In later years, centers were likened to "a mini-Manhattan Project," that which created the atomic bomb, because no single researcher, nor any single institution, can accomplish such very large tasks alone.[204]

Reagan's promise to double the NSF budget in 5 years had not yet been fulfilled and that, too, was causing tension in the community. But Bloch told his interviewer that a requested 19% rise in the budget for FY89 included $150M as a:

> single appropriation to be made at the start, in order to fund 12 to 15 [11, as it eventuated] Science and Technology Centers for their first full five years. Congress says it won't do that and instead proposes to fund the centers one year at a time. So without up-front funding to

---

[200] Ibid.

[201] Ibid. On page 48, Goodwin's article moves into the interview he held with Bloch during the first half of 1988 and which became the primary substance of the *Physics Today* expose.

[202] Ibid.

[203] Umminger interview.

[204] Ibid.

support the centers for five years, NSF is asking [instead] for a 13 percent increase for [FY89].[205]

While the audience for *Physics Today* was more interested in the traditionally termed "hard sciences" and the strongly quantitative areas, the interview with Bloch gave the director the opportunity to answer, in a most notable fashion, one of the first questions posed by Goodwin. Goodwin asked about Bloch's background as an engineer, from industry, and one without a doctorate at that; academic scientists were on Goodwin's mind. To it, Bloch was able to answer by saying that: "No longer is research at arm's length from *application* in many fields—*biotechnology*, computers, materials science and many others" are all within the purview of NSF.[206] Engineer Bloch chose as his *first* example biotechnology.

Of considerable significance in the *Annual Report 1989* was the announcement of the first STCs. Those 11 were to be found in California, Illinois, Michigan, New Jersey, Oklahoma, Missouri, Texas, and Virginia. The New England or Eastern bias of NSF, as claimed by numerous early critics (see first two chapters), was clearly *not* present in that selection of states. However, with the possible exception of Oklahoma of the time, the remaining states were leading powerhouses in the industrial-scientific areas specific to the STCs established. One such center, as an example, was the National Instrumentation Facility for Nuclear Magnetic Resonance of Biological Studies at Washington University of St. Louis: it probed the "structure of biological molecules and living tissue." Works in experimental polymers, biodegradable plastics, and drug research were all facilitated by the Missouri center's instruments and activities.[207]

3.9 A Case Study of One Biological STC

Bruce Umminger came into the DCB's predecessor, PCM, after a period in the academy, and that as an organismal biologist. Such biologists generally take the organism as the primary unit of study, but that often implies, especially after the molecular revolution began, that doing research at the "organismal" level may be much more along cellular or even biochemical lines. That was the case for Bruce Umminger, who obtained his bachelor's at Yale University in 1963 when biology was still rather "classical" in many respects; the genetic code was just in the midst of being worked out, for example. He then went on to UC, Berkeley for graduate work in zoology but, as it was the era of the "free speech movement," he "didn't

205 However, an overhead slide used at the BBS AC meeting held on May 5–6, 1989, showed a pie chart where the STC component was emphasized being marked with a figure of 7.3%, not 13% as Bloch stated. One or the other may be a tentative figure, but internal evidence is such that the 7.3% seems not to be the final figure. Alternatively, the larger percentage represented *all* STCs, not just those from a single directorate, in which case Bloch's figure would indeed be correct.

206 Ibid., Goodwin, pg. 48. Emphasis added.

207 Ibid., pg. 27.

particularly care for the atmosphere there" and so returned to Yale to work on two masters degrees: an MS in biology (1966) followed by a masters of philosophy and biology (1968). He finished his studies there in 1969 with a doctorate in biology. Soon after he began his academic career, Umminger obtained a grant (1971) from NSF for his research on the physiology of fish (temperature and salinity adaptation, etc.); it was renewed continually until 1979. It was in that latter year that he came to NSF on leave from the University of Cincinnati, Ohio, for what was to be but a single year as a rotator. As with so many before and after him, the move evolved into a career; first as a program, then division director.[208] He decided to resign from the university after a decade there and came to NSF for what would become more than 26 additional years in his discipline, and many other fields, retiring only in 2006. The first two decades were in BBS/BIO and his final 6 years were as leader of the STC program in the Office of Integrative Activities/Office of the Director, as previously noted.

Simply enough, he had found that he liked management, but he also sought more experience in science policy. He had a suspicion of this interest earlier through his activities with the American Society of Zoologists (ASZ)—later to become the Society for Integrative and Comparative Biology, which, under both names, has had a long and particularly close association with NSF. He felt that remaining at the Foundation on a permanent basis was a "good career move."[209] He wanted to represent his fields of comparative physiology and comparative endocrinology, which were covered under the more encompassing umbrella of the Regulatory Biology Program at BBS. He was formally a rotator for 2 years before becoming permanent with BBS, all of which he has called a "broadening experience." As a program director of regulatory biology from 1979 to 1984, Umminger read some 100 or more proposals per panel period, and there were then three panels per year for the program. He noted that he was much busier at NSF than at his previous university position, but he "really enjoyed it…it was fun"![210]

In 1984, when PCM was broken up into the Division of Molecular Biology and the Division of Cellular Biology (DCB), Umminger became Mary Clutter's deputy director of DCB until 1989. In 1985, Clutter had become a senior science advisor for Director Bloch; that lasted till 1987. During those 2 years, Umminger functioned as the acting director of DCB. When Clutter returned to BBS from the director's office, Kingsbury was on the verge of leaving NSF and so she moved into the AD position temporarily. Umminger once again found himself acting director of DCB for 1988–1989. His performance was such that Clutter appointed him full division director and that lasted until the completion of the second major reorganization, 1991–1992. He moved into the directorship of the then new Division of Integrative Biology and Neurosciences (1991–1999), or IBN, prior to leaving BBS to take command of the STC program until his retirement.

---

[208] Umminger interview.

[209] Ibid.

[210] Ibid.

In characterizing one aspect of his years in BBS, he pointed out that "budget time at NSF is all the time," as one is operating under the current budget, making a case for the following one, and preparing the one after that, all at the same time![211] But there were also more philosophical questions that occupied minds at BBS during the decade of the 1980s. Reductionism was a key factor in much of what made up not only the discipline of biology in the closing quarter of the last millennium, but also the nature of its primary patron, BBS/BIO. If that is so, then Umminger's statement that he came to NSF "with a strong bent on protecting organismic [organismal] biology"—not precisely a reductionist area—must be examined.[212] His training was in organismal biology and that more specifically in the comparative aspects of physiology, endocrinology, and biochemistry. Note that all three are centered upon or are mostly below the cellular level. It was those aspects of his education and experience that made him a good choice several years into his life at NSF as a director of a division of *cellular* biology. Physiology was lodged in the PCM for the first 5 years that Umminger served with BBS as program director for regulatory biology. And, too, the American Society of Zoologists was, as Umminger reminds us, "an organism[al] biology group" and so he continued to view himself as such. Indeed, during the 1991–1992 reorganization, Umminger would speak vigorously for organismal biology up to and later through his years in IBN.

Reductionism and organismal biology were not the only issues of concern to the new program director in his earliest years, however. Umminger reflected back to the beginning of the 1980s when Reagan came into power and said that he (Umminger) "woke up one morning and read the headlines in the *Washington Post*" and saw that the NSF had barely survived cuts in the President's budget. He was stunned and "thought…the whole agency's been zeroed out!"—and he had only just become a permanent employee! He saw, as had others, that NSF had actually been *forgotten*—literally—altogether in the Reagan/Stockman budget for FY82 (see earlier above). The error was quickly corrected. When Reagan appointed Erich Bloch as director (September 1984), the DCB budgets were such that the typical grant award had become some $50 k a year, double that of the mid-1970s. But it was clear that one could not continue to do "science by the yard" (that is, one grantee at a time) as there were more and more proposals coming to not only DCB, nor even only to BBS, but to all of NSF that suggested that big science, in the financial sense, was rapidly on the rise. Some NSF-sponsored centers were already extant and the STCs were imminent: if one wanted to get a major "bang for the buck" a great many bucks would have to be made available to get the science done.

When Umminger took over the directorship of the STC office in 1999, it was based upon his long history associated with the Centers' program going back to the 1980s in BBS. It was then, as Umminger had it, that Bloch said that NSF had to get more "interdisciplinary teams" to do the science that was arising in the period. Large multiyear grants made to places such as concentrations of specialization

[211] Ibid.

[212] Ibid.

where scientists of several disciplines might work together was the vision Bloch had promulgated and was a notion closely tied to the growth of biotechnology at NSF. One of those initial centers was DCB-sponsored when Umminger was the division's head, and it was in the first group that was funded at the inauguration of the STC program. When the request for proposals went out to interested parties across the country in 1987, one respondent was the Pittsburgh-based Carnegie-Mellon University's (CMU) department of biology in the person of Professor D. Lansing Taylor.[213] CMU Biology proposed what would become the Center for Light Microscope Imaging and Biotechnology (CLMIB). It represents a good example of the earliest centers supported under the STC program for BBS. It went on to operate into the first decade of the twenty-first century. Why a case study of the CLMIB is a good example is, as Umminger stated, because:

> they changed directions a couple of times…they had a lot of growing pains…but it's a good one from the standpoint of their success in spin-off companies and technology transfer; it was really good at that.[214]

There exists the possibility that Taylor was shrewd in adding the term "Biotechnology" to the proposed name of the center given the tie of STCs to that discipline at NSF. He was, as well, prescient when one examines his later career in the same field as an entrepreneur after leaving the academy. At least six panelists adjudicated the CMU proposal in the late fall of 1989 as the first round of STC bids were being examined. Not unlike so many such judgments of this sort, the review responses to BBS ran from as many as three pages to as small a document as two paragraphs. In sum, they were all complementary and dealt with the several benchmarks against which the 323 proposals received would be assessed: intrinsic merit; research competence; training plans; knowledge transfer; rationale for the particular center; research utility; and, institutional support. One panelist, in answering the first criterion, intrinsic merit, summed up well the basic nature of what CLMIB would be and do:

> This Center application proposes to develop a complete light microscope workstation that will be user friendly, yet capable of sophisticated image analysis including the rapid computer processing of complex data sets and the generation of graphic displays from them that adequately conveys the imaging data. Also proposed are a series of experiments focused on the development of new photochemical fluorescent and biological probes which will be required to extract molecular and chemical information from biological preparations. A small number of biological experiments, largely in tissue culture models, will be explored concurrently to guide the development of methods and instrumentation. The utility of the analytical instruments and materials which should result from this Center is vast. Their

[213] D. Lansing Taylor had been professor of biological sciences and the vice dean of molecular sciences in his time at CMU from 1982 to 1996. He was the founder of the CLMIB and then went on to found and become Chair and CEO of Cellomics, Inc. After that, and to the time of this writing (2010), he is president and CEO of Cellumen, Inc., and continues as an adjunct professor at CMU. "He was a pioneer in the development of instrumentation and fluorescence-based reagents to measure the functions of living cells" and as such holds more than a dozen patents. For the provenance of this quoted sentence and more on Taylor, see: www.cellumen.com.

[214] Umminger interview.

> successful development will likely foster a range of new engineering fields and technologically advanced products.[215]

Another reviewer noted that:

> [p]rogress in science is often limited more by lack of instrumentation than by lack of ideas. The instrumentation and chemical probes developed in this work should have direct utility to large numbers of biological scientists.[216]

Both reviewers, as well as others of the six panelists, were themselves prescient when recognizing the present and future value of the CLMIB. Taylor went on to found a fluorescent cell probe-based company, Cellomics, Inc. (followed by Cellumen, Inc.) in the later 1990s.[217] Technology transfer personified!

In order for the initial group of 48 finalists to move into the "top ten," a number of panelists would visit all the sites in the semi-finalist cohort. The Site Visit Team for Carnegie-Mellon submitted its Report in the latter part of April 1990. They noted that:

> [a] technological revolution based on modern advances in physical optics and computer science has enabled a new instrumental power in the classical tools of light microscopy… User friendly microscopy work [results can] be made available through technology transfer through industrial partners…[as] fluorescent probes [are] becoming more and more widely used to approach *fundamental* questions of cell structure, function, and metabolism…[218]

Clearly, "goal-directed basic research." The Center was a logical follow-on from the 6-year-long effort of Taylor and his colleagues and students in a research group known as the Center for Fluorescence Studies at CMU: they had a "distinguished history of research accomplishments and dissemination of information through publications…and academic training," averred the Site Visit Team. The Carnegie-Mellon provost at the time provided criteria for how the University would evaluate the Center. He pointed out the two major activities of the CLMIB as being technology development to include instruments, software, and reagents, all aimed at fluorescent probes, and biological research into cell biology (such as: contractile mechanisms in cells; cellular-level calcium signaling—an area that would become of vast importance to biology within a few years—and, mitochondrial metabolism), developmental biology (cell migration and patterns of gene expression during development), and general cytological research (in situ hybridization, ploidy analysis, cell phenotyping, etc.).[219] It eventuated, then, that NSF did fund CLMIB, but at a lower level than had been hoped for by CMU.

---

[215] Taken from a collection of review sheets associated with files made available to the author from Maryanna Henkart (see later chapters for more on Henkart). Reviewers' names, though the sheets evidence them, are not given here. Materials taken from this group of documents are known hereinafter as the Henkart Papers.

[216] Ibid. Taken from a second reviewer.

[217] See several footnotes above concerning Cellomics, etc.

[218] Site Visit Team (names not listed), "Site Visit Team Report, Carnegie-Mellon University, Taylor, 89–20, 118, April 23–24, 1990," 4 ppg. Quotation from pg. 1. Emphasis added. Henkart Papers.

[219] Paul P. Christiano, Provost, "Criteria for Evaluating the Effectiveness of The Center for Light

In January of 1992, a team from BIO (BBS was then extinct) did the first annual evaluation of the new Center at Carnegie-Mellon.[220] It was led by Gerald Selzer of the post-1991 Division of Biological Instrumentation and Resources (BIR; Chap. 6), and composed of two additional scientists, one each from the University of Massachusetts at Amherst and from Dartmouth College. While both the president and provost positions at CMU had changed occupants during the CLMIB's first year, the new senior administrators of the University and the continuing science dean were strong proponents of the Center's role in the promotion of both computational biology and biotechnology. There had been significant progress in the six programs of the Center during the inaugural year, but a seventh, molecular biology, was underfunded by the University due to the smaller start-up grant given by NSF. Had the immediate future of biology been seen more clearly at the University, they might have chosen to shortchange some other program rather than molecular biology! The state of the facilities impressed the visitors, as did a strong program in K-12 education tied to both a Supercomputer Center and an ERC, each of which were also NSF-funded. Summer laboratories and museum exhibits were part of the education programming, too. The undergraduate and graduate courses (Biology of the Brain, Molecular Biophysics, and Computational Biology) were presented to the assessment team. The Center did succeed, it was shown, in raising non-NSF funding totaling $712 k for FY92 and the State of Pennsylvania and CMU had delivered fully on their start-up funding commitments. Both film- and optics-giant, Kodak, and the German microscope-maker, Zeiss, also funded promised graduate fellowships, as did a number of other philanthropies (Keck Foundation, Markey Trust, etc.). The second-year budget would total $1.5 M with $475 k of that being NSF funds. NIH monies were sought for that next year, as well. While the Center did seem a considerable success, the site team noted that there was a lack of women (none) in the Center's faculty group and only one on their external advisory board. When the molecular biology component restarted, however, two faculty women would be involved thenceforth.[221]

The CLMIB is considered further later as it was successful into the third millennium and so remained a good exemplar, for present purposes, of the STC program.

3.10 Change of Command at BBS: II

Mary Clutter could move into the Assistant Director position at BBS because of a growing problem bubbling in the closing years of the 1980s. Late in 1987, Mark Crawford, a writer for the journal *Science*, penned an article that was to be one of

---

Microscope Imaging and Biotechnology," Carnegie-Mellon University, July 31, 1991. Henkart Papers.

[220] Memo from Gerald Selzer, BIR to File, "[First] Annual Site Visit to STC for Light Microscope Imaging," January 30, 1992. Henkart Papers.

[221] Ibid.

several that would chronicle Kingsbury's difficulties and that would lead, in part, to his eventual resignation.[222] It should be noted that Kingsbury's involvement in the many organizations mentioned earlier was quite independent of BBS and NSF. Nevertheless, the entire matter was deeply complex, as shall be revealed. He had become entangled in both industry and a number of federal Executive Branch committees, and those compromised his NSF position. In addition to his board position on the *Journal of Chemical Technology and Biotechnology*, Kingsbury had signed an agreement to sit on the board of a California medical diagnostics company, IGB Products Limited, in 1986. More so than a journal editorial position, membership on a biomedical corporate board was a matter that could be viewed as a conflict of interest for a federal employee.

Beyond that, Kingsbury had also recently taken over a White House-level committee, the crucial Biotechnology Science Coordinating Council (BSCC), from its departing head, Bernadine Healy.[223] The BSCC was looking into how the field of biotechnology was regulated by the federal government. It had been formed under the OSTP primarily "to deal with interagency disagreements over the release of genetically altered organisms into the environment."[224] The Department of Justice was called into the investigation because of Kingsbury's chairmanship of BSCC, though not because of his BBS AD position, as he had "personal finances and ties to the biotechnology industry."[225] Kingsbury had said, Crawford reported, that he (Kingsbury) had "severed all ties with the company, IGB," prior to coming to NSF in 1984 and "had never agreed to serve as a director" for the company. When he was shown a copy of a January 10, 1987, document that made him a director of IGB, in fact, he said to Crawford "[t]hat's my signature, I guess."[226] He further stated that "[m]aybe it's my fault for not paying attention to what I was signing," though in doing so he was agreeing not only to his *own* appointment to the board, but also that of his wife and several other directors. He stated that he "may have confused the document with papers related to the receipt of stock certificates" that eventually had been returned to IGB. Crawford noted, in the first of what would become three articles in the next 2 years, that there was no evidence that Kingsbury's position and decisions had in any way benefited IGB and its parent, Porton International PLC of London, England.[227]

---

[222] Mark Crawford, "Document Links NSF Official to Biotech Firm," *Science* (1987):742.

[223] Kingsbury interview. See also: http://www.nlm.nih.gov/changingthefaceofmedicine/physicians/biography_145.html and numerous other sources. Healy had been deputy director in Reagan's OSTP and she would, in later years, become the first woman director of NIH, along with many other major positions. Like Kingsbury's, her own career was not without controversy. She died in late 2011; see: http://www.huffingtonpost.com/susan-blumenthal/bernadine-healy-md-pionee_b_923193.html.

[224] Kathi E. Hanna, ed., *Biomedical Politics* (Washington, D.C.: Institute of Medicine/National Academies Press, 1991), pg. 136.

[225] Crawford, "Document," pg. 742.

[226] Ibid.

[227] The name Porton carries considerable baggage in the history of science and with regard to the

He was tied to other Porton subsidiaries, beyond IGB, and from which he had apparently withdrawn prior to taking his NSF position, but he remained with IGB beyond the beginning of his federal appointment, one of Porton's chief executives had said.[228] Kingsbury intended to return to Berkeley after what was to be a 2-year position only at NSF, though by then more time than that had passed. And IGB wanted merely his technical advice during his projected Foundation appointment, they had said. But the IGB executive, Zsolt P. Harsanyi,[229] told *Science* that Kingsbury was actually a director of *four* subsidiaries of Porton International. Porton supplied that information to the Department of Justice, which had started its investigation in April and which had been instigated by the House CST. The inquiry was kept under wraps for some period, however. Kingsbury had argued that he had no connection to the four subsidiaries, but when he saw the January 10 document, he said "you got a thing that said I agreed…I don't have anything else to add."[230] There were inconsistencies in whether or not Kingsbury would be paid for aid he may have proffered to the Porton subsidiaries, but such pay was not to come until *after* leaving NSF's service. He stated that "[t]here never was an agreement to that effect." By the time of the first Crawford article, no assessment had yet been made by Justice as to whether Kingsbury had broken any "civil or criminal statutes." In February 1987, Kingsbury severed all ties to Porton and its subsidiaries. He also left his chairmanship of the BSCC later in the year. But he was worried about his NSF job: "I am starting to have some reservations, I don't know what's going to happen."[231]

The BSCC was to be abolished on October 1, 1987, as originally envisioned by its creator, George Keyworth. That would, necessarily, also be Kingsbury's departure date from it. Keyworth, earlier OSTP head, had planned for the BSCC either to have its life extended in the same or some reconfigured form after that time, or be eliminated totally. It was resuscitated, in fact, and remained about the same in its

history of WWII (and beyond). Porton Down, near Salisbury in Wiltshire, was the home during the War for research in chemical and biological weapons. Porton International was a spin-off of certain aspects of those earlier days and research efforts. Porton Down, as a UK governmental research park, dated to 1916 and WWI, but during the Second World War "research concentrated on chemical weapons such as nitrogen mustard [gas] plus biological weapons including anthrax and botulinum toxin. In 1942, highly successful tests of an anthrax bio-weapon developed at Porton Down were held at [the test site of] Gruinard Island." Quotation from: http://www.absoluteastronomy.com/topics/Porton_Down. Further, "[i]n 1970 the Chemical Defence [British spelling] Establishment became the title of the senior establishment at Porton Down and remained so for the next 21 years. Preoccupation with defence against the nerve agents continued." Ibid. cited website. See also: http://www.mod.uk/DefenceInternet/AboutDefence/WhatWeDo/HealthandSafety/PortonDownVolunteers/ and, G.B. Carter, *Chemical and Biological Defence at Porton Down, 1916–2000*, Ministry of Defence, Great Britain.

228 Crawford, "Document," pg. 742.

229 A brief biography of Harsanyi can be found at: http://www.expobio.com/company-profile/management-directors.php.

230 Ellipsis in original *Science* article.

231 Ibid.

structure, but was limited to study of the five federal agencies then funding "research in biotechnology or overseeing commercial applications of genetically altered organisms."[232] With Kingsbury's departure from the BSCC, James B. Wyngaarden, then director of NIH, took the helm and John H. Moore, then still new in the position of deputy director of NSF,[233] became the Council's assistant director, interestingly enough. Crawford stated that:

> Kingsbury's departure [from the BSCC] appears to have been driven partly by infighting with the executive branch and by concerns about allegations that Kingsbury did not fully disclose his relationships with biomedical subsidiaries…of Porton International.[234]

Crawford went on to say that Justice was still investigating Kingsbury at the time (November). In October, Kingsbury had told Crawford that he (Kingsbury) had been involved in "a 'power struggle' over the future shape of the BSCC with Beverly J. Berger, assistant director for life science [of the OSTP]." The president's science advisor, and necessarily the head of OSTP, was by then William R. Graham, Jr., and he was "reported to have been worried about the Justice Department's probe of Kingsbury" in that it might damage the credibility of the BSCC, to say nothing of NSF.[235] Crawford went on to state that "[a]ccording to White House sources, several months ago OSTP urged NSF officials to find a replacement for Kingsbury on the BSCC."[236] The BSCC never handled policy, though doing so had been considered at its inception.

The body to take on policy issues was the then new Committee on Life Sciences (CLS), and which was chaired by, interestingly, Beverly Berger. Many federal agencies and offices would become members of that Committee, all of those having strong roles to play in biotechnology in one manner or another. The Committee on Life Sciences was welcomed on many fronts, including Congress, as it would balance the tilt that the OSTP was believed to have at the time toward the physical sciences, Crawford discovered.[237] While some groups were unhappy that the

---

[232] Mark Crawford, "Wyngaarden to Chair Biotech Council," *Science* (1987):1504–1505. That was in the December 11 issue of *Science.*

[233] Moore was announced to the public through the pages of the NSF *Annual Report 1988*, among other publications. See: http://www.acsh.org/about/pageID.110/default.asp for a short biography of Moore.

[234] Crawford, *Science* article, pg. 1504.

[235] Ibid., ppg. 1505–1506.

[236] In Kingsbury's interview with the author years later (2009), he said that "I became a political target…And they made stories up and put them in *Science* magazine." While he was not in reference to any single article or commentary by Crawford, he did see the matter in the sense of "made up" information. He was at the "tip of the spear," he averred in the interview, and became a "lightning rod." Congress' interest in him, he stated, began with a staffer in Representative John Dingell's (D-MI) office who said that the regulatory policy that BSCC had promulgated was not liked. Kingsbury told the author that organizations such as Greenpeace and the Sierra Club, specifically, did not like the policy framework.

[237] Crawford, *Science* article, pg. 1505. That bias has been discussed in the first sections of this history, but small incidents that exemplify it can be seen in various other documents. In the NSB "Minutes" for their August 16–17, 1979, meeting, there is mention of a film viewed by the Board

structure of the BSCC was strongly biomedically oriented, and short on the environmental sciences, Wyngaarden's (an M.D.) leadership was positively accepted widely in the scientific community. The BSCC, under new leadership, and the new CLS met for the first time in 1988.

As it happened, the BSCC was, in fact, temporarily suspended at its sunset date that October of 1987 by Graham, who planned some new variant of it in which Berger would have a major role.[238] BSCC survived, but the "power struggle" between Berger and Kingsbury came out in the scientific news. The struggle concerned Kingsbury's *not* "pressuring the regulatory agencies to ease up on proposals for field trials" of genetically modified organisms (GMOs), which Berger wanted to have changed to the contrary and which industry sought also. Graham did not consult Kingsbury, then still chair of BSCC, or any of the other six members of the Council about his (Graham's) reorganization plan. The future for Kingsbury was already clouded by the Department of Justice investigation and he had already ceased to participate in BSCC while he was under scrutiny, though he had not yet resigned from it. Whether he had agreed to do so under pressure from some direction, or other, was contested by Kingsbury, who held that his decision had been his own choice.[239]

In late 1988, "Kingsbury's political situation deteriorated quickly," one history has it, and other features of the entire set of intra- and extra-federal agency struggles during that "formative period" in the history of biotechnology and government relations would play into NSF's role in another area: plant science.[240] Kingsbury "emerged as a mediating [force]" in a Congressional imbroglio over the Department of Energy having been chosen as the lead agency in the human genome project over NIH, which was spending ten times more money per year on the subject than was Energy. (The National Science Foundation never played a *direct* role in the human genome project, but the BBS instrumentation program did fund Leroy Hood at Cal Tech with regard to the gene sequencer, as well as those machines that sequenced proteins and polynucleotides.[241] The company Applied Biosystems spun out of that

---

members that was taken by the Woods Hole Oceanographic Institution (WHOI) in its submersible ALVIN during a RISE ("Riviera Submersible Experiment") dive. A discussion of the dive mentioned physical features of hot water vents, recently discovered by researchers at Oregon State University, but made no mention of the spectacular life forms they typically support. Were there no organisms to be seen, or was the report so inclined to the physical sciences that aspects of those sciences were all that was considered worthy of placing in the Minutes? A request by the author to WHOI (summer 2010) to view the film was not answered.

[238] Yvonne Baskin, "Graham Shakes Up U.S. Biotech Panel," *The Scientist* (1987):20. That was in the November 2 issue.

[239] Ibid.

[240] Hanna, "Biomedical Politics," pg. 136–137.

[241] Eric Lander became the leading scientist with regard to the DoE/NIH Human Genome Project (HGP) and Kingsbury, in his interview with me, noted that it was Harriman and John Wooley (see main text above) who "discovered" Lander. Lander had gotten his first research grant from NSF to create a software package to do genetic map construction, Kingsbury told the author. The HGP will not be discussed broadly in the present study as it is not part of NSF's history; it will be mentioned

work. Much of that was applicable to the human genome work.) But if Kingsbury were to play such a mediating role, and that via his chairmanship of the BSCC, NSF would have to stand aside concerning the support of plant science research until the Congressional struggles were decided—a pivotal Foundation employee very much at the center of action. It was a period of considerable confusion and infighting quite independent of Kingsbury's own difficulties, but he was out of the broader genomics' war once the Justice investigation had grown and his withdrawal from BSCC came to the fore.

In late summer 1988, Bloch had received the Department of Justice's findings regarding Kingsbury. The investigation had been ongoing for nearly 18 months. Bloch had announced its existence to the press much earlier, in November of 1987, but took no "immediate administrative action" at that point, as it would have been "inappropriate."[242] Even late in 1988, Crawford was able to say that decisions made by Kingsbury in the name of either BSCC or NSF could *not* be shown to have "directly benefited" Porton International. Most of what Justice had been doing in the case, at any rate, was an investigation centering around the Porton subsidiary, IGB, specifically. The basic question was whether the public would conclude that its confidence in government had been damaged by Kingsbury's ties to IGB. A grand jury was also called to question the issue. The agreement between Kingsbury and IGB had been signed on July 24, 1985, and was to end on May 31, 1988. The AD was to have been paid $22,000 annually for 35 h per month of work for IGB, but with payments for same not to begin until *June* 1, 1987, the time when Kingsbury had expected that his initially agreed-to 2-year rotator position at NSF would

---

occasionally in the main text as necessity warrants, however. But, importantly, Kingsbury has argued (interview) that his role in the HGP was considerable, even though NSF's was not. His activities with the BSCC was the basis for that. He said that Charles Cantor of Columbia University and Sherman Weissman of Yale had been involved with the DoE and that the Department was trying to start up what became the HGP: Cantor and Weissman thought it a "good idea," Kingsbury related. The two men asked Kingsbury, due to his role at NSF, to tie the Foundation into the project. He said no as the "human [aspect] takes us out of it." However, he offered to "change to my other hat," that of chair of the BSCC, and would look for more "players" to begin the HGP. Since Kingsbury had a "very good relationship" with James ("Jim" to Kingsbury) Weingarten, Director of NIH, the two of them put together a "package" for the nascent HGP and ran it past the White House and Senator Ted Kennedy, as he was behind part of the thinking on the subject. The same was true for Senator "Pete" Domenici (R-NM; in Congress from 1973 to 2009), and in whose home state the Los Alamos National Laboratory was situated: it would become a major locus for the HGP, as would Oak Ridge National Laboratory in Tennessee. A series of hearings attended by Kingsbury was undertaken in Santa Fe, NM, and later at NAS; the HGP began in 1990. For more on Eric Lander, the interested reader is directed to his short biographies at http://en.wikipedia.org/wiki/Eric_Lander, and also at http://genome.wellcome.ac.uk/doc_WTD021050.html. For more on the HGP, see especially http://www.ornl.gov/sci/techresources/Human_Genome/project/hgp.shtml, and the publications page at that site: http://www.ornl.gov/sci/techresources/Human_Genome/publicat/publications.shtml. A listing of book-length historical studies can also be found at: http://www.google.com/search?q=history+of+the+human+genome+project&hl=en&client=safari&rls=en&prmd=b&source=univ&tbs=bks:1&tbo=u&ei=h5UWTNaTLYH98Abk7935CA&sa=X&oi=book_group&ct=title&cad=bottom-3results&resnum=12&ved=0CEsQsAMwCw

[242] Mark Crawford, "Kingsbury Resigns from NSF," *Science* (1988):28.

conclude.[243] While the agreement was later modified for him to work fewer hours and for the pay to be less than a quarter of the initial amount, Kingsbury "claimed the agreement was never in force," Crawford wrote. However, on *March* 13, 1987, IGB had paid Kingsbury $9201.67 by check for some tissue sample costs, it was said, but not as income. Harsanyi contested that by noting it was, *indeed*, "for consulting services," Crawford stated. Kingsbury had not only endorsed the check on the back, but cosigned it on the front, suggesting he had signing authority for IGB on March 13. Kingsbury did, however, return that amount to IGB on June 22, some 3 months later. The House CST, continuing to watch matters, felt that the money should have been paid to George Washington University (GWU), where the AD taught part-time, not Kingsbury personally, reported the writer for *Science*. It was becoming apparent that Kingsbury's days at NSF were numbered and, finally, on September 23, 1988, Erich Bloch had in hand David Kingsbury's resignation.[244]

After Kingsbury's departure, and though the inaugural semi-annual report of the NSF Office of Inspector General provided no details about the Office's founding on April 1, 1989, it was the Kingsbury affair that was the reason Congress required NSF to set up that new office.[245] NSF had for a long period in its existence continued in so many ways to function much as does the academy: more collegial than bureaucratic. But it was clear that, like other agencies of the federal government whether independent or not, time and circumstances required the establishment of an internal watchdog.

Also in 1989, Kingsbury, though no longer at NSF but rather a full-time professor at GWU, published the first of two articles on computational biology as applied

---

[243] Ibid.

[244] Kingsbury told the author (interview) that he (Kingsbury) and MPS director Richard Nichols, his close friend, had both become disillusioned with Bloch's policies and so each of the two had been looking for new positions outside the NSF some time before Kingsbury left: "it's time" to leave, he said to Nicholson, and also to the author ("it was time to leave."). Kingsbury further stated that both of them had "spent so much time trying to teach him [Bloch] what science was. First, he was an engineer and, second, he didn't have a PhD. Didn't trust faculty." How much of that failure to appreciate the nature of science was true is unknown, but Bloch's acerbic style, at least, has been discussed in the main text above. Nichols, shortly thereafter, became the Executive Director of the AAAS and preceded Alan Leshner in the position. Kingsbury went full-time to GWU in October of 1988 in medical microbiology. He later took a project leader position at Johns Hopkins University (JHU) in Baltimore, Maryland, in the Human Genome Project. Yet later, JHU offered him the associate deanship for research and information technology in the medical school and thereafter he also became the founding editor of the new *Journal of Computational Biology*. In later years, Kingsbury worked in prion (infectious proteins) research with both Stanley Prusiner and Carlton Gajduseck, who later won the Nobel Prize for his efforts in that new area of biology.

[245] See the historical collection of reports of the NSF OIG at: http://www.nsf.gov/pubs/2002/old-semiannuals/oigseptember1989.pdf. It should be noted that the NSB had promulgated a "[Board] Policy Statement on Conflicts of Interest" at its April 17–18, 1980, meeting. Well before the Kingsbury matter, that policy had placed limitations on the "[m]any research workers with [NSF] grants or contracts," and was not then aimed at Foundation employees. See NSB, "Minutes," NSB-80-192, 215:19 (document number NSB/Res-80-39/B).

to biotechnology.[246] Former NSF colleague, John C. Wooley, director of the new Division of Instrumentation and Resources (DIR; partially the subject of Chap. 6) of BBS, published the second in the series (below).[247] Kingsbury was finally able to speak of desktop computers as typical tools for the scientist. They were coming to be seen widely at NSF by that time, thanks to the interests of former IBM leader, Erich Bloch. Kingsbury argued, citing others, that computers had become the "third node in the experiment-theory-simulation paradigm of scientific research."[248] He addressed computer modeling and networking, shared data and supercomputing, as well as graphics and what desktop computers should be able to do in future. He spoke of the "electronic lab notebook" and "electronic mail" just then making its appearance, though invented as far back as 1971.[249] Kingsbury had also been lecturing on virology, from fall of 1986 onward, at the GWU School of Medicine while still at NSF full-time.[250] Wooley had followed Kingsbury's paper a month later in the same journal with the second part of their joint review. It was there that Wooley examined, in particular, needs and opportunities for *computational biology related to biotechnology*.[251] He considered dynamic simulations, prediction of functional properties of molecules, X-ray crystallography, nuclear magnetic resonance, and numerous other areas. Simulations were, as was once said, the "third paradigm of science that illuminates theory and drives experiment."[252]

A massive change, not only for BBS but also for the Foundation itself, was on the horizon, but its magnitude for BBS, in particular, was not entirely clear in the final year of the 1980s. A reorganization of some sort had been under discussion off and on since the NSB initiated mostly informal discussions on it in the middle part of the decade. The history of the change that, among other things, would split all of the social, and part of the behavioral, sciences away from the biological disciplines, is

---

[246] Kingsbury interview.

[247] DIR was new at the time and Wooley was much interested in biological computation, hence DIR had the effect of replacing, to some degree, the lost IST that had been incorporated into CISE. Recall that one of the first suggestions made by the new AC was that IST should remain in BBS; it did not, however. DIR had as its concern all sorts of instrumentation for biological sciences; computers played no small role. See also main text earlier in this chapter. Wooley and DIR are considered in more detail in the following chapter.

[248] David T. Kingsbury, "Reviews: Computational Biology for Biotechnology: Part I: The Role of Computational Infrastructure," *Trends in Biotechnology* (1986):82–87. Quotation on pg. 82. Kingsbury was in reference to H. J. Raveche, D. H. Lawrie, and A. M. Despain, *A National Computing Initiative: The Agenda for Leadership* (Philadelphia, Pennsylvania: Society for Industrial and Applied Mathematics, 1987).

[249] See Ian Peters' history of email at: http://www.nethistory.info/History%20of%20the%20Internet/email.html, as well as http://www.historyoftheinternet.info/history_email.html; and Naomi S. Baron, *Alphabet to Email: How Written English Evolved and Where It's Heading* (New York: Routledge, 2000), *passim.*

[250] Kingsbury interview.

[251] John C. Wooley, "Reviews: Computational Biology for Biotechnology: Part II: Applications of Scientific Computing in Biotechnology," *Trends in Biotechnology* (1989):126–132.

[252] Peter Gund, *Science* (1992): Vol. 256.

considered in Chap. 5. Nonetheless, hints became more and more clear for BBS at the close of the 1980s that restructuring could not wait much longer. Too, the value of the still young BBS AC was becoming ever more apparent to the undercurrent of felt need for change. Under the dynamic leadership of Peter Raven and the visionary replacement for Kingsbury, Mary Clutter, the AC and BBS, respectively, were both moving into a position of greater notice in NSF overall.[253] Biotechnology and the STCs, even though most of the initial group of 11 were outside the biological sciences, signaled an increased importance for BBS as the last decade of the millennium was dawning. That was to be seen in so many of the discussions held by the NSB and in interest demonstrated by Bloch himself. The promise of Watson and Crick and the biological revolution was being recognized by the rapid development of molecular biology and its applied features within BBS and the communities it supported.

3.11 The AC at the Close of the 1980s

The AC met in May of 1989 and considered several matters of business.[254] In the Foundation at large, the problem of shortages of women and minorities, as well as the disabled, in science had been under discussion for some time (as it had been governmentally and nationally): the NSB had shown strong concern. Those shortages were seen in all the sciences, engineering and technology, biology included. The AC was concerned with the matter and efforts were being undertaken to bring change to the issue, Raven noted. But there were other matters of organizational structure, as well as pure science, on the minds of the members of the AC, too. Structurally, BBS continued to grow in program numbers and organizational complexity, Mary Clutter explained.[255] She presented institutional and national statistics

[253] Mary Clutter did not immediately become AD upon Kingsbury's departure. Frank Harris had been a program director in DEB when Kingsbury came and so the latter appointed Harris deputy AD. When Kingsbury left, Harris took over as acting AD until Clutter was put into that same acting position by Eric Bloch and then, some months later, into the permanent AD's post. Harris moved on to the Oakridge National Laboratory in Tennessee and Clutter appointed James Edwards to the deputy AD position. More is said of Edwards in the following chapters. Kingsbury interview.

[254] Interestingly, Mary Clutter argued that the AC had been "inactive" in a memo of hers to M. Rebecca Winkler, the Committee Management Officer of the NSF Division of Personnel Management (February 27, 1989) when Clutter was Acting AD for BBS. She indicated to Winkler that she (Clutter) would move "rapidly to revitalize this important advisory body" once she became permanent AD. This clearly happened because the AC met several months later in May of 1989. Exactly what Clutter meant in her memo to Winkler remains unclear as there does not seem to be any suggestion, in the paper records, that the AC was "inactive" for any period of time prior to May of 1989.

[255] Kingsbury said (interview) that Clutter "paid more attention to process" than did he (Kingsbury). She was quite different in her approach and Kingsbury opined that she "struggled to find the relationship she desired with the scientific community." As well, she was "more organismal[ly oriented] and less molecular than I was," he opined. Later in the interview, Kingsbury held that he

informing the AC of the state of science leading into early 1989.[256] The 5-year plan for the period FY91–FY95 was also a matter of interest to the AC. And the bugaboo of the STCs taking money away from funding individual grantees was still alive and well, as concerns made by a new AC member, Margaret Davis (University of Minnesota faculty member), were evidenced at the Committee's meeting.[257] Mary Clutter adduced data that showed that such concerns were unfounded and even argued that the rise of the STCs themselves was tied to the increase seen in NSF's budget in the final year or two of the decade.[258] A fair balance was obtained for both centers and individual grantees, it was implied.

As to science itself, three major initiatives were under discussion by the AC: global environmental change (discussed in later chapters); the program in dynamic form and function (DFF) lodged in the DCB (below); and, plant sciences. John Wooley also presented a discussion on biological databases. Coming out of the discussion of global environmental change was the ratification of the notion that *interdisciplinarity* was essential to its understanding. The then recent EXXON VALDEZ oil spill in Alaska was given as an example; it required a broad collection of specialists to attempt to deal with the disaster. Even though the main programmatic leadership for global environmental change came from the geoscience directorate, many divisions across the Foundation were involved, and BBS was certainly an important one among them. Roberta Miller, director of SES (Chap. 2), discussed global change human factors while John Brooks, division director of Biotic Systems and Resources, and Frank Harris, who led the discussion, spoke mainly in terms of big science. But at least one individual present, Margaret Davis, continued to feel that interdisciplinarity was stymied by the basic organizational structure of NSF. Further, the very rise of big science itself, such as the Long Term Ecological Research (LTER) program and the STCs, would endanger individual researcher's funding, opponents claimed, Mary Clutter's argument to the contrary apparently notwithstanding.[259]

A lunch discussion followed the first morning's reports and three issues emerged: a need to return to emphasis on funding individual researchers (that would only ever be partly realized, even though such was the overwhelming percentage: big science was becoming well ensconced at NSF); improvement of communications to facilitate more "multidisciplinary research"; and, improvement in the educational arena.

"funded science and she funded scientists"; another of their differences in approach. By doing so, in his opinion, the amount of money for a given grant went down. That approach would be reversed by her successor, James Collins, Kingsbury stated.

[256] A large group of pages (paper versions of overhead projected transparencies) includes data pages and the then-current organizational chart for BBS. BIO Lateral Files.

[257] Margaret Davis, AC Member from the University of Minnesota, stated that she was concerned with the issue on more than one occasion at the two-day AC event and even felt the need to apologize for what may have been strong words (not included in the Minutes) to the ecologists about her remarks concerning LTER. See also main text and footnote further below.

[258] BBS AC, "Minutes," May 5–6, 1989, pg. 1. BIO Lateral Files.

[259] Davis was that individual. See earlier footnote above.

The first two of those issues were tightly intertwined in the direction of the whole two-day meeting. Of particular concern to the trend of the AC's discussion, two presentations on the second day of the meeting were of importance: division director James Brown's discussion of DMB's initiative in dynamic form and function (DFF), and Bruce Umminger's DCB initiative in plant science. The latter was noted as having many cross-divisions, thus to some extent interdisciplinary, features. From the viewpoint of the historian's task, it is one of many areas of biology at NSF that is not easily amenable to explication in but a single telling. In fact, a careful examination of Fig. 3.3 evidences no *single* program listed as "photosynthesis" or "plant sciences." At the time, virtually *all* plant science fell under the Cell Biology program of the DCB, much of that under Developmental Biology. However, photosynthesis research, *per se*, was primarily under the Biophysics Program.[260] Thus, such sub-organismal and organismal aspects of plant research must be discussed in this and several of the following chapters: if interdisciplinarity is a theme of this history, so must also be the matter of how an institution (NSF; BBS) treats levels of biological organization.[261]

When the AC returned for its second meeting in October, just a bit more was said with regard to science education, the third issue brought up 5 months earlier.[262] NSF Deputy Director, John Moore, presented a discussion on both pre-college and college science education, but as it was held desirable to have a member of the Science and Engineering Education directorate present for the discussion, further consideration was tabled. Nonetheless, the NSB had heard in March that NSF had given out its 25,000th graduate fellowship![263] Much more was looked at in regard to global climate change, biodiversity, and computational biology. Wooley's colleague from the DIR, Jean M. Lauder, noted that only 5 years previously, "biologists were hostile towards computation but [that] the attitude" had been changing. Lauder offered the "influence of molecular biology" as a *central feature* to the increasing interest in that quantitative area. Erich Bloch was present at the meeting and, when a concern was expressed that computational science may not be moving quickly enough, he asked if it was thought that "NSF was dragging its feet" on the issue. Raven responded that in his "dealings with the NSF for fifteen years or so, NSF is always the leader."[264] True or not, the issue was coupled to the rise of big science by the voluble Margaret Davis, who was still openly concerned about the ratio of big science versus little science funding and also the continuing lack of Foundation mechanisms that would invite more interdisciplinarity. She wondered if NSF was "too conservative" in moving toward yet more computerization, though NSF's record

---

[260] Clark interview, *op. cit.*

[261] Ibid.

[262] BBS AC, "Minutes," October 23–24, 1989, pg. 2. BIO Lateral Files.

[263] NSB, "Minutes," NSB-89-58, March 15–17, 1989, 3–89:6.

[264] Ibid.

would hardly have supported her worries. Bloch answered: "reorganization alone doesn't always solve problems."[265]

Bloch brought his own agenda to the meeting. He answered an inquiry concerning Congress' attitudes about and understanding of science. The director reminded the AC that, with the partial turnover of members in the House of Representatives every 2 years, it was "necessary to start the education process all over again": not an inconsiderable point to be borne in mind by everyone. On his own agenda were science education and its relationship to research, sharing of research data, conduct of research, and conflict of interest—the last item being a matter much in the front of Bloch's mind recently due to the Kingsbury matter, presumably.[266] Those issues and others (global change, biodiversity, long-term science projects) were also in the director's thoughts as he had begun 1989 by telling the NSB that NSF "now has a 5-year authorization" for its budget, vice annually.[267] Looking further into the future for the Foundation had suddenly become much easier, it appeared. And notable the budget had become as NSF approached its 40th anniversary: "for Fiscal Year 1990, the appropriation passed $2 billion for the first time."[268]

Later in the year, Bloch, supporting the general direction of the federal government as promulgated by Ronald Reagan, noted that NSF was in the mainstream of policy abidance by stressing international cooperation, intensification of efforts in competitiveness of America's industrial prowess, and keeping pace with other countries as the times seemed to be, in the view of some, "a new Sputnik era." That was in evidence in the *Annual Report 1989*, which contained a large picture of grantee and biotechnology revolution leader Leroy Hood and a view of the prototype of the DNA sequencer. The article featured his and his team's creation of a computer chip and associated software that sped up the sequencing effort manifold. While the title of the piece was "Genetic Chip," a computer chip on which *actual* DNA would be attached was still in the future. Regardless, Hood's new invention was a boon to the already inestimable value of the sequencer.[269]

The new directions for BBS's little biology and biology of the little in the later 1980s were strongly oriented toward *applied bioscience*, providing the places, *centers*, and conditions, *computerization* (in part) for that to occur, and for support and infrastructure to be optimized on a large scale, and *big science*, for the little end of

[265] Ibid.

[266] National Science Foundation *Annual Report 1989* (Washington, D.C.: Government Printing Office, 1990), ppg. 4–5.

[267] NSB, "Minutes," NSB-89-30, February 9–10, 1989, 2–89:4. The five-year authorization approach has been used for NSF, but not without difficulties. Even at the time of this writing (2010), Congress required three attempts prior to approving the America COMPETES Act that reauthorized a five-year budget for NSF and several other science agencies.

[268] National Science Foundation, *Annual Report 1989* (Washington, D.C.: Government Printing Office, 1990), "Director's Statement," pg. 4.

[269] Ibid., ppg. 6–7.

the biological spectrum. Indeed, the 1980s had been, by some, "characterized as the 'decade of biology.'"[270]

3.12 Photosynthesis First, Then Plant Science

The origins of interest in photosynthesis by various incarnations of the biological sciences unit at NSF dated back to 1951 when the early BMS selected four areas for research emphasis: protein synthesis, enzymology, marine biology, and photosynthesis.[271] Since the Foundation was keenly interested in supporting scientific conferences, it funded the NAS Committee on Photobiology in 1952. Working together, that Committee and NSF organized two conferences on photosynthesis (and one also on another area of photobiology, bioluminescence, in connection with the expressed interest in marine biology). That commitment by NSF to support research in photosynthesis continued over the decades and it became one of the areas that moved from support primarily to individual scientists (little biology) to what the Foundation would eventually term *initiatives*—examples of which have already been noted above. The initiatives notion was born in the 1970s as a focus to "gain favor with the NSF hierarchy, OMB, and Congress" and depended upon internal commitment by program managers and independent lobbying groups (here, biologists with specific interests), and "a cogent argument for social relevance." Given its potential importance to agriculture and human food production, photosynthesis along with work in nitrogen fixation, biological catalysis, and microbial conversion of natural products were the primary initiatives in which BMS/BBS were first involved.[272]

By 1976, in the *Annual Report*, NSF was able to place photosynthesis (and nitrogen fixation) in the first few paragraphs of the opening "Director's Statement." That commentary had been titled "The Frontier is Still Endless," and Richard Atkinson, in his first such Statement, gave early notice of the importance of BBS's work in how green plants create foodstuffs.[273] In its own section of the report, BBS took up the issue of the importance of plant sciences indicating that the Foundation was taking on "greater responsibility" for the area due to its "immense importance to the world's food supply." Indeed, work in plant science was equivalent to one-third of BBS's entire budget at the time![274] Even with that said, BBS noted that while some other agencies (e.g., USDA) provided more support for plant sciences, "many promising" areas within that field were "going unexplored," certainly in NSF's

[270] National Science Foundation, *Thirty-First Annual Report for Fiscal Year 1981* (Washington, D.C.: Government Printing Office, 1982), pg. 29.

[271] Appel, "Shaping Biology," pg. 85.

[272] Ibid., ppg. 267–268.

[273] National Science Foundation, *Twenty-Sixth Annual Report for Fiscal Year 1976* (Washington, D.C.: Government Printing Office, 1977), pg. vii, *op. cit.*

[274] Ibid., pg. 55.

bailiwick of *basic* scientific research. However, in many subsequent years, little was said in NSF publications, or even in unpublished internal documents at the Foundation, about either photosynthesis or other areas of the plant sciences. Even with the taking on of "greater responsibility" in the area, plant sciences seemed to have had the same invisibility that so characterized them in much of the historiography of biology for the twentieth century.[275] Still, in that *Annual Report*, BBS discussed the research that had been ongoing both at UC, Berkeley and the University of Rochester, New York. How plants capture sunlight and how the reaction centers where the light is utilized, were both being explored at the two universities.

Other locations where progress in plant sciences was "considerable," and funded by NSF, included UC, San Diego, the Universities of Washington, Pennsylvania, and the City University of New York, especially with regard to molecules involved and the time sequence of their reactions in the complex photosynthetic apparatus. Electron transfer was of special interest as were the proteins involved in the flow of photon-excited electrons. A special type of microscope at the University of Oregon aided in locating the chlorophyll pigment and a laser tracked electrons in the energized chlorophyll molecule. As Nobel laureate, Albert Szent-Gyorgi, once had it: "What drives life is…a little current, kept up by the sunlight." Understanding the flow of that current—the trail of an electron and the molecules through which it

[275] In his quasi-historical account of twentieth-century biology, Ernst Mayr, in *This is Biology: The Science of the Living World* (Cambridge: Belknap Press of Harvard University Press, 1997), provided two pages on photosynthesis and one page on plant ecology—in 270 pages of text. Admittedly, Mayr placed a strong and primary focus on animal evolution in most of what he wrote. That includes his magnificent *The Growth of Biological Thought: Diversity, Evolution, and Inheritance* (Cambridge: Belknap Press of Harvard University Press, 1982), in which photosynthesis is not listed; plant breeders get three pages, and plant classification 13—of 860 pages of main text. Lois Magner's *A History of the Life Sciences* (New York: Marcel Dekker, 1993), a textbook typically used by professors who are other than professional historians of biology, gives one page to photosynthesis and two to plant sexuality. The book does not limit itself to the twentieth century, albeit, but covers the history of biology from ancient times. Garland Allen's highly regarded *Life Science in the Twentieth Century* (Cambridge: Cambridge University Press, 1978) lists neither the word plant nor botany! In all the foregoing examples, of course, the taxonomy and physiology sections often treat plant science, especially the former. As well, the term "plant science" itself grew to replace the less specific "botany," especially after about 1969 and the growth of the five-kingdom concept of living things—though not entirely (see below in this footnote). Fungi are not plants, nor are most of the other organisms studied under the moribund epithet "botany." NSF/BBS were both prudent and prescient in the 1970s in choosing the term plant science as opposed to botany to describe its choice of research areas to support. However, in an interview with the author, systematist and long-time BBS/BIO scientist-staffer, Jim Edwards, said that he had "two or three meetings of a group" of botany department chairs from academic institutions and they "wanted to keep the term botany because they were mainly systematic botanists." That is common in the field. They were, he said, "afraid that plant biology [as a term] implied that everything was going to be done via molecular kinds of approaches and systematic botany would be frozen out." Their fears would not be realized at BIO, but that botany would become increasingly molecular, would very certainly come to pass at the Foundation. James Edwards Interview with the Author, February 18, 2009.

passed—was an early critical issue to understanding photosynthesis at a deeper level than theretofore.[276]

The broader area of photobiology, however, occasionally got notice as it did in the 1977 *Annual Report* in a discussion of the photoproduction of hydrogen gas by certain species of both bacteria (*Rhodospirillum*) and eukaryotic algae (*Chlorella*).[277] And individual grants were being awarded for research in photosynthesis, such as that for $70,000 to faculty members Feher and Okamura for their study of "Primary Processes in Bacterial Photosynthesis," at UC, San Diego during that year.

Within a year, BBS was examining its potential future, as noted above, with regard to "New Scientific Opportunities and Research Resources."[278] Under the aegis of the PCM, plant science was the first subject listed as a "new [*sic*] opportunity." It was followed by fertilization and reproduction, gene splicing, and macromolecular assemblies. Specifically, a "major research question" was "by what biochemical/biophysical mechanisms are the early events in photosynthesis accomplished?" But, as the five-page white paper covered *all* of BBS, no more was said on photosynthesis or plant science research, *per se*.

Atkinson's "Director's Statement" in the 1978 *Annual Report* only gave plant sciences a passing notice when he said that "[s]tudies last year at the cellular level… opened new vistas in both plant pathology and the neurosciences." It was not the specific duty of the director to address issues peculiar to only a single directorate, of course. Though one-third of BBS's budget in recent years had gone to plant sciences, the level of visibility in the scope of work done by NSF to support them did not offer the glitter that would be biotechnology within a short period. It remains somewhat unclear as to why plant sciences, even though strongly emphasized by both the NSB and the BBS in the record, seemed in those years still to be something of a stepchild. However, it may have been that the very *sudden* rise of biotechnology, and concomitant *heavy funding*, broad utility, and financial promise overshadowed the strongly voiced intent to increase research in the plant sciences at just that same time. Historically, too, botany was always a lesser player in university biology departments than were the animal sciences.[279] Additionally, due to the tough cellulose in the cell walls in plants, working with the protoplast (everything that makes up the plant cell inside the cell wall) remained difficult for a very many years. Yet further, the neurosciences were being heavily funded during the period. Finally, it may have been that David Kingsbury had little interest in plants and that was reflected in the actual support that the directorate provided to the community: his interview with the author hints at this conclusion.

BBS *did* note in its section of the *Report*, that "plant scientists are now trying to specify the molecular processes underlying the regeneration of whole plants from

[276] NSF, "26th Annual Report 1976," pg. 58.

[277] National Science Foundation, *Twenty-Seventh Annual Report for Fiscal Year 1977* (Washington, D.C.: Government Printing Office, 1978), pg. 65.

[278] BBS, "5 Page Issue Paper," January 30, 1978, pg. 1, *op. cit.*

[279] Both general and personal knowledge.

vegetative cells," however.[280] (Gottlieb Haberlandt, as early as 1902, was the first to suggest that whole plants could, indeed, be cultured from single cells as his own attempts, though unsuccessful, clearly suggested the possibility.[281] But it took most of the twentieth century to get to the point where that was successful.) That was a growing area of interest to BBS-supported plant scientists as it implied that single cells (from a leaf, say) seemed to be *totipotent*. In other words, their full genetic load was active, or could be activated, to function as if they were a zygote (fertilized egg) that could lead to an adult plant, but with no sexual reproduction necessitated. A complete plant (admittedly a clone), then, could be grown up from virtually any single adult cell of that species.

"Cell division factors" were being investigated by Fred Meins at the University of Illinois in the totipotency connection, as was the study of character selection within single cells by James Shepard at Kansas State University. Shepard was seeking virus resistance in Russet potatoes: the human food connection remained very much in place with NSF-supported plant science, but that was at a basic science level, as compared to projects awarded from the much more applied mission agency USDA.[282] Even in important farm states and those with leading agricultural universities, there did not seem to be major support visible in the plant sciences: Nebraska only received four total PCM grants in 1978, none in plant science, as just one, albeit isolated, example.[283]

The "Program Report" that came out from PCM in July of 1979 was when botanist Mary Clutter was still director of the program in developmental biology, a part of the DCB and a locus for much of plant science. The division director was still then Henry C. Reeves, but Clutter would assume that role later. She was able to pursue the matter of plant cell culture, one of the areas her program supported. As she noted,

> [f]or the past 10 years, every major study on food and nutrition has emphasized the need for the United States science and technology establishment to direct its efforts toward solving future global food needs.[284]

There was the bold hope that rDNA in plants allowed for accomplishing "within one plant generation" that which field agricultural would require a "minimum of 20 years" to complete. That matter was foremost in the mind of Clutter, and many others in the early period of biotechnology. While Clutter averred that Haberlandt's

---

[280] National Science Foundation, *Twenty-Eighth Annual Report for Fiscal Year 1978* (Washington, D.C.: Government Printing Office, 1979), pg. 61.

[281] A short festschrift on the occasion of Haberlandt's 80th birthday can be seen at: http://www.ncbi.nlm.nih.gov/pmc/articles/PMC439112/pdf/plntphys00322-0148.pdf. His birthplace is given there as Ungarisch-Altenburg, but the city was also named Magyarovar due to its Hungarian (Magyar) history. As an interesting aside, Carl Erich Correns, one of the three biologists who, almost simultaneously, rediscovered the work of Gregor Mendel, was a student of Haberlandt.

[282] NSF, "28th Annual Report 1978," pg. 66.

[283] National Science Foundation, "G&A, 1978," pg. 118.

[284] NSF, "Program Report: PCM," July 1979, *op. cit.*; "Organismic-Level Biology: Plant Cell Culture," pg. 33ff.

1902 predictions for plant cell culture were not too bold, the hope that an early start on genetically modified organisms (GMOs) where those creatures were plants and not animals *was* premature even three-quarters of a century later, as we have just seen. Clutter discussed the work of Shepard at Kansas State. But even though he was working with the naked protoplasts (cell walls removed), he was not attempting to create a GMO; that would come shortly, however. On the other hand, animals, lacking a cell wall around their cells, were much easier targets for roles as GMOs. Even though bacteria also possess cell walls, their nature is very different from those in plants so genetic manipulation with them was not so daunting; hence part of the reason for their central role in the history of biotechnology.

Plant cell culture continued to be the star of the BBS section the following year when the fusion of two protoplasts from the potato had become a line of experimentation in attempts to produce a "super potato."[285] But a nod was made in the *Annual Report* to plant molecular biology as one of the highlights of the past year's grantee accomplishments, too. The discoveries of Robert S. Bandurski at Michigan State University and Shang Fa Yang at UC, Davis were described with regard to new work with the long-known plant hormone, indole acetic acid, and the newly discovered hormone, ethylene gas—the first known gas to play a role as a hormone in a living thing. Further, nitrogen fixation research had been pursued by Timothy C. Hall at the University of Wisconsin, Madison. He made discoveries in protein quality and the genetic changes responsible for that in leguminous plants.

The work of others was also described and finally began to demonstrate that, in fact, a goodly amount of research in plant science *was* being supported by BBS during the 1970s, even though both the paper and electronic records have left this subject somewhat veiled for the historian. (Given the difficulties of grant data retrieval at NSF, the building of a broader historiographical picture assembled primarily from individual award stories remained difficult at this writing.) The 1980 *Annual Report* began to lift that veil, though, by noting important work on the genetic history of maize (corn) recounting Hugh Iltis' crucial discovery of *teosinte*, the ancient relative of the familiar crop. Iltis, at the University of Wisconsin, Madison, found that teosinte was "immune to several major corn virus diseases" and if a GMO could be created between it and choice maize varieties, it would be a signal accomplishment. Though listed under the header of "Environmental Biology" in the report the teosinte discovery, as well as research done by Norman Negus of the University of Utah on plant-animal relationships, indicated the growing interdisciplinarity seen at BBS. Iltis' work was tied to organismal biology, genetics—and archaeology. That of Negus examined field ecology, but that through macromolecules in plants and their effects upon animals at the organismal and community levels. A close reading of the *Annual Reports* shows that it was hard to separate the life sciences into neat taxonomic categories for reporting their successes: the overlap, interdigitation, and melding of subdisciplines and organizational levels within biology, and even some

[285] National Science Foundation, *Twenty-Ninth Annual Report for Fiscal Year 1979* (Washington, D.C.: Government Printing Office, 1980), pg. 52.

of the social/behavioral sciences, made it impossible. While PCM focused on biology of the little in plants, the Division of Environmental Biology (DEB) focused on "plant distribution, plant-plant and plant-environment interactions affecting growth and productivity."[286] Such separation of little and big became less so with time. That was just exactly what had been wanted in biology at the Foundation in the first place: choosing the *functional* organizational concept of Paul Weiss as far back as 1951.

Interest continued to focus on the plant hormone abscisic acid in its role inside maize kernels on the cob.[287] The matter was of concern in increasing production of maize in connection with the basic plant sciences' ties to world food needs. Nitrogen fixation remained a notable area of research, as well. The work of Frederick Ausubel at Harvard was exemplary. He had made a number of discoveries in regard to various genes in the bacterium *Rhizobium* that is found in the nodules of leguminous plants and which is responsible for converting nitrogen gas to usable forms of nitrogen for the green plant host. Though funded by DEB and not PCM, the translocation of photosynthate (primarily sugars) in plants was under study. Using radioactive carbon (C^{14}), the movement of that element had been traced in plant tissues earlier by a number of researchers, but the destruction of the plants was necessary to assess that movement. Instead, researchers at both Duke University and Texas A&M University found that using another radioisotope of carbon (C^{11}) they were able to measure the circulation of that element without damage to the living plant. The work was done at Duke's Phytotron, a facility for plant science research originally established by NSF in 1968 and managed by the Foundation for many years. It later became fully a part of Duke University's biology department.[288]

Board member John Hogness, in his role as chair of the BBS mini-group in the NSB, was charged with a number of issues to examine beginning in early 1982. One of those was "what the relative BBS emphasis on research in the plant sciences should be, and how it compares with that of other Federal agencies."[289] It was at that meeting that the nature of BBS was held to be "diverse and diffuse" (see earlier) and analyses of the directorate's performance was made difficult as a consequence. It will be recalled that that was the genesis for establishing the first, but abortive, AC for BBS. More came of the discussion than that, however. Plant sciences (along with the neurosciences) "were identified as examples of high priority thrust areas": but at what minimal level of support could a program remain viable and produce "quality science"? No answer was forthcoming at the time, but by the June 1982 meeting, the Board listed 14 topics for consideration, one of which was plant sciences and agriculture.[290] It was agreed that two areas, engineering and plant sciences:

[286] From the BBS 1984 Budget Request document, pg. BBS-2.

[287] NSF, "Annual Report 1981," *op. cit.*, pg. 30.

[288] See: http://www.biology.duke.edu/plantfacility/phytotron.

[289] NSB, "Minutes," NSB-82-103, March 22, 1982 (February18–19 meeting), 2–33:10.

[290] Ibid., NSB-82-243, August 23, 1982 (June 16–18 meeting held at UC, San Diego), 2–36:17ff.

> have been so frequently discussed by the Board that no further agreement was needed before moving ahead with discussion with the external community and the relevant Assistant Directors…We agreed to experiment in these two cases with the use of individual Board members, rather than committees [or mini-groups] to do the fact-finding and discussion…Gene [Eugene H.] Cota-Robles [was charged with the plant sciences.][291]

While the evidence in the NSB Minutes does *not* support the conclusion that plant sciences had been "so frequently discussed by the Board," it is immaterial to the decision that finally had been made. The text went on to state that:

> [i]n plant sciences…Cota-Robles will be looking beyond NSF, asking what opportunities exist in plant sciences for the United States, how they might benefit the country, and what kinds of research programs are needed (whether at NSF or elsewhere). The objective is for us to understand opportunities that exist in the plant sciences and how we might provide leadership for the community as a whole, in cooperation with others. Cota-Robles…will provide us with a status report.

In a table that accompanied the Minutes of that Board meeting, plant science and agriculture was given a "purpose": "Understand where this field, which NSF identified to OSTP in 1980 as deserving of a special initiative [above], is going and where we fit in, including relations to programs in NIH and Agriculture." Betsy Clark (still then AD) was listed as a possible NSF participant along with Cota-Robles as the "responsible" party.[292]

The plant sciences, even if flagged as important and as an area for strong initiatives, appear not to have been so clearly dealt with at NSF during the 1970s as was biotechnology—and plants played little role in the early period of that applied area. Biotechnology, once it had become apparent that the community of biologists would rapidly turn to it and develop it to such strength, was supported avidly by NSF/BBS in the later 1970s, and beyond. The plant sciences appear to have had a more irregular beginning in the 1970s within the Foundation, no matter the level or intensity of discussion about them. But interaction between BBS and the community appears to have gained both order and momentum by the beginning of the 1980s with regard to plant research and it can be said that they were then on a more solid foundation. Once botanist Mary Clutter assumed the leadership of BBS in the late 1980s, that foundation grew apace, as discussed in later chapters.

At that August 1982 NSB meeting, Cota-Robles made his report for the membership. He was then charged with pursuing the matter and bringing a group of plant scientists to the October meeting to discuss further the contents of the report; and at that meeting the discussion of plant sciences was extensive indeed.[293] Specialists included: Lawrence Bogorad, the Maria Moors Cabot Professor of Biology at Harvard and one of the leading figures in the study of photosynthesis and plant

[291] Ibid., 2–36:18.

[292] Ibid., Table 1, 2–36:20.

[293] Ibid., NSB-82-365, November 22, 1982 (October 21–22 meeting), 2–39:7–9. In a footnote on the beginning page (2–39:7), it was indicated that BBS staff associate, Thomas S. Quarles, wrote the summary for the presentation. Quarles is mentioned above in the main text and in footnotes in another connection.

genetics, and a member of the new BBS AC;[294] Robert H. Burris, professor of biochemistry at UW, Madison; and, Virginia E. Walbot, associate professor of biological sciences at Stanford and also an AC member. Bogorad's topic was plant molecular genetics. He indicated that science knew the positions on chromosomes of only some ten genes of 100,000 estimated to exist in plants! And little was known of what those ten coded for. However, successes were being seen in the introduction of foreign genes into protoplasts via an infectious bacterium (*Bacillus thuringiensis*; commonly termed Bt) as the vector (carrier agent). The production of genetically modified organisms was thereby facilitated. Burris spoke primarily on nitrogen fixation in the 10,000 known species of legumes. The general environmental requirements of the 17-gene nitrogenase system were becoming known, but the hope for introducing such a complex set of interlinked macromolecules into non-leguminous crop plants such as maize or wheat did not look promising—nor has it become a reality even at the time of this writing.[295] Chemical fertilizers could not be eliminated, as was the hope if plants *other* than legumes could use atmospheric (gaseous) nitrogen as fertilizer. The nitrogen fixation system also has a high energy requirement. Maize might produce over 100 bushels per acre, whereas the legume soybean produced only 40: the primary difference was how much energy goes to the nodules for nitrogen fixation.

Walbot discussed plant pathology with special attention paid to chemical signaling between pathogen and host and their evolutionary relationship, among other more general topics.

More to the point of what NSF could do, given the commentary by the three plant scientists, was discussion about the shortage of trained such researchers and a lack of "critical mass" of sufficient specialists on university faculties across the country (not necessarily "botanists," however). Since, as it was argued, the plant sciences had a close relationship to "chemistry, physics, and biochemistry," sophisticated equipment was needed and the universities were also short in that area, too, though NSF support to that time had been significant and was praised. More was needed, however. That plants had a close relationship to chemistry and biochemistry, as areas of biological study, had long been acknowledged. As the leading ethnobotanist in America, Richard Schultes, once had it: "Plants live by their chemical wits." The speakers did indicate that the majority of agricultural and plant science research occurred at land-grant colleges with USDA funding, as well as at USDA laboratories, and seed and chemical companies. In most cases, research was *applied* in the latter three of those institutions and training did not occur, except at the universities and colleges. A further problem was that most USDA funding was for a very few crop plants, many not amenable to laboratory study or as teaching materials. Continuity of funding was also unstable, especially from USDA. Europe and

[294] For more on Bogorad, see: http://www.mcb.harvard.edu/NewsEvents/News/Bogorad.html, several web pages there; and also *Plant Physiology* (2001):1345–1346.

[295] See, for instance, http://www.springerlink.com/content/l34121n818807122/, where biological nitrogen fixation is examined via implanting of whole bacterial cells into a host plant as opposed to the early hope of creating a GMO of a non-legume capable of acting as if a leguminous plant.

Australia were doing much better than the United States, the three guests asserted, and NSF's support was crucial to increase plant sciences on all fronts in this country.

There were other issues afoot at the time that would change, to some degree, the balance of funding sources for support in the sciences, overall. "In 1980, industrial sources of R&D expenditures exceeded Federal Government sources for the first time in the last two decades," reported *Science Indicators 1982*.[296] Furthermore, research and development was strongly tilted toward the "D" at a rate of growth from 1980 through 1983 of 5.4% by industry, whereas it was only moving upward at 3.1% annually via federal sources. And that had represented a decline, from all sources, of 2.4%.[297] Supporting science financially at that time would seem not to have been an easy task. Under Reagan's increase of support for the military, national defense R&D would represent 64% of the 1983 total federal R&D budget, where it had been only 47% in 1980! However, basic research was to see a 22% increase by 1984 as the government was being taken out of "commercial demonstration projects."[298] NSF could hope, then, to put more money into basic research in the plant sciences, as well as support education in the area.

Data showed that "almost all S/E's [scientists and engineers] who want jobs are employed," though not all of them within their respective areas of interest and training. By 1981, there were some 2.8 million persons in the S/E workforce.[299] So it was due to a combination of the assessment by the three NSB guests, the R&D environment, and the employment milieu, that one BBS scientist introduced the notion of a program for Postdoctoral Research Fellowships in Plant Sciences (briefly noted earlier). Carter Kimsey, associate program director in DCB's metabolic biology program, took her argument to her supervisors to establish such a fellowship program.[300] Her five-page memo, and attachments, contained extensive justification for the design of the fellowship program she proposed.

By the beginning of 1983, the BBS was able to note three major research areas of growth for the directorate: neurosciences, biotechnology, and the plant sciences. It was the third area that "promises enormous commercial payoffs in forest and field and exciting discoveries in the laboratory and greenhouse."[301] Beyond that, plant scientists had a long "wish list" of plant characteristics that they sought to improve for human benefit, and those scientists were "determined" to see that research through. Startlingly, "[t]he estimates are that plant genetics is about 20 years behind bacterial and animal genetics."[302] Catch-up would have to be part of the game plan.

---

296 National Science Board, *Science Indicators 1982*, pg. 40.

297 Ibid.

298 Ibid.

299 Ibid., pg. 62.

300 Carter Kimsey to Elijah B. Romanoff and William van B. Robertson, May 20, 1982. Provided to the author by Carter Kimsey.

301 National Science Foundation, *Thirty-Second Annual Report for Fiscal Year 1982* (Washington, D.C.: Government Printing Office, 1983), pg. 27.

302 Ibid.

Legumes were used to study the distribution of resin-producing plants across continents. It was an excellent example of interdisciplinarity in research as it lay at the juncture of taxonomy and ecology, as well as geology with regard to the history of plate tectonic movement (Gondwanaland distribution of the legumes), but little else in plant science was mentioned for that year.[303]

As has been seen, the *Status of Science Reviews 1984* came out with reviews of all areas under BBS's purview.[304] The then recent COSEPUP—Committee on Science, Engineering and Public Policy of the NAS[305]—review on the outlook for science over the coming 5 years found three significant areas for PCM to deal with; one of those was plant biotechnology.[306] While "fundamental knowledge" was clearly the first area to be attacked when the field was being considered, genetic engineering to create desirable crop plants was the main theme. Several attendant areas of interest were mentioned: plant development; plant and pest interactions; organisms in the rhizosphere (the soil-root zone); photosynthesis and carbon metabolism; mineral uptake; biological (nitrogen) fixation; and, the plant genome structure and gene expression. More will be said of these points in later chapters.

It was certainly clear from the list that a very great deal had yet to be understood about plant biology beyond the long classical foci on taxonomy, nomenclature, anatomy, and morphology. Indeed, even in the nineteenth century, French entomologist Jean Henri Fabre bemoaned the fact that:

> [h]istory celebrates the battle-fields whereon we meet our death, but scorns to speak of the ploughed fields whereby we thrive; it knows the names of the king's bastards, but cannot tell the origin of wheat.

Such a situation was in need of change and Carter Kimsey's suggested postdoctoral fellowships' program, accepted and instituted in 1983, was cited as one way through which new knowledge could be forthcoming. The OTA and OSTP reports on biotechnology (above) were anticipated and the American Society for Microbiology recommended supporting research on photosynthetic bacteria and Archaea (evolutionarily ancient bacteria-like organisms). The land-grant colleges issued a report on "Emerging Biotechnologies in Agriculture" at just that same time.[307] The authors of the *Status of Science Reviews* went on to point out the great importance of interdisciplinary research noting that research with plants "falls appropriately within the domain of each PCM program, but it may also be viewed as an interdisciplinary activity." The closely related program in the Chemistry of Life Processes also considered molecular biology and spilled over into the Chemistry

[303] Ibid., pg. 33.

[304] Division of Planning and Policy Analysis, Office of Planning and Resource Management, NSF, *Status of Science Reviews 1984*, November, 1983.

[305] For more on the history and purposes of COSEPUP, see: http://sites.nationalacademies.org/PGA/COSEPUP/PGA_044177.

[306] Authors of the "Status of Science Reviews 1984," citing, on pg. 78, the COSEPUP review.

[307] Ibid., pg. 79.

Division of the MPS directorate, as well as other programs in BBS, as borders separating these various units were "broad and arbitrary."

There was also work in photosynthesis where a connection to medicine showed "how unanticipated benefits can emerge from an esoteric fundamental study." Few examples were more powerful in regard to interdisciplinarity than the work done by Alexander N. Glazer at UC, Berkeley along with Lubert Stryer and Vernon Oi at Stanford when they noted natural light absorption behavior in the phycobiliproteins of the so-called blue-green algae (properly blue-green bacteria or, preferably, cyanobacteria) and saw a possible application to biology more broadly—and also to medicine. There are a number of differing types of those protein molecules and they fluoresce brilliantly in a variety of colors. Joining the phycobiliproteins to selected antibodies (making a conjugate) and then introducing them to cells, which would take up the antibodies, allowed the researcher to see various colored "lights" in specific locales within, say, a mixed tissue sample. Should some group of cells or another be responsive to a given conjugate, such molecular tools might identify an invisible tumor, for instance.[308] The oft-repeated phrase that no matter how long basic research findings may precede something other than intellectual satisfaction in the knowing, they someday may very well find application. Basic scientific research needed no better example of its value than the work of Glazer, Stryer, and Oi.

Under the heading of what had become, by 1982, "Biotic Systems and Resources" (formerly "Environmental Biology," DEB, and then again DEB in years to come), research in the mineral nutrition of plants was underscored with regard to carbon physiology by ecologists.[309] However, it was not just the BBS that spoke of its grantee's achievements of the time, but so also the NSB that continued to promote the need for more work in the plant sciences. It had been, and continued to be, one of just a few areas that the Board hoped to see expanded. The BBS mini-group reported to the full NSB in January of 1983 that three areas of significance to the directorate "transcend the various disciplines." They included the need for more young investigators, more instrumentation, and a "focus on plant biology."[310] Of particular concern to the Board was the fact that the plant sciences were so widely spread programmatically across BBS (and beyond), that AD Betsy Clark was charged with bringing back to the NSB additional information on how all of that was managed. That would never come to pass, however, as Clark would leave NSF by mid-year. The Board had also produced a skeletal 5-year plan that, in brief text points, stated the priorities for BBS from 1985 to 1989. For plants, the directorate was to "extend new techniques of molecular and cellular biology to plant systems," and to "[i]mprove biochemical understanding of plant pathways and enzymes" as research priorities.[311]

[308] National Science Foundation, *Thirty-Third Annual Report for Fiscal Year 1983* (Washington, D.C.: Government Printing Office, 1984), pg. 30.

[309] Ibid., pg. 34.

[310] NSB, "Minutes," NSB-83-58, February 22, 1983 (January 20–21 meeting), 1–83:16.

[311] Ibid., NSB-83-92, March 22, 1983 (February 17–18 meeting), 2–83:21.

By 1984, there appeared mention, as has been seen, of the use of thale cress, *Arabidopsis thaliana*, as a model system for the genetics of flowering plants.[312] One of its many virtues is that it can be grown easily in the laboratory aseptically from seed, thus uncontaminated by other organisms and their genomes.[313] That was one feature, which "led several scientists to believe that it may represent the simple model plant system for molecular and genetic studies" and, for the many reasons of its ideal nature, "several molecular biologists have begun research with this plant."[314] The *Annual Report 1984* said little else about thale cress beyond those two sentences and one benign photograph. Though known as early as the sixteenth century, thale cress was just coming onto the scene in molecular biology. Much more will be said of the species in later chapters as its use in the several subdisciplines of biology, to which it is so well suited, continued to grow at BBS and later at BIO. As Richard Fortey, arguably the leading figure in the study of the extinct arthropods known as trilobites, once quipped: "[t]he deep language of the genes is an Esperanto of biological design which can be understood by a Babel of organisms." The Babel had, by 1984, grown beyond the virus lambda, the fruit fly *Drosophila*, certain bacteria, maize, and several others, to include the lowly weed, thale cress.

That same year of 1984 witnessed the first *regular meetings* of the fully formalized BBS AC. One of the speakers, Mary Jane Osborn of the NSB, emphasized at

[312]While little of what follows in the first half of this footnote was known in 1984, thale cress was shown, over time, to be *the* ideal model system for the genetics of flowering plants because of its small genome (later shown to have 115,409,949 base pairs) of DNA in five chromosomes, so 2 N = 10; little of the DNA is "junk"; the total number of genes proved to be 25,498; it is easily amenable to GMO studies as it can accept vectored foreign genes with ease; it is small, 6–12 inches in height above a basal rosette, and fast growing (maturing in 5–6 weeks) in the laboratory; the seed production is prolific (up to 10,000 seeds per plant); mutations using radiation or chemicals are easily produced; and, since it is self-pollinating, the induced mutations become homozygous quickly. See: http://users.rcn.com/jkimball.ma.ultranet/BiologyPages/A/Arabidopsis.html. See also the official site, TAIR, for the plant at: http://www.arabidopsis.org/portals/education/aboutarabidopsis.jsp. There are numerous other websites devoted exclusively or partially to thale cress, as well, such as that at Ohio State University: http://abrc.osu.edu/. At the TAIR site, Elliot Meyerowitz provided a brief history of our knowledge of thale cress: "*Arabidopsis thaliana* was discovered by Johannes Thal (hence, thaliana) in the Harz mountains in the sixteenth century, though he called it Pilosella siliquosa (and it has gone through a number of name changes since). The earliest report of a mutant (that I know of) was in 1873 (by A. Braun). F. Laibach first summarized the potential of Arabidopsis thaliana as a model organism for genetics in 1943 – he did some work on it much earlier though, publishing its correct chromosome number in 1907. The first collection of induced mutants was made by Laibach's student E. Reinholz. Her thesis was submitted in 1945, the work published in 1947. Langridge played an important role in establishing the properties and utility of the organism for laboratory studies in the 1950s, as did Rédei and others (such as J. H. van der Veen in the Netherlands, J. Veleminsky in Czechoslovakia and G. Röbbelen in Germany) in the 1960s. One of Rédei's many important contributions was to write scholarly reviews on Arabidopsis, a particularly thorough one is in *Bibliographica Genetica* vol 20, No. 2, 1970, pp. 1–151. He wrote a more easily found one in *Ann. Rev. Genet.* (1975) vol. 9111–127. Both go through some of the early history of the use of Arabidopsis in the laboratory, though the longer 1970 one has all the details."

[313]NSF, "Annual Report 1984," pg. 20.

[314]Ibid.

the January meeting the importance of plant science, noting that the Foundation could point to successes in that area of its support and expected the plant sciences "to continue to be of importance for the indefinite future" both scientifically and agriculturally.[315] Nothing more was said of plant science in either that first meeting of the AC in 1984, or at the second, September, meeting. However, the FY84 budget justification spoke in terms of "molecular botany" and modern subdisciplinary studies in plants, including the now familiar areas of interest to BBS, plus also a newer interest in plant stress (tied to an early grasp of global environmental change). By that period the amount of funding provided to plant science by NSF had grown to be equivalent to that given out by USDA.[316] And the PCM promoted plant science grant applications heavily, stating that:

> [t]he Division encourages research in plant biology in all of its programs. Newly emerging areas of high interest such as the molecular biology of plant systems are emphasized [but so also are] plant biochemistry, embryo development, somaclonal variation and the basis for totipotency.[317]

"All of its programs" included then: Alternative Biological Resources; Biochemistry; Biological Instrumentation; Biophysics; Cell Biology; Cellular Physiology; Developmental Biology; Genetic Biology; Metabolic Biology; and, Regulatory Biology. In June of 1986, an NSF/USDA/DoE plant science program was drawn up in a joint statement after discussion with the OSTP.[318] That new competitive awards program called for interdisciplinary research and training centers at academic institutions. As American agriculture was "one of few major national enterprises that account for a positive trade balance," and as future competitiveness depended upon both strong agricultural and basic plant research, and also recognizing that the latter is "completely inadequate," the joint program sought to correct those difficulties. By use of cellular and molecular techniques and genetic engineering, as well as ecological techniques (ecological processes in agroecosystems, land management, natural resources use), increases in plant science faculties, and increases in industrial scientists, the program would move forward. Five multidisciplinary centers were proposed, each for 5 years, and with possible renewal, including: rhizosphere dynamics and microbial ecology; plant biotechnology; complex carbohydrates; ecological processes; and, systems research and analysis. New facility construction funding was *not* included, so the centers would not be under the

---

[315] BBS AC, "Minutes," *op. cit.*, pg. 2.

[316] Internal document for BBS for FY84 budget consideration; pg. BBS-3. "NSF is essentially the exclusive source of support for research that is not immediately related to the mission of the [USDA.]" The "Programmatic Topics," pg. BBS-I-3, emphasized the areas of research interest to be supported in the plant sciences; it did not differ from what has been discussed in the main text above, but went into some detail about plant disease genes (BBS-I-5), and plant community structure (DEB; BBS-II-5).

[317] A general program announcement for all of the PCM programs for FY84, pg. 1.

[318] "Planning for Enhancement of Plant Science Research in the Fiscal Year 1988 Budget," June 17, 1986, signed by the Assistant Secretary for Science and Education, USDA; by the Director of Energy Research, DoE; and by the Director of the NSF.

STC program also coming on line that year. The joint program centers would include all sectors of the economy with the first year's budget to be $10 M shared equally by the three participant agencies. However, the Gramm-Rudman-Hollings Act, which took effect March 1, 1986, would shorten NSF's FY86 budget by 4.3%: over $65 M.[319]

The *Annual Report 1987* saw the first significant recognition of the growing science surrounding and made possible by *Arabidopsis thaliana*. The "distant cousin to the common mustard plant," thale cress, was "creating excitement in the community of scientists" studying plant genetics.[320] Cal Tech's Elliott Meyerowitz was said to be the first scientist to study *A. thaliana* at the molecular level. Certainly, he was a pioneer then working on a complete restriction map of the cress' genome.[321] Restriction mapping is a procedure undertaken prior to cloning a gene, among other purposes. Meyerowitz' group was also studying flower development and one of his graduate students went on to become an NSF Plant Postdoctoral Fellow at the University of Minnesota working on male sterility in *Arabidopsis*.[322] Another previous Fellow, by then on the faculty at the University of Pennsylvania, was researching the endogenous synthesis of plant hormones. Yet other senior researchers, such as Gerald Fink and his Postdoctoral Fellow, Robert Last, at MIT were creating auxotrophs (specific nutrient-requiring mutants), highly difficult to do in plants, but long then common with bacteria. Such living "tools" were used to aid in the explication of biochemical pathways in an organism. "What we…learn about the basic genetics and biology of this weed will have a great impact on the potential for bioengineering food plants."[323]

Clearly, plant science was growing rapidly by the later 1980s and would continue to do so into future years under Mary Clutter (later chapters). The divisional funding for plant biology from FY88 to FY90 moved from over $18 M to more than $21 M, an increase of greater than 17%. The total DCB budget for plant research was running nearly one-third of its full budget and continued to increase in percentage to nearly 39% by FY90, a 12% leap in just 3 years.[324] The NSF/USDA/DoE plant research cooperative program soon included centers and was highlighted in the *Annual Report 1988* for the first time. The first of the centers was funded in September of that year at Cornell University, long one of America's leading institutions of plant science and agricultural research: the Center for the Experimental

[319] NSB, "Minutes," NSB-86-32, January 17, 1986, 1–86:2.

[320] National Science Foundation, *Annual Report 1987* (Washington, D.C.: Government Printing Office, 1988), ppg. 48–49. The article was unusually long for that period for the *Annual Report* and also contained four pictures, one was of a scientist at Elliot Meyerowitz's Cal Tech laboratory of plant genetics. Meyerowitz wrote the history of the use of thale cress in 1998, still in use in 2010 on the TAIR website; see footnote further above.

[321] A current (2010) tutorial on restriction mapping can be seen at: http://faculty.plattsburgh.edu/donald.slish/RestMap/RestMapTutorial.html.

[322] NSF, "Annual Report 1987," pg. 49.

[323] Ibid.

[324] Data taken from an ephemeral sheet dated by hand as "approximately" July 20, 1990.

Analysis and Transfer of Plant Genes was funded by NSF. Under the joint program, USDA simultaneously supported the establishment of Michigan State University's Center for Genetic and Biochemical Alteration of Plant Lipids and Starch. Finally, the DoE aided in the founding of the Center for the Study of Early Events in Photosynthesis at Arizona State University (ASU). The triumvirate of those agencies was an excellent example of federal cooperation. Additional funds from DoE went to ASU's already-established Plant Science Center for equipment.[325]

At the May 1989 meeting of the AC, Bruce Umminger, who had replaced Mary Clutter as director of the DCB as she moved to head the directorate, presented a short discussion on the plant science "initiative."[326] The so-called initiative had by then become a collection of well-established programmatic efforts in the plant sciences, both little and big, that included most visibly the three-agency centers' efforts. Umminger focused on: the cross-divisional nature of the initiative; the need for increasing students committed to the plant sciences; and, a comparison between the well-funded neurosciences over the previous two decades and the lesser amount available for the plant sciences. One part of the reason for the disparity was that the neurosciences received considerable support for training grants in a "front-loading" manner, Umminger averred, whereas plant science education support did not come right at the beginning period of efforts to improve those sciences nationally. AC member Walbot, who had joined Bogorad and Burris at the October 1982 NSB briefing on the status of plant sciences, suggested at the AC meeting that a "subdirectorate" for plant sciences be created.[327] Though that did not come to pass, the funding of plant research by all federal agencies involved climbed fairly steadily from 1978, when the total was about $30 M annually, to 1990 when it reached about $170 M per year. Only in 1985 was there a sudden increase; afterward it returned to a smoother slope.[328] In the various presentations given to the AC membership, overhead slides evidenced the involvements of BBS as a whole. But with regard to the plant sciences portion as presented by Umminger, one slide emphasized the 1983 plant postdoctoral program (Kimsey's proposal) as an early activity of the plant science initiative. In another line in the graphic, the plant science centers' program appeared for 1988. The plant science initiative for the long-range vision portrayed the coming period of 1991–1995 as including "collaborative research groups, research initiation awards, university/industry long-term research, [more] plant sciences centers, instrumentation and facilities, and a basic scientific research base

[325] National Science Foundation, *Annual Report 1988* (Washington, D.C.: Government Printing Office, 1989), pg. 31. In more recent years, the ASU Center has become the Center for Bioenergy and Photosynthesis. For more information, see: http://bioenergy.asu.edu/.

[326] BBS AC, "Minutes," May 5–6, 1989, *op. cit.*, pg. 3, BIO Lateral Files. It is significant to note that in most of the 28 interviews conducted by the author for this study, the importance of Mary Clutter to the rise of plant science at NSF was commented upon repeatedly.

[327] Ibid.

[328] Data taken from an ephemeral paper copy of a set of overhead slides that apparently came from an oral presentation on the state of BBS in about 1989. It is likely that the presentation was done for the May 5–6, 1989 AC meeting.

[individual grants]."[329] The plant sciences were indeed widespread divisionally within BBS.

At a conference at Cold Spring Harbor, New York, in the early 1980s attended by Bogorad and many others in the photosynthesis research community, the latest scientific information on the subject was heard. It was later gathered in a *Proceedings* that Bogorad and three colleagues edited. In their preface, they noted the major leaps made in their subject during the second half of the twentieth century and divided the era decadally. The 1950s had been the period of elucidating the carbon reductive cycle, better known as the Calvin (or citric or carboxylic acid) Cycle named for its primary elucidator, Melvin Calvin, who later won the Nobel Prize for his work. The early 1960s saw many discoveries on the electron transport system through the work of Hill and Bendall. In the later part of the 1960s and into the early 1970s, there was a "resurgence of interest" in the area of carbon metabolism, including photorespiration more generally, through the work of Hatch and Slack on the C-4 system. Overlapping those major trends, the two decades preceding the early 1980s and the conference saw more work with regard to chloroplast structure, the complexes between pigments and proteins, light-gathering mechanisms, electron flow, model systems, and use of rDNA to accomplish many of these tasks, especially with regard to chloroplast biology.[330] Much of this was supported by the NSF.

To conclude this chapter, it may be said that little biology and biology of the little can be seen as distinct from big biology and biology of the big in certain aspects at the National Science Foundation and its Directorate of Biological, Behavioral and Social Sciences in the decade and a half from the first major reorganization to the next. But, true to the earliest efforts at NSF's life sciences unit to look not at disciplines as well-bounded categories but to go beyond to Weiss' functional approach, it is not quite so easy to dissect and define the biological sciences by organizational hierarchy, as has been amply demonstrated in this present chapter. The line dividing little and big science and biology of the little versus biology of the big is impossible. Plant science at NSF is a good example of that. There was the necessity to study plants under both a heading of little science, here, and one of big science, as is demonstrated in the following chapter.

[329] Ibid. Another slide in the series.

[330] Katherine E. Steinback, Susan Bonitz, Charles J. Arntzen, and Lawrence Bogorad, eds., *Molecular Biology of the Photosynthetic Apparatus* (New York: Cold Spring Harbor Laboratory Press, 1985), pg. xvii.

Chapter 4
The Big End of the Spectrum

Ecology is physiology carried into the actual habitat.
Clements and Shelford
Ecology is physiology under the worst conditions.
Graduate student motto

4.1 Ecology Reduced…And Expanded

If it is true that "ecology is physiology carried into the actual habitat," it goes far in saying that even as early as 1939, Frederic Clements and Victor Shelford in their book, *Bio-ecology* [*sic*], saw reductionism as a central feature of the big end of the spectrum running from molecules to biomes and beyond. The interface of little biology and biology of the little, as well as that of big biology and biology of the big, could readily be found at the National Science Foundation's BBS in the years prior to 1991. After the reorganization, the many themes and threads that have been introduced in earlier chapters would be seen in ever-increasing importance. The reductionist theme, discussed throughout this study, fitted as well with the biology of the big—ecosystems, biomes, and the many other hieronic levels of biology near and above the "center point," that of the organism, as it did with the levels below: physiology, cell biology, molecular biology, genetics, and so forth. Indeed, a program director of the time, Jim Edwards, said: "[t]hat's one of the major things that happened, I think, while I was at NSF: that the movement of molecular tools, molecular thinking, crossed into the environmental sciences."[1]

The *size* of research grants, big or little, played alongside the Foundation's interest in support of biology at different scales. The National Science Board (NSB) began to consider that very issue in the second year of interest to this history when it took up the matter of "Larger but Fewer Research Grants" in its deliberations of

[1] Jim Edwards Interview with the author, February 18, 2009.

D. J. McGraw, *Millennial Biology: The National Science Foundation and American Biology, 1975-2005*, https://doi.org/10.1007/978-3-030-56367-7_4

June 1976.[2] As has been discussed in previous chapters, a task force of the Board undertook the assignment of "how to utilize more effectively the limited funds available" in order to reap the best science at the lowest reasonable cost. The how, as has been seen, was to offer block grants, multi-investigator grants, and variations on such patterns. This may have presaged the establishment of the Science and Technology Centers Program, the STCs, discussed at some length in the previous chapter, and played a role in the growth of big biology. Sufficient funding for large, expensive, multi-use equipment was also a concern for the NSB at the same meeting. In other meetings that same year, large databank support was discussed, as well.[3] But the issue of science of the big would come for the Board in 1976 in the form of the biggest area of study of all: the environment.

The NSB had numerous *ad hoc* committees over the years to examine particular issues, report upon them, and then have those committees dissolved once the problems were solved, the questions answered, or some new program institutionalized. One such group was a subcommittee of one of the then two standing committees: that for Programs. The Subcommittee on Environmental Programs was chaired by Board member David M. Gates, who was at the time the director of the University of Michigan Biological Station.[4] The subcommittee's report, "Strengthening Environmental Programs," focused on recommendations concerning "lake systems, urban hydrology, role of the Gulf of Mexico in relation to climate, and drought"—areas where the NSB had determined that "less than adequate effort is currently being made."[5] Gates and his two Board co-authors, Jewel Plummer Cobb and Saunders Lane (known as "Mac"), began their report noting that "[c]oncern for environmental issues appears to have reached a peak in the mind of the public during the 1960s" and, by 1970, the first Earth Day had taken place on April 22, having been celebrated across the nation.[6] It was, as *American Heritage Magazine* argued, "one of the most remarkable happenings in the history of democracy."[7] Earth Day was founded by U.S. Senator Gaylord A. Nelson, a Democrat from Wisconsin.[8] He served in Congress from 1963 until 1981. His initial election to the Senate came in November of 1962, just two months after Rachel Carson published her electrifying

[2] NSB, "Minutes," 178:38, June 4, 1976.

[3] Ibid., 180:16, May 24, 1976.

[4] See: http://www.lsa.umich.edu/umbs/alumni for some more recent information that notes that a new (2009) lecture hall at the University of Michigan was named for Gates and his wife. Gates was director of the Biological Station from 1971 to 1986.

[5] NSB, 'Minutes,' 180:30. See: David M. Gates, Jewel Plummer Cobb, and Saunders Mac Lane, *Strengthening Environmental Programs* (Washington, D.C.: National Science Board, 1976), pg. 1. Note that it is Lane, not MacLane.

[6] Ibid.

[7] Cited at: http://earthday.envirolink.org/history.html, pg. 1.

[8] See: Gaylord Nelson, Susan Campbell, and Paul Wozniak, *Beyond Earth Day: Fulfilling the Promise* (Madison: University of Wisconsin Press, 2002); and Bill Christofferson, *The Man from Clear Lake: Earth Day Founder Gaylord Nelson* (Madison: University of Wisconsin Press, 2004). See also: http://bioguide.congress.gov/scripts/biodisplay.pl?index=n000033.

book, *Silent Spring*, which not only facilitated the ban on the pesticide DDT in the United States, but is widely regarded as having launched the environmental movement in America.[9]

Nelson said that the idea for Earth Day had the roots of its inception as early as 1962 when he convinced then President John F. Kennedy to take a tour of the country to promote conservation. The tour covered some 11 states over five days in September of 1963, just two months before his assassination. While Nelson hoped to see the issue of conservation—his term, but soon to become known popularly as environmentalism—put into the public and Congressional consciousness, the presidential tour was not a success. Nonetheless, it was, as Nelson has averred, "the germ of the idea that ultimately flowered into Earth Day."[10] The actual idea for Earth Day came to Nelson in 1969 when he was on one of his many environmentally oriented speaking tours: he covered some 25 states during the 1960s speaking on the subject. By the end of that tumultuous decade, "teach-ins"—efforts at non-violent protest—were occurring all across the nation in opposition to the Vietnam War. Nelson said that "[s]uddenly, the idea occurred to me—why not organize a huge grassroots protest over what was happening to our environment?"[11] The rest is history and Earth Day is celebrated annually in later April throughout America to the present day.

By the time that Gates and his colleagues penned their NSB report and some seven Earth Day's later, they could cite their own Board's earlier and, by then, third annual report on the subject, "Environmental Science—Challenge for the Seventies" of 1971. In the 1976 study, the authors spoke of atmospheric degradation due to a fleet of supersonic transport (SST) airplanes, Freon © refrigerant gas injected into the air, ozone, dust increases from poor agricultural practices, something then already called "climate change," let alone damage to terrestrial ecosystems and environments: fresh and salt water pollution, drought, pesticide contamination, etc. But with all the areas that could be taken on by the NSF, the NSB chose to have the Foundation focus on "water resources and climate change."[12] In Chap. 9, this will be seen in greater detail as both take off more rapidly in the 1990s. Gates and his associates had been guided by the "Challenge for the Seventies" report, which had stated that

> [e]mphasis should be given to projects, manned by coordinate teams directed to intermediate scale or 'mesoscale' problems, that is, problems on the scale of lakes and estuaries, urban areas, regional weather systems and oceanic fisheries.[13]

"Coordinate teams," "mesoscale"—big science. The NSF had, in its Division of Environmental Biology (DEB), a well-established group to which to turn in 1976 to carry out the suggestions of the Board with regard to environmental research sup-

[9] Rachel Carson, *Silent Spring* (Boston: Houghton-Mifflin, 1962). See, among many sites and sources on Carson and her book, http://www.nrdc.org/health/pesticides/hcarson.asp.

[10] http://earthday.envirolink.org/history.html, pg. 1.

[11] Ibid., pg. 2.

[12] Gates, 'Strengthening Environmental Programs,' pg. 1.

[13] Ibid., citing the 'Challenge for the Seventies' report.

port, liaison with the cogent scientific communities, guiding big science, and a division with a long history of supporting environmental and ecological studies. Indeed, in 1975 in the Director's "Program Review"—the occasional internal report of a given program for select audiences—evidence indicated that interest in biology of the big had been growing rapidly in the later 1960s and into the first half of the 1970s.[14]

In 1955, the Systematics and Environmental Biology programs had been established under the earlier BMS division, but ecology (broadly defined), was a "weak and relatively minor biological science in the 1950s."[15] Other programs that had been established in the period from 1952 to 1964 included those in inland field stations and tropical biology.[16] By 1969, a program in Biological Oceanography was established, but two years later it was transferred to the Division of *Environmental Science*, not a part of BMS, but with a title easily confused with programmatic names in the latter.[17] Just one year earlier, the Environmental Biology program itself had been split in two and did not include further activities with regard to marine research. The NSB's interest in research with regard to "water resources," fresh or marine, could be handled in more than one directorate, clearly. The two new BMS programs issuing from that split were Ecosystem Studies and Ecology. Though considerably more funding went to the former, the latter was considered by those in the discipline at BMS to be, along with systematics in DEB, the "core" programs.[18] Such areas as Systematics continued on with little change in the critical growth period for DEB from about 1969 to, and beyond, 1976. By the early 1970s, the effects of Earth Day became apparent as the scientific community turned its interest toward environmental science and began to seek greater funding from NSF: some three-quarters of the "innovative [basic] research in the subject sciences [ecosystem studies, etc.] in the United States is supported through the Division of Environmental Biology," with only small fractions coming from such other agencies as USDA, NIH and a few others.[19]

By the time of the 1975 environmental biology report, there were four programs in DEB; another would be added the following year. Paul Risser, a rotator from the University of Oklahoma, was leading that in *ecosystem studies* and had had particular responsibility for the International Biological Program, the IBP (more below), then having just recently ended. John Magnuson, a rotator from the University of Wisconsin, was program director for *ecology* where aquatic studies were then placed and from whence Board member Gates, *et al.*, took some of their direction in making the study of freshwater one of their interests in their "Strengthening

[14] Program Review Office, Office of Planning and Resources Management, Division of Environmental Biology, *Director's Program Review: Environmental Biology, October 21, 1975.*

[15] Appel, 'Shaping Biology,' *op. cit.*, pg. 226.

[16] Ibid., pg. 380.

[17] 'Program Review: Environmental Biology,' pg. 2. For more on the history of biological oceanography, inland field stations, and tropical biology in the 1960s, see Appel, ppg. 188–205.

[18] 'Program Review,' pg. 2.

[19] Ibid., pg. 5.

Environmental Programs" report. William Heed, a rotator from the University of Arizona, was the program director for *systematics* which, at the time, was supporting study of the evolution of fruit fly (*Drosophila*) species in Hawaii, the archipelago being a "machine for making [them]." Finally, a new *Biological Research Resources Program* (BRR) was just beginning and was led by William Sievers. The fifth program, added in 1976, was Population Biology and Physiological Ecology (PBPE) and was led by Acting Program Director John Wright.

Each of these five programs will be considered in the following sections.

4.2 Ecosystem Studies: The IBP, Hubbard Brook, and the Early LTER

Charles Darwin is given credit for first envisioning something that would, only in the following century, become community and ecosystems ecology when he studied "entangled banks" near his home, Down House, in Kent, England. He noted the numbers and interactions of several organisms such as flowers, pollinating bees, mice, and cats but discussed the matter only to a limited extent in *Origin of Species*. The relationships of their respective populations and changes in their numbers suggested to him something of a "war of nature," yet still there was the presence of relatively stable systems even as the conflict raged. He did not take the leap to create the science of ecology, particularly at the community and ecosystem levels, but it is clear he saw some of its fundamentals in the field—his own extensive back yard.[20]

Joel B. Hagen has argued that the modern birth of *systems ecology* was a product of the early to mid-1960s, the same time that Nelson came to Congress and that Gates spoke of the beginning of environmental awareness in America. Appearing upon the scene then too was one Bernard C. Patten who, like a few others, trained in "traditional biological programs"—undergraduate zoology at Cornell, in his case. He was soon intrigued by more modern interests in biology: for him, its interface with mathematics. He then "caught the systems bug," he said, and studied information theory. A doctoral dissertation on measuring the diversity in a phytoplankton (microscopic algae) community assured his commitment to what was soon to become a new discipline, systems ecology[21] (the term systems, is not to be confused with the term systematics).

Patten taught for a short time at the College of William and Mary, Maryland, and assigned as reading material in his marine ecology course Ross Ashby's influential *Introduction to Cybernetics*—a statement as to where Patten saw the field progress-

[20] Joel B. Hagan, *An Entangled Bank: The Origins of Ecosystem Ecology,* (New Brunswick, New Jersey: Rutgers University Press, 1992), pg. 1–2. As a personal aside: Hagen dedicates his book to Norm Ford and Paul Farber. The present author was Farber's first doctoral student in the history of biology.

[21] Ibid., pg. 132. The 'bug' quotation is from Patten and cited by Hagen.

ing.[22] But in 1963, he moved to Oak Ridge National Laboratory in Tennessee. Though better known for its work in the physical sciences, the Laboratory also had the considerable computing power that Patten needed for his research. Two other biologists, Jerry Olson (University of Chicago doctorate in botany) and George van Dyne (UC, Davis, doctoral dissertation on cattle grazing in grassland ecosystems) were also employed at Oak Ridge; Olson there since 1960, Patten arriving in 1963, and van Dyne in 1964. The National Science Foundation was present at the beginning having supported the trio, along with other funding from the Ford Foundation and Oak Ridge itself, as the three men collaborated on a graduate course in systems ecology at the nearby University of Tennessee. While both Olson and van Dyne left the Laboratory to return to academe in 1968, the mold was cast.

Using the early computing language of FORTRAN, van Dyne began work in systems ecology using *digital* computers. However, though they would not remain of wide importance, *analog* computers were used more heavily for the early efforts in the new field as they could more handily model both very large ecosystems and rapid time simulations, thus offering "tremendous theoretical and practical implications" for prediction.[23] Such things as food chains and energy flow, both early interests in community ecology, could be elegantly done using analog computing. Systems ecology was "big picture" ecology, as Hagen had it, and was right up the NSF's alley, though it would become even more so in later years as big science support grew. NSF data evidence the rapid growth of the BMS' ecosystems studies program by and after 1969. Others quickly applied the van Dyne programs and methods to such things as population ecology, as well.

Further, Hagen has stated that the NSF-supported IBP boosted early work in both ecosystem and systems ecology.[24] However, the IBP has been blamed for a major "intellectual schism," Hagen has argued. From 1968 to 1974, the United States participated in the IBP, which had a strong orientation to *ecosystem studies* and which term was subsequently used for the program of the same name later at NSF in keeping with its employment by the IBP. The IBP had been conceived in 1962 taking its inspiration from the already highly successful IGY, or International Geophysical Year of 1957–1958.[25] The creative mind behind IBP was the brilliant British embryologist Conrad Hal Waddington, known to history as C. H. Waddington.[26] He had

---

[22] Norbert Weiner had said, as the subtitle to his book *Cybernetics* (New York: John Wiley and Sons, Inc, 1948), that it was "control in the animal and machine."

[23] Hagen, 'Entangled Bank,' pg. 133–136. Quotation on pg. 134. Hagen gives a clear explanation of the fundamental difference between digital and analog computers on pg. 134. However, much more detail and a fuller explanation of analog computers can be found at: http://www.cds.caltech.edu/~hsauro/Analog.htm.

[24] Hagen, 'Entangled Bank,' pg. 134.

[25] The IGY and *Sputnik* both helped this author and many others from his cohort choose science, of some stripe, as a profession.

[26] While there appears to be no full-length biography of Waddington, much has been written on him in smaller works. See his many obituaries, but also see: J. M. Slack, "Conrad Hal Waddington: The

proposed the theme for the IBP: "The Biological Basis of Productivity and Human Welfare." The Program began in 1964. As many as 1800 American scientists alone worked on the Program with the NSF being the primary sponsor pouring between $40M and $60M into it over the years (according to Hagen; but $39M, according to Appel).[27] Some 70% of the total went to five (seven, if one splits out certain subsystems) large biome studies, which were testing computer models for research, but not for modeling[28], and doing extensive field studies, etc., Appel recorded.

Appel also noted that the expenditures in ecology, *qua* discipline, and biology more generally at NSF, were primarily those made for the IBP at the time, but that led to its "bec[oming] the most divisive example of big biology at NSF."[29] The "intellectual schism" of Hagen and the "divisive example" of Appel require some elaboration. Hagen averred that not only did other biologists, such as those in the molecular end of science funded by BMS, feel cheated by such vast amounts of the division's money being directed to the ecological systems' work, but so also did many ecologists (e.g., evolutionary and physiological ecologists) whose work was typically more in the realm of little science. The feeling of these other ecologists was, Hagen stated, "hostile." Even after the IBP ended in 1974, criticism continued because the argument was made that little of value came of the whole program. Worse, those ecologists argued that IBP's most undesirable features had become "institutionalized in the new Ecosystems Studies Program at NSF."[30] However, Hagen went on to say that ecology had become "intellectually fragmented" during the 1960s and its practitioners did not meet the expectations required for the financing of big science that characterized the earlier Manhattan Project during WWII or the then contemporaneous U.S Space Program. Ecologists desiring to do large ecosystem studies during the years just before the IBP did not have a united front and could not lobby Congress for funding, nor could they excite the many interest groups, both scientific and the general public, necessary to support the effort. As well, few were tutored in the ways to seek funding from agencies such as the NSF. Finally, "[e]nvironmentalism lacked the glamour of space travel and the immediacy of cancer," then also high on the national radar.[31] Thus, Hagen's "intellectual schism." However, it should be noted that the fragmentation of the ecologists did not affect employment possibilities once the IBP began in earnest. But, it was said that "many" of them who staffed the biome projects were "mediocre scientists"

Last Renaissance Biologist?", *Nature Reviews Genetics* (2002):889–895.

[27] Hagen, 'Entangled Bank,' pg. 164; Appel, 'Shaping Biology,' pg. 227.

[28] Dave Coleman, personal communication. See footnote further below for details on Coleman.

[29] Appel, 'Shaping Biology,' pg. 226.

[30] Hagen, pg. 165.

[31] Ibid., pg. 166.

who could not have won individually funded grants on their own; a charge possibly open to question.[32]

Appel's "divisive example" was based upon the same notion, that is, that

> [b]efore, during, and after the IBP, biologists have battled with one another over whether the IBP was a highly innovative experiment that established systems ecology or *a politically motivated diversion of money that could have been much better spent elsewhere.*[33]

Appel cited numerous sources for her assertion and, along with Hagen's analysis (which she did not cite), provides a basis for understanding features to be discussed concerning big biology and biology of the big at BBS; especially NSF internal views of LTER in its early period.

It should be recorded, however, that a number of critical scientific contributions did, in fact, devolve out of the IBP for the benefit of science, especially ecosystem science. At a national meeting (2009) of the scientists involved in LTER, David C. Coleman offered eight contributions (later expanded to ten) that IBP made to that science:

1. Evidence of the importance of belowground processes in terrestrial ecosystems;
2. Roles of canopy processes in nutrient cycling, particularly nitrogen fixation;
3. The stream continuum concept;
4. Microbial processes are of paramount importance in ecosystem function;
5. The development of cross-biome and international collaborative studies of food webs (corollary of number 4);
6. The pulsatile nature of ecosystem function;
7. The absolute necessity to conduct whole-system experiments to better understand mechanisms occurring; and
8. The paramount importance of studying old growth forests, and the mature stages of other ecosystems, to fully understand ecosystem nutrient cycling and overall ecosystem resilience.[34]

The end of the IBP was by no means the end of big field ecology at NSF, but a transition to some other projects of that big science needed to become clear such that a leap from the IBP to some new, broadly unified approach was called for: ecosystem studies had become "institutionalized" at NSF, after all. Thus, during the years from 1974 and the close of IBP and the inception of the Long Term Ecological Research (LTER) program in 1980, there was no large international effort ongoing

---

[32] Ibid., 176.

[33] Appel, pg. 227. Emphasis added.

[34] I thank Dave Coleman for allowing me to cite these eight points from several of his overhead slides that were shown in a presentation he gave at the "2009 LTER All Scientists Meeting: Integrating Science and Society in a World of Constant Change," held in Estes Park, Colorado, September 14–16, and attended by this author. I also thank the NSF for trip costs, which were kindly provided by the Foundation as an extension to the contract under which this history has been written. Some months later, Coleman's (Emeritus Professor at the Odum School of Ecology at the University of Georgia) history came out: *Big Ecology: The Emergence of Ecosystem Science* (Berkeley: University of California Press, 2010), just as the present chapter was being written. In his book, Coleman added two more contributions to ecosystem science: No. 9, "Simulation models are tools, not ends in themselves," and No. 10, "Ecosystems are active all year round." Each had more substance than is given here. The full text of all ten is listed in his ppg. 76–77.

in systems ecology. Nonetheless, the *Annual Reports* evidenced an ongoing effort in big ecology even if not within a nationally or internationally unified network.

The seven biomes (some authors combine and reduce that number to five) that were studied during the decade of the International Biological Program included two that were of particular importance as the data, methodologies, and conclusions involved were notable in both heuristic and epistemological value, quality, size and other features—and also consumed nearly half of the IBP budget![35] The Grassland Biome Project and, most especially, the Eastern Deciduous Forest Biome Project then became the jumping off place for new directions at DEB. Both had ties to Oak Ridge National Laboratory (ORNL) as IBP project scientists widely employed radioactive tracers for the study of energy and nutrient flow in the biomes. Some 176 scientists worked on the forest biome project over time with nearly 25% coming from ORNL.[36] Patten, Olsen and van Dyne had been at Oak Ridge, of course, and had been supported, in part, by NSF; as well, human connections had been made. And, too, NSF was the prime sponsor of the IBP. It was van Dyne who directed the grassland biome project, for instance, and Coleman has given notable historical consideration to that study and to its leader.[37]

Further, then new AD Betsy Clark, said that

> [t]he biome projects are having substantial impacts in advancement of basic biologic [*sic*] science, management of ecological and environmental research, and production of knowledge of immediate use in resolving environmental and energy problems[:] environmental scientists...have embraced a systems-oriented interdisciplinary approach.[38]

However, she went on to say that

> [d]espite the substantial progress...the support of basic ecosystem science must continue. A great deal of work remains to be done in placing these largely empirical programs on a theoretical basis [and] attaining...synthesized research results.

Clark's latter points could only come to pass in a ***unified*** network: the LTER of the immediate future. So it was clear in the mid-1970s that no matter how well funded the *micro* end of the biological spectrum was to be at BBS, so also should be the *macro* end. And while it is true that considerable funding did come to DEB, the big end of biology would still have very much less in total than the little end, and that by nearly two to one: PCM was to be funded at the level of just over $67M as compared to the next highest figure, that for DEB, at somewhat less than $38M in

---

[35] Hagen, pg. 175. All authors examined list five biomes, but in the 'Program Review: Environmental Biology,' pg. 8, it is stated that IBP examined six biomes. This is the only known (to this author) use of the number six. They include: grassland, deciduous forest, coniferous forest, tundra, desert, and tropical forest, and the 'Review' is organized around those six.

[36] Ibid.

[37] Coleman, 'Big Ecology,' ppg. 29–44.

[38] Betsy Clark, "What payoffs have resulted from biome programs in the IBP?", pg. 4. This emphemeron is a typescript of some 15 pages in length and signed by Clark. It appears to be an overall assessment of BMS/BBS near the time she became BBS director. She provided the author with copies of these pages (4–15), but could not recall the provenance of the document.

1979[39]; comparable ratios had been in effect for a number of years, and would continue to be so. Even once LTER began, little budgetary change was seen (below).

The Board continued to show interest in research in environmental/ecological areas when it reapproved a multi-year, multi-investigator grant for what has been a central piece of history in American ecology. The NSF tie to the famous Hubbard Brook Experimental Forest study in New Hampshire commenced in 1963 when the Foundation began funding the work of plant ecologist F. Herbert Bormann and freshwater ecologist Gene E. Likens who had begun small watershed studies in order to better understand aspects of a whole forest ecosystem. The two scientists were together at Dartmouth College in 1961 where they had "developed, refined and perfected the technique for controlled watershed experiments."[40] They sought yet another round of funding, by then long continuing, from the Foundation years later in 1976 under a proposal entitled "Collaborative Research on Hydrologic-Nutrient Cycle Interactions." That grant was aimed at summarizing and synthesizing their many years of results. By then, they had been supported, in total, by just under $2M over 13 years by an uninterrupted string of Foundation grants. That was some of the best money the NSF had ever spent as evidenced by, among other things, an astonishing 250 publications (to 1976) emanating from one of the truly great field investigations of all time. If one takes only some of the *human* data from the deciduous forest project, for example, it eventuated that 47 students (by then) had earned doctorates working within the biome. There had been four foreign students who studied at the several forest biome sites and five foreign scientists collaborated, as well. Finally, ecologist W. Frank Harris told fellow scientist, Dave Coleman, that "[i]f you now [2009] look at a lot of the leadership in American ecology today, these folks cut their teeth on IBP."[41] It should also be noted that the support for the Hubbard Brook research and that of IBP were almost simultaneous during the 1960s and 1970s coming, first, from the BMS program of Environmental Biology and, after its split in 1969, from the Ecosystems Studies Program. Other grants from the Ecology Program supported associated work to that of ecosystems study, too.

What eventually came out of the Hubbard Brook Experimental Forest work guided vast areas of research in field ecology in forests and other types of ecosystems and is both currently ongoing and very significant, even in today's science. That grant request for synthesis work in 1976 was funded for Yale University, to which Bormann had moved from Dartmouth in 1966 and who would eventually become the Oastler Professor of Forest Ecology, for $207k for FY77 and for $160k for FY78, for example. Likens had also moved, not to Yale but to Cornell University in 1969. Cornell was awarded just under $218k and about $165k, respectively, for

[39] See the discussion in Chapter Four. The source for these data is in: NSB, "Minutes," NSB-80-58, February 25, 1980, Appendix A, 212:31.

[40] See the Tyler Prize site at: http://www.usc.edu/dept/LAS/tylerprize/bormann_likens.html for the provenance of the quotation.

[41] Dave Coleman, presentation slide contents, *op. cit.* Coleman later repeated this phrase in his book.

the same two example years.[42] What is particularly significant about the Hubbard Brook project in this current history is that it was one factor in the establishment of the NSF's LTER, famous in its own right and considered both in this chapter and again in Chap. 9. In fact, Hubbard Brook Research Forest eventually became one of the early LTER sites. Bormann and Likens won, among others, the Tyler Prize in 1993 for their work in "bringing *fundamental order* to the science of ecology and for creating the *premier model for ecosystem studies* in the world."[43] A large body of literature, both of an historical nature and a purely scientific one, exists on the Hubbard Brook story and its two central players, and a host of others. That corpus can be consulted by the interested reader.[44]

Of the four programs in DEB starting in 1969, the first of these, the Ecosystems Studies Program, began by supporting two major efforts: IBP and Hubbard Brook. When this present history formally takes up at 1975, IBP had already ended and Hubbard Brook continued on and was, in part, a seed for LTER. Indeed, Coleman has asserted that

> [p]erhaps the greatest legacy that the IBP has produced in North America and elsewhere in the world is the development of the LTER network, and soon afterward…with the I[nternational]LTER network.[45]

The Long Term Ecological Research Project was still a few years into the future, however, when NSF published its *Annual Report* for 1976. The authors of the Environmental Biology section of the text averred that "[c]ommunity ecology has constituted the intellectual core of the science of ecology since its beginnings early in this [20th] century."[46] Of special interest to BBS at the time was the interaction between plants and insects at the organismal level and below: "secondary substances," plant chemicals (often metabolites); population genetics, etc. The *Report*'s authors spoke also of disciplinary offshoots including population biology and ecosystem science, and the interdisciplinarity arising from these, as well as genetics

---

[42] NSB, 'Minutes,' 181:13, July 15, 1976.

[43] 'Tyler Prize' website, *op. cit.* Emphasis added.

[44] It is recommended that the interested reader begin with Sharon E. Kingsland's *The Evolution of American Ecology, 1890–2000* (Baltimore: The Johns Hopkins University Press, 2005), ppg. 224–228 for a brief overview of the significance of the Hubbard Brook project; an important contribution is Hagen's work (*op. cit.*) at pages 181–186; see also Stephen Bocking, *Ecologists and Environmental Politics: A History of Contemporary Ecology* (New Haven: Yale University Press, 1997), *passim*; see the Tyler Prize site; as well, see http://www.hubbardbrook.org/overview/HBEF_establishment.htm, and the Home and other pages there; and view http://www.ecostudies.org/people_sci_likens.html as Likens is as of this writing director of the Cary Institute of Ecosystem Studies at the University of Wisconsin, Madison; see also http://www.uvm.edu/~cmncmnt/commencement2005/?Page=Boormann.html for more on Bormann, among many other publications, both paper and online. Most especially, see Bormann and Likens' own *Pattern and Process in a Forested Ecosystem: Disturbance, Development and the Steady State Based on the Hubbard Brook Ecosystem Study* (New York: Springer, 1994), *passim*.

[45] Coleman, 'Big Ecology,' pg. 83.

[46] National Science Foundation, *Twenty-Sixth Annual Report for Fiscal Year 1976* (Washington, D.C.: Government Printing Office, 1977), pg. 62.

and behavior. Coleman, too, stressed the importance of "interdisciplinary/multidisciplinary" aspects.[47] As the *Report* stated:

> Ecosystem science…has made substantial progress through the U. S. biome programs under the auspices of the [IBP, which] completed its existence as a formal organization several years ago, but the support of some of these large integrated projects has continued [, the biome projects being] in varying stages of completion, and a second generation of integrated ecosystem research projects, smaller in scale, [has] begun.[48]

Such a new project was the taiga (the Russian word for the coniferous forests south of the tundra) ecosystem study that had been a spin-off of the IBP Tundra Biome Project. The post-IBP taiga study was then examining the white and black spruce (*Picea glauca* and *P. mariana*, respectively) and their relationship to permafrost, as just one feature of the larger whole.[49] Indeed, the taiga project was termed "second-generation" when comparing the IBP to post-IBP periods.[50] Parts of the 140M acres of taiga, bog, and fen in Alaska were the sites for that project and provided information on plant community *succession* (one group of species replacing another group over time). The notion of succession was then and had long been a powerful key concept in ecology that would come under challenge only in the first decade of the new millennium. Several wetlands' projects in the Okefenokee Swamp of the Georgia-Florida border region was yet another of those second-generation undertakings, and would become an LTER site on the second round (1981) of grant competition. The research team was led by none other than Bernard Patten of the University of Georgia, he who had "caught the systems bug." They focused on nutrient dynamics and plant community succession on several levels from the whole ecosystem to habitat components, that is, systems ecology.

The argument has been made several times in this study that there is a very close tie between science of the little and that of the big. It is reiterated once again in the words of the *Annual Report* for 1977 when its authors pointed out that "[f]or the past several years, evolution-oriented research in population genetics and research in population and physiological ecology have been converging."[51] That was also true for ecosystem science. Why should that have been so? Because, "[t]o a considerable degree, this converging trend is a consequence of the availability of the analytical technique of electrophoresis,"[52] and, for purposes of this history, the *technique*, along with its associated *instrumentation*, allowed for a deeper

---

[47] Coleman, 'Big Ecology,' pg. x.

[48] NSF, 'Annual Report 1976,' pg. 62.

[49] Ibid., ppg. 65–66.

[50] National Science Foundation, *Twenty-Seventh Annual Report for Fiscal Year 1977* (Washington, D.C.: Government Printing Office, 1978), pg. 70.

[51] Ibid.

[52] Electrophoresis is a "process by which molecules (such as proteins, DNA or RNA fragments) can be separated according to size and electrical charge by applying an electric current to them. Each kind of molecule travels through the medium at a different rate, depending on its electrical charge and molecular size." See: http://www.medicinenet.com/script/main/hp.asp for the rest of the discussion on electrophoresis and a good definition of it.

understanding about what was happening in the *field* aspects of biology (an ecosystem, for example). It was not just a technique and instrument tied to purely laboratory, little, biology alone. Electrophoresis and its employment (it being done in a laboratory setting) can stand as just one exemplar of the fusing of little and big biology. Many others could and will be adduced later in this text, but electrophoresis was centermost at the time.[53] This already well-established tendency of convergence would do nothing less than grow such that little and big biology would find themselves ever closer by the end of the last millennium: *integrated*. Much more evidence has yet to be offered in this study to continue to demonstrate this clear trend.

In the case of both of the Alaskan taiga and another second, and third, generation site, the Okefenokee wetlands, *biogeochemical cycling* (often termed "nutrient dynamics," as in the case of the swamp research) was of concern. Such cycling examines living nature at the molecular and even ionic level, but here in ecosystem level settings: hence, once again, the fusion of little and big biology. By the period of the 1977 *Annual Report,* some "seven of the more than two dozen projected volumes in the Synthesis Series" of the IBP had been published.[54] The IBP, as a supported endeavor of the NSF, was rapidly providing sites for continuing, or recently established, research projects that would be funded by the Foundation, either again or newly so. In fact, sites such as the taiga would fall under the notion that a "primary criterion for support of an identified facility is the strength of the ongoing research program." Those universities and other institutions that were part of the Tundra Biome Project of the IBP were naturals for support once the taiga project began under the same academic aegis (e.g., the University of Alaska, Fairbanks, in this case). The taiga project did, in fact, become a third round LTER site.

The period, then, from 1974 to 1980 was, but only in historical retrospect, an interregnum of sorts as there was not a large national or international set of project sites, IBP, or a single administrative structure, as there would soon be with LTER. *So ecological research, especially in the ecosystems studies area, was done more by way of completing the goals that had been set for IBP and concluding many projects while beginning others logically related; for example, Tundra to Taiga.*

Scientists at Colorado State University, Fort Collins, were examining the "impact of a variety of vertebrate and invertebrate herbivores on plant metabolism and productivity, seed production, and community structure" as noted in the *Annual Report 1978*, for instance.[55] Community ecology was being funded, as were other studies, in the four programs of DEB (below). More than that, many papers were coming out

---

[53] This assertion was further confirmed in an interview with retired NSF systematist Jim Rodman (done by the author May 11, 2009; more below in main text) in which he said of the growing field of biochemical systematics in the later 1970s and early 1980s that it was "changing over from so-called natural products industry…into…I'll say, electrophoretic and eventually the DNA sequencing technology."

[54] NSF, 'Annual Report 1977,' pg. 70. See also: http://www.nsf.gov/awardsearch/showAward.do?AwardNumber=7721903 for the grant award for the "Synthesis Series."

[55] National Science Foundation, *Twenty-Eighth Annual Report for Fiscal Year 1978* (Washington, D.C.: Government Printing Office, 1979), pg. 71.

in the journal *Ecology,* which had grown rapidly in pagination once the IBP had concluded and results were eventually published over a number of years.[56] And with regard to project funding sources noted in the period from 1978 to 1982 in that journals' articles, there were some 27 issues with a total content of 637 articles with nearly 80% citing their sources of support. Of those, just over 80% received federal dollars: of the federally supported research, 75% came from the National Science Foundation. One-half of all the 637 articles' authors had received NSF support *exclusively*: a testament to the Foundation's signal role in the rise of ecological sciences in America preceding the era of the LTER.[57] Biology was certainly being "shaped" in this country, as Appel had it.

In the BBS "Five Page Issue Paper" of January 1978, cited in the previous chapter, DEB was looking at research opportunities that included the "[b]iology of the humid tropical forests, [c]onsequences of plate tectonics for evolution of vertebrates, [and improving] ecosystem analysis."[58] Under the header of research resources for supporting aspects of each of the three areas they were considering, DEB scientists could offer, respectively, "long term ecological data sets, [field s]ites to develop…[and s]ome mode of support to maintain continuity for long term data sets." Clearly, Betsy Clark continued to be keen on the fact that a "great deal of work remains to be done in placing these largely empirical programs on a theoretical basis [and] attaining…synthesized research results." *Large database support* was obviously more called for by the later 1970s than ever before in BBS (or BMS) history, to judge by the repeated requests for funding. Such calls were seen in the Minutes of the NSB and numerous places within the paper record of the BBS. The IBP work had certainly left a rich data legacy for NSF. For instance, the ELM model (Intermediate Level Model, ILM; but termed ELM) of the Grassland Biome Project, indeed its crown jewel, was not only the *product* of much data, but had the capacity to *generate* vast amounts of data, too.[59] The "Issue Paper" also pointed out a major research question seeking to understand the "role of the terrestrial biomass as a net sink or source for atmospheric CO_2"—an issue that would grow over the following decades and reach a high pitch in world interest in the earliest years of the twenty-first century.

Late in the year of 1979, not so many months prior to LTER's initiation, the BBS added a significant new member to the Directorate in the person of Frank B. Golley, a rotator who had taken leave from his positions as Professor of Zoology and Environmental Design and Executive Director of the Odum Institute of Ecology at

[56] See: http://www.esajournals.org/loi/ecol. In 1975 there were a total of 1,473 pages in the six annual issues. That number jumped to 1,992 by 1982.

[57] Data found in a one-page ephemeron dated July 28, 1983 and attributable to the Population Biology and Physiological Ecology Program (noted at the bottom of the page). From the Courtney Files held by the Historian of the NSF as of this writing.

[58] "Biological, Behavioral, and Social Sciences: 5 Page Issue Paper," *op. cit.*, ppg. 1–2.

[59] See Fig. 2.2, pg. 33 in Coleman, 'Big Ecology.' The ELM diagram is remarkably complex, but highly informative.

the University of Georgia, Athens.[60] Golley already had a distinguished career by the early 1960s having established the concept, with several others, of *energy flow* in an ecosystem, when he joined with pioneer ecologist Eugene P. Odum, "Gene" or "E.P.," and his brother Howard Thomas Odum, who was known for his work in general systems theory, ecosystem ecology, and thermodynamics. The three had established big science in tropical Puerto Rico, and also at Eniwetok Atoll (a major site for atomic weapons testing in past years) in the Pacific, prior to the IBP years.[61] Golley was, in the early history of the development of the eco*system* concept, very pleased with the word itself, as it seemed to connote something of a machine in nature, an idea that he supported.[62]

Golley would later be one of the five scientists, along with the primary founder, James Thomas (Tom) Callahan, of the LTER project. Callahan was Associate Director for the Ecology Program of DEB. Paul Risser (noted earlier) and three others played central roles. One of those three was John L. Brooks, a Yale limnologist and Deputy Division Director (and for several years, Acting Director) of DEB, which Golley eventually took over from Brooks. LTER had appeal to Golley because, as Gene Odum noted, there was *no central theory to test in the IBP* and to that unworkable reality, Golley "drily remarked that the IBP carried 'bricolage to a new level of activity and scale.'"[63]

Brooks', and then Golley's, DEB would have plenty to do in the late 1970s as

> [o]n the biological macroscale, it becomes increasingly important to understand how ecosystems work and populations function. If the information base can be enlarged sufficiently…[the data] could guide regional development efforts as well.[64]

Echoes of Clark and the drive to expand database activities! The *Annual Report 1979* went on to say that the tropics were a "particularly pressing problem" and this interest fit with the BBS's desire to work on humid tropical forests as deforestation there was moving apace and local extirpation of many species was occurring. The DEB was also supporting work on soil nematodes and continued with its interest in nutrient cycling in ecosystems that year.[65] Nematodes are a massively important group of worms, which, among other features, are major players in food webs.

---

[60] NSB, "Minutes," 209:6, November 19, 1979.

[61] Coleman, 'Big Ecology,' pg. 12. See also Hagen, 'Entangled Bank,' pg. 106 for something more about Eniwetok—"an early benchmark for ecosystem studies"— then to pg. 110 as Hagen follows Golley to his directorship of the Savannah River Ecology Laboratory, South Carolina, a part of the SRNL (like Oak Ridge, one of the national nuclear research laboratories), and to pg. 180 and his work on the IBP.

[62] Kingsland, 'Evolution of American Ecology,' pg. 215.

[63] Ibid., pg. 222. Kingsland was quoting a line on pg. 139 of Golley's own *History of the Ecosystem Concept in Ecology: More than the Sum of the Parts* (New Haven: Yale University Press, 1993), to which the interested reader is heartily commended.

[64] National Science Foundation, *Twenty-Ninth Annual Report for Fiscal Year 1979* (Washington, D.C.: Government Printing Office, 1980), pg. 56.

[65] Ibid., ppg. 56–58.

Leading up to the birth of LTER, NSF had funded three workshops, one each in the years 1977 through 1979, which addressed the issue of the "feasibility of collaborative long-term ecological research."[66] As well, they focused on the *philosophy* of collaborative research and developed a "centralized working hypothesis" to manage such a research style.[67] For the purposes of this history, that became in later years another excellent example of both *big science* and *interdisciplinary* research. The IBP projects had been smaller, were set up much more quickly, and ran a shorter period of time compared to what the LTER projects were to do. Indeed, by 2010, LTER projects were three decades old and still very much alive. The IBP projects had, as well, only "limited applicability to the practical problems of environmental management."[68] Only two, the LTER sites known as H. J. Andrews and Coweeta (below), had offered *management* knowledge in their earlier IBP incarnation. After the end of the third workshop, NSF was ready to engage in a pilot study. In later 1979, the Foundation announced a call for proposals for pilot sites with two major goals:

- Initiating the collection of comparative data at a network of sites representing major biotic regions of North America, and
- Evaluating the scientific, technical and *managerial* problems associated with such long-term comparative research.[69]

So in 1980, six sites were selected for NSF funding: North Temperate Lakes (Wisconsin)[70], H. J. Andrews Experimental Forest (Oregon)[71], Coweeta Hydrological Laboratory (North Carolina)[72], Konza Prairie (Kansas)[73], North Inlet Marsh (South Carolina; the site withdrew from the LTER in the early 1990s)[74], and Niwot Ridge (Colorado).[75] With the exception of the Konza Prairie and North Inlet Marsh locales, the remaining four had been IBP sites. The initial funding for the six pilot sites was $300k apiece. Grants were to extend for six years—a long period in NSF grant types. That same year, the lead scientists from each site met in Washington, D.C. to establish a steering committee for "coordination and accommodation of mutual goals." NSF announced a second round of proposals in 1980, as well.[76]

The history of the LTER, beyond NSF *per se,* took on its own independent life, as to management, but not financially; that remained with NSF. The LTER story has been partially covered by Coleman in his excellent 2010 book, and the Internet site,

[66] Kingsland, 'Evolution of American Ecology,' pg. 239.

[67] See: http://www.lternet.edu/about/history.html, pg. 1.

[68] Coleman, 'Big Ecology,' pg. 90.

[69] See: http://www.lternet.edu/about/history.html, pg.1. Emphasis added.

[70] See: http://lter.limnology.wisc.edu/.

[71] See: http://www.fsl.orst.edu/lterhome.html/.

[72] See: http://coweeta.uga.edu/.

[73] See: http://www.konza.ksu.edu/.

[74] There was no Internet site for North Inlet Marsh at the time of this writing.

[75] See: http://culter.colorado.edu/NWT/.

[76] See: http://www.lternet.edu/about/history.html, pg. 1; see also Coleman, pg. 91.

www.lternet.edu, contains a wealth of information about the history and science of the whole LTER venture discussing all of the eventual 26 sites of the network. Each has its own website, as well; the interested reader is directed to them. However, the NSF organizational history *per se*, a central concern of this study, is not closely examined by either Coleman or LTERNET. That requires comment.

It is an interesting fact that the NSB did not consider LTER in its discussions in the year of its birth given just how much the proposed network reflected interests of the Board: expansion into ecology; big science; rich database creation; and, interdisciplinarity. It may be that having to deal with three different "revolving" door directors of NSF in that year of 1980, Atkinson, Langenberg, and Slaughter, and one deputy director's (Pimentel) leaving, was enough to keep the NSB busy. The discussion concerning reorganization of the NSF overall, and BBS in particular, was on the Board's plate at that time, too, as was a variety of other issues: then current OSHA (Occupational Safety and Health Administration) expectations of NSF; numerous retirements (Atkinson, Pimentel, Board Chair Hackerman—they "occur in a batch"[77]); conflict of interest policy; the Women in Science Act; the Bayh-Dole Act; etc.[78] Possibly more notable yet, was Director Slaughter's plan to move applied research into each directorate and also to create the engineering directorate.

DEB was also an active place with Frank Golley releasing his program status and update in 1980 on the "Decade of the Environment" in response to a request by the NSB.[79] The first Earth Day having been in 1970, Golley was able to examine ten years of success in the growth of environmental biology at and beyond the NSF. There was the establishment of the Council for Environmental Quality, which had been created under the then also new National Environmental Policy Act, NEPA (January 1, 1970), and the founding of the Environmental Protection Agency (December 2, 1970). NSB's own report, "Challenge for the Seventies" (1971) spoke in terms of the urgency of developing a comprehensive national program to forward the efforts in environmental science.[80] That Board report emphasized team approaches, global scale effort, and training. Overall, Golley noted that research in the area must be "treated as system problems" and in a "hierarchical pattern" from the smallest portions of an ecosystem to the largest, the biosphere. The "systems" issue was clearly on the minds of those in DEB and that carried over to the *Annual Report 1980* when its authors stated "[a] key to effective progress in environmental biology is to view the objects of our research as systems."[81] As well, the report addressed the issue of arranging matters in a hierarchical manner. Use of similar terminology in widely separated types of internal and external publications indi-

---

[77] National Science Foundation, *Thirtieth Annual Report for Fiscal Year 1980* (Washington, D.C.: Government Printing Office, 1981), pg. vii.

[78] NSB, "Minutes," for the year 1980, *passim*.

[79] Office of Planning and Resources Management; Division of Strategic Planning and Analysis, Program Review Office; Frank B. Golley, "Program Report, Vol. 4, No. 7, October, 1980: Environmental Biology,"; pg. 1.

[80] Ibid.

[81] NSF, 'Annual Report 1980,' pg. 55.

cated a strong unity of approach that NSF took in adhering to thematic structures established in one or another directorate across the Foundation. And such unity was evidenced in many publications: for example, the statement that "NSF provides about 90% of the support for basic environmental biology research at colleges and universities," as alluded to above.[82]

The notions of reductionism, big and little science, models, interdisciplinarity and several other unifying themes discussed in this history can now be joined by the concept of *systems*. In future chapters, it will be seen that systems and systems theory, long established in the physical sciences, engineering, and computer science, gain ever more in importance in biology as the second millennium morphed into the third.[83]

The systems approach, in part, was on the minds of those in BBS in 1980 in a rather different connotation too, as DEB became the Division of Biotic *Systems* and Resources, BSR, in FY81 (though it would return to the title DEB in 1992 after the reorganization). The use of the new name of BSR did not "resonate with a number of the PIs" submitting proposals and that was why, at least in part, it went back to DEB in later years: "we always had a lot of explaining to do" with PIs and the "negatives with the name BSR grew too great."[84] The original choice of BSR seems to have been based upon the desire, "frankly, to get away from the words environment and evolution" as the former was not well liked by biologists and the word "evolution" which, as has been seen, was unacceptable on many fronts, especially in the Reagan years that began at that same time.[85] Nevertheless, it took 11 years to make the change back to DEB.

BSR (DEB) continued its interests in tropical biology with interdisciplinary research in Central America as it pursued, through support of the work of Edward S. Deevey at the University of Florida, the paleoecology of Mayan urbanization of the tropical forest; a very interdisciplinary study, indeed! As well, the genetic diversity of maize was studied with regard to its ancient progenitor, *teosinte*, in the so-called cloud forests of the Cordillera Orientale of Mexico. But BSR also supported the work of well-known paleontologist J. William Schopf of UC, Los Angeles as he pursued microfossils in 3.5 billion-year old rock formations in Australia.[86] During that same year, of course, LTER was gearing up.

---

[82] Ibid.

[83] It should be noted that Paul A. Weiss, discussed in Chaps. 1 and 2 herein, had been strongly influenced by Karl Ludwig von Bertalanffy upon their first meeting in Vienna in 1924. It was Bertalanffy who, during the early and middle parts of the 20th century, developed general system theory (GST). See http://www.bertalanffy.org/c_26.html for a more detailed, yet brief, history of Bertalanffy, and also a mention of Weiss. It is clear how Weiss was influenced by Bertalanffy and was thus led to suggest functional versus disciplinary organization to the young NSF and its biologists in the 1950s.

[84] Edwards interview.

[85] Ibid.

[86] NSF, 'Annual Report 1980,' ppg. 56–58.

Actually, picking out LTER in the then premier publication of NSF, the *Annual Reports*, presents a historiographical conundrum when one considers the magnitude of the project as a part of the support given out by BSR: there was an inexplicable lack of use of the very term LTER in the *Reports*. In that publication for the second year of LTER's existence, the authors noted that

> in 1981 NSF began initial support of a network for long-term ecological research. Six active projects, originally funded in 1980, were joined under a steering committee charged with overseeing the development of network characteristics, such as data comparability and joint experiments.[87]

Further, the ratio for the annual budget for PCM, versus BSR, was much as it had been for a number of years—almost. In actual FY80 figures, PCM stood at just over $72M, while BSR received not quite $40M, as has been pointed out several times above. In FY81, the figures were nearly $78M and just over $41M, respectively: BSR was losing ground as LTER was coming online! Still, those figures would see a notable increase as the House CST boosted the NSF's requested $172M for BBS to $198M for FY82. Only the science Foundation's education directorate and cross-directorate programs got more than BBS. The actual percentage change for BSR itself, however, was just under 7% up between FY81 and FY82. (That same change for BBS's Behavioral Sciences was a stunning 38% *decrease*, while Social and Economic Sciences dropped by a staggering 57%, after several other previous years of decline; victims of the *early* Reagan era.) Support for LTER did not come from BBS alone, it is important to note, as biological oceanography and geosciences contributed, also.[88]

Several conjectures may be offered as to why neither the Minutes of the NSB nor the pages of the *Annual Reports* seem informative. The first of these may be in a comment made by Coleman, in his book *Big Ecology*, as he pointed out that Risser, Callahan, Brooks and one other, all of BSR, had to *convince* the Board of the value of engaging in what would become LTER.[89] It was the three workshops and the call for proposals that finally led to support of the six pilot sites, yet little was said in any internal or external NSF or NSB documents prior to the minimal statement for 1981 given above. Nevertheless, the second round of competition lead to five more sites beyond the initial six. Another reason why so little is to be found in Board Minutes is that during those same years other areas of BBS's purview, biotechnology, and plant sciences were of considerable interest and concern to NSB members (previous chapter), and were heavily discussed. Dialogues concerning the social and behavioral sciences, too, were constantly a part of the Board's records during much of the 1970s and 1980s. Finally, to return to the arguments of both Hagen and Appel and the divisiveness that IBP expenditure's caused in the community, it may be that that was the ultimate reason that extensive mention or discussion of LTER was so muted

[87] National Science Foundation, *Thirty-First Annual Report for Fiscal Year 1981* (Washington, D.C.: Government Printing Office, 1982), pg. 34.

[88] Coleman, 'Big Ecology,' pg. 91.

[89] Ibid., pg. 92.

in the earliest period. Buried as part of one sentence in the *Annual Report 1982*, for instance, the phrase was much the same as that of 1981: "the Foundation is supporting a series of integrated, long-term environmental studies."[90]

However, by the following year, LTER became more visible when, in the pages of the *Status of Science Reviews 1984*, the project was shown to be investigating several important issues. Included in those was the mention of acid rain (North Temperate Lakes site)—which was catching national interest at the time[91]—increasing levels of CO2 in the atmosphere, and a few others. Still, the total length of that first LTER-headed text was but a single, short paragraph. Although, in the more critical (because of its wider spread) *Annual Report 1983*, there was the phrase "[a] long-term ecological research initiative begun in 1980 continued." More conspicuously, that was followed several paragraphs later by the comment that

> [e]cosystem scientists have long recognized the need for systematic, long-term, and coordinated investigations. But individual projects often produced incomplete results. Studies suffered from insufficient knowledge of ecosystem variability, unsystematic monitoring of long-term trends, and the conversion of natural ecosystems to uses incompatible with research. And the *lack of a coordinated network of sites* prevented monitoring and comparing data at different locations.[92]

Clearly, a reference to the then concluded IBP. The *Report* went on to say that the solution was to create a network, designated as the LTER, including both field sites and regional research facilities, thus making up a "national resource." Data were to be deposited in a manner usable by future scientists. Further, LTER had become a "magnet for researchers" and the data its scientists were producing were "integrated and cumulative." Beyond that, a conference took place in 1983 at NSF, the purpose of which was to review issues of the first few years of the network's efforts.[93]

Other significant administrative activities had preceded that conference: in 1980, Richard Marzolf of the Konza Prairie site was elected chair of the newly created LTER Network Directorate (not an NSF directorate) with the first meeting in December of that year. In 1981, the first LTER All-Scientists' meeting was held in Las Cruces, New Mexico.[94] By 1982, a Steering Committee was formed with NSF funds and "created a policy for workshops." Marzolf was succeeded, in 1983, by Jerry Franklin of Oregon State University, host of the Andrews Forest site. At that same time, the LTER *Network Office* was formed. Franklin served in his new capacity until 1995, with the Office moving to the University of New Mexico permanently.

[90] National Science Foundation, *Thirty-Second Annual Report for Fiscal Year 1982* (Washington, D.C.: Government Printing Office, 1983), pg. 32.

[91] The National Park Service (NPS) was busily installing acid rain monitoring devices in select parks, such as Sequoia and Kings Canyon National Parks, at the time. Personal experience of the author, an NPS Ranger-Naturalist (Interpreter) there during the 1970s and 1980s.

[92] National Science Foundation, *Thirty-Third Annual Report for Fiscal Year 1983* (Washington, D.C.: Government Printing Office, 1984), pg. 27. Emphasis added.

[93] Ibid., pg. 28.

[94] Coleman, 'Big Ecology,' pg. 98.

In that year, an executive director was hired: Bob Waide.[95] Finally having a network-wide office, the NSF was better able to liaise with and coordinate its funding and data transfer through a centralized location.

A significant statement of the period was that the "science supported by this division [BSR], unlike that of many subdisciplines of biology, draws upon all aspects of biological knowledge." That was to say that ecologists required knowledge of not only their own primary specialties, *but also that of the micro end of the spectrum* of the life sciences—as well as mathematics, computer science, geology and other areas—a very rigorous form of *interdisciplinarity*. Therefore, the BSR supported work in 1983 on such issues as the mass extinctions of the geological past, human use of patchy ecosystems by the African peoples of the Lake Turkana region, Kenya, and such other research as nitrogen flux in a mosaic of patchy landscapes at Maryland's Rhode River.[96]

Much occurred with regard to grants by BSR beyond the LTER efforts, of course. The Board was thinking in concert with BBS when it considered the directorate's priorities for the period of 1985–1989 in its February meeting.[97] LTER was fourth in its list (general plant biology was third) and priority number five emphasized "longer-term (five-year) research programs in plant biology"—giving plant science double billing—thus combining the thinking of both BBS leadership with that of the Board in regard to the successes of LTER and its influences on rapidly expanding research in plant biology. The latter had been a keen interest of Betsy Clark, but it was the same year in which she left NSF. BBS would soon be led by David Kingsbury (previous chapter). Late in that year, too, BSR established its Postdoctoral Research Fellowships in Environmental Biology.[98]

When Kingsbury became AD of BBS, a new type of biology would be undertaken: there was a distinctively different milieu to be sensed in the directorate. The Long-Range Planning (LRP) document for the period 1984–1988 evidenced this stylistic and topical redirection.[99] It may well be such that one could not term it a full paradigm shift, but it was clearly an intellectual redeployment in how biology was to be conceived in Kingsbury's BBS as compared to his predecessor's. Allusions along such lines have already been cited. For instance, the LRP spoke in terms of "[e]cological studies in [an] expanded Space-Time framework." This, along with "Integrative studies," marked an altering of how biology was conceived and what terminology was to be used in the newer versus the previous BBS: Clark's style was

---

[95] Ibid., ppg. 98–99. There is a minor discrepancy in what constituted the "first" All-Scientists' meeting as the official LTER website states that the "first" meeting was held not in 1981, but in 1985, and that in Lake Itasca, Minnesota. See: http://www.lternet.edu/about/history.html, pg. 2. The latter year is likely correct as Coleman does say "[i]n 1981, the first All-Scientists' meeting, *in effect*, was held…" in his history. Emphasis added.

[96] NSF, 'Annual Report 1983,' ppg. 33, *passim*.

[97] NSB, "Minutes," NSB-83-92, March 22, 1983, 2–83:20 (new pagination style).

[98] Ibid., NSB-83-373, February 16, 1984 (but the November 17–18, 1983 meeting), 11–83:9.

[99] BBS LRP 1984–1988, "Topics for Increasing Support," *op. cit.*, pg. 1. The LRP is a typescript document of seven pages.

more classical, whereas Kingsbury's was more systems-oriented, mathematical, and computer-based, in form. In a word, the swing was from a more qualitative style of biology to a more quantitative mode; it was certainly in keeping with the directions of the community—the *intricate dance*. As Michael Levitt, a British biophysicist and later professor of structural biology at Stanford University, once said: "computers have changed biology forever, even if most biologists don't know it yet." *Integrative* approaches to biology were already underway, but would become something near dominant in the following decade and after. As one of integrative biology's founding thinkers, Marvalee Wake, said much later (2001), "'[i]ntegrative biology' suddenly has cachet.'"[100] It was a slow process, to be sure, and the very phrase had been in coinage for much of the last half of the twentieth century. In a more formal version of the LRP, Kingsbury spoke in terms of the "[i]nitiation of new research emphases[, for example,] chemistry and biology of life functions," but also noted the importance of continuing other emphases, such as LTER, through at least FY90.[101] In the event, it continues well into the present century. The redirection toward more quantitative biology could be seen not only in such documents as LRPs, but in the *Annual Report,* also: "[f]inally, ecologists and systematists are making extensive use of computers in approaching a variety of problems."[102] Erich Bloch would come from the IBM Corporation to take the directorship of NSF in September of 1984, as has been seen, and the move into a more computerized world was already underway by then.

It will be recalled that just after Clark had left, her deputy, Robert Rabin, was heading BBS temporarily. By the time that Kingsbury had arrived, the first meeting of the newly established BBS Advisory Committee had occurred. In its Minutes for that meeting, penned by Rabin, it was noted that a detailed review of the BSR division had been prepared for the director and considered by the NSB. It was discussed in a broader fashion by the AC, as well. The "general consensus [of the AC] appeared to be that the Foundation historically had been rather passive [about how] decisions were made about areas of emphasis," that is, research directions and subsequent funding, within the NSF.[103] That passive approach to budgeting for newer and sizable initiatives was found to be a poor one, so better ways were established for providing for such things. The LTER was an example of interest in their discussion. That, along with "extensive interactions" with the community of ecologists, "provided a [more] firm basis" for development of both large new initiatives for plant biology, as well as for the LTER.

[100] Marvalee Wake, "Integrative Biology: Its Promises and Its Perils," *Biology International* 2001:71–74.

[101] BBS, "Long-Range Planning FY 86-90," a seven-page document dated January 4, 1984. This LRP appears to be an updated version of the one cited in a footnote further above.

[102] National Science Foundation, *Thirty-Fourth Annual Report for Fiscal Year 1984* (Washington, D.C.: Government Printing Office, 1985), pg. 24.

[103] BBS AC, "Minutes," January 26–27, 1984, pg. 9. All AC Minutes are found in the Lateral Files of the present BIO.

The "Director's Program Review" of BSR was extensive with regard to funding aspects, among others, but many details were provided to the director. Technically, that person was then Acting Director Edward Knapp, but Bloch also read it after he came later in 1984.[104] A background paper, a synthesis of the review, was presented in a meeting of January 1984. BSR had several programs ongoing, of course, but two that were strongly emphasized were those associated with LTER and those tied to "Classifying the World's Plant and Animal Life," that is, systematics. The "Review" was retrospective for the previous five years, beginning with FY79. BSR staffing from then to FY84 had doubled, but that was figured upon a small base: so permanent scientist positions went from four to eight, but rotators remained unchanged at five for all of the division, not just the Ecosystem Studies program under which LTER fell. Full-time support staff went from seven to 11. Total BSR division funding went from \$34M to over \$53M with average grant award size going from nearly \$50k to \$71k. Actual awards changed very little: in FY79 they were 685; they had gone up to just 755 in FY84.[105] By that time, the percentage of support for environmental biology, once at 50% of the BSR budget, was then just over 56% (including monies from all directorates of NSF, not just BBS). The Smithsonian was providing something over 13% and USDA about 18%, with several other agencies in the single digits. The biome projects (started under the IBP) "were very large by NSF standards" and had had funding running from \$1 to 3M per project. One of the LTER projects (11 total by then) was averaging about \$1.3M over a five-year period.[106] Small, non-LTER grants were running from about \$2 to 10k each: LTER was the elephant in the room. That the "environment is in fact a single entity, a gigantic system" was becoming ever more clear and big projects were fully justified, and had been obvious since the Board report of 1971 ("Environmental Science—Change for the Seventies").[107] It was even more relevant in 1984 with LTER's rapidly expanding size.

The next year, BSR was able to define its program areas in a call for proposals for each of its five divisions.[108] That for Ecosystem Studies came first, a likely statement concerning the over-arching importance of the program to BSR.

> The Ecosystems Studies Program supports mainly multidisciplinary studies of the structure and function of complex biotic-abiotic associations. Processes, mechanisms, and system behavior comprise a major research focus, as in the study of energy and nutrient transfer through ecosystems.

---

[104] Internal evidence from various source documents.

[105] BBS (Syl McNinch), "Director's Program Review: Biotic Systems and Resources Division: Background Paper," January 10, 1984, pg. 1.

[106] Ibid., pg. 3.

[107] Ibid., pg. 4.

[108] Of course, annual competitions had been issued to the communities for years, but I am choosing the 1985 document to use throughout the present chapter. It was OMB 3145-0058, December 31, 1985 (aka NSF-83-75), pg. 1.

The outline went on to consider scale ("landscape" level), spatial and temporal aspects, ecological theory, "mathematical modeling for analysis and integration," and long- and short-term studies with varying foci.[109]

It is interesting that even though much was said in Chap. 2 about how NSF's social and behavioral science projects were attacked by the conservative press and certain Congressmen, Clark had noted that many of the more purely "hard" science projects were, too. In the background paper for the "Director's Program Review," NSF author Syl McNinch provided an example of a newspaper article's judgment:

> Woodenheaded government bureaucrats are wasting $75,000 of your hard-earned tax dollars so a college professor can measure trees on a tropical island off the coast of Panama.[110]

By the AC's second meeting in later September of that year, Bloch had arrived and was a guest of BBS to address the AC on a number of broad topics, the FY86 budget being primary among them. Kingsbury spoke in terms of two major initiatives, Chemistry of Life Processes and supercomputers: the BBS reflected not only Kingsbury's more quantitative vision, but the NSF's under Bloch, as well. BSR and the LTER, however, were not topics of the discussions. And, for several years, major publications of NSF had little more to say about the LTER as, for one, the *Annual Report*s had changed greatly in their design under Bloch's influence and lacked many of the useful features of directorate activities seen in earlier issues. It was not until 1988 that an article appeared on the LTER, and then it was a very short one. It did reflect, however, some of the research that was then ongoing with regard to established sites and noted that several new sites were added, for instance.[111] Nonetheless, much had been happening in the previous several years both within the network sites and at NSF. The Foundation announced its third call for new LTER site proposals in 1986, which led the following year to selection of sites in: the Alaskan arctic; an experimental forest, also in Alaska; the renowned Kellogg Biological Station, Michigan; and, a coastal reserve in Virginia. But the most notable addition was that of the Hubbard Brook Experimental Forest, long a site needed in the LTER. Now a part of its own system with its own management, but still funded by NSF/BBS, the LTER began to publish its internal "Network News" in 1987.[112] That same year, Robert Robbins was hired by NSF to fill an LTER technology development position created *within* BSR.

In 1988, three more sites came into the system, including one in northeastern Puerto Rico, another in New Mexico, and the Harvard Forest, one of the most intensively studied natural areas in the United States. That year also saw two sites withdraw, one of which was the Okefenokee location. At NSF headquarters, the "Shugart Report" came out. Named for its chair, H. H. Shugart of the University of Virginia,

---

[109] Ibid.

[110] McNinch does not give the provenance of the quotation.

[111] National Science Foundation, *Annual Report 1988* (Washington, D.C.: Government Printing Office, 1989), pg. 33.

[112] See: http://www.lternet.edu/about/history.html, pg. 2.

the report was the product of two meetings held in January of 1988. The topics under consideration included a number of issues:

> research topics in spatial variability; documenting and interpreting long-term temporal variability in ecosystems; developing interbiome comparisons; developing and validating simulation models; acquiring [a] geographical information system [the then new GIS] capability across the LTER network; developing a network remote-sensing analysis capability; and, augmenting wide-area and local-area computer networks for the LTER system.[113]

The recommendations were greatly detailed. The report functioned in a sense as a set of guidelines for yet more unification of the full network, by then approaching a score of sites. And in that expanding unification effort, NSF added another position. It was filled by Caroline Bledsoe who became Research Coordinator in BSR. The post was created in order to encourage inter-site research efforts. By the end of the year, Robbins and Bledsoe authored the first comprehensive directory of LTER personnel.[114] The two BSR positions were independent of the LTER's permanent office in New Mexico, but worked closely with it.

As the decade drew to a close, NSF engaged in a 10-year program review of the LTER chaired by Paul Risser in 1989, but did not see publication until 1993 (see Chap. 9). As the network continued to expand in the United States, an exchange and collaborative program was established with the People's Republic of China, only the first of many foreign connections and sites under the ILTER yet to come. BBS had a budget of some $285M by FY89 and was the third most well financed directorate at the Foundation. It would break $300M in FY90. BSR, as a division, had become the largest budgetary area of BBS, by far. In FY89, what was once a single division, PCM, had been split into two divisions, DMB and DCB (previous chapter). The *individual* budgets for each of those two, over $44M and approaching $56M, respectively, hardly rivaled BSR. As it was, though, the *combined* figure was just over $110M, a statement as to how strong the micro end of the biological continuum continued to remain—and the disparity would grow in favor of little biology in coming years as the genomics era came upon the scene at the change of the millennium.[115]

The BBS AC, when it met in May of 1989, was chaired by Peter Raven. The BBS was now headed by Mary Clutter, David Kingsbury having departed in September 1988. At that meeting, as has been seen, three major scientific initiatives were before the directorate for the five-year plan period to cover FY91 to FY95. Along with plant science and dynamic form and function, global environmental change was discussed. Frank Harris of BSR spoke about that subject as it related to the NSB Task Force on Biological Diversity, cross-directorate activities, and the global change's home area at NSF in the Geosciences directorate. John Brooks, director of

---

[113] See: http://www.lternet.edu/technology/background/shugart_report/shugart.htm. NB: The actual wording has been changed only slightly in this quotation in order best to reflect grammatical necessity for the present rendering.

[114] Ibid.

[115] Data extracted from a set of overhead transparency slide paper copies from a variety of sources across NSF and used at the BBS AC Meeting of May 5–6, 1989; Lateral Files at BIO.

BSR, spoke specifically about the LTER. Concerns were voiced that the "LTER sites were NSF's answer to the need for research on global change [, and so w]hat of the individual investigator?" that is, small grants/little biology. The questioner was AC member Margaret Davis (met in the previous chapter), whose strongly activist voice favored grants to individuals. It would appear that such, in her view, was almost to the total exclusion of multi-investigator grants, including LTER, the STCs, and other large projects. Davis did indicate on the second day of the meeting that she had not meant to have been so critical of LTER, it must be said, and she did aver that NSF's "structure needed to become more accommodating to multidisciplinary research."[116] Possibly so, but LTER had two positions in the BBS that were aimed specifically at administrative unification of the Project, and LTER had its own Network Office. And furthermore, Davis got a taste of an LTER site that she visited later in the year. She presented the results of her trip at the second AC meeting in fall, 1989, when she held that she was, in fact, "impressed with the way that participation in the network was being handled."[117]

The great reorganization of the NSF, and BBS cum BIO, was on the horizon for 1991–1992, but LTER would continue on into the future much as it had been developing. The directions of LTER will be considered again in the closing chapter.

The BSR overall remained much the same in its major structural configuration from 1976, when Population Biology and Physiological Ecology (PBPE) was added to the four extant program areas present in 1975, to the biological sciences directorate of 1992. Each of those other smaller (not an entirely apt word, though) programs will be considered below. Their funding in the first decade and a half considered in this three-decade history typically ran from about one-third to one-half, respectively, of that of Ecosystem Studies with its behemoth LTER. But after 1992, LTER would be moved to a new division.

The Board, in early 1989, forwarded a resolution on "International Cooperation of Global Environmental Change Research," thus fulfilling an intent stated years earlier in 1976. Since the "global environment is indisputably changing, and some of the change is clearly man-made," the NSB therefore indicated that NSF would seek to engage more in international research efforts, provide leadership to interagency coordination in the United States, "establish effective mechanisms for planning the science agenda" internationally and, in a myriad other ways, play a central role with regard to world-wide efforts to understand global environmental change.[118] And as the NSF was celebrating the close of four decades of support of American and world science, LTER was a *featured* programmatic aspect in the pages of the *Annual Report 1989*.[119]

---

[116] Ibid., pg. 4.

[117] BBS AC, "Minutes," October 23–24, 1989, pg. 5.

[118] NSB, "Minutes," NSB-89-58, Appendix NSB-89-55, March 15–17, 1989, 3–89:8–9.

[119] National Science Foundation, *Annual Report 1989* (Washington, D.C.: Government Printing Office, 1990), ppg. 42–43.

Ecosystem Studies, the LTER, and many other issues and programs, continuing and newly established, will be discussed again later after a consideration of the reorganization of 1991–1992 (following chapter).

4.3 Systematic Biology: Ordering the Living World

"As early as 1952," said Appel, "NSF became the primary federal patron of systematic biology."[120] Within several years, that area provided as much as 80% of all such funding nationwide. The Systematic Biology Program had reached a budget of some $5.6M by 1967. Though individual grants were typically quite small at the time, around $10k, the Program was responsible for awarding many of them. However, it also aided in the creation of important organizations: the Association of Systematics Collections was founded in 1972, for instance.[121] "Collections" suggests a central work of systematics: classification—an absolute essential in doing biology of any type. Biological systematics (sometimes given as biosystematics), the study of the *diversification of life over time*, should not be confused with taxonomy, the organizing of living and extinct organisms by type, nor with nomenclature, the giving of names to both closely and distantly related organisms, living or fossil, nor even with evolutionary biology. Both of the first two cover all hieronic levels of classification: taxa. However, biological systematics must make extensive use of these two fields; they are integral to it. After 1859 and the publication of Darwin's *Origin of Species*, systematics has also been inextricably tied to evolution, the fact, and evolution, the discipline. By the mid-1970s, biosystematics was a highly developed field as evidenced by commentary in the *Annual Report 1976*: ecology and systematics *together* had become, the authors averred, the "core disciplines of NSF's environmental biology programs." Further, the programs of DEB had "undergone profound transitions during recent years."[122] Those transitions referred to the ever-increasing importance of systematics to the findings made in biology of the little: the multifaceted significance of DNA playing a role, most especially.

Appel spoke of the work of NSF and the early biological sciences directorate (BMS) as *shaping biology* in America in her book of that title, and Jim Edwards (detailed further below) would later opine that systematics was central to that, in at least one sense. "[R]ight from the beginning [that is, 1955], the actual names of programs that were given to fields within the [NSF] *already were shaping biology*."[123] One of those first programs was systematics. "At the time that [the title was chosen],

[120] Appel, 'Shaping Biology,' pg. 216.

[121] Ibid., pg. 217. Appel erred in referring to it without the final "s" on Systematics. The ASC's founding records from 1972 to 1988 are a part of the Smithsonian's collections. See: http://siarchives.si.edu/findingaids/FARU7459.htm.

[122] NSF, 'Annual Report 1976,' pg. 62.

[123] Edwards interview. Emphasis added.

there was this gigantic fight [within] the community about whether to call it taxonomy or systematics." As Edwards went on to say, to many scientists it did not matter, but to others it was "a really, really big deal" and for NSF to choose to "use the term systematic biology was an important signal."[124] In the *intricate dance*, this decision on the part of NSF very probably had a definite influence on the community.

Edwards felt that biology in this country was *shaped* by NSF in four different ways. He believed that the *rotator concept* was of great significance in that it continued to bring to NSF new ideas and current thinking from the scientific community. (It should be added that the reciprocal was also true.) He felt that the rotators brought on a "random walk," but not all that random because "various rotators ensure [that] program[s move] in different ways."

> [The] pseudo-random walk influence of the rotators [would] counter [the possibility that] a single permanent program officer...would get a lock on a particular idea and the program [wouldn't] change[, thus] leading to stultification within that particular field.

Still, he did "see several instances where he felt the narrowness of the program officer was not in the best interest of the fields covered." Nonetheless, Edwards averred that to have a single permanent officer and a group of rotators in a given program was the "best of both worlds."

Second, he felt that *panels* "play a big role" due to the community feedback they provide to NSF. In BBS, a mixture of live and mail panels was used, so even more views flowed in. Third, Edwards felt that the biology directorate was, especially under Mary Clutter, "a leader in several different fields." Even with the NSB often worrying that NSF was too reactive and not proactive enough, Edwards found BBS/BIO to be otherwise. He pointed out plant genomics as a prime example. Thanks to Clutter, he believed, the biology directorate was quite the reverse of reactive. BBS went beyond shaping just pure biology, as the plant science community was "highly fractured" and a great deal of political effort was required on the part of the AD for biology to bring them together. While "NSF officials cannot work directly with politicians...they can work through OMB and...OSTP in ways that can help get particular programs going and in appropriating funds to move those programs forward." That Mary Clutter did.

In his fourth, and final, point, Edwards thought that later Foundation Director Rita Colwell "played an important role in getting the community of scientists to think larger and broader." We shall come back to Colwell, the first woman biologist to lead NSF, as well as the founder of the concept of *biocomplexity*, in later chapters.[125]

Returning to the 1970s, the work with products of DNA—proteins—such as hemoglobin, albumin, and cytochrome C (those three were heavily studied in the 1960s), and the differences in those molecules over various species, particularly the

---

[124] See also Appel's comments on her page 65, in particular, and other references to systematic biology in 'Shaping Biology.'

[125] The quotations for several foregoing paragraphs are those of Edwards from his interview with the author.

vertebrates studied in the earlier years, was being used in a new form of biosystematics: molecular systematics. The "differences in amino acid sequence of the [protein] molecules can be attributed to differences in…genes, that is, the DNA sequences that code for these proteins."[126] DEB supported the research of Allan C. Wilson who was then seeking to demonstrate that the "rates of organismal change would provide information on the mutation rates of genes" associated with a variety of characteristics of given experimental species. However, he showed that the rates of genetic change and that of, say, expressed anatomical characteristics examined, showed an unexpected independence. That is to say, albumin in frogs and that in mammals, fairly distant relatives, had evolved at about the same rate. In a slightly different research direction, he also found that some 44 proteins of humans and chimpanzees were found to be 99% identical; early steps in demonstrating, in later years, the fact that chimpanzees are human's closest living relatives. Somewhat baffled, Wilson then turned to "regulatory genes" in search of an explanation for obvious evolution in organisms but which showed minimal change in "structural genes"; the former controls the activation or expression of the latter. Wilson and his colleagues became founders in the area of molecular evolution when he, especially, proffered the "Mitochondrial Eve" hypothesis.[127] The entire area of molecular evolution became of interest to NSF, as it was also in the community: the *intricate dance*. The subject was considered in detail in one of the prime publications of the Foundation at that time: *Mosaic*.[128] The article's author noted that evidence of evolution "has come traditionally from two major sources: comparative morphology and the dating of the fossil record." But over the previous two decades, the author went on, the emerging field of molecular evolution brought new data into the equation.[129]

---

[126] Ibid.

[127] A short biographical note concerning Allan Charles Wilson can be seen at: http://www.ncbi.nlm.nih.gov/pmc/articles/PMC1682534/. He "revolutionized the study of evolution," by, in particular, the use of molecular approaches.

[128] *Mosaic* was a product of the public relations group within NSF and was a slick magazine published from its first issue in Winter 1970, until its last in Fall 1992. The editor was Warren Kornberg at the time of the 1979 issue (see main text above), with some six issues per year then, though the publication began and ended its life with four issues per year. Arthur Fisher in his "An Appreciation of *Mosaic* and the Devoted People who Made it All Possible," *Mosaic* (1992):52–53 (reprinted from a then contemporary issue of *Sciencewriters*, the newsletter of the National Association of Sciencewriters), said that "I write not to bury *Mosaic*, but to praise it." The article is the best history of *Mosaic* in existence (personal communications between the author and Ellen Y. Weir, Christina Bartlett-Whitcom, and Susan E. Olmsted, all of NSF, and to all of whom the author offers his thanks). The founding editor, Jack Kratchman, named the publication as he did because of the " 'nature of the scientific effort, which relies so very much on many individual contributions to knowledge coming together to form a pattern of science theory.' " Quotation from the Fisher article, pg. 52. This is another example of how one can view the rise of interdisciplinarity at NSF overall. All of the *Mosaic* issues and articles can now be found at an online site and provide an excellent historical overview of much of the NSF's work for over two decades; see: http://www.mosaicsciencemagazine.org/.

[129] Author unknown, "Molecular Evolution: A Quantifiable Contribution," *Mosaic* (1979):12–21.

Over the next several years, use of DNA and other molecules proved invaluable in helping to understand the evolution and, necessarily, the systematics of organisms such as flightless birds, ratites, as one example of the new directions in systematic biology. Charles G. Sibley and Jon E. Ahlquist of Yale were supported in work on the ratites and established the notion that all the world's living ratite species (emus, cassowaries, rheas, tinamous, kiwis and ostriches) derived from a common ancestor; a situation known as monophyly.[130] (That idea remained as a given for many years until work done in 2008 contradicted it and demonstrated that parallel evolution was at work and the ratites were, in fact polyphyletic—deriving from multiple ancestors.[131]) Sibley and Ahlquist continued their work (1986) in reclassifying not just ratites, but *all* of the bird group, the Aves. Over more than a decade, the two compared the DNA of 1600 widely related bird species and found that many aspects of the then standard systematics were in error.[132] Even with the later reversal of their ratite findings, much of their research stood the test of time.

But support in systematics was by no means limited to the molecular end of the spectrum: whole organisms and their important roles in nature were of interest, as well. Nematodes, noted earlier, are to be found in soils worldwide and play a significant part in a number of processes, including as major plant pests (pathogens) in agricultural settings. It has been said that if one were to take all the nematodes on the planet and have them outline everything visible from buildings to blades of grass and then remove all but the nematodes, the world would look familiar to us, so numerous are they.[133] Since the reproductive modes of nematodes are quite unusual, their classification was highly problematic: a systematist's nightmare. At North Carolina State University, Chapel Hill, Anastasios Triantaphyllou was a leading nematode biologist having worked out many of the problems with both their reproduction and classification, especially the plant pathogenic ones. Many of the most damaging species seemed to be *polyploids*, creatures having more than a single full set of chromosomes (diploidy); the condition is common in plants, but rare in animals. Those nematode polyploids were poor survivors in nature, but did well in cultivated crop situations—artificial environments. "Triantaphyllou's research is an excellent example of systematic biology—study of a group of organisms that offers a challenge as a taxonomic group."[134] Such *applied research* was in keeping with the

Quotation from pg. 12.

[130] National Science Foundation, *Twenty-Eighth Annual Report for Fiscal Year 1978* (Washington, D.C.: Government Printing Office, 1979), ppg. 71–73.

[131] See J. Harshman, E.L. Braun, M. J. Braun, *et al.*, "Phylogenomic Evidence for Multiple Losses of Flight in Ratite Birds," *Proceedings of the National Academy of Sciences* (*PNAS*) 2008:13462–13467. The monophyly view held even as recently as 2005: Allan J. Baker, *et al.*, "Reconstructing the Tempo and Mode of Evolution of an Extinct Clade of Birds with Ancient DNA: The Giant Moas of New Zealand," *PNAS* 2005:8257–8262.

[132] National Science Foundation, *Annual Report 1986* (Washington, D.C.: Government Printing Office, 1987), pg. 14.

[133] Personal knowledge.

[134] NSF, 'Annual Report 1979,' *op. cit.*, pg. 57. Director Richard Atkinson also found the nematode

direction of NSF at the time and both the nematode and ratite grants were also typical of the smaller ones of the period.

Even though grants were small, averaging around $10k for much of the 1970s and into the 1980s, the names of researchers could sometimes be very big, indeed. The late Stephen Jay Gould of Harvard University, one of the most well-known names in science during the closing years of the last millennium, along with colleague David S. Woodruff of UC, San Diego, were being funded to study the West Indian land snail, *Cerion*, in 1982. The species of that genus were unusual because they had evolved into some 600 morphologic forms (gross appearance) without being reproductively isolated and with little genetic differentiation.[135] Gould and Woodruff sought to determine what the snails could reveal about the "processes of adaptive radiation and speciation in general." The team sampled over a thousand localities in the Bahamas islands finding more than 250 *named* species. However, after careful study, they believed there to be only about ten *biological* (actual) species present! The 250 found showed little genetic or anatomical diversity, but their morphologies were widely varied, especially shell color and shape. That variation was shown to be based upon the "snails' adaptation to thermal stress and predation by land crabs."[136] It may be said, "morphology is a poor indicator of species status" and that reproductive isolation had been overstated by earlier investigators; later, most of the so-called species were shown to hybridize. Just what caused the genetic variation present remained unknown at the time. An active Internet website on *Cerion* hosted by the Smithsonian's National Museum of Natural History exists today due to NSF funding to Gould and other colleagues.[137] It is a small, but fitting tribute to Gould, a master of science communication, let alone several fields in geology, ecology, biology and their histories.

The year 1982 saw a change at BSR when Jim Edwards came as Program Director of Systematic Biology, for the first two years as a rotator and then as the permanent director until his move in 1985 to BRR, the biological resources division, as director. At that time, and for many years both before and for a few years thereafter, program directors in smaller programs were rotators, some programs having no permanent scientific personnel at all. Individual secretarial staff members were often long associated with a given program, however. Edwards, with his doctorate from UC, Berkeley, had come from a tenured position at Michigan State University, East Lansing, to direct the systematics effort at BSR. He had been at Michigan State since 1976 and tenured since 1979, and he had been funded by NIH until 1982. He was receiving NIH funding as his work in that period was in functional morphology, not systematics. With an undergraduate degree in paleontology,

---

study of interest and mentioned it as one exemplar of NSF's support of scientific research in the country in his "Heading for the 1980's" Director's Statement in the 'Annual Report 1979'; see page vii. He called it "a 'flora and fauna' kind of study."

[135] NSF, 'Annual Report 1982,' *op. cit.*, pg. 32.

[136] Ibid.

[137] Grant number EAR 0106936 was awarded to Gould, G. Goodfriend and M. G. Harasewych. The website for *Cerion* at the Smithsonian is http://invertebrates.si.edu/cerion/.

he was working on the evolution of terrestrial locomotion of salamanders and got his support from the neuroscience program at NIH as they were interested in his work on the spinal chord, its neurons, and its potential application to humans. NSF did not fund him because of the human connections; one of the fundamental bases for either NIH or NSF choosing to support a given proposal when submitted to both agencies. Not long thereafter, it became impossible to submit to both agencies at once due to newer BBS, but not NSF-wide, rulings under Mary Clutter, as noted earlier. Some program directors in the BSR division (as well as the cellular and molecular areas) did not like the decision, as they wanted to fund "superstars," said Edwards, and not let NIH have all the credit in such matters. As it happened, his attempt to renew his NIH grant (1982) did not materialize. Further, his department at the University was in turmoil at the time. "Out of the blue," he recalled, he was offered and gladly accepted the rotator job at NSF.[138] He had been known to the Foundation since the 1970s as a reviewer of proposals in systematics. When he arrived in August of 1982, he was the senior of the three new rotators coming into the program. Very soon thereafter he found that he liked the position greatly and enjoyed "doing research vicariously," a sentiment that may well fit many another rotators who stayed on when a permanent position was offered. He and his wife liked the capital area and she, a sign language interpreter, found work at Gallaudet University (for the hearing impaired) in Washington, D.C.

John Brooks was the BSR director at the time and felt that the systematics program should have a permanent director and not merely a rotator. Edwards applied for the new position and got it; in the process he gave up his tenured professorship at Michigan State. He held the systematics position until his move to the BRR directorship. The three systematists in the program divided up the work based upon, not surprisingly, their specialties. In FY82, those one-year rotator systematists included Grady L. Webster and Harold K. Voris. Edwards, working with vertebrate proposals, joined as director later in that year. The trio in FY83 had included: Wallace E. LaBerge (University of Illinois) for invertebrates; W. Hardy Eshbaugh (Miami University of Ohio, Oxford) for botany; and, Edwards for vertebrates.[139] Edwards recalled that at the time he came, NSF was just getting desktop computers—one was on his desk when he arrived—and he and his colleagues realized the potential for systematics in using the new tools; the secretaries were still creating all documents on typewriters using carbon paper for copies.[140] In fact, in 1984, when Erich Bloch became director, he placed a desk top computer on his secretary's desk and she "almost quit[:] 'I can't work with that, I need my typewriter.'"[141] Further, BSR over the next several years, developed programming for a database of reviewers and, with Mark Courtney of BSR's PBPE program, developed a standardized

[138] Edwards interview.

[139] Personal communications with James Rodman, May 7, 2012. Rodman maintained a roster for those years.

[140] Edwards interview. It was the beginning of the era of desktop computers; this author saw his first such device in 1978, a Radio Shack brand TRS80, and took his first programming course in 1981.

[141] Erich Bloch Interview with Marc Rothenberg, Historian of the NSF, February 11, 2008.

questionnaire for potential reviewers. Edwards noted the special nature of the database as it allowed him and his colleagues to no longer merely count on persons they knew in their respective fields as potential or realized reviewers, but to have a system in which specialists could be chosen on more rational grounds. In the process, many more potential reviewers could become actual ones and bring both more talent and knowledge to the process, as well as provide for possible future rotators. Edwards used the database to choose three reviewers that he did not know for a given proposal—he called them "wildcards"—and three that he did, typically from a circle of colleagues. Edwards believed that the other divisions within BSR did not have a similar system at that time; a lack of evidence to the contrary suggests he was correct.

As the 1980s progressed, the systematics program began to grow in numbers of personnel due to the increasing number of proposals that demanded such growth. A fourth systematist would often manage the BS&I program, but was a formal member of another cluster, not systematics[142] In 1983, Jim Rodman joined the systematics program as a rotator. He considered himself an "NSF baby" as he had attended a Foundation-sponsored summer science camp when he was in high school.[143] It was held at Michigan State University and, lasting some six or eight weeks, it emphasized biology and mathematics. He was encouraged to go on to college there where he worked on an hourly basis for his last two years on an NSF grant to the University's Herbarium. He had his own NSF Graduate Fellowship in his doctoral years at Harvard that was made available, he believed, due to the increased funding for science across the nation, "probably courtesy of *Sputnik*." His graduate work included work in biochemical systematics and he had knowledge of numerical taxonomy.[144] He took a faculty position at Yale for the following ten years and had NSF funding throughout, two grants of which came from the systematic biology program. Other funds came from the Ecology Program and supported a project on plants and insects. In his last few years at Yale, Rodman was approached by his undergraduate mentor, John Beaman, who had become a rotator in the systematics program, and was asked to become a panelist for the program. Rodman put in three years on panels and then left Yale to take his own rotator position, which later led to his career at NSF.[145] Rodman and Edwards overlapped for a short time before Edwards' move to BRR.

---

[142] Rodman communication of May 2012.

[143] Jim Rodman Interview with the Author, May 11, 2009.

[144] A highly mathematical form of taxonomy, numerical taxonomy may be defined as "(1) A type of taxonomy using numerical similarity values, e.g. cluster analysis, to rank organisms into categories based on the degree of overall similarity, [or] (2) A way of grouping and ranking organisms into taxa according to the numerical evaluation of the affinities and similarities between taxonomic entities or units." http://biology-online.org/dictionary/Numerical_taxonomy.

[145] David Schindel came in 1986 replacing Kohn as a rotator in invertebrate zoology (more on Schindel later in the main text above). Rodman overlapped with Eshbaugh for a few weeks replacing him in summer 1983. Finding a full roster of all persons in all years in the biology directorate did not prove possible at the time of this research, but may be so. The *Annual Reports*, for instance, listed the more senior personnel at NSF, but only down to the level of division, not program, direc-

Rodman averred that he came into a program in which there was a considerable amount of team spirit with team-based skills, and overall "a well-oiled machine already; the staff was held to high standards," apparently by Edwards as its senior scientist. In that "very task-oriented" group, Rodman noted that one of the only aspects that was not positive for him was that all choices for panel members had to be vetted by John Brooks. Brooks had knowledge of systematics and knew a great many of the people proposed as panelists, Rodman allowed. After some time, Rodman reevaluated his feelings and concluded that Brooks had much to offer—"I sort of appreciated his knowledge and skills."

The early and mid-1980s was a time remembered by Rodman as the "great cladistics revolution." Cladistics is a more modern form (since the early 1950s) of taxonomic analysis and phylogeny-building for any type of organism.[146] Since the team had some "very strong personalities around[, there were] shouting matches at meetings and other shenanigans probably through the review process," Rodman recalled.[147] But he went on to say that "I don't think any of the panels were ever jeopardized." All of that generated heat was based upon the *methodologies* of systematics chosen by a given grant seeker, some of which fell into the area of cladistics (that number increased markedly with time). The whole approach of using cladistics often engendered considerable animosity amongst many classicists in the community, at least in the early period.[148] Even before the early and mid-1980s, there was much heat and little light shed over the cladistics' resistance movement, as one might term it.[149]

Rodman explained the apparent lack of anything other than grants to individuals in the systematics program during much of the 1970s and 1980s, by saying that the Program did not offer any notable *initiatives*: "the first years I was there, things were quiet" with regard to large, or even small, initiative-type offerings. The paper historical record, if not flawed, seconds Rodman's assessment. The manner in which

---

tors until 1988, and then only down to the directorate level from then onward.

[146] "Cladistics is a particular method of hypothesizing relationships among organisms. Like other methods, it has its own set of assumptions, procedures, and limitations. Cladistics is now accepted as the best method available for phylogenetic analysis, for it provides an explicit and testable hypothesis of organismal relationships. The basic idea behind cladistics is that members of a group share a common evolutionary history, and are 'closely related,' more so to members of the same group than to other organisms. These groups are recognized by sharing unique features which were not present in distant ancestors. These *shared derived* characteristics are called *synapomorphies*." http://www.ucmp.berkeley.edu/clad/clad1.html.

[147] Rodman interview.

[148] See, for instance: http://www.palaeos.org/Cladistic_controversies; Ernst Mayr, "Cladistic Analysis or Cladistic Classification?" *Journal of Zoological Systematics and Evolutionary Research*, 1974:94–128; http://golab.unl.edu/teaching/phylobio/papers/BrowerCladistics.pdf: the journal is that of the Willi Hennig Society, *cf* http://www.cladistics.org/; the German botanist Hennig created cladistics in the 1950s. One of the best examinations of the cladistics controversies is to be found in David L. Hull's *Science as a Process: An Evolutionary Account of the Social and Conceptual Development of Science* (Chicago: The University Press, 1988), Chap. 5, "Systematists at War," *passim*.

[149] Personal experience.

individual proposals were sought from the community would have been such that the

> encouragement in various fields would have been informal…potential PIs calling on the phone…[and program officers] chatting them up about where things were going [and] who was working on what.[150]

Such was a typical method of inviting proposals across the entire Foundation and remains a major methodological aspect of how scientists and the NSF come into contact for the vast majority of individual grants over the continuing life of the agency, the *intricate dance*. Program calls for proposals did exist then too, of course, it should be said. Rodman began to promote his own interest area in the community by such individual contacts "chatting up" molecular systematics. "Not everyone was doing flavonoids, or sunflowers, or…a particular group of [chemical] compounds for a particular group of families [of flowering plants]," he stated. For that reason, "whole chunks of the Tree of Life [were] being ignored."[151] Rodman pushed molecular, or chemical, systematics to the community.

Why did the Systematic Biology Program not offer *initiatives* in the earlier period of interest here, and why did it start to do so in the later 1980s? Rodman offered that it may have been a simple matter of funds available to the program. The budgets for all areas of DEB/BSR from Fiscal Year 1976, the year that the Population Biology and Physiological Ecology Program was established, until 1984 show that Ecosystem Studies (the initial home of LTER) was always the leading program with regard to budget in DEB/BSR overall (Fig. 4.1). Still, within the Systematic and Population Biology subset, broken out from Ecological Science from 1981 onward, Systematic Biology was always the leader within its group in those years. It is not surprising that PBPE should be second to Systematics in funding in 1976, as it was a new program. However, it would remain so throughout all that period. While PBPE generally led over BRR, always a smaller program, in some years they were nearly or precisely the same in their respective annual budgets. Both of those programs remained lesser in size than Systematics throughout, however. The ratio between Ecosystem Studies and Systematics over the years tended to be very roughly 3:1 or 2:1. By the time that Jim Edwards came on the scene in 1982 the ratio was under 2:1 and did not again fall into the 3:1 range during his tenure. Coupled to what Rodman had to say about the taut ship that he saw in 1983 upon his arrival, it seems that the systematics program was on a general, but still slow, upswing and so more likely to begin to offer initiatives and competitions in future years. Indeed, two workshops to promote fields that would benefit by computerization were offered under the guidance of David Schindel who had joined the program in 1986.[152] But Rodman recalled that it would not really be until the early 1990s that initiatives would finally be seen. Given that the total for the Systematic Biology Program had risen to $11M by 1984, one might ask if initiatives were not fiscally possible by

---

[150] Rodman interview.

[151] Ibid.

[152] Ibid.

BIOTIC SYSTEMS AND RESOURCES
FY 1976 - FY 1984

Rev. 12/02/83

Fiscal Year	Program Activity, Subactivity, etc.	Amount
1976	Environmental Biology :	
	Ecology	$ 3,030,000
	Ecosystem Studies	11,903,481
	Systematic Biology	4,375,000
	Research Resources	3,915,000
	Population Biology and Physiological Ecology	3,615,000
	Total	26,838,481
1977	Environmental Biology :	
	Ecology	4,169,401
	Ecosystem Studies	10,791,968
	Systematic Biology	6,901,633
	Research Resources	4,596,230
	Population Biology and Physiological Ecology	3,714,310
	Total	30,173,542
1978	Environmental Biology :	
	Ecology	4,240,707
	Ecosystem Studies	10,811,992
	Systematic Biology	7,429,629
	Research Resources	4,980,506
	Population Biology and Physiological Ecology	4,026,844
	Total	31,489,678
1979	Environmental Biology :	
	Ecology	4,942,244
	Ecosystem Studies	11,663,199
	Systematic Biology	7,550,647
	Research Resources	4,696,561
	Population Biology and Physiological Ecology	5,100,132
	Total	33,952,783
1980	Environmental Biology :	
	Ecology	6,772,484
	Ecosystem Studies	12,996,059
	Systematic Biology	7,868,948
	Research Resources	6,608,744
	Population Biology and Physiological Ecology	5,462,572
	Total	39,708,807
1981	Environmental Biology :	
	Ecological Science :	
	Ecology	6,504,939
	Ecosystem Studies	14,204,658
	Subtotal, Environmental Biology	20,709,597
	Systematic and Population Biology :	
	Systematic Biology	8,413,992
	Population Biology and Physiological Ecology	6,111,664
	Biological Research Resources	5,816,988
	Subtotal, Systematic and Population Biology	20,342,644
	Total Environmental Biology	41,052,241

Fig. 4.1 Biotic Systems

Fiscal Year	Program Activity, Subactivity, etc.	Amount
1982	Environmental Biology :	
	Ecological Science :	
	Ecology	6,239,000
	Ecosystem Studies	15,027,875
	Subtotal	21,266,875
	Systematic and Population Biology :	
	Systematic Biology	7,990,674
	Population Biology and Physiological Ecology	6,082,774
	Biological Research Resources	6,649,851
	Subtotal	20,723,299
	Total	41,990,174
1983 Actual	Environmental Biology :	
	Ecological Science :	
	Ecology	7,291,433
	Ecosystem Studies	15,080,243
	Subtotal	22,371,676
	Systematic and Population Biology :	
	Systematic Biology	8,612,172
	Population Biology and Physiological Ecology	6,782,750
	Biological Research Resources	7,739,999
	Subtotal	23,134,921
	Total	45,506,597
1984 Current Plan	Biotic Systems and Resources :	
	Ecological Science :	
	Ecology	8,030,000
	Ecosystem Studies	16,770,000
	Subtotal	24,800,000
	Systematic and Population Biology :	
	Systematic Biology	11,000,000
	Population Biology and Physiological Ecology	8,800,000
	Biological Research Resources	8,800,000
	Subtotal	28,600,000
	Total	53,400,000

Fig. 4.1 (continued)

then. The answer to why such were not offered, while not definitively known, may lie in the fact that the program had long been under nothing but rotators, until Edwards become its first *permanent* program director in 1985 following his own two years in a temporary position. Furthermore, LTER was taking off at that time and had followed both a long period of an active IBP and then big ecosystem project completion, under Clark, for several years of the IBP/LTER interregnum. The mind-set of DEB/BSR seemed to be strongly oriented toward the Ecosystem Studies end of things. Finally, many NSF publications, though not all, stated over time that the ecology and ecosystem studies programs—not systematics—were the "core" of DEB/BSR.

David Schindel arrived at NSF to take the rotator position in invertebrate zoology in July of 1986. Edwards had moved to BRR for FY86 when the trio included

Rodman, Alan J. Kohn (University of Washington), and Mark S. Hafner (Louisiana State University). Schindel knew NSF as a grantee just the year or two before in his professorial position at Yale. He was a paleontologist with orientation to evolutionary theory in systematics having done pre-doctoral work at the Smithsonian in the later 1970s. He began at Yale as a new faculty member in 1978 and he was awarded his first NSF grant in the mid-1980s. That project was to set up a sort of multidisciplinary center at the University that was aimed at the "computer-aided analysis of shape and shape-change" over time in organisms: the field of morphometrics.[153] But it was also used to study archaeological specimens and other objects that could be defined quantitatively based upon morphology. Schindel came to the Foundation and the Systematics program with a strong computer orientation, which had immediate effects upon the nature of the program. Being the program officer for the invertebrate aspects of systematics meant that many proposals would deal with insects. Since Schindel was a paleontologist, therefore a great deal of his training was in geology, research that was to deal with living insects was a "bit of a stretch" for him, he stated. There was a concern in the community, too. It was held that his lack of training in the biology of living animals would be problematic for fully informed judgment of entomologists' proposals. He had stated in his job interview for the NSF position that he could "recognize a good argument [in a proposal] from a bad [one]". This, but more so his strong computer-aided expertise in morphometry, was more than enough to satisfy his interviewers; in fact, his computer knowledge was much desired. It was, of course, the time of Erich Bloch.

Since BSR was more than pleased with his background, they were also willing to allow him to promote programmatic ideas beyond merely maintaining a program officer's typical duties. He began contacting the community and encouraging them to submit proposals for short courses that would support training for biologists in the use of computers to aid their research. Soon thereafter, it became possible through that effort to start courses (each of two week's duration) in morphometrics for taxonomists, evolutionary biologists and others.[154] Those efforts led to the development of software that was later "distributed on a nonprofit basis" to those desiring it. The short courses not only trained those who attended, but raised the awareness of a more quantitative biology for practitioners long accustomed to mostly qualitative study. The first of the courses was offered in the year Schindel began at NSF and was presented at the University of Michigan. The second was at the State University of New York, Stony Brook, and others followed thereafter. Schindel designed the courses to be somewhat in the nature of those that had been offered for a number of years in molecular biology at Woods Hole Oceanographic Institution/Marine Biology Laboratory, long renowned for short courses in many fields of biology; with some of those earlier efforts supported by the NSF, incidentally.

---

[153] David Schindel Interview with the Author, May 11, 2009. Morphometry is simply the measurement of the outside of something, and is a sub-disciplinary area in biology, both of living and fossil creatures.

[154] Ibid., the interview.

Schindel was happy that he could "introduce new ideas, new directions [and] new possibilities" into the program. In so doing, he felt that he could "move the field [into] productive lines of research." His vision had been affected greatly by the morphometry center he established at Yale just before his move to NSF. That center was not a part of the Science and Technology Centers Program, but Schindel did arrive at the Foundation in the inaugural year of the STCs. He was "intrigued [and] tremendously excit[ed]" by the STC program, and remained so even in an atmosphere of mixed feelings about it at BBS early on. He found it "kind of ironic and sad" that BBS did not get many of the earlier STC awards even as late as the turn of the millennium when the second big round of STC proposals were coming in. That was long *after* the initial 11 years committed to by the NSB and was evidence of a highly successful concept and program. Schindel was of the opinion that the "top management of biosciences did not like the STC program," as the "idea of making [those] very large investments in one place [was] taking the bread out of the mouths of [individual] researchers." Whether he meant by that sentiment a dislike of Mary Clutter or others, or some combination, he did not elucidate. Going "head to head [in a] competition between [*sic*] biologists, geoscientists, physicists" and others, was inappropriate for biologists in particular, he said: it seemed they had less chance at winning STCs.[155] The BBS personnel were "starkly divided," he opined, over whether STCs were good for biology or not. Some, Schindel felt, held that the Centers were "hiding places for academic deadwood," but yet others, including him, held that they were places that "creat[ed] matches" among scientists and that they allowed for easier tracking of funds, and that the "whole would be greater [in scientific productivity] than the sum of the parts." Indeed, Schindel became so much involved with the STCs that, in the 1990s and into the present millennium, he headed the STC program in the Director's Office, having left BBS to do so.

It should be considered that only modest things in the way of big science were done anywhere at BBS in the 1980s, with the obvious exception of LTER and the still youthful biotechnology. Mary Jane Osborn, member of the NSB, noted that "[i] n general, the BBS…is viewed as a directorate of 'small' science."[156] It is interesting to ponder whether BBS equated big biology with ecology only and did not yet envision little biology in a big format, even though biotechnology was becoming just that. More compellingly, Acting Director Knapp said much the same thing about BBS being a locus of small science at the AC meeting in January of 1984:

---

[155] Ibid.

[156] The NSB was made up of a broad cross-section of scientists representing many fields, certainly not biology alone. Nonetheless, there were a number of biologists on it at all times, of course, with the few exceptions when the Board was unusually short in its total membership; such did occur. Mary Jane Osborn had a bachelor's from U.C., Berkeley and a PhD in biochemistry from the University of Washington. She did postdoctoral work in microbiology and a decade later became a professor at the University of Connecticut Health Center School of Medicine in Farmington from 1968 and its head from 1980 onward, as of this writing. She was an NSB member from 1980–1986.

> there might be large projects within the Directorate in the future, driven by the need for expensive instrumentation, in contrast to the 'small science' mode characteristic of the Directorate in the past.[157]

Clearly, Knapp (a physicist) equated large projects with the need for "expensive instrumentation" (telescopes maybe?) and not with what might come to be large projects (genomics, for example, yet in the future). Indeed, a real-life variation on such a presumed theme was already brewing (following paragraph). In FY87 and FY88, Rodman became Acting Program Director and permanent program officer with Tim Lawlor, invertebrate biology, and Schindel as Associate Program Directors. In FY89 and FY90, William S. Moore (Wayne State University) replaced Lawlor.[158] Schindel later followed Edwards again in 1990 when the latter left BRR to become Deputy Director for BSR; Schindel thus became director of BRR from 1990 until 1992. In 1993, Edwards became the Deputy Director of the still new Directorate for Biological Sciences (BIO) under Mary Clutter.[159] As something of an aside, it should be noted that rotators typically came to NSF in the summer after spring graduation at their home universities (if they were academics) and so saying that they were in a PO position in a given fiscal year is something of a misnomer: they preceded the beginning of a fiscal year by several months.[160]

During Schindel's time at BSR, there were three panels per year in the systematics program and the awards were running at about a 30% success rate. This would have been judged, at that time, a sign of good panel member efforts and equally good program officer work. But it became clear to the three program officers that proposals coming in to them were falling out in a "multimodal" way, Schindel stated. That is, some were very molecularly oriented, especially in molecular phylogeny as determined by genetic data. Those were "expensive, high-tech [projects with] a lot of visibility." Others coming in focused on "traditional systematic revision studies"—those that generate monographs on various taxa—"and then there was a group of proposals that came to be known as Biotic Surveys and Inventories." That latter group was not faring well in competition with the other two, especially the molecular ones, simply because the more chemical ones were what was *hot* at the time—and would remain so for the foreseeable future. For that reason, the three systematists (Moore, Rodman, and Schindel) decided to set one of the three annual panels aside for the surveys and inventory proposals exclusively. It would "level the playing field," Schindel held.

---

[157] BBS AC, 'Minutes,' January 26–27, 1984, pg. 1.

[158] Personal communication with Rodman in May 2012

[159] Data derived from a combination of the transcripts of both the Edwards and Schindel interviews.

[160] Rodman communication.

Some of those proposals in the inventory category required the use by American investigators of the collecting of specimens in foreign countries. Simultaneously, the NSF was preparing for the imminent likelihood of an international Convention on Biological Diversity (CBD), which came to pass and is still in effect as of this writing.[161] The Foundation was thus emphasizing the importance of international agreements and its intent to adhere to such. CBD's central *raison d'être* is based upon three main objectives: the conservation of biological diversity; the sustainable use of components of that diversity; and, the fair and equitable sharing of the benefits from use of a signatory nation's genetic resources (diversity).[162] "Many countries saw gold mines in their biodiversity," especially the potential for pharmaceuticals derived from natural resources, such as plants. Therefore, it was imperative to Schindel and his colleagues in the Program, to assure that grantees who intended to collect in foreign locales have the proper permits. That became crucial after the final singing of the CBD in late 1993.

The hot areas of research that were making things difficult for biotic surveys and inventories' proposals yet to be funded were part of the *reductionist trend* argued for in this history and that Edwards' himself seconded as standing out as a theme at NSF: he stated that he saw reductionism "over and over again." Even in his own area, "a lot of proposals in systematics…were using genetic sequences"; they being valuable for population biologists as well systematists.[163] Indeed, the coming genomics revolution would define *Millennial Biology* in many ways.

The call for proposals document of 1985 solidified the brief wording that explained BSR's systematics program at the time by noting that it

> supports basic research on the characterization, affinities, adaptations, and evolutionary histories of all groups of the earth's biota. The dominant research emphasis is evolutionary, involving such interrelated disciplines as comparative morphology-physiology, biochemical and numerical systematics, biogeography, floristics, faunistics, and elements of paleobiology except studies that deal with man.[164]

It continued for a few more words, but made it clear just what would be funded and what would not: the long-term taboo of studies on man. Those were funded by the Anthropology Program in the social sciences' group then under attack again by, what Rodman termed, the "Reaganauts." Studies on humankind then, was left to NIH or some other agency, along with the Anthropology Program at NSF.[165]

---

[161] See: http://www.cbd.int/. The UN Environment Program began work on what would become the CBD in November of 1988. By June of 1992 at the "Rio [de Janeiro] Earth Summit," the Convention was signed and came into effect for the 168 original signatory nations by December 1993. The U.S. signed the CBD on June 4, 1993.

[162] This is a close paraphrase of the material at http://www.cbd.int/convention/about.shtml.

[163] Edwards interview.

[164] NSF 83-75, *op. cit.*, pg. 2.

[165] Rodman communication.

4.4 Population Biology and Physiological Ecology

In 1976, PBPE joined the other four programs in DEB/BSR discussed in this chapter. But its own history was somewhat better recorded than that of those other four, at least for its beginning years. Figure 4.1 shows data for many parts of the following discussion. Starting out at just under $4M for its inaugural year, it increased in funding by about 11% per year for the next few years and, by 1984, stood at nearly $9M, on a par with the longer established BRR.[166] PBPE provided support to grantees in the four areas of: population genetics; and of population, behavioral, and physiological ecology. The PBPE had come into existence in 1976 due to the moving of the contents of those four areas either from Ecology or from Systematics into the new program. In that first year (FY77), the Program awarded 112 grants and, by FY80, they had gotten up to 147. PBPE was not unlike the Systematic Biology Program in that some 95% of its funds went to individual researchers with the small remainder going to such things as "symposia, equipment, supplements, and dissertation research."[167] Nothing went to large initiatives then. There was a mix of one- to two-year awards and some three- to five-year "continuing" grants. Those continuing awards increased over the first three years of the program from 20% of the total to 37%: one aspect of big biology expanding. It was the intent of those first program officers to keep funding for *new* awards at or above 50% of the annual budget for PBPE. This varied over BBS but was in keeping with a general philosophy at NSF.

Of the four disciplinary areas funded by the program, population genetics was foremost in awards made. This was in concert with the increasing interest in the ecologically oriented research community in genetics, more generally. Figure 4.2 shows the distribution of funds to the four disciplines for the first four fiscal years of the program.[168] The data are broken down by major taxa in Fig. 4.3: over 70% of the dollars went to animal projects; vertebrates, *per se*, received a three-year average of 36% of PBPE funds. Work on fish typically emphasized population genetics, that on birds either behavioral or physiological ecology. Over 70% of invertebrate funding was for insects, with but 5% or so being for behavioral work for invertebrates, more broadly. Plant research requests, already in single digits, decreased for a period, but then began to pick up consonant with a growing interest in plant biology, partially promoted by Mary Clutter from 1988. Virtually all those funds given out by PBPE focused on terrestrial herbaceous angiosperms (non-woody flowering plants). An increase in interest in tropical studies also became apparent and was in keeping with what was a focal point in the view of the NSB, and then also of the NSF.[169] Tropical research amounted to about 20% of all PBPE funds. While the program officers of

---

[166] Donald W. Kaufman, Mark W. Courtney, and Penn R. Chu, "The First Three Years of NSF's Population Biology and Physiological Ecology Program," *BioScience* (1982):51–53.

[167] Ibid.

[168] Table 1 of ibid., pg. 52. Used with permission of the American Institute of Biological Sciences.

[169] Ibid., the article's text, pg. 52.

	FY 1977		FY 1978		FY 1979		FY 1980	
	% tot. $ (3.6 mil.)	$/Grant-Year	% tot. $ (3.8 mil.)	$/Grant-Year	% tot. $ (5.0 mil.)	$/Grant-Year	% tot. $ (5.2 mil.)	$/Grant-Year
Population Genetics	53	30	36	28	47	37	38	41
Drosophila	14	33	7	35	11	38	10	43
Other Animal	29	31	16	29	27	42	18	43
Plant	9	28	9	25	6	32	8	44
Theoretical	2	20	3	20	3	21	2	21
Population Ecology	22	26	28	28	24	31	23	32
Animal	17	26	21	28	20	32	14	31
Plant	4	26	5	27	4	35	8	36
Theoretical	-	-	2	30	<1	10	<1	40
Behavioral Ecology	12	20	14	31	12	35	10	33
Physiological Ecology	14	23	22	33	17	36	29	35
Animal	3	25	15	33	12	36	13	32
Plant	11	22	8	33	6	35	15	38
Avg. $/Grant-Year for All Awards		27		29		35		36

Fig. 4.2 Summary of research support by subelement of the Population Biology and Physiological Ecology Program, NSF. Taken from "The First Three Years of NSF's Population Biology and Physiological Ecology Program," *BioScience*, 32(1):52. Used with permission of the American Institute of Biological Sciences

TAXONOMIC GROUP	FY 1977					FY 1978					FY 1979				
	Percent Dollars	Percent of Support to: PG	PE	BE	PHE	Percent Dollars	Percent of Support to: PG	PE	BE	PHE	Percent Dollars	Percent of Support to: PG	PE	BE	PHE
ANIMALS	73	58	22	16	4	72	31	29	20	20	82	47	24	15	14
Vertebrates	30	42	27	27	5	40	13	20	31	35	38	36	23	23	18
Pisces	7	79	17	4	0	5	100	0	0	0	10	64	15	21	0
Amphibia	3	51	49	0	0	8	7	41	0	52	4	30	46	0	24
Reptilia	1	19	81	0	0	8	0	30	32	38	2	10	54	36	0
Aves	10	10	16	63	11	10	0	16	45	39	13	7	12	38	43
Mammalia	9	49	32	16	3	9	0	9	59	32	9	52	31	11	5
Invertebrates	43	69	19	9	3	32	53	39	6	1	44	56	24	8	11
Noninsecta	13	80	11	0	8	8	55	34	11	0	8	50	15	0	34
Insecta	30	65	22	13	1	25	52	41	5	2	36	58	27	10	6
Drosophila	14	90	10	0	0	9	84	16	0	0	15	77	23	0	0
Other	16	43	31	24	2	16	35	55	8	3	21	44	30	17	10
BACTERIA	1	100	0	-	0	2	29	71	-	0	0	0	0	-	0
PLANTS	25	36	18	-	46	21	45	17	-	38	16	40	23	-	36
Terrestrail	25	36	18	-	46	21	45	17	-	38	15	42	25	-	33
Lower Plants	1	100	0	-	0	1	100	0	-	0	1	0	0	-	100
Gymnospermae	2	100	0	-	0	4	80	0	-	20	1	37	63	-	0
Angiospermae	22	30	20	-	51	16	33	22	-	45	13	45	23	-	32
Herbaceous	19	30	23	-	47	14	34	25	-	41	12	39	25	-	35
Woody	3	24	0	-	76	2	24	0	-	76	1	100	0	-	0
Aquatic	0	0	0	-	0	0	0	0	-	0	1	0	0	-	100
THEORETICAL	1	100	0	0	0	5	66	34	0	0	2	82	18	0	0

Fig. 4.3 Summary of research support by taxonomic group for the Population Biology and Physiological Ecology Program, NSF. Relative allocation of funds within each taxonomic group is listed for population genetics (PG), population ecology (PE), behavioral ecology (BE), and physiology ecology (PHE). Taken from "The First Three Years of NSF's Population Biology and Physiological Ecology Program," *BioScience*, 32(1):53. Used with permission of the American Institute of Biological Sciences

PBPE had only three years worth of data and were not willing to extrapolate too greatly from it, they did make an interesting conclusion:

> *Scientists are opportunistic* concerning new ideas and developments and, therefore, future short-term changes are not necessarily predictable from long-term patterns of funding.[170]

The *sentiment* about opportunism being a characteristic of scientists is one that few of them could deny with regard to one very good reason: funds are needed for research, the bread and butter of science, and NSF was in the lead as the source. However, the *claim* of opportunism has been heavily debated, especially around the philosophy of Paul Feyerabend (see references in below footnote). Further discussion is left to others.

Not unlike Jim Rodman and his early contact with the Foundation, Mark Courtney was also an "NSF baby," having gone to a summer science camp just after ninth grade.[171] He continued in that vein with two more such camps in following years. Even as an undergraduate, Courtney did NSF-sponsored research. He had completed his bachelor's degree at Indiana University, Bloomington, and had started on his PhD at the University of Arizona, Tucson, but received a military draft notice just three months into his doctoral studies. It was late in the Vietnam War era. He converted to a master's program to shorten his time at Tucson and, when he went in to the Army, he did so as a biological sciences assistant in a hospital company where he worked in preventive medicine. After his two-year service commitment, he was able, in 1973, to pick up just where he had left off at the University on the same fellowship under which he had begun his degree work. For the next nearly five years, he worked on his doctoral research at Arizona, but just before beginning to set it to writing, he received an offer from the NSF as a scientific assistant for a one- to two-years' rotation. That had come, not from all his earlier experience and his being known at NSF, but as a consequence from his blind mailings in job hunting. That effort, however, eventuated into a 31-year career at the Foundation! He, therefore, had to integrate his dissertation writing into his life along with his new job at BSR, which began in July of 1977. It took three more years to complete his dissertation, so demanding were the two simultaneous 'jobs.' The degree was finally awarded in 1983 and Courtney had already become an assistant program director by then. Once he had the degree in hand, he moved up to associate program director. In FY93, all POs became Program Directors without distinction as assistant or associate, in part as a result of the great reorganization of the directorate.[172]

---

[170] Ibid., pg. 53. Emphasis added. Wernher von Braun was often considered an opportunist, financial and otherwise. See: http://bak.lernen-aus-der-geschichte.de/resmedia/document/document/C006T05E.PDF. See also: http://www.catholicintl.com/noncatholicissues/miller.htm and http://www.newscientist.com/blog/environment/2007/10/al-gores-inconvenient-truth.html, but most especially, see Paul Feyerabend, *Against Method* (London: New Left Books, 1975) and http://plato.stanford.edu/entries/feyerabend/. Finally, the "anything goes"/opportunism notion is considered synoptically at: http://www.marxists.org/reference/subject/philosophy/works/ge/feyerabe.htm.

[171] Mark Courtney Interview with the Author, February 12, 2009.

[172] Ibid. Courtney retired from NSF in March 2008 and moved to New Mexico State University, Las

When Courtney first arrived at BSR, he aided in the computerization of the lists of potential reviewers and panelists finding that "[a]t that time there weren't that many people who had dealt with large databases in biology." However, one of the two major changes that Jim Edwards said he witnessed in his years' as director of the systematics program (1982–1985) was the "incorporation of large-scale databases into environmental biology," an area where he opined that he had played a "catalytic role."[173] The database established for BSR lasted, Courtney remembered, about 15 years until overtaken by a more modern system. Working at the NSF at the time was "a secure source of income, which…was not all that common at that point for a young biologist."[174] His colleagues, he recalled, were not a cohesive group in regard to views about the recent reorganization of BMS into BBS, but were more focused upon their disciplines and their jobs than on bureaucracy; the academic-like atmosphere at NSF. Some were "quite compartmentalized in their view of biology…the stovepipe philosophy." Others were "quite synthetic and broad in their approach," thus making PBPE somewhat different in its outlook than were the more team-oriented members of the Systematics group: some in PBPE "saw bridges, where others saw moats surrounding the castle." There was, then, a "mosaic of perspective," he stated.

The concept of a mosaic applied to more than just personnel, as award levels, funding, proposals arriving, and other data points for measuring programmatic activity were mixed. By 1980, a funding scare arose. Associate Program Director of PBPE, Donald Kaufman, was communicating with Frank Golley, director of DEB, asking that an impending reduction of $200k for PBPE's budget *not* occur as commitments had increased for the program by $2M.[175] Fewer than two months later, both Kaufman *and* Courtney were pleading in a co-authored memo to Golley making their concerns known as to what was, by then, an imminent budget cut. They spoke in terms of how much "science per dollar" (a term well established throughout the Foundation) could be gotten through the program. They made several arguments: that no reduction (of dollars) scheme could fit all proposals; that quality of proposals was paramount (a view that was true Foundation-wide over its existence); and, the ideal of an "'optimal foraging strategy' for NSF, DEB or any individual program."[176] Awards were climbing, from 112 in 1977 to 132 in 1979, for instance.

Cruces. He has continued, as have others, to serve on panels, to function as a site reviewer, and in other capacities for the Foundation to the time of this writing. Jim Rodman also reminded the author of this 1993 change in his communication of May 7, 2012 thus leading to an error correction concerning dates.

[173] Edwards interview.

[174] Ibid. See this author's comments (footnote, earlier chapter) about employment for biologists at just that point in history.

[175] D. W. Kaufman to F. B. Golley, February 5, 1980, Courtney Files.

[176] Donald W. Kaufman and Mark Courtney, PBPE, to Frank B. Golley, Director, DEB, April 1, 1980, Courtney Files. This second memo was more formal in its use of personal names and that was reflected in strong arguments presented in the text and in the two tables and a hand-drawn graph of expenditures and budgetary expectations.

However, Golley had good reason to consider a budget decrease for PBPE: while the number of "actions" (proposals considered and actions taken) had gone from 25 in 1977 to 49 in 1980, the total dollar levels in awarded grants went from a figure of $2.*933*M in 1977 to $2.*903*M in 1980. Clearly, the views, as seen by the PBPE and BSR leadership, were not in concert. More notably, the carry-over of *unspent* funds went from about $330k to $750k in the same period!

Golley apparently did not reduce PBPE funds as, in another round of financial threats to the program two years later, the officers would argue that PBPE was the only program in BSR that was still "undergoing rapid growth."[177] That was true, as by then some 250 new proposals were piling up on the program officers' desks annually. The four areas covered in PBPE in 1976 had transformed somewhat to include three of the originals: population ecology, physiological ecology, and population genetics. But a newcomer was population structure, an area in "its infancy" but which disciplinary discoveries "will profoundly alter our views of how populations are organized" and so was needy of a strong budget. PBPE did not suffer during the period as its share of BSR funds climbed steadily from 1979 right through 1984 (see Fig. 4.1).

By the call for proposals issued in 1985, PBPE was able to articulate its program in the following way:

> The [PBPE] Program provides support for research which seeks to identify the most important ecological and evolutionary factors responsible for shaping the adaptive response of organisms to their environment, and how these adaptations at the individual level translate into interactions at the population and species levels. The focus is on research using theoretical, experimental, laboratory and field approaches to the understanding of populations of single species with regard to genetic relationships, population dynamics, and behavioral and physiological adaptations to specific environments.[178]

At the end of the 1980s, BSR was asking about the "consequences at higher levels" of biological organization from the "emerging fields [of] molecular evolution and genetics."[179] PBPE was being well served by such program officers as Mark Courtney, who regularly attended scientific meetings and who conscientiously provided brief but informative trip reports. In July of 1985, he had gone to Brighton, England for the Third International Congress of Systematic and Evolutionary Biology meeting where, he said, "there was little especially remarkable" at the conference except for a debate on the mythical Big Foot. In that regard, he noted that everything the presenter had to say was "immediately and thoroughly refuted" and

---

[177] Author unknown, "[PBPE]: Past, Present and Future"; a four-page typescript ephemeron; Courtney Files. The draft document appears to be an appeal to not reduce funding to PBPE; the Program had already instituted a 66% reduction in the fall, 1982 panel as requested (by BBS?). Should a 10% increase in the planned $6M budget occur, it would lead to one of $6.6M that was desired, but a 10% reduction to $5.4M would be highly damaging, it was argued.

[178] NSF 83-75, pg. 2.

[179] From a collection of paper copies of overhead slides used in a presentation at BBS; an ephemeron; *op cit.*; Courtney Files.

that in his view the presenter's arguments were "sas-squashed"![180] However, he did note that Lynn Margulis gave a "spirited presentation" on the origin of sex in the protists.[181] In mid-1988, as another example, but of a more valuable meeting, he attended a conference at Ohio State University on the "Foundations of Evolutionary Biology," one attended equally by both biologists and philosophers. Having addressed "currently hotly debated questions," he felt that future collaborations between the two disciplines should "likely result in new and insightful research."[182] Those in attendance were a who's who of leaders in the field including Stephen Jay Gould, Niles Eldredge, Douglas Futuyma, David Hull, William Wimsatt, Michael Ruse, and others; a total of 22. David Kingsbury had asked what would be the "consequences at higher levels" for biology and philosophy. Of this, more later. Just one example of the possible, the consequences of interdisciplinary intercourse, fit very well with the directions NSF and its directorates were taking.

4.5 The Ecology Program

It may be true that Ecology, along with Systematics, were the two "core" programs of the BSR division in the minds of some, but it is equally clear that funding levels did not suggest that the Ecology Program was to be equivalent in that arena, at least.[183] A close examination of Fig. 4.1 demonstrates that not only was Ecology subordinate to Ecosystem Studies as early as 1976, but it was, in fact, at the budgetary bottom in the list of the five programs of BSR! It remained in fourth or fifth position right through 1984.

A reading of what the Ecology Program funded, as stated in the 1985 call for proposals, exhibits certain features that would almost guarantee its rather lowly fiscal standing.

> The Ecology Program provides support for ecological research on relationships within and between communities of animals and plants in terrestrial and aquatic habitats. This includes studies of interspecific competition, predator-prey relationships, and symbiosis between

---

[180] Mark Courtney, "Trip Report: Third International Congress of Systematic and Evolutionary Biology, University of Sussex, Brighton, England, July 4–8, 1985," typescript of one and one-half pages in length; Courtney Files. See also: http://onlinelibrary.wiley.com/doi/10.1111/j.1502-3931.1984.tb01624.x/abstract.

[181] Protists are mainly unicellular organisms considered by some biologists to be in a taxonomic kingdom of their own. They include a great range of unicellular algae-like organisms, protozoans, unicellular fungi-like creatures, slime molds, and others.

[182] Mark Courtney, "Trip Report: Foundations of Evolutionary Biology, July 11–15, 1988," typescript of July 21, 1988; Courtney Files; quotations from pg. 3. See also: http://philosophy.osu.edu/news/logos/pdf/logossummer1988.pdf

[183] The *Annual Report* authors, influenced by the materials supplied by the division and directorate leadership, continued over the years to state that environmental biology was composed of ecology and systematics. See, as just one example, the National Science Foundation, *Thirty-First Annual Report for Fiscal Year 1981*, pg. 33. It was routinely repeated in slight variations annually.

> species, as well as studies of species that comprise communities. The program funds, for example, research on co-evolution of animals and plants, microbial ecology and paleoecology. Emphasis is placed on multi-species interactions at the community level.[184]

It should be noted, however, that the program was used to fund single researchers and/or projects more or less in the little biology category. The Ecology Program sought to answer many questions, but to co-opt philosopher Jose Ortega y Gassett: "Tell me the landscape in which you live, and I will tell you who you are." Ortega may have liked the foregoing program statement: it was highly classical in nature, almost an early textbook definition of ecology. While that is not problematic on its surface, if one compares the activities in the other programs in the division and examines the trends in biological research during the 1970s and 1980s, it is clear that little in the way of more classical studies was being undertaken in ecological science generally. If it is true that "scientists are opportunists," then the hot areas of research that combined little and big biology—molecular genetics as applied at the population level, for instance—was what many scientists sought research support to undertake. Much of the work being done on the IBP, then later completing the "biomes projects" spoken of by Betsy Clark (the IBP/LTER interregnum, above) and, finally, gearing up LTER, contained many aspects of classical ecology. However, the funding for that work was coming from the Ecosystems Studies Program (ESP) and so ecological research done independently of the big science supported through ESP would be as many smaller, individual grants that would come from the Ecology Program.

Over the years, the *Annual Report*s, admittedly aimed at a broad audience not always scientific, such as Congress, did not break down their stories by programs, but only by directorates and often, indirectly, by divisions. With often only a paucity of primary historical sources having been saved and archived by BMS/BBS/BIO, the historian may be forced to consider tertiary literature at a more intense level. Perusing those information sources, such as the *Annual Report*s, through time it is obvious that unless a given *program* is specified or singled out for discussion, it is difficult to discern which of a related series of programs' research awards may be under consideration. Closely related programs in the biology directorate were not at all unusual through the years. That is particularly true for the Ecology Program because of the close relationship of at least four of the programs in BSR, they all being very similar in content. While Systematics may be a bit different than Ecosystem Studies and Ecology, and even PBPE, it is only BRR (below) that is notably different: it dealt more with the instruments and infrastructure of science rather than the intellectual content of the science itself. The text of the *Annual Report*s, then, typically made it a challenge to determine which of BSR's programs had funded a given project. That was especially true when comparing Ecosystem Studies' projects with those of Ecology. And, as BBS became continually more interdisciplinary, such separation in programs, both with regard to reports upon their successes and with how the actual science and funds were divided program-

[184] NSF 83-75, pg. 2.

matically, became progressively more opaque. The work on plants was widespread over BBS, for example, and the emerging field of microbial ecology, one of the areas supported by the Ecology Program, was adding another level of complexity to what BSR, and even BBS, were doing with regard to supporting their communities of investigators.

The "core" of ecology, it was held at BBS/BSR, was community ecology, and had been so since the beginning of the twentieth century.[185] The work on secondary substances in plants, for example, often had ties to co-adaptation between those plants and certain insects. Falling, as it did, under the heading of community ecology in the 1976 *Annual Report*, it was also noted that not only was co-adaptation of two organisms under consideration, but so also was the *little science* of "microchemical analysis" of the secondary substances, and even the "population genetic aspects of the relationships" was part of an undeniably interdisciplinary project. As referenced earlier, "two disciplinary offshoots of systematics and ecology have developed: population biology and ecosystem science." It is not surprising, then, that the historical record of the Ecology Program should be so much integrated with that of the rest of BSR.

Community ecology "persist[ed]" in the activities of BSR in the 1970s as research interest remained high in the areas of "interspecific interactions."[186] One small part of it was just then beginning to emerge into what, by the first decade of the present millennium, would become one of the most notable, and possibly least foreseen—but long pondered upon—areas of biology: microbial ecology. "Despite long-term recognition of the essential function of decomposition in the environment" by microbes (bacteria, fungi, and select other organisms) the "heterogeneous" nature of the microbial community in soils, for instance, made their study especially challenging: methods were lacking. However, a breakthrough was made by E. L. Schmidt of the University of Minnesota in NSF-funded research on fluorescent antibodies tagged to soil microbes in his investigation of both the distribution and metabolism of previously cultured (laboratory) creatures, but here viewed in natural systems.[187] Schmidt, already known for his laboratory manual in soil micro-

---

[185] National Science Foundation, *Twenty-Sixth Annual Report for Fiscal Year 1976* (Washington, D.C.: Government Printing Office, 1977), pg. 62. Interestingly, Sharon Kingsland, in her 'Evolution of American Ecology,' does not consider specifically community ecology. On the other hand, Joel Hagen considers it in some detail in his 'Entangled Bank.'

[186] National Science Foundation, *Twenty-Seventh Annual Report for Fiscal Year 1977* (Washington, D.C.: Government Printing Office, 1978), pg. 70.

[187] It has been said by some, that as few as 1% of all suspected microbial inhabitants of the soil have ever been cultured. While the number may still represent a mere guess, it has only been in recent years that molecular techniques have allowed for an expanding science of microbial ecology to consider not only the biota and its functions in soils and natural waters, but in other microbial ecosystems, such as the human skin and, especially in the present era, the human intestine. See, for instance, a brief overview of the history of microbial ecology at: http://www.eeescience.utoledo.edu/Faculty/Sigler/COURSES/Microbial%20Ecology%20Lecture/01%20-%20History%20of%20Microbial%20Ecology.pdf, as well as the many works by Thomas D. Brock, a modern founder of the discipline.

biology, gained additional notice for his work with the rhizobia and their role in the nitrogen cycle—long an interest area of several BBS divisions. Microbial ecology grew over time to play an especially notable role in biology and is considered further in later chapters.

Grantees continued to do work in the areas of "interspecific interactions" in following years, one project being aimed at resource partitioning amongst competing species in an ecosystem. The question had long been open as to "whether these differential efficiencies [of competitive interactions] in exploitation of resources that were in short supply were an important restrictive mechanism in nature."[188] Earl Werner and his colleague's work at the Kellogg Biological Station at Michigan State University examined both freshwater fishes and old-field plants to determine the answer to the mystery of partitioning mechanisms. Spatial segregation during feeding periods by the fishes indicated that, with mixed species in limited locales, broadly consumed food types were replaced by limited, but preferred, food types for each species. Werner received the Ecological Society of America's Mercer Award for his work.

In other studies that year, James Detling and his associates at Colorado State University did work on the impact of herbivory on plant growth, metabolism, and productivity.[189] They discovered that not only do herbivores function as "'mowing machines' but also as 'inoculators' of salivary biochemicals." That is to say, the inoculating saliva contained plant growth-regulating compounds that accelerated re-growth as compared to just simple mechanical damage to the plants. Always in the background, however, was the role of molecular biology and its "remarkable discoveries." Those findings were having a profound influence on many areas in biology, not least ecology. In fact, "the stage is now set for successful *syntheses and reconstructive* approaches" in the biological sciences.[190] *Ecology, like the rest of biology, would be very much affected by the molecular revolution.* Indeed, it had been so already, as seen in the exemplars stated above. Since a more clearly articulated systems' approach was being taken during those years, the notion of a reconstructed biology meant that moving toward a greater internal interdisciplinarity of biological subdisciplines was to be expected in future years. *Biology would be newly rearticulated after a long period of specialization had led to its then more diffuse character.* In essence, the century following the 1880s when specialization in biology (and other) sciences began. In fact, that would be a very slow process—on the multi-decadal level—into the twenty-first century, as shall be seen in the rise of integrative (re-integrative?) biology.

Work in herbivory continued to be an area strongly supported by BSR. Norman Negus of the University of Utah, doing work not unlike that of James Detling but in a reciprocal situation, showed that a plant biochemical could have decided effects

---

[188] National Science Foundation, *Twenty-Eighth Annual Report for Fiscal Year 1978* (Washington, D.C.: Government Printing Office, 1979), ppg. 69–71.

[189] Ibid., pg. 71.

[190] National Science Foundation, *Thirtieth Annual Report for Fiscal Year 1980* (Washington, D.C.: Government Printing Office, 1981), pg. 49. Emphasis added.

on a mammalian herbivore. An estrogen-like secondary substance in plants affected reproduction in field mice, reducing their populations and hence the pressures of herbivory.[191] Years' later, estrogen and similar compounds, having become environmental pollutants, would be of broad interest as agents possibly interfering in human growth and reproduction.

Always alert to the fact that a study in ecology also implies a study of abiotic features of an ecosystem, the work of the first six LTER sites was referenced, but without mentioning LTER, in the 1981 *Annual Report.* Readers were reminded that geological aspects such as soil types, parent rock, atmospheric gases, etc., were all necessary of consideration in a proper research protocol.[192] Indeed, it was that very same publication that headlined an intriguing notion: "Genes, Physiology, Ecology—The Grand Connection." The subdisciplines in biology were shown to be aggregating into a more unified whole in future to become part of the view known as *integrative biology* (later chapters). Once again, in reference to the great significance of electrophoresis to a biology becoming ever more molecular—and equipment-dependent—in its orientation, the *Annual Report* authors reminded their readers that the technique in the "1960s [more so in later years, however] allowed biologists to inventory the genetic variability" of populations of organisms.[193] A great deal of such variability had been seen, but the explanation for it had eluded the research communities. Dennis Powers' work at Johns Hopkins University used a highly favored *model system* amongst vertebrates, the killifish (*Fundulus heteroclitus*), as they are noted for their abilities to withstand great variations in salinity in the mixed estuarine and marine littoral areas of the Atlantic shore and riverine systems they inhabit. Their reproductive behavior leads to limited geographic dispersion and so populations are highly adapted locally. For that reason, Powers determined to study the killifish at the genetic, physiological and whole animal levels—what would later be considered very much in the integrative biology tradition. Examining certain genotypes and ecotypes, and genes at the molecular level along with environmental variables, he was able to demonstrate that certain enzymes differed in populations along a continuum in Chesapeake Bay coordinate with a continuum of salinity values. His research had just begun, but by using research approaches from several subdisciplines in biology, Powers was doing a small part in unifying the science. The authors of the *Annual Report* stated that biologists had found "surprising links between [*sic*] genes, physiology, and ecology." In many ways, such findings were not truly surprising, but having used the term was not ill-advised in the sense that it may have taken a *group* of specialists to attack the problem in the way a *single* biologist, Powers, chose to do it. In so doing he took on a much more *synthetic* approach, an *integrative* approach.

---

[191] Ibid., pg. 59.

[192] National Science Foundation, *Thirty-First Annual Report for Fiscal Year 1981* (Washington, D.C.: Government Printing Office, 1982), pg. 33 ff.

[193] Ibid., pg. 36.

The Ecology Program continued to favor supporting research in herbivory. Jean Langenheim of UC, Santa Cruz, was doing work with the terpenoid compounds of the leguminous algarrobo[194] tree (*Hymanaea* spp.) of the New World Tropics and their function as a deterrent to insect attack. Langenheim was doing tropical biology at the same time that Gould was working on the tropical snail *Cerion*—a period of considerable NSF support of work in that area of the world and consistent with NSB interests. Research in the plankton communities of Lake Tahoe (California/Nevada) and nutrient cycling in soil heterotrophs was also among the studies reported in the early 1980s.[195] The continuing breadth of research interests in awards made by the Ecology Program was very much as could be expected—and they were in keeping with the statements made by BSR in its portion of the *Status of Science Reviews 1984*.[196] That section began with five bulleted synopses that matched with the five programs within the division. While genetics or molecular aspects of ecology were not much mentioned in the five single sentence bullets, the introductory text that followed focused almost entirely on the micro end of the biological spectrum and its value to ecological science. It was almost remarkable in how intense that commentary was in that regard. It ended with pointing out the three main areas of keen interest to NSF for BSR funding: tropical ecology; LTER; and plant biology. The statistical changes in the environmental biology community were notable, too, in regard to numbers of ecologists: in 1970, the report authors stated, there were some 3500, whereas by 1980, that number had risen to 6000. Adding to that the number of systematists, the total was estimated at 11,000—some "20% of the total labor force of bioscientists."[197] (That rise had taken a drastic plunge by the new millennium, however.[198]) Furthermore, requests for support from NSF rose 216% from 1970 to 1982; that roughly paralleled the growth in the numbers of environmental biologists, about 6,000 (coincidentally) of whom were employed in the academy. Some 50% of all federal funding for basic research in environmental biology came from the NSF; that aimed at the academic world, specifically, was 95%! Finally, the reports emphasized the increasing importance of microbial ecology and its special

[194] Also spelled as algarroba.

[195] National Science Foundation, *Thirty-Second Annual Report for Fiscal Year 1982* (Washington, D.C.: Government Printing Office, 1983), pg. 33–35.

[196] Division of Planning and Policy Analysis, Office of Planning and Resource Management; National Science Board, *Status of Science Reviews 1984*, November 1983, NSB 83-360. See ppg. 68–70.

[197] Ibid., pg. 69.

[198] It was being argued, by the end of the first decade of the new millennium, that not only were many species going extinct, but so was the class of specialists that study them, taxonomists. Why? "Because we have devalued their contributions, both monetarily and scientifically," said Craig McClain in "The Mass Extinction of Scientists Who Study Species" at www.wired.come/wired-science/2011/01/extinction-of-taxonomists/. It would not be the first time that such a dire prediction was made, to be sure, as McClain also noted that NSF had begun a program in 1994 (see main text in later chapters) "to enhance taxonomic research [that] provides training, [but] does not create job opportunities."

role in explaining plant community structure: microbes were rapidly being shown to do much, much more than merely decompose organic matter in the soil.

Betsy Clark left NSF in 1983 and by the following year Kingsbury was in her place. His quantitative leanings, his facile use of computers, and the nature of both his own experience and training, let alone the fact that he came from the University of California at Berkeley, titular home of the term "integrative biology," must have played a direct role in him being the first to use the phrase at BBS. It is assumed that the term was quickly picked up by the leadership in BSR. The historical record shows that it was used in the long-range planning (LRP) document for the BSR for the period 1984—1988: the LRP had three major points, the third of those was "integrative studies."[199] The subpoints under that head noted the need for research at the population level examining "genetic variability with population structure and physiology," exactly what Dennis Powers was doing with killifish. The next subpoint was at the community level and emphasized "species interactions and community energetics," the second topic of which involved extensive physical science: *interdisciplinarity*. Another point focused on global levels and emphasized funding research in environmental factors and "system control processes"—a term not seen before in the records of the BBS. In a word, the BSR leadership saw a more quantitative division with "integrative studies" as including the PBPE, Ecology, and Ecosystems Studies programs, in particular, hence unifying BSR all the more.[200] The overall vision at BSR specifically, was of a highly integrated nature, anyway. Ecology, in the broadest sense, seeks that very thing, but like all of biology its subdisciplinary practitioners had gone the way of specialization. However, into the 1970s and 1980s especially, that diaspora was beginning to reverse. It was 1984 in which the very quantitative High Performance Computing Program began at NSF and BBS was clearly to profit, as will be examined in a later chapter.

As the decade came to a close, the general BBS initiatives included not only the familiar plant science and the newer dynamic form and function, but also Global Environmental Change. The latter would become a source of funds for the ecology community in the new decade when BBS became BIO. In the BBS AC meeting of October 1989, the Biodiversity Task Force of the NSB was discussed with Jim Edwards, by then Acting Director of BSR, leading the presentation. There were three issues that had emerged from a luncheon roundtable of the AC. The first was that "ecology" education[201] needed to begin at the grammar school level. The second was that "ecologists have not 'tapped' into molecular biology…rapidly," and, though while not entirely a true statement, it was clearly still a problem in the community. Finally, college freshman may need a course in the introduction to environmental sciences.[202]

---

[199] BBS LRP, 'Topics,' pg. 1.

[200] Ibid., pg. 2.

[201] Emphasis in the original 'Minutes.'

[202] BBS AC 'Minutes,' October 23–24, 1989, pg. 1.

4.6 The Biological Research Resources Program

The BRR Program was unlike the others that made up BSR, as it was more of a *service* program in that it provided grants to individuals and organizations, both in the academy and to a variety of non-profits, for *infrastructure* for the biological sciences. Content-oriented research was not excluded, however. The 1985 call for proposals explained it best.

> The Biological Research Resources Program provides for the refurbishment and operation of research facilities in the biological sciences that are considered essential at the national level. Included are systematic research collections and publications of long taxonomic monographs, living organism stock centers of known genetic characteristics, and biological field research facilities.[203]

By the mid-1980s, some grants were already long-lived, such as that supporting the *Escherichia coli* Genetic Stock Center at Yale University. The CGSC, as it was termed, was founded in 1971 with NSF support and the funding has been continuous since that time: a facility "essential at the national level." The Center began with a strain collection that had been amassed by E. A. Adelberg, a leading bacteriologist, co-author of the heavily used textbook, *The Microbial World,* and a founder of microbial genetics.[204] NIH also has funded the Center and, beginning in 1987, user fees have supplemented federal support.[205] From 1971 to 1993 the collection was curated and directed by Barbara Bachman. She was succeeded in that latter year by Mary Berlyn, who remained in the position until 2007. The Center was assembled with the aid of the NSF's "expert advice [and] encouragement," and that of the Genetics Society of America. The Center both supplied strains of *E. coli* to the research community and typed (identified) them, as well. It continues to do so to the time of this writing. Many of the most famous names in microbiology from the era of the founding years of the CGSC have contributed yet more organisms to the Center from their own collections over time. As early as 1972, collections from some 40 laboratories around the world had been contributed. Nobelist Joshua Lederberg (earlier chapters) was among those donating his strains when he became president of Rockefeller University in 1978. The collection was computerized from the earliest period using software of the 1960s: SAM, a program co-developed by Adelberg. That effort evidenced the fact that many strains were improperly typed and Bachmann, supported by NSF, went to laboratories in several foreign nations to study original stocks and so to rectify that issue. Thus, by the early 1980s, some

---

[203] NSF 83-75, pg. 2.

[204] Adelberg is still at the ECGSC as of this writing. The well-known textbook is by R. Y. Stanier, M. Doudoroff and E. A. Adelberg, *The Microbial World* (Englewood Cliffs, N. J.: Prentice-Hall, Inc., 1964). On a personal note, the author used different editions of the book as both an undergraduate and later as an assistant professor teaching the subject. The Center can be found online at: http://cgsc.biology.yale.edu/index.php.

[205] Some strains of *Escherichia coli* are strongly pathogenic to humans and have, in recent decades, caused a number of nationwide epidemics in hamburger meat and other food products, thus the NIH funding.

95% of all the strains had full and correct pedigree information. One of the most important strains of *E. coli* is the renowned K-12 for which the Center's stocks allowed for the publication of the genetic map of the bacterium. Both the strain and its map were and remain of immense importance to a number of research communities. By the end of the 1980s, a new relational database had taken form based upon a card file system also in use at the CGSC. So useful had that software become that it was a model for the later genetic database, MaizeDB, used by the plant research community.[206]

The use of computers in the collection management of living, as well as preserved, organisms had attained a real standing by the early 1980s when BRR supported both workshops on the subject and the collections themselves. The BRR program director from 1980 to 1985 was James Chase Tyler II, an ichthyologist and, from 1985 to 1987, Associate Director of the National Air and Space Museum. He was its Acting Director from 1986 to 1987. He later became Senior Scientist at the Smithsonian's National Museum of Natural History.[207] Tyler was enthusiastic about a workshop report of the time authored by Alan E. Leviton, the Curator of Herpetology at the California Academy of Sciences, and his colleagues—and well he should have been. Leviton, *et al.*, brought together

> scattered or otherwise obscured [*sic*] older information into a coherent whole [thus allowing] curators to properly analyze the cost-benefit ratio of computerization in collection data base management.[208]

The report noted that computerization ("automated data processing" in the parlance of the time) of museum collections had been under discussion for more than two decades prior to 1981, but its successful employment was rife with "tales of woe [and] project failures."[209] The collections' database management program of the 1960s, SELGEM, took many hours to process data, while that of the early 1970s, TAXIR still required a large mainframe computer with attendant specialized personnel, high costs, and to which only one major museum (the Smithsonian) committed itself. The two programs competed for a few years, but with the rise of small desktop computers and dispersed data management just coming about at the end of

---

[206] I thank John E. Wertz, Director, ECGSC (2007 to the time of this writing) who supplied the majority of this information; email from John Wertz to the author, May 21, 2009. In that same email, Wertz supplied the author with a "running narrative" (functioning as a bibliography) from each of their grant renewal requests from NSF of the papers published in one connection or another with the Center. Some 26 are listed. The list is held by this author.

[207] His papers from 1950 to 2005 are held at the Smithsonian. See: http://siarchives.si.edu/findingaids/FA05-082.htm. He had interests in both living and fossil fishes: http://openlibrary.org/authors/OL391294A/James_C._Tyler. See a short biography of Tyler at: http://www.pwrc.usgs.gov/resshow/perry/bios/WBFC_booksm.pdf.

[208] James C. Tyler to Alan E. Leviton (and Committee: Robert H. Gibbs, Jr; Robert K. Johnson; and, Roy McDiarmid), October 18, 1982. I thank Michelle L. Aldrich for making copies available to the author of this letter and the report (following footnote) from Leviton's personal files.

[209] Alan E. Leviton, *et al.* (see note above), "Computer Applications to Collection Management in Herpetology and Ichthyology: A Survey and Report Submitted to the [NSF] and Supported by the Biological Research Resources Program at NSF," September 15, 1982, Washington, D.C.

those two decades, large mainframes could be dispensed with. The report authors did not "advocate a single solution," to database management for the collections' and museum communities, but did present an "overview of the issues and problems involved" in computerization as it stood during the period.

Such post-workshop reports had been appearing for a number of years supported by the BRR program and several important data collections were supported by it. But not all such support came from the BRR, and indeed, some famous collections that had direct applicability to BBS interests were more properly funded by, for instance, the Biophysics Program in the chemistry directorate. Such an example is the Protein Data Bank. As early as 1971, a database concerning protein structure was created at the Brookhaven National Laboratory.

The year 1985 was notable for other reasons, too. In February of that year, Jim Edwards became the new program director for Biological Research Resources. He would stay (until March 1990) into the early period of discussions about the next major reorganization of NSF, and BBS in particular. His view of the rise of reductionism for environmental biology (above) was such that he saw the use of large databases for *genetic* information as a model for such repositories to be built for *organismal* biology because they acted not only for data storage and retrieval, but also as "discovery…and organizing tools."[210] Though that realization came when he was BRR program director, it was not until he had moved into the position of Deputy Division Director for Environmental Biology, where he stayed until 1993, that he was able to pursue more along the lines of information technology as a formal part of his duties. It was during the reorganization of the early 1990s when he felt that BBS was going through a "speciation event." That happening was the split in the old BBS structure and the giving rise to two new "species [directorates]," social sciences and biological sciences that would each follow their own respective evolutionary trajectories. It was possible to focus more easily on large-scale databases for biology alone then. The social sciences had long used them (Chap. 2).

In a very different vein, the BRR project partially funded some meetings of the Association of Systematics Collections (ASC), a group that BRR had helped found. The ASC's meeting in Los Angeles in May of 1985, for instance, focused on establishing a National Biological Survey, a notion long in discussion within the biological communities of North America.[211] The "NBS" never came to lasting fruition in

[210] Edwards interview.

[211] ASC Annual Meeting for 1985, Schedule of Events, Los Angeles, California, May 23, "Community Hearings on a National Biological Survey [NBS]." A cover letter was attached from Stephen R. Edwards, Executive Director of the ASC to David Kingsbury inviting his participation. Kingsbury was, however, otherwise committed and could not attend; Kingsbury to S.R. Edwards, March 13, 1985. It was the ASC itself which first formally suggested an NBS for America in 1985, but see the primary publication on it: Committee on the Formation of the National Biological Survey (Peter H. Raven, Chair), National Research Council, "A Biological Survey for the Nation" (Washington, D.C.: National Academy Press, 1993). From the Preface: "The National Biological Survey will produce the map we need to avoid economic and environmental 'train wrecks' [such as the then current spotted owl controversy, among others] we see scattered across the country. NBS will provide the scientific knowledge America needs to balance the compatible goals of eco-

the form many envisioned for it, but aspects finally did come to exist as a function within the U.S. Geological Survey.[212] Such support by BRR indicates its wide participation in *service*-based activities for the advancement of the life sciences in the widest communities of biologists in all parts of the nation. Not all funding engaged in by NSF was for bench or field research.

In late 1987, the BRR awarded 20 grants to provide "sophisticated instrumentation to Biological Facilities Centers" (Chap. 3) for the purposes of "investigating basic problems" in a variety of fields, such as biotechnology, environmental biology, and neuroscience.[213] Instrumentation support was a major area for the BRR.

The 1980s ended with survey and workshop results from the ASC and with the National Museum of Natural History as cosponsor. The Executive Summary stated that the time was one of both "great promise and great danger for systematics collections." Biodiversity mapping and comparative studies were about to "assume a central role in biology," the authors held. "[A]ll biological research is validated" by use of systematics collections and those resources were in need of considerable financial support. NSF, through the BRR especially, had long supported collections, and would continue to do so into the future. The service to biology and to society of those collections was crucial in a time of "global change, the growth of natural products for pharmaceuticals and other human uses, and the [NIH's] human genome project." The specific "danger" that the ASC and others feared was the notion that collections were "old [or] 'out of date'" sorts of things in a time when such new

system protection and economic progress," by Bruce Babbitt, Secretary of the Interior; pg. vii. President Bill Clinton had appointed Babbitt as Secretary just a month before the NRC report came out, but the "train wrecks" (Babbitt's words prior his becoming Secretary) were a major challenge for Clinton just as he took office. For more on this, and a broader historical consideration, see a history of the failure to establish a lasting NBS in Frederic H. Wagner, "Whatever Happened to the National Biological Survey?" *BioScience* (1999):219–222. Wagner pointed out that the Canadians had first suggested such an agency as early as 1977 (his pg. 219), but the idea was quickly taken up in the U.S. and, after a long and tortured history, was established on October 1, 1993 and staffed by 1,200 Department of the Interior employees, all of whom were transferred from within DoI's other agencies to the new NBS, though most came from the Fish and Wildlife Service. That staffing plan created severe early policy problems, but when "Republicans swept both houses of Congress" in 1994 (pg. 220) and introduced the Conservative Agenda, environmental agencies came under attack because of the belief that private property rights would be trammeled by biologists hunting for endangered species (the term "survey" was not properly understood by many in Congress). Morale dropped at NBS and by October of 1996, it was transferred to USGS as the BRD—see following note.

[212] The current incarnation of the "NBS" is expressed in the U.S. Geological Survey (USGS) NBII program: http://www.nbii.gov/portal/server.pt?open=512&objID=236&mode=2&cached=true. The USGS has been the location for biologically oriented offices and agencies for a long period of time. In 1996, the "National Biological Service" (not "Survey") was moved into the Biological Resources Discipline (BRD) at USGS. For a brief history, going back to biological agencies as early as 1885 having a place at USGS, and an introduction to the BRD, see: http://biology.usgs.gov/pub_aff/welcome.html.

[213] National Science Foundation, *Annual Report 1988* (Washington, D.C.: Government Printing Office, 1989), pg. 32. See the commentary in Chapter Four on Margaret Lindorfer, University of Oregon, and her use of the BRR-funded peptide synthesizer.

directions as that characterized by biotechnology were hot and funds poured in to support the area. "The very age of the collections and the buildings housing them" was a threat to the whole enterprise. The report provided a list of trends with regard to use of collections and, more importantly, a statement of "Priority Actions." The authors addressed issues such as specimen backlog, lack of knowledge about collections, infrastructure, education, personnel, computerization, and gaps in collections, libraries. In conclusion, the authors noted that our society was one "whose imagination is set on the stars" and so billions of dollars were being spent on physics and astronautics but only about "[ten cents] per year for each species of the world's biota." Astronaut Sally Ride, they averred, had commented on the need to "focus some attention on the Mission to Planet Earth."[214]

Such attention would be so focused in years to come, but the NSF and the BBS were on the verge of a major reorganization and two new directorates would come out of BBS during the next two years' deliberations. The split of the social sciences out of BBS would leave, for the first time, a directorate for the biological sciences alone: BIO. It would not come easily and the struggle for the birth of BIO would be demanding, but it would have a steady hand at the tiller for that process in the person of Mary E. Clutter.

Clutter was a native of Pennsylvania and attended Allegheny College in Meadville where her undergraduate degree was in biology.[215] Her family did not live far from Pittsburgh, so with scholarships in hand, Clutter went to the University of Pittsburgh. The AIBS synopsized her career thus:

> Dr. Clutter has also served [beyond her career at NSF] as the U.S. Chair of the U.S.-European Commission Task Force on Biotechnology, a member of the Board of Trustees of the international Human Frontiers Science Program, a member of the Board of Regents of the National Library of Medicine, a member of the National Agricultural Research, Extension, Education and Economics Advisory Board, chair of the Biotechnology Subcommittee of the Committee on Science of the National Science and Technology Council (NSTC), co-chair of the Subcommittee on Ecological Systems of the Committee on Environment and Natural Resources/NSTC and co-chair of the NSTC Committee on Science's Interagency Working Group on Plant Genomes. She is a member of numerous professional societies and has served on the Board of Directors of the American Association for the Advancement of Science (AAAS). She is a Fellow of the AAAS and the Association for Women in Science. Dr. Clutter received her B.S. in biology from Allegheny College and her Master's and Ph.D. degrees from the University of Pittsburgh. She also holds honorary doctorates from Allegheny College and Mt. Holyoke College and has received the Medal of Distinction from the University of Pittsburgh.[216]

Returning to earlier times, she had been a co-PI on an NSF grant in her first few faculty years (Yale) and later on a review panel in the early 1970s through an invitation from Melvin "Mel" Spiegel. He was the rotator Director of the Developmental

[214]The Association of Systematics Collections and the National Museum of Natural History, Smithsonian Institution, "Systematics Collections Resources for the 1990s: Results of a Survey and Workshop," January 30, 1989, Grant No. BSR-8809937.

[215]Mary Clutter Interview with the Author, May 14, 2009.

[216]See: https://www.aibs.org/about-aibs/mary_clutter.html.

Biology Program at BBS and a zoologist from Dartmouth College whose interests were in the molecular area. At the time, Clutter recalled, the program director for the developmental program alternated between a zoologist and a botanist with no permanent person, not uncommon, as has been seen; and more examples shall yet be seen. The panel invitation was to see if Clutter was interested in a professional involvement with NSF: "I found it fascinating," she said. She agreed to come for one year as a rotator in the program. She returned as the faculty at Yale and some years later was to return to NSF when courted by Herman W. Lewis, who, as Appel has said, "came from Michigan State as a rotator for a year and stayed for twenty."[217]

Lewis was the head of the Genetic Biology Program from 1962 to 1973 and, at the time Clutter arrived, Lewis was head of the Molecular Biology Section after BMS had become "more bureaucratized in 1964," according to Appel[218] (the term Section was in use at the time and was and would continue to be until it was dropped with the 1975 reorganization). Lewis had a strong ethical bent and a keen view that there were insufficient women professionals at NSF. Clutter believes that he was instrumental in her hiring. Both she and her husband felt that a move from Yale to NSF was a wise one and so she took the offer. Upon her arrival in the early 1970s, she set herself to improve the situation for the plant sciences and during 1991–1992, by then AD for BBS after Kingsbury left in 1988, she led the reorganization efforts. We shall return to her involvement with plant sciences for her whole tenure at NSF in later chapters (she retired in 2005; that is the basis for that year as end to the present study). And after consideration of the second big reorganization in the period covered by this history, we shall return to the new incarnation, the Directorate for Biological Sciences, and follow the directions of BIO as a new structure.

[217] Appel, 'Shaping Biology,' pg. 158.

[218] Ibid.; and also partially drawn from Clutter interview.

Chapter 5
A Second "Time of Tumult" and a New Home for Biology

If this century is the age of physics,
the twenty-first century will be the age of biology.
D. Allan Bromley

5.1 Revolving Door Directors, et al., Redivivus

If Allan Bromley, Science Advisor to then President George Bush, *pate*r, was correct, then NSF, for its part in making the first century of the third millennium the "age of biology," would *have* to place biology into its own house. The necessity for the separation had become almost universally held among those in BBS, higher in the Foundation and in the relevant communities. That is precisely what came out of the reorganization of 1991–1992: two new homes for the divisions of a dissolving Directorate of Biological, Behavioral, and Social Sciences. It had not been since 1975 that the life sciences were last reorganized at NSF on a large scale but such was in the offing for BBS, let alone for several offices, other directorates, and the senior leadership of the Foundation, during the transition to the Directorate of Social, Behavioral, and Economic Sciences (SBE) and that of biology (BIO).

Mary Clutter had been on a 2-year assignment as Science Advisor from BBS to the Director's Office from early 1985 to early 1987 when much occurred at the director's level that signaled the broader restructuring to come. In later years, a then BBS behavioral scientist rotator, Kathie Olsen (who would by 2005 become Deputy Director of the NSF), was asked who *initiated* the reorganization of 1991–1992. She noted that it was Bloch who was primarily responsible.[1] He did not stay on at NSF

[1] Kathie L. Olsen Interview with Marc Rothenberg, Historian of the NSF, October 10, 2007. Olsen had first come to the NSF as a rotator in 1984 from the State University of New York (SUNY), Stony Brook, and stayed 2 years in her position in the Division of Behavioral and Neuroscience of BBS. She returned to her SUNY laboratory for the following 2 years but then once again came to NSF in the fall of 1988 and stayed permanently, rising up through the ranks, an uncommon situation at the Foundation, to become deputy director.

D. J. McGraw, *Millennial Biology: The National Science Foundation and American Biology, 1975-2005*, https://doi.org/10.1007/978-3-030-56367-7_5

to carry it out; however, but he did begin it in the form of full-blown computerization at the agency.[2] Clutter had at least two major assignments in her time in the Office of the Director: to help create a computer science directorate and to reexamine the status of all of BBS.[3] Clutter, in her temporary position, became a very close associate of Bloch's—*simpatico*, it might be said—and thus played a significant role in those imminent changes. She had been asked by Bloch to "assemble the pieces to establish a new directorate [for computer science]," as she put it. Bloch's style was new and different at the Foundation, and the in-house rise of "office automation" was already bringing about a changed environment in the form of the creation of secretarial pools: no small matter in shifting institutional demeanor.[4]

It should be recalled that Bloch had become director when Acting Director Edward Knapp, in the office from November of 1982 until August 1984, became the last of the "revolving door directors." That is, the *first set* of revolving door directors. Those early senior personnel changes would hardly be all at the end of the Foundation's fourth decade. By the later 1980s, Clutter had originally taken on the science advisor position for 1 year only as a test, and then, should it work out, she would stay for the second year, she told Bloch.[5] It did work out, and that is how she came to be in the Director's Office at a very notable time in the history of NSF: a total commitment to both full computerization of the Foundation, as well as that of the research community of the nation; the early days of the Internet and NSF's role in it; and, the first time that the discussion of a major reorganization since that of 1975 would actually bear fruit. But because Clutter had been an academic prior to her arrival at NSF in the 1970s, she stated in her own colloquial style many years later that she knew "zilch about management." So NSF sent her to a number of courses on the subject prior to Bloch's period, including one at the Brookings Institution in Washington, D.C. Another round, at the Wharton School of Business at the University of Pennsylvania, she found to be "an invaluable experience."[6] Indeed, those learning experiences would provide a good deal of theoretical management knowledge. She needed it when she found herself in the midst of the practical aspects of the reorganization of her familiar BBS: a restructuring she was destined to lead. After her 2-year rotation as science advisor with Bloch, she returned

[2] Bloch's valedictory to the NSB is given in the Minutes of the Board of its June 14–15, 1990, meeting, NSB-90-139 (subpart NSB-90-93). It is an excellent commentary and is recommended to the reader. The Board commended him at its August 16–17 meeting.

[3] Erich Bloch Interview with Marc Rothenberg, Historian of the NSF, February 11, 2008, *op. cit.* Bloch asked how many computers (desktop types) the Foundation needed and received the answer, as he recalled, "around 50 or 150, or something like that." He said, "well, I want you to order 1200 for next year." "Well, we don't have 1200 people, we only have 1000," he was told. He responded: "That's why I want you to order 1200, because [personal computers] are useful for other things than replacing typewriters"!

[4] Ibid.

[5] Clutter interview.

[6] Ibid.

to divisional and programmatic work at BBS for the next couple of years and was there at the time David Kingsbury left as AD (September 1988).

Other personnel changes appeared around the corner when the Foundation's deputy director, chemist John Moore, left his position in December of 1989; engineer John White would not fill that vacancy until September of 1990. And even so, White was only in an acting capacity until March 1991. Most notably, in August of 1990, Bloch himself would depart and another complex period of not only "revolving door directors" but also revolving door *deputy* directors would ensue. Frederick Bernthal would take on the acting director's role the next month and would, from September of 1990 until March of 1991, hold the position.[7] He would thus parallel the time of John White's role as acting deputy director. That March, Bernthal would replace White as permanent deputy director; but that would only last until April of 1993 when Bernthal, *again*, had to step into the acting director's place. Why had that occurred? It was Walter Massey, a former member of the NSB, who would soon be appointed by Congress as Director of the NSF after Bernthal's first, short acting role. Massey had been expected to stay on for his full 6-year Senate appointment, but he, too, became a revolving door director when he left in April of 1993, thus throwing Bernthal back into the role of acting director for the second time in only two years. Massey and Bernthal led the Foundation during the reorganization period of 1991–1992, which had been envisioned by and commenced with Erich Bloch. It was a period characterized by Byzantine musical chairs.

The director at the actual time of the reorganization was, then, Walter Eugene Massey. He was born in Hattiesburg, Mississippi, in 1938 and was one of three brothers. He became the second African-American to lead the NSF. His bachelor's was from Morehouse College, Atlanta, in 1958 and his doctorate, in physics, from Washington University, St. Louis, in 1966. He did a postdoctorate at Argonne National Laboratory in Chicago (1966–1968) and later returned to become its director (1979–1984) after having been a faculty member at Brown University, Providence, Rhode Island, then professor, dean, and Vice President for research, all at the University of Chicago (1975–1991). He remained at Chicago until he became director of NSF. After leaving the Foundation in 1993, he went first to the University of California and then became president of his *alma mater*, Morehouse, in 1995.

One of the most crucial personnel actions at the time was the choice of the person to follow David Kingsbury as AD for BBS. Thus, to add to the tumult of the later 1980s, and that which had most direct effect in the biological and social sciences' realms at NSF, was Mary Clutter's return to BBS. She came first as acting and then

[7] Frederick M. Bernthal had been a professor of chemistry and physics at Michigan State University from 1970 to 1978 when he joined the staff of Senator Howard Baker, one of the most powerful senators during the later half of the twentieth century; Bernthal was a native of Tennessee, Baker's home state. Bernthal remained with Baker until 1983 when he joined the Nuclear Regulatory Commission and, that same year, he had been made Assistant Secretary of State for Oceans, Environment and Science (1983–1988). From there, he went into the position of NSF Deputy Director from 1990 to 1994, from which he was thrust into the acting director's position twice, as has been seen. He left NSF that latter year to take up the presidency of the Universities Research Association.

fully the assistant director. Already-established Deputy Assistant Director, Robert Rabin, held the acting AD post for several months after Kingsbury's departure and until Clutter filled it. At that point in time, ADs were commonly presidential appointees confirmed by the Senate, such as both Betsy Clark and David Kingsbury had been. The law allowing the NSF Director to appoint ADs was passed in 1986.[8] Mary Clutter was appointed AD then by Bloch, after a major search. "Presidential Appointments with Senate" confirmation (known as PAS positions) were never again the employment track followed by any senior NSF personnel save the director and deputy director.[9]

From 1989 into 2005 and the close of this history, Mary Clutter was AD of first BBS and then BIO. Clutter had the necessary experience to be that leader: it had been won in the BBS over a number of years; it had been won with Erich Bloch; and, it had been won with her learning of theory at Brookings and Wharton. Most notably, though, it was due to her very substantial *inherent* leadership abilities. During her period as Bloch's science advisor, she was pleased that "Erich gave me assignments that were really very interesting and challenging."[10] Those had included an examination of the Department of Energy (DoE) national laboratories' budgets with the hope that some of the facilities might be closed to save funds under the Reaganomics' approach to federal fiscal management.[11] It was one of the expectations of the Director of NSF, as authorized in the Foundation's organic act, to advise the Congress on science activities *government-wide*, even though that duty was rarely carried out fully. The DoE effort "fizzled," said Clutter, but it was nonetheless another valuable learning exercise for her. It failed as the NSF director's authority to advise on federal science activities hardly sat well with other departments' leadership.[12]

Much more successful would be Clutter's "assembl[ing] the pieces" to create a directorate of computer science. Clutter literally "walked around all of the directorates looking for bits and pieces [and] programs that would fit into that kind of directorate." That, too, was remembered by her as a "valuable experience" and clearly gave her a portion of the necessary background knowledge to play a large role in bringing about the split of BBS within the next several years. Another very notable aspect of Bloch's eventual creation of the Directorate of Computer and Information Science and Engineering, CISE, was that Clutter, its lead architect, was able to bring

8 Personal communication with Rothenberg, June 3, 2011.

9 See: http://en.wikipedia.org/wiki/List_of_positions_filled_by_presidential_appointment_with_Senate_confirmation.

10 Clutter interview.

11 When Bloch was asked by his interviewer what it was like to work under Presidents Reagan and Bush (*pater*), as he had done, he spoke highly of Reagan because a number of the NSB members at the time were close to the White House: Simon Ramo, Glen Campbell, Anneliese Anderson, as well as Presidential Science Advisor George A. "Jay" Keyworth. They all had the betterment of the Foundation foremost, he averred. "My relationship with the Bush White House was a little bit cooler, at a lower level, but I would say constructive," he went on.

12 Clutter interview.

it off without any extra funding; it was totally an "internal reorganization." Such structural changes do not require congressional approval as the organic act permits the director to organize the Foundation as he or she deems fit. Therefore, no new funds had to be requested. Did the creation of the computer science directorate provide an "inspirational" feature in Clutter's mind with regard to taking on a similar effort for BBS? "Oh, yes" was the answer.[13] And that not surprisingly, the notion of creating both a directorate for the social sciences and a separate one for biology alone had been on the minds of all those concerned for many years, as has been demonstrated in earlier chapters. It took Erich Bloch to initiate, officially, that crucial change in the history of NSF and its relationship to biology and to the social sciences, but Massey would complete it. Still, it was the work of many persons that prepared the way for the final decision, as this chapter demonstrates.

CISE and NSF played a central role in the origin of the Internet, while Clutter was science advisor. The history of the military structure known as DARPA (Defense Advanced Research Projects Agency) has been considered by many chroniclers, but much of a specious nature has also come into common parlance. The official brief history can be seen on the very Internet that grew out of DARPA's work.[14] It was Bloch at that time, according to Clutter, who saw "well beyond" just connecting military sites, and so NSF:

> made some awards to universities to start putting [what would become a portion of the Internet] together. What was it called? Well, NSF Net. After that, NSF made awards for browser development such as MOSAIC [the web browser that popularized the World Wide Web]. Another award [under SBIR—Small Business Innovation Research program] established domain names to register URLs [Uniform Resource Locators] and email addresses. In the end it became the Internet.[15]

Clutter recalled that she was involved in "all of the meetings" in 1985 and 1986 and was "privy to all of [the] excitement" at a time when "everything we take for granted today—email, online—did not [then] exist." Mary Clutter was, so it happened, at a propitious place at a very gravid moment, and the broader learning she was accumulating was clearly of immense value when it came time to take on the reorganization of the BBS. She felt "very lucky to have been around to observe a lot of this," she averred.[16]

In the year or so after Clutter left the Director's Office, but before she became AD, "several significant actions were taken." The first of three was the revision of the position titles and descriptions at the BBS program level: all doctoral level professional scientists who had formerly been *program* directors, associate directors, or assistant directors (the latter two categories unknown outside of BBS) became Program Officers (POs). One PO would be Program Director, as well, but the other

[13] Ibid.

[14] http://www.darpa.mil/history.html. But especially see the page at that site where Mitch Waldrop presents the history of the Internet, *per se*: http://www.darpa.mil/Docs/Internet_Development_200807180909255.pdf.

[15] Clutter interview.

[16] Ibid.

titles were eliminated, thus streamlining the staffing charts and bringing BBS more in line with general NSF practice. At first, the Program Officer positions could be filled by either permanent or rotating individuals. A second effort, and closely allied to the first, was Clutter's success in establishing the Science Assistant program in which "very smart college graduates" were brought in to aid the POs on a limited term basis.[17] Finally, the third, somewhat unrelated effort, was all of the work that Clutter had put into the Foundation-wide committee that sought to increase the number of women scientists in senior positions in the academy:

> The main recommendation was to initiate a competitive program for junior women faculty. Universities nominated women they wanted to groom for professorships. Five-year research support was provided by NSF. All 100 women winners became full professors within the five-year award.[18]

It was an astonishingly successful program. Clutter's concern for advancing women in the academy, in science, and at the NSF was a *major hallmark* of her time as AD for BIO. Kathie Olsen had even helped her "develop and write two brochures, 'The Planning Grant for Women,' and then [the] 'Research Opportunities for Women.'"[19] It is widely acknowledged that Mary Clutter was an aggressive advocate for women at the NSF and beyond.[20] At just that time, 1989, Clutter became formally the AD for BBS, and with the creation of CISE behind her, she could apply her considerable toolbox toward the restructuring of BBS.

How did the reorganization effort begin, given that it had so long smoldered in many quarters both within the social and the biological sciences, as well as at the director's level and also within the NSB—to say nothing of the various communities? In earlier chapters, we have seen that the issue bubbled to the surface repeatedly, but the final political will appeared not yet to have been present to take such an extreme leap as to free the two sets of life sciences one from the other. The issue of "protective coloration," or the "riding of coattails," has been discussed as a central reason that the social and behavioral sciences were maintained within the cloak of the bigger sibling biology. As we have seen, too, not all agreed that was a necessity. Biologist and senior administrator, Bruce Umminger, in retrospect stated that, "I don't think that we really needed to protect the social sciences. They were pretty politically astute, they knew their strengths and weaknesses."[21] Over time, though, the *safety* of the social sciences was of progressively less concern than it had been from the founding year of the NSF into the earliest 1980s. Even the Reagan administration soon realized the need for social data (e.g., voting behavior patterns),

[17] Bloch interview. Christopher Platt has said that the Science Assistants arose out of the previous Program Assistants once the computer became common across the Foundation and overall paperwork was reduced and Program Assistants were eliminated. Christopher Platt Interview with the Author, February 19, 2009.

[18] Clutter interview.

[19] Olsen interview.

[20] The author heard this oft repeated by NSF employees, and it is corroborated in a number of the oral interviews cited in this study.

[21] Umminger interview.

though it was Reagan appointee David Stockman, Director of the OMB, that came close to eliminating all social and behavioral sciences, as well as science education, from the National Science Foundation. However, even into the present day, there are individuals in Congress who still seek to have certain areas of funding, such as political science, eliminated in the NSF budget (Chapter 2). This attitude continued right up to the time of this writing!

Nevertheless, it is likely that one single feature, person, stands out as the precipitating mechanism in the 1991–1992 reorganization: Erich Bloch. It was widely held that the director was no shrinking violet by any stretch and his views of moving forward with matters was such that political delicacy or, worse, simple fear to bring large-scale change to a federal agency was no road block. Unlike other directors then to date at NSF, it will be remembered; Bloch did not come from either the academy *or* from the federal establishment but rather from industry. Inherently, Bloch was both aggressive and a doer and his previous life at IBM only sharpened that approach. He was a pragmatist, and he believed that "criticism comes with the job" when taking on risky decisions.[22] One might say that he had nothing to lose by coming into a government institution and making the changes he believed necessary: he ran the Foundation on a more business-like model than it was theretofore managed. A CISE was called for and Bloch incisively made it happen. And from his close ally, Mary Clutter, it is possible that he heard the argument (though she did not accept it!) that not only did the social and behavioral sciences need their own directorate but so also did biology:

> And so Erich Bloch is the one that said to Mary…'Okay, if this is an issue, why don't you [create] a big task group and then have the task group look at biological and behavioral sciences for the twenty-first century?' Okay? So that's how, at least [to] my understanding, it emerged.[23]

Kathie Olsen had remembered it simply enough; and so did Mary Clutter, herself:

> So when I became AD, I decided that we really needed to be looking to the 21st century. And so I appointed a committee representing all of the various areas of BBS.[24]

It sounded straight forward enough. It was not. Far from the comparative simplicity of the 1975 reorganization, an apparent top-down *fait accompli* by Director Guy Stever that of splitting BBS into SBE and BIO was a significant undertaking that was hardly *sans* challenge or calumny.

---

[22] Bloch interview. Bloch was clearly pragmatic. When asked by the Historian of the NSF in the interview about the NSB and if its guidance was the best way to operate the NSF, Bloch responded: "You can ask the question, is that the smartest way to run an organization like the NSF? It doesn't matter, that's how it is."

[23] Olsen interview.

[24] Clutter interview.

5.2 A Representative Committee?

Mary Clutter had been aggressively promoting plant science within BBS from the earliest days of her directorship and boosting the role of women in science and at NSF, as well. Olsen opined that Clutter was very much a "goal-driven" AD and her promotion of plant science "really riled many communities."[25] The behavioral science group at NSF, and beyond, included Olsen as part, so she was present on the scene when Clutter began the reorganization. Olsen held that her social science colleagues were saying " 'Wait a second. I don't have a voice.' " That the biological sector of BBS would get all the funding as plant sciences, in particular, expanded, was a stated fear in the community, in the opinion of Olsen. But that may have been unfounded in that Kingsbury has said that plant scientist Clutter and BBS biologist Machi Dilworth, when dealing with the early emphasis on plant science, did not follow the Human Genome Project model: there was no "big center…they did a bunch of little things here and there." It was, he said, not unlike work that had begun on the bacterium *E. coli* at about that same time (previous chapter)—no "economies of scale."[26] But doing a "bunch of little things" was less expensive than the centers' approach. Still, it took 6 years to complete the *E. coli* genome: it would have taken considerably less had the "big center" model been used, Kingsbury believed. The implication was that surely the same thing must also apply to plants: the drive to reveal the *Arabidopsis* genome, and other work. We shall return to that story, specifically, later.

As Mary Clutter began to look toward what biology and the social and behavioral sciences might look like in the coming millennium, she sent out a number of documents to members of the BIO AC for reading in preparation for their planned meeting in March of 1990.[27] They included white papers on Beyond *Arabidopsis*; Structure of the Plant Science Initiative, Phase II; Biological Diversity; Human Dimensions of Global Change; Cognitive Science—An Interdisciplinary Initiative; and Computational Science for BBS. All of those issues had been discussed at a certain level at the previous October's AC meeting. What may have been conspicuous by its absence was any mention of reorganization. Peter Raven had been Chair of the AC for some time, but Clutter told the Committee in her meeting memo that President Bush had just nominated him for the National Science Board, so he had to resign from the AC.[28] His boost onto the Board may have been a propitious one, as matters turned out later. John G. Hildebrand, professor of neurobiology at Arizona

[25] Olsen interview.

[26] Kingsbury interview. Kingsbury also opined that Dilworth was "sort of Mary's person by history and association."

[27] Memorandum to the BBS AC from Mary E. Clutter, March 14, 1990. A cover memo precedes the several white papers (author's term) and the appended Minutes from the previous meeting. The Agenda for March 19–20 was also included and noted that Clutter would speak on the "Planning Environment" for FY92 and beyond.

[28] See also: NSB "Minutes," NSB-90-51, 3–90:4, March 15–16, 1990, where it is noted that Raven's term was set from March 1, 1990 to May 1994.

State University, Tempe, would take over the chair position, and he, like both Bloch and Raven, was not a reticent personality.

Kathie Olsen was one of the individuals that helped form the reorganization study committee for Clutter, Olsen recalled. She pointed out that there had been an earlier ad hoc group, formed at the request of the AD, of six persons. That group had been chaired by AC member Linda Smith, a psychologist on the faculty of Indiana University. Smith and her colleagues had noted that since the social and behavioral sciences' divisions within BBS had certain "concerns," the need for a larger study committee suggested itself, and so Clutter decided to include all parts of the directorate in the study. At any rate, Bloch had already directed Clutter to do just that. Smith had made her first report to the members of the AC at the March 1990 meeting and cited testimony of Herbert Simon (of the Simon Report noted earlier) to the House Science and Technology Committee where he called for a separation of the BBS into two new directorates.[29] Smith reported on the results of a single, 2-day meeting of the ad hoc committee of six social and behavioral scientists the previous February. Smith's final report would not come until June. Two recommendations came from the discussions, as she reported to the AC, and later in her final report. The first was that "resources must be radically increased in order for the opportunities in new knowledge to be realized" for the future. Second, reorganization was a necessity "in order for the NSF to play its leading role." Indeed, "[w]e specifically recommend consideration of a separate directorate" for the social and behavioral sciences.[30] To the AC, she spoke of "benign neglect" of those sciences. To that, Clutter took strong exception and responded that funds were short in biological areas as well: in plant sciences and biodiversity, in addition to other areas, for instance.[31] The second day of the AC meeting closed on continued discussion of the Smith report with her adding that a reorganization was a necessity for the social and behavioral sciences to have a "larger voice." Again, Clutter was defensive but took the opportunity to say she was going to form a larger committee to examine all of BBS.

The American Psychological Society's newsletter, the *APS Observer*, picked up on the story for its May edition and stated that:

> [t]he six-member ad hoc committee recommended that—in view of NSF's mandate to fulfill a leadership role in supporting basic science, the phenomenal growth in knowledge, and the potential for additional accomplishments, in the behavioral and social sciences—NSF

[29] BBS AC, "Minutes," March 19–20, 1990. See also next footnote.

[30] "Report on the <u>Ad Hoc</u> Committee on the Organization of Behavioral and Social Sciences within the National Science Foundation; February 26–27, 1990, Bloomington, Indiana." The Final Report was dated June 15, 1990. Note that the title did not say "within BBS" but rather spoke of those sciences "within [NSF]," thus suggesting that, even before getting to the final two recommendations, the reader might conclude that the social and behavioral sciences should/would have a new home at the Foundation.

[31] BBS AC, "Minutes," March 19–20, 1990, pg. 5 of a draft copy.

gave serious consideration to housing the behavioral and social sciences within a separate NSF directorate.[32]

The APS article used the past tense "gave" when it is clear that the word should have been "give." One wonders if it was one of the infamous "Freudian slips" so often spoken of in situations of hopefulness! Whatever the motivation, if it was not the more likely case of just a simple compositional error, no decision at NSF could be forthcoming until Clutter's larger task group concluded its work, the better part of a year into the future.

When that broader group was formed, Olsen noted that, initially, there was no one representing the neurosciences on it.[33] Kingsbury has argued that "one of the things [Clutter] didn't like was neuroscience, which she basically all but got rid of." If that had been the actual case, it may have explained the lack of a representative from that discipline. BIO, *after* the reorganization, however, can be seen in Figs. 5.1 and 5.2 to have a very substantial set of programs devoted to neuroscience and with that epithet still very prominent.[34] It will be shown, though, that against many opponents, Clutter's reorganization did heavily restructure the neuroscience programs.

Kingsbury also stated, in a slightly different vein, that since Clutter was neither a microbiologist nor a biochemist, she "hadn't followed what was going on in molecular genetics very carefully." Molecular geneticist Delill Nasser of BBS was particularly bothered that Clutter did not choose to follow the centers' model for the expanding work in plant genetics, alleged Kingsbury.[35] These two opinions held by Kingsbury—the second possibly also held by Nasser, though that remains unknown (she died in 2000)—and their relationship to the thinking of Mary Clutter, while intriguing, only remind us of the very strong personalities native to both Kingsbury and Clutter. Still, it is a matter to which we shall return.

There were other rising tensions in the directorate that were more severe than had been the case in 1975 and which played right into the matter of the split. Roberta Balstad Miller (Chapter 2), Director of the SES division within BBS, was a figure in leading the behavioral scientists to consider a bolt from the directorate altogether. That was possibly a part of the "concerns" alluded to by Smith. Kingsbury left NSF abruptly in late 1988. He has stated that the behavioral scientists' contingent had told him that, "'as long as you're [Kingsbury] here, we're going to be quiet, but when you go, we're going to try to get out.'" But what's more is that the social and behavioral scientists "were starting to write letters to Congress. And they didn't feel that Clutter would represent their work. Whether they were right or wrong, I don't

[32] *APS Observer*, "The National Agenda: A Separate NSF Directorate for Behavioral and Social Sciences?" (1990):19–20 (May).

[33] Olsen interview.

[34] While Fig. 5.1 is quite complex, a careful attention to the lines connecting any given program or group of programs can be followed as the directorate split took place. The illustration was a single page ephemeron associated with the June 26, 1990, Task Group charge draft document (footnote further below). It is known to have been created by Mark Courtney.

[35] Kingsbury interview.

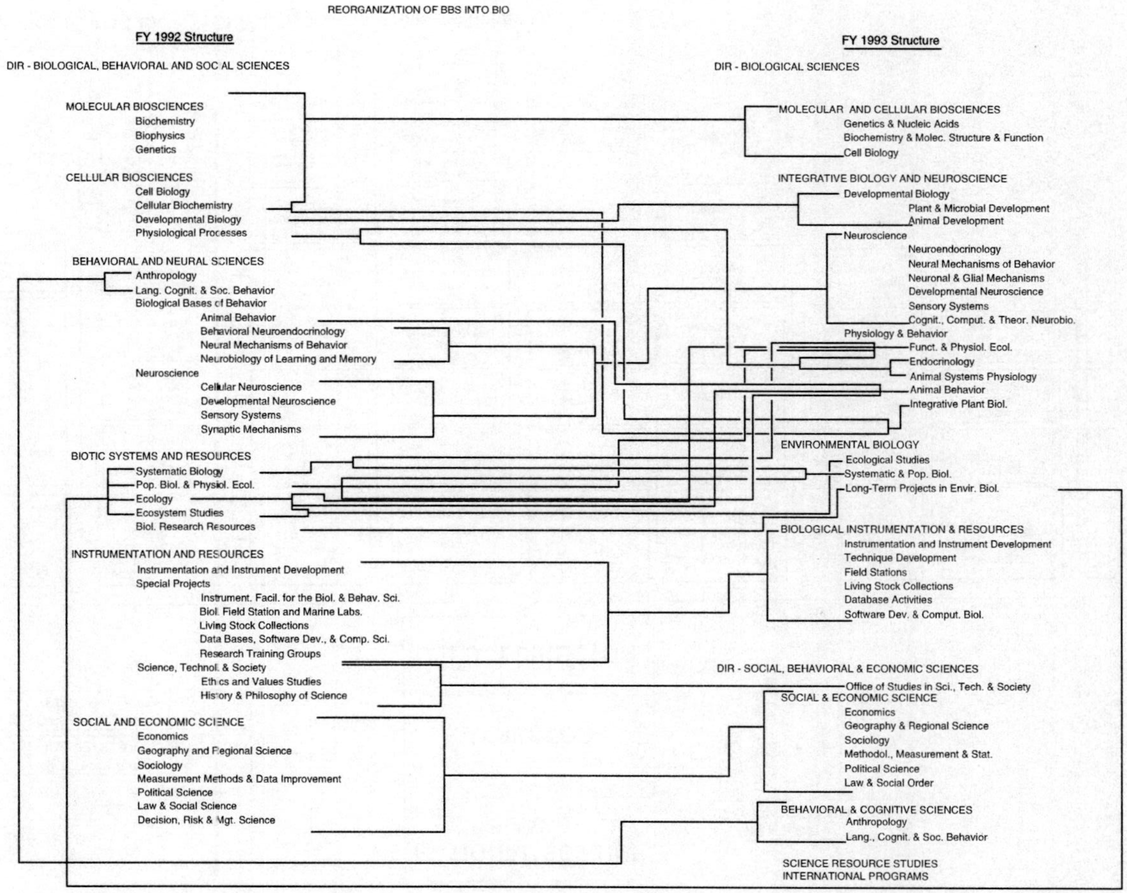

Fig. 5.1 Reorganization of BBS into BIO. NSF. Specific provenance unknown

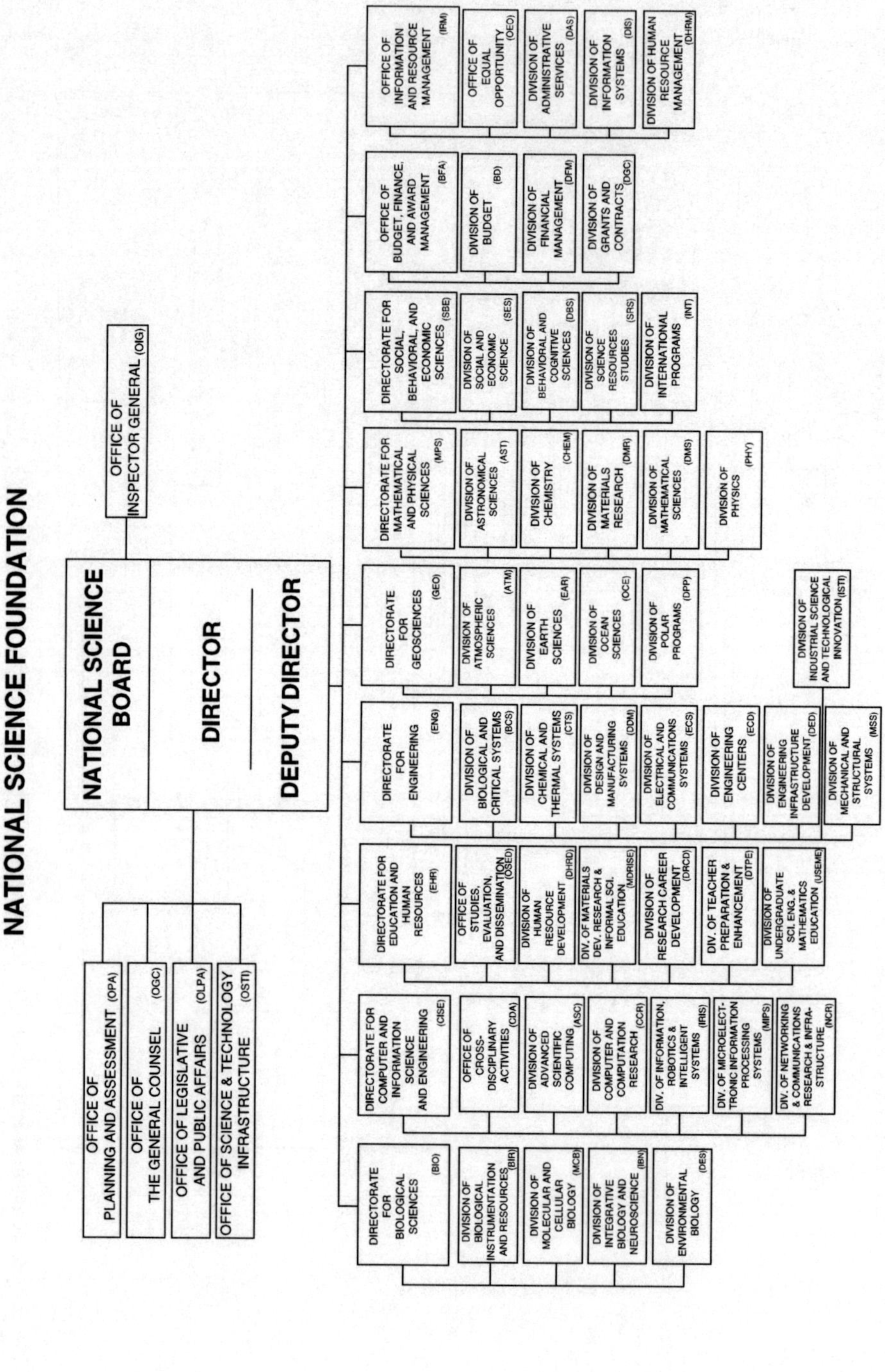

Fig. 5.2 Organization chart of the NSF, 1992. Specific provenance unknown

know," Kingsbury has asserted.[36] So the situation in which Mary Clutter found herself was not a simple one: politics, some of it decades old, had been simmering, but this time with the threat of boiling over. That may have eventuated had it not been widely known that Clutter was just then taking action to start the process of examining organizational issues for the twenty-first century: she had formed her committee for the purpose. Still, many other anxieties would lie ahead once the greater community was brought into the discussions.

What then was exemplary of those tensions appearing at the time? A bill was brought before the House of Representatives and may have played a role in the final outcome of the reorganization, if in no other way than simply shining a very public spotlight on the whole issue. Representative Doug Walgren (D-PA), along with his House colleague and long-time supporter of the sciences, George Brown (D-CA), wrote a letter to Acting Director Bernthal in early September 1990 commending him for establishing the task force to examine the status of the BBS. But they further stated that they:

> would like to take this opportunity, however, to express our concern that the charge to the task force seems broad and duplicative of past reports and specifically does not include consideration of a separate directorate for the behavioral and social sciences. This is particularly important since the BBS most likely would not be the subject of inquiry were it not for the concerns raised in Congress and elsewhere about funding for these sciences.[37]

The "letters to Congress" mentioned by Kingsbury were possibly the basis for the comments by Brown and Walgren. They had introduced H.R. 5543 in early August of 1990 as a bill to amend the organic act in order "[t]o establish a Directorate for Behavioral and Social Sciences within the National Science Foundation," because, as they said in their letter to Bernthal, that:

> if adopted, this legislation could begin to address the persistent inequities in funding which have characterized the NSF's treatment of behavioral and social science research for at least a decade. By creating a separate directorate…that parallels…[other disciplines at NSF] our bill would provide these sciences with a voice at the Assistant Director level [and alleviate the] disadvantages these sciences have worked under in budget after budget.[38]

The bill died in Congress.[39]

Clutter's committee—it would be a "Task Force"—that would look into the fundamental aspects of BBS's structure did not see its first instructions until mid-1990, though Bloch told Clutter during 1989 to form it. Possibly simple inertia explains why the gap of almost a year intervened, but it would seem to belie the assumption

---

[36] No hard evidence of this matter has been found, so it must remain an opinion on the part of Kingsbury.

[37] George Brown and Doug Walgren to Frederick Bernthal, September 7, 1990, pg. 1.

[38] Ibid.

[39] See http://frwebgate1.access.gpo.gov/cgi-bin/TEXTgate.cgi?WAISdocID=Bx4nMW/0/1/0&WAISaction=retrieve. The *COSSA Washington Update* of August 10, 1990, mentioned on page 1 that the American Psychological Society, "a COSSA Affiliate[,] has actively supported the measure [Brown/Walgren bill]," but went on to say, that "action seems unlikely in the near future." There never was a future for the bill.

that Clutter was moving aggressively on creating the study group. The committee was first termed the "Task Group on Structural Review of BBS." Their assignment was to "review and to recommend options for optimal organizational structure."[40] The process section of the charge memo noted that there would be an internal group of persons, in addition to "external scientific experts" of up to 20 individuals to be chosen by Clutter from a list to be proffered by BBS scientific staff. She went on to note that a hearing would be held in mid- to late fall of 1990. It would include not only NSF internal individuals, and single members of the community, but also representatives of "Professional Scientific Societies and Organizations." She expected that the task group would meet four to five times altogether. The first two meetings before the hearing would be to develop the agenda and a working relationship within the task group. Two more meetings would ensue after the hearing for the writing of recommendations and a report. A retreat might make up the final meeting, all of which would occur in or very near the national Capitol area; NSF was still located at 18th and G Streets.[41] The timeline she provided broke down the duties in a detailed fashion and with the expectation that the final report would appear by April 1, 1991.[42] Tied to this effort, Clutter had hired not only a management company for assistance in bringing all those task group's efforts to fruition, including such other things as accommodations and meals for meetings, but also a science writer for rendering the final report.[43] Clearly, the reorganization of BBS bore little resemblance to the simple perfunctory memo from the director to BMS (and the rest of NSF) in 1975.[44]

---

[40] "Task Group on Structural Review of BBS: Outline of Task and Meetings," stamped as a "Draft," of June 26, 1990; cover memo by Mary Clutter. A mentioned "Attachment A" is missing. The "Process" portion of the memo noted that there would be an internal group of persons in addition to "external scientific experts" of up to 20 individuals. See later footnote.

[41] It was a time, as we have seen, in which many files "stored under the stair case" (literally) at 18th and G were lost, or disposed of (?), as the move to Arlington progressed. But the activity of disposing of files, and thus history, was not limited to moves alone. Bruce Umminger, long a senior leader within the biology directorate, recounts the story of one program director, a rotator (name withheld here), from a major northeastern university who said there was going to be a "catharsis" in the then PCM directorate of past award files including all the comments on the assessment documents, temporary and permanent personnel names, etc.—the "jackets"—because "'we're worried about what we're funding today [not] what we funded ten years ago.'" To which Umminger said, many years later: "my god, what a purge of information!" Further, Umminger said that "[PCM] was probably the biggest division in all of NSF at that time, that's why it split up…[it] just got too big. In that division, *everything got chucked*." Umminger interview; emphasis added.

[42] That had been set in NSB, "Minutes," NSB-90-161, 10–90:5, October 11–12, 1990 by the Board. The Charge to the Task Force was an Appendix B to NSB-90-161.

[43] "Task Group" charge memo. The science writer was Courtland S. Lewis.

[44] While one might be tempted to wonder if the reorganizations of 1975 and 1991–1992 were so different because they reflected the *managerial styles* of Stever/Atkinson versus those of Bloch/Massey/Bernthal, the answer is clearly no. The primary reason is that the biology directorates of those two eras were themselves very different in how they approached matters. In the 1991–1992 period, the BBS run-up to the reorganization had a long and involved history of revisioning itself as opposed to that of the BMS prior to 1975. Even though I demonstrate that the BMS was clearly considering a restructuring in the early 1970s, Mary Clutter, who was far more responsible for the

Clutter has said that the methodology for the whole exercise was fashioned upon the model of a "congressional hearing" wherein the task group would give "people five minutes to summarize [their views/case]," but that their "longer testimony [would be] submitted for the [written] record."[45] She was pleased with the technique—"it was a wonderful time"—but "what I didn't expect, and I think precious few of us [biologists representing professional societies] did, was that when the social scientists testified, they came extremely well prepared—the biologists were stunned"![46] Why should they have been? One must wonder at that given the many, many years during which the social sciences had chaffed under "protective coloration." The answer appears to lie in the fact that, for Mary Clutter, "as a biologist, I *really believed that biology was saving the social sciences.*"[47] So convinced was she, that she pointed out to the author that, in earlier years, Betsy Clark had "managed to keep the social sciences as part of BBS, [and] *she deserves all the credit for that.*"[48] In Clutter's mind, it seems, keeping all of those disparate disciplines under the banner of biology was a wise decision. However, the secretary for the Task Force was Sonya Mallinoff, who was very close to Clutter. She has said, very clearly to the author, that Clutter *did* want to see the separation occurs, that she *did* want two directorates.[49] The real views of Mary Clutter may not be knowable.

In Betsy Clark's day, maintaining the two rather disparate sets of sciences together almost certainly made sense, as we have seen earlier in this history. But times had changed. It would seem, Mallinoff's argument notwithstanding, that Mary Clutter did not appreciate the fact that even with some then-waning criticism from congressional members[50] on and off about the federal government supporting

reorganization than any of the five named directors above (possibly save Bloch), involved the *community* in her restructuring: that did not happen with Clark in 1975. In addition, I would argue that biology itself had grown much more complex in the intervening 16 years and demanded a more complex approach to establishing BIO from BBS than had BBS from BMS. It is my opinion that Clutter's very aggressive leadership was yet another factor in the nature of the 1991–1992 activities.

[45] Clutter interview.

[46] Ibid.

[47] Ibid. Emphasis added. It should be noted that Bruce Umminger sent Mary Clutter an email sometime prior to May 29, 1990, in which he must have told Clutter that Carter Kimsey (an associate program director in DCB; see Chapter Four) saw the social and behavioral sciences as the "*driver*" (emphasis added) for the formation of the Task Force, in Clutter's words to Umminger. Clutter further told Umminger that Kimsey was "wrong" about that supposition. This stands as another thread of evidence that many persons in the BBS felt that the entire reorganization exercise was *meant* to separate the sciences in BBS into two directorates.

[48] Ibid. Emphasis added.

[49] Mallinoff interview, *op. cit.*

[50] In an email (March 21, 1990) from one NSF worker to a number of persons in BBS, an attempt was being made to trace several grants about which a complaint had been made by a constituent to his Congressman, Representative Wes Watkins (D-OK; later R-OK). The complaint concerned "waste" over awards that included, as the constituent said, "$98,000 on why rats stomp their feet," "$174,000 on how wild radishes make love," $100,000 on how bugs nab spiders for dinner," and a number of others. The constituent wondered if that was "all that [NSF] has to do to justify their

the social sciences within one of its agencies, there were new and mitigating circumstances calling for the BBS split—and they were quite plain. First, Senator Proxmire, the most visible and vocal critic, had been gone from his congressional seat since early January of 1989; second, Clutter had the very astute, and not so shy, Roberta Miller leading the largest group in the social and behavioral sciences, the Division of Social and Economic Sciences (SES), along with her "concerned" subordinates ready "to get out"; and, third, COSSA had been on the scene and vocal for many years about raising the visibility and stature of the disciplines covered within the two divisions, SES (primarily) and BNS (tangentially), and keeping them from calamity arising through defunding. "Protective coloration's" time had passed: those disciplines had risen to their new statures and had even survived, in healthy condition, the mostly unrealized "threat" of Reaganomics. That the split was on the way to a *fait accompli*, regardless of what Clutter has said in more recent times, and from all the evidence available from the period, should have been seen then to be the case. Clutter's statement after the hearing and the appearance of the final report (1991) was that "the major recommendation was that the social science[s]...needed to have their own directorate...we were just stunned [again!] by that." Was Mary Clutter being disingenuous with the author during our interview, or was she truly "stunned"? The answer seems to be clear in her own words, when she said that:

> I went to NSF Director, Walter Massey, at that time. And I have to tell you, that guy is one of the gutsiest people I have ever met. Absolutely. I made a report to him on what happened...[a]nd he said, 'Well, Mary, you know, how can we, how would we ever make this [the committee's recommended splitting of BBS] happen?' I said, 'I don't know, it's going to be difficult.' He invited some leading social scientists and me to discuss the issues. We also went out to California.[51]

It was under Massey and via the invited, and visited (California), social scientists that the two new directorates, SBE and BIO, came to exist, at least in part. But we leap ahead of the story. What occurred in the meetings that preceded the appearance of the report? What did the "Adapting to the Future: A Report of the Task Force Looking to the 21st Century" state? And what did the professional societies and all the individual scientists, of all disciplines concerned, who were either asked to or who indicated their desire to speak up, actually say?

---

existence?"

[51] Clutter interview. Many months later, Jane E. Buikstra, President of the American Anthropological Association (see also main text) wrote (November 8, 1991) to, by then Director, Walter Massey, thanking him for meeting with some five anthropologists and social scientists in "northern California" (San Francisco) to obtain their views on the potential split.

5.3 Convening a Task Force

Mary Clutter announced to the BBS Advisory Committee in a memo of August 1990, that the "Task Force," the term finally settled upon, had been established.[52] The earlier, small ad hoc committee under Smith, along with the echoes of Bloch's suggestion to Clutter to form a task group, led to the decision: the small group's suggestion to form a larger committee was noted in the memo. Bloch's recommendation of nearly a year before did not play a role in the decision. But in her later, final communiqué of August 1990, the AD had appointed and named the 20 outside advisors. They represented 14 US states and included leading scientists from such disparate areas as economics, botany, geography, zoology, and the humanities, as well as much else. Further, she provided a list of executive directors of 120 scientific societies to whom invitations to the projected hearing were being sent. They covered the social, behavioral, and biological sciences: from the American Association of Anatomists to Sigma Xi: The Scientific Research Society and from the big to the small AAAS (the world's largest general scientific society) to the Linguistic Society of America (new in 1983). Also, the charge had been refined by late June of that year and included three major areas: examine the current structure of BBS; evaluate its adequacy and effectiveness in meeting current and future needs; and recommend options for an organizational structure to optimize the ability of NSF to serve the community in the future.[53] It did not mention *separation*, as Congressmen Brown and Walgren had pointed out. The three main charges were further fleshed out in the guideline questions proffered (not a limiting list, it was stated) and comprised a number of issues. They included asking if the NSF's role was unique; what fields were served; which of those were not well served and how to address that; do disciplinary areas get the necessary encouragement; is interdisciplinarity provided for; is the structure flexible enough to respond to emerging areas; and are infrastructure needs of the community sufficiently addressed, etc.[54]

Energy was building toward the first meeting of the "BBS Task Force Looking to the 21st Century" that was to be held on September 10 and 11, 1990, at the Embassy Row Hotel in Washington, D.C. Emails were coming and going to Clutter concerning critical features of the actual hearing, which was set for late November. Bruce Umminger had been Clutter's Deputy Director of BBS from 1984 to 1989 and Acting Cell Biology Program Director for Clutter earlier when she was science advisor for Bloch. He was, therefore, very familiar with her and so sent a memo on

---

[52] Mary E. Clutter to Margaret B. Davis [one member of the AC; similar memos went to all the others], August 16, 1990. There was a cover letter and three numbered attachments. The AC at the time had ten community members, including Chair Peter Raven (Attachment 1); the second attachment was the list of 120 society directors; and the final attachment was the list of the 20 community advisors for the "looking to the twenty-first-century" advising committee.

[53] These points are given here as a very close paraphrase of the original, but not set in quotation marks as they are not precisely the same words.

[54] BBS to Concerned individuals, June 22, 1990; ephemeral single sheet with two areas of emphasis only: Charge and Guidelines.

major issues for the "reorganization," as it had for some time routinely been called. How one might define "reorganization" would vary with the speaker, however. Umminger noted that the first concern was finding a home for the social and behavioral sciences—as well as dealing with the neurosciences, that borderline area that had often caused so many questions of organizational placement. He noted also that "organismal biology" was an issue ("role and placement"), and so was the matter of the BBS's "interface" with oceanography, polar biology, and biotechnology—all of which were tied through interdisciplinary connections from several other directorates to biology.[55] Umminger, thinking strongly about organizational issues coupled to programmatic size, pointed out that such large cohesive units as Cell Biology, with proposal loads of more than 400 per year, and small but internally cohesive programs with very few, presented challenges: does one further divide the one type and coalesce a number of the second, he wondered?

It was not his first memo to Clutter on the subject, though, as he had written her 3 days earlier suggesting that the whole Task Force effort should be aimed at dealing with placement of the social and behavioral sciences *only*.[56] It is likely she answered him in the negative, but such response is missing, if ever in written form. There is good reason to think her presumed response would have been nay, however, as there is strong evidence of her own particular predilection for the definition of "reorganization." That definition became almost entirely clear, as considered below, and it was *not* in concert with nearly everyone around her. Whatever prompted Umminger's second missive, it is clear that he was giving considerable thought to the entire matter, not just the locus of the social sciences.

Virginia Walbot, a strongly active member of the BBS AC from Stanford University (not chosen to be among the 20 for the task group, though), who we met earlier in this history, addressed herself exclusively to review panel structure and function.[57] The potential reorganization had to deal with not just what would be the fate of the social sciences, behavior, and the neurosciences but *all* aspects of how BBS carried out its *mission* within NSF and for the community. One contemporary document listed pros and cons with regard to maintaining the BBS or separating it into, at least, two directorates.[58] It could even be three, if the neurosciences were found to require a completely separate unit. That was an issue not widely seen in the historical record, but it was clearly on the minds of some unknown number of BBS personnel at the time and became one of the final sticking points in the whole effort. The three units could be subcomponents of yet another variation, as others suggested.

---

[55] Bruce Umminger to Mary Clutter, "Major Issues for 'Reorganization' Task Force," August 8, 1990.

[56] Umminger to Clutter, "Follow Up to April 4 DD [Division Director's] Meeting," August 5, 1990.

[57] Virginia Walbot to Mary Clutter, "Task Force on BBS Charge to Examine BBS Structure," August 26, 1990.

[58] The ephemeron, from the Courtney Files, had attached to it two more pages titled "Other Issued Highlighted in Hearing"; note the past tense. Neither the author of nor the date of the pros and cons' lists is given. The author matters less than the fact that there were *many* issues that had been brought to the fore in one way or another within BBS and needed close consideration.

The list of pros in connection with retaining a single BBS included such matters as continued maintenance of cross-disciplinarity, animal/human "bio/sociology," avoiding damage by splitting disciplines, continued "protective coloration," maintaining low overhead (two directorates would cost more than one; another area of considerable dispute), avoiding "unnecessary 'balkanization,'" etc. The cons included only two: maintaining the social sciences as "'second-class citizens'" and pointing out that the numbers of programmatic areas in the social sciences had grown considerably and thus they might not each be fairly represented in a continued single directorate.

The second option, the one that would lead to the separation of BBS into two (or three) directorates, had nearly as many cons as pros. The pros were the obvious value of an independent "home" for the social sciences, etc.; it would give them their own voice at NSF; it sent the signal that NSF saw them as having matured "on a par with the biological sciences"; all of their sub- and disciplinary areas would rise to a higher level; it would eliminate the "no [natural] overlap with biology" problem existing for economics, political science, etc., but not seen with such areas as sociology and behavior; it would improve proposal handling; and, it would allow better infrastructure management for the included fields. But the many cons came back, once again: to a variant of "protective coloration"; higher administrative costs; possible hampering of cross-disciplinarity; "painful rending of the fabric of science in some areas"; near-duplication of some programmatic areas (neurosciences); a fear of reduction by about 15% of the pool of research funding available to BBS broadly—a crucial issue; competition for those pools; and possible infrastructure funding redundancy.

It is obvious that the matter of separation of all those sciences long located in BBS was not a simple, clear-cut issue: it would not be like creating CISE. AC member Linda Smith had made suggestions to Mary Clutter that August, too.[59] Smith brought up four questions about what the Task Force, the "TF" as it became known to those involved, should consider. The first was the issue of simply having the TF consider the split from the social sciences exclusively—just as Umminger had suggested. The theme was an old one, and many persons must have felt that it alone would be sufficient in any restructuring; Clutter obviously felt it was not all that had to be laid out to do a proper job of considering a reorganization. Smith's second thought was that a separate subcommittee of social and behavioral scientists should be formed to assess the direction they would take if they did *not* continue in "riding the coattails of biology." Third, why was funding not mentioned in the charge, Smith wondered? And finally, big and little science, top-down directives versus PI-initiated grant-seeking, and the dynamics of those issues interacting together, was of interest to Smith. This, she noted, intrigued her as the world had just seen the Soviet Union collapse and Smith noted that the methodology of "big, inflexible… direct[ion] from the top" did not work. A response indicated that Clutter thought

[59] Linda Smith to Mary Clutter, August 21, 1990.

that all of Smith's concerns would come up in the TF meeting set for September "as surely as day follows night."[60] But Clutter:

> wanted to keep the discussions *centered on science*...In other words, does it still make sense to be organized along disciplinary lines?...For example, BNS was established to foster neurobiology more than 15 years ago. Has it accomplished its goal? Remember, at best, NSF can only serve as a catalyst. I hope this effort does not degenerate into a disciplinary fight over scarce resources...Linda [Smith], I urge you not to politicize this sincere effort to address crucial scientific issues facing our disciplines and NSF. Some people think I am insane for inviting community participation. If we see the whole thing collapse my colleagues will be convinced that scientists are unable to help in making policy decisions.[61]

The attitude and desires of Mary Clutter in setting up such a large and complex study of the structure of BBS and its future could not have been more clearly stated! It would not be 1975 repackaged.

Acting Director Bernthal was given a set of "talking points" to use as he addressed those gathered for the first Task Force meeting in September. Hearing his set of expectations may have unsettled the social scientists present as his three points never mentioned them but focused instead on biology only: capitalizing on the current ferment in biology (the molecular revolution); the critical needs in infrastructure; and education and training (an area then no longer taboo in the federal establishment). He went on with a variety of expected commentary that dealt with such issues as limited funds at NSF, innovative interdisciplinarity, moving on "to the next plateau," and the like. But one of his points harkened back to Clutter's statement to Smith (Clutter may have composed the words for Bernthal):

> It needs to be stressed that if the task force retreats into limited disciplinary domains exclusively, the results will be of limited usefulness. People were selected for [their] excellent scientific insights as well as their ability to think broadly.[62]

Bernthal's talking points went on to highlight some hot research areas, "scientific opportunities," which included the now-familiar economics, biotechnology, plant science, biodiversity, "Global Change including the Human and Economic Dimensions," computational life sciences, and cognitive sciences. Where the rest of the social sciences were to be lodged was not clear in the "talking points" script.

For the first day of the Task Force meeting, besides hearing their charge from Bernthal, the TF was supplied with a large number of documents. A number of historical issues were presented: the former Division of Biological and Medical Sciences was considered; the long struggle with the state of the behavioral and social sciences (B&S below) in society and at NSF was given in synopsis; the period

[60] Mary Clutter to Linda Smith, August 25, 1990.

[61] Ibid. Emphasis added.

[62] "BBS Task Force Looking to the twenty-first Century: Expectations for the Task Force: September 10–11, 1990, Embassy Row Hotel," a four-page ephemeron upon which were penned the words "Talking Points for Dr. Bernthal." A second and more clearly circumscribed set of talking points (one page) dated October 11, suggest Bernthal, apparently, kicked off another meeting of the TF where he mentioned the one that had concluded on September 10–11 but spoke of the upcoming November 29–30 hearing.

of 1957–1976, when most BMS organizational change was primarily at the programmatic level, was discussed; the 1971 founding of the Neurobiology Program was mentioned; and the 1975–1976 reorganization of BBS that established the Division of Population Biology and Physiological Ecology ended the early history.[63] The 1984 split of the Division of Physiological, Cellular and Molecular Biology into the Division of Cellular Biology and the Division of Molecular Biology was also considered. The establishment in 1988 of the Division of Instrumentation and Resources was yet another topic of discussion (see following chapter). A valuable point for the TF membership at the time—and to this history now—was the comment that "[f]or the most part, organizational changes in the BBS *have occurred to facilitate administration of proposals*."[64] A most salient assertion, but it was only noted in retrospect. For that reason, the statement should be accepted with some care by the reader if one wishes to conceive of it as any sort of historical driver.

One of the most notable statements to be found in the handout for the opening efforts of the TF was the following:

> It is no secret that a significant segment of the B&S community see[s] as the ONLY ACCEPTABLE outcome of this task force THE recommendation that the B&S sciences have their own Directorate. That may well be one viewpoint. In crafting a charge to the Task Force, we have stressed the need for OPTIONS and a[n] elaboration of PROS and CONS for each. In so doing, our intent is to retain control in order to make optimal use of their advice. There may indeed be structural arrangements that facilitate the visibility and development of the B&S sciences *short* of the notion of a separate directorate.[65]

If there had been any doubt issuing from the comments made by Mary Clutter in her interview with the author that she "believed [that] B&S" should stay under biology in some form or another of a BBS, her words above to the Task Force in the fall of 1990 make it abundantly clear that that was her thinking. It was not all of the scientists of BBS that had been "stunned" later, not at all, but primarily its AD.

The September meeting, in the event the only one prior to the hearing, prepared all involved to take on the challenge of November 29 and 30, 1990, the hearing itself, which was held at the Key Bridge Marriott Hotel in Arlington. A third day, December 1, was a recap meeting for the TF membership and selected BBS personnel, only. The plan for the hearing had been spread in the community some time before when the invitations went out to all of the 120 societies that Clutter had chosen. The obligatory announcement in the *Federal Register* had been made, as well.[66] The knowledge of the impending hearing was already causing waves in the community as C. Richard Taylor, the Alexander Agassiz Professor of Zoology and Director of the Concord Field Station of Harvard University, had already written a

[63] The historical points and some others were in an additional four pages (unnumbered) beyond the first four cited in the footnote just above but were almost surely part of the handouts of September 10; the title was the same, but not so the contents.

[64] Ibid. Emphasis added.

[65] Ibid. The capitalized emphases were in the original, except for the word that I have italicized.

[66] *Federal Register*, 55, no. 163 (August 22, 1990):34358.

letter of "concern" to Clutter in early November.[67] The concern was the seeming pending loss of physiology (below). The letter was a follow-up to a telephone conversation that he and Clutter shared yet earlier. Taylor cited a number of well-known zoologists and physiologists (Jared Diamond, George Somero, Donald Griffin, and others), all members of the National Academy of Sciences, as they asked to visit Clutter the following April or May (1991) when all of them would be in Washington, D.C. during an NAS meeting. They assumed from what they knew was to occur in late November at the hearing that *physiology as a science* might "get 'lost in the cracks' of a restructured BBS." They felt that physiology "will be at the 'cutting edge' of biological research for the next 10-20 years." For that reason, they expressed their concern. The NSF should "consider a structure that is most beneficial to the pursuit of science, rather than to the establishment of administrative domains."[68] That view had been Clutter's, as well.

Internally, individuals within BBS were also very much alert to the possibilities brought about by the expected reorganization. Patricia A. Werner, Division Director of Biotic Systems and Resources (BSR), took the opportunity to write to Mary Clutter and suggest that BSR allow David Schindel, then still a program director in the division, to spend half of his time with the Directorate for Education and Human Resources in liaison with elementary science education. Werner provided Clutter with a thick sheaf of papers in support of the idea, one of which came from the Ecological Society of America (ESA). The ESA's then embargoed draft interim report, "The Sustainable Biosphere Initiative: An Ecological Research Agenda for the Nineties," was created by a large committee that included Paul Risser and was chaired by Jane Lubchenco.[69] The report was not unlike other similar documents that were circulating in the ecological community of the time, one of which would end up in front of the Task Force at the November hearing in an odd form, as we shall see.

More feelings like those expressed by Taylor and his confreres in the community were to come even before the hearing had taken place. By November 20, Clutter was writing a memo to the members of the Task Force to tell them that BBS had received "over 69 requests from organizations to provide oral as well as written testimony."[70] That is to say that more than 50% of all of those 120 societies invited not only planned to come but also wished to convey their feelings on the subject of a reorganization. Such a notable number stands in evidence as to *just how powerful was the influence of the BBS*, and by extension NSF, to the scientific communities of the social, behavioral, and biological sciences in America. After all had been contacted by telephone, a final number of 60 societies held to their plans to testify orally; though that number dropped slightly to 54 by the close of the hearings.

[67] C. Richard Taylor to Mary E. Clutter, November 6, 1990.

[68] American Physiological Society, "APS Testifies on NSF Reorganization," *The Physiologist* (1991):17 (February).

[69] Lubchenco would later become the Administrator of the National Oceanic and Atmospheric Agency (NOAA) under President Barack Obama.

[70] Mary Clutter to BBS Task Force, November 20, 1990.

Clutter had already gathered their earlier written testimony into a "massive document" that she asked each TF member to read, or read "at least the executive summaries," prior to the hearing, then only 9 days away. Finally, Clutter assembled a set of "Instructions to Witnesses" as a necessity to "accommodate all organizations requesting time to present oral testimony." So many were they that they had been divided into some 12 panels of 3 to 6 societies each. The 5 minutes given to each witness (a society representative, generally) "must end" at the appointed times, she insisted.[71] At least 23 witnesses represented various of the social and behavioral sciences; the remainder were biological/social science societies or multidisciplinary organizations such as COSSA and AAAS.

One document of the period was a listing of the programs from all the divisions of the BBS placed into three suggested plans of reorganization. The first of them retained all of the social, behavioral, and biological sciences in a single directorate. The second split them into a "Biology and Behavioral Sciences" directorate and another for "Cognitive, Economic and Social Sciences." But it was the third that was the most curious: it split the BBS sciences in the same manner as had the second plan for the non-biological sciences but included a third structure termed the National Institutes of the Environment (NIE). The "Biology and Behavioral Sciences" directorate would have included *only* behavior, molecular and cellular biology, and instrumentation and resources: all of the organismal and ecological level biosciences would have been placed in a proposed NIE structure. Biology, in the broadest understanding, would have been deeply separated into two directorates! The NIE structure would have included a number of "Institutes," to wit: Ecological Systems, Biological Adaptations, Conservation Management, Biotic Resources, Sustainable Development, and some number of unnamed Intramural Centers. However conceived, the NIE, as a *division* (essentially) of the new biological directorate for NSF, must have been very odd to biologist's eyes. It is very possible that Richard Taylor and his biologist/NAS colleagues saw the handout earlier or learned of its contents, and it was that which led to their "concern" as physiology, *per se*, was not obviously present in *any* of the lists of programs/divisions/directorates.[72] The NIE has its own history and little more is said of it beyond the just given footnote.

---

[71] Magee was even more forceful, yet humorous, on the day of the hearing: "After four minutes, [the green light] will turn yellow. When that amber light goes on, you are supposed to slow down, not go through the intersection. And when the red light goes on, you are supposed to stop. If you don't stop, a beep will come on. After that, I believe there is a trap door underneath the witnesses' table." From Magee's introduction in National Science Foundation, "Biological, Behavioral, and Social Sciences Task Force Looking to the twenty-first Century [Minutes]", Volume I (of two), pg. 6–7, November 29, 1990, verbatim Minutes prepared by Diversified Reporting Services, Inc., Washington, D.C. Volume II contains the verbatim testimony given on the second day of the hearing, November 30.

[72] The National Institutes of the Environment had not been an incautious creation of the TF, or even of BBS, for the purposes of discussion at the hearing. Rather, the NIE had been an idea that was in currency at the time. In an eight-page, professionally designed circular, leading ecologists Stephen P. Hubbell (then at Princeton; at this writing at UC, Los Angeles), Henry F. Howe (University of

5.4 Heard at the Hearing

The public hearing occupied 2 days, November 29–30. The next day, December 1, a Saturday, the Task Force alone considered what had been heard during the previous 2 days and how to deal with what was learned. Due to the "Sunshine Act," the Saturday meeting had to be open, but apparently the TF and the NSF "resource" persons met with no others present on that day.[73] The summary minutes that was issued after the hearing testimony and the interpretation by the TF was a document of two score pages that provided a synopsis of every one of the panel testimonies and indicated how each society came down on the relevant issues for the reorganization.[74] Some terms were used in multiple or various ways by the several communities, so the Minutes provided a definition of "social and behavioral" sciences by having them, "for the sake of brevity[, include] the social, psychological, cognitive, and economic sciences and their respective subdisciplines." The place of the cognitive sciences, then, was clearly in the social and behavioral camp. The neurosciences, on the other hand, were not defined as a part of some greater heading, social *or* biological. For the panels, discussion was had of funding, of single investigator awards and big science versus little science, and of peer review, increased communication among the NSF directorates, databases, instrumentation, duplicate submission of proposals (especially to NSF and NIH), and other matters—all issues by now familiar to readers of this present study. These were discussion points about which Mary Clutter hoped to hear. However, the major topic of dialog was the one thing most expected: splitting the social and behavioral sciences away from biology. *For many of the social sciences, economics, political science, and much of sociology, "biology is irrelevant," the second panel had concluded.*[75]

The third panel included five organizations: American Speech-Language-Hearing Association, Society for Complex Carbohydrates, Society for Research on Biological Rhythms, American Anthropological Association, and the Society for American Archaeology. Chair of the TF overall and author of the summary minutes was Paul Magee, Dean of the College of Biological Sciences, University of

---

Illinois, Chicago), and David E. Blockstein, Director of the "Committee for the NIE" in Washington, D.C., stated the need for and purposes of an NIE. It did not come to fruition, exactly, but in that same year the Committee, the CNIE, went on to establish itself as an organization that, in 2000, changed its name to the National Council for Science and the Environment (NCSE). The relationship among the NSF, NSB, and the CNIE of the 1990s is of considerable importance but beyond the scope of this study. For a fuller view of the NIE, see Henry F. Howe, "Report: Towards the National Institutes for the Environment," *Global Environmental Change* (1990):71–74.

[73] Mary Clutter had sent a memo to the TF members, and, among other purposes of her communiqué, she pointed out that the December 1 conclave could *not* exclude the public. Mary Clutter Memo to [the Task Force], October 5, 1990. Nonetheless, no member of the public seems to have come.

[74] "Summary Minutes of the BBS Task Force Looking to the twenty-first Century; November 29–December 1, 1990; Rosslyn, VA," 35-page typescript. The Blockstein quotation, see earlier footnote, is on page 12.

[75] Ibid., pg. 2.

Minnesota, St. Paul. Panel 3 was notable because, he said, it was likely the closest to the "cleavage plane" where a collection of disciplinary societies would likely come down unambiguously on one side or the other on separation. In fact, all five were *opposed* to separation! Indeed, later, the anthropology community, doubly represented on Panel 3, would be, along with psychology, the most opposed of any disciplinary area to the split, both before and once it had been decided upon.

The fourth panel brought something of a surprise. All five societies were oriented toward social and economic sciences: four of those, all economics organizations, were in favor of the split, as expected, but the fifth society in the panel remained undecided. Most surprisingly, it was the powerful, international, and broadly representative Social Science Research Council.[76]

Possibly no less strange was the outcome for Panel 5, all biological societies. Two were opposed to the split, and one was undecided. That organization was the important American Society for Zoologists—later to become the Society for Integrative and Comparative Biology—already discussed earlier in this history and to be considered again later. Two others did not even address the matter.

The last panel of the first day of the hearing included four social science societies, three of which favored the split, but one, the American Statistical Association (ASA) would not take a position. The ASA's two representatives on the scene, however, were personally in favor of separation.

Six more panels were convened on the second day, and responses were as mixed as had been seen the previous day. The History of Science Society (Panel 10) took a seemingly apathetic position. It "expressed no opinion," but it "made clear its preference to remain with the biological sciences in the event of a split (not because of any greater affinity with that group, but simply because there is no reason to move elsewhere)"![77]

By later Friday evening, the 54 societies' representatives had dispersed. But on the morning of the third day, the 20 members of the Task Force and a number of BBS personnel took up *the* most notable issue: the separation. Magee indicated that at the TF dinner the evening before, he had "observed that a near-consensus [amongst the members of the TF] seemed to have been reached already in favor of a new directorate." How that was concluded, given what the TF members heard based upon all the rifts across virtually all the panels, is quite opaque indeed! The "consensus" may have been present for the TF members, but it was hardly the case for the actual societies' votes, nor yet for some number of NSF personnel.[78] And even if Clutter's various statements made over then recent times had indicated her strong leaning *not* to split BBS, it seemed that the Task Force felt that was *exactly*

---

[76] For more information, see http://www.ssrc.org/.

[77] "Summary Minutes,' pg. 11.

[78] Still, in an article in the APS *Observer* ("NSF Directorate Now Proposed in Senate," (1991):1; 16, [May]), the author stated that "[t]he task force's recommendation [for splitting BBS] is believed to have come as a surprise to NSF, and many observers believe that the agency will continue to resist a separate directorate." How anyone could have been "surprise[d]" at that point in time remains a mystery.

what was called for. In fact, months later (April 1991), the presidents of the American Anthropological Association and of the Society for American Archaeology wrote a strongly worded letter to the editor of *Science* noting several salient points:

> We would like to correct the impression [in a newsbrief in an earlier issue of *Science*] that there [was] unanimity among the organizations testifying on the issue of a separate…directorate for the social and behavioral sciences…Not all organizations 'disagreed' with the doubts expressed by Mary Clutter and others within NSF.[79]

They went on to point out that of the 54 organizations that testified, 37 "voiced definite opinions." Of that group, 20 were in favor of a new directorate, but 17 were opposed! Hardly "unanimity." The two presidents noted that their respective organizations were opposed, as were all those on Panel 3 (above), they said. They had highly placed allies, too. E.A. Hammel chaired the Section of Anthropology at the National Academy of Sciences and wrote to Mary Clutter earlier that same month stating that the "members [of the Section] were unanimous in their opposition to changes that would violate the intellectual integrity of the discipline."[80] He employed even stronger words when he said that the "members…view most unfavorably any reclassification that would divide or relocate the discipline within the Foundation." Reason for pause? Maybe.

Why had this "consensus" come about in the Task Force? Upon Magee's polling the morning of December 1st, 19 of 20 did in fact agree on the idea of a split; though some were only "mildly" in favor. The one clear holdout, however, remained adamant in his opposition: Harold Morowitz, Robinson Professor of Biology and Natural Philosophy, of NSF's neighboring George Mason University. His argument, according to Magee, was that the split would undermine the intellectual trend "toward convergence of the disciplines" then part of BBS, and it was merely a concession "to political and managerial considerations."[81] "[T]he disciplines" must have referred to the social and behavioral sciences, it is assumed, because it is not abundantly obvious that a "trend toward convergence" was ongoing in BBS overall; certainly not in any great sense in regard to the social and the biological sciences "converg[ing]." Efforts at interdisciplinarity had been pursued for some time, of course. The question remains: Why did Magee state that a "consensus" seemed to have been drawn? Circumstantial evidence and the general feeling of so many groups and so many individual scientists of so many stripes—and congressional pressure—it seems, all combined to lead the 19 members of the Task Force to propose separation, even if the data from the previous 2 days did not, *sans* considerable ambiguity, support the concept. In concluding as they did, the Task Force must have known that they were clearly going against Mary Clutter's wishes. Clutter was very obviously in a small minority even after decades of the social and behavioral scientists chaffing under "protective coloration."

---

[79] Jane Buikstra and Prudence Rice to the Editor, *Science* (1991):April 19 issue.

[80] E.A. Hammel to Mary Clutter, April 30, 1991. He copied Jane Buikstra, which may have suggested that she (and Rice?) had written to him seeking his intercession in the matter.

[81] Ibid., pg. 13.

Morowitz, a very plain-speaking person, may have been the only other member of the "stunned" camp of Mary Clutter's when he made it quite clear to her what he thought was afoot at the hearings:

> First, it came as a surprise to me when the social and behavioral scientists appeared with such a well defined agenda to form a separate directorship. I had assumed our charge was to take a broad look at the [directorate] and then make recommendations. The intensity of *preformed secessionalist* views from task force members and witnesses really precluded carrying out our charge...The almost unanimous decision to go with a new directorship came from the relentless campaign of the social and behavior scientists and the [additional] view of the biologist[s who, in essence, said]: 'To hell with them, if they want out so badly, let them go.'[82]

He went on to say, quoting the philosopher Jose Ortega y Gasset, that he had "rarely seen such an example of 'the barbarism of specialization,' [as e]veryone showed up representing a guild."[83]

There was one group of disciplines, and it is hardly surprising that was causing challenges for the Task Force if it were to recommend the split: that was cognitive science, animal behavior, and, "to some extent neuroscience." No "breaking of the continuity" or "parceling out" of those areas, which were covered from the molecular to the cognitive in the BBS, would be countenanced by the relevant community, the "mild" members, of the 19, held. Much has been made in this history of these very areas that came back to cause concern within the ranks of those TF members less strongly committed. Controversial areas lead to contentious responses. Some proposal was needed to satisfy that contingent's uneasiness. The first step along those lines was taken by Task Force member Michael Greenberg, Director of the Whitney Marine Laboratory at the University of Florida, St. Augustine. Using cell biology as a metaphor, Greenberg suggested that the split for those areas be based upon two groupings. The first one of which—experimental psychology, human cognition and perception, linguistics, and social psychology—would be placed in a new directorate of social sciences, but they would maintain close relations with a new biological sciences directorate. As in the metaphor of cell division, he saw those fields (then current BBS programs) as having "sticky ends" (of chromosomes) such that they would not easily pull apart in the later phases seen in the dividing cell and so would be stretched over the two new directorates.[84] (The term "sticky ends" became something of a catch phrase at BIO and would resurface in future times and for other purposes.) The second group, more nearly non-human areas, included

[82] Harold J. Morowitz to Mary Clutter, December 11, 1990. Emphasis added. That was one of a great many letters in the BIO Lateral Files from Morowitz, a most productive commentator.

[83] Months later, on April 29, 1991, Morowitz wrote to then newly-confirmed Director Massey reiterating that he, Morowitz, had been a "minority of one" on December 1, 1990. His letter was very sharp and spoke of the separation as a "shotgun divorce." Massey's response (May 16, 1991) was very diplomatic and non-committal, as might be expected.

[84] The term "sticky ends" specifically refers to the double helix of DNA when it is "unzipped" and replicating: its open ends "attract" nucleic acids—they are "sticky." It is not so much used in the sense that Greenberg chose, but what one might term the parametaphorical value was clear, nonetheless.

neurobiology of learning and memory, animal learning and behavior, and psychobiology. Those would be placed in the new biology directorate. Interdisciplinary programming, joint review panels, shared reviewers, and maybe even joint oversight at the AD level would be mechanisms used to maintain the closeness of the "sticky end[ed]" programs; it was suggested.

Stephen Anderson, of the Johns Hopkins University Cognitive Science Center in Baltimore, Maryland, "presented a tentative formulation of the prospective new directorate" of social sciences by outlining a somewhat detailed vision of its structure but which details were never recorded in the Minutes.[85] Discussion ensued, and some TF members felt that possibly it might be better left to NSF organizational experts to delve so deeply, but the "countervailing" argument was that it was important to convey to NSF the thinking of the Task Force. For that reason, the first two of five Task Force working groups (WGs), which had been set up prior to the hearing, were sent off to consider suggestions for a structure for each of the two potential directorates. Note that these groups had been set up *before* the hearings: one was aimed at considering what a social sciences directorate might look like and the second what a biological sciences unit might be. Why would there have been such a working group arrangement *in advance* of the hearings if there *was not already* a strong indication that the separation was a foregone conclusion? That seemed clear enough to some in the community as Warren E. Miller, Regent's Professor of Political Science at Arizona State University told Clutter a few days into 1991 that he "appreciate[ed] the independence afforded the Task Force—all the more because of the a fortiori nature of their switch in sentiment following the hearings."[86]

Once the WGs reassembled, they announced their suggested structures. That for the social sciences, WG-A's findings, were as most had already surmised they would be. There should be divisions for, respectively, social sciences, psychological sciences, and economic sciences and a new division for measurement methods and resources. They also suggested, but then dropped, the idea for a division of cognitive sciences.[87] The group did not get into deeper layers of structure. That working group, as well as all the others, was made up of not only TF members but also BBS division directors most likely to be of help in the deliberations of a given group: the directors were dubbed "NSF Resource[s]." For instance, Roberta Miller (SES), James Brown (DMB), and Nathaniel Pitts (BNS), all mentioned earlier in this history, were aiding group A as "resources."

The working group for the biological sciences, WG-B, on the other hand, saw their task somewhat differently as they felt that biology at BBS, broadly speaking,

---

[85] "Summary Minutes of the BBS Task Force Looking to the twenty-first century"; footnote further above.

[86] Warren E. Miller to Mary E. Clutter, January 7, 1991. It was a four-page missive that suggested issues of strategic planning for the new social science's directorate.

[87] That area would later gain standing in the final version of the new directorate as the Division of Behavioral and Cognitive Sciences, though; see Fig. 5.1. The figure was created by Courtney in 1991 as a tool to help him keep track of all that was ongoing for the 2-year period. The final, official organization is seen in Fig. 5.2.

required a *strong reassessment* in a number of areas. That group was aided by the "resources" of Bruce Umminger (DCB), John Wooley (DIR), and Patricia Werner (BSR). Their report noted that the apparently dissolving BBS's Division of Behavioral and Neural Sciences (BNS) should be taken apart such that those mostly human-oriented features would move into the suggested social sciences directorate. Behavior, *per se*, would remain with biology, but the "sticky ends" approach would be used as a device to cover certain programmatic areas that bordered along the human/non-human divide. They also suggested a new program in *behavioral ecology* to be placed somewhere in a new biological sciences directorate.

Besides the BNS area, it had long been felt that *organismal biology* had not been well served in what was fast becoming the disappearing BBS. Umminger had suggested that notion to Clutter (above), but he was hardly alone either in time or sentiment within BBS. For that reason, the WG-B discussed *organismal and integrative biology* as a program.[88] They envisioned that:

> [s]uch a program would facilitate work on both animals and plants [allowing] further development of important biological concepts such as adaptation, integration, homeostasis, optimization, and segmentation.

Therefore, they suggested a finer division of programmatic features that should be included in a new segment of an apparently growing directorate for biology, to wit: physiological ecology; comparative physiology; integrative aspects of development; organismal endocrinology; chronobiology; biomechanics; and, symbiosis.[89] Tellingly, Marvalee Wake of UC, Berkeley, a founding figure of integrative biology, was not only on the Task Force but was deeply involved in the discussion by WG-B of better recognizing organismal and integrative biology: she was a notable player, then, in the new directions that the eventual BIO would take as the reorganization progressed and as BIO became the future for the biological sciences at the Foundation. We shall meet her again.

The WG-B members also considered other aspects of a new directorate that should include long-term studies (e.g., LTER); global and regional studies (an interface between BSR programs and the geosciences directorate); conservation biology (for biodiversity efforts); a "refocused DIR" to bring in more on methodology such as "experimental statistical techniques"; and establishment of a program for studies in science, technology, and society. That latter field, widely known as STS, once the split had been accomplished, became an "Office" that would be set up in the new *social* sciences' directorate. STS would *not* be lodged within biology, contrary to the somewhat uncommitted desires expressed earlier by the History and Philosophy of Science program officers—and the History of Science Society. It was NSF personnel who determined not to place STS with biology. The Office is shown in Fig. 5.1 (lower right quadrant), but, as it was not a division, does not appear in

---

[88] A header in ibid. on page 28, Attachment F (biological working group report) reads "Organismal and Integrative *Behavior*" (emphasis added) but is surely a mistake as no evidence within that area of text indicates that behavior was the intended word; rather it seems it should have been Biology.

[89] Ibid. The terms are as given in documents, though rearranged here for convenience.

Fig. 5.2. Further, the WG-B spoke for the need for the environmental sciences ("perhaps [including] systematics") to remain with a BSR division; efforts to address the continuing disappearance of taxonomists in the American scientific labor pool (the efforts at reversing the trend would become a notable undertaking in years to come); and making sure that at least one permanent and one rotating program officer was present in every program in the directorate. The NIE was not given any consideration (see earlier mention).

Once working groups A and B had made their reports, Magee reassembled the participants—TF members and NSF "resources"—into other permutations for three earlier-designated working groups (C, D, and E) to consider large topical areas. Infrastructure—education, equipment, and resources—would be examined by group C, while D was to deal with organization to facilitate scientific change. Group E would define a "Unique Role for the Directorate" (implying both, if two were to materialize). Though not its chair, Magee himself made the report for WG-C. The D group was headed by Wake and the E group by Greenberg. Those reports were placed as attachments to the summary minutes and were envisioned to constitute an "Introduction to the [final] report of the Task Force," Magee said.[90] As the end of the third and last day was approaching, the TF members were sent back to their respective home institutions to continue "fleshing out the issues they had identified" and to report back by January 7, 1991.

The attachment which group D provided considered such points as structural issues concerning the *program level* (details the social scientists never undertook): more interactions among program directors to avoid matters "falling between the cracks"; shorter proposals (now a long established requirement at NSF[91]); increasing the number of rotators; a pool of reserve panelists for peak periods; subpanels convened by telephone for cross-disciplinary proposals; careful selection of all players in the review process for "breadth of expertise"; and so forth. Other considerations were made at the division, directorate, and interagency levels. Cross- and interdisciplinary and barrier removal matters were discussed, as well. Working Group E, by dint of its assignment, was very broadly philosophical in its recommendations, addressing NSF's role in federal support to science and to research, *sensu lato*.

Marvalee Wake went back to Berkeley. She had been enthused with her time in the East leading to the accomplishments of the "Task Force Looking to the 21st Century." So committed must she have been that, along with her colleague Albert F. Bennett, then of the Department of Ecology and Evolutionary Biology at UC, Irvine (later Dean of the School of Biological Sciences there), they sent a three-page document to the secretary of the TF proposing the establishment of "A New Program in Organismal and Integrative Biology" for NSF.[92] Clearly, Wake wanted to see her

[90] Ibid., pg. 14.

[91] Clutter and others at the NSF had already issued a new policy on grant proposals requiring biographical sketches to list only ten publications: "NSF Stresses Publication Quality, Education with New Grant Format," *The Scientist* (1989):2.

[92] Paper copy of a Fax from Bennett and Wake to Janell Richardson of the TF; December 13, 1990.

suggestions made at the TF discussions on December 1 solidified in a formal proposal. In their summary, they said they hoped to see the program:

> study the form and function of organisms in view of their evolution and environmental interaction…[how creatures] are [evolutionarily] designed and work and how their different components integrate in organismal function.

The hearing was having the effect of generating what may have been its first programmatic proposal and the first drive to use the term and concept of "integrative biology" at NSF, at least in program nomenclature. The term would be incorporated into the new BIO and that at the highest level: the Integrative Biology and Neuroscience division (see Fig. 5.2). Neuroscience then would also be a part of that division, but that came about only with considerable struggle.

Marvalee Wake's return to *her* institution may have been done with a light heart, but a fellow TF member from elsewhere was still having "reservations about [the separation decision] being the right conclusion" many months later. That member had been reminded of some unpublished words of the great environmentalist Aldo Leopold, who had said:

> One of the anomalies of modern ecology [applied ecology and conservation] is that it is the creation of two groups, each of which seems barely aware of the existence of the other. The one studies the human community almost as if it were a separate entity, and calls its findings sociology, economics, and history. The other studies the plant and animal community, [and] comfortably relegates the hodge-podge of politics to 'the liberal arts.' The inevitable fusion of these two lines of thought will, perhaps, constitute the outstanding advance of the present century.[93]

5.5 Contrary Voices Lost in Transition

Nothing was lost in translation as translation was not required: everyone understood the language of the hearings. But the voices of those opposed to the separation, while heard at the hearing, were not heeded. The transition to two new directorates was firmly ratified by 19 of 20 Task Force members at the poll that Pete Magee took on Friday night. But it was many months and much work from actually coming to be. Congressmen Brown and Walgren received letters from Acting Director Bernthal the previous mid-October, more than a month after their letter of concern *for* a separation, assuring them that Mary Clutter was seeing to it that the "review was deliberately inclusive of organizational change." And more than that, "very importantly," the review sought the advice of the social and behavioral sciences' communities. Bernthal was "confident" that the Task Force would be "thorough and effective." It was a very mollifying letter, if nothing else.[94] But, in the event, neither Brown nor

[93] Robert Goodman (Member of TF) to Pete Magee, April 23, 1991. Goodman attached Leopold's words, which he said in his letter to Magee had never been published. Like Leopold, Goodman was from the University of Wisconsin, Madison.

[94] Frederick M. Bernthal to Honorable George Brown, October 17, 1990. Although not found, a

Walgren nor anyone else who sought the separation needed to have worried as it would appear that such persons entering upon the hearing along with the work of the TF were predestined to recommend the split of the BBS Directorate. Almost everyone, except Mary Clutter and a very few others, that is. Who was opposed and why?

The American Anthropological Association (AAA) was the "oldest and largest" such organization in the country. They noted in their written testimony prior to the hearing that with the 1975 reorganization, the scientists they represented—archeologists, sociocultural anthropologists, linguists, and biological anthropologists—were placed in the BNS. They saw that division as a "'bridge'" between the social and biological sciences. They felt that the BNS, as such, was:

> more beneficial to the discipline as a whole than either moving anthropology to a new social science directorate, or splitting the unity of anthropology by placing socio-cultural anthropology in one directorate and archeology and biological anthropology in another.[95]

Over the intervening years since 1975, anthropology, in its most inclusive definition, had profited by "riding the coattails" of biology and, thus, had remained safely in the camp of funded disciplines. Apparently, the communities that the AAA represented saw themselves as very much of an interdisciplinary one, and they were not willing to see their structure rent by parts going to two different directorates: they hoped to see the BNS continue on as one unit. From the moment that the two working groups set off to discuss matters that Saturday morning, the unity of BNS was not likely to survive. Indeed, the spokesperson for AAA, Kenneth Weiss of Pennsylvania State University, said "tightly compartmentalized research by discipline severely impedes cooperative interdisciplinary ventures" that were needed to deal with "complex problems confronting human societies."[96] If the split occurred, such "tightly compartmentalized research" would be the norm, Weiss said, and that would destroy the value of interdisciplinary interaction. While NSF was keenly on the side of interdisciplinarity and had worked ceaselessly to bring it about, the salvation of it for AAA and its component fields would have to be sacrificed, as Morowitz would have had it, "to political and managerial considerations." Indeed, that was to be the case as the desires of virtually all parties involved, save AAA and a few others, could not halt what was not only becoming inevitable but desirable in most eyes.

Another source of opposition to separation was former Director Erich Bloch. But *former* was the key word in his case as Bernthal, and then Massey, most especially, would be the ones who would have to deal with the issue. Bloch "did not look fondly on the creation of new entities within NSF," stated an unnamed writer in *Science & Government Report*. "[N]or did he accept as accurate the claims of fiscal

similar letter is presumed to have gone to Representative Doug Walgren.

[95] "Testimony from Kenneth M. Weiss (Penn State), on Behalf of the American Anthropological Association for the BBS Task Force Looking into [*sic*] the Twenty First Century; Submitted November 15, 1990." Quotation from the Executive Summary.

[96] Ibid.

neglect in the social and behavioral sciences," the article went on to say.[97] The writer argued that Bloch had "insist[ed]" that those sciences had "made a good comeback from Reagan-era cuts," but H.R. 5543 of Brown and Walgren "reflects the contention that they have fared poorly." They needed "high visibility" that would come with greater financial support. One of the most arresting features of both the oral and written testimony given by the social and behavioral sciences' societies and representatives throughout the entire hearing was the pervasive claim that, as the Smith committee had said, a "radical" increase in funding was a necessity. Such an issue had been a worry of Clutter's, but, with separation, some other assistant directors would have that particular anxiety on her or his head; Clutter, the biologist, would have only funding for biology to concern her. And biology was in the throes of a very positive revolution at the time and not particularly in need of "radical" help in the fiscal arena.

There were other lost voices. Serge Lang, Professor of Mathematics at Yale University, wrote to Representatives Brown and Walgren in early September saying that he had just discovered their H.R. 5543 and "object[ed] very strongly" to it.[98] Lang was from a rather radical camp that was long known to exist (earlier chapters; see also last sentence of this paragraph): the "so-called social 'sciences' merely make up what they call theoretical models having no basis in fact…and pass off their political opinions as science, illegitimately." His second argument was that the fiscal/intellectual review by "other" scientists of the social scientists would not come to pass if H.R. 5543 did. Of course, that did not happen, but Lang feared that "over the last two decades," such review was in the very "hands of precisely those who transgress the scientific empirical approach." Lang may have been extreme, but he may have represented some number of fellow intellectuals and others who would

[97] "Social, Behavioral Sciences Seek Upgrade at NSF," *Science & Government Report* (1990):6 (October 1). That was a *claim* made in the newsletter, but actual proof for that can be seen in a letter to Robert A. Roe, Chairman of the House Committee on Science, Space and Technology from Bloch of March 16, 1990, in which Bloch stated, among other things, that he had been "disappoint[ed]" with the NRC's 1988 report (see earlier chapters), "The Behavioral and Social Sciences: Achievements and Opportunities," the topic of the exchange. The Report failed, as Bloch saw it, in helping to "[set] priorities," an issue closely tied to organizational structure, of course. In regard the financial issue, Bloch told Roe that the report concluded that "more resources are needed" for the social sciences, "but this is a concern across all scientific disciplines." Bloch went on to say (pg. 2 of the letter), that "I believe a streamlined organization with fewer, not more, standalone entities encourages efficiencies as well as maintains short communication lines": Bloch, the industrialist.

[98] Serge Lang to George Brown and Doug Walgren, September 2, 1990. On November 4, 1990, Lang wrote to Bernthal including a copy of the Brown/Walgren letter. Not wanting to be repetitious of the contents of that letter, Lang chose to "add a few more remarks" to his initial two-point complaint: social scientists were "parasitic on the natural science community" at NSF, was one; he did not have "quarrels" about anthropology, but political "scientists" were unacceptable; and, the peer review system had broken down with some "peers" having too much power in both the NAS and NSF. He saw as critical the issue that the review of the social scientists *by* natural scientists would not occur should H.R. 5543 come to pass. He copied his letter to a number of other senior federal science leaders, too.

have supported his view.[99] He was not concerned so much with the separation, of course, but instead held the more Proxmirian views of the recent past: eliminate social sciences from the NSF *altogether*.

The American Society of Zoologists, the ASZ, as we have seen, had some mixed views on the subject of the separation. That was so as some members were animal behaviorists and neurobiologists, and the split could have negative effects upon their research, funding, features of interdisciplinarity, etc. That was pointed out by Lynn M. Riddiford of the University of Washington, Seattle, who had testified for the ASZ.[100] Riddiford concluded her testimony by saying that the neurobiologists and animal behaviorists were "adamant that they should be included in the biology directorate" if the split did come about. In the event, they would be.

The question about the placement of the neurosciences in any new structure was of interest to the Society for Neuroscience as their two representatives, Thomas Carew and Michael Davis, both of Yale, noted that they were "deeply concerned" about funding issues.[101] They had both been trained in psychology departments and that had a considerable effect upon their thinking, they stated. They pointed out that (contrary to what Bloch had said) support in constant dollars for the social and behavioral sciences had fallen by 38% in the previous decade, while the total NSF budget had grown by 27%. There was corroborating evidence of that, and it was widely published in the newsletters of a number of scientific societies of the period. The two neuroscientists concluded that it would be "crucial" for NSF to increase the "status, voice and funding level" for the social and behavioral sciences.[102] They supported a new directorate for those sciences and proposed a structure for it, mentioning H.R. 5543.

While, for *this* part of BBS/NSF existence, the resources available to the historian are truly vast, the foregoing provides an overview of, primarily, the state of affairs at BBS about 1990.[103] But the opinion of the Task Force to split the directorate was just the beginning of a year that truly could be called, as Guy Stever might have said, one of "tumult." Splitting a directorate at the National Science Foundation was no simple task.

---

[99] In a copy of a letter from Henry L. Bretton to Lang (n.d., but likely fall of 1990), Bretton said that, "I have profound reservations regarding claims of 'mathematical rigor' in political analysis." Bretton was the author of books on political analysis and the well-known work, *The Power of Money: A Political-Economic Analysis with Special Emphasis on the American Economic System* (Albany, NY: State University of New York Press, 1980).

[100] Lynn M. Riddiford to Mary Clutter, November 15, 1990, with attached "preliminary draft" testimony.

[101] "Testimony of Dr. Thomas Carew and Dr. Michael Davis [both of Yale University] to the Biological, Behavioral and Social Sciences Task Force, November 29, 1990," representing "the governing board of the Society for Neuroscience."

[102] Ibid. Emphasis in original.

[103] The history of the biological communities' responses to the proposed split is extensive, and it is recommended that a close consideration of these materials, beyond that feasible for the present study, be undertaken to better understand the *zeitgeist* of biology at that period.

5.6 A Diary of 1991

The community was well aware of the recommendation of the Task Force to split BBS by January of 1991, once academics returned to their respective campuses following the holiday break. The American Psychological Association (APA) was quick to announce the news to its membership that month. Their representative said:

> This is an historic event for the behavioral and social sciences…[f]or these disciplines, the task force's proposal indicates a victory in their long struggle for recognition at the federal level.[104]

The "proposal" was still just that and would not become fact until much later in 1991. Still, the APA was very much pleased by it, yet feeling the necessity to say that both the BBS Directorate and its Advisory Committee were "biased" toward biology. Indeed, the APA article said that the BBS AC had eight biologists among its members but only one social and one behavioral scientist. The author noted that while the director's position at NSF was vacant (Bernthal was acting), President Bush had appointed Walter Massey. The latter would take the helm in February 1991, and it would be he who would have to formalize the "proposal"—*if* he chose to do so. The article ended by quoting Richard Louttit, director of BNS, when he said that "[y]ou have to hand it to the behavioral and social sciences…[t]hey certainly were organized."[105]

Announcements and commentary were heard from many scientific societies over the next several months as their respective newsletters arrived on the desks of their members. Some articles were short; others quite extensive. In general, the biology organizations made mostly brief announcements, but the social and behavioral scientists had much to say, almost universally in favor of the notion of a new directorate for them. Wallace LeStourgeon of Vanderbilt University, Tennessee, was typical of biologists when he represented the American Society for Cell Biology. He stressed what Mary Clutter was seeking to hear: views of societies on issues of *science*—that which she had told the Task Force to focus upon. LeStourgeon spoke for the directorate "to support more individual investigators" in basic research and to raise the level of science education.[106] The expected split was merely stated as news. On the other hand, an article headlined page one of the *Observer*, the newsletter of the American Psychological Society (APS), where the first sentence read: "Christmas came early for research psychologists…with [an NSF] task force playing the role of Santa."[107] APS Board Member Nancy Cantor was the only psychologist on the BBS Task Force and she helped it understand her field. Indeed, "the task

---

[104] American Psychological Association, "Task Force Recommends New Directorate at NSF," *Science Agenda* (1991):1–2.

[105] Ibid., pg. 2.

[106] The American Society for Cell Biology, "ASCB Testifies on Reorganization of BBS Directorate," *Public Policy Briefing* (1991):4.

[107] American Psychological Society, "NSF Panel Recommends Behavioral Directorate," *Observer* (1991):1; 10.

force has really moved beyond disciplinary turf issues," the APS article quoted Cantor as having said.[108]

On the other hand, AAA's Executive Director, Eugene L. Sterud, wrote to Pete Magee in February reiterating the anthropologist's stand in "continu[ing] to support maintaining the existing structure" of BBS. If the restructuring was to become effective, though, the AAA wished to be placed in what would finally be termed SBE (Fig. 5.2; lower right corner).[109] "Under no circumstances, would the AAA wish to be split in order to accommodate the diversity of disciplinary activity," Sterud opined. He went on to say that "[g]iven our somewhat unusual disciplinary interests, it is truly a shame that anthropology was not directly represented on the Task Force." That particular loss was not mentioned by Kathie Olsen when she noted the lack of a neuroscientist on the originally conceived TF.[110] But for the social sciences broadly, Allan Bromley, Bush's Science Advisor and head of the OSTP, would appoint a specialist in the area in the person of Pierre Parrole, a China scholar and former US science counselor in Beijing during the late 1980s.[111] Such an appointment accorded both wider standing and greater visibility to the social and behavioral sciences. The announcement article in *Science* concerning Parrole went on to say that the social sciences "may be getting their very own directorate."[112] But citing Mary Clutter's feelings, the article quoted her and:

> others hav[ing] expressed doubts about such a reorganization, saying that it could operate to the detriment of the behavioral sciences by further isolating them from their biological underpinnings…but witnesses from a range of professional societies…disagreed.[113]

It was further noted that while Bloch had, in the past, "firmly rejected" a new directorate, Massey was likely to be more "sympathet[ic]." Massey went before the Senate Labor and Human Resources Committee, chaired by Senator Kennedy, on February 7. He was confirmed by the Senate on February 28 and sworn in on March 4, 1991.

By March 9–10, the social science working group of the TF had met in San Francisco to create its interim report but was behind schedule and "had nothing in hand," said an article in the COSSA newsletter. Clutter appended an indignant note to her copy of the article averring that COSSA Executive Director, Howard J. Silver, who had been at the meeting, "did not listen" to information from the January meeting: the group was *not* "behind schedule."[114] Clutter was under pressure at the time

---

[108] Ibid., pg. 10.

[109] For a short period, some newsletters were using the term SEPS—social, economic, and psychological sciences directorate. Pete Magee had used the term, too, but it was soon discarded. See the COSSA footnote below.

[110] Footnote 35 above.

[111] *Science*, "Social Scientists Get a Man at OSTP…" (1991):742.

[112] *Science*, "…and an NSF Directorate?" (1991):742–743.

[113] Ibid., pg. 743.

[114] Ibid., "With Report Unfinished, BBS Task Force Looks to Future," (1991):4 (March 18). There was a Post-It© note on the page indicating Clutter's disagreement with "Howard" [Silver's] words

as the Task Force's January 13–14 meeting had led to discussions of revising their interim report such that findings would *not* lead to an expensive separation of the sciences at BBS. This was because of fear that the upcoming budget of President Bush would likely be austere. Indeed, Clutter asked the TF to make "short-term solutions" that would not include the split; her estimate of when separation could occur was not "before 1993."[115] The federal budget was such that Clutter need not have worried, as the funds were present for the creation of the two new directorates.[116] Given her opposition to the split, it may be that she hoped to stave off the separation by her suggestion to find "short-term solutions."

It was not until July 22 that the March 9–10 meeting notes from San Francisco became available. The BIO group (originally Working Group B) was working on its version of the final report chapter, but the "SEPS" directorate (known by that title only a short while and later to be SBE) group was not, as it did not have a draft in hand. Contrary to what Clutter said in her indignant note, Howard Silver seems to have been correct: the social and behavioral scientists were not prepared. Discussions in San Francisco involved the federal role in science, K–12 science education, science journalism, STCs, and many more topics, in the BIO session. One of those topics was based upon one of two "major themes" of the meeting; that the "biological sciences are moving away from the era of reductionism…to integration."[117] (The present study will refine that vision for the history of biology.) The second theme was concerned with the fact that integration would bring "people and information" together because "systems understanding cannot be integrated in isolation." The BIO group was to write its version of the final report with the social sciences to do the same; the meeting content of the latter group had been rather more generalized since there was no final chapter draft text to discuss. Clutter explained to all members of the Task Force that she would brief the BBS AC, BBS division directors, the NSB, and Presidential Science Advisor Bromley about the results of the March meeting.

Upon Clutter's return from the West Coast to NSF, she found that ongoing discussions among many of the program areas in BBS were aimed at what NSF, broadly, might look like in the year 2000.[118] They were, after all, "looking to the twenty-first century." The report that came out of the BNS division's POs, for

---

concerning "nothing in hand."

[115] American Psychological Association, "NSF Task Force Prepares Report," *Psychological Science Agenda* (1991):2–3 (March/April).

[116] See http://www.usgovernmentspending.com/downchart_gs.php?year=1990_2010&view=1&expand=&units=p&fy=fy11&chart=F0-fed&bar=0&stack=1&size=m&title=&state=US&color=c&local=s.

[117] Courtland S. Lewis, "Summary Minutes of the BBS Task Force Looking to the twenty-first Century; March 9–10, 1991, San Francisco, California, Draft 1, July 22, 1991, Prepared for the [BBS]." Quotation on unnumbered page 5.

[118] Richard Louttit et al., "NSF in the Year 2000," March 27, 1991, a four-page report prepared for Mary Clutter's use in an Executive Council (of the NSF) retreat. Transmittal memo attached; dated March 28, 1991.

instance, assumed increases in neither funding nor personnel: a wise baseline from which to work, except for what it led to. The BNS, under Richard Louttit, produced a report that considered issues common to many areas of the Foundation such as duplicate proposals, numbers of POs, workloads, and other matters, but which text suggested that the old BBS would remain *as it was*! At least among those in the BNS, or possibly only Louttit himself, there seems to have been no firm belief that BBS would be broken up with BNS, conspicuously, suffering among the greatest restructuring of all sections of the directorate (see Fig. 5.1). Such major reorganization seemed imminent to most, but it was not assured, however, as Massey remained uncommitted upon his arrival at NSF.

As Clutter and many on the Task Force were working hard to meet the April 1 deadline set earlier for the final report, it became obvious that such was not possible: the social and behavioral scientists were, indeed, "behind schedule." In fact, even the Interim Report, which had been set for January 31, was never issued. Accordingly, for the final report, a new set of "milestones" was promulgated and sent out by letter to certain members of the TF in the last days of March.[119] The new completion date was aimed at "mid-June." The scientific community, however, was becoming anxious to see the document, as Magee had been receiving requests for some time.[120] And scientists were not the only ones keen to see the report: Representative Rick Boucher (D-VA) had asked Director Massey for a copy of it.[121] Massey had not yet made a decision on whether he would support a new directorate when he testified before a congressional hearing of the CST, chaired by friend of science George Brown.[122] Massey told the Committee that he had "an open mind" on the matter and had "no preconceived notion" but was "willing to hear the arguments."[123] Massey was found by some to be a person with a "softer approach" to leading the NSF, especially compared to his predecessor Erich Bloch. The question had been bandied about whether Massey would follow in Bloch's footsteps:

> Most hoped he wouldn't. The voltaic Bloch, [scientists] felt, tramped with hobnailed boots through the manicured gardens of academic science…Although Bloch did the right thing[s] politically, he succeeded in alienating much of the basic science community.[124]

---

[119] Mary E. Clutter to Joan Huber (TF Member), March 28, 1991. Only five of the 20 members received the same letter. Why is not definitively known, but internal evidence suggests they may have been an executive committee in charge of the completion of the final report.

[120] Gerald Holton to Paul Magee, February 6, 1991. A Post-It© note from Magee attached to the letter directed that a final copy of the report be sent once available.

[121] "BBS Task Force Report Delayed, Massey Tells Science Committee," *COSSA Washington Update* (1991):1 (April 15).

[122] Brown became chair of the Committee in January 1991.

[123] Quotations as given in the *Update* article (second footnote above. However, they were derived from the HST hearing transcript, an ephemeral three pages which were in an unaccessioned group of documents at NSF. When Massey ended his comments (quoted in the main text above), he indicated with regard to pros and cons that "we will evaluate them." Boucher returned with: "That makes two of us." See transcript lines 1026 through 1048.

[124] Wil Lepkowski, "Walter Massey Takes Over Helm of [NSF]," *Chemical & Engineering News* (1991) (April 22).

By May 1, some 70 letters had been received in Massey's office in favor of the separation. A cover commentary concerning those letters, apparently written by a staffer and aimed for the director, attached a "sample [typical] letter," and offered the assessment that *all* of them had been sent by members of the American Psychological Association.[125] The staffer found "stock" phrasing in every one of them. The most notable wording spoke of a decade of "lost ground" with the social and behavioral sciences being a part of BBS and not an independent directorate. That arrangement had actually been ongoing for over 15 years by then, of course. In addition, a brief letter from David L. Featherman, President of the influential Social Science Research Council, to Massey included a copy of his "Special Report" published in the newsletter of the Council. The editorial spoke of "drastically alter[ing] the conduct of social science in the next decade."[126] In fact, "[w]ere NSF to become more narrowly structured along disciplinary lines, it could send a message to scholars that [the Foundation] is not prepared to adapt its structure to changing science," but rather the opposite. Mary Clutter responded for Massey saying that she was sympathetic and that the draft final report was due shortly. But she did add a comment that might have indicated a shift in her feelings about separation:

> While I feel positive about the importance accorded interdisciplinary research, I can not see how any NSF organization alone can lessen the *balkanization already extant* in the social and behavioral sciences. These changes have to come from the leadership of the involved scientific community.[127]

With the final report delayed and the interim report never having materialized, the Senate became agitated as well as had the House, but the Senate's unease had come about with a particular scientific society's instigation. While Brown and Walgren had introduced H.R.5543 in August of 1990, Senator John F. Kerry (D-MA) did much the same thing in the upper house with S.1031 of May 9, 1991, both of which called for a separate directorate for the social and behavioral sciences.[128] Kerry's came at a very crucial time in the discussions, however. In the case of the Senate bill, the American Psychological Society (APS) "worked closely with

---

[125] Dorothy Fragaszy (Associate Professor of Psychology, The University of Georgia, Athens) to Walter Massey, April 16, 1991. This typical letter was preceded by an unsigned, single page assessment of the group of such letters, which memo author is presumed to have been an O/D (Office of the Director) staffer.

[126] David L. Featherman to Walter Massey, April 19, 1991. Attachment: Featherman, "New Directorate: A Cautionary View," *Items* [of the SSRC] (1991) (March). These two documents were under cover of a letter to Featherman from Clutter; see following footnote.

[127] Mary E. Clutter to David L. Featherman, May 8, 1991. Emphasis added. See also footnote above.

[128] See: http://thomas.loc.gov/cgi-bin/query/z?c102:S.1031.IS:. An email was sent to Mary Clutter, and others, by John Chester (NSF Office of the General Counsel) on May 13, 1991 and contained the words of Senator Kerry as he introduced his Bill. One statement in the email was underlined by hand by an unknown person (Chester?) where Kerry's words read: "Unfortunately, NSF has been slow in publishing the task forces' [*sic*] report, and has taken absolutely no steps toward implementing their recommendation." Chester had titled the email: "The 'take the BS out of BBS bill.'"

Senator Kerry in the introduction of the legislation."[129] That may have been with Alan G. Kraut, Executive Director of the APS, as he had not only represented the APS at the Task Force's hearing the previous December but also gave testimony concerning the FY 1992 appropriations (and, again, called for a separate directorate) before the subcommittee that had charge of NSF's budget. That was on May 2, 1991, just 1 week before Kerry introduced S.1031. Indeed, Kraut had been chair of the BBS AC in past years and so was a very well informed individual. He was aggressive, too, as he told the Committee that the arguments for retaining BBS *as it was* needed "dispelling."[130] He said that the "'vulnerability'" issue, "perhaps the most alarming" such matter, is that "without biology for protection" the social and behavioral sciences would be "unable to defend [themselves] against funding cuts." To that, he said: "Our response is something along the lines of 'with friends like these…'" One might say that the era of "protective coloration" met a sudden end that day. He also listed two other allegedly specious arguments, the second being "increased administrative costs"—one given by Mark Courtney and others in then recent times, let alone in the past—and noted that NSF had made no estimates of what it would cost to create a separate directorate. It was a factor on Massey's mind, too (below). Kraut said that, "informally…others" had estimated it at about one million dollars, but "[f]rankly, that appears to be a scare tactic." He did not name "others" nor had he a real fiscal answer to offer to the issue himself. But with all these actions ongoing, it is clear that extreme pressures from many fronts were being placed on NSF, BBS, and Mary Clutter to bring the issue both before Massey and for Massey to decide in favor of the split.

June 20 and the next meeting of the National Science Board was fast approaching, but newsletters from many societies were still coming in to NSF with thoughts on the future, either pro (mostly) or con (fewer) with regard to the split. COSSA was especially on top of matters for its many represented societies. In their June 10 newsletter, they implored their members to write to their Senators in support of the full NSF appropriation's bill. COSSA could point out the strong decision of the House to provide the Foundation with all its requested funds for FY 1992.[131] Indeed, in a vote of 240–173, the House had passed the Foundation's budget in May and had sent it on to the Senate. NSF was accorded its full $2.72B request, being short only $1 M for funds not tied to BBS concerns. The NSB had met by the time of COSSA's next newsletter, and so the Consortium had much to say about what the Board considered at its June 20 meeting.

The members of the NSB had entertained Mary Clutter, who spoke on the topic of "Special Focus on BBS Disciplines." She gave a brief history of BBS and then introduced Paul Magee. The TF chair provided an overview of the Task Force's findings and recommendations, including those 19 of 20 members in favor of the split.

---

[129] APS, "NSF Directorate Now Proposed in Senate," *Observer* (1991):1; 16 (May).

[130] Alan Kraut to "Interested Parties," May 13, 1991; APS letterhead. A memo to which was attached a copy of Kraut's testimony to the House Committee on May 2.

[131] COSSA, "Space Station & NSF Funding Increase Survive in House Vote," *COSSA Washington Update* (1991):1 (June 10).

Discussion ensued among the Board members on the matter. The "consensus of the Board was to further investigate" the feasibility and implications of the separation, while Massey (also present) told the NSB that he would speak to "key NSF staff" to consider the "full impact" on the budget, primarily, and would report back to the Board "in the near future."[132] It would, in fact, be some *4 months later*, leaving the question open as to whether a split was to occur or not. While no other details existed in the Minutes of the NSB's meeting, COSSA's newsletter of *4 days later* was well informed. Peter Raven was by then a member of the NSB, having had to give up his role as chair of the BBS AC. He told fellow Board members "that there was 'no real reason to continue the marriage' between the biologists and social/ behavioral sciences," and thus supported the separation, as COSSA had reported.[133] The Board had a "mixed reaction" to the notion of a new directorate for the social and behavioral sciences with some having expressed "concern over the administrative costs," stated the article. Others "asked questions about congressional reaction," and one, Charles Hosler, "wondered whether the Congress would earmark increased funds for the SEPS [SBE]."

Only one social scientist was then on the NSB: W. Glen Campbell, of the Hoover Institution of Stanford University, and he wanted to know about the breakdown of funds for groupings of divisions within BBS. Clutter explained that SES received just over $39 M, BNS got $32 M, and all biology divisions had a combined budget of just over $244 M. The NSB members who questioned the "protective coloration" issues were told by Magee that the matter "was 'not compelling.'" The Board also showed interest in where anthropology would be placed. Raven said it belonged to SEPS, but Hosler "was not so sure." Massey himself raised questions, even though still of an "open mind." Because interdisciplinarity might suffer, Massey thought the split "would not be the logical conclusion." Raven's response to Massey was an intriguing one.

> Raven echoed [interdisciplinarity] in his remarks, noting that there is probably more interdisciplinary work in bio-chemistry than in bio-SEPS [SBE], and biology and chemistry are currently in separate NSF directorates.[134]

A powerful statement! By the time that the APS had its annual meeting in July that year in Washington, D.C. and with over 2000 psychologists in attendance, the matter of the creation of a social and behavioral sciences directorate at NSF was still open.[135] But Kraut was in the fray again. With the suggestion at NSF that yet another group (the third) should be formed to consider the Task Force's recommendations, he saw the Foundation's musings as nothing "other than a stalling tactic."[136] That

[132] NSB, "Minutes," NSB-91-113, 6–91:4, June 20–21, 1991.

[133] COSSA, "Science Board Reviews SEPS Directorate," *COSSA Washington Update* (1991):1–2 (June 24).

[134] Ibid., pg. 2.

[135] APS, "Over 2000 Attend APS3 in DC," *Observer* (1991):1;18 (July).

[136] APS, "NSF Panel Recommends Separate Directorate, But Status Still Uncertain," *Observer* (1991):3;25 (July).

was an especially cogent statement as Massey had met with AP*S* President Gordon Bower and Executive Director Kraut 3 days prior to the June Board meeting, and Massey had told the two APS officers that he was still "not yet convinced that a separate directorate was the way to go."[137]

The AP*A* had also been informing its membership concerning the inexorable moves at NSF that might predestine the appearance of the new directorate. When the Task Force's recommendations were presented to the BBS AC in late May, there was some resistance to separation, but finally the AC agreed not to oppose the TF and sent the report on to the NSB with its support, though with concerns indicated.[138] A second hurdle cleared, as APA had it, was that the Board could have, on June 20, dealt the TF recommendation "a deadly blow," but thanks to Raven's "impassioned presentation" the NSB did not deliver that coup.[139] And contrary to the need for Massey to worry about budget issues should the directorate be created, "[k]ey Congressional staffers have assured [APA's] Science Directorate that they will seriously consider the recommendation of the BBS Task Force," claimed the APA's newsletter. Clearly, the APA and the APS had different views.

Not only was Raven's support "impassioned," the report given by Magee was, Mary Clutter held, "smashing"![140] In her letter of thanks to Magee, she noted that the "internal feedback has been very positive." She was, though, in the dark as far as anyone other than the one that really mattered, Walter Massey, was concerned, she told Magee. Even if members of the Task Force had been silent for some time near the beginning of July, Clutter could tell her TF chairman that "Massey has asked for an informal briefing." It was to contain three "case studies," one each on SES, BNS as related to biology, and another on BNS that could go "either way"—BNS as related to biology *or* as related to the social and behavioral sciences. While the Massey request was a private correspondence to Clutter, Roberta Miller of SES learned of it. To that, Clutter averred to Magee that she was "no longer outraged by the leaks. I am now simply fascinated by the rapidity of the process"![141] But the more significant view that she expressed to Magee was the critical issue that physicist "Walter [Massey], as a true scholar, is learning about the science involved before making any decisions." And, she was pleased to say that the letters from psychologists had dwindled to only a few each week and there "has been no pressure whatsoever from other disciplines." For her, and all concerned, the next few

---

[137] Ibid., pg. 25.

[138] APA, "NSF Moves Two Steps Closer to New Directorate," *Science Agenda* (1991):2 (July/August). Massey spoke at the May 28–29 meeting of the AC having been given a set of talking points on the 24th. That was his first chance to meet with the key players in the reorganization effort who were also present at that meeting.

[139] Raven, a botanist, apparently impressed Howard Silver, the Executive Director of COSSA, as they offered Raven a seat (one of four at-large members) on the Consortium's Board. Raven thanked them but begged off as too committed already. Silver to Raven, October 15, 1991, and Raven to Silver, October 22, 1991.

[140] Mary E. Clutter to Paul Magee, July 3, 1991.

[141] Ibid.

months would turn out to be the decisive ones with regard to the erecting, or not, of two separate directorates.

Kraut, though, kept up his drumbeat by writing to significant players, including NSB member Raven and Board Chair Mary L. Good, who had missed the critical June 20 meeting but remained well up on the issue.[142] More quietly, the various materials that Massey had requested of a number of staffers were coming in to his office. A "History of Social Sciences at NSF" had been prepared by Ron Overmann of BBS in which he had interviewed many leaders of BBS, BNS and even SES over time, such as Otto Larsen, Betsy Clark, Richard Louttit, Howard Hines, and others we have met in Chapter 2.[143] Within a month, the final product had come from the printer and the 106-page *Adapting to the Future: Report of the BBS Task Force Looking to the twenty-first Century* was made publicly available. As COSSA had it, "the key recommendation" was:

> The Task Force recommends that two distinct directorates be established to replace the present directorate of Biological, Behavioral, and Social Sciences. One of these is to be devoted to research in the biological sciences (BIO), and the other to the social, economic, and psychological sciences (SEPS) [SBE].[144]

To the social and behavioral sciences, that were the primary matter of interest, but in 106 pages, the report had much else to say.[145] So much so, in fact, that a separately bound "Executive Summary" was also published. That 19-page booklet repeated the Foreward by Clutter and the Preface by Magee but reduced the larger texts' main sections to easily digestible pieces that, broadly, included discussion on the *divergence* of disciplines then in BBS, issues of infrastructure, and the future of the sciences involved. For our purposes, the primary recommendation, the split, remains now as it was then: the most important matter decided by the Task Force, among many other points they made. But what the report had to offer were recommendations, not realities: *Massey's decision was still being awaited.*

Finally, on October 10 and 11, the National Science Board met with numerous items of business to discuss but, arguably, none more important than to hear Director Walter Massey report on his decision with regard to restructuring.

> The Director announced a major reorganization of the NSF, the details of which are spelled out in [the attached documents]. [He] further stated that an implementation group had been

---

[142] Alan G. Kraut to Peter Raven, June 27, 1991; Kraut to Mary L. Good, July 29, 1991. Each letter had extensive attachments such as a copy of Kerry's bill, material representing the Brown/Walgren bill in the House in 1990, etc. The fact that Goodman was not present, but that the June meeting of the Board was chaired by NSB Vice Chair Tom Day, is seen in: NSB, "Minutes," NSB-91-113, 6–91:4, June 20–21, 1991.

[143] Ron Overmann to Marta Cehelsky (in the Office of the Director), August 29, 1991; cover memo with 11-page history, plus attachments.

[144] COSSA, "BBS Task Force Report Released: SEPS Directorate Key Recommendation," *COSSA Washington Update* (1991):1 (September 23)

[145] National Science Foundation, *Adapting to the Future: Report of the BBS Task Force Looking to the twenty-first Century,* NSF-91-69 (Washington, D.C.: NSF, 1991), 106 pages. See also the online version at: http://www.nsf.gov/pubs/stis1991/nsf9168/nsf9168.txt.

established to develop guidelines, time lines, schedules, and budgets for implementation of the reorganization.[146]

One of the two attached documents was a Staff Memorandum from the Office of the Director dated October 11 that informed the entire NSF.

> I am pleased to announce my plan to establish a Directorate for Social, Behavioral and Economic Sciences (SBE)...The Directorate for Biological, Behavioral and Social Sciences will be renamed the Directorate for Biological Sciences and include the biological and neural science programs. Dr. Mary E. Clutter will continue to serve as Assistant Director.[147]

Other features of reorganization went on at the same time, and Frank Harris was appointed Acting Assistant Director of SBE until a national search could bring forth a permanent director for the new home for the social sciences. "SEPS" was dropped as a title and the important place of economics was assured in a titular fashion, but the psychological sciences were now buried deeper in the infrastructure of the new SBE.

In the opinion of Nathanial Pitts (who was considered in Chapter 2)

> [w]e ended up with a reorganization of the biology directorate, and that wasn't anything Walter [Massey] was interested in. [Mary Cutter] knew enough about the politics of science and where NSF fit in the Washington scheme. She also knew what an Assistant Director's authority could be if they [*sic*] wanted to exert it, and she decided to exert her authority and...the Director's Office really wasn't inclined to worry about [it], 'Well after all, it's only Biology and Social Science.' Why should they worry about it? They were too busy worried about the things that NSF generally worries about, which is the next telescope or the [next] fleet of ships...I mean, the big things is what the Director's Office used to tend to. They don't pay much attention to how [you are] rearranging the deck chairs in the organization, if you will.[148]

There had been a BBS Retreat in early September that had begun examining the structural elements of what would become the new biology directorate, as well as that for the social and behavioral sciences; Clutter still had the latter in her portfolio until all the dust settled, of course.

---

[146] NSB, "Minutes," NSB-91-173, 10–91:3, October 10–11, 1991.

[147] Staff Memorandum from Walter E. Massey, October 11, 1991, O/D 91–27. In the same documents, Massey also created an Office of Planning and Assessment, an Office of Budget and Financial Operations, and an Office of Administrative Management. Those offices became official on November 17, 1991 (Walter E. Massey to Administration and Management, O/D 91–29, November 19, 1991). He also disestablished the Directorate for Scientific, Technological and International Affairs (STIA; earlier chapters) and transferred its programs to other locations throughout the Foundation. An embargoed news release went out at 9:30 a.m. local time on October 11, as well. "Strong and very important advantages will be gained for research as a result of this reorganization," Massey was quoted as having said. Massey also presented a memo to the NSB (NSB-91-174) on October 11 in which he said that the BBS' "scope...is far too broad to give sufficient attention" to all the changes that had occurred in both the biological and the social and behavioral sciences since 1976. The split could be "accomplished with no appreciable increase in operating costs," he said.

[148] Nathaniel Pitts interview, *op cit.*

It later came to pass that four divisions would be formed within BIO (see Fig. 5.2). However, these four divisions did not just simply appear by mid- to late October but had been the products of a number of working groups' considerable efforts over the previous weeks; and much longer—years—in the thoughts of those involved. This has been demonstrated in earlier chapters. These internal organizational labors will be considered in the following section of this chapter. The four divisions for BIO would include those of Biological Instrumentation and Resources (BIR, the word biological was added to the older BBS version of DIR), Molecular and Cellular Biosciences (MCB, a combination of the two in BBS; again, very much a return to the past), Integrative Biology and Neurosciences (IBN), and Environmental Biology (DEB, a return to an older title with the elimination of BSR). The newest division was IBN, at least in the expansive sense of including neurosciences. These four also provide a basis for the structure of the remaining portion of this historical study; each division is considered in a separate chapter.

It hardly needs be said that the scientific community, both biological and social/behavioral, was caught up in the new directions at NSF. Less disciplinary-oriented news sources also picked up the story, as *Science & Government Report* (S&GR) stated that:

> [i]n the venerable Washington quest for a name on the door if not on the building, the social and behavioral sciences achieved victory last week in their protracted campaign [to acquire an] enclave of their own at [NSF. The] SBE, in NSF-ese…is to be extracted from the back-room of its longtime home…leaving the old shop simply titled the Directorate for Biological Sciences.[149]

The S&GR's concluding statement read: "Massey's announcement received virtually no attention outside the science press." That was not entirely the case, however, as the same writer, Daniel S. Greenberg of both S&GR *and* the widely popular *Washington Post*, penned a strong article about the "little-noted administrative shuffle" at NSF.[150] He averred that the reorganization leading to the "elevation of the social sciences tells a good deal about the evolving relationship of knowledge and politics." He reviewed the history of the social and behavioral sciences and their difficulties at NSF over the years from its founding in 1950 and then their slight rise in 1976 "despite their poverty" when it came to the budgetary pie. As others would also conclude, Greenberg held that the new directorate "can be *attributed to effective lobbying*," as well as to "the intellectual growth of the social sciences," and to "maturity" on the part of biological and physical scientists "who still dominate the [F]oundation."[151] And, he said, the "dwindling of confidence" of the country to deal with its "economic and social problems" demonstrated the need for social research.

[149] Daniel S. Greenberg, "NSF Gives Social Sciences Their Own Directorate," *Science & Government Report* (1991):4 (October 15).

[150] Daniel S. Greenberg, "'Soft' Sciences Grow Up," *Washington Post* (1991), November 13 edition. It should be noted that editorial commentator Greenberg was also the editor and publisher of the S&GR; see note just above.

[151] Emphasis added.

Still, the "trips to the moon were kid's stuff compared to the problems in the domain of the social sciences."

The disciplinary press was not nearly so subdued, as Greenberg held the general press was, however. As if a new child had entered the world, the Federation of Behavioral, Psychological and Cognitive Sciences' newsletter chose to banner the split with the words: "It's A Directorate!"[152] A special edition of the COSSA newsletter was also issued with banner headlines.[153] In the story, it spoke of the TF report as one titled "*Advancing* the Future": the error was picked up by many other society publications over the following months. The article noted that the Foundation's intent was to have a permanent Assistant Director "on board" by March 1, 1992. The restructuring of the directorate was already well underway. Beside the expected program moves, the International Programs of the Foundation were also placed in the new SBE; that done by Deputy Director Bernthal.

The influential *Chronicle of Higher Education* immediately brought up the notion that the separation might well bring more funds for all the sciences involved. Massey was quoted as having said that it was "certainly possible" that biology might well profit from the split and expand faster than the physical sciences, especially due to its "explosive growth" of the period.[154] He also said that he would "try to win more support" for the social and behavioral sciences.

The headline in large type that appeared on the November newsletter of the APS read: "New Directorate: Yes!" "It's finally going to happen," the first line fairly screamed.[155] As the APS might have been expected to say, "[a] directorate for *psychology and other* behavioral and social science disciplines" was to be created at NSF.[156] Interestingly, they did not comment on their titular loss in the dropping of the term SEPS, though. With a multi-page timeline of events from early 1990 to the formation of SBE, the newsletter provided a large one-page spread showing seven members of Congress with lines from their names—and their portraits—leading to the center picture of Walter Massey. The article stated, in no uncertain terms, that it was "Congressional Pressure" that made the difference. Later in the article, Representative George Brown of California was quoted as having said, in regard to the "protective coloration" issue, that "behavioral and social science researchers need to continue to develop and improve their scientific methods and objective data collection" as it is done by "biology and the rest of the physical sciences."[157]

---

152 FBPCS, "It's A Directorate!" *Federation News* (1991):2 (October).

153 COSSA, "National Science Foundation Creates Separate Directorate for Social Science," *COSSA Washington Update* (1991):1 (October 15). That issue of the *Update* was a "Special Edition," so important to COSSA was the creation of SBE.

154 Colleen Cordes, "Social and Biological Sciences Could Benefit From Reorganization of Science Foundation," *Chronicle of Higher Education* (1991): Section A (October 23).

155 APS, "NSF Directorate: Yes!" *Observer* (1991):1; 28 (November).

156 Emphasis added.

157 Ibid., pg. 28.

If someone were to have a last word on the separation, it most likely should be given to Sonya Mallinoff, present through it all, when she said to the author many years later that Clutter was

> successful in devising a means to finally let them [the social and behavioral sciences] get out on their own. Mary was a superb negotiator [though] people became very frustrated with her method of always insisting on consensus. It was a strategy that I think was very successful…She always had the highest level of ethics and values of anybody I worked with.[158]

Both SBE and BIO were born, then, on October 11, 1991, and began separate journeys into the next millennium. It is from this point that this history no longer deals with the social and certain of the behavioral sciences (not including animal behavior and neuroscience, both of which became part of BIO) as part of what has been considered in the last 15 years in the life of the Directorate for Biology and Behavioral Sciences. The remaining four chapters will deal with biology as it was interpreted at the National Science Foundation's Directorate for Biological Sciences from 1991 until 2005.

5.7 A New Home for Biology

The purely biological sciences, it might be assumed, were already rather well delineated in the former BBS and would require only minor reorganization, with the conspicuous exception of the more biological areas covered by the deconstruction of Behavioral and Neural Sciences. But that was rather far from the case, as Fig. 5.1 demonstrates. The disestablishment of BNS and the rise of a division for integrative biology and neuroscience in BIO and of what came to be Behavioral and Cognitive Sciences in SBE were just one of several complex adjustments.

How the biological sciences reformed from BBS into BIO was not without considerable and spirited discussion by the major players: PDs, POs, division directors, and various associates, let alone BIO AD Clutter. It was she to whom all these persons provided reports in the weeks after the September retreat. The commentaries would be crucial to how BIO chose to organize itself, even if the final four division titles seemed mostly familiar from the former BBS structure. *And it would make a difference as to how the history of biology progressed in America in coming years.* Certainly, researchers in the behavioral and neuroscience areas would now approach the Foundation seeking funds in a very different way than in the past, given the new BIO (and SBE) structures.[159]

---

[158] Mallinoff interview.

[159] The short form name "BIO" was chosen to correspond to the earth sciences directorate, "GEO," and to specifically avoid the shortening *biological sciences* to "BS," said a writer in an article about the reorganization! See American Society for Microbiology, "NSF Retailoring Biology Programs," *ASM NEWS* (1992):7 (58:1).

The various program officer-led working groups were given their charges on September 23.[160] As Massey had not yet made his announcement of October 11, of course, the biologists were tasked with the job of "look[ing] positively toward scientific opportunities and how best to foster such advances, rather tha[n] 'fixing' current problems," that is, attempting fully the layout of a separation. That included how the biology programs might interact with the "proposed 'SEPS' directorate." Of special note, recommendations were made to the "BIO Division Directors [to seek] possible options for restructuring biology activities at the program level and above." It was clear that a simple creation of four more or less unchanged divisions that would become BIO was not all that was in the thinking of the time: *programs* and other parts of the directorate's structure needed examination, as well.[161] The long-considered desire to eliminate duplicate proposals (see earlier), among other matters, had been accepted at the Retreat, but Massey's approval was still being awaited, so issues not directly organizational in nature were also on the table.

The charged working groups were reminded that one of the Retreat's decisions was that BIO would have four divisions, one of which would deal with infrastructure. Also, none of the new divisions could retain *precisely* the titles used in BBS, Clutter required. Finally, no increase of personnel or budgets was anticipated. October 11 was set to consider budgets for FY93, something Massey would be doing for all of NSF that same day(s). Other aspects of restructuring were only slightly less pressing: final *draft* reports were called for by October 18; they were to be discussed at the BBS AC meeting on October 31. The final reports were due November 15. All in all, little time was available to the working groups—after more than 15 years of one major structure having held sway. The matter of their charge was taken quite seriously, and so, for instance, four division directors (DDs) signed the second page of the memo committing to the work to be undertaken by each group. One, as an example, included Phil Harriman for Molecular Biosciences; Richard Louttit for BNS; Bruce Umminger for Cellular Biosciences; and, Patricia Werner for BSR. Similar signature pages were present on the reports from the other PO working groups. Each of the groups would be made up of five or six POs, one of which would be a "leader," or two of which would be "co-leaders." None of the 21 persons spread over the four working groups came from SES, but several naturally came from the now nearly disestablished BNS. (SES was finally represented by Thomas Baerwald, as it happened.) The names of the working groups were Molecular-Cellular Biology Continuum, Relation of Neuroscience to the Rest of Biology, Integrative Physiology and Functional Ecology, and Molecular and

[160] Memo "Establishment and Charge to PO Working Groups," from Biology Division Directors to PO Working Groups in Biology, September 23, 1991.

[161] There is a large collection of emails in paper format that survive from the period and represent some portion of the person-to-person and person-to-group discussions that played into the reorganization scenario. It is beyond the scope of the present work to consider them fully, but a future historical study should examine them in depth to understand all of the interpersonal dynamics of the period; such consideration will help better to define the state of the history of biology at the time.

Cellular Evolution. The names tell a great deal about the thinking of the time, and when one compares the final four divisions of BIO (Figs. 5.1, 5.2, and 5.3), parts of what the working groups were considering can be seen in the outcome. Others, however, cannot. The use of the term *evolution* had long been a sticking point in the federal establishment, and it could not end up in a formal title for a division of the NSF. The BIO scientists were dealing with the quite new notion and term (certainly in modern times) of *integration* in biology and had to find the best way to use the epithet in a final title and organizational structure, as another example. *Integrative biology had to be defined at NSF by dint of the structural features it assumed in the reorganization.* It would then affect the nature of the specialty in the community into the future. Clearly, neuroscience had to make itself fit an incipient BIO once BNS was disassembled. *More so than any other area, neuroscience would be the most contested in the reorganization efforts.* Animal behavior did not offer the same difficulties as did psychology or anthropology/archaeology. It should be pointed out, too, that a parallel effort was ongoing at the same time in regard to the social and behavioral elements of BBS, but their findings will not be considered here.[162] Nonetheless, Mary Clutter had to preside over the formation of the new SBE, as well as BIO.

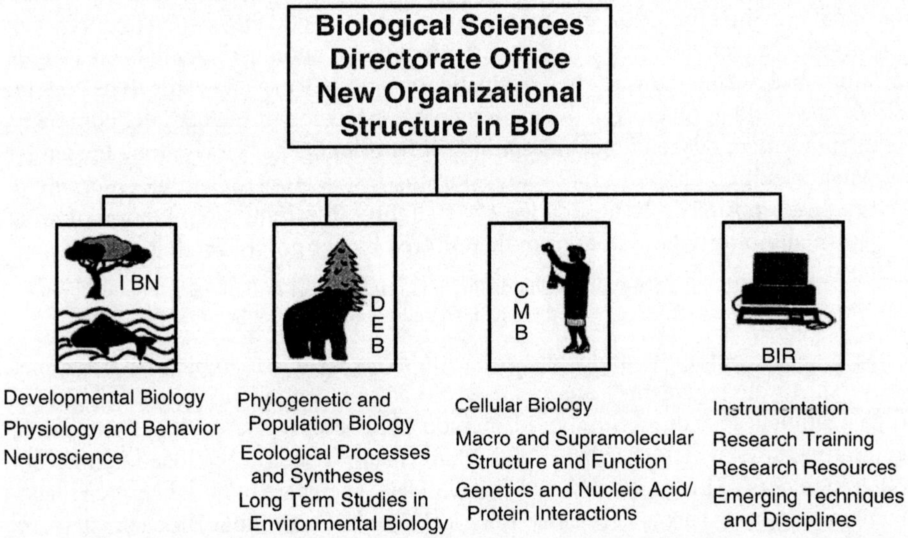

Fig. 5.3 New organizational structure for BIO. (NSF. Taken from a visual of the BIO directorate meeting of November 18, 1991. See main text for details)

[162] However, the documents that make up the working groups for the social and behavioral sciences of the rapidly ending BBS of the time would make an excellent starting point for any future history to be written on the SBE from 1991 onward.

Reports began to come in to the AD from October 15 onward for several days. Bruce Umminger, of the BBS Division of Cellular Biology, was prepared well in advance, as he said in his opening sentence: "In response to your request…*today*…I am providing several options…[for] the new BIO Directorate."[163] As Deputy AD, he had long been considering the matter, and many of the restructuring ideas were also, of course, widely shared within the BBS—and had been for many years. His three *scientific* divisions (as opposed to a more mechanical and infrastructural DIR; soon to become the BIR) were named and constructed much as they would finally turn out, even though he offered Clutter three options on his basic plan. Werner's plan came in 3 days later and was a take-off on Umminger's borrowing both his division names and ideas, but with "some differences."[164] She and her colleague, Jim Edwards, would have, as with Umminger, avoided beginning a formal program in molecular evolution, for instance, as "[c]ontent, not techniques, should drive our restructuring." She offered *four* content divisions plus one more for infrastructure; her views were based upon how integrative biology was to be interpreted.

The five-person group that undertook Integrative Physiology and Functional Ecology had included Machi Dilworth, who would later submit an "addendum" report, and animal behaviorist Fred Stollnitz of BNS, plus three other biologists. Their thinking was strongly along functional lines, and therefore they decomposed older programs placing parts into newly proffered locations but did not actually suggest divisions to accommodate the widely restructured programs. Many would have been put into the group's *organismal biology* structure, one they called for, however. Dilworth's "addendum" report was occasioned based upon her having been away at a conference during the week her group did most of its work and came to its conclusions; she could not then sign the main report.[165] She did not see her later comments as a "minority or descending [*sic*] opinion." She oriented her views along the nature of what constituted organismal biology, turning instead to words in the *Adapting to the Future* report. The Report called for a slightly different interpretation of what organismal biology should be, from that offered by her group, she felt.

> Organisms represent a particularly interesting and crucial level of biological organization, one clearly separate from cells on the one hand and populations on the other.[166]

As they are not at the molecular and cellular level, where "biological principles universal to all organisms" are studied, she argued, organisms needed a distinct organizational structure and should include plant biology, a personal and prime commitment for which she also called. Plants had not been so included by the four other members of the group, but had been left as "taken care of" by their then placement in scattered aspects of the Physiological Processes, Cellular Biochemistry, and developmental biology programs. She noted that, in the end, one or more of those programs might not look the same in the new structure. She was prescient in that, as

[163] Bruce L. Umminger to Mary E. Clutter, October 15, 1991; six ppg. Emphasis added.

[164] Patricia A. Werner to Mary E. Clutter, October, 18, 1991.

[165] Machi Dilworth to Don Jackson (DCB), et al., October 16, 1991.

[166] Cited in ibid.

seen in Fig. 5.1 where, for only one example, Physiological Processes were split up rather conspicuously into three new programs: endocrinology, Animal Systems Physiology, and Integrative Plant Biology. Clutter had called for looking deeply into the program level, and Dilworth and her group, in their two respective reports, did exactly that. Others would do the same with neuroscience.

The molecular-cellular continuum group provided very carefully thought-out structural suggestions that led to a highly integrated MCB for the future and retained the older terms Division of Environmental Biology and Division of Biological Instrumentation and Resources as much the same—not surprising given that they were specialists in the little end of the spectrum.[167] They dealt with an "Organismal/ Integrative Biology" division such that it fit well against their proposed MCB by handling various areas within physiology logically over the two suggested directorates. However, they were not unanimous in their decisions. Edward M. Lieberman, PD of the Cellular Neurosciences (CN) program, provided a true minority report.[168] He objected to placing the current CN program into the newly proposed MCB in the manner the majority had suggested because he held that it would "destroy both the identity and integrity of the [CN] and will weaken the entire Neuroscience continuum." He preferred a single division level structure for all the neurosciences and was, by no means, the only one to suggest that. In the end, however, the CN was broken up (as can be seen in Fig. 5.1), and the neurosciences became a group of several newly restructured program level entities but still under the heading of the neurosciences, a term used also as part of the new Division of Integrative Biology and Neuroscience. In many organisms, of course, a nervous system is the primary *integrative* mechanism for those creatures, but the word neuroscience had strong standing by itself and could hardly be eliminated, Kingsbury's opinion of Clutter's views on the subject notwithstanding.

The "Relation of the Neurosciences…" group was a different entity than the MCB one, and so Lieberman was ill-placed in the latter group, *if* one merely takes his expertise and minority report into consideration. However, examining all four of the groups that Clutter composed, it is clear that she sought a cross-section of disciplinary and programmatic representation in each. It may well be that she was seeking a variety based upon personalities, as well, an area difficult to judge at this remove. Demeanor and style during oral interviews for this history suggest that possibility. For instance, with regard to discipline specifically, Jim Reynolds of BSR was on the "Relation of the Neurosciences…" group with only two of the five

[167] Marcia Steinberg, Leader, et al., to Mary E. Clutter, n.d. (likely October 18), 1991; 11 pages plus minority report given separately; see following footnote. This and several following communiqués did not actually have dates or "to" lines on them and are gathered under a single cover sheet, itself without dates or other identifying marks with the exception of a handwritten note ("Sonya") indicating that it had either been a copy of Sonya Mallinoff's, or was gathered by her and, presumably, presented to Clutter. I have placed "to…Clutter" in a couple of the following citations, though that has been done for convenience alone.

[168] Edward M. Lieberman to Chair of the MCB Continuum Group (Marcia Steinberg); October 18, 1991.

members actually closely tied to the subject area of that committee.[169] However they, like others, suggested a new BIO of five divisions, one being neurosciences, just as Lieberman had called for, even though Clutter called for a total of only four. They took a line from the Retreat that read: "fragmentation of neuroscience…would be counterproductive." They held that such would be the case on several planes and so chose not to place it with their, also suggested, integrative biology division. One of their "rationale[s]" for maintaining a unified and separate division for neurosciences was the statement that it was

> strongly interdisciplinary in its methods, but its content is focused on understanding the brain and behavior. **Consequently, it is distinct from other disciplines within biology, although the methodologies are complementary**.[170]

It was due to the fact that a "combination of approaches" required by the very nature of the neuron and the "organ" of the nervous system that molecular biology, genetics, biochemistry, physiology, and behavior had to be used in carrying out research in neuroscience. They also provided discussion on the other divisions they proposed; those were not unlike ones proposed by other groups. But, as it was with the MCB group and Lieberman's minority report, Susan Porter Ridley, Assistant PD of the Genetics Program in DMB and member of the "Relation…" working group, also provided such a report. She argued that all of the then current eight programs in neurosciences should be kept *intact* and that plant science (also examined by the group as necessitated by the directive that all groups consider *all* possible new divisions) should not be isolated, as the majority would have had it. An important historical point arises here as the dominant *zeitgeist* at NSF, overall, was such that organizational structure within the directorates should be of a nature that strong communication between persons and disciplines, beyond just fields and program officer membership, should be promoted: *interdisciplinarity* in program planning and in discussion of grant proposal reviews, in particular. Any sort of intellectual isolation should, at all costs, be avoided. That had been the case for a long time, and, philosophically, it was so inbred into the minds of those permanently at NSF—and quickly taught to those present only temporarily—that such free interchange was one of the most sacred of goals.

The group working on Molecular and Cellular Evolution noted that BBS had been dealing with molecular evolution for "about 5 years."[171] Studies in the area had been situated in BBS programs in Population Biology and in Genetics. Indeed, the Society for the Study of Evolution (SSE) had hoped to meet with BBS division directors (DDs) in 1990 to pursue issues of mutual interest, but the internal activities of reorganization then ongoing were such that the meeting never occurred.[172] However, a 1-day workshop had been conducted in late September of 1991 (under PI Morris Goodman of Wayne State University, Detroit, Michigan) supported by

[169] Kathie Olsen, et al., [to Mary E. Clutter], n.d., 1991.

[170] Ibid., unnumbered page 2; emphasis in original.

[171] Peter Arzberger (DIR) et al. to Philip D. Harriman et al. October 21, 1991.

[172] Ibid.

NSF and the Sloan Foundation, on the subject of molecular evolution. That growing discipline had been funded by Sloan for 6 years in order to "catalyze" the field. Since the field was "exploding" and given that it was a costly area for research, there was a general feeling in the community that not only was NSF's "nurturing" of the area called for, but its presence must be very obvious at the Foundation, as well. The evolution society was writing a report on the workshop at the time, but the version to be circulated in the community would not mention NSF! The reason was not stated.[173] However, "[t]he statement from the workshop serves to define the field [of molecular evolution]," they said.

John Wooley, DD for DIR (following chapter), sent four memos—parts of a "set," he averred—to Mary Clutter all on 1 day in later October reflecting on the place of DIR in the future organization of BIO. In that complex of memoranda, he addressed: issues of proposal reviews, "straw" structures (a heavily used term at the time in BBS) for a new BIO, alternative straw structures, and, finally, retaining "special projects competition" for field stations and marine laboratories (known as FSML)—a program of considerable significance, he argued, and one which should be kept in the new BIR. He chose the rather odd approach of four separate memos on four subjects so that they could be "explicitly developed or abandoned," one-by-one.[174] The FSML matter was visited by several of the four working groups as a whole and also some of their 21 members individually. Patricia Werner provided a very brief history of where the FSML had been placed organizationally (including cross-directorate activities with Biological Oceanography; not part of BBS) and funding history. She supported the need for more involvement from NSF personnel and retention of FSML competition in a new BIR. Appel has considered the subject of that competition in earlier years.[175] It was a long-continuing program and remains in existence at the time of this writing.

By October 23, Chairs Louttit, Umminger, Werner, Harriman, and Wooley reported to Clutter their distillation and synthesis of all of the working groups' suggestions and came back to three scientific divisions and an infrastructure/resources division. The one notable difference in what they had to say, as opposed to what finally came to pass, was the notion of a Cellular and *Structural* Biology division. It would, they said, represent the "continuum" of molecular to cellular, reflecting the working group's title, but followed a more strictly "so-called 'reductionist' approach to biology."[176] However, the matter of how many scientific/content divisions there should be came up again at the October/November meeting of the BBS AC and the discussion moved, once again, to the notion of having four such divisions with neurosciences alone being one. This satisfied some, but not others within both the BBS itself, as well as the AC. Mary Clutter was "quite pleased" with the AC meeting:

---

[173] Ibid.

[174] Division Director, DIR (John Wooley) to Assistant Director (Clutter), October 22, 1991; a set of four different memos; subjects covered as described in main text above.

[175] Appel, "Shaping Biology," see especially ppg. 93 and 169, among some ten references to the subject of MS/FL (=FSML) in her history.

[176] Richard T. Louttit, et al., to Mary E. Clutter, October 23, 1991.

BBS "received good advice…and the level of discussion was simply outstanding."[177] And, while some at BBS were anxious to say that neuroscience would have its own division, Clutter told one of her colleagues that she first wanted to hear feedback from the neuroscientists "before I make my decision about the structure of BIO."[178] It must be kept in mind, too, that David Kingsbury had recently opined (earlier chapter) that neuroscience was not an area of interest to Clutter, and as the final decision about BIO's structure was looming, Clutter *may* have been affected by her antagonism, *if* Kingsbury was correct in that opinion. Among those very much worried about damage to the discipline's integrity at NSF was John Hildebrand, a neuroscientist and then chair of the AC. Hildebrand was quoted in *Science* as being fearful of BBS making such a move as BIO's structure was on the verge of being decided.[179] But that issue of the journal was dated the second day of the AC meeting (likely coming out about 3 days later), so it is clear that the interview with Hildebrand had occurred some time before.

Hildebrand wondered, said *Science*, why NSF would, after having been such a major force in building modern neuroscience and doing so much to unify it, now consider taking it apart in order to satisfy aspects of reorganization. "'It would be unthinkable,'" Hildebrand asserted, were NSF to "dismantle such a successful program."[180] Neuroscientists William Greenough, University of Illinois, Urbana-Champaign, also quoted in the article, along with the peripatetic Thomas Carew of Yale and Robert Wurtz of the National Eye Institute, went to visit with Clutter to discuss the apparent threat. "The three weren't alone," said the article. It went on to quote Clutter, in her inimitable style, as having said she heard from "beaucoup neuroscientists," along with many others opposed to "dismantl[ing]" neuroscience. While welcoming these comments, Clutter noted that she was not "swayed by them." *She felt that the time had come to "mainstream" neurobiology* and so place, for example, cellular neurobiology with other cell biology programs, and so forth. After all, the integrity of neuroscience had been "protected" at NSF for the "last 15–20 years," she was quoted as having said. The issue, as far as the article in *Science* went, had to be left hanging until Clutter made her decision. But Clutter had "beaucoup" comments to think about, including that of her own AC chair, Hildebrand, who had written her very near the time he must have been interviewed for *Science*. He told Clutter how worried he was about the possibility of breaking up neuroscience: "You can bet that I will be vocal if not eloquent on this subject" and

---

[177] Email from Mary Clutter to unknown recipients (BBS and AC?); a copy of the full document has not been found. The partial text is tied to an email from Judy Finkelstein, BBS, to Clutter of November 11, 1991, in which Finkelstein informs Clutter that she will be attending an annual meeting of the Society for Neuroscience (SfN) on November 11–15, and hopes to inform them that a full division of neuroscience was about to be established.

[178] Ibid. Clutter wanted to wait until after the annual meeting of the SfN to which Finkelstein was going, assess the thinking of the neuroscience community as discovered/reported by Finkelstein, and then decide on the structure of BIO.

[179] Marcia Barinaga, "Neuroscience at Risk at NSF," *Science* (1991):643 (November 1).

[180] Ibid.

that he "felt VERY threatened by what is happening"; it was, he held, "going to be the shoot-out at the OK corral."[181] "The sociologists got their way [with the new SBE], but I cannot roll over and watch as neuroscience suffers disintegration and neglect!"

On November 18, 1991, Mary Clutter presented a quarterly review with overhead slides showing the process that had been used over the previous 14 months or so to bring about the reorganization of BBS into BIO. Figure 5.3 is a reproduction of the slide she used to introduce the new BIO organization. *The neurosciences were, indeed, reorganized* (as shown by the lines in Fig. 5.1) with some programs from BBS coming into BIO intact (e.g., animal behavior, though under a new heading of Physiology and Behavior) and others having been restructured rather heavily: Cellular Neuroscience moving, in part, to Neuronal and Glial Mechanisms, as but one example. Clutter's "mainstreaming" had occurred for neuroscience, but the division title of IBN still gave a unified title for the field: that must have been at least a sop to some of those in the community.

And, thanks to the establishment of BIO, the parts of the new directorate "will emerge phoenix-like from the ashes," said its AD.[182]

---

[181] John Hildebrand to Mary Clutter, email of October 15, 1991. Emphasis in original.

[182] Mary E. Clutter to BBS AC, October 24, 1991.

Chapter 6
Technology and the "Fearless Biologists"

The twenty-first century is going to see a cohesion of the sciences and a disappearance of their borders
David Baltimore

6.1 Foundation for Cohesion

The years 1991–1992 were ones of considerable activity in the finalizing of the National Science Foundation's reorganization, and particularly in the realm of biology by year's end for 1992, BIO was clearly established. But Leon Lederman, then President-Elect of AAAS, was fearful at the beginning of that year. He worried that science in the United States was becoming moribund. He found, through a rather eclectic survey he conducted, that even though the "federal investment in basic research" support was almost the fastest growing area in "non-defense government spending," the scientific community was evidencing "low morale."[1] Why was that? Lederman sent out a survey that asked four questions and sought, in return, a letter from each survey participant elaborating on the questions asked. Those included one about research funding availability, another about the ease of acquiring such funds as compared to past times, a third inquiring about how a given scientist's "plans and expectations for the future" might be influenced by then current funding issues and, finally, "other factors" that might influence a scientist as he or she conducted research in the "current setting." He found, in both quantitative and qualitative data generated from 250 respondents, that three factors were at play with regard to the "low morale" matter. One finding was that the cost of regulation was increasing, especially in the life sciences. Guidelines that must be met in "animal care, human subjects, low level radioactive waste, and hazardous substances" inevitably "reduce the amount of research that can be done for a given amount of money."

[1] Leon Lederman, "Science: The End of the Frontier?," Supplement to *Science* (1991); 18 ppg. Quotation from the Preface.

D. J. McGraw, *Millennial Biology: The National Science Foundation and American Biology, 1975-2005*, https://doi.org/10.1007/978-3-030-56367-7_6

Another discovery was that the cost of overhead was rising rapidly: doing science in 1968 incurred overhead costs of some 16%, but that had risen to 28% by 1986. And, "[c]harges equivalent to 70 percent of salaries are not unusual [in 1991]"! He said that many researchers held that it was "tax on research." Even though it was a "legitimate component," he conceded, it meant "less money is available to the laboratory scientist for the direct costs of research."[2]

A third, and very notable, finding was what he called "complexity" and what others, he said, referred to as " 'sophistication inflation.' " That is, the questions scientists were asking nature were becoming ever more complex and thus required increasingly complex—and expensive—*equipment* to obtain answers. He gave the example of the price of a "state-of-the-art dye laser" that cost $19,000 in 1974 and, by 1990, it was $160,000. Even correcting for inflation, he held, a researcher had to pay three times the price for that instrument. Indeed, it cost *ten* times more to equip a laboratory for a new assistant professor in 1990 as compared to 1968.[3] "*The complexity factor is a direct cost imposed on research by increasing sophistication in science.*"[4]

For these reasons, Lederman told the readers of *Science* that "[c]olleagues have told me of senior researchers about to 'throw in the towel' " and of bright young assistant professors unable to get started in their fields due to the lack of research support.[5] He knew his survey "may perhaps reveal indications of passion and advocacy"—it was informal and highly qualitative, after all—but his concern was for the state of science in America and for the "profound cultural and economic benefits that science brings." Indeed,

> American leadership in scientific research, as measured by published papers and Nobel prizes, is threatening to go the way of the automotive, tire, machine tool and consumer electronics industries.[6]

To "go the way" was literally to go away, the comparison suggested.

The early 1990s brought other views of the future, as well. The coming change of the millennium was greeted by some as a time of apocalypse on many levels. Mary Clutter had been asked to write a guest opinion piece for the newsletter of the Federation of American Societies for Experimental Biology (FASEB), an organization of considerable importance to the work of many researchers who were regularly funded by NSF/BIO.[7] She pointed out in a commentary, however, that "The End of Biology?" was hardly the case and that the "prognosticators" who were then "speculat[ing]" on the future were incorrect in certain of their assumptions.[8] John Horgan, author of *The End of Science: The Limits of Knowledge in the Twilight of*

[2] Ibid.

[3] Ibid., pg. 10.

[4] Ibid. Emphasis added.

[5] Ibid., Preface.

[6] Ibid.

[7] See more on FASEB at: http://www.faseb.org/

[8] Mary Clutter, "The End of Biology?," *FASEB Newsletter* (1996):2–3 (December).

the Scientific Age, had claimed, as she stated, that "all of the big questions of science that can be answered have been."[9] Furthermore, all that was left to do was a sort of "clean-up action…a filling in the blanks," she told the FASEB readers that Horgan had argued. His views were somewhat comparable to the, perhaps apocryphal, story of the inventor who, *in 1900*, hurled himself off a building as he held that everything that could be invented had been. Though Horgan had a strong science writer career and was respected in the field, Clutter showed why he was in error in his thinking, with regard to biology, at least.[10] Her FASEB opinion piece spoke of the "emergence and melding of disciplines" both within NSF and in the community. "Revolutionary breakthroughs" were on the horizon, she said, and gave computational biology as an example; it was an area that not only had been championed by David Kingsbury, but one for which Mary Clutter took up the cudgel to continue and advance. Support for the biological sciences "can be found in every Directorate" at NSF: she gave 11 examples of such cross-Directorate projects and programs; interdisciplinarity was broadly established. The Directorate for Biological Sciences "itself has a number of special responsibilities."

> It provides about 95 percent of available federal funds for biodiversity research, 75% for evolutionary physiology, 90 percent for long-term ecological research, 75% for environmental biology, 50% for plant biology, 65% for biological informatics, and 45% for computational biology.[11]

The funding picture that worried Lederman in 1991–1992 *was* justified on one level: the value of then-current dollars had dropped. For the sciences, including the life sciences, the two preceding years (1989 and 1990) were affected by that drop and that may have contributed to a part of the "low morale" he was seeing in the community. The purchasing power of those dollars had fairly crashed over the long run: the rate of inflation change from 1975 to 1990 was just under 143%! But, for the period of those two fiscal years of FY88 and FY89 (Fig. 6.1) preceding Lederman's survey where NSF funds had decreased, the inflation rate was just over 10%; not a serious burden for the community, comparatively speaking.[12] On the other hand, *NIH funding* available to life scientists had dropped radically in that same two-year period. From FY81 to FY88, the funding of projects by NIH varied only about 2% above or below the mean, but for FY89 it had dropped 29% and for FY90 the drop was about 24%. Since such rates tended to skew awards to those investigators with proven track records, it was no surprise that young and aspiring new researchers were heavily disadvantaged, as Lederman discovered.[13] (It is also likely that Lederman's admittedly biased survey did not include the types of

---

[9] John Horgan, *The End of Science: The Limits of Knowledge in the Twilight of the Scientific Age* (New York: Broadway Books, 1996).

[10] See: http://www.johnhorgan.org/

[11] Clutter, "The End," pg. 3.

[12] See: http://www.usinflationcalculator.com/ as a tool to create the figures given in the text above.

[13] From a draft ephemeron; two pages of typed text entitled "The Problem" and commentary in handwriting; author unknown.

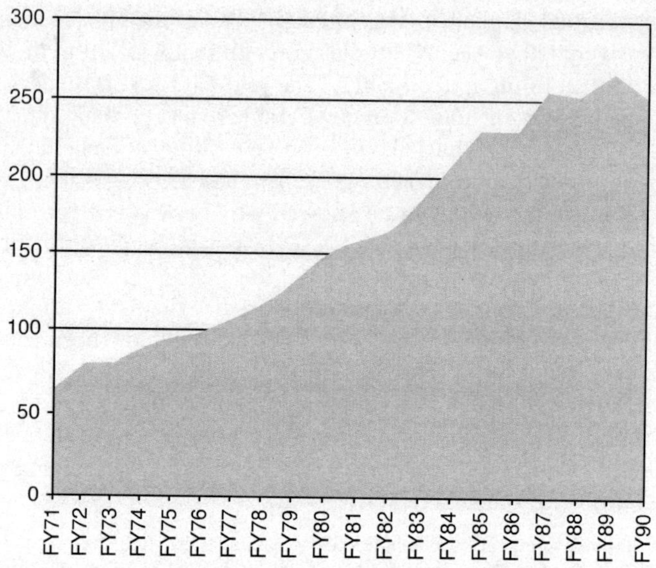

Fig. 6.1 Life science funding at NSF from fiscal year 1971 to 1990 (millions of dollars). Specific provenance unknown

researchers who would seek funds from NSF primarily, but included those investigators who typically went first to NIH.[14] Note that Lederman spoke of laboratory science, but did not address field science; above.)

The biological sciences at the National Science Foundation had begun their independent existence as a dedicated directorate on October 11, 1991, among a short downturn in NSF funding, a relatively new president—George Bush, *pater*, was inaugurated in January of 1989—and a new war that had begun for the United States in the Middle East in early August of 1990. The latter would be over quickly, however (February, 1991), as would the downturn at NSF. The funding for the biological sciences at the Foundation would rebound in FY92 and begin another upward climb, albeit with a regressive beginning (FY92: $293 M; FY93: $286 M), but rising then without a break through FY99. By that latter year, it had reached $393 M, a full 100 million dollars above that of just seven years before![15] Even with spiraling inflation, the growth was remarkable—for NSF. But a struggle over fewer grants supported with greater funding versus more grants at lower dollar levels was also brewing and would create tensions in the community.

Michael Hahn's work in signal transduction in plant cells at the University of Georgia was a case in point. He first obtained support from BBS in 1987, at which

[14] This conclusion by this author is merely based upon Lederman's own interests in science, but likely correct.

[15] National Science Foundation, Division of Science Resources Studies, *Survey of Federal Funds for Research and Development: Fiscal Years 1970–2001*, (Arlington, Virginia: NSF, 2001): Table 2; pg. 48.

time his two years of grants totaled $40,000 per year. That was followed by three more years of $75,000 each. Those levels had allowed him to do excellent research for five years. But by 1993, he stated that even had he been offered $100,000 per year, it was equivalent to only about $70,000 in real spending power: no longer "enough to do cutting-edge research."[16] That same year, his young colleague in plant science at the University, Debra Mohnen, was seeking her first support from NSF. Having been unsuccessful, she argued that, "if they cut down the size of grants, they could afford to take a few more chances on a good idea." The Foundation was doing almost the opposite, however, and so "the difficulty of rewarding those already in the system," such as Hahn's research, "is that new people are hurt," she stated.[17] The small size of grants at NSF at the time and their short lengths forced scientists onto a "grant-writing treadmill." The late 1980s had seen the average "size and duration of…grants" increase, but by 1993, both size and numbers "plummet[ed]." Even though there was a shift aimed at supporting younger researchers and minority scientists, with both of those groups at undergraduate institutions, the Foundation was still "moving toward longer, bigger grants." And BBS (to be mimicked by BIO shortly) had made "larger grants a priority," having increased their value by 13% since 1989 and concomitantly—and consequently—lowering the award numbers by 8%. Bloch had supported the notion of larger grants, but could not in his period carry it out. Massey, however, repeated Bloch's earlier pledge and, given that the NSB had said in its newly released report by its Commission on the Future of NSF that "we favor research grants sufficient to do the work for which the grant is awarded," Massey would see the suggestion enacted. Finally, Director Neal Lane (permanent from October, 1993), while also in general agreement, noted that there was a lot of variation of belief from one discipline (directorate) to the next on what was the best size and length of award.[18]

The period saw three successive US presidents promise double-digit increases for the Foundation, but Congress balked each time. More funds were approved during the period for education at NSF than for any of the six pure science directorates; a rather stunning turn-around from the days of MACOS. And interest in supporting training and other educational efforts within BIO spiraled upward into the new millennium. What Debra Mohnen saw was what Jim Edwards, then executive officer of Clutter's BIO, reminded a science reporter: BIO was giving out fewer awards and at higher levels of support. It had been just the year before, 1992 that Mary Clutter decided to ban dual applications to such other agencies as NIH, for example. That caused the number of applicants to NSF to drop by 12%, but it also allowed BIO to maintain its success rate at 27% while increasing its awards by 4% in total dollars.[19]

---

[16] Jeffrey Mervis, "At NSF: Fewer, Longer Grants," *Science* (1993):1636–1638; quotation on pg. 1636.

[17] Ibid.

[18] Ibid.

[19] Ibid.

Notably, too, the Soviet Union collapsed at the same time the restructuring was occurring at BBS and that would mean a rather rapid change in direction of support within the national budget: away from heavy orientation toward defense spending and, with a new global economy about to arise, more basic and applied research toward global *competitiveness* would be the new order. "Superpower tensions have declined," said the NSB: in fact, the Board was very much alert to these issues and thus called for the creation of the Commission on the Future of the NSF.

> The new strategic plan will determine the direction of our future evolution in an environment that is very different from the one we were in only five years ago…[The plan must] recognize the changing nature of research and the changing context in which research is conducted.[20]

In the fall, their report came out and one recommendation stated that "[t]here is a *convergence between science and technology* arising from technology having a stronger basis in theory and data [than in earlier years]."[21]

Overall, calendar 1991 had been a highly productive year across the new BIO. The very last day of the year, for instance, saw the *Annual Report* for the Population Biology and Physiological Ecology Program come out.[22] PBPE was the one division added in 1976 after the 1975 reorganization of BMS into BBS. Its history had been a strong one with 218 awards made from 616 considered in FY91, for example. They totaled over $13 M in funds distributed. The program "participated in several major initiatives" during that year, including global change research, biodiversity, plant science, and molecular evolution. Aspects were spread over systematics, population biology, genetics, biochemistry, developmental biology, and the important and rapidly growing *computer science in data analysis*.

Peter Raven, as a member of the NSB, reminded the Board that its own Task Force on Global Biodiversity set up in the late 1980s had accepted its first report in the year 1991, too.[23] However, due to the funding constraints brought about by the Gramm-Rudman-Hollings Act (usually referred to as "Gramm-Rudman") in Congress,[24] the budget for biodiversity work had been "sidetracked." Reassessing the issue, Raven said that the FY92 budget should add $15 M to the earlier base budget of $20 M and that within five more years, the total ought to be $72 M! That was a pet area for Raven, but one that also was both very important and much in need of support. Biodiversity research would flourish in the age of highly computerized biology just then on the horizon, as it eventuated.

---

[20] See especially: NSB, "Minutes," NSB-92-146, August 14, 1992, 8–92:10ff, Appendix A, "Discussion Paper on the Future of NSF." By November, the committee charged with creating the report had completed it and it was incorporated into the "Minutes," of the 311th meeting (NSB-92-194) of the Board that month as "A Foundation for the twenty-first Century."

[21] Ibid., pg. 11–92:14 (Appendix A). Emphasis added.

[22] Carol B. Lynch, Mark Courtney, et al., "Annual Report FY 1991: Population Biology and Physiological Ecology Programs, Division of Biotic Systems and Resources," December 31, 1991.

[23] NSB, "Minutes," NSB-91-96, 5–91:8–9, May 8–10, 1991.

[24] "Balanced Budget and Emergency Deficit Control Act [Gramm-Rudman-Hollings Act; GRH]," P.L. 99–177, December 12, 1985.

Accomplishments in biological research were given in the Foundation's *Annual Report 1991*. Its authors had been for many years noting that bioprocessing and environmental biotechnology were expanding thanks to NSF support of the Center for Microbial Ecology at Michigan State University, East Lansing, and on biofilms at the engineering directorate-funded Engineering Research Center at Montana State University, Bozeman.[25]

Also of considerable import was the expanding work on mapping the *Arabidopsis* genome where the research community's first year's successes included expanding the number of known genetic markers in thale cress by some 36%. The multinational effort, begun in 1990, also found 200 new mutations in important developmental and metabolic control areas of the genome. The *Arabidopsis* story is a significant one and will be developed in more detail in Chap. 7.

The *Annual Report 1991* also reflected the changes that would establish the new BIO by providing the four division titles and their award numbers and dollar amounts in the financial tables. MCB had awarded 891 grants totaling nearly $77 M; IBN 998 at just over $72 M; DEB gave out 678 for a total of over $68 M; and BIR awarded 21 for nearly $38 M. That came to some 2784 grants totaling $255 M plus—and in a year in which reorganization seemed to swamp all.

So, the restructuring effort was generally complete: the path the four newly constituted divisions would take now needs consideration. As the NSB had said, a convergence of science and technology was neigh and it is that *rise of technology for biology* that is foremost in what this chapter explores. Clearly, the "End of the Frontier" was not imminent, nor was the "End of Biology" looming in the early 1990s: indeed, quite the contrary! The decline and brief plateau for life science funding at NSF as the last decade of the passing millennium dawned and as the rapid changes being seen in the disciplines of biology grew apace led to much that was new and that would make the next century that of the science of life. The twenty-first would be the "century of biology," as Bromley had predicted—and as Clutter had reiterated to the FASEB membership.[26] In addition, and notably so, biologist David Baltimore's sense of the new directions for science—cohesion and the end of borders—would begin to be seen more clearly than ever in the second decade and a half—1990–2005—covered in this study.

---

[25] From the download at: http://www.nsf.gov/pubs/stis1992/msf921/nsf921.txt. See ppg. 8ff of the 56 ppg. at the site. The *Annual Reports* became available that year online and all since that time have been widely accessible by going to the NSF website.

[26] Clutter, "The End," pg. 3. Clutter paraphrased Bromley's comment seen at the opening of Chapter Six above.

6.2 Founding a Division of Instrumentation and Resources

Choosing to begin this section of this history with a discussion of the division in BIO charged with supporting infrastructure and resources—the "convergence of science and technology"—versus any one of the three research divisions, is deliberate. That is so because, in every way, the charge to that division was to *support the needs* of investigators as they turned to one or another of those other three, *content* or *pure science,* divisions to obtain project funds specifically. And, as we shall see, the *convergence* foreseen by the National Science Board would lead directly to a number of completely new disciplines within biology. BIP/DIR/BIR/DBI: though changing its name over time, it was always charged with the funding of research *tools*, in the very broadest sense of the term, including fellowships and other forms of graduate support—"human resources." The coordinate divisions of DEB, IBN, and MCB are disciplinary and interdisciplinary, but provide little or no *infrastructural* needs, at least on a large scale: the instrumentation, student aid, and other resources to *do* biology. There are exceptions, such as the doctoral dissertation support grants in DEB. While the physical sciences directorate chose to have its various divisions handle infrastructure at that level (the astronomy program in charge of observatories; the physics program in charge of particle accelerators; etc.), BIO chose to set up a *special division* for the same sort of thing.[27] The Research Collaborations Networks was also created and placed in the division. That was the location for "glue money" that allowed for the running of workshops and many other types of collaborative activities—a "sort of living infrastructure," it had been said.[28]

The BIR that came out of the 1991–1992 reorganization, then, was a modification of the DIR of but a few years before, and the BIP of even earlier. As the DIR, it was the youngest of the divisions within the old BBS and, like all others that were children of AD Mary Clutter's restructuring, the division name was banned from being the same in the new BIO as it had been in the former BBS. But what had been the DIR of the late 1980s had begun years earlier as the Biological Instrumentation Program (BIP) and was lodged within the PCM division.

BIP was, in December of 1983 when John C. Wooley interviewed at BBS, without a program director. He filled that opening when he was hired by the Foundation on February 6, 1984. The PCM division was then led by Jim Brown and his deputy director, the still relatively new biologist at BBS, Mary Clutter. The BIP had been placed in PCM as there was a strong relationship between instrumentation and the little end of the biological spectrum, Wooley has stated.[29] Furthermore, BIP would not play a notable role in the *big end* of biology until 1986 when Wooley was given "authority" to support environmental biology, neuroscience, and behavioral science—the rest of "the whole biology directorate." Wooley also developed a

[27] David Schindel Interview with the Author, February 8, 2009.

[28] Ibid.

[29] John C. Wooley Interview with the Author, May 28, 2009.

relationship with Richard Louttit of the behavioral sciences area and aided his grantees in the instrument line. As well, funding was made available to the economics program for awardees needing the types of support peculiar to BIP.

Future BIP director Wooley first encountered the NSF when in high school having taken courses at a Foundation summer institute at the University of Kansas. Later, he became a graduate of the Honors College at Michigan State University, which did not require of its students that they have a disciplinary major; all that was needed was the right number of academic credits to graduate. That arrangement provided its students with considerable freedom, as Wooley recalled, and so he was able to take a wide variety of science courses.[30] From MSU, he went on to the University of Chicago to do his doctoral work in applied physics and biophysics. He did a short post-doc period there and then was a G.D. Searle Fellow at the pharmaceutical giant's research branch in England for a time. He moved on to become a research associate in molecular biology at Harvard, then to the Marine Biological Laboratory at Wood's Hole, Massachusetts, and finally became a professor of biochemistry at Princeton, where he taught a variety of courses in biology.

When he came from Princeton to BBS, Wooley filled the BIP leadership position as a rotator. He remained in that post until he was made a permanent NSF employee in 1986 and then, after a period of executive training in 1988, joined the federal government's senior executive service in 1989.[31] It was in 1988 that Kingsbury, then in his last year as BBS AD, came to Wooley's office one day and proclaimed: "I am going to make you a division." Wooley responded, "that's nice, David."[32] It took two years to get the plan through the many hurdles, including Congress, but Kingsbury did manage to begin the process; Mary Clutter finished it after Kingsbury's departure. Sonya Mallinoff recalled that Kingsbury had been pressing the BBS divisions to "do something with computational biology instrumentation, [but] they would take the money year after year and do nothing," she said. So, Kingsbury placed a levy of 10% on funding that went to all the BBS divisions in order to build the budget for BIR. There was unhappiness in the divisions over the technique, not surprisingly, but Mallinoff held that Kingsbury did not care: "He was quite an advanced thinker [who] wasn't afraid to take risks," for which she "admired" him.[33] The new division was funded at a very respectable $60 M—a long way from the first $4 M for BIP that Wooley had encountered just four years before. It also put Wooley and two new hires "all in one room." The division became the rather short-lived DIR. "It has morphed a couple of times," Wooley reminisced.[34] Crucially, too, the establishment of a division provided territoriality where budget was concerned: Wooley now

[30] Ibid.

[31] The SES is described at: http://www.transitionjobs.us/federal-civil-service-outlining-three-categories

[32] Ibid.

[33] Mallinoff interview, *op. cit.*

[34] Wooley interview.

controlled fully the funds that previously Kingsbury had held for him and released as necessary under BIP.

Wooley brought specific background and experience with him that made him an ideal fit for BIP and then for the new DIR. Such was the case, because while he had been working on his doctorate in earlier years at the University of Chicago, Wooley did graduate work in biophysics at Chicago's Fermi Institute.[35] A major fact in the history of biology is that many physicists of the 1940s and 1950s turned to biology in the post-Atomic Age in America.[36] By the time of Watson and Crick (the latter of whom also started out as a physicist) in 1953, a great many leading names, or persons soon to become great figures in biology of the time, were immigrants from physics to biology. Stories of their conversions, one might say, were legion at both Chicago and Fermi, Wooley enthusiastically recalled. And while some Chicago researchers, such as Nicolas Rashevsky, pioneer of mathematical biology, were developing the field in the late 1940s, well before Wooley's time there, Wooley averred that "biology wasn't ready for math." That attitude would live on widely into future decades.[37] Nonetheless, it was Wooley's introduction to both mathematical and computational biology—central features of programmatic activities of the DIR—that Kingsbury, and Wooley in particular, would later fashion. The latter's introduction to those disciplinary areas had been in the 1970s and so, upon his arrival at NSF in 1984, he was able to say that he was "someone who had used computing thinking…it was like any other tool." Much the same thing could have been said by or about Kingsbury. Indeed, Wooley has stated, "with apologies to [musician Bob] Dylan…the times they were a changing…and biology had matured a lot."[38]

By that comment, Wooley implied that biology was then becoming much more mathematical and any future directions it took would necessarily have to be, accordingly, more quantitative, let alone facing "sophistication inflation." Upon Wooley's arrival at NSF, Robert Rabin was acting AD for BBS, filling in during the interim between Betsy Clark's departure and David Kingsbury's arrival. Clark had strong interest and training in the highly mathematical area of biophysics, but Herman Lewis, PCM's chief, filled that gap once she had gone. And, according to Wooley, Lewis "created eukaryotic molecular biology"—an area strongly dependent upon biophysics—at the directorate. As the NIH "empire" grew, Wooley averred that the institutes took over the field from NSF. Wooley described Rabin as a very cautious

[35] For a brief historical introduction to the Fermi Institute, see: http://efi.uchicago.edu/about/

[36] See especially Gunther S. Stent's Norton Critical Edition (New York: W.W. Norton, 1980) of James D. Watson's, *The Double Helix*; *passim*. See also the classic work by Erwin Schroedinger, *What is Life?* (Cambridge: University Press, 1967) for his important commentary about the relationship of physics and biology.

[37] Introductory university physics courses, to this day, often come in two forms: "college physics," with minimal mathematical rigor; and, "general physics," much more quantitively based and hence often termed "calculus-based physics." It was common nationwide for undergraduate biology majors to take the former variety. Personal experience; personal knowledge. See also: http://nature.berkeley.edu/getzlab/Reprints/BiocalculusP.pdf for additional commentary on this matter.

[38] Wooley interview.

bureaucrat ("it's the safe way"), but one who listened to the community. Of Clark and Clutter, Wooley argued that both were "very organized, [and were] organization, rule and process centered"; but also, they were "risk averse,"[39] a view that may well be contested in the case of Clutter, as foregoing evidence suggests! Clutter had been very much influenced by Clark, he held, and Clutter's later role in raising the status of women at BIO, and beyond—as others have also said—was considerable. But under Rabin, the BBS was a "relatively nice place to be…[however, it] wasn't going anywhere." After all, though, Rabin was only *acting* AD and the search for Clark's replacement was about to bring the highly dynamic and very quantitatively oriented Kingsbury aboard. Furthermore, the dedicated computer scientist Erich Bloch was the director at the same time, of course. This confluence of interests that were shared by Kingsbury, Wooley, Bloch, and shortly Clutter would have a *profound influence on the direction of biology in America*, I contend.

Computational science, *per se*, as well as the rise in the use of computers *in* science, including biology, was becoming *de rigueur*. That, too, was the period when information science at NSF was still part of BBS, though soon to see its independent efflorescence under Bloch and Clutter with the establishment of CISE. Computation was rapidly pervading the Foundation. Wooley was fully in the thick of things from the very time he arrived. He had been supported both by NIH and DoE earlier in his career in his "own instrumentation research," but NIH cared more about what instruments could accomplish, he said, than *developing* them for whatever purposes they could serve. Overall, he argued, "there had been a lack of attention in government to new instrumentation." For that reason, Wooley "multiplied by a factor of 20 the level of commitment of the [BIP] in…dollars to instrument development."[40] He held that, outside of computing, supporting *instrument development* had been his "most profound impact" at NSF.

One particular example of making instrumentation available, he recalled, was in providing mass spectrometers (MS) to anthropology, still then part of BBS.[41] The BIP brought in experts on the instruments and also a group of anthropologists who did *not* use the machines and "we put them together in a room so they could analyze" the grant applications in which their colleagues were requesting funding to purchase the complex devices. Panels were reconstituted mornings and afternoons for several days by moving the MS specialists from one area to another, not only at BBS, but to other directorates to aid in doing the same type of thing in regard to MS machines. It was to manage the new BIP that put its director and his newly hired secretary (Laura Williams; who later became active in the group Women in Cell Biology; later chapters) and a single PO (Gerald Selzer, to remain at BIO into the new millennium) "all in one room." The comradery was strong: people across NSF were "very, very positive about their jobs." That was especially appreciated by

[39] Ibid.

[40] Ibid.

[41] For more on the field of mass spectrometry, see: http://www.asms.org/whatisms/p1.html. The site is that of the American Society for Mass Spectrometry.

Wooley, as he was fighting cancer at the time and had gone bald from the effects of chemotherapy. Pleasant coworkers made his travails more bearable. The disease relapsed in later years, but he continued to survive it to the time of this writing. The instrumentation initiative for BIP was, then, the "nucleus" for both the program and the division soon to come.

6.3 A Home for Protein and Other Macromolecular Data

In that period, Wooley "simply took the liberty of the authority vested in the program officer at NSF to take on new activities." Computing was "the first and foremost, and most visible, of those."[42] There were two "singular events," he said, that led to the expansion of computerization of biology at NSF and in the community, and for which BIP/DIR was partly responsible: funding the earlier established Protein Data Bank (PDB), and the Supercomputer Initiative. The second of these two has been discussed to some degree in earlier chapters, but there is more to tell. The first, the PDB, was the largest database supported by NSF in biology and has been the longest funded.[43] An allusion to it was made in the previous chapter, but a more detailed account is necessary. The story began at Brookhaven National Laboratory in 1971.

Three young crystallographers, Helen M. Berman (later to become a member of the BIO AC), Gerson Cohen, and Edgar Meyer, had first discussed the idea of creating a protein data repository, open to all, during the American Crystallographic Association (ACA) meeting in Ottawa, Canada in 1970.[44] (Proteins can be crystallized and so in that form they had been studied for decades.) The following year's ACA meeting in South Carolina saw a greater number of interested researchers join in the repository discussion. That led to a plan, which was subsequently submitted to the US National Committee for Crystallography for consideration; they later graced it with their imprimatur. At another meeting at Cold Spring Harbor (CSH), New York, later in 1971, the matter was pushed far enough such that one investigator, David Phillips, considered that the field was " 'coming of age' " and that it "heralded a new era in biology."[45] In many respects, that was true given the central role proteins play in cellular and, hence, all life. What the group, including such leading lights as Nobelist Max Perutz, was suggesting was the establishment of a dedicated location to accept, store, and share protein data (typically the numerically stated geometric point coordinates of crystal forms) among all interested scientists. Theretofore, individual researchers merely shared those data among themselves,

[42] Wooley interview.

[43] Ibid.

[44] Helen M. Berman, "The Protein Data Bank: A Historical Perspective," *Foundations of Crystallography: Acta Crytallographica Section A* (International Union of Crystallography, 2008).

[45] Ibid., pg. 1 of the online version of Berman's article. See: http://scripts.iucr.org/cgi-bin/paper?S0108767307035623

and that by the then best method: punched (IBM style) computer data cards. Myoglobin, the first protein fully described down to its component atoms and their positions within the structure, would require about 1000 cards to elucidate. It was Max Perutz and John Kendrew who shared the Nobel Prize in 1962 for demonstrating myoglobin's three-dimensional structure—and thus laid the foundations for the discipline of *structural biology*.[46] That would be an area of considerable interest to the NSF and the community in coming decades. But in the early 1970s, it was clear that a central data repository was a foremost necessity to move the field into the future.

One of the chemists, and leading crystallographers, present at the CSH meeting was Walter Hamilton of Brookhaven National Laboratory (BNL). The interested crystallographers approached Hamilton with the idea of setting up the repository at the Laboratory, and he agreed. At the same time, it was decided to set up a second, identical, set of files in the United Kingdom. Perutz, Vienna-born but long established in England, along with Olga Kennard, founder of the Cambridge Crystallographic Data Center (CCDC), met with Hamilton in the summer of 1971 in England and worked out the plan for how the centers would archive and share the protein crystal data with one another and the community. The BNL-CCDC arrangement was established and even with the sudden and unexpected death of Hamilton in 1973, the BNL connection was assured. Tom Koetzle, a former graduate fellow of Hamilton's, became the first director of the Protein Data Bank—the PDB.[47]

By January of 1976, various investigators had deposited structural solutions to some 23 proteins in the Bank, and 375 data sets had been shared with 31 laboratories within that year. The data were open to sharing in the form of the, by then, more advanced magnetic tapes instead of the earlier stack of cards. The NSF became involved for the first time, via BIP, with a grant of $33 k. PDB funding decisions were not without a history of some calumny at BIP, however, as a panel of the late 1980s felt that NSF could not award funds for the management of the data set itself more than what one could give an individual PI. That logic required change, clearly, but the panel determined to set a limit of $90 k per year for support of the PDB, which the newly arrived Wooley found "ludicrous."[48] Their logic was that the organization was a data*bank*, and not a data*base*. In essence, they were opposed to supporting an open repository as compared to funding individual PIs: they saw data set support as an undesirable precedent. Further, they did not want persons who did *not* deposit a given data set to be doing error checking on it. That was just one example

[46] For information about Perutz and his Nobel Prize, see: http://nobelprize.org/nobel_prizes/chemistry/laureates/1962/perutz-bio.html,

and for John Kendrew, see: http://nobelprize.org/nobel_prizes/chemistry/laureates/1962/kendrew-bio.html.

Structural biology has become a major subdiscipline of biology and has led to the formation of societies and journals devoted to it. See, for example: http://www.nesba.net/, the website of the New England Structural Biology Association.

[47] Berman, "A Histor[y]."

[48] Wooley interview.

as to why the desire on the part of the panelists was to favor a closed repository. The panelists were all crystallographers. Wooley was appalled: the panelists wanted a "pure *savings* bank," not an open community resource, he railed.[49] He went to his colleague in chemistry at NSF, Art Kowalsky, for advice. "I learned a lot from Art," he said. Wooley argued that the PDB may be *called* a databank, but it is a database, nonetheless, and should be open to everyone (as it had been), not just to those who, like his panelists, were part of the crystallographic community. The more notable aspect of that small brouhaha to this history was that Wooley held that the interchange was the beginning of his involvement in *bioinformatics*.[50] The PDB was a "knowledge resource" and must remain so.[51] The issue was resolved, finally, and the Foundation continued to support the PDB; it still does as of this writing. As of early 2011, the Bank contained nearly 71,000 structures!

That first (1976) NSF award to PDB was accompanied by the formation of an Advisory Board for the Bank at BNL and the PDB remained at the National Laboratory for the following 22 years. In 1999, it became a part of a yet greater organization worldwide, including such banks as the original cofounding CCDC and others. It then moved to Rutgers-The State University of New Jersey, New Brunswick, and is there called the Research Collaboratory of Structural Bioinformatics, the RCSB PDB (see also the end of this section). Other organizations that joined in the RCSB included the UC, San Diego Supercomputer Center (following section). The National Institute of Standards and Technology (NIST) also became a player in the RSCB PDB.[52] Today, the PDB is co-managed by Rutgers and UC, San Diego and is funded by seven US agencies, NSF first among them.[53]

But in the initial decade of the existence of the PDB as NSF support grew, the Bank was making considerable efforts to entice the community to submit voluntarily (still the approach as of this writing) those data on the structure of various macromolecules which scientists in that community had solved. Protein data were submitted (such as that for hemoglobin and lysozyme), but so also were the structures of such things as tRNAs, or transfer RNAs, which are not proteins at all, but nucleic acids.[54] The structural biology community was so small at the time (1970s and early 1980s) that virtually everyone knew one another. By the mid-1980s, though, rapid growth was seen in the field because of the concomitant rise of molecular genetics, and the soon-to-be-realized possibility for cloning genes. Once that had occurred, one could force such genes to make the products for which their nucleic acid sequence was coded. Thus, there was a supply of expressed proteins (at that time the most commonly recognized products of genes) which structures lay

[49] Ibid.

[50] For a definition of bioinformatics, see: http://www.ncbi.nlm.nih.gov/About/primer/bioinformatics.html. The first sentence there reads: "Bioinformatics is the field of science in which biology, computer science, and information technology merge to form a single discipline."

[51] Wooley interview.

[52] Berman, "A Histor[y]"

[53] For more on the RCSB PDB, see their home website at: http://www.pdb.org/pdb/home/home.do

[54] See one aspect of this fascinating history at: http://www.ias.ac.in/jbiosci/oct2006/453.pdf

open to assault by the crystallographers.[55] However, BIP funds also made possible, in part, the development of new and different types of instruments with which to visualize the three-dimensional structure of proteins. While the basic form of X-ray crystallography had been the method in use for more than three decades, advances such as nuclear magnetic resonance and synchrotron radiation came into play. In later years, other methods became available, too.[56]

During that period, computerization was rapidly growing with regard to both speed of processing and storage capacity. The appearance of graphic computing was a dramatic development for structural biology as it allowed for visualization of the molecules.[57] As computerization techniques, both software and hardware, permitted greater ease in solving structures, more and more of them were published. "Soon it became commonplace to see pictures of structures on the covers of journals such as *Nature* and *Science*."[58] Structural biology was growing apace and "definitive information" was appearing that seemed to be the "key to understanding the molecular basis of biology and medicine." Committees of the crystallographic unions, societies and associations were set up to consider the matter and to determine if it had not finally become necessary to *require* data submission to the PDB (reminiscent of the Asilomar decision discussed earlier). By 1989, such agreement was formulated and published by the International Union of Crystallography.[59] At the same time, the PDB was able to accept and share data fully electronically over the early Internet; no longer dependent upon punch cards or even magnetic tapes. As well, nuclear magnetic resonance (NMR) became available as a tool to determine structures; and flash freezing and use of synchrotron (a more modern update of the classic cyclotron[60]) radiation, made it possible to solve the structures of such complex entities as viruses and ribosomes; structural data for them began to arrive at the PDB and was archived.

Since the PDB has been funded by NSF continually, reports on its progress over time were necessarily submitted to BIR, and its later incarnations, and were received by PO Gerald Selzer there. By 1996, PDB's 25th anniversary, the Bank had some 691 structures newly archived that year of a total 4050. Not only were proteins and nucleic acids in that grouping, but also some 12 carbohydrate structures had been deposited. Techniques to determine the structures on file had grown from the primary use of X-ray diffraction (3865 of the total) to include the more newly established NMR (651), and also some structures were listed that had been created using

---

55 Berman, "A Histor[y]."

56 See, for instance: D. Blow, *Outline of Crystallography for Biologists* (Oxford: University Press, 2002), *passim.*

57 Berman, "A Histor[y]." See also the PDB website at the following page for the example of an animated graphic applied to the structure of hemoglobin, with and without its oxygen atom: http://www.pdb.org/pdb/static.do?p=education_discussion/molecule_of_the_month/pdb41_2.html

58 Berman, "A Histor[y]."

59 Ibid. See also: *Acta Crystallography* (1989) A45:658.

60 For an explanation of a cyclotron, see: http://education.jlab.org/glossary/cyclotron.html

theoretical modeling (136). Selzer's hand-written notes on the PDB/BNL annual reports indicate a keenly interested PO.[61]

By the first decade of the new millennium, the PDB archived structural data with weights of over 2 M atomic mass units (amu; also known as Daltons)—very large structures, indeed. By comparison, myoglobin only weighs just over 17,000 amu.

By the mid-1990s, too, the NSF had already issued a grant competition to form a "Macromolecular Structure Database" (MSD) that would go beyond PDB.[62] BIO stated that "the design, development, implementation and use of biological information resources [was] a priority." Therefore, the MSD was needed and NSF was joined by the Office of Health and Environmental Research, DoE, NIGMS, and NLM, as sponsors. Any grant application made under the MSD solicitation must then have also shown a plan to incorporate data from the PDB; and the Bank's data were increasing at 45% per year! The MSD competition was won by Rutgers, and hence the RCSB came into existence.

The Protein Data Bank and its cousin, the RCSB, then, were responsible in some degree for the rise of several new disciplines in biology. But they depended upon the simultaneous developments in computerization of data of all sorts—and the BIR/DIR was there to move that issue along. Therefore, with regard to the major themes of this history, it is posited here that the areas of structural biology, bioinformatics, and computational biology (below) represent excellent aspects of both interdisciplinarity and, as will later be considered, the significance of reductionism to Millennial Biology.

6.4 Supercomputing

Why had Hamilton and the Brookhaven National Laboratory played a part in the history of biology? After all, the work of the national laboratories was aimed primarily at physics, especially atomic physics in those earlier years, and later focused on energy and other areas in the later middle-third of the last century. Hamilton was a crystallographer, but by no means was that discipline relegated only to those interested in the crystallized form of organic compounds, such as proteins, or even nucleic acids—long an interest in England, for instance.[63] Those acids, when combined, formed certain intriguing macromolecules and it had been Rosalind Franklin whose X-ray radiographs of such a macromolecule, DNA, were heavily used by

---

[61] For example, see the "Annual Report of the PDB, July 1996."

[62] National Science Foundation, "Macromolecular Structure Database: Proposal Solicitation," NSF-94-3.

[63] The Australian, (later Sir) William Lawrence Bragg, was a founder of x-ray crystallography in England from as early as 1915 and central to the DNA story of the early 1950s in several senses. See a brief biography at: http://nobelprize.org/nobel_prizes/physics/laureates/1915/wl-bragg-bio.html

Watson and Crick to solve the three-dimensional structure of the genetic material.[64] The PDB accepted data on not only proteins, but nucleic acids, too—and many other structures over time. The answer as to why the PDB was situated at a national laboratory was due to the *computing capacity* of those institutions. Why BNL specifically is simply because Hamilton was present at the right moment in history to offer the Laboratory's services. As has been seen, several new areas just then arising in biology were entirely dependent upon large computers that, in the 1970s, were to be found almost exclusively at the national laboratories. Indeed, the first true supercomputer, the Cray 1, was installed in the national laboratory at Los Alamos, New Mexico, but not until 1976.[65] So even when the PDB was founded five years earlier, true supercomputers were still in the future. Nevertheless, large, high capacity machines, needed for dealing with macromolecular architecture, were present at BNL and just a few other laboratories in 1971.

By 1984, all that was to change, however. The rise of what would later be called a cyberinfrastructure for the biological sciences was likely at the center of the efflorescence of a number of new fields. Supercomputers played a *seminal role* in the establishment of the disciplinary areas of structural biology, bioinformatics, and computational biology, let alone seeing DIR appear shortly thereafter. They were all birthed at generally the same time in connection with the computerization of biology. As early as the 1984 NSF *Annual Report*, computers for systematics were reported to have been seeing use.

> Computers have contributed to recent gains in the evolutionary meaning of shape and in image analysis, two areas of systematics…Karl J. Niklas [of Cornell University, has done computer simulations of] modern-day tree ecology, as well as the evolution of plant form. Millions of years of evolutionary change can be simulated [and] comparisons between simulations and the fossil record offer means to test evolutionary hypotheses.[66]

It was that same year that the Foundation's High Performance Computing and Information Technology initiative began and the BIO AC met to "evaluate if and how biologists could use supercomputers."[67] The answer to the first question, they said, was a "resounding yes." That to the second "led the Instrumentation Program [BIP] to undertake entirely new directions." The little biology of the PDB, in

---

[64] Much has been said about Franklin in recent years and the failure to have given her sufficient credit in the 1950s, let alone in following years. The literature has grown and her place is now better recognized by most in the history of biology community, and beyond. There has been a sort of apologetics community arise concerning Franklin. Though not always a reliable source, Wikipedia's article and bibliography on Rosalind Franklin is excellent. See: http://en.wikipedia.org/wiki/Rosalind_Franklin

[65] See: http://www.osti.gov/accomplishments/nuggets/historynatlabs.html; Seymour Cray's first supercomputer was delivered to Los Alamos National Laboratory in 1976. For a timeline of supercomputing milestones, see: http://www.supercomputing.it/What/chrono.htm. It was only after the early 1980s that supercomputing really began to burgeon.

[66] National Science Foundation, *Thirty-Fourth Annual Report for Fiscal Year 1984* (Washington, D.C.: Government Printing Office, 1985), pg. 27.

[67] See: "Historical Background: Building a Cyberinfrastructure for the Biological Sciences," at http://research.calit2.net/cibio/bground.htm, pg. 1 of 3 online pages.

particular, was a central feature of the rise some years later of CIBIO—the Cyberinfrastructure for the Biological Sciences Program, discussed below. (Even later, CIBIO would be applied to programs in big biology, such as LTER, equally successfully.) The funding provided in the CIBIO effort was the first within the federal government and it was BBS that was at the forefront.[68] The appearance of computer-based biology, the rise of supercomputers, and the birth of a linking device, the Internet, were all coming together to interface almost simultaneously. CIBIO was to be, and became, a highly democratic device and a critical program in the history of modern biology in America. It was once humorously said by Alan Bleasby of the European Bioinformatics Institute at Cambridge University that, after all, "two months of hard work in the laboratory can easily save an afternoon on the computer."

After computer scientist Erich Bloch arrived at NSF and when the supercomputer centers, institutions other than the national laboratories, were becoming established through Foundation support, it was Wooley's BIP that was deeply into the funding efforts.[69] The center at San Diego then became one of the premier locations for the rise of computational biology, close relative of both structural biology and bioinformatics. NSF's biologists became involved in all of this because the information science program was still part of BBS, and Bloch and Kingsbury needed to support compute cycles (time use) at supercomputer centers for grantees, BBS and otherwise. BIP funds during that time came to the supercomputer center in Columbus, Ohio, for *generic* compute cycle use, but a center and a machine aimed at biological uses *in particular*, quickly became focused at San Diego. That Center opened in early December of 1985, and saw the first of over 100 already approved projects come under study immediately.[70]

Wooley had great concerns with the use of funds for compute cycles as, just after his arrival at the Foundation, he discovered that half of his budget for BIP went to computer service charges only—it was hard to see the science coming out of that. That first year, his program was given $8 M, but $4 M was instantly lost to such use. Similar compute cycle problems were being faced by most directorates at NSF at the time, so Wooley went to Rabin (Kingsbury had not yet arrived) to get permission to do something about the drain on the budget. Wooley was able to arrange a meeting with DoE, ONR, and the crystallographic community, and thus gathered about 90 persons together for his conference. He arranged the participants into four groups that emphasized various areas of the aborning field of computational biology. They held a three-day meeting to talk about the best uses of NSF funds—the dilemma that Wooley had stepped into. They solved the issue such that BIP was able properly to direct funding to *individual awards* that produced good science, and did so by employing supercomputers. The grants *included* costs for calculation cycles as part of biological *project-based* research, not just cycle service costs. By the late 1990s,

[68] Ibid., pg. 2.

[69] Wooley interview.

[70] See: http://ucsdnews.ucsd.edu/graphics/images/2010/10-18timeline_web_.jpg

Wooley has stated, biology was the biggest user at NSF of supercomputer time at the centers: some 28% of cycles going to that area alone. He went on to say that Kingsbury had been a major promoter of such use and began that upward trend during his tenure at BBS. Further, Wooley held, BBS Deputy AD, Alan Leshner, along with Kingsbury and himself, all contributed to the early growth of computational biology: they all had the "just do it" attitude, he said.[71] And Leshner's promotion by Kingsbury to become the latter's deputy AD was a "profound" decision as Leshner had been to that time working, Wooley said, "in the trenches": Kingsbury saw him as a leader and so thus the promotion. The three were something of an idea factory in that the early growth of DIR was due to input from each and it would later become difficult to determine whose ideas led to what, Wooley averred.[72] It was a "very proactive period." But it wasn't

> just about NSF being proactive, it [was] about NSF having failed to be responsive adequately before…there [was] no proper infrastructure [as in the sense of a needed DIR], there was inadequate support [for the community]…science [had] changed and new opportunities [had arisen]…the community was struggling…running around getting private donations in order to do instrumentation [and computerization].[73]

The growth of computerization in biology is basically bioinformatics and the first person to lead the program in that area under Wooley was Bob Robbins. Also playing a role in the early expansion of that area was Peter Arsberger, an applied statistician who was the director for the statistics program in the mathematics division when it was still part of the physical sciences, mathematics, and engineering directorate. Arsberger worked with Wooley part of the time in the period of 1986 to 1988 before BIP became the DIR. The former helped create a bioinformatics grant competition; and providing an example here of cross-directorate—interdisciplinary—interaction.

6.5 "Fearless Biologists"

Once the DIR was established fully, the programmatic activity extended beyond the two major programs of databanks/bases and supercomputing to include not only bioinformatics but also graduate training programs. The first PO that had been hired into the new DIR had been Jerry Selzer (above) and he was charged with, among other things, graduate education programming. It was now clear that something beyond the rather limited, but highly productive, BIP could be offered to the community in the form of educational programs under the new DIR. The community would benefit by the new division not only providing instrumentation and its development, as well as supercomputer-based projects, but also by providing for a cadre

[71] Wooley interview.

[72] Ibid.

[73] Ibid. The interview with Wooley was best characterized as wide-ranging: mixing subjects was typical throughout its length. It was one of the most valuable interviews I did for this study.

of new young biologists adept in the use of the new technologies then appearing. Leon Lederman, it might be imagined, would have seen the value of "sophistication inflation" in all of that. And, again, interdisciplinarity was given yet another face through the training programs. John Wooley argued that the DIR was building the infrastructure, and the training of generations of users of it was an obvious corollary need. A committee was created to help guide the establishment of the training program. They saw the future.

> What you are trying to do, John, is to create a generation of *fearless biologists* who are willing to use computer-centric tools, computational tools, bioinformatics, physical instrumentation, chemical instrumentation.[74]

That training program led to others, such as the Interdisciplinary Graduate Research Fellowships.

The 1989 appearance on the scene of Mary Clutter as the new AD played a crucial role for BBS, and later BIO, by helping distinguish NSF from NIH in a very critical way. During the period while DIR was coming into existence, Mary Clutter had replaced David Kingsbury, both of whom can be said to have been *fearless biologists*. As we have seen, quantitatively oriented Clutter had already helped create CISE and, in addition, was a vigorous spokesperson for plant science: she saw it as a keystone that was the "ticket that describes us as different than NIH…a boundary with NIH," as Wooley interpreted it.[75] And he was not alone in recalling the importance of Clutter's push to get out from "living under the shadow" of NIH. Sonya Mallinoff remembered it the same way. "I've been through every kind of strategy to deal with that shadow, and none of them worked."[76] Nonetheless, Clutter moved with alacrity into a much greater commitment in the plant sciences than was theretofore the case at BBS. From the period 1988 to 1992, the directorate held a series of retreats to build up the plant sciences, concomitant with the great reorganization of BBS into BIO. At the same time, another area of less direct concern to NIH and the casting of shadows was environmental biology: it was also growing with support from Kingsbury and then Clutter, Wooley averred.

Thus, by the end of the early 1990s, BIO would not only exist, but the creation of the four divisions that Mary Clutter had called for from the various working groups she had assembled to frame the new directorate, would become established. As per instructions, no old title for a division could remain in the group of four for BIO: thus, the former DIR of BBS had become the BIR of BIO, though it would often be called the DBI (Division of Biological Instrumentation and Resources) and, later, officially so. The division drafted what would become its *Opportunities*

[74] Wooley interview. Emphasis added.

[75] Ibid.

[76] Mallinoff interview. It was also Mallinoff's belief that Clutter did not only seek more funding for the plant sciences alone, but to raise the entire level of funding for BBS, later BIO, and so all, including social and behavioral sciences, areas of the directorate would profit. The argument to highlight plant sciences to distinguish BBS (NSF) from NIH, then was only part of Clutter's thinking, according to Mallinoff.

brochure for the "fearless" biological community for the first years of the decade and explaining its role, thus:

> [BIR] is to support the necessary investments in people and their research capability for the broad spectrum of research supported by the research divisions of the Directorate for Biological Sciences. The various kinds of investments made by BIR all directly support the research activity in DEB, IBN and MCB.[77]

The BIR of the early 1990s told the community about some of their "program perspectives…[and] priorities." They had just begun funding 15 interdisciplinary Research Training Groups (the RTG Program) in FY90 and FY91 for undergraduates, graduate students, and postdoctoral individuals. The seminars and workshops and equipment that were funded under the RTGs supported many students and were often inter-institutional and multidisciplinary in character. The RTG program was suspended for a year (the funding decrease; above) and then the plan to return it in FY94 was forwarded as a high priority area identified by the Foundation.

The REU Program, Research Experience (later plural) for Undergraduates that had begun in FY87 with some $1.2 M, had supported 300 students under 33 awards for all of BBS's social and biological sciences. It grew to $2.7 M in FY93 for BIO alone and so funded 860 students in biology, specifically, scattered over 58 university sites nationally: 8–15 students per locale.

In addition to the two student-oriented funding lines, the new BIR supported the more clearly infrastructural and resources' areas such as database work in biology. Since biological sciences were "becoming increasingly data rich, not just in the volume of data, but also in their complexity," the division was then focusing on three major areas: search algorithms based on images, three-dimensional structure, or pattern recognition (of the brain, for instance; or Landsat satellite imagery for ecological studies); methods for searching databases, even if distributed (over the growing Internet); and, techniques for long-term (multi-decadal) storage of large databases. The Protein Data Bank was just one example of funded projects.

Other areas that BIR supported included: high performance computing and communications; modular instrumentation software; high capacity instruments; equipment for biotechnology; BIR's portion of the Plant Science Initiative (a very large area of commitment in BIR, as well as the other divisions); and, the field research facilities that included the biological field stations and marine laboratories. A significant infrastructural effort of BIO was that of those Science and Technology Centers, STCs (see following section), which were biologically based.

Concomitant with the change from BBS to BIO, Wooley stated that more oversight was coming into NSF and that Clutter had to deal with more program monitors, more rules of a variety of kinds and, in general, more tension. But in the period 1991–1992, Wooley's cancer had relapsed and he was under stress from his health. In the fall of 1992, he left the NSF and moved to the DoE, where he stayed for another seven years. The DoE site was closer to the locale where Wooley received

[77] Typed draft of "Division of Biological Instrumentation and Resources: 'Opportunities'" likely 1993, though the date is not clear; a seven-page document.

his cancer treatment than was the Foundation's headquarters. Crucially, though, after he departed from Energy in 1999, Mary Clutter asked him to serve on the BIO AC. He did so and stayed on as a member until 2006, by which time he was in the senior administration of UC, San Diego. In the years after he left NSF and during those in which he was on the AC, he was still able to keep close track of happenings at BIO and at the Foundation: thus, he played major roles in the rise of cyberinfrastructure for biology led by BIO. He watched the early development of the NEON project and the growth of microbial ecology (both considered in Chap. 9), as well as the tenure of biologist Rita Colwell as Director of NSF and her promotion of the concept of biocomplexity. That subject is considered later in this chapter and was of great interest to Wooley as he saw in it many of the efforts that the early DIR had engaged. Thinking along the lines of biocomplexity seemed to be "out there": Kingsbury and Colwell had gone to graduate school together in time and space, at least, for instance. Wooley held that DIR was creating the infrastructure to examine "the complexity of biology."[78] "The 21st century is about biocomplexity," he has stated.

> The dimensionality is extraordinary. The living world has degrees of intricacy and n-dimensional space in terms of understanding how it works. If you look across time, space, and organizational complexity, it's hard to have anything that compares. It is naturally bounded, so you can go from atoms to ecosystems [over] billions of years of evolution. You've got a new level of experimentation…where computing and bioinformatics really comes in…knowledge repositories…models…remote instrumentation. Biocomplexity: trying to put together a large historical spectacle of the environment.[79]

Not the language of a classical biologist!

Wooley was replaced in 1992 temporarily by Frank Harris, who was Clutter's first permanent Deputy Director of the new BIO at the time. Wooley had also held together the infant SBE until it had its own AD—the final act of the great reorganization. James L. "Jim" Edwards would follow Harris in a few years. Harris was later succeeded by James H. Brown, by 1996, as permanent director of BIR, though Brown had been permanent director of MCB in 1992, and of PCM years before that. Machi Dilworth (below) would succeed Brown. The situation was one less of "revolving doors" for division directors than an internal "musical chairs."[80]

Looking back as the founder of the DIR, Wooley said that "[e]verything I funded paid off. But there are things I sure as hell wish I would've funded. There's only so much money, and you have finally got to make a choice."

---

[78] Wooley interview.

[79] Ibid.

[80] These data are to be found in NSF telephone directories of the period. Some of those, during the 1990s and beyond, are available on the NSF website.

6.6 Centered

The specifically biologically oriented Science and Technology Centers (STCs), discussed at some length earlier, were a responsibility of the BIP when the first of them appeared in 1988; they then fell under the new DIR after the reorganization. In either case, they were another part of Wooley's rapidly growing division and were overseen by Jerry Selzer. So large a project did the whole series of them become Foundation-wide—there were 25 by 1992—that they came to be managed through an office of their own, as has been seen. The Board approved the creation of a new entity, in August of 1987, called the Office of Science and Technology Centers Development (OSTCD) as the *coordinating* device for the coming growth of centers NSF-wide. That included biology STCs, as well. But it was an administrative office, not a science content unit; the science was left to BIP/DIR, at least for those centers tied to BBS/BIO. By the early 1990s, however, the OSTCD was superseded by the Office of Science and Technology Infrastructure (OSTI), a larger and broader entity covering many aspects of the whole Foundation, STC management and coordination included.

With the collapse of the Soviet Union, so "sudden and unanticipated," and the quickly foreseen subsequent redirection of national budgetary priorities, the directors of the 25 STCs met in Michigan in early October, 1992, and created a document that they submitted to the NSB's Commission on the Future of the NSF. They were particularly concerned that the Foundation should expand its efforts in connection with a specific recommendation of Director Walter Massey: that was not only to continue to support basic research, but to "bridge the gap between [*sic*] academia, industry and society."[81] There had been fear of a reduction of funding for the STCs, given the just-ended two lean years (see earlier). And there would be the unknown impact of likely cuts in military funding and what that may have portended for emphasis elsewhere in the annual federal budget. It was that same situation that had also excited those scientists polled by Leon Lederman. While the economy was to turn around just at the time the STC directors wrote their letter, such was not yet obvious. And there were other, possibly more important, reasons the directors were concerned. Since the STCs "represent a very successful paradigm," increased linkages to *societal needs* were of great concern to the directors, but it may well have been that the worried directors felt the need to shine a light primarily upon the broad directive from Massey.

The qualms over finances by the STC directors were important as, by late 1994, there was discussion at NSF of phasing out support for the centers. There had been from the inception of the STC program the rule that after 11 years any centers begun under the umbrella of the NSF would no longer be funded (see an example from Virginia below). An algorithm had even been created to slowly wean the STCs away from NSF support, and a "thorough evaluation is being initiated" at the six-year

[81] A seven-page letter, with attachments, from the Directors of the STCs to the NSB, October 14, 1992. Quotation from pg. 3.

mark for the original 11 centers (evaluated every three years) and of the program—its first assessment.[82] It was extremely thorough, indeed.

The review was closely tied to the then new Government Performance Results Act of 1993 (GPRA)—another reason for some level of fear on the part of the center directors. GPRA would turn out to be immensely important across the federal establishment, not least at NSF. Assessments done under the Act found that there was "waste and inefficiency" in federal programs across the agencies and that their managers were "seriously disadvantaged in their efforts to improve them." That was so because of poor articulation between program goals and actual performance. Furthermore, due to "insufficient attention" having been paid to program performance and results, Congressional policymaking was "seriously handicapped." Given that, findings under the Act sought to "improve the confidence of the American people" by: holding "federal agencies accountable for achieving program results"; measuring programs against their set goals; promoting goal focus; "help[ing] federal managers improve service delivery"; and, several other points.[83] *Accountability* had come to government in a very big way.

"For decades, the Cold War was the indispensable pretext for national investments," *The Washington Post* stated in early 1994.[84]

> The National Defense Highway Act had nothing to do with security, but the name allowed Congress to dodge a debate about whether government could make sound economic investments. Even improving education had to be camouflaged as the National Defense Education Act.

While the *Post* writer was focused primarily upon the growing idea of "economic competitiveness" as a replacement for Cold War "camouflaged" policy-making, there was the question of how best to guise support for basic research. A Senate Appropriations subcommittee under Senator Barbara Mikulski (D-MD), concluded that NSF should "devote at least 60 percent of its funds to 'specific national goals' [via] 'strategic research,'" an amorphous and undefined matter. "Could Darwin or Mendel have set 'annual, quantifiable, performance milestones'?" An uproar led Mikulski to clarify the issue that strategic research could only be interpreted to mean applied research and that, no, basic research could not be performance-driven nor accountable in the same manner as could applied research—biotechnology and

---

[82] Memo from Nathaniel Pitts to STC Technical Coordinators, which represented a large number of persons at NSF, October 26, 1992. Pitts was, at that time, director of the Foundations' Office of Science and Technology Infrastructure, OSTI. David Schindel was by then gone from the systematics program leadership having then moved to OSTI. See later footnotes.

[83] See the "Findings and Purpose" section of GPRA at: http://www.whitehouse.gov/omb/mgmt-gpra/gplaw2m#h2. See also a history at: http://www.acf.hhs.gov/programs/cse/pubs/1996/reports/gpra_implementation.html. GPRA had become a very crucial tool in federal management and it continued (and still does as of this writing) to become more and more important to all agencies, not least NSF. The bill had originally been introduced by Senator William V. Roth (R-DE) in the fall of 1990. After much work in committees, the Act was finally signed by President Clinton in August of 1993.

[84] Jessica Mathews, "Could Darwin Have Set 'Performance Milestones'?," *The Washington Post*, March 2, 1994.

high performance computing were given as examples of the latter.[85] That environment was a new one in a number of ways, one in which GPRA had seriously to be taken into account.

The Office of Management and Budget, using GPRA, focused at the Foundation on the Science and Technology Centers as one of about 70 "Pilot Projects" marked for special attention. And it was the OMB that had an "issue" with the Center's program: there was a "need for measurable quantitative performance indicators."[86] However, OMB conceded that quantitative measures might not have been fully appropriate to the STC program and that options needed to be explored. From 1988 to 1994, NSF had spent about $200 M on the STC program. The program, as with virtually all features of the US government, according to GPRA, was required "to define and communicate its goals, objectives, and results in terms of performance" to be understandable by OMB, Congress, and the American people. David Schindel, discussed earlier, then in the broadly inclusive Office of Science and Technology Initiatives and specifically in charge of STCs, wrote a white paper on performance measurements for the Center's program. He noted that:

> This approach to program management is new to many federal agencies, and the use of quantitative performance indicators is foreign to the academic community in which NSF invests public funds.[87]

Academics were, he reminded his readers, "generally assessed using some form of qualitative peer evaluation." The task of responding to the intent of GPRA was to move from qualitative to quantitative measures. As fundamentally an "investment broker," NSF did not have a clearly defined " 'provider-customer' relationship" with the American public. Quick returns on investments were not characteristic of the activities in which the Foundation engaged. But since OMB had already signaled that the STC program (and others at NSF) *may* not lend themselves to purely quantitative evaluation, Schindel turned to Sect. 1115(b) of GPRA. There, allowance was made for indicators that were a *mix* of both qualitative and quantitative factors, and the section also recognized that time lags existed in inputs and outcomes for programs such as those engaged in by the NSF. New weapons' development would have been a sterling example. Schindel crafted an "alternative approach to GPRA" that measured quantitative input indicators, such as numbers of researchers supported by an STC program, the proportion of females and minorities at the centers, and the amount of research instrumentation required. His *output* indicators would be a mix of quantitative and qualitative factors including expert opinions on articles published, numbers of students graduated, placement of those students in academe or industry, and new instructional materials created. Out*come* indicators included

---

[85] Ibid.

[86] Draft Memo from David E. Schindel to Anne C. Petersen, Deputy Director of NSF, n.d., but sometime in 1994.

[87] David E. Schindel, "Performance Measurement and the Science and Technology Centers Program," a white paper attached to a series of memos that comprised a growing package of STC evaluation materials. Undated, but likely later 1994.

evidence of industrial use of a given STC's research results, new areas of inquiry opened, application of discoveries or new instruments developed, success of graduates, etc.[88]

By 2003, the NSF was hosting some *seven different categories of centers*, many of those oriented toward engineering and materials science, some in chemistry, a few in education, and four in the area of biological analysis and synthesis. The STCs were the second oldest of those categories, with the Engineering Research Centers being two years older. BIO had become centered in not only STCs, but the new category of Centers for Analysis and Synthesis (CAS) that had been established in 1995. The CASs were

> designed to continue development of new tools and standards for management of biological information and meta-information, support data analysis capabilities with broad utility across the biological sciences, host [interdisciplinary] workshops…and begin to host and curate databases.[89]

In a word: bioinformatics. Four of them had come into existence and included the: National Center for Ecological Analysis and Synthesis (NCEAS); National Evolutionary Synthesis Center (NESCent); and, a Plant Science Cyberinfrastructure Collaborative (iPlant; formerly PSCIC) for "grand challenges in the plant sciences." Also established was the National Institute for Mathematical and Biological Synthesis (NIMBioS; formerly CIMBS)—all strongly interdisciplinarity!

Among the 25 STCs, *per se*, those that were biologically oriented included the: University of Washington's Center for Molecular Biotechnology; Center for Microbial Ecology at Michigan State University; University of Illinois' Center for Magnetic Resonance Technology for Basic Biological Research; Carnegie-Mellon University Center for Light Microscope Imaging and Biotechnology; UC, Davis Center for Engineering Plants for Resistance Against Pathogens; and, the Center for Biological Timing at the University of Virginia—the CBT.

The CBT can stand as an exemplar for the biological STCs. The Center survived for the 11 years promised each center by NSF. After the Foundation's start-up and maintenance for that decade plus, a center was to become self-supporting in some manner, if it wished to continue. The CBT had not done so by 2002, *qua* center, but represented a successful effort, nonetheless. It began in that very active year of 1991 and lasted until 2002, leaving behind a distinguished history of discoveries. The University of Virginia (UVA) set the beginning date for the Center at 1991, but it was 1993 when the CBT was formally constituted as a multi-institutional, multidisciplinary collaboration being supported by NSF. It had a precursor known as the Biodynamics Institute (BI) dating back only to 1989. When, in 1991, the Institute sought STC funding from NSF, its research role was expanded from biological timing alone to also include all forms of oscillations in living systems. When first considered by NSF, UVA was alone in its proposal and the grant was not made. In 1993,

[88] Ibid.

[89] See: http://www.nsf.gov/about/budget/fy2009/pdf/37_fy2009.pdf. The CASs are described on the first two online pages.

however, UVA had added Northwestern University, Evanston, Illinois and Rockefeller University, New York City to the collaboration. That consortial arrangement was approved by the Foundation and the CBT became an official STC. In later years, the University of Virginia entered into additional relationships with Scripps Research Institute, San Diego and Brandeis University, Waltham, Massachusetts.[90]

In the first year, the CBT leadership did not have a fully clear idea of its direction, but with the addition of an advisory board, the Center "was forced…to become more focused." That was accomplished by the creation of a "core project," termed the Clock Genome Project, in which the novel clock mutants of *Arabidopsis, Drosophila*, and the common house mouse (*Mus musculus*) were sought in isolation. The project had biologists creating mutants of those three organisms; after that, a molecular group would characterize them. Even further beyond, physiologists would look at how the brain had changed with the mutations and, finally, endocrinologists would study how the biological clock(s) affect the circadian (approximately 24 h) clocks of the body. That process of research was sequential and multidisciplinary so that the Center was often viewed as a "factory."[91] At the time of the final NSF review of the CBT in 2000, it had 29 PIs, 14 of whom were receiving Center (that is, indirectly NSF/BIO) support and the remaining group had other funding. There were some 38 postdoctoral fellows and an additional 38 graduate students. The Center was the primary locus for the production of scientists involved in biological timing studies in the United States for much of a generation: a superb example of the fruits of the *intricate dance*. Undergraduates (there were 20) were funded, as well, and there were some 105 visiting scientists. Even precollege students—223 in the year 2000—had experiences at CBT. The NSF review team of the period spoke in terms of the "smashing success" of the Center, its investigators having published no fewer than 40 articles in *Science* journal alone, a mark of considerable prestige, over the decade of its NSF-funded existence and, in the year 1999 as an example, over 112 publications came out in total.

The scientific impacts were "profound and far-reaching." The "most significant scientific discovery was the identification of the biological clock genes," the *raison d'être* of the "core project." Beyond that, the first *mammalian* clock was among five they had found, and the second major impact was linking those findings to molecular genetics. As well, researchers at CBT came to understand the "frequency modulation of…gene expression."[92] There were methodological achievements, too. The employment of luciferase (a molecule that lights up under certain conditions and, hence, known as a *reporter*) in gene expression in transgenic mice was of considerable note. A colored protein from jellyfish was also used in a reporter capacity to demonstrate certain protein-protein interactions.

The 11-year mark had been reached for the Center, so NSF STC funds had come to an end in 2002, but the $1.4 M per year accorded UVA for its center had been a

[90] NSF, "Site Visit Report," April 21, 2000.

[91] Ibid.

[92] Ibid.

superb investment when the discovery of biological clock genes is alone considered—a major achievement in science. One of the founders of the CBT, Michael Menaker, Commonwealth Professor of Biology at the University, noted that funding was being sought in 2002 for continuing the efforts of the PIs and others at the Center and that the Foundation's seed support over the years had been invaluable.[93] But for the purposes of this history, one of the other very notable accomplishments of the Center was stated by Gene Block, UVA Provost and director of the CBT, when he noted that the multi-university effort "taught us how to work effectively across institutional boundaries."[94] Insofar as our concern here goes, Block's statement *can stand as an example of the considerable value to the community of the* intricate dance *and* also *as another variant of the face of* interdisciplinarity. It is a particularly important point that the comment came from a member of the *community*. And yet further, Block stated that the CBT had done some "high-risk research that has paid off greatly." One of the intents of the move into *applied research* at NSF in earlier years had been to take risks. The discoveries made at the Center for Biological Timing could not be a better exemplar of NSF's entire STC program in not only the biological sciences, but in science more broadly.

6.7 Revolving Door Deputy Directors Redivivus…and a "New Era"

Two periods of "revolving directors" of the NSF have been considered so far in this history. Another had earlier occurred for the *deputy* directors: and it would not be the only one. The early 1990s were not only marked with the expansion of centers of various types, the promulgation of GPRA, and the youthful period of BIO and BIR, but also the arrival of a new director for the National Science Foundation. Walter Massey had left the agency in April of 1993, leaving Fred Bernthal to return for his second period as acting director. Like his first such assignment just two years earlier, Bernthal would find himself sitting in for about six months until Massey's permanent replacement took command. Neal Lane became the Foundation's next director in October of 1993 and would remain so for almost five years (see especially Chap. 9). As a consequence, Mary Clutter left as acting deputy director of the Foundation and Bernthal returned permanently to that post. But the next year, Bernthal left the NSF altogether.

So, in July 1994, Clutter would again function as acting deputy, but that time for Lane and for only one month. Behavioral scientist, Anne Petersen, would take the deputy director's helm, but would stay on for only slightly over two years. Revolving

---

[93] Fariss Samarrai, "Clock Stops on NSF Biological Timing Center, But the Momentum Carries On," *Inside UVA Online* (2002):1–3 (online pagination). See: www.virginia.edu/insideuva/2002/11/bio_timing.html

[94] NSB, "Minutes," NSB-93-243, October 14–15, 1993, 10–93:2.

door *deputy* directors *redivivus* that time. More of the musical chairs seemed to be coming when engineer Joseph Bordogna would function as acting deputy director from September of 1996 to June of 1999. But, breaking the mold, he then became permanent. He stayed on in the position for a full six years until June of 2006, nearly ten years total—and also became one of the best and most beloved persons to fill that post.

As it was with 1991, 1993 was also an event-filled year. That was the time period when the move from 18th and G Streets in Washington, D.C. was underway with some 500 NSF staffers already having relocated to the new headquarters buildings in the Ballston District of Arlington, Virginia.[95]

In other connections, January of 1992 saw the official end of the reorganization process and within a year, Mary Clutter could report that up until that time, BIO had maintained some 17 advisory committee/review panels. In an effort to save funds and streamline operations, Clutter established the "cluster mode for programs"—the first NSF directorate to do so.[96] Clutter's clusters provided BIO with a group of individual researchers who could be "on call" for reviews and that also allowed for more flexibility during each review cycle. Clusters were more than that, however, as they represented a somewhat less formal structure *within* BIO paralleling the Program Director (PD) level to an extent. For instance, in 1996, James Brown was Director for the Division of Biological Infrastructure, DBI (BIR had morphed into DBI by then), and under that there existed an Instrumentation-Related Activities Cluster composed of two PDs, a Program Assistant, and a clerk. There was also a Research Resources Cluster (RRC) for the *Arabidopsis* Genome composed of a single PD and an assistant. But also under the RRC, there was another subcluster (my term), the Plant Biology Partnerships, with the same PD and assistant. Those two subclusters were composed of the same personnel in order to liaise with the community on closely related aspects of the RRC that they managed. As an aside, one might wonder if some programs in BIO were too closely related to warrant their continued individual existences. Indeed, over the decades, there were occasions in which programs were consolidated; some are discussed in this history. The clusters were reorganized over time, as necessary, and they were evaluated by the Committee of Visitors (CoV or COV) approach described below.

Returning to the committees and panels in 1993, the new MCB and IBN divisions that had been birthed in the recent reorganization, had their numbers of those committees and panels reduced from 17 to 12. All three content divisions (DEB was the third) operated three committees/panels each, and BIR had two. The twelfth committee was the overall AC for BIO itself. BIR, specifically, had an Advisory Panel for Instrumentation and Instrument Development (IID) with 3–10 members, numbers varied as required by the types of instrument proposals and numbers of

[95] Ibid., 10–93:5. The NSF buildings' address at 4201 Wilson Boulevard, has become familiar to a whole generation of scientists for nearly three decades now.

[96] Mary Clutter to Constance McLindon, Director of IRM, "BIO Advisory Committee Reduction," February 25, 1993. An extensive set of materials describing all current, as of this writing, 17 committees/panels and their planned retention, termination or establishment was attached.

projects to be reviewed. In 1993, BIR was receiving about 500 proposals annually and was funded at over \$46 M.[97] Two older BIR panels were terminated: the one with responsibility for field stations and marine laboratories was combined with IID; the second was a "special emphasis panel" (SEP) on instrumentation and resources. However, one new panel was formed, and used that same term of SEP, but was fully titled Special Emphasis Panel on Biological Sciences, making it much broader than the previous SEP during the BBS years and the only such SEP for all of BIO beginning in 1993; though intended for the whole directorate, it was lodged within BIR. The SEP would undertake grant proposals that included research training, small business features, and infrastructure and databases; all had to be decidedly *interdisciplinary,* and many would be cross-directorate, as well.

Additionally, FY93 was well underway and the Foundation sought, for the first time, a \$3B figure for FY94 with more than a quarter of a billion dollars going just to the High Performance Computing initiative across all directorates. Biodiversity and Global Change (Chap. 9) were other major programmatic efforts. A special request beyond the base budget was that for \$213 M to complete the move to Arlington. Congress not only accorded the hoped-for new figure, but an unexpected 11% more than the FY93 request level brought the actual to \$3.2B and allowed for the conclusion of the relocation of the headquarters.[98]

So, it was onto that very active stage that the new director would step. Neal Francis Lane was born in Oklahoma City in 1938 and earned all three of his degrees in physics at the University of Oklahoma, therefore making another physicist Director of the National Science Foundation. He had a distinguished career in several universities, but finally came to be most closely associated with Rice University, Houston, Texas. He moved upward from assistant professor of physics eventually to full professor and then to Provost of the University. He took leave from Rice on several occasions, but one was to head NSF's directorate of physics, mathematics and engineering in 1980. When he came to the Foundation as its overall director in October of 1993, he brought with him strong knowledge of the institution. Taking that knowledge after leaving NSF in August of 1998, he was able to apply it to his new position of Assistant to the President (Bill Clinton's second term) for Science and Technology, and also as Director of the OSTP through January of 2001 when George Bush, *fil*, assumed the presidency.[99] While Lane did not quite fulfill the six years of his directorship, it would herald the first time since the early 1970s, and

[97] Clutter used the figure of \$46 M, however, the actual budget for BIR in FY92 was \$41.38 M and the originally proffered request for FY93 had been \$50.17 M, but the revised FY93 budget saw a drop down to only \$43.97 M. See NSF, "Justification of Estimates of Appropriations to the Congress, Fiscal Year 1994," pg. 45. The BIR request for FY94 was \$46.71 M and that may have been the figure she was using in her memo to McLindon (footnote above). That amount was divided over the two programs then extant: IID at \$14.85 M, and SEP at \$27.41 M.

[98] Wil Lepkowski, "NSF Director Neal Lane Faces Challenges on Homefront, in Congress." *Chemical and Engineering News* (1993):45–48.

[99] Lane had been nominated by Clinton on July 13, 1993, and confirmed by the Senate on October 7 of that year. NSB, "Minutes," NSB-93-243, 10–93:2 *op. cit.*

before, at the Foundation that a period of considerable constancy would mark the highest position there. From his arrival in 1993 until the close of this history, the National Science Foundation would experience only three directors: Lane, Rita Colwell, and Arden Bement. More will be said of Colwell in later commentary.

When Lane came aboard, there was an ongoing sense of "reeling" among the employees over a lack of leadership at the top (Bernthal was only acting); and the Congress, likely inspired in part by the mood set by GPRA, was calling for NSF (and all agencies) to "prove [their] strategic worth to the American people," as one science reporter had it. Could the

> mild-mannered physicist from a Texas college known for its strong academics and weak Division I football teams cure a government science agency of the worst identity crisis in its history?[100]

The article in *Chemical and Engineering News* the month after Lane became director may have been stating matters rather too dramatically, as it went on to say that "[m]any NSF officials say the agency really ought not to be experiencing an identity crisis, [but rather] be crowing not quaking," so many successes had it had.[101] Nonetheless, the situation at the Foundation was such that one staffer there had been quoted as saying that "[w]e need someone to come in and define us."[102] There had been the latest rounds of musical chairs with both directors and deputy directors for several years, by then. And it seemed that Congress' new concept of something called "strategic research" was causing concerns at NSF: clarification was called for. When Senator Mikulski, who chaired the subcommittee that controlled the appropriations for NSF, first stated the concept for the press, it was not at all clear what it meant to many both within and without the government, not least those at the National Science Foundation. Apparently, there was a feeling that the popular "curiosity-driven" research, more or less another phrase for basic research, might be somehow *overdone* at the Foundation, even though that had been its primary purpose for over four decades! It may not have had the right *sound* to political ears, either. Further, Mikulski was " 'a socially-minded, up-from-the-streets Baltimore politician' " and she had already been on a year-long campaign to get NSF to " 'focus more clearly on the transfer of knowledge and technology for broader national goals and objectives,' " and so

> Lane's agenda is to show Congress that NSF can make a difference in the social and economic fabric of the country and not just appear as *a feeding trough for curiosity-driven individual investigators.*[103]

Mikulski wanted the Foundation to " 'identify that which is specific, immediate, and realizable in pursuit of this broader mission.' " Rethinking the mission was already underway, however, at the Board level and with the new "strategic" notion spreading

[100] Lepkowski, "Challenges."

[101] Ibid., pg. 45.

[102] Ibid.

[103] Ibid. Emphasis added.

through the federal establishment, the NSB's future plan for NSF would come out in a year and rectify the causes of fear about the matter of the Foundation's image and definition, a reason for worry to many, it seemed. Mikulski was hardly alone either in her thinking on strategic research as John H. Gibbons, then head of OSTP, and Leon Panetta, OMB director, issued a statement to the effect that the terms, at least for budgetary purposes, of basic research, applied research, development, and facilities be eliminated! They were, the two men said, " 'artificial distinctions'," and were not of use when it came to making policy and budgetary decisions. They proposed replacing them with new categories to include "manufacturing, communications and information, natural resources and the environment," and others of such ilk.[104] Lane told Congress that "everything we do is basic research" and that curiosity-driven and strategic research were essentially one and the same. Even with some question among the politicos about Lane's stance, the "Foundation's future, as far as the Clinton Administration is concerned, is secure."[105] Technology policy was a major set piece for Bill Clinton, and Gibbons had also pointed out that a "protective line" had to be drawn around basic research to shelter it.[106] In essence, the long-adrift tenets of RANN, research applied to national needs, of the 1970s, had returned and required Lane's attention.

In November of 1993, only one month after Lane's arrival and Mikulski's public comments, the national election had brought Clinton to the presidency. The Democratic Party had control of both Houses of Congress *and* the presidency for the first time since the Carter administration.[107] And just a year later, in mid-October of 1994, the NSB's Commission on the Future of NSF reported out its strategic plan, which the Board entitled "NSF in a Changing World." It was heavily oriented toward not only the many features for which NSF had become known as the United States' agency for basic scientific research, but strongly reiterated and promoted education, as well.[108] This would have profound effects at BIO over coming years. Neal Lane, in his annual message pointed out that 1995 would witness the 45th anniversary of NSF and chose to highlight, in his first few words, the "astonishing advances" in science and arranged those in an order that gave evidence of the rising role of biology. He mentioned first the stunning success of the Internet and NSF's role in its early period, and quickly followed that with biotechnology, but "especially

[104] Ibid., pg. 46.

[105] Ibid., pg. 48.

[106] See: "Testimony of John H. Gibbons: Statement of the Honorable John H. Gibbons, Director of Office of Science and Technology Policy before the Subcommittee on Veterans' Affairs, HUD, and Independent Agencies Committee on Appropriations, U.S. House of Representatives," April 30, 1996, at http://clinton4.nara.gov/textonly/WH/EOP/OSTP/other/ts043096.html. For Clinton's orientation toward technology policy, see John S. DeMott, "The Clinton Push on Technology—President Bill Clinton's Technology Proposals," *Nation's Business* (1993):cover story. Vice President Al Gore was also strongly committed along those lines and played a major role with Clinton on promoting the issue, as DeMott notes.

[107] James P. Pfiffner, "Partisan Polarization," pg. 37, *op. cit.*

[108] NSB-94-183.

the recent advances in the mapping of plant genomes" and the possibilities that particular work would ordain (following chapter). He emphasized the "NSF in a Changing World" notion and promoted the social aspects of science that reflected the tenor of the history of the previous several years: making "strategic research" central to American economic competitiveness and leadership in a world no longer held together by a Cold War framework.[109] And 1995 was a productive year for the biological community, too: BIO had given out 2877 awards during FY94 totaling almost $288 M, and would grant 2905 the following year.

But the change at the top of the US government, the collapse of the Berlin Wall, the break-up of the Soviet Union, and a new Foundation director, brought about aspects of "turbulence [to] NSF" during the previous nearly two years, reported one member of the BIO AC at their November 1995 meeting.[110] The budget was not expected to pass during the then current Congressional session. But, as we have seen, after the leaner years at the turn of that decade that had worried Leon Lederman, the fiscal climate for NSF began a continual upswing through 1999. Budget aside, and for a variety of reasons, Clutter saw the time as a "New Era." And on that view, Jim Edwards, still then Executive Officer for BIO, updated the AC on the state of affairs broadly across NSF, there having been a Foundation Senior Management Retreat some months before.

Of matters of more direct concern to the AC for biology, however, the committee was treated to a discussion of how BIO set priorities during the time of "turbulence" (reminiscent of an earlier time of "tumult"!), as Mary Clutter explained to them the many inputs she sought in order best to do such setting (Fig. 6.2).[111] One such technique was the use of the Committee of Visitors. COVs were a new concept at the time and each of the divisions would be examined by a COV (also seen in the record as CoV) appropriate to the disciplinary areas for which each was responsible. The assessments would be done once every three years for each division. The COVs were another spin-off of the expectations seen in GPRA and so such outside groups' opinions were seen to be one source of accountability information that would satisfy the Act's requirements, beside the obvious and more immediately valuable end of adding more cerebral power, even if temporary, to BIO. Another was the use of decadal evaluations, such as that being done at about the same time for the LTER Program (Chap. 9). Further, Clutter reminded the AC that the reorganization efforts of 1991–1992 had led to the establishment of SBE and BIO, the new IBN division of BIO, the cluster concept, and "sticky ends" (previous chapter), all of which played into her view of a "New Era."

For BIR and its role in education, the first of "four core strategies" for the Foundation that were stated in the Board's plan for the NSF's future, was that for

---

[109] See: http://www.nsf.gov/news/frontiers_archive/nsf961/7annrep.jsp

[110] BIO AC, "Minutes," November 2–3, 1995 meeting. The member who gave the "Congressional Update" was Joel Widder. See pg. 2 of 9 in the online version at: http://www.nsf.gov/bio/bioac/meetings/minutes/9511.jsp. Paper copies in BIO Lateral Files.

[111] The illustration is one of the overhead transparency slides utilized by Clutter at the November, 1995 BIO AC meeting (footnote above)

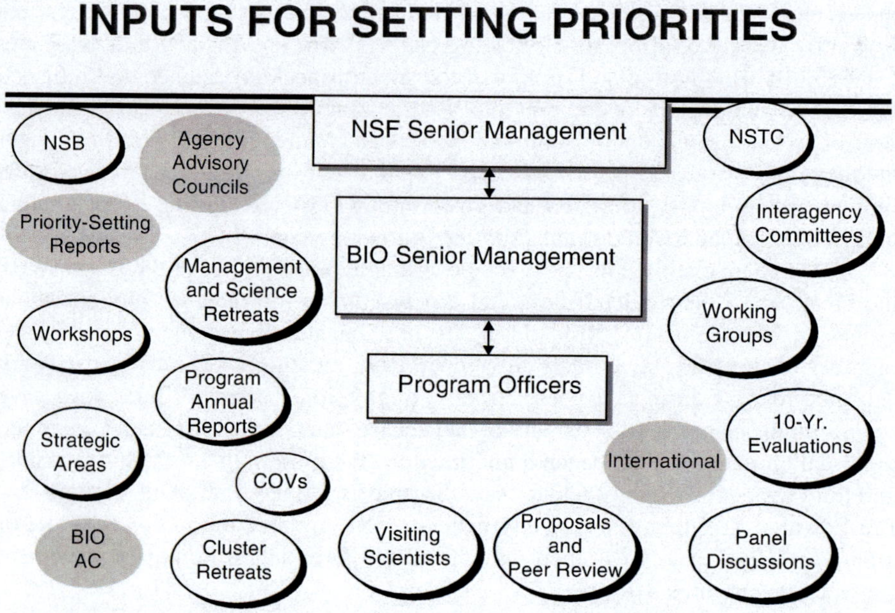

Fig. 6.2 Inputs for setting priorities. NSF. Taken from a visual of the BIO Advisory Committee meeting of November 2-3, 1995

integrating education and research. The hue and cry for "hand's on" education in science would come to characterize the field in the 1990s.[112] In extensive discussions about education, Nina Federoff, eminent molecular biologist and former member of the NSB as well as then chair-elect for the BIO AC,[113] reported on BIR's recent Committee of Visitor's assessment of the division. It had found that BIR had no fundamental flaws and it was doing a "quite adequate" job in its many charges, education being but one.[114]

[112] See, for instance, the National Academies Press statement on the subject at: http://www.nap.edu/openbook.php?record_id=4962. Much of the movement had its roots in the long-running HOSO (Hands On Science Outreach) program of Montgomery County, Maryland, efforts from 1982 to 2007. Many programs later developed across the country based upon that model.

[113] See: http://www.state.gov/r/pa/ei/biog/89337.htm

[114] BIO AC, "Minutes," pg. 5.

6.8 BIR's Many Charges

6.8.1 Education and Training

The budget for BIR was nearly $46 M in FY93 and $51 M for the following year. For its budget request for FY95, BIO had asked for nearly $56 M for BIR—an increase of $10 M over just two years.[115] The whole request for BIO was $314 M, the first time over $300 M. The requests for BIR emphasized education via the Research Training Grants ($11 M plus) and Research Experience(s) for Undergraduates ($15 M plus) programs and yet others leading to a total of nearly $34 M for education and training.

By way of comparison, High Performance Computing came in at $17 M, with STCs and LTER sites seeking $29 M combined. The LTER sites would make a considerable leap of $4 M over FY94, once approved. A long way from MACOS.

6.8.2 Centers

By the early 1990s and the youthful days of BIO, the Science and Technology Centers supported by BIR included those noted earlier above. They were spread across the BIO landscape: Molecular Biotechnology (BIR), Microbial Ecology (DEB), two in IBN—Magnetic Resonance Technology, and Biological Timing—and the two centers under MCB, Light Microscope Imaging, and Engineering Plants for Resistance Against Pathogens. Under *facilities* handled through BIR, some $100 k would go to its share of a nanofabrication network along with the physical sciences and engineering directorates. The STC program finally got the first major six-year review for the Centers and, for the STC program itself, the first-ever major review, both in 1995. The massive review and resulting document were done *by* the National Academy of Public Administration (NAPA) *for* the Senate Appropriations Committee's subcommittee on science.[116] The review covered five selected centers of the then 24, two of which were biologically oriented: Molecular Biotechnology and Biological Timing. The conclusions of the NAPA teams were that the "STCs provide[d] a good return on federal funds which is in the national interest," and recommended that Congress continue to fund them and to use the same approach in management as was then ongoing. NAPA did suggest that a subcommittee be formed with appropriate representation by industrialists or other stakeholders as monitors and as aids in helping replace weak or departing center directors.

---

[115] National Science Foundation, "Biological Sciences: Justification of Estimates of Appropriations to the Congress, Fiscal Year 1995."

[116] National Academy of Public Administration, "National Science Foundation's Science and Technology Centers: Building an Interdisciplinary Research Paradigm," July 1995; 128 pages.

At the BIO AC meeting of late April 1996, Jim Edwards made a presentation on the future of the STCs relying, in part, on the recently completed NAPA report. He asked the AC to consider several questions: are STCs valuable or essential in furthering biology; do they enhance education and interdisciplinarity; is their scope appropriate; and should another competition be held? As to the last question, another competition would be opened three years later as evidence was clear that the STCs, Foundation-wide, were of decided value and the BIO AC was "[o]verall…very supportive of the STC program" as realized in BIO.[117] There were feelings that: the 11-year period may be too long; that internal projects at a given STC were not as rigorously peer-reviewed as compared to little biology grant proposals were at NSF; that ethnic diversity was not great at STCs; and, that there was a worry over centers not continuing outreach activities once NSF funding ceased. Still, the STCs were well accepted by the AC "overall."

6.8.3 Tools and Technology

In the BIP of the early 1980s, even BBS could not easily cover all the large instrument needs of a nation of biological researchers, so NSF entered into an agreement (a Memorandum of Understanding; MOU) with the NIH under the Shared Instrument Grant Program (SIG) at the institutes.[118] The MOU grants required that PIs had to make parallel proposals to both NSF and NIH, *if* they wanted to use the SIG approach. By 1996, the National Center for Research Resources, a section of the NIH with roots going back to 1962,[119] became partnered with NSF and thus expensive and shared equipment was purchased via the SIG program in concert with NSF's own special fund in cases where the costs for a given item would exceed $500 k. That special fund was the NSF's Major Research Equipment (MRE) account. It was established that same year of 1995 (and renamed the MRE and Facilities Construction, MREFC, fund in FY 2002.) NIH had primary responsibility for solicitation and the reviewing of proposals. NSF participated in NIH Study Sections (somewhat comparable to panels at NSF) where proposal adjudication occurred. That first year of 1983, the only equipment types sought from the community were nuclear magnetic resonance machines, but within two years, high-end microscopes and even computers were being requested.

NIH's SIG had suffered a financial challenge in FY92 when its budget was cut from $32 M to only $9 M and so the MOU was a saving device for the community in that NSF could cover a greater part of the costs per award. By FY96, the SIG budget had moved up to $20 M, but it was still a long way from what it had been

[117] BIO AC "Minutes," April 25–26, 1996, pg. 8 of 9 in the online version; see: http://www.nsf.gov/bio/bioac/meetings/minutes/9604.jsp

[118] Gerald Selzer to Machi Dilworth, August 14, 2001, "MOU with NCRR." See the first page, "History," of the four-page memo with attachments.

[119] See a synopsis history of NCRR at: http://www.nih.gov/about/almanac/organization/NCRR.htm

prior to the "hit" it had taken in 1992. By 2001, it had risen to $44 M, but without accounting for inflation.[120] By that time, too, the SIG proposals in which PIs sought a *joint approach* to funding were very minimal: nine out of 337. The remainder was all aimed at NIH alone as most of the proposals were for instruments that would be used exclusively for *biomedical* or *medical* purposes and so could not be funded by NSF. For that reason, it was felt, in 2001, that the MOU had served its purpose and should be concluded. The portion of SIG proposals that had come in under the MOU as opposed to those strictly to NIH alone, was always a small percentage of the total. In 1996, only four of 285 were dual proposals and the awards from NSF totaled under a million dollars that year. The high came in 1999 with 17 of 294 proposals being joint in nature, but the funds awarded from NSF still barely crossed the million dollar mark.[121] The MOU had been managed under the BIR's IID program and it was no longer clear by the first year of the third millennium that it "play[ed] a useful role" within a slightly reorganized DBI, so it was eliminated.

That MOU is one example of how NSF partnered with another federal agency to accomplish purposes not entirely within the capability of a single agency alone, particularly in dealing with the high costs of modern research. The NIH "empire" shared so many features with NSF in the research world milieu of the life sciences, as has been abundantly stated in most chapters in this study, that many types of MOUs were crafted over time between the two agencies. Others with USDA and many more federal agencies, NGOs, and similar institutions, were equally created from the earliest days of NSF.

In 1995, also, the notion of "emerging technologies" was thoroughly on the minds of a broad sector of the scientific community, biological and otherwise, but it was a group of interested biologists that gathered in late June at a workshop on the subject. Chaired by Erich Bloch, then with the Council on Competitiveness[122] in the nation's capital, there were members of the BIO AC (Helen Berman, for example) present, as were directors of STCs (for instance, Gene Block of the Center for Biological Timing), long-term grantees of NSF, such as Leroy Hood at the University of Washington and Paul Magee (then still Chair of the BIO AC) of the University of Minnesota, and many others from industry and academe. In the Report that emanated from the Workshop, Bloch averred that

> Compared with physics, chemistry, and other physical sciences, biology has not drawn as consistently, as quickly, or as thoroughly on the capabilities afforded by emerging technologies traditional imaging, data collection, and analysis techniques have remained the foundation for progress in the biological sciences.[123]

[120] Selzer, "Memo," pg. 1.

[121] Ibid.

[122] See: http://www.compete.org/about-us/

[123] National Science Foundation, "Impact of Emerging Technologies on the Biological Sciences: Report of a Workshop," June 26–27, 1995. Quotation taken from the Preface by Erich Bloch. See: http://www.nsf.gov/bio/pubs/reports/stctechn/stcmain.htm

Still, he recognized that many recent advances in biology were directly due to the "creative application of new technologies." The Workshop had been organized under the leadership of the STC directors along with BIO. Lansing Taylor, director of the Center for Light Microscope Imaging and Biotechnology at Carnegie Mellon University, BIO's James Brown, and Jerry Selzer, head of the BIR SEP, spearheaded the workshop. They stated that:

> Increasingly, biological research problems drive technology development while, in turn, new technologies stimulate advances in science [so the two aims of the workshop were:] identifying emerging technologies that can have an impact on biological research, and finding ways to develop and harness those technologies more efficiently through changes in the science infrastructure and support mechanisms.[124]

The latter, of course, was the purview of BIR. The highest priority areas were deemed to be five: bioinformatics; computational biology; functional imaging tools using biosensors and biomarkers; transformation and transient expression technologies; and, nanotechnologies. "Biology is at a crossroads," they said, and it had "lagged" behind the physical sciences in the application of technology. For that reason, they made a series of recommendations that included the comments that: "NSF should promote and facilitate efforts" to develop tools and methodologies; NSF and NIH should partner in "interdisciplinary training in biology"; the Foundation should broaden K-12 and undergraduate education, and BIO should partner with the Education and Human Resources (EHR) directorate; NSF should reexamine databases for possible commercial application; and, a host of others. Concern was expressed about such matters as integration of education and research (was it optimized?) and whether technology transfer offices at universities have the right "perspective," as well as "socioeconomic" issues with interdisciplinarity at STCs and similar institutions. The authors commented that:

> The exciting new possibilities in biology are just starting to be realized, led by advances in medicine, pharmaceuticals, agriculture, ecology, and forensics. This is only the beginning.[125]

Most notably, they stated that "[i]t is evident to the workshop participants that we are undergoing a *major paradigm shift* in the way that science and technology development are pursued."[126] To make such a statement was not glib, not with the level of scientists involved at the workshop, not with a brilliant former director of NSF chairing the meeting. It must, then, be taken quite seriously, historiographically. To suggest that *truly major changes were afoot at the end of the last millennium in biology broadly* appeals as very much the case and is central to the present study. The section on "The New Biology" later in this chapter will develop this theme yet further. At very minimum, a shift in *zeitgeist* was surely a truth of the times. That can best be viewed in the fact that to do biology during that period, and afterward, the use of highly advanced technological tools became part and parcel of

---

[124] Ibid.

[125] Ibid., pg. 8 of 23 in the online version.

[126] Ibid., pg. 11. Emphasis added.

research in biology. The tenor of the times with regard to *methodological* approaches to doing biology was new. What had long been seen as classical biology was no longer at the forefront of the discipline by the latter few decades of the millennium and, along with the many other themes developed in the present study, this *methodological shift* must, too, be taken into account. *Indeed, I will summarize in the Epilog that the years of the mid- and later 1980s and into the 1990s was a critical period in this shift of* zeitgeist.

One of the practical matters to come out of the Workshop was the notion of developing "cross-disciplinary enabling tools" for development of new instruments: that is, methods and support within various directorates to encourage forward movement in such creative arenas, Jim Brown told the BIO AC at its April 1996 meeting. At that meeting, the still- new Deputy Director of the NSF, Anne Petersen, would be present, but by the fall meeting, she was no longer at the Foundation and Joe Bordogna, then acting deputy, would represent senior leadership of NSF to the AC members. Bordogna had spoken at the November meeting on GPRA, but other matters were of greater interest to the committee's membership. Jim Edwards addressed the FastLane approach to proposal submission. In 1994, a Foundation-wide Task Force was established to look into the electronic submission of proposals, disbursements on grants awarded, etc., that would work between NSF and the community.[127] By early December 1996, the new online method was established and allowed the over 400 organizations (mostly universities) then communicating with NSF to do so more conveniently via the Internet. Within two more years, that number grew rapidly to over a thousand. In its first 18 months, the proposal numbers submitted electronically went from about 250 in the first quarter of 1996 to about 4300 in the second quarter of 1997. While use of FastLane was optional at its beginning, it became mandatory by 2000. Once again, though a bit slow in coming, NSF had committed to a highly automated system, which BIO, and all other directorates, would use. Eventually it would be the, by then former BIO program director, Dave Schindel, who would be accorded a meritorious service award for having pioneered an all-electronic proposal review, no paper in use.[128]

While the 1993 GPRA may not have been of compellingly great interest to the AC, at least at first, it called for an annual performance plan from each agency beginning on October 1, 1997.[129] But that would be reduced to every other year from March of 2000 onward. Because those performance plans were required, NSF Director Lane created a five-year strategic plan (FY97-FY03) to account for GPRA's mandated features; the Lane concept would continue to be renewed. The NSB report of early 1995, "NSF in a Changing World," had been, the Director stated, "a touchstone for all our activities, providing an overarching sense of purpose and

[127] See: http://www.nsf.gov/news/special_reports/history-nsf/1996_fastlane.jsp

[128] Schindel interview.

[129] The annual responses to GPRA by future historians of the directorate are commended to them as a very useful source of information.

direction."[130] Now that GPRA requirements were coming due, the NSB statement needed to be put into operational reality. And so, Lane noted that NSF wanted to assure that it

> reap[ed] optimal benefit from the revolution in information, communications, and computing technologies [because they are among the] many factors contributing to rapid change in the character of academic institutions.[131]

The document was designed for a broad readership and took the approach of placing interesting research accomplishments as sidebars in scattered locations throughout the report; many were associated with specific goals of the implementation plan. Several were from BIO: Life in Extreme Environments; Small Business Innovation Research (SBIR) using Envirogen, Inc., as exemplar, and early career development using DBI as an example case.

By the fall 1997 issue of the annual NSF "Guide to Programs" for BIO, DBI had already assumed official status as the new name for BIR.[132] That would remain the case to this writing. DBI continued to then maintain its three familiar clusters: Instrument-Related Activities; Research Resources; and, of course, Training. Of significance to the almost tectonic shift of biology into a cyberworld, Computational Biology was lodged in the Research Resources cluster, as it was so closely associated with databases; plant sciences were also included under the NSF/DoE/USDA continuing program—Collaborative Research in Plant Biology—and the *Arabidopsis* genome project. Living stock collections and research efforts in systematics' collections were also supported. In the training arena, undergraduate, graduate, and post-doctoral education was all funded, along with special programs in minority post-doctoral research and other initiatives. In essence, DBI had changed little over time, programmatically, but it played its particular role in new initiatives that arose over the years.

By the following year, however, the term training would be dropped as a program title and two remaining areas would be combined. DBI, then, would offer just two major programmatic areas, Research Resources and Human Resources, the latter containing all the education and training programs. The level of funding sought for the two major areas was just over $46 M and a little over $18 M, respectively.

More broadly at NSF, the Foundation announced the first major change in peer review methodology since 1981, when, in early 1997 it adopted a set of *merit* review criteria. Though the long-standing system had a "track record of success," said Neal Lane, he looked forward to the new approach.[133] Acting Deputy Director Bordogna said that the new system was "clearer and easier to apply," its establishment recognizing changes with regard to a much greater commitment to education by the

---

[130] National Science Foundation, "GPRA Strategic Plan: FY 1997-FY 2003; September 1997," To the Reader, pg. 2. See the online version at: http://www.nsf.gov/od/gpraplan/gpraplan.htm?org=BIO

[131] Ibid.

[132] National Science Foundation, "Guide to Programs: FY 1998," December 1, 1997.

[133] News release, "NSF to Adopt New Merit Review Criteria," April 8, 1997. See: http://www.nsf.gov/pubs/1997/pr9728/pr9728.txt

Foundation and more partnered research ventures. There had been four criteria in the older peer review system, but only two in the new one, and those more all-inclusive and qualitative, it was said.[134]

6.9 A more Powerful AC…and a New Director

The BIO AC of the later 1990s became more and more of an essential element in the management of the directorate. That trend would continue into the next millennium, as well.[135] By the April 1999 meeting, Frank Harris reported that the "caliber of people attracted to the BIO AC means that the AC could be of greater assistance and have greater influence on NSF."[136] Indeed, that had been true for most of the history of the AC, but only became more visibly the case when examining the "Role of [the] Advisory Committee in the 21st Century"—Harris' report topic. Mary Clutter had formed the modern AC in 1992 after coming out of the great reorganization and it was a permanent feature of the way BIO would be operated from that time to the time of this writing. Chairs changed more or less annually and new members were almost always introduced at each session, so the dynamics and idea-rich environment led to great productivity. As features of GPRA came into greater play, the AC would take on yet more impressive responsibilities with regard to crucial issues of general management, along with duties more specific to biology, *per se*. The AC was a valued partner of the AD.

As an example, Director Lane addressed the AC at its April 1997 meeting and explained issues of how NSF would respond to GPRA while working within a rather flat budget. He also considered the new merit review system and the concerns of the NSB with regard to several major new efforts within the Foundation: Knowledge and Distributed Intelligence (KDI); Next Generation Internet; and, research education. NSB, recognizing the impact of that flat budget, would drop back to the earlier plan of four meetings per year from the six to which it had become accustomed; a cost-saving decision. A new program in CISE was the Partnerships in Advanced Computing Infrastructure. Matters were circling around the one major area of great interest NSF-wide: computing and the many aspects associated with it.[137]

---

[134] See also the Board's "Final Recommendations" from its Task Force on Merit Review, NSB/MR-97-05. The issue came up again in front of the Board as recently as 2011 when the two criteria, intellectual merit and broader impacts, were under consideration. The NSB's Task Force on the matter was to look at all options for the system: keeping the criteria as is, "completely rewriting them, or anything in between." See: http://www.nsf.gov/nsb/publications/2011/01_19_mrtf.jsp

[135] As just one example, the OAD of BIO noted the need for changes in COV reporting mechanisms called for by several issues, one of those being "BIO's desire for COVs to address higher order issues." BIO AC, "Minutes," November 13–14, 2003.

[136] BIO AC, "Minutes," April 22–23, 1999.

[137] BIO AC, "Minutes," April 9–10, 1997.

KDI had been the topic of a BIO directorate retreat and Bruce Umminger, Director of Integrative Biology and Neuroscience, gave a presentation that considered a number of features of the initiative. Among other things, the program announcement (call for proposals) for KDI encouraged technology development to solve research problems. For BIO's response to it, it was clear that work on the *Arabidopsis* genome, in order to accelerate, needed to be sure that the necessary analytical tools were developed, as just one example of KDI implementation. Other matters under that initiative were: creation of new technologies altogether; use of virtual laboratories; and, "development of biological information standards and architecture" for databases—the matter of standards was one that would haunt the community and NSF for years to come. More on KDI is given in the following chapter.

By the October meeting, Machi Dilworth, long a program director in several plant-related areas in DBI, etc., had become acting director of the division, and permanently so by 1999. At that same time, Nina Federoff yielded the Chair position of the BIO AC to ecologist and long-time BIO employee, Frank Harris. By the close of 1998, a very major change for the entire NSF had come about that must have pleased the collective memory of the AC: former member, Rita R. Colwell, had become the eleventh Director of the National Science Foundation.

Colwell was born in Beverly, Massachusetts and earned her bachelor's degree in bacteriology from Purdue University, Indiana. She did her masters there in genetics and followed that by a doctorate in oceanography from the University of Washington, Seattle, thus allowing her to specialize: a career in marine microbiology, and much, much else. In her academic life before coming to NSF as director on August 4, 1998, Colwell was president of the University of Maryland Biotechnology Institute (earlier chapters) and is, to the time of this writing, a professor of microbiology and biotechnology there. She has also been chair of the Board of Governors of the American Academy of Microbiology; president of the AAAS, the American Society for Microbiology, Sigma Xi–The Scientific Research Society, and of the International Union of Microbiological Societies. She is a member of the National Academy of Sciences and several other such organizations. An honorary member of microbiological societies in at least six other countries, she also holds 40 honorary degrees. The Colwell Massif in the Antarctic was named after her due to her work in polar science. She became the first woman to hold the directorship of the NSF and the first true biologist; biochemist William D. McElroy (director from 1969 to 1972) was her only predecessor to come from a similar discipline. She has some 600-plus scientific publications, including authorship or co-authorship of 16 books. Finally, she would essentially complete her six-year term, ending her NSF service in February of 2004.[138] She went on to become Distinguished University Professor at the University of Maryland and also at Johns Hopkins University Bloomberg School of Public Health (she long held research interests in the cholera bacterium, both in and out of marine environments) and, upon retiring from NSF, she took the Chair of

[138] See: http://www.nsf.gov/news/speeches/colwell/colwell_bio.jsp

Canon US Life Sciences, Inc., a subsidiary of the internationally-known photography firm.[139]

Colwell succeeded Neal Lane as Director. Lane stated that Bill Clinton's choice of Colwell was an "outstanding" one.

> Her record of accomplishment speaks for itself…[as] an internationally-recognized researcher [and] admired educator…she has consistently stepped forward as a leader in science and technology.[140]

Interestingly, Colwell was originally to be the Deputy Director, but with Lane's move to the OSTP and presidential Science Advisor posts, Colwell was asked to take the top position at the Foundation. In September of 1997, Colwell had been approached by John H. "Jack" Gibbons, science advisor to President Clinton and director of the OSTP from 1993–1998 (he was followed by Lane). In response to her being asked if she would consider being *deputy* director of NSF, she told Gibbons, "Gee, not really…I would go to be director [, however]"![141] In December she got a phone call.

> My [Colwell's] secretary says, "the Vice President's on the phone," and I said, "the vice President of what?" And she said "the United States." So I went into my office and I took the call. And the secretary, his secretary, said, "Is this a bad time, do you want him to call back?" I said, "It's fine; no, I'll talk to him."

In answer to Vice President Al Gore's question as to whether she wanted to be *director* of NSF, Colwell said: "Absolutely. It would be an honor and a privilege."[142]

If the third millennium was not already well on the way to becoming that of biology, the appointment of the first biologist to head the NSF certainly secured it. *That was especially true as Colwell brought a biologically-driven agenda with her to the position. Among other matters, her notion of* biocomplexity *was a central thread during her tenure as director.* More will be said of that below and in following chapters, as will other features of her biological bent as the Foundation's leader.

At the April 1998 meeting of the BIO AC, Colwell's nomination had been mentioned, as was Lane's intended new directions at the White House. Lydia Villa-Komaroff had succeeded Frank Harris as chair of the AC. Numerous matters of more significant administrative aspects of BIO became commonplace in the AC discussions: such things as COVs; the continuing, but never realized, interest in a National Institute for the Environment; diversity of the scientific workforce; the rotator program—the number of rotators in professional positions in BIO was hovering at 50% at that time—and the ever-present GPRA. The brand new National Plant Genome Initiative (following chapter) was introduced and KDI and related initiatives were discussed at length by Bruce Umminger and the Committee membership. Planning for FY 2000 was equally of concern and other new initiatives

---

139 See: http://www.cbcb.und.edu/~rcolwell/

140 NSF, News Release, February 13, 1998.

141 Rita R. Colwell Interview with the Author, May 12, 2009.

142 Ibid.

(following chapters), such as microbial biology, for instance, were topics considered.[143]

Colwell, as a former AC member, represented a good exemplar for Harris' statement about the caliber of the Committee's membership over time. As occurred at most AC meetings, Clutter had, at the October 1998 meeting, the occasion to introduce a new member to the AC, James P. Collins of Arizona State University, Tempe. He was another remarkably high caliber member. By late 2005, Collins would himself move into a high leadership position at NSF: he would succeed Clutter as Assistant Director for Biological Sciences.[144] Three others were also new to the AC that fall and Rita Colwell was able to address her old familiar Committee as the Foundation's new director. She pointed out the successful overall budget, having moved upward by 7% over the previous year, with the dollars for research specifically going up by nearly 9%. She promoted both *information technology* and her concept of *biocomplexity* as two major areas of concern for coming years. Biocomplexity was to be examined on the "grand scale" and the thematic area of Life and Earth's Environment would provide the gateway to the study of the new area. The funds for attacking that Foundation-wide initiative and all the other aspects of what BIO wished to accomplish, would come with their requested FY98 budget of just under $331 M for BIO alone, a figure that included nearly $72 M for DBI.

The impending change of the millennium—and the 50th anniversary of NSF—was permeating the Foundation as the NSB produced its report titled "Toward the 21st Century: The Age of Science and Engineering." It read:

> The 20th century will be remembered for ushering in an era of dazzling advances in science and engineering. In an extraordinary chronicle of scientific and technological progress, the systematic pursuit and exploitation of new knowledge during the last half-century has, through small steps and transcendent leaps [essentially created a new world.][145]

Colwell was busy during that same year creating the first "Accountability Report," essentially a fiscal audit, required under GPRA.[146] In her "Message from the Director," she spoke to the dual revolutions in information technology and biotechnology and opined that "at least half of the nation's productivity growth over the last 50 years" could be attributed to innovation and discovery brought about by science and technology. Indeed, from the period of 1986 (Kingsbury and Bloch's time) to 2007, it has been argued that worldwide computing capacity *grew by 58% every year*![147] A truly astonishing statement, but one that was not without difficulties:

---

[143] BIO AC, "Minutes," April 6–7, 1998.

[144] Ibid., October 22–23, 1998.

[145] National Science Board, "Toward the twenty-first Century: The Age of Science and Engineering," November 19, 1998. See the online version at: http://www.nsf.gov/nsb/documents/1999/nsb98215/nsb98215.htm

[146] National Science Foundation, "FY1998 Accountability Report," NSF-99-114. See also the online version at: http://www.nsf.gov/bfa/dfm/stmtpg.htm

[147] Josh Fischman, "Dumped on by Data: Scientists Say a Deluge is Drowning Research," *The Chronicle of Higher Education* (2011):online page 1. See: http://chronicle.com/article/

many aspects of the "data deluge" and incommensurability of disparate data sets had been of concern to DBI, as well as NSF and the science community more generally, for years. Solutions had been sought along the way, but were often stymied.

6.10 Biocomplexity: First Appearance

The concept and term were both coined by Rita Colwell, but much of what she included in her thinking with regard to biocomplexity would be similar to systems biology. However, and very importantly, she would meld in the social and behavioral aspects of various species and would include human biology and our social structures into the mix, as well. "It may not be in our lexicon right now, but it is where we need to go in our quest to understand the Earth's biosphere," she stated in an address before the national meeting of the AIBS in early August of 1998.[148] She was just then assuming the directorship of NSF and thus was able to bring the notion of biocomplexity to one of the scientific societies most prepared to appreciate her creation.

> Much like the Chinese definition of 'crisis,' there is both opportunity and responsibility for the science community. This is where biocomplexity takes shape as a research direction, as well as a key to social understanding. To my mind, biocomplexity reaches beyond biodiversity…the phrase 'understanding biocomplexity' speaks of a deeper concept. It is not enough to explore and chronicle the enormous diversity of the world's ecosystems…but also [to] reach beyond, to discover the complex chemical, biological, and social interactions in our planet's systems.[149]

It is almost a curiosity that while for so many years there was the felt need to separate the biological sciences from the social and behavioral, yet with a different director of the Foundation the concept of biocomplexity appears on the scene and sought to reunite them! Indeed, it easily can be argued that the new notion takes into account not only these sciences, but the Earth and physical sciences, as well as mathematics and computational biology, and suggests that, as it was with Paul Weiss in the NSF of the early 1950s, a *functional* approach was viewed as more desirable than a *disciplinary* one. Interdisciplinarity flew in on silver wings with biocomplexity like it had never before been seen at the National Science Foundation.

Underlying her concern that the broader picture of nature must be understood at a grand level, she argued that "from personal experience…the [NSF] grants were too small and they were too short in time."[150] In a word, they were "just insufficient." But more philosophically,

---

dumped-on-by-Data-Scientists/126324/

[148] See the online version of the whole speech, as delivered, at: http://www.nsf.gov/news/speeches/colwell/rc80802.htm

[149] Ibid.

[150] Colwell interview.

> [f]or the biosciences, I knew that we had to be in a time of holistic science in that the highly reductionist science that had been practiced successfully, most valuably for the last 50 years[, had now to be brought] together.[151]

Big science was called for in both its definitions. Her view of the role of reductionism was in keeping with its accomplishments, but in order for her to have aspects of biocomplexity understood, reductionist science would have both to continue and to contribute to the holism, the grand view, she sought. She cited Simon A. Levin at Princeton as one of the early grantees under the biocomplexity initiative.[152] Levin later stated his research interests in such a way that his own fascination and contributions to biocomplexity could be summed up by remarking that

> [m]y principal interests are in understanding how macroscopic patterns and processes are maintained at the level of ecosystems and the biosphere, in terms of ecological and evolutionary mechanisms that operate primarily at the level of the organism…The work integrates empirical studies and mathematical modeling, with emphasis upon how to extrapolate across scales of space, time and organizational complexity.[153]

His contemporary statement of his personal research interests is virtually a working definition of biocomplexity.

By the April 1999 meeting of the AC, data had been accumulated on the first, and very successful, Phase I of the Biocomplexity Initiative. Joanne Roskoski, of the BIO AD's office, reported upon work supported by BIO over the previous decade that had features of biocomplexity in it, though it was not so termed at the time.[154] But under Phase I, with emphasis on microorganisms in the environment and held as a restricted competition, she stated that there had been 113 pre-proposals; they would involve 525 researchers who sought some $325 M in grants. Of the 113, only 34 PIs were asked to submit full proposals and, once done, they would compete for $11 M available under the initiative. That was strong evidence of very considerable interest in the community, to state the obvious! Phase II was already under way in early planning and its emphases would include all types of organisms and be in the areas of *modeling* and the *properties* of complexity, such as emergent ones, feedback mechanisms, hierarchies, sensitivity to initial conditions, etc.[155] One of the discussants from the BIO AC on the subject was the new member, Jim Collins, who led the AC group; Roskoski led from the federal side. (Roskoski would later work directly for Collins once he became the AD of BIO a half-decade later.)

A number of points were made as to why biological systems afforded such a special opportunity for understanding complexity in natural systems more broadly. Among other reasons given, was the fact that

---

[151] Ibid.

[152] Ibid.

[153] See: http://www.princeton.edu/~slevin/

[154] BIO AC "Minutes," April 22–23, 1999.

[155] Ibid.

> [u]nlike physical or chemical systems, each biological system has some unique properties…[that include such things] as self-replicat[ion]…are scalable…and multidimensional across all levels of hierarchy…[are composed of] individuals…[and are modelable.][156]

The aspect of modeling was significant to Colwell and others, as she said to the BIO AC:

> Biocomplexity is an important issue in that it brings together the complexity that has been fermenting in all disciplines. Building prediction models will help reach decision making goals in environmental policy.[157]

She went on to point out the importance, too, of the new initiation in information technology (her other major agenda item as the new director), termed IT2, which the Foundation was then embracing through both BIO and CISE. Her interests also included the value of the STCs and of graduate education that BIO and the education directorate were working on together.

The final month of the second millennium closed with the BIO AC meeting of December 1999 and with three new members joining the Committee, including the founder of the BIR/DIR/DBI, John Wooley. As we have seen, Wooley played a major role in the history of infrastructural features for the biological community in America through his work at BIO. Now he would have the chance to remain of value to the directorate with his service on its AC for over six more years. As a working group co-chair with AC colleague Claire Fraser, Wooley was able to make his first report to the Committee. In the case of Fraser and Wooley, theirs was on the NSF's role in regard to genomics: developing a list of higher organisms that needed to be *sequenced* (determining the order of nucleotide bases in DNA); facilitating the establishment of infrastructure and databases (among Wooley's special interest areas); and providing both student and faculty training opportunities.[158] Wooley would become as much of an asset to the AC, and hence to BIO, as he had been in previous years employed directly at NSF.

Indeed, that meeting included a number of discussions on a vital subject: the "Unmet Opportunities" for 2000 and beyond. They included:

- NEON—the National Ecological Observatory Network (Chap. 9).
- Genomics (following chapter).
- 2010 Project—An extension of the work to date on the functional genomics of *Arabidopsis* (following chapter) which goal was to determine the functions of all 25,000 genes in thale cress
- Collaboratories/Networking Activities—Multi-investigator opportunities that would ask broad conceptual questions.
- Nanotechnology and its relationship to biology.
- Portfolio Balance—the matter of core programs versus special competitions, staffing, workload, etc.

---

[156] Ibid.

[157] Ibid.

[158] Ibid., "Minutes" of December 2–3, 1999.

As such things are wont to do, other more significant topics edged out some of the above from standing, but most especially something that would soon come to be called "the New Biology." Among significant personnel changes in BIO, Mary Clutter reported at that same meeting that long-time figure, Bruce Umminger, had moved from being the director of the Division of Integrative Biology and Neuroscience to the Office of Integrative Activities, the Foundation's unit that handled all of the cross-cutting programs of NSF, including Umminger's special interest, the STCs.[159] Also that December, a number of major COV reports were discussed, as well as the FY 2000 budget. Colwell had managed to get a 6.6% increase ($240 M) for the NSF budget for the year in order to start up her new initiatives (and some already established ones) of biocomplexity, IT2, plant genome research, and terascale computing systems.[160]

Colwell's arrival upon the scene of the National Science Foundation as it ramped up to its 50th anniversary, and also as it found itself at the beginning of a new millennium, offered a striking contrast to how the NSF began in 1950. Then, as we have seen, it was oriented strongly toward physics and it was led for many years by physicists, but with the molecular revolution in biology that had begun before 1953, and then was sparked by Watson and Crick's discovery that year, and with a concomitant waning of atomic physics in the century's latter quarter, the rise of biology was fully assured by the year 2000.

Something "new" seemed to be on the horizon. Inspired by the directions biology as a science was taking, inspired by the concept of biocomplexity, or just simply an historical shift in the very nature of biology, things were clearly changing.

6.11 "The New Biology"

For the historian, the year 2000 is like yesterday, at least with regard to the cherished notion of *perspective* in historiography. Gaining perspective in analysis of what is and is not of note to the historian obtaining from just a mere decade before he or she writes is not fully possible. Nonetheless, it can be argued that definite trends seemed to be appearing at and after the first year of the third millennium that floated around something that the BIO directorate thought of as "the New Biology." Indeed, throughout this entire study, the historical record shows that biology was very definitely moving toward a *renaissance that would be one of integration* and "the New Biology" may be just what it would become. Over the next several chapters, that centrality will be developed further. But it was, at least in a titular sense, taking a place in the Minutes of the BIO AC by the close of 2001: "Biology has reached the point where new tools and information allow it to embrace an *integrated as well*

[159] See the current description of the OIA at: http://www.nsf.gov/dir/index.jsp?org=OIA

[160] "Minutes," December 2–3, 1999. Terascale computing refers to computers running tens of billions of calculations per second.

[as] a reductionist approach," averred Mary Clutter to the Committee. Reminiscent of David Baltimore's assessment (head of this chapter), Clutter went on to say that:

> As borders between disciplines have blurred, a New Biology, which is multidisciplinary, multidimensional, education oriented, and information driven, has emerged.[161]

DBI's role had been very great in that shift with regard to computerization, "sophistication inflation" in instrumental design and capabilities, guiding and supporting the flourishing of databases and their uses, and other special efforts in which the division had been a leader. The emergence seen by Mary Clutter led her to appoint AC Chair Ellen Goldberg to a breakout group on the subject: "[t]he charge was to consider how to bring about the *integration of reductionistic* and *holistic biology*."[162] While that charge was weighty, to say the very least, at the second day of the two-day AC meeting, Goldberg addressed the issue of facilitating and enabling "the New Biology." Four issues arose in the breakout group's discussions: to support informatics; to train biologists during their mid-careers (to shift into a possible new paradigm?); to train biologists to manage multidisciplinary teams; and, to develop "infrastructure and institutional framework to support interdisciplinary work." An additional subcommittee on the subject was appointed to report back to the AC at a later date.

Indeed, actions had already been afoot to establish inter-directorate methods by which interdisciplinary research could better be accomplished. Specifically, BIO had partnered with the Directorate for Mathematical and Physical Sciences and its team had made its study report in February of 2001; it was obviously incorporated into Clutter's thinking by the time of the AC meeting that November. The intent was not to

> describe a new, short-term initiative, but rather to set out a number of steps that, taken together, will encourage and sustain cutting-edge interdisciplinary research by providing mechanisms for both internal and external community interactions.[163]

Some 30 highly specific recommendations were made in the report with such suggestions to: "[i]nclude a module on interdisciplinarity in all Program Officers training programs"; and, to "[p]romote an entrepreneurial approach to solicit novel proposals by encouraging PIs to submit proposals that can be co-funded and by linking PIs who would not normally collaborate."

The 30 recommendations were remarkable as a set of tools by which to guide interdisciplinary project planning and management. They went far to answer basic questions about how one *does* interdisciplinary research: long a question in an agency devoted to the very idea of promoting such research. The observer may marvel that the *mechanisms* for doing interdisciplinary program creation and

---

161 "Minutes," November 8–9, 2001. Emphasis added.

162 Ibid. Emphasis added. For more on Goldberg, see: http://www.salzburgglobal.org/2009/includes/FacultyPopUp.cfm?IDSPECIAL_EVENT=381&IDRecords=13482

163 An Internal Report: To Mary Clutter, Director, BIO and to Robert Eisenstein, Director, MPS, "Establishing Inter-Directorate Mechanisms to Foster Cross Disciplinary Research: Report of the MPS-BIO Working Group, February 15, 2001."

management had not already been long-established at the Foundation. It may be, however, that just the right vision by just the right group of individuals had to come together in time and space to truly move one crucial facet of interdisciplinarity forward. Persons such as Colwell, Clutter, and Wooley, and surely others, came together at NSF at that time and were at the center of a creative nexus. It may be that their time at the *fin de siècle* was similar to the critical period of the mid- and later 1980s when Kingsbury, Clutter, Wooley, and Bloch led the charge to computerize the biological sciences.

The "New Biology" was not just a passing issue, then, as by the BIO AC meeting of spring 2002, Clutter was reporting that the Senior Management Team of BIO had met and was "revisioning" the directorate, because it was now "driven by the characteristics of the Biology of the 21st Century." That is, it was "multidimensional, multidisciplinary, data-driven, education-oriented and internationally engaged."[164] The latter point was newly added from the list seen in the November 2001 AC Minutes. It may have been something of a political statement, but it remained present in more formal publications. It was certainly of importance, however, given the great significance of the National Plant Genome Initiative (NPGI), as one notable example and which was ongoing in many countries. *The revisioning process was offering up its first plan: the establishment of a new "virtual division" of BIO to be called "Emerging Frontiers."* That virtual division exists to the time of this writing and augments the well-established divisions of DBI, MCB, IBN (eventually also to change its name; Chap. 9), and DEB as part of the greater structure of BIO.[165] Indeed, a set of five "scenarios" came out for the revisioned BIO directorate that considered different numbers of divisions, but the thinkpiece remained essentially that and, barring only the addition of Emerging Frontiers (and a name change for IBN), the directorate remained fundamentally the same from 1991 to 2005.[166] The BIO budget for FY 2003 noted, under the rubric of "The New Biology," the notion that the biological sciences were undergoing a "profound transformation" due especially to "recent advances in genomics, proteomics, informatics, computer science, mathematics, physics, chemistry, engineering and the earth and social sciences." The Clutter mantra was taken up by science journalist David Brakke, as just one example, in his article in *AWIS Magazine,* as he emphasized experimentation and observation across a "wide range of scales in time, space and organismal and system complexity."[167]

Not unlike Mary Clutter's major achievement of reorganization of BIO in the early1990s, she promoted "the New Biology" heavily in the first years of the

---

[164] BIO AC, "Minutes," April 25–26, 2001.

[165] Ibid.

[166] An ephemeron of two typed pages that listed the five scenarios (first sheet) was appended to the front of a page that provided the information details to the FY 2003 BIO budget request and expanded upon the use of the terms "multidimensionality; multidisciplinarity; information-driven; and, education-oriented."

[167] David Brakke, "Biology for the twenty-first Century," *AWIS Magazine* (American Women in Science) (2002) 31(2).

twenty-first century. For example, she gave an illustrated presentation at the Santa Fe Institute, New Mexico, on "Revisioning the Biological Sciences at NSF," in which she saw the process as coming in two phases: the first of those was the addition of the virtual division of Emerging Frontiers (EF) and, for 2004, she saw the thorough evaluation of EF and a generally realigned BIO. How would that occur? EF was to have an initial portfolio that oversaw integrated research areas in biology, research coordination networks, and the Tree of Life project, just beginning, as one example of a named undertaking. EF's main goal was to "foster 21st century biology."[168] Along with EF, she listed Plant Genome Research coordinately with the four actual divisions of BIO, so major had the whole area become (following chapter). Her processing approach employed two simultaneous time lines of activities that merged at the planning stage for the budget for FY 2003. By *calendar* 2003, "long-range planning" was once again a matter of consideration at the AC, as well as at BIO itself.[169]

By the BIO AC meeting of April 2002, the directorate was set for a minor internal reorganization. While several scenarios were considered, some quite radical, a conservative approach was taken for three of the four divisions. IBN would see a name change to Integrative Organismal Systems, IOS, with other changes, too (Chap. 8). The status of all four, less Emerging Frontiers, and "IOS" (neither were formalized by February 2002, the date of this organization chart) is shown in Fig. 6.3. Note the significance given the Plant Genome Research Program. The IAR, in the figure, was a directorate level (front office) function.

The BIO AC was told in April 2003, that Emerging Frontiers was expected to be "an incubator for 21st Century Biology," and that Congress authorized another plan to double the NSF budget over the five years: Congress or the president had made such plans before and they were not always carried through.[170] At any rate the final FY03 appropriation saw a 12% increase for BIO despite the fact that the Senate had recommended 3.5%.

6.12 DBI in the New Millennium

As 2000 dawned, the substructure of DBI returned to the three familiar clusters: instrument-related activities; research resources; and, training.[171] "Human Resources" had been dropped. In annual budget proposals to Congress, however, the first two were usually combined for simplicity. The first of the three included: the familiar multi-user equipment, usually large and expensive; the development of

[168] Mary Clutter, set of overhead slides for presentation at the Santa Fe Institute, New Mexico, February 25, 2002.

[169] BIO AC, "Minutes," April 24–25, 2003.

[170] Ibid.

[171] National Science Foundation, "Guide to Programs, FY2000 NSF Funding Opportunities," pg. 40ff.

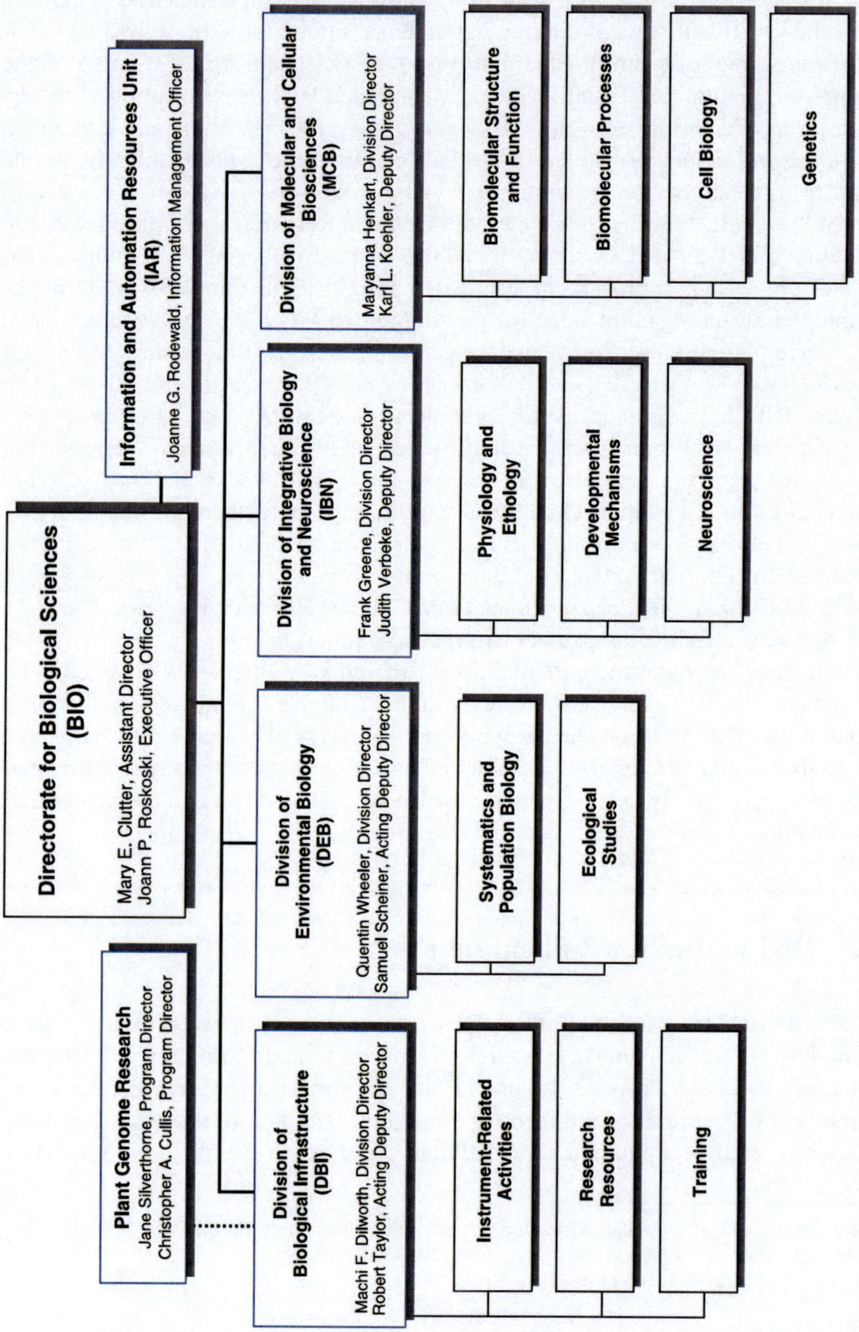

Fig. 6.3 Organization chart for BIO as of February 2002. NSF. Specific provenance unknown

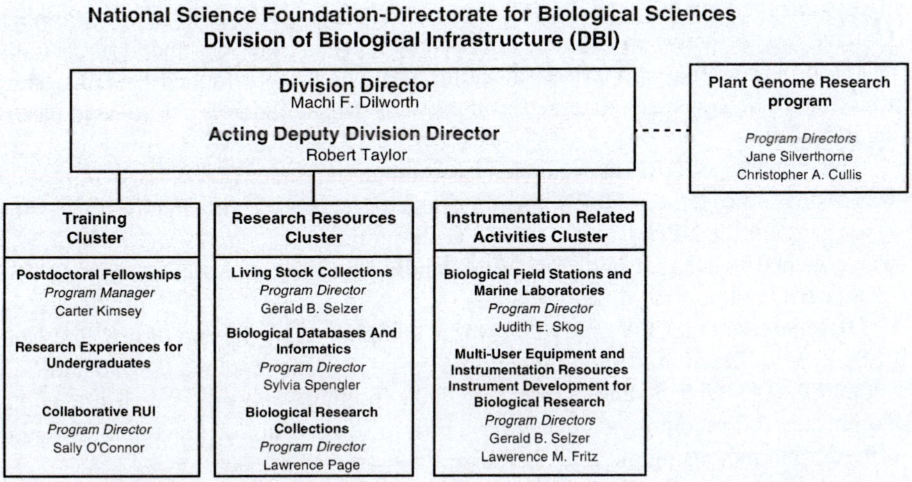

Fig. 6.4 Organization chart for Division of Biological Infrastruction (DBI) of BIO. NSF. February 2002. Specific provenance unknown

such equipment; and, improvement in facilities, communications and equipment at field stations and marine laboratories—the long-lasting FSML program. In addition, the Research at Undergraduate Institutions (RUI) program, begun originally in 1994, continued to be available under that first cluster.[172] Fig. 6.4 provides details of DBI's structure in February of 2002.

The research resources' cluster focused upon biological databases and informatics, plant genomes, living stock collections, and biological research collections; once again, all areas long-familiar to BIR/DIR/DBI. The plant genome program had been initiated along with the National Plant Genome Initiative (NPGI) of 1998 that had been established by the OSTP.[173] (Its initial budget was $40 M, but would be over $94 M by FY 2005.[174]) The 2000 call for proposal's focus was such that "[t]he program will consider research on the genomics of plant-associated microbes including fungi, if addressed within the context of host-microbial interactions." That orientation was consistent with the rising importance at BIO of microbial ecology (Chap. 9).

The training cluster included: Research Experiences for Undergraduates (REU), something of a companion program to the RUIs, though housed separately; the graduate program; several others, such as the minority postdoctoral fellowships, and that in microbial biology—in all, some seven different programs. A subcommittee of the BIO AC, co-chaired by Marvalee Wake and Laura Hoopes, both major players

[172] See Program Announcement (PA) NSF 94–79.

[173] See PA NSF 99–171.

[174] Figures taken from a paper copy of an overhead slide from Machi Dilworth's presentation at the BIO AC meeting of November 17, 2005, just a month after Mary Clutter's retirement and her replacement by James P. Collins.

in biological education at leading California institutions (UC, Berkeley and Pomona College, respectively) and having effects nationally via the AC, undertook a study of the programs that supported education and which were funded by BIO. They found a high level of success in those covering kindergarten to mid-career scientists.[175]

The structure of DBI two years later was much the same (Fig. 6.4). So important was a single project, the NSF's portion of the plant genome research (BIO was the lead group for the NPGI), that it existed as something of a *virtual program* itself and was couched in DBI. It held a similar standing on the organization chart for the greater BIO at the same time.

DBI underwent a COV study that was accomplished by the end of that first year of the new millennium,[176] and the authors of the report concluded that, in addition to other foci, "DBI was ranked favorably in the funding of new discoveries." But the Committee did recommend reviewing the ratio of rotating to permanent program officers, and increasing the total staff size.[177]

At the time, about one-third of NSF POs were rotators, a number that had fluctuated upward and downward over time from that very generalized mean. But it was found within the AC that other matters were more pressing for DBI to consider: long-term management of databases, for one. Leonard Krishtalka,[178] a very active member of the AC during the period, pointed out to the Committee that "whenever possible, informatics standards should precede research."[179] So crucial had data use and management become in an era of highly computerized biology, that Krishtalka argued that "[i]deally, 40% of any research budget should be for informatics." The matter of establishing standards that would allow interoperability (data commensurability) was crucial to avoid "balkaniz[ation of] information," and the unacceptability of *lost data.* That issue had been bubbling for quite some time, as has been alluded to earlier, and would continue to be a problem with which the scientific community would have to grapple into the early years of the present century, not finding all-pervasive solutions even at this writing.[180]

DBI continued to support those areas for which it had become well-established in the biological community, but was "investing increasingly in research on

---

[175] Marvalee Wake and Laura Hoopes, Co-Chairs, "Report of the Education Subcommittee to NSF BIO Advisory Committee, April 6–7, 2000." Six ppg.

[176] The argument as to whether one should consider 2000 the end of the second millennium, which it clearly is, versus the first of the third such era, is left to semanticists. Here, I choose to use it as the beginning of the third millennium simply because it begins with a 2 and not a 1.

[177] BIO AC, "Minutes," November 16–17, 2000. Member Ralph Quatrano's report to the AC on the findings of the COV. Other suggestions were made, as well.

[178] See his University of Kansas professorial website at: http://www2.ku.edu/~eeb/faculty/krish.shtml

[179] Ibid., April 26–27, 2001 meeting.

[180] There is a large literature on biological database management and the challenges inherent in differences in types of data computerized from disparate disciplinary areas. See, for example: http://www.engineeringletters.com/issues_v13/issue_3/EL_13_3_13.pdf

biological databases and bioinformatics tools," such as the project known as SPECIFY.[181] That open license (i.e., *gratis*) online software supports herbarium and museum research data processing and is hosted by the University of Kansas. One use of the SPECIFY program allowed researchers to produce a genome map of the fungus *Magnaporthe grisea* (rice blast fungus), which is a significant parasite. Not only has it killed enough rice to feed as many as 60 M people in a single year, but could well be used in agricultural terrorism as a biological weapon.[182]

The budget for DBI hovered around $70 M during the first few years of the new millennium with the great majority of that (over $55 M) going to Research Resources and the remaining (about $16 M) being directed toward training. The BIO real budget for FY02 was the first one over a half billion dollars. The Research Resources support at that time included: the 2010 Project; Biocomplexity in the Environment (BE); Information Technology Research; Nanoscale Engineering; Genome-Enabled Science; and, Systems Biology. Funding for SPECIFY, for instance, came from the Information Technology program, and that for infrastructure support from BE.

As biology became increasingly more data-oriented and concomitantly more dependent upon computerization, the AC arranged a "small, community workshop" in mid-July of 2003: "Building a Cyberinfrastructure for the Biological Sciences," dubbed CIBIO ("sea-bee-yo"; alluded to above). With 28 participants—some being from the AC, but mostly others—it was chaired by John Wooley. Given his enormous background with the rise of BIR/DBI, his then presence as Associate Vice Chancellor for Research at UC, San Diego, and as a computational biologist, he was an ideal choice. More than that, however, the very idea for the workshop was Wooley's; certainly, he provided the rationale and direction for it. Evidence for that was to be seen in a letter to Mary Clutter the previous May. Wooley pointed out the fact that NSF/BIO had done more than any other agency to promote computational biology and bioinformatics and was the natural leader in establishing a cyberinfrastructure for the biological sciences.[183] Because biology had "entered an era characterized by data intensive research observations," building the infrastructure to manage it all was vital, Wooley argued. It would be "democracy in action for biology" and would lead to all community users, from K-12 to professional scientists, to "participate more fully in the power and joy of discovery and the impact of its consequences." It would "accelerate [the] access" brought on by the Internet and would be both open and inclusive.

> Beyond the great success of reductionist approaches of the past five decades, biology is moving into an environment to consolidate these gains through *information integration*.[184]

---

[181] See: http://specifysoftware.org/content/welcome-specify-6

[182] This commentary combines information from the 2002 budget request to Congress, section on BIO, pg. 107, and the National Science Foundation, "Management and Performance Highlights: Annual Report, FY 2002," NSF 03–024, pg. 16.

[183] John C. Wooley to Mary Clutter, May 29, 2003; a highly detailed letter of some three and a half pages of single-spaced type (with attachments). It is a masterpiece of argumentation for building the cyberinfrastructure for biology, and that to be done by NSF's BIO.

[184] Ibid. Emphasis added.

It could not have been better stated. Biology was clearly becoming "new," it was to be a "biology for the 21st century." Wooley noted that NIH had created a network[185] in 2001 for neurosciences that was moving rapidly in the area of *information integration* and that same thinking could be incorporated into NSF's LTER, NEON, and other activities of BIO and the biological communities it served; indeed, all of biology.

What led to the workshop was Wooley's spirited suggestion that the BIO AC "engage to facilitate a small work group or planning session," followed by a "public workshop, leading to white papers, a web presence," and a report. While he saw the workshop to come about by August of 2004, it occurred less than two months after Wooley's letter to Clutter, thanks to rapid planning at BIO's front office—and a long period preceding that in which Clutter had been thinking about the nature of "the New Biology." For the infant twenty-first century, Wooley envisioned "Biology 21: An Endless Frontier – Enable Biology Through Information Technology," such that "[t]he language for understanding biology at a systems level will be bioinformatics/IT [just] as calculus/math has been the language for understanding physical sciences."[186] While the web presence did not develop, the workshop report continues to be an Internet icon for the new biology of the twenty-first century and remains worthy of reading.[187] Fig. 6.5, an overhead slide Wooley used at the workshop, goes far in showing, for instance, the scalar features of information integration into widely disparate areas of both biology of the little and that of the big. *Reductionistic biology of the genome serves holistic biological study of the biosphere.*

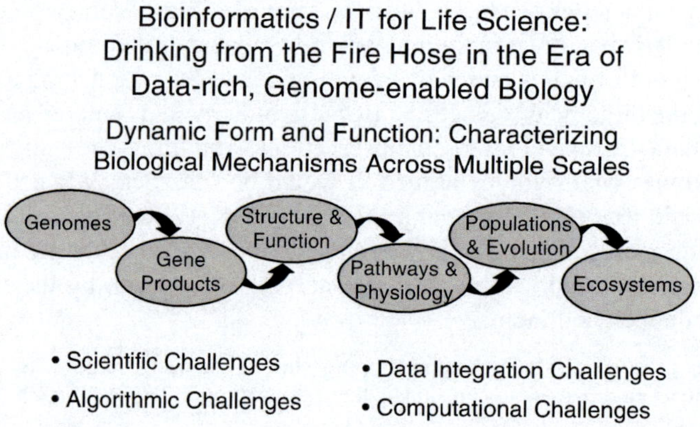

Fig. 6.5 Bioinformatics/IT for Life Science: Drinking from the Firehose… NSF. Taken from a visual used at a meeting in August 2004. See main text for details

185 See: http://www.birncommunity.org/about/history/

186 Wooley to Clutter letter attachment, and possibly an overhead at the workshop.

187 See: http://research.calit2.net/cibio/archived/CIBIO_Overview_Report.pdf

There was a strident push made to get the community to embrace the new biology that CIBIO would provide, and that was spearheaded by its primary proponent, Mary Clutter. The AD gave a presentation on "Meeting the Challenges," in which she specifically spoke in terms of "overcom[ing] 20th century barriers" in moving forward.[188] By barriers, she had reference to disciplinary ones; her overhead slide proffered a graphic of "silos and smokestacks" with the classic belching chimneys of early industry. Whether "the New Biology" (integrative biology?) has edged aside disciplinary biology will be left to the consideration of future historians of science: both the author's personal experience and the lack of perspective at this short time remove, makes it impossible to detect the end of silos of knowledge in biology. To borrow from a comment in the BIO AC Minutes from early in 2003, there are "challenges and natural tensions created when technological change is more rapid than institutional change."[189] Information integration through technological change could not have been moving faster.

The AC heard reports from various division personnel during early 2003 on just how well matters *were* progressing, however, when the 2005 budget theme for BIO was considered. "Transforming the Core: Research, Education, and Infrastructure Across Scales" appeared to be successful already, well before the actual planning stages for that fiscal year was upon BIO. Quentin Wheeler, then division director for DEB, noted that the Tree of Life project, large-scale international collaborations, and biodiversity inventories were all "obvious signs of 21st century biology and [the] transformation of the core." NSF/BIO had been the incubator for much of this. And Maryanna Henkart, director of MCB, dealt with aspects of microbial biology with regard to interagency efforts and also practical applications of metagenomics[190] (following chapter). Machi Dilworth, continuing director for DBI, reviewed the NPGI and pointed out how it had transformed the core. As a heavily invested effort in *information integration*, and that at an international level, the plant genome project was an ideal example of transformative research. Judy Verbeke, then acting director for IBN (Chap. 8), added comments concerning the need for synergy between annual budget themes and the actual budget documents and budget levels, while Judy Skog discussed the Frontiers in Integrative Biological Research program's successes.

The "fearless biologists" that John Wooley was said to have been trying to create in the late 1980s became a term of special significance to him as he used it again in his discussion before the AC that November concerning the then recently concluded workshop on CIBIO.[191] The AC membership discussed the already growing "success and far-reaching significance of [the] workshop" and

[188] Mary Clutter, "Meeting the Challenges," February 28, 2003; a paper set of overhead slides. To whom the talk was addressed and what the venue was are unknown.

[189] BIO AC, "Minutes," April 24–25, 2003.

[190] Metagenomics is the study of all the genetic material obtained from a *mixed sample* of environmental material and does not take into account individual species present, *per se*. This subject is considered in Chapter Nine.

[191] BIO AC, "Minutes," November 13–14, 2003.

noted that a cyberinfrastructure could "truly integrate natural and social systems"—Rita Colwell's biocomplexity and Mary Clutter's biology for the twenty-first century. Indeed, the LTER network had held a workshop of its own just a month after the AC's CIBIO conference and determined to "forge a bold new decade of synthesis science."[192] It remains ongoing as these words are written. After all, "today's new initiatives are often tomorrow's core."[193]

6.13 The National Science Foundation at 50

The British journal *Nature* stated it thus: "As the NSF approaches its fiftieth birthday…its prospects have seldom looked brighter."[194] President Clinton's proposed $675 M increase for NSF in FY2001 was but one example of the success of the Foundation during the period. It was the largest such suggested increase in its history—by twice that of any former figure. But with a budget then of about $4B annually,[195] Director Colwell's desire to see it doubled in her tenure turned out to be frustrated; still, "[w]e are guardedly optimistic," she had said. The budget would not double in coming years. NSF had always been a small part of the nation's R&D portion of the budget, which was about $80B at the turn of the millennium. The author of the *Nature* article made a good point noting that there had been a five-year "wrangle" (really more than five years, as we have seen in Chap. 1) *prior* to the establishment of the Foundation in 1950. With the already established Naval Research Laboratory, the Atomic Energy Commission, and especially NIH by 1950, NSF was destined to begin small and stay that way, at least as a proportion of total US R&D expenditures. However, it became the giant for *basic research* support, of course. Though *Sputnik* in 1957 did spur a rise in the NSF's budget, it rose little again until the early 1980s, "when economists and politicians began to associate basic research with economic growth."[196] And, while Eamon Kelly, then chair of the NSB seconded that conclusion, it should be pointed out that, unlike NASA, for whom Senator Mikulski "will protect [its] budget to her last breath…no one will go to the wall for NSF." Still, Colwell and Nobelist Leon Lederman, long a supporter of and a funded scientist through NSF, taught the school children of Washington, D.C. how to do science as a kick-off event of the 50th anniversary celebrations.[197] While he may have wondered about the possible "end of the frontier" in 1991, it was

[192] Ibid., William Michener's report to the AC in August 2003 on the LTER workshop "Environmental Cyberinfrastructure Needs for Distributed Sensor Networks."

[193] Ibid., in reference to aspects of several COV reports.

[194] *Nature*, "NSF Celebrates Half a Century with Big Plans," *Nature* (2000):8.

[195] National Science Foundation, "Fiscal Year 2001 Budget Request Overview," pg. 2. The FY00 current plan was then $3.897B and the FY01 request was $4.572B, an increase of slightly over 17%.

[196] *Nature* "NSF Celebrates."

[197] Ibid.

that self-same Leon Lederman who was quoted as having said, a decade later, that "[s]cience education is a serious business, but it's also great fun." Science was hardly at the end of its frontier.

Into that milieu, the National Science Foundation's Directorate for Biological Sciences stepped across the threshold of the third millennium determined to make "fearless biologists."

Chapter 7
Genes and Beyond

The metabolizing cell is more than
the sum of its genes.
Holgar Jannasch

7.1 Expanding Upon Genetics

Holger W. Jannasch, a marine microbiologist, was more than a little aware of what is now termed epigenetics,[1] in the modern sense of the word, and realized that the great complexity of life could not be explained solely by the encoded directions in genes alone. As a grantee of the National Science Foundation, Jannasch was "one of those rare microbiologists," because, among many other traits, he "was always ready to describe and explain his work in fascinating detail and with wonderful humor."[2] Holger Jannasch represented very much the ideal of an organismal biologist, albeit his organisms were of the unicellular ilk. His research moved up and down the scale from biology of the little to biology of the big, a necessity for those doing microbial ecology. In that guise, Jannasch might be an example, even an epitome, of a biologist at the end of a millennium poised on an innovative redirection for the science of life, that of Mary Clutter's "New Biology." Jannasch died in 1998 and thus closed that millennium's final century, the most rapidly flourishing and creative 50 years in the history of biology.

[1] R. Holliday, "Mechanisms for the Control of Gene Activity during Development," *Biological Reviews of the Cambridge Philosophical Society*, (1990):431–471.

[2] The quotation about his rarity was made by the great microbial physiologist, Cornelius van Niel, in a recommendation letter to Woods Hole Oceanographic Institution, Massachusetts, upon Jannasch's job application there in 1961. The second quotation in the main text above derives from Jannasch's 1998 memorial tribute by WHOI. See: http://www.whoi.edu/page.do?pid=10934&tid=282&cid=817&ct=163. Van Niel was quoted in the memorial. Exemplary of his funded research were 1975 grants DES 75–17,017 and GH 29665, that led to a Note in a leading journal of the field: "Microbial Dark Assimilation of CO_2 in the Cariaco Trench," *Limnology and Oceanography*, (1979):746–753.

D. J. McGraw, *Millennial Biology: The National Science Foundation and American Biology, 1975-2005*, https://doi.org/10.1007/978-3-030-56367-7_7

The half-century of biology from the near-contemporaneous dual events of the founding of the NSF and the discovery of the nature of DNA, to the early years of the new millennium, has been the subject examined by both Toby Appel in her *Shaping Biology* and by the author of the present study. Few areas within biology better characterize the massive changes in the discipline than do those of molecular and cellular biology, the subject of the present chapter. The work supported by the MCB, the Division of Molecular and Cellular Biology of the Directorate for Biological Sciences, from the post-reorganization period of BBS/BIO, goes far in demonstrating the growing *integrative* nature of modern biology. Such integration was, and remains, a consequence of the division's primary area of responsibility, biology of the little, because it blended into biology of the big more and more over several decades. The science of the macro end of the spectrum was, of course, done in the Division of Environmental Biology, but MCB's charge verged into indistinction between the two units as biology continued to integrate at NSF. The term "leaky borders" had been used at BIO for many years and may have been coined by Mary Clutter; certainly it was used by her. Cross-divisional integration and leaky borders will be one theme of the present chapter and will be shown through, especially, the major programs that were fully or partially a responsibility of the MCB.

In 1991, when MCB was formed, the three basic areas of interest (programs) were termed: Genetics and Nucleic Acids (GNA); Biochemistry and Molecular Structure and Function (biochemistry); and, Cell Biology (CB). That organization appeared to be suitable as 4 years later the same programs were still extant. However, it could not persist as just a simple three, as each of those would see a continuing parsing and increasing specialization of focus after mid-decade. The leadership just after the reorganization, some of whom we've met in earlier chapters, included: James Brown, MCB division director with four divisional staff; Phil Harriman and Delill Nasser were two of the four program directors (rotators counted among them), along with two assistants in the GNA; biochemistry, and other areas in that program, were staffed by six program directors (PDs) and two assistants; while CB was composed of three program directors and two assistants. One of those PDs, Maryanna Henkart, would by 1997 become the division director (DD) for MCB and would remain so through the period covered by the present study. By 1995, microbiologist Julius H. Jackson[3] had replaced Brown as DD and Henkart had moved from a program director to the position of deputy DD, a post not in existence under Brown. As it had been for some years, the titles of program officers were streamlined to include only Program Directors and program assistants, some of whom were designated as senior such assistants. The appellations of Associate and Assistant Program Director had been out of coinage for some time.

Also, the *cluster* mode had become fully established by the mid-1990s with the biochemically-oriented programming spread over Biomolecular Structure and Function as one cluster and Biomolecular Processes, as another. Cell Biology was

[3] Jackson maintains a faculty webpage at Michigan State University; see: http://www.mmg.msu.edu/jackson.html.

one unified cluster and Genetics yet another. By the final year of the past millennium, the cluster numbers would increase in the biochemistry and molecular areas, particularly, to include: Molecular Biochemistry; Molecular Biophysics; Biochemistry of Gene Expression; and, Metabolic Biochemistry. In the cellular area, Cellular Organization and Signal Transduction were the two groupings, while genetics had been further clustered into Eukaryotic Genetics and Microbial Genetics—all reflective of the directions the new biology was taking in both the community and necessarily at BIO. Persons from the division rosters were organized among the clusters based upon their individual disciplinary specialties, whether permanent or rotator.

For many years prior to the formation of clusters, in both BBS and, later, BIO some two open competitions were held per year that were for *unsolicited proposals*: their subjects fit no particular program/cluster. The PDs, and more staffers (such as rotators) in later years, would gather to determine whose program a given proposal might best fit and distribute them accordingly during what could be quite chaotic meetings.[4] The meetings alone would have been insufficient as they may have led to favoritism in funding a given proposal over another. For that reason, a "routing sheet" had to follow each proposal and had to be signed off by *two* persons in a division to indicate more than just one PO had read it.

A salient point arises when one examines the growth in numbers of programs, clusters, and levels of specialization for MCB during the 1990s: the place of *reductionism* was clearly increasing, but so was *integration*. The Signal Transduction in Cells program indicates the rise of interest in cell-to-cell communication, a major research involvement of the community at the end of the millennium and one of increasing size and significance during the twenty-first century. The twin features of reductionism and integration are well exemplified by that program, but even more so with the genome sequencing of *Arabidopsis,* as shall be recounted. The microbial genetics cluster's appearance and expansion parallels that of DEB's role in microbial evolution and, especially, microbial ecology—termed "microbial systems in the biosphere" in the later 1990s. By the first decade of the present century, microbial ecology would become of compelling interest not only to the environmental sciences, but also to human biology—a renaissance of interest in the role of microbes in all sorts of "environments," from soil, to deep-Earth rocks, to the human gastrointestinal tract, was dawning brightly. The same would be true with The Microbe Project in which microbial genomics, particularly metagenomics, would play the center position from the turn of the millennium.

The foregoing examples, and others, will be critical to the story covered in this chapter and those following. But there is more to consider in the present discussion. With the rise of subspecialties in genetics, new terms and areas for study were seen during the 1990s and have since become commonplace in biology—the so-called "omics."[5] Genomics (the study of the entire DNA sequence of an organism),

---

[4] Maryanna Henkart Interview with the Author, February 17, 2009.

[5] Satya P. Yadav, "The Wholeness in Suffix *–omics,* *–omes*, and the Word *Om*," *op. cit.* Yadav noted

metagenomics (below), proteomics, metabolomics, and other exotic creations would, by the present century, make up a very large fraction of research done in the science of life. Yet another area of concern to the present chapter is an examination of the state of biotechnology at NSF during the second decade and a half studied here. Further, the same may also be said about the STCs.

Finally, and of signal importance, was the establishment of, first, the study of the genetics and genomics of the thale cress, *Arabidopsis*, and later the expansion into the genomics of plants more broadly—the National Plant Genome Initiative (NPGI). While both the cress program and that of plant genomics were formally lodged in DBI for management purposes, I have saved the discussion of them for this chapter to both balance the length of chapters and to tie genomics more closely to other features of genetics/genomics handled by MCB. At any rate, it was becoming very difficult to separate programmatic features by division with the increase of interdisciplinarity and integration.

7.2 An Institutional Overview of MCB

One individual, Maryanna Henkart, was present at BBS from well before MCB's inception in 1991–1992 through the close of this study in 2005. She knew well the division's history. Henkart had first encountered the NSF during her undergraduate studies at Reed College, Portland, Oregon when she had a summer scholarship under her advisor in the field of plant physiology.[6] She went on to do her senior thesis supported by Foundation funds. She also had another NSF-funded summer as an undergraduate fellow at UC, Berkeley. She earned her doctorate at Harvard within the then still unified biology department (which later split along specialty

that he made a then recent (2007) visit to the online search engine Google for words ending in –omics and found some 829 of them! The word genomics was first coined by Thomas H. Roderick, a geneticist at the Jackson Laboratory, Bar Harbor, Maine, Yadav has asserted. Yadav provided a short story concerning the event that is said to have occurred at a since closed eatery, McDonald's Raw Bar, in Bethesda, Maryland during a conference where Roderick and two colleagues "[a]fter a bit of beer," saw Roderick create the word for a "yet-to-be-published journal" in need of a name; it became the journal *Genomics*. That was in 1986. However, the word genome (not genomics) is believed to have been coined by Hans Winkler in 1920 in a German language book where he used the German spelling "genom" of a Greek stem and had reference to a haploid set of genes. There is no Greek root "-ome." Some hold, Yadav averred, that –ome was actually created as a borrow from chromosome. The suffix –ome *may* have arisen from the Sanskrit intonation "om" signifying wholeness or fullness in Indian mantras; hence a *full* set of chromosomes, Yadav offered. But see also Joshua Lederberg and Alexa T. McCray, "'Ome Sweet 'Omics—A Genealogical Treasury of Words," *The Scientist* (2001):8, who tells the story somewhat differently and cites some other significant sources. Which of the two articles is the more authentic is left to others to argue. Finally, see the Oxford English Dictionary under *genome* for yet more information as well as the online Web of Science site. Incidentally, Lederberg coined the term *microbiome* (let alone such major words in biology as plasmid, exobiology, and prototroph); see Lederberg and McCray.

[6] Henkart interview.

lines in 1969). Her post-doctoral work was undertaken at UC campuses in both San Diego and Los Angeles. Additional post-doctoral studies were done at NIH in 1971. By 1974, she had become a junior faculty member at the Institutes and had her own laboratory. Later, moving just across the street, she worked at the Armed Forces Radiobiology Research Institute for a short time, then returned to NIH. Unexpectedly, Henkart received, "out of the blue," a telephone call from Mary Clutter who was recruiting rotators for her PCM division in BBS. Henkart, who had started out in plant physiology and had moved to cellular physiology (covering all types of organisms), electrophysiology, and electron microscopy, was able to take up the rotatorship as program director in cellular physiology at NSF beginning in January of 1984. She stayed on for nearly two and a half decades. While the early cellular physiology program was strongly animal-oriented, some plant cell research was also being funded, Henkart recalled. Given that she had a "weird combination of expertise and background," she said, Clutter found her to be ideal for the position. As her career advanced at BBS/BIO, she had various programmatic leadership assignments, though staying close to the cellular aspect, until she became deputy director of MCB, and then (1997) its director. She averred that: "there's no perfect way to distribute scientific matter" in regard to such issues as where the NPGI was placed, for example. That comment was a correlate to her statement to the author: "you have to have bite-sized pieces so that you can handle [the] work."[7] As BIO continued to integrate, the boundaries between divisions continued to blur ever more strongly.

Henkart recalled that Clutter's creation of BIO required that each division have a director and a deputy director, either one of which would be permanent and the other a rotator. The premise behind that was that one would provide "institutional memory and the other [would bring] new ideas [from the community]," according to Henkart. "[D]ifferent divisions had different combinations" in filling the two posts over time. Julius Jackson, successor as DD of MCB to James Brown, was an example of a rotator director; he stayed 2 years. It was Henkart that succeeded Jackson and remained in that role for a decade, leaving it only in 2007 and retiring from the Foundation the following year. Henkart remembered that Jackson had fervently promoted greater study of the *microbial biosphere* in his short tenure at BIO. His excitement about the area seems to have taken root at the directorate in several ways (see the end of the present chapter and especially Chapter Nine). Classically, microbes had been studied primarily in the laboratory, and then only those that could be cultured. This had been a long-standing maxim in bacteriology, but it had also been long believed that many species were being missed under those circumstances.[8] With the rise of certain areas within molecular biology, she said, and especially metagenomics,[9] unearthing of an entire undiscovered world became

[7] Ibid.

[8] Personal knowledge; knowing that one was missing most species in laboratory culture had been a dictum of bacteriology for much of the discipline's existence.

[9] "Like genomics, metagenomics is both a set of *research techniques*, comprising many related approaches and methods, and a *research field*," according to the National Research Council's, *The*

possible. MCB and DEB met at that boundary—leaky borders—besides many others. Metagenomics offered a "huge opportunity," Henkart averred, and it was promoted the more so by Rita Colwell's biocomplexity initiative—Colwell was, after all, a microbial ecologist.

But this discussion of the later 1990s leaps somewhat ahead of the earliest years of MCB. Henkart pointed out in an interview with the author that the interdisciplinary feature brought about by microbial ecology was only one aspect that she had either participated in as a program director in the first of that decade, or actually led as DD after 1997. Other features, she remembered, included the continuing need for NSF to distinguish itself from NIH (previous chapters), a part of what Mary Clutter's efforts had addressed. For Henkart, it was to bring together biology with physics (*biophysics*), and biology with chemistry (toward a more modern understanding of *biochemistry*), to an ever-greater degree than theretofore at BBS/BIO. Creating "that interface was a big part of what my responsibilities were in MCB."[10] The cross-disciplinary field of biophysics was well known to Henkart, she having worked with its methods as applied to biological questions. She was an early member of the Biophysical Society and later joined the Society for Neuroscience, the latter of which was also inclined toward biophysics; her interests in cellular physiology were very broad.[11]

MCB held a "number of interdisciplinary workshops [over time] that…led to the development of programs that had relatively short lifetimes," such being "indicative of new directions"; a *significant point* in considering BBS/BIO's institutional history. In the early 1990s, she like all program directors had long done, sought to "promote the advances [of new] frontiers," and not just to fund research requests that "came over the transom." Therefore, the workshops for her program, and for the division both before and later, were designed to discover the thinking of the community, to promote the *intricate dance*, and to fund the frontiers: standard practice at the Foundation. Those included ones at the interface of mathematics and biology, not unlike those sponsored by DBI under John Wooley, and also of network analysis (such as must occur to understand, say, the nervous system). *Modeling* was central to both of those areas. As she opined, "the development of tools parallels the development of questions, and they feed on each other."[12] The *Arabidopsis* story may be

---

New Science of Metagenomics: Revealing the Secrets of Our Microbial Planet (Washington, D. C.: Nation Academies Press, 2007), pg. 3. "Metagenomics overcomes the twin problems of the unculturability and genomic diversity of most microbes." In practice, individual species in an environment of any type are not sought out nor isolated; only the genetic material of a *mixed population* is examined. Unlikely as it seems, vast new knowledge can be and has accrued by employment of the technique.

[10] Henkart interview.

[11] More information can be discovered on the Biophysical Society, founded in 1957, at: http://www.biophysics.org/AboutUs/Overview/tabid/2238/Default.aspx. The Society for Neuroscience web presence is to be seen at: http://www.sfn.org/. SfN was founded in the late 1960s.

[12] Henkart interview.

an ultimate expression of the matter of *model systems*, not unlike that of *Drosophila* in earlier years.

In all of that, she found that there was a "false dichotomy" between the "reductionists and the integrators"—an argument supported throughout the present study. "[W]hat we need to do is [to] understand the parts and [to] go back and forth between parts and then, when we have tools to integrate [them], we can do that." Thus, I would argue that, *reductionism prepares the way for integration.*[13] The tools would have included aspects central to computational biology, she held, among other growing disciplinary areas. Then, "being clever as you can be, you try to use informatics' tools and trace the combination of…what this all means to the whole system." That statement reflected the move of reductionism into a new holism greatly dependent upon discoveries in biology of the little—and which matter will come upon again in the following chapters.

In addition to cleverness in regard technological tools, management of funding had matured a long way by the middle 1990s, and onward, in regard the many values of the *cluster concept*. Besides offering leaky borders on the intellectual plane, such permeability also allowed for greater freedom in how funding could be directed to the best ends: often interdisciplinary and inter-unit. That is, a proposal, the contents of which were found to lie across more than one programmatic or divisional or even directorate line, could be funded by several budgets from the program through the directorate level. However,

> the cluster structure as it actually was practiced wasn't always what people hoped it would be and wasn't even what some people tried to believe that it was…I may be describing it in a more…idealistic tone than it actually played out,

said Henkart in her opinion of the system. Not surprisingly, new institutional structures may have flaws and the long-time DD of MCB saw several reasons why, in this case. For one thing, the cluster system "varied a lot from one division to another and from one cluster to another"—for several reasons. One of those was that "when you ask people to start behaving in teams when they haven't been accustomed to doing that," difficulties arise. That was especially true in the early period of constituting clusters. "Some people are more willing to try [to work in teams] than others." Another cause of those variations was due to "personalities of individuals." For instance, "it just takes one poisonous person in a cluster to cause a lot of grief," Henkart argued.[14] Moving persons from one cluster to another could often improve the "chemistry" of the group and was an approach taken.[15] Further, different DDs

[13] While the statement is mine, see a comparable comment in Camille Ripoll, et al., "Defining Integrative Biology," Essays and Commentaries, *Complexity* (1998):19–20, where they say (pg. 19) "[integrative biology may be] the natural extension of a reductionism, which has been highly successful until now, or [it may be] of a globalism, alias a holism." Further, they ask, "[o]r could integrative biology transcend both reductionism and globalism?" (The French preferred "globalism" to the American choice of "holism.")

[14] Ibid. Many references to personalities are to be found in Henkart's interview transcript.

[15] The author interviewed one DD (name withheld here) who had needed to repopulate a given cluster to seek better internal relations, as one known example.

were committed to the cluster concept to different degrees and, in some case(s) would only create them to the letter of Mary Clutter's expectation, but not to her intent, in the opinion of Henkart. (As with other features of the 1991 reorganization efforts, Clutter consulted heavily and broadly amongst her personnel prior to reaching the decision to initiate clusters.)

Little more needs be said about human nature and group work on the negative side. On the other hand, Henkart also felt that, with time, the cluster system became much better, primarily as persons who knew only the pre-cluster model left NSF. New individuals who came to BIO during the 1990s were introduced to the necessity of working in teams; they did not have the prior experience at the Foundation of working mostly alone. Indeed, then-current cluster scientists would interview potential new employees—cluster members to be—so accommodation to one's workmates began even before hiring.

While much of the foregoing was extant in MCB, it was just as common in the other divisions. That was so, because inclusive staff meetings at BIO had long been established by its predecessor administrative units and ADs. Interdivisional communication was typically open and engaging.[16] And, further, much that has been said just above can be seen to have had a direct relationship to the personalities leading the various units from AD to DD to permanent PDs. We have seen the strong and intellectually powerful nature of both David Kingsbury and Mary Clutter, in particular, as ADs. Others, DDs, met in this history include John Wooley, for example, and many another predecessor of the divisions and programs of BIO, and BBS earlier. More such leaders in the biological sciences at NSF are still to be introduced.

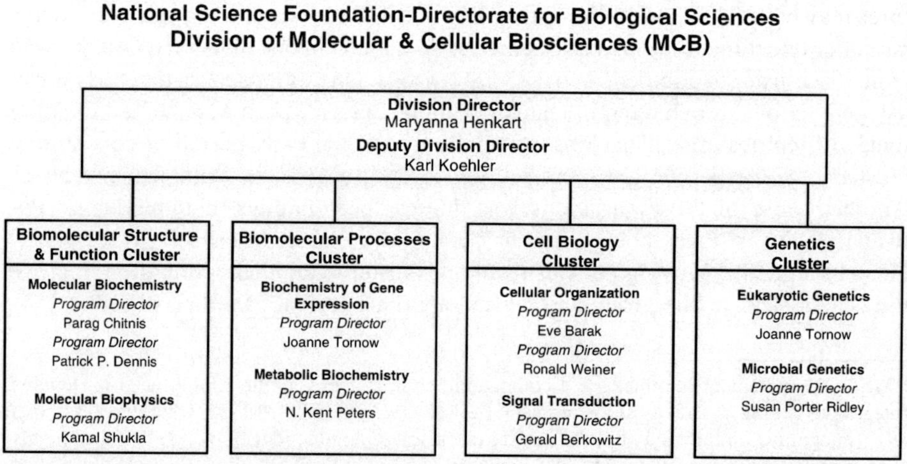

Fig. 7.1 Organization chart for the Division of Molecular and Cellular Biosciences (MCB). NSF. February 2002. Specific provenance unknown

[16] Ibid.

The organizational chart for MCB by February 2002 is shown in Fig. 7.1. A minor reorganization occurred for all of BIO within a few months, however. The budget for the division was, at that time, over $121 M and requests for the 2010 Project (below), Biocomplexity in the Environment (more in Chapter Nine), Genome-Enabled Science (below), and Systems Biology, were made. Mary Clutter gave a talk at the Santa Fe Institute in late February in which she discussed the "revisioning" of BIO.[17] She provided extensive details for both NSF and BIO concerning where matters stood at the time noting, among other things, the appearance of "Emerging Frontiers," and the establishment of "virtual programs" with other Foundation directorates, Mathematics and Physical Sciences (MPS) directorate being one with regard to mathematics and biology.

7.3 Biotechnology at Millennium's End

Biotechnology (earlier chapters) was *institutionalized* at NSF/BBS in the mid-1980s with Erich Bloch's founding of the Office of Biotechnology Coordination and David Kingsbury's aggressive direction of BBS. And when Mary Clutter assumed its leadership at the end of that decade, she continued the NSF commitment to the field. It passed safely through the great reorganization into the new BIO.

But before that restructuring, Phil Harriman had taken a sabbatical as a Congressional Fellow in 1981, as we have seen, being then the director of the prokaryotic genetics portion of the Genetic Biology program, which he co-directed with Delill Nasser. Again, and pertinent here, Harriman was able to take a second paid sabbatical in 1987 to work in the laboratory of Hamilton Smith at Johns Hopkins University Medical School.[18] (Smith was just then starting to study the genome of the bacterium ***Haemophilus influenzae***, which is considered more in a later chapter.) In 1978, Smith, who was a grantee of NIH but not of NSF, had shared the Nobel Prize with Werner Arber and Daniel Nathans for their work in the discovery of restriction enzymes, so well applied by Boyer and Cohen to initiate the biotechnology revolution.[19] Harriman, as it eventuated, became highly informed about the laboratory practices and techniques under Smith and which were central to the revolution. He was, then, able to bring back to his position at BBS considerable insight into an area so heavily funded by the directorate, and particularly by programs in his division. As the reorganization approached, biotechnology was well supported across many sectors of NSF and had been since 1984, and in a less formalized sense before that.

There was, though, unhappiness in BBS, recalled Harriman, when it was noticed that he had had two sabbaticals in the 10 years he had then been at the Foundation.

---

[17] Paper copies of a set of overhead slides of the February 25 presentation.

[18] Harriman interview, *op. cit.*

[19] http://nobelprize.org/nobel_prizes/medicine/laureates/1978/

Because of that discord, he said, a move was made at the directorate "making it more formal" with regard to persons being selected for sabbaticals; that internal program having not existed prior to that time.[20]

After the reorganization, Harriman continued as one of the four directors of the newly named Genetics and Nucleic Acids Program under the infant MCB, along with Nasser and, by 1992, Richard Eberle and Todd M. Martensen. (Martensen was also a PD in MCB's Biochemistry and Molecular Structure and Function Program.) Two assistants completed the program roster. Jim Brown had been on sabbatical at Michigan State University during the crucial years of 1991–1992, but then returned as DD for MCB.

Prior to his year's leave-taking, Brown had already appointed Harriman to a newly forming biotechnology self-study working group at BBS. Joann Roskoski (previous chapter; and still at BIO as this history was being written) was another representative, she for the BSR/DEB division. Finally, Bruce Umminger of DCB assigned Judith Plesset (though she was involved in the evolution and development program areas, she had a "long-standing interest" in biotech) and rotator Ed Leadbetter of his division to the group that October.[21] That was the first strong effort to bring biotechnology, as it was seen at BBS/BIO, into the last decade of the century.

During those same years of spiraling growth for biotechnology in the community and at BIO, the effort saw notable publications appear from the Foundation, and other government agencies, academe and industry; the latter always a part of the picture. NSF broadly considered its own place in the report of its combined self-study efforts with a university/industry advisory panel, "Biotechnology Opportunities: The NSF Role."[22] The Biotechnology Research Subcommittee (BRS) of the Federal Coordinating Council for Science, Engineering, and Technology (FCCSET[23]; somewhat disingenuously pronounced "fix it"), produced its first report. By the early and mid-1990s, biotech was one of the six initiatives

[20] Harriman interview.

[21] This information from a series of ephemeral printed versions of three emails: Shirley E. Hill to Harriman, Umminger, et al., October 9, 1990; Roskoski to Umminger, October 10, 1990; and, Umminger to Peggy Weber (BBS OAD staffer of the time), October 9, 1990.

[22] National Science Foundation, "Biotechnology Opportunities: The NSF Role; A Self-Study and University/Industry Advisory Panel Report," NSF 91–56 (Washington, D. C.: NSF, 1991).

[23] FCCSET had been established under Title 42, Chap. 79, Subchapter IV in 1976 and chaired by the OSTP head; the OSTP itself was established under the same act. FCCSET and OSTP began as separate entities under the act, but see below. FCCSET included cabinet members, or deputies, from 12 federal departments, various science agencies, and OMB. It was charged with providing more effective planning in the subject areas, identifying research needs, using facilities and agencies more effectively, avoiding unwanted duplication, and furthering international cooperation. The following year, FCCSET was abolished but reconstituted *under* OSTP in a different form, though retaining its name. See: http://govinfo.library.unt.edu/npr/library/reports/NSF1.html, and other sites. Mary Clutter, in addressing the NSB about "emerging opportunities" in mid-1994, stated that "every Presidential initiative that arose through the [FCCSET] process originated at NSF," to that time. NSB, "Minutes," NSB/AO-94-21; August 10, 1994, 6–94:8.

under FCCSET, all memberships were voluntary by the agencies involved: FCCSET did not have a mandated membership requirement. NSF had interests in the biotech effort, in particular, along with others of FCCSET's six main efforts, such as High Performance Computing.[24] FCCSET was not considered effective by 1994, however, and efforts were afoot to improve its efficacy. Such was later accomplished.

Bruce Umminger's report to Mary Clutter in April of 1991 on the findings of the self-study group noted that the committee found seven areas that benefited from NSF support of biotechnology; they covered a number of subject areas and were often cross-directorate.[25] Running from agriculture to biomedical sciences to engineering aspects, the seven areas were examined with regard to how to implement a formal national initiative for biotech. An ad hoc advisory panel for NSF was assembled early in that same month from members of the self-study group along with representatives of the academy and industry. It was chaired, appropriately, by then NSB member Rita Colwell, herself long deeply involved in biotech in Maryland. The panel utilized the findings of the report and promoted extensive *interdisciplinarity* and sought mechanisms for *technology transfer*—a particular interest of Colwell's.[26] A number of opportunity areas were examined with those approaches in mind while numerous other features were additionally discussed. Most notable was NSF's concerns that certain areas were underfunded in biotechnology broadly and so the Foundation sought to emphasize those areas in which it had been successful in past efforts and for which other federal subsidy was weak. They included such subjects as bioremediation, bionetworks, connections between BIO and CISE, bioelectronics, and plant biotechnology. With the, by then, well-established commitment to science education at the Foundation, that became another area of special concern, as would eventually the social ramifications of the field; that issue was finally undertaken by the new Directorate of Social, Behavioral and Economic Sciences (SBE).

Implementation of the entire enterprise had not yet matured by the time of the initial report. Nonetheless, the FY90 budget of the previous year had been a vibrant one for the subject as the total had exceeded $1.7 M from NSF alone. At the same time, however, the NIH "empire" was putting into the field nearly $2.3B! DoD was second only to NIH in funding commitments, and NSF was a distant third in a list of 12 federal agencies.[27] Figure 7.2 provides an overview of the federal commitment at the time. The report noted that the "New Biotechnology Initiative" of the NSF, as but one in that multi-agency project, would focus on *five* areas: environmental,

---

[24] Ibid., the website.

[25] Bruce Umminger to Mary Clutter, April 17, 1991: handwritten memo attached to a draft version of the report on the work of the self-study group. See also a memo concerning an earlier meeting on the subject: Umminger to Clutter, March 14, 1991. Finally, see the report, 'Biotechnology… NSF Role,' Appendix A of seven parts (pp. 29–44).

[26] Purdue University, Indiana, has considered the subject at: http://www.prf.org/pdf/The%20Technology%20Transfer%20Process.pdf. Colwell had keen interests in technology transfer; author's personal experience with her.

[27] Umminger to Clutter memo attachment, Tables 1 and 2.

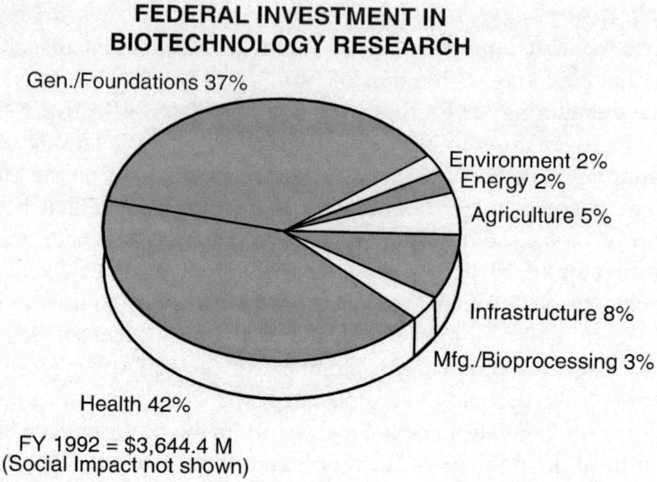

Fig. 7.2 Federal investment in biotechnology research. "Biotechnology Opportunities: The NSF Role; A Self-Study and University/Industry Advisory Panel Report," NSF 91-56. Appendix A

bioprocessing, bioelectronics/bionetworks, agriculture, and marine biotechnology—a refinement of earlier discussions and suggested directions for up to seven areas.[28] More refinement in that number was still to come. Foreign competition was noted within the very first paragraph of the Executive Summary and the intent to keep America remaining in a "preeminent international role" was an imperative; but international involvement would soon be both sought and supported. The proposed beginning for funding of the initiative was FY93.

The fall of 1991 also saw Senator Al Gore's comments on the subject in a paper published in the *Harvard Journal of Law & Technology*.[29] He pointed out, quite brilliantly, that

> Pioneers and planners are, by nature, opposites. Pioneers must rebel and revolt against society to renew it. Planners try to relate the novel to the normal to provide continuity and growth.

Citing the successes of Watson and Crick, Henry Ford, the Wright brothers, and supercomputer's Seymour Cray in history, he wondered who might be remembered for planning biotechnology policy.[30] He said, in another memorable statement, that the "genetic code is the organic arm of the information revolution," and in so saying neatly related the two great revolutions of the end of the last millennium—and much of what was central to the funding efforts of the National Science Foundation then and for years to come. Since "biotechnology is developing faster than any other

[28] 'Biotechnology…NSF Role" Appendix A.

[29] Senator Al Gore, "Planning a New Biotechnology Policy," *Harvard Journal of Law & Technology* (1991):19–30.

[30] Ibid., pg. 19.

technology and, in the process, creating a wider gap between practice and policy," establishing the latter clearly was of central concern to the Senator. He told his readers that, due to the Asilomar Conference of 1975, "policy and pioneering" were put on equal ground at the same time then. But, 15 years later, the growth of biotechnology was moving well ahead of policymaking and the Recombinant DNA Advisory Committee was "relax[ing] itself out of a job." Who would police biotechnology's super highway? At that time, Congress was considering biotechnology in relation to: capital formation, patent protection, drug pricing (a perennial issue right up to the present), and research funding—but each one independently of the others. Nothing was being done about "ethics or [the] future direction" of the biotechnology juggernaut, Gore bemoaned.[31] Indeed, an answer would not be forthcoming at the time.

Such concerns as expressed by Senator Gore came right at the moment that a ratcheting up of biotechnology could be seen coming from all quarters. That was so because, concomitantly, FCCSET was assuming the leadership role for the new efforts of academe, industry, NSF, and the 11 other federal agencies involved. Thus, FCCSET seemed to be the organization that would play the leadership role that Gore wondered about: FCCSET brought out its own report in 1992: "Biotechnology for the 21st Century." The report was very wide-ranging being a statement from two groups: the Committee on Life Sciences and Health (CLSH), chaired by James O. Mason of the Department of Health and Human Services and the Biotechnology Research Subcommittee (BRS) of FCCSET. The CLSH subcommittee's *co-chair* was David J. Galas, of DoE, who was also the *chair* of BRS, the second of the two subcommittees authoring the report and the one to take on the actual coordination of biotechnology in the fullest sense for FCCSET. The area of medical biotechnology was so much larger than any other, at 42% of the federal investment (Fig. 7.2), that it was CLSH that was central to much of the report's content.[32] But basic research (NSF) and non-medical applications (for example, environmental at 2%) still represented a substantial portion of the national picture when all those other areas were tallied to 58%—and most of those areas fell under NSF partially or totally. Six broad priorities were stated that included the matters of: "[maintaining] momentum" in the face of growing international competition; filling gaps "in support for research in critical areas"; increasing use of a "broad range of organisms" as research subjects; strengthening work on macromolecular structures; expanding interdisciplinary research; and increasing the level of training.[33]

Additionally, the formalization of the national scheme for biotech became a Presidential Initiative in 1993. By then, "all quarters" were involved, clearly enough. By mid-decade, too, another report, "Biotechnology for the 21st Century: New Horizons." It had appeared via the BRS under the chair of its parent Committee on

---

[31] Ibid., pg. 29.

[32] "Biotechnology for the twenty-first Century: A Report by the FCCSET Committee on Life Sciences and Health," (Washington, D.C.: OSTP, 1992), pg. 7, Fig. 7.

[33] Ibid., Executive Summary, pg. 1.

Fundamental Science, M.R.C. (widely known as "Marcy"[34]) Greenwood of OSTP, and co-chaired by Neal Lane of NSF and Harold Varmus of NIH.[35] The results of that second and even more major report allowed for it to become a supplement to the FY 1994 budget for the biotechnology "strategic" Presidential Initiative.[36] It was pointed out that the federal commitment to biotechnology was primarily focused on human health and that it was having a "profound impact." However, the time had come to "bring the power of biotechnology to bear in other fields." If the 2% for environmental sciences of the previous report 3 years earlier was any indication, non-medical biotech clearly could profit by a substantial increase in support from federal agencies, NSF not least among them, of course. Agriculture, manufacturing or bioprocessing, and marine biotechnology and aquaculture were the final *three* principal priorities having been slowly whittled down from seven, then five, of a few years earlier. Indeed, the Executive Summary of the budget supplement averred that "[b]y the year 2000, the biotechnology industry is projected to have sales reaching $50B in the United States" alone. (It actually reached that mark, $50.7B, by 2005, in fact.[37]) It would create "thousands of jobs and promote renewed economic growth." Clearly, NSF was strongly tied to the economic well-being of the country, but basic research remained necessary at the foot of the biotechnology *industry* and it was basic research that fed applied research. The former was, of course, the primary focus of NSF/BIO then and on into the future.

The previous year's budget for all of that increase in forward motion had been a good one, as President Bush, *pater*, had unveiled his FY93 proposals in January of 1992.[38] NSF Director Walter Massey was "grinning" when he saw the figures as the Foundation was to get an 18% boost overall and a 21% increase in basic research funds, specifically. Presidential Science Advisor Bromley was equally gleeful as he felt that the administration was clearly committed to investing in science and technology that would keep the nation "prosperous and secure." While military R&D dropped, the civilian equivalent in basic research rose 8% across the agencies.[39] Special mention was given to five initiatives, biotechnology being first among them. It would be given just over $4B.[40] That was the promised 18% increase over the previous 2 years.[41]

The letter's writer, BRS Chair David Galas told Lederberg that the work of FCCSET was the first "systematic compilation and analysis of Federal programs

---

[34] Personal knowledge.

[35] "Biotechnology for the twenty-first Century: New Horizons; A Report from the Biotechnology Research Subcommittee of the Committee on Fundamental Science, National Science and Technology Council," (Washington, D.C.: NSTC, 1995). It can be seen online at: http://www.nal/usda.gov/bic/bio21, as one of many sites.

[36] See: http://www.accessexcellence.org/RC/AB/IE/In_the_National_Interest.php

[37] See: http://www.soc.duke.edu/NC_GlobalEconomy/biotechnology/overview.shtml

[38] Colin Norman, "Science Budget: Selective Growth," *Science* (1992):672–675.

[39] Ibid.

[40] David Galas [DoE] to Joshua Lederberg [Rockefeller University], June(?) 16, 1992.

[41] 'Biotechnology for the twenty-first Century,' pg. 1.

and budgets on biotechnology research and infrastructure." Biotech at NSF had been funded for three-quarters of a decade without being part of any national infrastructure, such being lacking. But the signal importance of the field by the end of the "science century" had become such that neither the Foundation nor any other federal agency, let alone the academy or industry, could long continue to go without that needed cross-organizational integration. Biotechnology was after all "the organic arm of the information revolution" and the two would be at the forefront as science and technology moved into the third millennium.

Susan McCarthy said that same year that "[b]iotechnology is on the move."[42] She was the coordinator for the Plant Genome Data and Information Center at USDA's agriculture library in Maryland. She noted that there were some 1100 companies in the US, at the time, that were engaged in biotechnology in some fashion: 76% of those were small, each with fewer than 50 employees. But their revenues belied their size: $4B in product sales in 1991. The Presidential Initiative was well received in Congress, too, with a bipartisan Biotechnology Caucus that had come into existence.[43] And, within the following year, biotechnology could be considered, along with America's space policy, for example, as a *major area of concern to the country*.

The highest-level overview of science and technology was undertaken by the new National Science and Technology Council, the NSTC. Established in November of 1993 by President Clinton, it was modeled after the long-successful National Security Council. Unlike George Bush's 1990 President's Council of Advisors on Science and Technology, PCAST, or the older FCCSET, the NSTC was made up entirely of federal officials. The President was the chair of NSTC (as he was on PCAST) and the Council is composed of the Vice President and Cabinet members along with department heads of primarily science- and technology-oriented units within the government. (Even as late as 2008, however, NSTC seemed to be less well-known than its predecessors, according to some.[44]) Along with the Congressional Biotechnology Caucus, the burgeoning field had gained a place in American policy discourse, let alone in American industry and funding the basic science that supports it.

Within several years of the 1991 initiative in biotechnology that had focused on seven areas of specific interest to NSF, the matter had become more refined. At the end of 1994, funding for a special competition was requested in a memo to the Foundation's Deputy Director by Mary Clutter and William Harris, AD for the Mathematics and Physical Sciences (MPS) directorate. The Program was to be aimed at three areas where the Foundation "has a unique and critical role" in biotechnology: environmental, plant/agricultural, and bioprocessing (a slight variant of the other group of three mentioned above, marine aspects having been subsumed

[42] Susan McCarthy, "The Federal Biotechnology Commitment," *Probe* (1992)(2):1.

[43] Members included: Senators Frank Lautenberg (D-NJ) and Hank Brown (R-CO); and Representatives Tom McMillen (D-MD) and Tom Bliley (R-VA).

[44] 'Beyond Sputnik,' *op. cit.*, pg. 30. PCAST was the successor to PSAC, sacked by Nixon in earlier years. NSTC was the successor of PCAST.

under environmental).[45] As noted earlier in this history, the Foundation was virtually the only federal agency that supported work in environmental and plant sciences at the basic level, whether such research might be in the otherwise applied area of biotechnology or not: the whole of that rapidly expanding field was still dependent upon a steady stream of basic research knowledge. The basic science for bioprocessing was, as well, not heavily funded by others than NSF; bioprocessing, too, being very reliant on more pure science at its core. Clutter and Harris pointed out that human health-connected biotechnology had been well supported. That funding came through the NIH "empire," of course, but other areas of concern primarily to NSF remained "poorly developed." Marine biotechnology and social and economic dimensions of the field needed consideration, and they were areas where NSF excelled and NIH had little interest. The request memo was peppered with commentary about the Foundation's particular expertise in devising *interdisciplinary* mechanisms to support basic science, not unexpectedly. A multimillion-dollar investment was sought per year for each of 5 years. Three other directorates had already agreed then to contribute $1 M each; BIO and MPS would offer $2.5 M apiece. Clutter and Harris were strongly committed to the new competition. SBE would contribute even more than the other three due to the fact that *all of NSF's international scientific efforts* had been placed under that directorate's purview during the great reorganization: multinational aspects were one area to be emphasized in the proposed program. The total investment of $10 M in the first year would permit 6–10 grants being made, each for three to five years duration.[46] Multinational efforts in the *Arabidopsis* genome sequencing effort—where a biotechnological use could be envisioned—would be an exemplar as, by the end of the century, some 30 countries were involved in that particular project.

The request was positively received in the NSF Director's office and BIO/MPS created the program announcement the following year. Referring to the findings of various earlier reports, the announcement stressed the fact that "two major generalizations" could be made:

- Health-related biotechnology research has represented by far the major component of the Federal investment; this must remain a high National priority.
- This pattern of research support has left important gaps in the knowledge base underlying future applications in critical areas beyond health: in biotechnology related to the environment, to manufacturing-bioprocessing, agriculture, in research on the social impact of biotechnology, and in the infrastructure and training specific for biotechnology research in these areas.[47]

---

[45] Mary E. Clutter and William Harris to Anne Petersen (DD of the NSF), "Opportunity Fund Request," December 1, 1994. A Draft Program Announcement was attached to the original of the memo: "Collaborative Multidisciplinary Research in Biotechnology."

[46] Ibid., pg. 2.

[47] Typescript ephemeron of the proposed program announcement (including several draft variants), "Biotechnology for the twenty-first Century, Phase Two: Beyond Health," 1995.

The "New Horizons" subtitle was an apt soubriquet because, as John Gibbons said in its introduction, the industry sector was then contributing $11B to the national economy. However, the actual amount was calculated—Susan McCarthy had declared $4B just 3 years before—it was obvious that the business of biotechnology was "about more than jobs and the economy; it provides the means for a vastly improved quality of life for all Americans and the world," Gibbons offered. The report represented all 12 of the federal agencies that had been involved since FCCSET's BRS took on coordination of the biotech initiative nationwide. MCB's Maryanna Henkart was the BRS Vice-Chair, a position formerly held by Bruce Umminger.

What the heady report did not consider, however, was a simultaneous movement within the biotech industry that had certain real or potential negative effects. "[C] onsolidation [and] restructuring" in the industry sector were just then ongoing and the rise of larger firms, such as Amgen, Chiron, and Genentech, were acquiring many smaller biotech companies. The economy was not necessarily impacted, but many other measurable features of the industry were being "skewed" by those changes.[48] A "new space" for economic growth had been created in the mid-1970s and had accelerated massively by two decades later; but with such activity came such features as "consolidation" of the map-points of an industrial landscape.[49] The concept of "new space", and related issues, have been considered in other ways in the history of science and technology: NSF, however, continued to make funding available for basic research in whatever manner the industry filled new spaces in the economic milieu.[50]

The international feature was doing well also, as by the opening of 1997, the US-EC Task Force on Biotechnology Research had just concluded its sixth meeting the previous year in Brussels, Belgium. Maryanna Henkart and Jim Edwards (at the time Clutter's Executive Officer for BIO) were the NSF representatives. Henkart presented an overview of US federally-funded research which was coming, by then, from 13 government agencies in America. An area of special note was the large amount of support going to research on the *Arabidopsis* genome, as well as toward bioremediation. She also noted that training funds, too, were coming from NIH. At a meeting just two days before in the same city, the US and its European colleagues had discussed marine microbial biotechnology. Though the Europeans held that they had been slow to adopt the technology with regard to the marine microbiota, the NOAA representative offered that America had not moved fully into the breach

---

[48] William O. Bullock and Mark D. Dibner, "The State of the U.S. Biotechnology Industry," *Tibtech* (1995), no pagination; "Feature Article" of five pages; pg. 1.

[49] Joseph Schumpeter had originally spoken of the "new space" in his early work: *Business Cycles: A Theoretical, Historical and Statistical Analysis of the Capitalist Process,* abridged by R. Fels (New York: McGraw-Hill, 1939). The term and concept were borrowed by Martin Kenny in his "Biotechnology and the Creation of a New Economic Space," in Arnold Thackray, ed., *Private Science: Biotechnology and the Rise of the Molecular Sciences* (Philadelphia: University of Pennsylvania Press, 1998), ppg. 131–143.

[50] Ibid. See other papers of the rather polyglot collection in Thackray (endnote above).

yet either. But Rita Colwell, who was present and representing the University of Maryland's involvement in that field, provided a long list of areas for research opportunities with marine microorganisms. Other areas related to biotechnology, *per se,* included bioinformatics, biosafety, metabolic engineering, and structural and computational biology, which were addressed in a presentation by Jim Edwards.

Plant biotechnology was an area that NSF declared to be of particular interest to it and simultaneously with the biotech work, plant genomics, and then functional genomics, would grow allowing for greater commitment by NSF of support to the world of plant biotechnology. Plant science is considered later in this chapter.

7.4 STCs in Biology: The Number Grows

The Science and Technology Centers were products of the very active period at NSF and BBS of the mid- and later 1980s, as has been seen in earlier chapters. Biotechnology was a central reason that the STCs came into existence in the first place. The first round of competition had produced 11 of them across all directorates at the Foundation. In the biology area, though, Carnegie-Mellon University's (CMU) Center for Light Microscope Imaging and Biotechnology, CLMIB, was one. Bruce Umminger, so closely tied to the STCs over time, and their leading officer at NSF from 1999–2006, suggested that CLMIB was a good example of how the STC program was modeled and managed (Chapter Four). But by the early 1990s, more biologically-oriented STCs were coming on line.

Also in 1991, the University of California at Davis was awarded over $21 M to begin the Center for Engineering Plants for Resistance Against Pathogens (with the none-too-euphonic acronym of CEPRAP) with George Bruening as the PI.[51] It was biotechnology applied to an area that had significant potential and would later profit by knowledge generated in the various plant genomics programs of NSF, USDA, and other agencies both within and without the US. The goal of the center was to "develop an integrated molecular and biochemical understanding of plant responses to diverse pathogens." The Center opened with the tomato (*Lycopersicon esculentum*) as the *model system.* As with other centers in the second-round competition, the Davis center was a multi-university effort that involved both the Davis and Berkeley campuses of UC, Cornell in New York, and Washington (Missouri) universities, along with industrial scientists from Calgene, Inc. (California). One of the several areas of interest in the major research objectives of the center was signal transduction in cells, a program that would come into existence at BIO in 1995 and is a good example of the *intricate dance.* As the "molecular basis of host–pathogen interactions is one of the major unknowns in plant biology," the center was to be a pioneering effort in the plant sciences. NSF funding came from a number of

[51] See: http://www/sciencestorm.com/award/8920216.html

programs in BIO, in education, and in other areas, but the management was via MCB under PO Susan Porter Ridley.

The tomato model proved a good choice as within 5 years the molecular mechanism of bacterial speck disease of that fruit was characterized.

> This is the first demonstration that there is a lock-and-key mechanism[[52]] at the molecular level involved with the plant's ability to recognize and mount a resistance response to a pathogen,

said Steven Scofield, a research geneticist at the Center.[53] An immune response from an antigen–antibody combination in the plant was not unlike that which was seen in animals. Exactly what explained the phenomenon had come from the molecular level studies of the "gene-for-gene" concept that had been investigated from the early 1940s by H.H. Flor of the USDA. After three decades of such work, Flor defined the concept as

> for each gene that conditions reaction in the host there is a corresponding gene in the parasite that conditions pathogenicity. Each gene in either member of a host-parasite system may be identified only by its counterpart in the other member of the system.[54]

The Flor Hypothesis, as it was known, was a valuable one to the growth of work done at CEPRAP, and the selection of the tomato as a model system continued to be of the greatest significance. By the later 1990s, members of two academic departments and the Center, all at UC, Davis, were able to show that a gene responsible for resistance against a nematode worm, also conferred resistance against the potato aphid, an insect. That gene was the "first example of a plant resistance gene [shown to be] active against two such distantly related organisms." Beyond that, the gene (termed *Mi*) was also the first "isolate-specific insect resistance gene to be cloned."[55] Both tomatoes and potatoes are in the same botanical family, Solanaceae, and so molecular structural relationships might be expected. What had been believed to be two entirely different genes providing resistance, one to a worm (*Mi*) and the other to an insect (formerly *Meu-1*), was the same gene (both now termed *Mi*) in the two related plants.

Across the decade and into the next millennium, the CEPRAP was a success story of the first order for BIO, NSF, and the STC program itself. For example, following up on the earlier work on tomato speck disease, Greg Martin of Purdue University (Indiana), was awarded MCB funding to work with CEPRAP to study

---

[52] A mental construct to understand the manner by which an enzyme and its substrate interact. The idea was first proposed by chemist Emil Fischer in 1894. See especially: http://www.tulane.edu/~biochem/lecture/gbch701/koshland2.pdf

[53] See: http://www.nsf.gov/news/news_summ.jsp?cntn_id=101830

[54] H.H. Flor, "Current Status of the Gene-for-Gene Concept," *Annual Reviews of Phytopathology*, (1971): 275–296; pg. 275.

[55] Magdalena Rossi, et al., "The Nematode Resistance Gene *Mi* of Tomato Confers Resistance Against the Potato Aphid," *Proceedings of the National Academy of Sciences* (1998):9750–9754; pg. 9750.

the cause of the disease, a bacterium named *Pseudomonas syringae* pv. *tomato*. What Martin found was that

> disease resistance in plants is not triggered by the pathogen itself. [Rather,] they found that the tomato's gene produces a protein that recognizes the protein produced by the bacterium's gene. The tomato's protein locks onto the bacterium's, and their interaction alerts… the plant's defense mechanisms.[56]

By the new century, signal transduction had become a major field and several researchers from CEPRAP, two other Davis departments, and a colleague from Canada considered resistance not only with regard to the interaction between pathogen and host, but also brought in the undeniably important issues of environmental pressures upon the host plants. Again, using the model system of tomatoes, R.M. Bostock and colleagues examined the complex interactions of the biological aspects of disease tied to environmental aspects and, in so doing, discovered a ceramide signal that complemented the already known salicylate- and jasmonate-mediated responses in infected plants.[57]

Only 3 years after the establishment of CEPRAP, the Center for Molecular Biotechnology (CMB), that had been founded originally at Cal Tech in 1989 during the first STC round, was moved to the University of Washington, Seattle. That was through an agreement with the NSB and was newly funded at over $16 M for 5 years.[58] Though some staff remained at Cal Tech and still participated as members of the center, most moved to Seattle. From 1989 through 1994 support of well over $20 M came from NSF and more than $33 M in additional funding from other federal and non-federal sources for CMB. (The center was funded at NSF by DBI, but is discussed here under MCB.)

The PI at CMB was Leroy Hood, a major figure (earlier chapters) in the history of biotechnology and primary inventor of both the gene *and* the protein sequencers.[59] He had moved from Cal Tech establishing himself at UW, Seattle and thus arranged with NSF to have the center moved there. Though the CMB no longer exists at UW, having reached its NSF-mandated 11-year life, Hood went on to become head of the Institute for Systems Biology (founded in 2000) at the University. The CMB had been affiliated with the then (1992) newly established Department of Molecular Biotechnology. As with other STCs then, and as of this writing, the CMB had an extensive educational program, both internally and as outreach. The center made particular use of a classical model system in biology, the sea urchin (part of the history of embryology/developmental biology in the late 1800s onward). The CMB collaborated with some 25 commercial firms, a very remarkable number

---

[56] See: http://www.nsf.gove/news/frontiers_archive/11-97/11plants.jsp

[57] Rossi, 'Gene *Mi*.'

[58] Neal Lane to Members of the NSB, November 4, 1994; NSB-94-198.

[59] Michael Fortun in Thrackray, 'Private Science,' has considered Hood in regard to the "speed" of processing gene sequences; see: Fortun, "The Human Genome Project and the Acceleration of Biotechnology," ppg. 182–201. See specifically pg. 194. See also Daniel J. Kevles and Leroy Hood, eds., *The Code of Codes: Scientific and Social Issues in the Human Genome Project* (Cambridge, MA: Harvard University Press, 1992), see ppg. 72ff.

among the STCs, likely as Hood himself had been involved in the founding of some 14 biotech companies, including one of the giants in the field, Amgen.

As the last millennium drew to a close, the NSB had approved five new STCs. In July of 1999, the Board announced that the five would share in a total outlay of $94 M for 5 years of support.[60] In-kind contributions at the various universities would raise that figure yet more. Those five were the first in 8 years to be newly established. Of the 25 in existence from the first two competitions in 1989 and 1991, 23 were still operating in 1999. The STC program had been modified in 1997 to bring in additional partnerships for the centers such that by the 1999 round, 36 US colleges and universities, federal laboratories, and some international academic institutions, would be involved in the five newest centers.

Specifically, the new centers included one on nanobiotechnology at Cornell that aimed at tissue engineering to create new tissues and organs for implantation and to examine neuronal cell-to-cell communication. As well, another biologically oriented center was that for Behavioral Neuroscience at Emory University in Atlanta, Georgia. Three others were in the physical and environmental sciences.

CLMIB at Carnegie-Mellon University continued to prosper having, by 1994, invented a light microscope that could render cells in three dimensions, among a string of other patented microscopes and technologies.[61] By the mid-1990s, too, more than 90 scientific papers had emanated from the center, and a major exhibit mounted by it for the Carnegie Science Center in Pittsburgh, Pennsylvania, opened in the fall of 1995. The CLMIB success level was such that Carnegie-Mellon University organized a new Division of Life Sciences as part of its academic structure. An unknown author, possibly from BIO, said that "[t]here is no doubt that this STC is fulfilling its specific goals and the general goals envisioned for the STCs."[62] Indeed, that was so.

About that time, the National Academies' COSEPUP, Committee on Science, Engineering and Public Policy, had signed an agreement with NSF to do a study on the status of *all* of the STCs. An expert panel was sought, and NSF awarded Abt Associates, Inc., of Cambridge, Massachusetts a contract for the evaluation. The PI for the effort was Steven Fitzsimmons of Abt; he was at the same time under contract to the Foundation to evaluate engineering's ERCs. He had a long record with NSF having had assessed many of the programs of the Directorate for Education and Human Resources.[63] CLMIB offered its annual self-report in late 1995 and was

[60] National Science Foundation, Office of Legislative and Public Affairs, "National Science Board Approves Five New NSF Science and Technology Centers," July 29, 1999, Press Release NSF PR 99–45 (NSB 99–141).

[61] Byron Spice, "3-D Microscope," *Pittsburgh Post-Gazette*, January 24, 1994, A-5,6.

[62] Ephemeron of one page titled "Diary Note: Recommendation," in a file on CLMIB. It appears to be part of a document regarding an informal visit to CLMIB. Its unknown author requested that a then-planned Site Visit be postponed until the center director, D. Lansing Taylor, could organize a workshop at NSF for BIO.

[63] David Schindel to numerous persons, April 13, 1995, paper copy of an electronic memo; Ibid., forward by Schindel of open memo from Mary J. Osborn, COSEPUP to all interested, June 13, 1995, paper copy of an electronic memo.

formally evaluated in August of 1996 by six panelists from various universities working under the aegis of NSF—a typical assessment effort for the centers. Julius Jackson found that the Carnegie-Mellon center was "FIRST-RATE" and felt that it should have its support renewed.[64] It was recommended by Maryanna Henkart, Jackson's still then MCB Deputy Director, to Mary Clutter for continued (sixth year) funding that would last until year nine (of the 11-year limit).[65]

While Jackson's view was that CLMIB was doing first-rate work, the work of Abt Associates overall was not viewed in the least well by the National Academies. The COSEPUP panel was to "review and interpret the data gathered by an outside contractor to NSF [i.e., Abt]" and to "reach conclusions regarding the progress of the STCs program."[66] The "study was not to critique the accomplishments of individual STCs," said the panel.

> The use of Abt *as well as the Academy* was an experimental effort to have a contractor develop some of the empirical data desired by NSF while maintaining the role of expert guidance from an [A]cademy panel…the panel did not view the experiment as successful.[67]

The panel went on to specify in detail how Abt performed its assessment using Abt's own words from its report. The panel then stated that:

> In other words, Abt collected primarily positive data from the various participants in the [STC] program, as opposed to collecting information and guidance from those outside the program who might have a more balanced and more negative view (such as members of the [NSF] site visiting committee)…All these factors [more were given] affected the panel's belief in and interpretation of the results of the Abt report. As a result, the panel does not recommend that this serial process whereby one contractor carries out data collection and another group reviews the results should be used in the future.

The COSEPUP panel ended up making "guarded and selected use" of Abt's data, but only then when they could find supporting evidence for Abt's statements. COSEPUP concluded its own extensive review of the STC program by saying, foremost, that it should be continued; it was. They did recommend that the 11-year period be reduced to 10 and, in future, "avoid redundant data collection."[68] As for CLMIB, it was replaced at the end of the 11-year term by the current Molecular Biosensor and Imaging Center, which is supported by NIH.

In the 1999 round-three competition, the Cornell nanobiotechnology center was one of the two biologically oriented STCs to become established, officially, in

[64] Julius Jackson to several persons with cc to Henkart, "Comments on the Review and Recommendation for Continued Support," November 13, 1996.

[65] Maryanna Henkart to Mary Clutter, "Recommendation or Renewed Support for the [STCLMIB] at Carnegie Mellon University," December 24, 1996; one year after CLMIB turned in its Annual Report to NSF.

[66] COSEPUP, "An Assessment of the National Science Foundation's Science and Technology Centers Program," (Washington, D.C.: National Academy Press, 1996), pg. v.

[67] Ibid., pg. vi. Emphasis added.

[68] Ibid., ppg. 2–3.

January of 2000. Its story is well told at its website.[69] The behavioral neuroscience center in Atlanta had a somewhat different founding history in that it was originally established in 1998 by a grant from the Robert W. Woodruff Foundation and, with the third round activity at NSF in November of 1999, it became an STC. Its history is also told at its website and considered more in the following chapter.[70]

7.5 Some Smaller Programs

7.5.1 Metabolic Engineering

A Metabolic Engineering (ME) Working Group, an oversight body providing and maintaining a project inventory of that discipline's research funded by the eight federal agencies in the group, included NSF. The group was begun in 1995 as a spinoff of an OSTP committee and also as a result of their second report, the "New Horizons" document. Through "targeted and purposeful alteration of metabolic pathways," it was the expectation of the grantees of the multi-agency program to find better ways toward "chemical transformation, energy transductions, and supramolecular assembly."[71] The Project Inventory was the product of the second of two workshops and produced annual reports on funded projects. Grants were from one to three years in length and typically provided substantial sums of money, generally over $100 K and, occasionally approaching a half million dollars. As with the vast majority of funds, they went to universities, but NGOs were also included: for example, the American Type Culture Collection.[72] In the earliest years, the total annual awards came to over $3 M and, later, to over $7 M. Charles Roth and his colleagues at Rutgers University, New Jersey, for instance, had a grant for just under $1 M to study the "Molecular Network Controls of Hepatocyte Metabolism."[73]

Metabolic Engineering was never a major effort at NSF, but MOLE was rather larger.

---

[69] See: http://www.nbtc.cornell.edu/program_mission.htm

[70] See: http://www.cbn-atl.org/

[71] See: http://www.metabolicengineering.gov/project.html. Another area of bioengineering, tissue engineering, though handled entirely through the NSF's Directorate of Engineering, has been studied historically by Jessica Viola, et al., "The Emergence of Tissue Engineering as a Research Field," and can be seen at http://www.nsf.gov/pubs/2004/nsf0450/start.html. It is recommended to the interested reader.

[72] See: http://www.atcc.org/About/tabid/138/Default.aspx

[73] For more on Roth's research interests, see: http://lifesci.rutgers.edu/~molbiosci/faculty/roth_c.html

7.5.2 Molecular Evolution: MOLE

When BIO opened its first year of existence after the great reorganization, numerous programmatic efforts were under study for development and were placed widely across the divisions. One of those was the product of the Working Group in Molecular and Cellular Evolution, a cross-divisional group including MCB. That creation would later be known as MOLE.[74] Though not directly tied to MCB, I have chosen to place it here more for convenience, as later commentary in this chapter addresses the *related* areas of genomics, particularly of plants and the *Arabidopsis* genome project. Indeed, the latter was also managed under DBI, but had many ties to MCB. Too, genetics was the domain of MCB.

While the Working Group's first report came out in 1991, the discussion had been ongoing for more than 5 years. It had reached something of a climax by 1990, and so, at the urging of the executive council of the Society for the Study of Evolution, a meeting between them and BBS division directors was planned. Given all that was occurring with regard to reorganizations at both NSF and the Society too, as it happened, the meeting never came to pass. The Working Group, though, had been growing from its starting point in the Population Biology Program, to include individuals from the systematics and genetics programs. Judy Plesset, still then of the dissolving Division of Cellular Biology, represented what was to become MCB. MOLE was, as the authors of the memo said,

> truly a 'grass roots', or 'bottoms up' initiative, because it stems from identification by NSF POs of increasing interest and concern from their scientific communities.[75]

That statement could not have been more clearly indicative of how the *intricate dance* proceeded at the Foundation in response to the lead the biological community took across the dance floor. A workshop, held in late September in Washington, D.C., was organized in response to the scientist's "increasing interest and concern." It was sponsored by both NSF and the Sloan Foundation. It was not the first time that particular NGO aided in conference support (see earlier chapters). The PI for the workshop was Morris Goodman of Wayne State University, Detroit, Michigan. The Sloan would mount one scientific initiative at a time and would maintain that for, typically, 5 years. Molecular evolution had been their interest for the preceding nearly 6 years and so support would drop off from Sloan shortly thereafter. It was under those circumstances that NSF took up the cudgel in response to the community's urging via the POs.

The Working Group was able to rapidly assemble a "distinguished group of scientists" including three who were members of the National Academy and the rest of whom were journal editors or officers of relevant scientific societies. That rapidity of action was due, they said, to the "exploding interest" in molecular evolution.

---

[74] Peter Arzberger, et al., to Philip Harriman, et al., "Program Officer Working Groups Report[:] Molecular and Cellular Evolution," October 21, 1991.

[75] Ibid., from the lengthy memo's "Introduction and Background" section.

With MOLE becoming a priority at NSF/BIO, the great cost of doing work in the field (highly expensive technologies were necessary) became more manageable. And, after all, "the promise of molecular technology is tantalizing."[76] The workshop would produce a report that stated NSF's interests, offering a definition for the field, and explaining its importance as an "interdiscipline." An edited version of the report would be published as an article in the newly founded (1992) *Journal of Molecular Phylogenetics and Evolution*, edited by Morris Goodman.[77]

The group defined molecular evolution as

> an exciting new, synthetic field science that brings to evolutionary biology the powerful methods of both molecular biology and the computational sciences. Molecular studies generate comparative data that can most effectively be interpreted within the framework of evolutionary theory, leading to new insights into the operation of evolution.

The goals of the field included "reconstruct[ion of] the genealogical history of all the ramifying lines of life," and to understand the processes underlying the evolution of genotypes and phenotypes. Indeed, "[t]his truly represents a new synthesis, and a culmination of the Darwinian revolution."[78] Powerful words. Al Gore's analogy fitted the MOLE situation well given the Working Group's repeated references to information science and its related areas. MOLE also "illuminates the broad sweep of biological history through the reconstruction of genealogical histories."[79] For political reasons (the matter of evolution), that part of the memo that included a short statement on the role of the NSF would not appear in the more public venues of the workshop findings, but it did state, not surprisingly, that the concerns of MOLE were not a mandated research area of any federal agency so NSF was "uniquely placed to foster expanded activities in this developing area." The same may have been said about so much of what NSF did, and does, over time with regard to new basic research fields. Further, and aimed at more senior leadership at the Foundation, let alone Congress, the memo authors noted that "[n]o new bureaucracy" was needed to support MOLE. The effort, though not suggested as a formal *program,* was to be an initiative that would span four directorates at NSF. Special funding based in each area would be drawn upon for grant proposals reaching any division or directorate involved. Interdisciplinarity was actually "forc[ed]" by the initiative.[80] No budgetary figures were initially suggested by the Working Group,

---

[76] Ibid.

[77] See: http://www.sciencedirect.com/science/journal/10557903

[78] MOLE memo section on results of the workshop itself.

[79] One cannot help thinking the phrase had been inspired, at least mechanically, by the famous (some would say infamous) phrase of Ernst Haeckel: "ontogeny recapitulates phylogeny."

[80] The term "forcing" was used in the original memo. In a single sheet ephemeron attached to a file of information about the beginnings of the MOLE program, there is a commentary from scientist and writer Conrad A. Istock (not to be confused with Conrad A. Stock in Google web searches) of Cornell University, which considered the position and nature of molecular evolution. In fine, Istock saw its rise as a third era of synthesis for Darwinism, the second era being neo-Darwinism. Istock is the author of a number of books, but most notably *Storm Over a Mountain Island: Conservation Biology and the Mt. Graham Affair* (Tucson: University of Arizona Press, n.d.).

however, but possible figures arose within a month and the first year's financing was to be just under $3 M as contributions came from various units within BIO.[81]

In 1994, a Post-Doctoral Research Fellowship program in MOLE was opened and co-sponsored by NSF and the Sloan Foundation. It would cover the 4 years from FY95 through FY98. Both BIO and SBE directorates were involved in that educational effort and the fellowships were set at $80 K total for a two-year period for each award.[82] The program was placed in DBI where science education was lodged under Carter Kimsey (previous chapter).

It was typical of Mark Courtney, of DEB, to go to a variety of scientific conferences and then provide a brief report of one or two pages on his experience at the given meeting. A Keystone Symposium[83] held in Santa Fe, New Mexico in 1997 on MOLE, was an excellent example of an *intricate dance* amongst once widely disparate scientific disciplines because

> [e]volutionary biologists and molecular and cellular biologists actually seemed to be speaking the same language in addressing both the proximate and ultimate explanations for the particular phenomena they were examining. [Moreover, i]t seems clear that many of the practitioners in this field are beginning to think about their work at levels ranging from the molecular to the ecological, in both empirical and theoretical contexts.[84]

Courtney could not have better stated how NSF/BIO had been successful in its programmatic efforts to: assist in the creation of a new field (MOLE); promote an *intricate dance* at another level which might be considered in the present study (that of two disciplines coming together); and, aiding in bringing together, to a degree, polar ends of the biological sciences continuum, that is, molecular and ecological.

In moving from aspects of MOLE into the genetic science that ultimately underlies it, genetics and genomics at MCB can better be appreciated in light of Courtney's commentary.

7.6 Genetics and Genomics: An Overview

MOLE-related research was becoming established at BBS just before the great reorganization when two PDs in two of the old directorate's programs sent a memo to "Interested Parties" with the introductory comment stating that

---

[81] C.B. Lynch to J.L. [James] Edwards, "Molecular Evolution Set-Aside," November 19, 1991. Carol Lynch, of the University of Colorado, was in BSR (earlier chapters) and ended her memo saying that she believed that the majority of grants in the subject area over time came from either the systematics program or PBPE.

[82] Original typescript of the Program Announcement NSF 93–151 (revised July 1994).

[83] See: http://www.keystonesymposia.org/AboutUs/Mission.cfm

[84] Mark W. Courtney, "Report on Keystone Symposium: Molecular Mechanisms of Evolution, Santa Fe, New Mexico, February 16–21, 1997." A one-page report dated March 12, 1997. Courtney Files.

> NSF currently has no formal mechanism to review and encourage research in comparative genome evolution. This oversight will become more conspicuous as the human and other genome mapping projects progress and comparisons of genome organization among related species become a primary focus of research. We believe that NSF should play the major role in synthesizing the information generated by the diverse genome projects.[85]

The two NSF biologists saw the importance of relating the findings of the Human Genome Project (HGP) to many other species of interest, but the HGP was the perquisite of the NIH and, since it was out of bounds for NSF to support areas of research into human biology, *sensu stricto*, focus on the genomes of non-human species would be the Foundation's central interest. "Since no creature fascinates us as much as ourselves," interest in human genetics began shortly after the rediscovery of Mendel's laws in 1900.[86] By 1907, eye color in humans had been shown "convincingly" to be genetically transmitted in our species. That was also true for the inborn error of metabolism (a genetic disease) known as alkaptonuria, discovered that same year.[87] Widely known as the search for the "Holy Grail" of biology, the human genetic assembly—the human genome—was underway in 1989–1990 and would not relent even to the present day.

In 1995, the first genome of any living thing, the bacterium *Hemophilus influenzae*, was published.[88] At the time of this writing, more than 180 organism's genomes are known.[89] The HGP published the first level sequence of *Homo sapiens* in 2000 and a complete one in April of 2003, but with many more milestones occurring after that time and through to the present. The HGP was led by Ari Patrino of the Department of Energy when efforts began in 1989, but Francis Collins of NIH became the primary leader and is given historical credit for the multi-agency US *federal* effort to sequence the human genome. Simultaneously, a *private* effort, led by J. Craig Venter of the Celera Corporation, was ongoing.[90]

> The human genome project will greatly accelerate progress in understanding the diagnosis, treatment, and prevention of disease. *Comparable benefits are likely to result in basic biological research.*[91]

---

[85] Fred W. Allendorf and William S. Moore to Interested Parties, "Comparative Genome Evolution Projects," January 17, 1990. An ephemeron memo of more than one page, but with only the initial page found.

[86] Daniel J. Kevles, "Out of Eugenics: The Historical Politics of the Human Genome," in Kevles and Hood, 'Code of Codes,' pg. 3.

[87] Ibid.

[88] R. Fleischmann, et al., "Whole-genome random sequencing and assembly of Haemophilus influenzae Rd," *Science* (1995): 496–512.

[89] A listing of those species is given at: http://www.genomenewsnetwork.org/resources/sequenced_genomes/genome_guide_p1.shtml

[90] A timeline for major events can be seen at the official website http://www.genome.gov/10001763 for the HGP and a picture of the three men is at http://www.ornl.gov/sci/techresources/Human_Genome/project/privatesector.shtm

[91] Allendorf and Moore memo. Emphasis added.

Such was the opinion at NSF/BBS in 1990 and as the HGP geared up, Mark Courtney was invited to a series of meetings held at Stanford University and other venues during those early years as an interested party to represent, in part, the non-participating NSF.[92] As well, Courtney reported that a "fairly impressive range of organisms and ages of biological materials are now being analyzed for DNA sequences" in his short meeting report from attending a conference on ancient DNA. "Most of the specimens" were organisms trapped in amber and some were "even from 'Jurassic Park'—or at least 'Cretaceous Park.' "[93] (It was also Courtney, the wag, who spoke of Big Foot having been "sas-squashed.")

During that same period of the first half of the 1990s, "emerging areas of excitement" for the Genetic Biology program of MCB included newer areas of research such as how the genes of a diploid individual were activated and/or repressed, DNA methylation,[94] and epigenetics. Genomes and the environment was also of interest as laboratory in vitro manipulation of single parameters left much to be desired when trying to understand the complexity of ecological domains such as soils. It was shown, for example, that certain genes were activated or inactivated in some bacteria in specific situations in soils: successes in bioremediation, say, were directly tied to how such organisms behaved in the environment rather than the laboratory. Heat shock genes were being discovered in all sorts of organisms and there was a need for them to be understood. Extreme environments, such as the perennially fascinating hot springs of Yellowstone National Park, presented special situations for life and they, too, opened areas of "excitement" for research in genetics and genomics.[95] As well, metagenomics would become a yet newer field of endeavor: it will be discussed in Chapter Nine.

The costs could be borne for those emerging areas as MCB's budget in FY 1995 was the largest of the four divisions of BIO at well over $88 M actual dollars. IBN was second at just over $80 M and DEB was virtually equivalent: $79 M. DIR, not unexpectedly, came in at the much lower figure of just under $53 M. As the FY97 budget was under preparation in that year, MCB was seeking about $92 M after having an estimated drop to $87 M in FY96. The total in actual dollars for BIO overall in FY95 was just under $301 M with a request for $326 M in FY97.[96]

At that time (1995), more than 85% of the budget for BIO went to fundamental research, investigator-initiated, and which was located primarily at universities and

---

[92] For example, Courtney attended a Human Genome Diversity Project meeting, a spin-off of the HGP, in early 1993. Letter from Marcus W. Feldman of Stanford to Courtney, February 11, 1993. Courtney Files. Cora Marrett was also in attendance (see Chapter Two herein), as was Mary Clutter.

[93] Mark Courtney, "Meeting Report: Second International Conference on Ancient DNA. October 7–8, 1993; Smithsonian Institution," October 29, 1993. Courtney Files.

[94] For more on this critical subject, see: A. Weissbach, "A Chronicle of DNA Methylation (1948–1975)," EXS, (1993)64:1–10. Review. PubMed PMID: 8418945; and, Rakesh Singal and Gordon D. Ginder, "DNA Methylation," *Blood* (1999):4059–4070.

[95] A two-page typed ephemeron likely intended as text for a solicitation document.

[96] NSF Budget Request for FY97, BIO section, ppg. 77–91; see especially the table on pg. 77.

colleges. Such had long been the practice and would remain so into the future. Much of that was in the areas of plant and environmental biology and on research support for beginning investigators, because those were the areas where NSF/BIO had "[p] articular emphasis."[97] As well, projects that focused upon education and training of future scientists were also of central concern. The average new projects in BIO were at the level of about $250 K over 3 years for FY96. MCB was particularly involved with the Microbial Biology initiative, begun in FY96, and sought $1.2 M to bring the total to $2.2 M. That cross-directorate activity that focused on "non-model system microbes and microbes from diverse habitats" was shared with the Environmental Geochemistry and Biogeochemistry programs in the GEO directorate; other units were also involved. The FY97 budget request totaled over $92 M for MCB. Figure 7.3 (termed a "crosswalk") depicts the request for all divisions of BIO for major programmatic areas for the FY96 estimated budget. The three programs of MCB are seen at the far left. Within 2 years, however, the FY 1999 budget request had gone from $326 M to nearly $418 M, a very significant jump.[98] All four divisions of BIO had increased, DIR notably so, and MCB continued to remain at the top, in total dollars, at nearly $108 M—plant genomics being a major aspect of MCB's funding portfolio. Also, molecular biology was a central concern, as was microbial biology, within the LEE and LExEn programs. More will be said of these programs in Chapter Nine.

By FY 1998, awards under the Collaborative Research on Functional Genomics rubric were oriented entirely toward plant research for both MCB and BIO, more broadly—and included agencies beyond NSF.[99] Since Mary Clutter had become AD, the emphasis on the study of plants had continued to increase, and with the simultaneous rise of genomics, both the *Arabidopsis* project and the NPGI were very large programs that continue to the present.

The activities of the COVs were increasingly more common by the later 1990s and into the next millennium, too. For instance, the COV that examined the MCB Genetics Cluster and its Biochemistry of Gene Expression program found that future COVs should be involved in strategic planning for the division.[100] They were disturbed that the gene expression area had moved about within MCB, first from the Genetics Cluster, then to Biomolecular Processes, among other homes. But MCB argued that such moves allowed for "flexibility" in rapidly expanding fields of inquiry.[101]

It was the rapid expansion of genomics during the decade of the 1990s that made it obvious to Maryanna Henkart that there was a "need to develop a long-range plan for NSF's role in genomics research." That was so as

---

[97] Ibid., pg. 78.

[98] NSF Budget Request for FY99, BIO section, ppg. 69–85; note the table on pg. 69 and the MCB discussion at ppg. 77–78.

[99] See: http://www.nsf.gov/bio/pubs/awards/pgr.htm

[100] BIO AC Minutes, December 2–3, 1999, unnumbered pg. 3.

[101] Ibid.

Crosswalk of the FY 1996 Estimate Directorate for Biological Sciences

(Millions of Dollars)

	Old Structure											
	Molecular & Cellular Biosciences (MCB)			Integrative Biology & Neuroscience (IBN)			Environmental Biology (DEB)			Biological Instrumentation & Resources (BIR)		
New Structure	Biochemistry & Molecular Structure & Function	Cell Biology	Genetics and Nucleic Acids	Physiology & Behavior	Neuro-science	Developmental Biology	Systematic & Population Biology	Ecological Studies	Long-Term Projects in Environmental Biology	Instrumentation & Instrument Development	Special Projects	Total, New Structure
MCB Molecular & Cellular Biosciences Research Projects	$34.36	$20.69	$32.15									$87.20
IBN Integrative Biology & Neurosciences Research Projects				$30.66	$33.24	$16.05						$79.95
DEB Environmental Biology Research Projects							$24.43	$27.76	$26.83			$79.02
BIR Research Resources Human Resources										$13.58	$23.95 $16.30	$37.53 $16.30
TOTAL, Old Structure			$87.20			$79.95			$79.02		$53.83	$300.00

Fig. 7.3 "Crosswalk" of the FY1996 budget estimate for BIO. NSF. Specific provenance unknown

> [s]everal workshops, an Interagency Report on Microbial Genomics, and an Interagency Genomics working group have identified gaps in areas of genomics. Some of the gaps…or needs include developing techniques for culturing novel microbes, establishing NSF's role in selecting or prioritizing potential non-model organisms for sequencing, encouraging the vertical integration of genomics into biology through the training of students, and providing support for infrastructure and database development.[102]

Much of that was already happening within the two big genome projects. Those "gaps or needs" included that for a Plant Genome website as BIO was entering the third year of competitions for the program.[103] As the Editor of *Nature* said: "Like it or not, big biology is here to stay."[104] All of biological research, he said, "has for decades been on an ascending slope of increasing systems complexity. The availability of genome sequences heralds a sharp acceleration in the upturn of that slope." More emphasis on computation was called for, the editor held, and high throughput sequencing of genes and proteins and their study in microarrays[105] was such that "the social change required to make this shift painlessly should not be underestimated."

BIO's involvement in plant science, *sensu lato*, had a history going back well before Mary Clutter—Betsy Clark recognized the need for more such support—but with Clutter's orientation and academic training, the focus at just the time that the human genome project's beginning was all the buzz in the scientific, and broader, communities could not have been better for NSF. NIH had ownership of the HGP, which left NSF to consider its own directions: plant genome research (PGR) was an obvious answer.

7.7 *Arabidopsis thaliana*: "A Simple Weed"

It is difficult to separate the general matter of plant genomics broadly, the NPGI, and the *Arabidopsis* genome-sequencing project, in particular. However, chronologically, the latter was the first effort for plant science in the realm of Big Biology to be undertaken by BIO. Part of the organizational management fell to MCB, but other divisions also played roles, and agencies and organizations beyond NSF and beyond America were deeply involved.

Figure 7.4 provides a view of how the efforts began and progressed to both the center of the timeline diagram, and to the actual beginning dates of US engagement in this massive international project.[106] During the initial year, the Multinational

[102] Ibid., unnumbered pg. 5.

[103] Ibid., unnumbered pg. 4. Comments by Jane Silverthorne, who headed up that programmatic area.

[104] Editorial, "Post-Genomic Cultures," *Nature* (2001):545.

[105] See: http://en.wikipedia.org/wiki/DNA_microarray for a shorter, but excellent definition. For more detail, see: http://www.mrc-lmb.cam.ac.uk/genomes/madanm/microarray/chapter-final.pdf

[106] The timeline was an overhead slide used by Mary Clutter in a talk, "Epigenetic Predictions," she

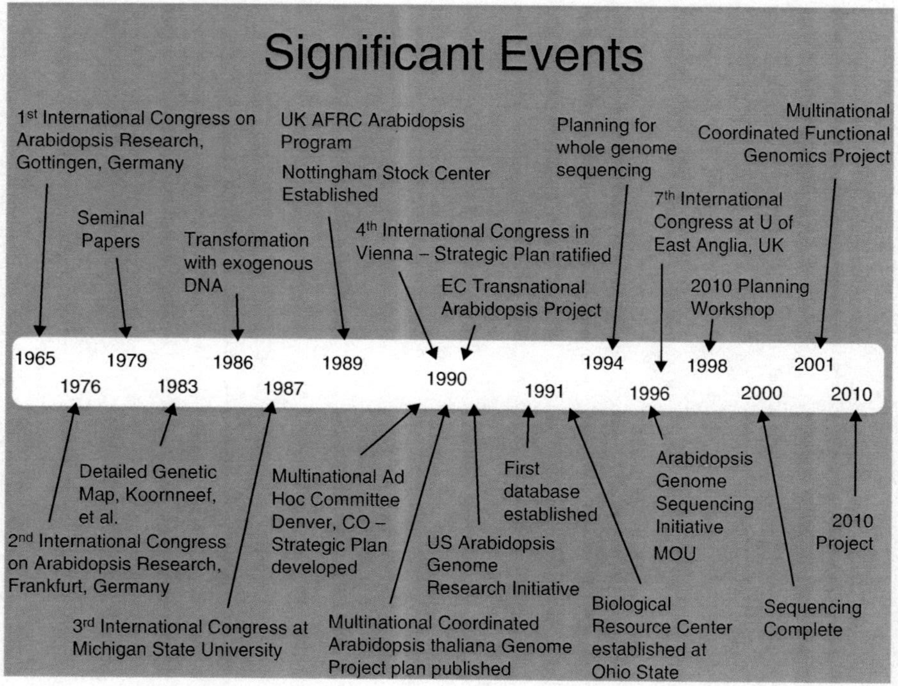

Fig. 7.4 Significant events in the effort to sequence the *Arabidopsis* genome. Taken from a visual used by Mary Clutter in a talk, "Epigenetic Predictions," she gave on September 28, 2008. Specific provenance unknown

Steering Committee was formed. Elliot Meyerowitz, of Cal Tech, was made its chair.[107] He was not only one of the earliest investigators to receive a grant from BIO to study the thale cress, but has also written the first, but very brief, history (2001) on this effort and I rely heavily upon his study for the early years of interest in the weed.[108]

It appears that *Arabidopsis* first became known, in a non-taxonomic sense, when Alexander Braun published a paper in Germany in 1873 concerning a mutant individual he found in a field in Berlin. (The mutant gene involved, *AGAMOUS*, was finally sequenced in 1990.) But it was not until a student, Friedrich Laibach, in Bonn published on thale cress in 1907 that the plant returned once again to the literature. His was a chromosome size and number investigation in the early years after the Mendel rediscovery. Again, in the scientific press for 1935, the Russians discussed their search for a plant that could be the *Drosophila* of botany and published on *Arabidopsis*, but incorrectly stated the number of chromosomes as three; Laibach had been correct earlier with the haploid number at five. The small size of the plant offered both advantages and disadvantages as a model system, but the short time to flowering (sexual maturity) was an ideal feature. Laibach took up the species once again in 1943 when he explored it further as a model system noting many features that made it so desirable, not least of which was its ready susceptibility to mutagenesis. Laibach's student, Erna Reinholz, created a collection of mutant strains. The title of her doctoral dissertation used a German phrase that caught the eyes of Nazi leadership in that she spoke of x-ray mutation and her findings were believed to be of possible use to them in their efforts to create an atomic bomb. Nothing came of it, of course, and the thesis was published only after the war (1947). Many other articles came out in the following decade and beyond and a number of mutant collections were known, but it was not until 1964 that Laibach (d. 1967) and others joined to produce the newsletter *Arabidopsis Information Service*; it finally ceased publication in 1990. Two international congresses (the first was a *festschrift* for Laibach's 80th birthday) were held on the plant in those earlier years (Fig. 7.4), but they preceded what Meyerowitz termed the "molecular biology era." However, it must be said that the second congress fell at just the time that molecular biology was aborning (1976).[109]

---

gave on September 28, 2008; venue unknown.

[107] More on Meyerowitz can be seen at: http://biology.caltech.edu/Members/Meyerowitz

[108] Elliot M. Meyerowitz, "Prehistory and History of Arabidopsis Research," *Plant Physiology* (2001):15–19. Corrections to several of his figures (photographs) appeared later that year in the journal at page 2203. He was a leading figure in the entire genome enterprise as chair of the multinational committee. See also: Michael Bevan, et al., "Sequence Analysis of the *Arabidopsis* Genome," *Current Opinion in Plant Biology* (2001):105–110. Bevan was a major player in the project in Britain. Also see the endnote further below for an article by Athanasios Theologis who presents a very different view of the early history.

[109] Ibid., pg. 15.

Laibach continued building a mutant (or, ecotype[110]) collection during those years prior to and following the two widely-spaced congresses. The collection finally became the foundation for a large collection known into the present.[111] In 1975, George Redei proposed using thale cress over petunias or tomatoes, both of which had many good characteristics as model flowering plant systems and were leading contenders for the role. A number of younger researchers took up the cress as the most ideal of the group and soon its value in certain types of genomic research won the day as the model system.[112] The first complete linkage map[113] (not a gene sequence) of the plant was published in the early 1980s and its use for studies on plant embryology, just one area of research, was realized. But crucial to commitment within the community for the use of *A. thaliana* as the flowering plant model system were articles in the journals *Science* in 1985 and *Trends in Genetics* in 1986 and again in 1987, Meyerowitz argued. Even more publications came out in the next couple of years and provided *methodological* reasons, beyond the already generally desirable characteristics of thale cress, to use it as *the* model system. The plant was certainly taken up community-wide and international congresses have been occurring ever since the third of them in 1987 (Fig. 7.4).[114]

As early as 1982, the Eukaryotic Genetics Program in the MCB of the time had "highlighted the potential of *Arabidopsis* as a model plant…and actively encouraged this emerging field of research." Indeed, the "importance today of that early leadership by NSF is widely evident."[115] Finally, by 1990, NSF became the lead agency for the *Arabidopsis* genome-mapping project in partnership with NIH (specifically, the National Center for Human Genome Research), USDA, and DoE.[116] An ad hoc committee composed of some nine scientists from the US, Europe, Japan, and Australia prepared the first report on the plan. That finally led to a simple,

[110] Ecotypes are geographically adapted variants that often show different phenotypes than others of the variants. One ecotype can breed easily with any other ecotype (one type of species test).

[111] See Meyerowitz's endnote 22 in 'Prehistory.'

[112] 'Prehistory,' pg. 15.

[113] "A genetic map showing the relative positions of genetic markers along a chromosome that is determined by the recombination frequency during crossover of homologous chromosomes." See: http://www.biology-online.org/dictionary/Linkage_map for the provenance of this definition. A linkage map should not be confused with a gene sequence map, the latter being a much more detailed analysis and one which could not yet have been created prior to the first such sequencing of any organism in 1995. See earlier endnote above.

[114] Meyerowitz, 'Prehistory,' pg. 16.

[115] Christopher Platt to Bruce Umminger, "Neuroscience Opportunities for FY95 re Infrastructure," May 18, 1993.

[116] While it has been stated that NIH does not typically do plant biology, Christopher Platt (neuroscientist in BIO; see Chapter Nine) told the author (interview, February 19, 2009) that he did an online search in 2002 via the NIH's CRISP system and found that the National Institute of General Medical Science (NIGMS) had "over 50 [active] proposals" on *Arabidopsis*, potato, and other plants: "In short, I found as many proposals active at NIH as there were new proposals coming into NSF in the Plant Biology area"! The CRISP system was internal at NIH until 1998 when it became public online. See: http://grants.nih.gov/grants/guide/notice-files/not98-099.html

one-page interagency agreement which was signed in June of that year by Mary Clutter for NSF, James D. Watson for NIH, and two others from Agriculture and Energy.[117] The model plant feature that made thale cress so useful to genomics, vice merely genetics, physiology, or other biological specialties, was the small number of chromosomes discovered eight decades before by Laibach.[118] The genome, with its five haploid chromosomes, had *30 times less* DNA than, say, corn, which had been so valuable as a model plant to Barbara McClintock's work with jumping genes.[119] The overall objective of the project was "to sequence the entire 100 megabase[[120]]...genome by the year 2004...[and to closely coordinate] with the European and other *Arabidopsis* projects."[121]

More specifically, the objectives included the intention of completing a physical map[122] of the plant's genome by 1997 and sequencing the first 10 megabases (MB) was to be completed in the first 3 years. In about 1994, it was estimated that approximately 2000 scientists would be participating in the project across the globe. The North American *Arabidopsis* Science Steering Committee met that year and, after discussions completed through an electronic newsgroup resulted in planning suggestions, the US proposal was developed. The European Community began a three-year pilot project on sequencing 2 MB of chromosome IV and another half MB spread across the genome.[123] Baker's yeast (*Saccharomyces cerevisiae*) had been the first eukaryote in which sequencing had been completed (1996), so early-period technologies were well established. Indeed, thale cress would be fully sequenced by the year 2000, the same as that of the human genome, and 4 years ahead of expectations![124] However, by as early as 1997, it became clear to NSF that the

[117] "Joint NSF, NIH, USDA, and DOE Agreement on Cooperation in Support of *Arabidopsis thaliana* Genomic Analysis," June 12, 1990.

[118] Very large chromosome numbers are common in the plant kingdom (notably ferns), sometimes due to polyploidy, but whatever the cause, finding plants with many good research features, and especially the thale cress with small numbers of chromosomes, is invaluable to the investigator. See in particular: http://en.wikipedia.org/wiki/List_of_organisms_by_chromosome_count

[119] National Science Foundation, *Annual Report, 1990* (Washington, D.C.: Government Printing Office, 1991) pg. 3.

[120] A megabase (MB, or Mb) is a fragment of DNA equal to one million nucleotides, or base pairs of the four building blocks of DNA.

[121] Mary Clutter to Members of the National Science Board's Committee on Programs and Plans, "Information Item for an *Arabidopsis thaliana* Genome Sequencing Project," n.d., but likely about mid-1990.

[122] A physical map is a "map of the locations of identifiable landmarks on DNA (e.g., restriction-enzyme cutting sites, genes), regardless of inheritance. Distance is measured in base pairs. For the human genome, the lowest-resolution physical map is the banding patterns on the 24 [*sic*] different chromosomes; the highest-resolution map is the complete nucleotide sequence of the chromosomes." See: http://www.ornl.gov/sci/techresources/Human_Genome/glossary/glossary.shtml#P

[123] See a more recent commentary on the subject at: http://tuebingen.mpg.de/en/news-press/press-releases/detail/sequencing-thousand-and-one-genomes.html

[124] The Arabidopsis Genome Initiative, "Analysis of the Genome Sequence of the Flowering Plant *Arabidopsis thaliana*," *Nature* (2000):796–815. The complete article can be seen online at: http://www.nature.com/nature/journal/v408/n6814/full/408796a0.html

project would likely be done, in fact, by the *fin de millennium*.[125] Still, the work leading up to that was extensive, to say the least.

The first announcement to the press of the Foundation's intent to take the lead on the sequencing project in *Arabidopsis* came in June of 1990 when the cooperative agreement was signed—the same year the 15-year long project (done in ten) was signed to establish the HGP.[126] The relationship between the cress project of NSF and the human project of NIH, while not particularly interrelated, were still seen as two sides of a coin being referred to as "the genome initiative"; two initiatives, in fact. Synergy was expected in that the

> research effort will require an inter- and multi-disciplinary approach that combines expertise in anthropology, population biology, sociology, computational biology…as well as the health sciences, ethical studies and international competition. This initiative has potential interactions with the proposed *Arabidopsis* sequencing initiative, with various efforts in database development.[127]

The estimates at the time for the thale cress project costs were made on the basis of $1/base of DNA sequenced. 50 million dollars was estimated for the years of 2000–2004 alone, which turned out to be unnecessary since the work was completed by 2000. Actual costs for the program would average about $50 M per year by the later 1990s for the cress work. The HGP, on the other hand, has been called "biology's first foray into 'big science,'" but that is *not quite correct,* as both the HGP and that on thale cress were simultaneous. They both began in 1990 and their respective genome sequences were clarified in 2000. That same year both enterprises were published.[128] The former was by far the better-known effort, however. (The LTER, Chapter Nine, was clearly a set of projects in environmental science well suited to the epithet "big biology" at the time, too.) Knowledge from two eukaryotes, both important and both good examples of "big biology," came from the human *and* the thale cress efforts.

A series of annual reports came out from NSF/BIO over the years of the project, including those beyond the 2000 sequencing completion date. The first of those, April 1991, was a simple typescript document, but indicated excellent progress for the infant project. The Fourth International Conference on *Arabidopsis* Research had been held in Vienna, Austria, the previous year with some 458 participants with

[125] National Science Foundation, *Annual Report 1996–1997* (Washington, D.C.: Government Printing Office, 1998); there was no annual report in 1996, so two years were covered in the issue noted here and the "Message from the Director" was Rita Colwell's first. She mentioned a completion date of 2000 and noted that the research had already shown considerable value with regard to "mechanisms of hormone action in plants, interactions of plants with environmental factors, plant pathology, developmental pathways, and complex metabolic pathways."

[126] See: http://www.ornl.gov/sci/techresources/Human_Genome/project/timeline.shtml

[127] A five-page ephemeron, likely a draft, titled "NSF/BIO Genome Initiative," n.d., but about 1995.

[128] Barbara R. Jasny and Leslie Roberts, "Building on the DNA Revolution," *Science* (2003):277. That issue (V. 300; April 11) of the journal contains a special section on the topic with scattered other articles of interest to that history. See also the references on the history of the HGP at: http://www.ornl.gov/sci/techresources/Human_Genome/project/hgp.shtml

no fewer than 73 papers presented orally, let alone a large group of posters. Japan held its first meeting and the UK its second at that same time.[129] The thale cress was the topic of papers presented at many other conferences internationally, as well. The second conference of those was held at Keystone, Colorado, home of the famed Keystone Symposia series. That report and many following can be seen on the Internet.[130] By the fourth annual report, the document itself had graduated from a simple typescript into a well-edited, tractable one in four colors: it had attained the status of a serial publication. Summaries of the recent progress, for one, were typically made available to the interested audiences. In 1994, for example, that fourth report noted at least eight major successes including "identification and cloning of genes needed for root and flower development," as well as building a genetic map that recognized some 400 markers, and the identification of 11,500 expressed sequence tags (ESTs; more in list below).[131] Goals for the following year were set in each of these reports. For instance, one planned for 1995 was to expand the markers from 400 to 500; the goal was met with even more than 500.[132] Between January of 1994 and November of 1995 alone, some 1235 articles on thale cress were published globally. The year 1995 saw the annual meeting of the rapidly growing project group held in Madison, Wisconsin, exceed 600 researchers.

Project planning was based on units of half-decade increments, not so much with regard to finishing the sequencing project, *per se*, but in recognition that efforts would continue well beyond 2000 in the study of the *new model organism*. *Arabidopsis* had clearly become what the Russians sought decades earlier: the *Drosophila* of plant biology. By the sixth annual conference, for example, the nucleotide sequence portion of the project was seen to be "two years ahead of the current timetable," yet five more years were projected for completion: only four would be needed, as it eventuated. Decadal goals were set, also: "[e]mphasis will shift from gene identification to protein function" was just one such new objective and it represented one avenue for continuing research once the full sequence was available to the community. Successes included not only the technical aspects, but also the originally intended educational ones. A Cold Spring Harbor Laboratory course on *Arabidopsis* was already training a "new generation of plant biologists familiar" with the molecular biology of flowering plants. Numerous stock centers (collections),

[129] National Science Foundation, "The Multinational Coordinated *Arabidopsis thaliana* Genome Research Project: Progress Report: Year One," NSF 91–60; 17 ppg + iii; two appendices.

[130] See: http://www.nsf.gov/bio/pubs/reports/index.jsp, which includes not only the original sequencing/genome project reports, but those on the functional genomics of thale cress, a later project, as well as those on the NPGI; both of the latter discussed at greater length in the main text below. Report One does not appear at this site.

[131] See the fourth annual report as given at the NSF/BIO reports website noted in the previous endnote. The matter of expressed sequence tags (ESTs) as a specific area of investigation is a "unique stretch of DNA within a coding region of a gene that is useful for identifying full-length genes and serves as a landmark for mapping. An EST is a sequence tagged site (STS) derived from cDNA." See this and more at: http://www.medterms.com/script/main/art.asp?articlekey=3328

[132] Fifth annual report at Executive Summary section; see http://www.nsf.gov/bio/pubs/reports/index.jsp, *op. cit.*

which had been established several years previously at Ohio State University and the University of Nottingham in England, were evermore closely involved with the international project. A half dozen websites were also in operation.

Annual reports for 1997 and 1998 (seventh and eighth years) were combined and could be found on the new project website: TAIR, The *Arabidopsis* Information Resource, an outgrowth of an earlier site, AtDB, the *A. thaliana* Database.[133] That combined annual report provided a useful historical perspective of the project then to date:

- 1983—Publication of the first genetic map of the thale cress;
- 1988–1989—Publication of RFLP maps[134];
- 1990—Initiation of the Multinational Project;
- 1991—Stock centers in Ohio and Nottingham established;
- 1991—YAC libraries established and YACs tied to RFLPs[135];
- 1992—Publication of first chromosome walk (or local contig[136]);
- 1993—Recombinant inbred map;
- 1994–1998—Expressed Sequence Tags linked physical maps to genetic ones;
- 1995–1996—Numerous data libraries established; and
- 1995–1998—Physical maps of all five chromosomes delineated.

Several other milestones occurred in later 1998, as well. The number of new websites spiraled upward and China was added to the consortial arrangement. Mary Clutter, Machi Dilworth, and Delill Nasser were observers at the international meeting of the project group held at NSF in July of that year. Not entirely sure of the completion date of 2000 confidently expressed by Rita Colwell at that same time, BIO issued a solicitation for research awards in 1998 seeking investigators to finish the initial project.[137] Within one more year, Clutter told *Science* that a new initiative had been previewed before the NSB, Project 2010, and would cost an additional $50 M per year above that same amount already being spent on the sequencing project annually.[138] Project 2010 would continue the use of thale cress once the

---

[133] The AtDB website was inoperable before the time of this writing (2011) and does not appear on the WWW finding device, waybackmachine.org

[134] An RFLP is "a difference in homologous DNA sequences that can be detected by the presence of fragments of different lengths after digestion of the DNA samples in question with specific restriction endonucleases. RFLP, as a molecular marker, is specific to a single clone/restriction enzyme combination." See: http://www.ncbi.nlm.nih.gov/projects/genome/probe/doc/TechRFLP.shtml

[135] A YAC is "a vector used to clone DNA fragments larger than 100 kb [kilobases] and up to 3000 kb. YACs are useful for the physical mapping of complex genomes and for the cloning of large genes." The term derives from "yeast artificial chromosome," as explained at: http://en.wikipedia.org/wiki/Yeast_artificial_chromosome

[136] "In shotgun DNA sequencing projects, a contig (from *contiguous*) is a set of overlapping DNA segments derived from a single genetic source. A contig in this sense can be used to deduce the original DNA sequence of the source." See: http://en.wikipedia.org/wiki/Contig

[137] NSF 98–52 with a deadline for submission in April 1998.

[138] Sciencescope (a brief news note), *Science* (1999):25.

sequence was complete, and the actual role of each and every gene in *A. thaliana* could be revealed over the following decade, it was hoped.

Anticipating the completion of the genome sequence within a year, NSF's *Annual Report 1999* announced that the project's purpose had been, "in other words, to create a genetic road map to flowering plants."[139] Some 70% of the entire genome's information had been deposited in GenBank by August of 1999 and at the end of the year two of the five chromosomes' sequences were totally complete. The importance of *Arabidopsis* as a *model* plant could not be overstated; as was said: "[w]hen you have seen one flowering plant (at the molecular level), you've seen them all": a play on Ronald Reagan's comment about redwood trees![140] But that was neither an expression of wit nor fundamentally incorrect as the gene(s) that trigger flowering, for instance, in the mustard weed or in a rose are essentially the same; such genes are retained in evolution within the flowering plants across taxa. The concept of a model system could not have been better represented than by the lowly thale cress, and Project 2010 would go on to demonstrate its many outstanding values as such. The research that would be made possible for flowering plants generally would be like that which had been made possible with the sequencing of one species of bacterium such that it would become a model for a great number of other bacterial species.

A perfect example of the value of a model system for angiosperms was the flowering mechanisms found in *Arabidopsis* when applied to the study of the same process in aspen trees, quite distant relatives. An aspen typically reaches sexual maturity—flowers—when it is about 30 feet tall. That can take 20 years. By molecular manipulation based upon findings in the cress model, aspens were made to flower within six months of seed germination and at a height of only two inches! Since about "80 percent of the world's food supply is made up of flowers or flower parts: fruit, grains, or seeds," the applicability of this sort of gene manipulation is profound.[141] Truly, by 2000, science was on the threshold of a " 'golden age' of discovery in plant biology."[142]

The genomics community bid "[g]oodbye to 'one [by] one' genetics" (above) the next year, at least in print. That was so because during the 1990s computer science's growth and its ever more tight bond with biology, especially genetics, helped create genomics and eliminate, in part, the need to discover the role of one gene at a time.[143] Laying out the entire genome after a sequencing program would permit

[139] National Science Foundation, *Annual Report, 1999* (Washington, D.C.: Government Printing Office, 2000) pg. 64.

[140] Ibid., pg. 65.

[141] Ibid., ppg. 65 and 67.

[142] Elliot Meyerowitz and C. Somerville, eds., *Arabidopsis* (New York: Cold Spring Harbor Press, 1994).

[143] Athanasios Theologis, "Goodbye to 'One by One' Genetics," *Genome Biology* (2001):Opinion section. Theologis presents a *strikingly different early history* (first half of the 1990s) than that given in the main text (above) and endnote further above. See the second of ten pages of the Theologis article, in particular. The view is a European one. This merely further indicates the need

discovery of individual genes in rapid order. The sequencing project began with believing that possibly 20,000 genes would be found in *A. thaliana*. By its end, however, it found some 25,500: a good initial guess, especially in view of the wild collection of estimates for the human genome. The full history of this massive project has yet to be written and space does not allow a more extensive discussion of it here.

However, all the while the concept of the gene was undergoing continual refinement and by the turn of the millennium, the definition was much, much less simplistic than it had been earlier in the twentieth century.[144] Indeed, genomics had become "functional genomics"—a field that uses the data gained from sequencing projects in order to understand the interactions of genes and proteins and other gene products (e.g., RNAs). It became a vehicle to transport genomic discoveries into what has been called a "Genomics Revolution 2"—that began in an "*annus mirabilis.*"[145]

The Steering Committee set forth its new direction in 2002 with a report, which subtitle was crystal clear in its intent: "Beyond the Whole Genome."[146] It would be a new mission statement for the project, indeed something of a new project, yet a continuation of the one just successfully completed: to determine the function of every gene in *A. thaliana* by 2010. The project called for the "complete study of the physiology, biochemistry, and development of plants"; that is, flowering plants in general based upon the newly sequenced *model system*. The Committee already spoke in terms of there being "a *tradition* of *Arabidopsis* research."[147] Indeed there was, even if a very recent tradition, and over the next several years annual reports appeared for either the functional genomics or 2010 Project until 2005 when a "mid-course assessment" was undertaken.[148] In the half-decade preceding the meeting that led to that report, NSF had funded 86 projects in the functional genomics of the thale cress. A web-based survey to determine the next five year's program led 580 investigators to respond—the vast majority of the *Arabidopsis* community. The 11 recommendations on what research was needed included: the development of reagents; improvement of databases and other tools; understanding non-protein coding genes; localizing gene products; seeking additional models; and enhancing the international collaboration.[149]

---

for the full history of this important genomics' project to be written.

[144] See the human gene guidelines definition at: http://www.medterms.com/script/main/art.asp?articlekey=3560; a complex explanation.

[145] See an advertising article by Peter Gwynne and Gary Heebner at: http://www.sciencemag.org/site/products/fungen.xhtml. They refered to 2000 as an *annus mirabilis*.

[146] National Science Foundation, "The Multinational Coordinated *Arabidopsis thaliana* Functional Genomics Project: Beyond the Whole Genome," June 2002; no document number; no longer available online; hard copy at the NSF Library as SB 608 .A7 2002.

[147] Ibid., pg. 5. Emphasis added.

[148] Mid-course assessment at: http://www.nsf.gov/bio/pubs/reports/index.jsp

[149] Ibid.

Three years after leaving as BIO AD in 2005, Mary Clutter gave a presentation she entitled "Epigenomic Predictions."[150] She ended her talk with ten such predictions including such things as the "Millennials will be the new leaders," and "[a]ll omics and omes will coalesce with a multinational epigenome project." Most notably, however, she predicted that the *Arabidopsis* epigenome would be decoded and the effort would lead to a Nobel Prize. She further saw *epigenomics as leading to a new unifying theory of biology.* Indeed, *the matter of* reductionism *could be seen ever so clearly in genomics, and* integration *is characterized in functional genomics,* it may be said.

7.8 Plant Science at NSF: Precursors and Partners to the NPGI

Early in 1990, discussions at USDA's Agricultural Research Service (ARS) led to that agency's plan to spend the following decade and approximately $500 M on a vaguely stated "program to study the genes that control traits like drought resistance and nutrient use in economically important plants."[151] Jerome Miksche of USDA, and head of the nascent program, pointed out that they would not attempt to "chart all the DNA in one plant." Indeed, NSF/BIO had already been supporting the work on *Arabidopsis* aimed at defining every gene. Its role as a new model system, the first for flowering plant biology, was sharpening. Agriculture hoped to bring *its* plan about as part of the next federal budget (circa FY93), but "supporters of the project said these diverse efforts have not been well focused and financing has been erratic."[152] ARS was not alone. Congress, too, had been pressuring USDA to do more "biotechnology research, including gene mapping," but Agriculture appeared to be dragging its feet on the issue.[153] Clearly, something such as the NPGI was called for, but other programmatic initiatives were also needed to expand the numbers of young PhDs that would work in plant science. For that reason, in part, NSF/BIO joined with several other agencies to establish several contemporaneous programs to study plants and to bolster graduate education in the area. Mary Clutter's

[150] Paper copies of overhead slides from a talk by Clutter given September 28, 2008; venue unknown.

[151] Warren E. Leary, "Genetic Maps May Point the Way to Healthier and Sturdier Crops," *The New York Times*, January 9, 1990. Charles Hess of USDA was a visitor to the BIO AC for its meeting of May 28, 1991 and reported that his agency wanted $500 M for the project, which was authorized, and so sought $100 M for the first year, but received only $73 M from Congress for FY91. In an attempt to make up the difference, they requested $125 M in FY92. BIO AC, "Minutes," May 28–29, 1991. The permanent Clutter version of the BIO AC was first established in the early part of 1992. Peggy Weber to M. Winkler, "AC Chairperson," February 10, 1992.

[152] Ibid. 'Minutes.'

[153] Ibid.

strong botanical orientation played no small part in generating interest on the subject and, in many respects, led the charge in the multiagency efforts.

In March of 1992, an interagency program announcement for the "DoE/NSF/USDA Joint Program on Collaborative Research in Plant Biology" (CRPB) was issued and called for proposals by the next month, a very short lead-time. It seemed there was suddenly not a moment to be lost. It is possible that Mary Clutter, still early in her post as BIO AD, was keen to see plant science moved forward with all due rapidity. The CRPB effort was for both research and training needs and was a "continuation of the collaboration that was begun by the 1988…Plant Science Center's Program" involving the same three agencies.[154] Noting that new frontiers in plant biology had appeared in just the previous few years and citing efforts by other USDA programs and the Foundation's early work on the *Arabidopsis* genome, several potential areas for research under the CRPB presented themselves. They included training highly qualified students, supporting innovative and interdisciplinary research, and increasing coordination and collaboration in the discipline. Unlike the separate NPGI (below), the CRPB did not emphasize plant genomics specifically, but it did add another route to bolster plant science broadly within the community.

Simultaneously, the NRC published its "Plant Biology Research and Training for the 21st Century" in which it spoke of a "major revolution" over the previous quarter century in the "understanding of fundamental life processes." That was especially true in human health-related areas, but it noted that there was also an ongoing "biological revolution [in the] agricultural and environmental disciplines." They even cited a letter of October 1989 from Mary Clutter to Frank Press (then president of the NAS) in the opening paragraph of the booklet's Executive Summary in which she pointed out the "disparity" between biomedicine/human health areas and that of plant science.[155] The NRC committee was none too shy about its declarations:

> [t]he members of this committee are convinced that the US research effort in plant biology is not keeping pace with biomedically related fields because of the defective *mechanisms* used for support as well as the small financial commitment to research and training in plant sciences. [There is a] patchwork system of support for research and [an] incomplete system of support for career training in the plant sciences.[156]

Those were words that packed a power punch, as the plant sciences within just the next 2–3 years would skyrocket in the consciousness of not only NSF/BIO, but other agencies, as well. Not least among those was the National Aeronautics and Space Administration, NASA. While the NRC committee went so far as to recommend the creation of a National Institute of Plant Biology to be shared by USDA

---

[154] National Science Foundation, "DoE/NSF/USDA Joint Program on Collaborative Research in Plant Biology," http://www.ibiblio.org/pub/docs/nih-nsf/nsfplantbio.920313

[155] National Research Council, Committee on Examination of Plant Science Research Programs in the United States, "Plant Biology Research and Training for the twenty-first Century" (Washington, D.C.: National Academies Press, 1992).

[156] Ibid., Executive Summary, ppg. 2–3. Emphasis in original.

and NSF, it did not materialize.[157] But in its second recommendation, the NRC held that all agencies then involved in plant science, NASA being one, "have provided valuable support for plant-biology research, and their continued—and increased—commitment is needed to fulfill the new institute's objective." Though there was no "new institute" as it eventuated, NASA and NSF would join hands in yet another plant science initiative of the period: by the opening of 1994, the two agencies had created a draft program.[158] While the funding management would fall to Machi Dilworth in the IBN directorate (the Integrative Plant Biology Program, IPB) both it and MCB (the Cell Biology Program) were involved in the joint effort within BIO.[159] The new publication was online on the then STIS system (pre-Fastlane) before the print version had become available.[160] By the first of May, there were already 18 proposals at the BIO office and it was believed that a "pile of proposals are sitting somewhere in this building" to add to that group.[161] At the end of June, the first panel would be held to adjudicate the "pile." Nine panelists were assembled to examine the by then 35 proposals. All were placed under IPB and Dilworth's management. IPB was a good moniker, because, as she said, "[h]ow plants perceive and respond to environmental signals is one of the major unanswered questions in the biological sciences."[162] Integrative biology was still a new phrase at BIO, but would become a standard not unlike computational biology and others that would begin then to characterize Clutter's "New Biology" at millennium's end. And the biologists funded by the new joint program would form a "research collaborative network…a center without walls" by which to make strides in the integrative study of plants. Slightly over $1 M was earmarked for the program that first year with intents to replicate about that amount over each of the succeeding 4 years. NASA had contributed $500 K and IBN and MCB a quarter million each to begin the effort. The eventual awardees included ten five-year long projects averaging $113 K each in year one. Signal transduction and cell differentiation were the primary areas supported. One of those grants was for a coordinator ($60 K/year) to manage the other nine, thus making the full program very much a collaboratory.[163] Plant biology was moving very rapidly within BIO in the first half of the decade of the 1990s and would see no cessation into future years.

---

[157] Ibid., pg. 6.

[158] The agreement was signed on October 7, 1993 by Mary Clutter for BIO and Harry C. Holloway for NASA.

[159] Machi Dilworth to Sonya Mallinoff, "Interagency Program Announcement: NASA/NSF Joint Program in Plant Biology," January 7, 1994.

[160] For more on STIS, see: ftp://athos.rutgers.edu/resource-guide/chapter3/section3-27.ps. The publication number was NSF 92–29.

[161] Machi Dilworth to Thora Halstead (NASA), "Joint Program Proposals," May 1, 1994.

[162] NSF News Release, "Scientists Study How Plants Respond to Environmental Signals," NSF PR 94–71, November 22, 1994.

[163] Data and other information in the several foregoing sentences in the main text above was drawn from Program Staff, "Annual Report—FY 1994: NASA/NSF Joint Program in Plant Biology," a typescript report of six unnumbered pages. The award success rate was 26%.

By the NASA/NSF program's third year, emphasis was formalized upon plant sensory systems, such as gravitropism[164] and others. A research network had become established for such sensory aspects in plants and it held annual meetings. Some of the researchers were employing *Arabidopsis* as it had become the primary model system for many, if not most, of the studies undertaken in flowering plant biology. The distal elongation zone[165] of the roots of the thale cress was being investigated at the time, for instance, and represented a "mega-project," as it was termed, covering all the investigators involved in the network. That activity was added to already established projects in various areas being done by the nine PIs. The plant sensory network was centered at Ohio State University and posted information on the Internet.[166] The education of graduate students was highly integrated in that the students routinely went for specialized training to various of the nine university laboratories across the country within the system.[167] The students particularly liked the "center without walls" approach, they averred. The third annual report of the program indicated that the intent to keep grants at about $1 M per year was holding steady and exciting work was being accomplished.[168]

Into the milieu of this variety of plant science efforts at BIO, there arose the notion of what would become the National Plant Genome Initiative. NSF had spent $87 M on plant science in FY95 alone, and all federal support came to $195 M that year (Fig. 7.5).[169] The NPGI was an independent project, *sensu stricto*, from that on the thale cress and the NASA/NSF project, as well as others, but it would be difficult to separate any of them entirely. Knowledge feeds upon knowledge.

---

[164] Gravitropism is one of many types of growth responses in plants. In this case, it is the tendency of plant roots to grow downward and stems to grow upward.

[165] That portion of the root where cells enlarge lengthwise to extend roots into the soil.

[166] See: http://www.google.com/search?client=safari&rls=en&q=plant+sensory+network&ie=UTF-8&oe=UTF-8 for an example of one of the network's documents. Many others are online, as well.

[167] In 1994, changes were made in the graduate education programs in BIO. For some time, there had been a program for post-doctoral training in plant science, but the directorate dropped it in favor of placing a number of such education programs under a broader heading of Environmental Biology. Carter Kimsey would direct the new program. Paper copy of an email from Sally Assman to NASA/NSF plant biology PIs, March 21, 1994.

[168] Information in this paragraph has been derived from a paper copy of an email from Mike Evans (Ohio State University) to Machi Dilworth and B. Zain at NSF and Tom Scott at NASA of October 15, 1996, and also from the "Report on the Annual Meeting of the NASA/NSF Plant Sensory Network Meeting," a three-page typescript, R:\NASA\RPTSTL96.DOC of October 18, 1996, and, finally, the "Annual Report—FY 1996," by Machi Dilworth, a four-page typescript, n.d. (1996 or 1997)

[169] A five-page document on "Plant Science Support" apparently prepared by the Office of Legislative and Public Affairs, OLPA, at NSF (by Catherine Corson?) for Kristen Kovalsky of NSF in connection with an exchange of letters involving Senator Paul Sarbanes (D-MD) and one of his constituents who was concerned that federal support for plant science was inadequate. Some $6B was spent in FY95 on all federal life science research support, the document noted. See several following endnotes for more financial details.

FEDERAL SUPPORT FOR BASIC RESEARCH IN PLANT BIOLOGY AT ACADEMIC INSTITUTIONS COMPETITIVELY AWARDED GRANTS FY 1978 - 1998 (EST) ($$m)						
	NSF	USDA	DOE	NIH	NASA	TOTAL
FY 1978	22	10	19			51.0
FY 1979	26	10	19			55.0
FY 1980	28	12	24		0.9	64.9
FY 1981	34	13	28		1.2	76.2
FY 1982	41	13	21	20	0.9	95.9
FY 1983	43	15	22	20	1	101.0
FY 1984	52	15	24	20	1.4	112.4
FY 1985	61	27	25	21	1.5	135.5
FY 1986	58	26	25	25	1.6	135.6
FY 1987	62	26	28	27	1	144.0
FY 1988	64	25	31	29	1	150.0
FY 1989	68	23	31	32	1	155.0
FY 1990	71	24	32	15	3.2	145.2
FY 1991	76	40	34	16	3.3	169.3
FY 1992	82	45	37	17	3.4	184.4
FY 1993	83	46	34	24	3.4	190.4
FY 1994	85	46	34	27	3	195.0
FY 1995	87	45	34	27	5.3	198.3
FY 1996	91	43	35	28	5.3	202.3
FY 1997	91	60	35	29	5.8	220.8
FY 1998						

Fig. 7.5 Federal support in awarded grants in plant biology from FY1978 to FY1998 (estimated). NSF. A five-page document on "Plant Science Support" apparently prepared by the Office of Legislative and Public Affairs (OLPA). July 2, 1997

7.9 The National Plant Genome Initiative: NPGI

There existed in the mid- and later 1990s a movement that was termed the National Corn Genome Initiative, which was being forwarded by the combined efforts of both the public and private sectors. That consortial group was seeking some $143 M over 5 years to complete the maize (corn) genome map and tried, unsuccessfully, to earmark the USDA budget in 1996 to do so. One leader of the consortium was Inverizon International, Inc., a consulting and business services company located in St. Louis, Missouri.[170] However, the National Corn Growers Association (NCGA) was also deeply, and very aggressively, involved in promoting the need for the sequence information. It was somewhat of a tangled web.

Senator Christopher "Kit" Bond (R-MO), for whom the NCGA was a powerful constituency group, came to NSF to visit with the BIO leadership and promote the

[170] In 2000, Inverizon sent a cease and desist letter to the telecommunications company, Verizon, for infringement of mark. Details of the case may be seen at: http://openjurist.org/295/f3d/870/verizon-communications-inc-v-inverizon-international-inc. Clutter mentioned Inverizon as a notable player in her "talking points" memo: see following endnote.

corn initiative plan in February of 1997. Mary Clutter issued a "talking points" document to prepare her colleagues for the meeting.[171] As it happened, maize gene sequencing, at an albeit much smaller scale, had already been an area supported by BIO funding for some years through PI Ed Coe at the University of Missouri, Columbia. In connection with Coe, BIO had also been supporting the work of Myron Neuffer, Coe's colleague at the University, for his collection of mutant maize seeds. Indeed, Neuffer had provided his information to the then already extant Maize Genome Database that was growing and that maintained an Internet presence.[172] Aspects of interest to the growers went back to the 1930s, but had turned strongly technological only in more recent times. Further, BIO was supporting other researchers, also at the Columbia campus, during that same period: Kathleen Newton's work on corn mitochondria, and also her Interdisciplinary Research Training Group. The graduate students in her Group worked with Monsanto, Inc., through internships to learn biotechnology via corn genetics, all under project structures in which NSF was a central player.[173] The students did need to learn, after all, that "the ethos of commerce is thoroughly enveloping academic life."[174] The University of Missouri was typical of academic institutions across the country as NSF alone supplied some 45% of all federal competitive grants for plant science in the academy.[175] Clutter and BIO had much to tell Kit Bond.

The talking points of the day included such issues as why BIO, and the biological community—and many another—would be interested in sequencing plant genomes. How it was done was explained to Bond and what problems the highly complex maize genome presented researchers was also made clear, let alone the great costs. Indeed, of the $91 M spent by NSF on basic plant research in FY96 (of the total federal outlay of $202 M), $20 M of that was specifically for plant genome research. *Arabidopsis* claimed $4 M, other plants and fungi some $3 M, genome-based research broadly about $11 M, and the remaining $2 M went for botanical infrastructure (databases, training, collections, etc.).[176] But none of the challenges of either difficulty or cost occurred for *Arabidopsis,* unlike maize, and so the thale cress was shown to Bond as a logical alternative—a *model system* that could aid a maize sequencing effort somewhat later in time. At the point of Bond's meeting, it was still believed that the thale cress would be done 2 years ahead of schedule, that is, 2002 instead of 2004. The actual completion date of 2000 was not yet foreseen, even as late as 1997.

---

[171] Mary Clutter, "Talking Points for Meeting with Kit Bond, 2/27/97," a two-page typescript document.

[172] See: http://www.agron.missouri.edu/

[173] Clutter, 'Talking Points.'

[174] Daniel Lee Kleinman, *Impure Cultures: University Biology and the World of Commerce,* (Madison: University of Wisconsin Press, 2003), pg. x.

[175] 'Plant Science Support,' *op. cit.*

[176] J. Widder, NSF/OLPA to Mary Clutter, "NSF Expenditures on Plant Genome Research," August 6, 1997.

The maize genome is billions of DNA base pairs in length so, in 1997, the full set of tools and techniques had not yet reached sufficient maturity to take on its sequencing. Once *Arabidopsis* was done, much of that information and technology would be ready to use on corn. In fact, the president of the NCGA was present when the thale cress genome was announced in December of 2000 at NSF headquarters, so keen was he on seeing the corn genome profit by what was learned with *Arabidopsis*. Indeed, what would become NSF's Plant Genome Research Program, the Foundation's part of the multi-agency NPGI, was fundamentally being framed in connection with Senator Bond's urging and BIO's already committed interests and support of sequencing corn, thale cress, and other plants and fungi. (Fungi were of particular importance due to their role in plant disease, maize not least among the victims.) In FY98, NPGI was established, but even then, it was clear that maize would have to wait a while: it would not be until 2005 that NSF would take on the challenge joining with USDA and DoE, once again. A total of $32 M was set aside for the project and 3 years were planned for its completion; it took four. Maize was not thale cress! The leader of the effort was Richard Wilson who directed the Genome Sequencing Center at Washington University School of Medicine in St. Louis. Other team members came from the University of Arizona, Tucson and Cold Spring Harbor Laboratory in New York, as well as scientists from Iowa State University, Ames. It was, after all, Patrick Schnable at the latter who had been working on maize genomics since the early 1980s—well before the "omics" were used as a set of terms—and pointed out that the "corn genome is as big as the human genome but more complex."[177] In the event, the maize genome would only be completed and published in November of 2009.[178]

When the NPGI began, it did so with a strong boost from Congress through the efforts of Kit Bond, he having secured $40 M (for which the corn growers association took considerable credit, whether earned or not) in later 1997.[179] As well, Bond was joined by Senator Mikulski in winning the funding. That amount was under negotiation from July/August of that year and involved Bond and the OSTP, but that Office's head, Jack Gibbons, had to convince a recalcitrant Bond that taking on maize at that point was just as difficult and would be as expensive as BIO had told him several months before.[180] Further, Gibbons said, USDA at least had to be involved along with NSF. Even more, the Interagency Working Group on Plant

[177] Jessica Ebert, "The Genetics of Ethanol Production," at: http://www.ethanolproducer.com/articles/4147/the-genetics-of-ethanol-production

[178] See: http://www.sciencedaily.com/releases/2009/11/091119193636.htm and also http://www.seedquest.com/News/releases/usa/Ncga/n3201.htm. The president of the NCGA in 2000 when the *Arabidopsis* genome was announced was Lee Klein, a farmer from Battle Creek, Nebraska. Evidently somewhat uninformed, he hailed the achievement by saying that the announcement "wouldn't have been possible without NCGA leading the effort to get major funding for the Plant Genome Initiative. Thanks to NCGA this sequencing project is four years ahead of schedule"!

[179] Ibid., the second webpage listed immediately above.

[180] Paper copy of an email memo from J. Widder at NSF/OLPA to a large number of in-house recipients at the Foundation, which copied an article in *Life Science* for August 6, 1997.

Genomes (IWG-PG) was on record as having said that at the rate that *Arabidopsis* was then (1997) being sequenced, maize would take about 138 years to accomplish! The IWG-PG was a creation of Gibbons some months earlier and included OSTP, Management and Budget, NSF, Agriculture, NIH and Energy, though not NASA, even though the latter was clearly a player at the time in plant science with NSF. Oversight of the IWG-PG was handled by the National Science and the Technology Council's Committee on Science.[181] One issue that the IWG-PG noted was that the state of bioinformatics was still not sufficient to handle a corn genome project, even though BIO had been putting considerable funding into bioinformatics for quite some years (previous chapter). The work on *Arabidopsis* was already straining current bioinformatics' capabilities, they said.

In order to have a strong and experienced leader for the new plant genomics program at NSF that the Kit Bond funding infusion allowed, BIO brought in Judith Verbeke from the University of Arizona, Tucson, in November of 1997. She already had a history with the directorate over several years as an "expert scientist" and rotator, but primarily in the development area within the IBN division.[182] Indeed, she was still not permanent in 1997, but became so the next year and returned to the development cluster in IBN, leaving the plant genome project. More will be said of Verbeke in the following chapter.

Japan was just at that time attacking the rice genome and the US should join in, IWG-PG suggested, so at least one major cereal crop genome would soon be known; a boost for maize, it was reasonably presumed. The rice genome (below) would be done by 2005, the year the corn effort began. *Arabidopsis* data *did* provide a boon to that effort in that 71% of the cress' protein-coding genes, specifically, were also found in rice; and 90% of thale cress proteins seemed to have a homolog in the rice proteome.[183] Rice's genome, at 389 megabases of DNA, was smaller than other cereal grains and so that made it ideal as the first such important food crop to sequence. The cereal grains are the primary food source in the world and thus genomic knowledge of them has great value with regard to breeding programs and other aspects.

The IWG-PG had made recommendations for the running of the NPGI and noted that since USDA's "management ability and commitment in the area of plant genomics 'is not impressive'…Bond placed the $40 million in…the NSF appropriation [line, instead]." Further,

---

[181] Ibid.

[182] Judith Verbeke Interview with the Author, February 25, 2009.

[183] Takuji Sasaki, "The Map-Based Sequence of the Rice Genome," *Nature* (2005):793–800. The paper is available online at: http://www.nature.com/nature/journal/v436/n7052/abs/nature03895.html where a fuller copy yet may be gotten via associated hotlinks. Proteomics is the study of the complete complement of proteins in an organism. The term proteome was coined in 1994 by doctoral student Marc Wilkins, but the term proteomics was not created until 1997.

> Congressional committees [associated] with USDA's oversight have a 'notorious reputation' for earmarking and pork barreling and 'would likely not provide good stewardship' for funds designated for this [maize] genome initiative.[184]

Indeed, Colin Macilwain, the Washington, D.C. correspondent for Britain's *Nature* journal, made the comment that NSF "appears to have succeeded in framing the [NPGI] on its own, scientifically rigorous, terms" bucking pressures from the National Corn Growers Association.[185] Thirty of the $40 M would be awarded within the year and would be directed toward economically significant plants, but maize, though listed, was not highlighted in particular. The remainder would devolve to the ongoing work with thale cress. The best science, in the long tradition of both NSF and BIO, would gain the awards because, as Mary Clutter said, it was the "ordinary way of doing business."[186] One thorn that arose, however, was the reference in the *Nature* news article to the effect that

> [o]thers worry about the principle of the NSF—an agency whose primary mission is to support science being conducted for its own sake—running errands for an influential senator from an agricultural state who happens to chair the appropriations subcommittee that funds it.[187]

A valid point no doubt, but the Foundation could hardly argue that they were not attuned to political needs as they often engaged in and generally won battles with Senator Proxmire in earlier years, let alone many other incidents not covered in this history. The NPGI was to be *big biology* in that a given proposal could seek as much as $3 M a year for one- to five-year long projects. Virtual collaboratories were sought in the proposal guidelines and such were not inexpensive endeavors. Should an investigator seek to work on an international effort such as the rice genome, for instance, they would be required to coordinate their work with the Japanese: big biology, *and* on the multinational scale. While sequencing was the main point, computational biology, bioinformatics, and other areas, when tied to genomics, could also be funded. So significant was the view held by BIO about the importance of the NPGI that Mary Clutter said, "[w]e're talking about a major effort. What NSF is doing is jump-starting something that will set the [direction] for agriculture of the 21st century."[188] The latter part of that statement, whether it would come to pass or not, could not erode the fact that the project was, indeed, a "major effort."

Under the NPGI, the US joined into the effort to decode the rice genome beginning in 1999. NSF, USDA, and DoE, the big three of similar efforts, estimated it would take 3 years and about $12 M to sequence the genome. It was very clear by 1999 that *Arabidopsis* was likely to be done the following year, those 4 years ahead

184 Widder memo that copied the *Life Science* article.

185 Colin Macilwain, "$40 M Plant Genome Sequencing Effort Targets the Best Science," *Nature* (1997):539–540.

186 Ibid.

187 Ibid.

188 Ibid.

of schedule.[189] The rice genome, then, was likely to be the second plant to be sequenced. In fact, 3 years was a good guess as it was exactly that. But it would be a private sector success in that *Oryza sativa* ssp. *japonica* was sequenced by 2002 in an effort led by Myriad Genetics and the Syngenta corporations. That same year, *O. s.* ssp. *indica,* the other of the two subspecies used in agriculture, was sequenced by the Beijing Genomics Institute, Zhejiang University, and the Chinese Academy of Sciences. The US corporate effort, using the Nippon-Bare variety of the subspecies *japonica*, chose not to have its genome published in a scientific journal or on the Internet, but made it freely available to any researcher that wished a copy of the sequence: a magnanimous decision on the part of industry for the sake of science.[190] Though not investigated for this history, there may have been other aspects underlying that magnanimity. At the time, Syngenta was a new company arising out of the merger of Novartis and AstraZeneca and it became the largest crop protection company and third largest in the "market for developing important commercial seeds." The corporate sequencing effort ran parallel to the federal agencies' triumvirate led by NSF.[191] At about 50,000 genes, rice was on the order of twice the size of the thale cress, at 25,500 in five chromosomes. But in base pairs, rice is massive at 430 MB spread over 12 chromosomes. (The fruit fly, *Drosophila melanogaster*, long *the* animal model system, had its sequence published in 2000 and had only 180 MB.[192]).

In the last 2 years of the millennium, no fewer than three plant genome competitions had been announced by NSF.[193] They were distinct competitions, albeit, but it was a bit difficult to tell one from the next. The whole area of higher plant genomics was blossoming at such a rapid pace that multiple competitions with slight variations seemed warranted. In that beginning year of 1998 for the NPGI, the IWG-PG issued its first five-year plan.[194] Working Group members included Mary Clutter as Co-Chair, but also Francis Collins, of the National Human Genome Research Institute of NIH, though he was represented by another individual, as Collins was so deeply involved in the HGP. Clutter's opposite co-chair was Eileen Kennedy of USDA. Others from USDA, DoE, OSTP, etc., were also members. The goals included:

---

[189] The Arabidopsis Genome Initiative, "Analysis of the Genome Sequence of the Flowering Plant *Arabidopsis thaliana*," *Nature* (2000):796–815. It was in the December 14 issue. Some websites erroneously give the date as December 16 (there *was* an issue of *Nature* on December 16, but in *1999*).

[190] That was a rough draft of the rice genome map done by Monsanto contractors in May of 2000. The contractors included Gregory Mahairas of the University of Washington at Seattle and Leroy Hood of the Institute for Systems Biology (ISB) in the same city; see also early chapters for mentions of Hood and also the ISB. Elizabeth Pennisi, "Underground Genome Project Pleases Rice Researchers," *Science Online* at http://sciencenow.sciencemag.org/cgi/content/full/2000/406/1

[191] See: http://news.bbc.co.uk/2/hi/science/nature/1138390.stm

[192] See: http://www.sciencemag.org/content/287/5461/2185.long

[193] NSF 98–30, NSF 98–52, and NSF 99–13 that replaced 98–30 in 1999.

[194] National Science and Technology Council, "National Plant Genome Initiative," 1998; see: http://www.whitehouse.gov/WH/EOP//OSTP/html/OSTP_Home.html

- completion of the *Arabidopsis* genome;
- sequencing rice as part of an international effort;
- developing ESTs, physical maps, and mutants for studying the cereal grains;
- increasing gene structure and function knowledge;
- bioinformatics;
- maximizing training; and,
- making data generated broadly available to a wide audience.

Considerable attention was also given to plant diseases (phytopathology), intellectual property rights, and relationships with the private sector. The amount of $20 M was earmarked to "speed up" the thale cress effort, while $40 M was aimed at doing 20% of the rice genome (the rest to come from the efforts of others, including Japan and Europe). Some $30 M was for ESTs in a purposefully undefined group of 20 major crops and up to $100 M from plants to include the specified crops of maize, wheat, soybean, sorghum, tomato, barley, pine.[195] An additional $60 M would go for infrastructure and $150 M for functional genomics. Those figures were to be spread over the 5 years of the plan. The Bond/Mikulski $40 M was not included in the IWG-PG total, which larger amount was stated as a desired $320 M total—but the list just above comes to $400 M. Explaining that, the report noted

> flexibility in this budget distribution will be essential because relative costs of plant genome research are changing constantly with rapid advances in technologies.[196]

Such bold planning led the public relation's office at NSF to say that the "Century of Biology" (the 21st) had been given a head start by all the plant genome work ongoing and planned by BIO. Funds were directed to Ed Coe at the University of Missouri for continuing his maize studies and Thea Wilkins at UC, Davis for cotton research, as but two examples. As well, Nina Federoff at Pennsylvania State University, doing research into environmental stresses on plants, was also supported. Federoff was a very prominent member and chair for several years (beginning in 1995) of the BIO AC, as has been said earlier.[197]

The flurry of plant genome research led to a workshop in November of 1998 at Stanford University in anticipation of the completion of the thale cress sequence that, as it eventuated, would be published just 13 months later. "New Directions in Plant Biological Research" saw 11 main participants along with an additional four observers discuss the fact that it was plant biologists who were at the forefront of genomics research because thale cress would be the first higher eukaryote, plant or animal, genome to be published. (The eukaryotic yeast genome had been published in 1996.) Actually, in some senses, *Arabidopsis* shared that place with the human

[195] Pines, and other conifers, are not considered flowering plants, but are not unlike them in many ways, not least of which is because they are extremely important economically.

[196] Ibid.

[197] Cheryl Dybas, "Upcoming 'Century of Biology' Given Head Start by NSF Plant Genome Research Program Grants," September 28, 1998, NSF PR 98–57.

genome. But the full human genome was truly completed 2 years later. Still, the DNA of both plants and peoples did appear in great part in print in 2000.

The Stanford meeting noted the revelation of some intriguing discoveries emanating from *Arabidopsis,* including the fact that, genes for glutamate receptors, known only in animal neurons up to then, had been found in the mustard weed. They seemed to regulate cellular channels and have a tie to light perception in the cress. Further, a number of enzymes called kinases were found in *Arabidopsis* and, in that plant, appeared to play a role in cell division and differentiation; they, too, were unexpected. Other oddities appeared and suggested a much broader role for many genes at first believed to have more limited functions and be native to other kingdoms of life. It was at that meeting, also, that considerable movement toward functional genomics as a field in need of expansion was pursued. Many specific goals were laid out for what needed to be understood about the function of the various genes and the proteins for which they coded.

The second Progress Report for the NPGI appeared in the fall of 2000.[198] Former Director of the NSF, Neal Lane, then Assistant to President Clinton for Science and Technology, addressed the second report to the scientific community by saying that "impressive scientific progress" had been made on IWG-PG's goals and chose certain of those upon which to report. The *A. thaliana* genome sequence was complete and had been placed in the GenBank public database.[199] Some 26 megabases of rice DNA had also been placed on the database; 9.4 MB of those coming from US efforts. A very unexpected discovery was that the centromeres[200] of plant chromosomes contain functional genes—totally unknown in any other eukaryotic organism. The quantitative trait locus (QTL) for fruit size in tomatoes was discovered and cloned, the first such QTL ever replicated in the laboratory.[201]

While biological discoveries had been made, significant technological leaps had also occurred. The NPGI had been experimenting with microarray technology[202]

---

[198] Interagency Working Group on Plant Genomes, NSTC Committee on Science, "National Plant Genome Initiative: Progress Report," November 2000, Executive Office of the President.

[199] See Bruno J. Strasser's intriguing new arguments concerning the nature of natural history, experimental science, and GenBank in "The Experimenter's Museum: GenBank, Natural History, and the Moral Economies of Biomedicine," *Isis* (2011):60–96.

[200] "(1) The constricted region joining the two sister chromatids that make up an X-shaped chromosome. (2) The site where [the] kinetochore[centromere is the more used term in American biology] is formed." See: http://www.biology-online.org/dictionary/Centromere

[201] NPGI 'Report,' Executive Summary. Quantitative traits in organisms are those that have a variable expression, such as height in humans, and are not in discrete units, such as white or red blossoms in *some* flowering plants.

[202] "A DNA microarray is a multiplex technology used in molecular biology. It consists of an arrayed series of thousands of microscopic spots of DNA oligonucleotides, called features, each containing picomoles (10^{-12} moles) of a specific DNA sequence, known as *probes* (or *reporters*). These can be a short section of a gene or other DNA element that are used to hybridize a cDNA or cRNA sample (called *target*) under high-stringency conditions. Probe-target hybridization is usually detected and quantified by detection of fluorophore-, silver-, or chemiluminescence-labeled targets to determine relative abundance of nucleic acid sequences in the target. Since an array can contain tens of thousands of probes, a microarray experiment can accomplish many genetic tests

and found it highly viable as a tool. Further, investigators within the program found that "chromatin charting" was also a useful tool. That technique allowed for "real-time visualization of movement of chromosomes in living cells within a plant."[203] At least eight more advances in science and education were made, too. Indeed, the NPGI had had a very successful second year.

The Plant Genome Research Program (PGR) was the proper name for NSF/BIO's portion of the much broader NPGI, but both terms were used somewhat interchangeably. So, by 2000, 16 new grants were funded under the PGR section of NPGI and totaled $48 M over the following 5 years. Research on maize, wheat, and rice would be included and functional genomics and research tools would be the emphases. Areas of strong interest included chloroplast biogenesis, plant nutrition, whole genome scale functions, host-pathogen interactions, and stress responses, Mary Clutter pointed out.[204] The extremely complex genome, actually multiple genome sets, of white bread wheat (*Triticum aestivum*), was studied by investigators at UC, Davis. They gave special emphasis to one of the four sets of chromosomes in the plant, the D genome. (Only three sets occur in any of the cells of the varieties/species of wheat at one time, however, as the most complex of those are triploids.) The D set is where most of the important genes for humans' use are located. It was the largest genome to be undertaken up to that time: the D is 2.5 times that of the whole maize genome, and ten times the size of that in rice.[205] Work was supported both at the University of Wisconsin, Madison, and the University of Georgia, Athens, for these studies. Other grants went to universities in Florida, Oregon, New Jersey, and elsewhere. NSF's strong early bias toward funding New England ivy-league universities at the expense of others across the nation had long-since diminished.

Polyploidy—multiple sets of chromosomes present in cells (such as in wheat)—had been well known in the plant kingdom (rare though not entirely absent in other kingdoms) for at least a century; ferns often having massive numbers. But an understanding of *plant evolution* could be, once large genomes were sequenced, better grasped as a basic biological reality. Indeed, the very next year saw a superb example of the use of understanding very large genomes for such evolutionary research: the closest living relative of land plants was found among the algae.[206] Jim Rodman, DD of DEB, working with the Tree of Life project (Chapter Nine) noted that the story about higher plant evolution had just become much more interesting as the actual algal group that gave rise to land plants had been demonstrated through genomic understanding. Though long suspected of being the ancestors, the group of

---

in parallel. Therefore, arrays have dramatically accelerated many types of investigation." See: http://en.wikipedia.org/wiki/DNA_microarray

203 NPGI 'Report,' Executive Summary.

204 Cheryl Dybas, "NSF Boosts Research for Understanding Structure and Function of Complex Plant Genomes," NSF PR 00–60, September 21, 2000.

205 Ibid.

206 Cheryl Dybas, "Researchers Find Closest Living Relative of First Land Plants," NSF PR 01–104, December 13, 2001.

green algae known as the Charales, were shown to include, in fact, the ancient progenitors of the land plants. The discovery was made by Charles Delwiche and his student, Kenneth Karol, at the University of Maryland, College Park, and Richard M. McCourt at the Academy of Natural Sciences in Philadelphia with his Maryland student, Matthew T. Cimino; all supported by NSF. The significance of the finding was of extreme importance to basic biology and was *an example of the interaction between basic and applied research*, but this time it was a case of applied research on large genomes for forwarding agriculture that fed crucial information and techniques to basic research in biology.

The PGR program, specifically, was assessed by members of a COV that same year of 2001. They found that BIO/PGR "has succeeded in bringing plant and agricultural biology into the *genomic era.*"[207] *Millennial Biology* may have many partial synonyms, not least of which could be "the genomic era." The "development of ideas, tools, and human resources" for the era could not have come to pass without the PGR, the COV averred, and it was "nothing short of revolutionary." They praised "large-scale projects" and reaffirmed the values inherent in big biology. Individual researchers would benefit by the mechanisms and structures established, the COV members held. They recommended that "project managers" be a routine part of such approaches to biological research. Indeed, that need had been foreseen by BIO for the first year of the NASA/NSF project and now had become a clear necessity, in the eyes of the COV. Cooperative research agreements had risen to new heights in solving issues that previously challenged management strategies. Bioinformatics was still a difficulty and much more had to be done in that area: massive collections of data were being produced, and rapidly, but informatics' management efforts were not keeping up—a challenge that would continue to plague biology well into the present century. That matter was the "most important of our concerns," said the COV.[208]

Indeed, the BIO AC had held a retreat in 2001 and determined areas for development by the Directorate to include "Genome-enabled Science, Biology of Complex Systems, and Beyond the Genome."[209] Clearly, genomes and what they had already and would yet spawn was centermost in the minds of many of those persons within and without BIO; and the concept of *genome-enabled science* had a firm grip upon the research community and its primary supporter. (Might future historians of biology call the 21st the century of genomics and its related sciences?) The AC noted that genome-enabled science had distinct impacts in the study of microbial genomics (below), the evolution of development (widely known as "evo-devo") and the Tree of Life—all of which will be discussed later in this history.[210] Members of the

[207] Committee of Visitors Report, Andrew Binns, Chair, "Plant Genome Research Program, [DIB, BIO]," May 9–11, 2001, 24 ppg. Emphasis added.

[208] Ibid., pg. 2

[209] BIO AC "Minutes," April 26–27, 2001, pg. 2.

[210] Ibid.

AC did call for "making [the areas] crisper and more distinctive," however, and so wanted BIO to be mindful of "overlap."

In 2003, an NSF/USDA cosponsored workshop addressed the wheat genome. Some 63 investigators, 18 of those from abroad, gathered in Washington, D.C. to consider the state of the sequencing project. Nine steps to completing the genome were among the conclusions that came out of the meeting. None was surprising nor unlike the other big undertakings of the time, but bioinformatics did get special mention and "integrating [it] at every step" for each end that it could serve was clearly stated.[211]

Nor was that the only meeting that year for examining the direction of large-scale genomics efforts. The NPGI planned its next 5 years (2003–2008) then, too. John H. Marburger, III, had become director of the OSTP under President George Bush, *fil*, and held that it was "critical to continue and even accelerate research efforts in plant genomics" in order to "maintain [the research community's] scientific momentum."[212] The objectives for the next 5 years were:

- Continued elucidation of genome structure and organization;
- Functional genomics—understanding the biological role of plant genomic sequence, including gene sequences, regulatory sequences, and repeated sequences;
- Translational plant genomics—application of genomics tools;
- *Bioinformatics in every plant scientist's research tool box;*
- Education, training, and outreach; and.
- Consideration of broader impacts (bioethics).[213]

Bioinformatics was very clearly presenting a major obstacle and was not keeping up with the pace of a new, much more rapid sequencing technology. Data (mostly as ESTs) were being produced faster than they could be consumed, it might be said. This challenge would continue.

Independently of the NPGI's own IWG-PG report, NAS had come out with a study several months before that covered the same years and also offered its own set of objectives that would "leverage…the translation of basic discovery to agriculture."[214] The NAS report had been requested by NSF/USDA/DoE and it stated that the NPGI should focus on a small number of "key plant species…in-depth," translate findings from those to related crop species, home in on the evolutionary history of plants, expand investment in bioinformatics, and create more interdisciplinary training opportunities for doctoral and post-doctoral students.

[211] Bikram S. Gill, et al., "A Workshop Report on Wheat Genome Sequencing," *Genetics* (2003):1–12.

[212] Interagency Working Group on Plant Genomes, NSTC Committee on Science, "National Plant Genome Initiative: 2003–2008," January 2003, Executive Office of the President.

[213] Ibid. Emphasis added.

[214] NAS, "The National Plant Genome Initiative: Objectives for 2003–2008," (Washington, D.C.: National Academies Press, 2002)

With all of that forward momentum and keen interest on the part of many audiences and participants, BIO only requested $75 M for plant science in FY 2004, the same amount it had in actual dollars for FY 2002 and then-estimated dollars for FY 2003! The text of the budget request for FY 2004 seemed to say that since "significant progress" *had* been made in the first several years of the PGR program, NSF/BIO had fulfilled its intentions of the late 1990s. That may well have been true, but the field was burgeoning and it would have appeared that NSF should remain as more than just a status quo member of the federal triumvirate.[215] Nevertheless, momentum certainly continued with NSF Director Rita Colwell signing the NSTC Committee on Science letter that preceded the NPGI 2004 annual report. So also did Kathie Olsen, by then Associate Director of the OSTP and, of particular note, so did Elias Zerhouni, Director of NIH. Clearly, plant genomics stood essentially equivalent to the HGP because it was at the "forefront of plant biology and its application to solving global problems in agriculture, *public health* [hence NIH's interest; was that a sop to involve NIH?], energy, and environmental protection."[216] And the clear enthusiasm regarding the success was well justified, because, during 2003: a high resolution genetic and physical map of maize had appeared; all of the genes necessary for plant nutrition had been identified; new wheat breeding strategies were defined; "Gramene," a comparative cereal grain genomics database with rice as the referent species, was established; a good deal of activity in education at all levels hoping to excite students to enter the study of plant science occurred; and, there was increased international collaboration.[217]

On August 11, 2005, the rice genome sequence was published, the first crop plant in which the total DNA codes became known. In November, Machi Dilworth addressed the BIO AC reviewing the PGR program and NPGI to date. She provided an overhead slide that showed the financial commitment to the PGR from 1998, $40 M, through 2005, over $94 M, and called the original investment promoted so strongly by Senator Bond a "rare gift" for plant biology. Summarizing the many achievements of the program, Dilworth also noted that the "omics era" was opened for plant science.[218]

[215] "NSF Budget Request for FY97," BIO section, ppg. 197–198.

[216] Interagency Working Group on Plant Genomes, NSTC Committee on Science, "National Plant Genome Initiative: Progress Report," January 2004, Executive Office of the President. A "Dear Colleague" letter. Emphasis added.

[217] Ibid., pg. 1.

[218] Machi Dilworth, "The Plant Genome Research Program," a paper copy of a set of overhead slides used at the BIO AC meeting of November 17, 2005.

7.10 The Microbe Project

Mary Clutter had been the sparkplug for plant science at NSF beginning in the 1980s and was in a position to do a great deal about that once she became AD and led the transformation of BBS into BIO. She was not as involved in promoting microbial gene sequencing as she was for plants, but others at BIO clearly were. Julius Jackson was an early advocate and urged the BIO AC to consider the importance of the area in an address to them in early 1996. Both microbial biology and *microbial genome sequencing* were discussed as the genome sequence of the bacterium *H. influenzae* had been published just the year before—the first gene sequence of *any* living thing. A seed had certainly been planted in a number of minds, but while they saw the microbial connection as an exciting opportunity, the AC wanted to focus on *Arabidopsis* as the cress had "more focused and tangible benefits" at that juncture.[219] Jackson again reported to the AC a year later on the Life in Extreme Environments (LExEn) program that was an "NSF commitment to study microbes in extreme geophysical environments" and the strategies to do so. The AC had many penetrating questions, mostly negative, at least with regard to aspects of possible management structures for the program.[220] Finally, by mid-1998, the BIO AC supported microbial biology as an emphasis for FY 2000. An inventory of international activities in the area would need to be done, they said, to avoid duplication of effort, especially in microbial genomics, and an education plan would have to be established.[221]

It was not before 1999, then, that the Committee on Science, Subcommittee on Biotechnology of the NSTC, along with OSTP, created an interagency task force that made an "informal inventory of Federally supported research" into microbial genomics.[222] The findings suggested more work was needed and so The Microbe Project was created under an Interagency Working Group on Microbial Genomics (IWG-MG) in August of 2000. The combined involvement of some 12 federal agencies finally came to pass, NSF senior among them; and the BIO AC well on board.[223] Rita Colwell was the lead co-chair among the three and, by 2001, the plan for the

[219] BIO AC "Minutes," ppg. 3–4, April 25–26, 1996.

[220] Ibid., pg. 2 of the meeting of April 9–10, 1997.

[221] Ibid., pg. 8 of the meeting of April 6–7, 1998.

[222] See: http://www.microbeproject.gov/. A few sentences only are devoted to the historical underpinnings of the Project at the second page of the introduction beginning on the home web page. A full history is needed.

[223] Aristides Patrinos of DoE (and a central figure in the HGP) was present at the BIO AC meeting of April 1997 and mentioned the already extant Microbial Genome Project at Energy that had begun in 1994 noting that DoE was seeking a $7 M increase in support from Congress. They had just then undertaken a new initiative on bioremediation research. BIO AC "Minutes," pg. 3, April 9–10, 1997. See also: http://microbialgenomics.energy.gov/pubs.shtml; and the first page at the website http://microbialgenomics.energy.gov/brochure.pdf

project was issued as an NSF publication under the imprimatur of the Subcommittee on Biotechnology of the NSTC.[224]

Exactly what role BIO would play in the NSF portion of that broadly constituted group was left to a BIO AC Workshop to produce a white paper for guidance. It was convened in August of 2000 by Claire Fraser of The Institute for Genomic Research (TIGR)[225] and John Wooley, by then gone from BIO and established at UC, San Diego.[226] Along with the two conveners, 16 other scientists participated directly. They included "a cross-section from the fields of microbiology, environmental microbiology, evolutionary microbiology, microbial genomics, and bioinformatics" from the academy, industry and the non-governmental (NGO) communities.[227] Four observers, including Maryanna Henkart, NSF Director Colwell, Mary Clutter, and Matthew Kane (then PD for BIO's Systematic Biology Program) were also in attendance. The findings of the committee were in many respects as might be expected in the milieu of the rise of genomics in that period, and of NSF/BIOs responsibilities generally to the community. The committee recommended graduate training and educational outreach, both typical of most programmatic/project efforts of BIO. But other suggestions were more specific to genomics of microbes. "[R]eadily affordable technologies for genome-enabled science" were called for, as were "tools for handling ever-increasing amounts of functional genomics data." Automated genome analysis tools were to be created by funding from BIO/The Microbe Project. The "complete genome analysis of cultivated prokaryotic species to achieve phylogenetic breadth" was sought as were rDNA "genome inventories" from various environments where individual species could not be cultivated (metagenomics). Viruses and eukaryotic protists were to be included, as well. Finally, four questions were to be answered:

How are genes distributed among organisms, and why?
How do genes define the interactions of organisms with the environment?
How is…diversity between and among organisms achieved and maintained?
How do organisms interact with each other in the environment?[228]

Already, by the turn of the millennium, over 120 microbial genome sequences had either been completed or funded, the report noted, but there are estimated to be

[224] See: http://www.nsf.gov/bio/pubs/reports/microbe_project/mp_2001.pdf

[225] See: http://www.jcvi.org/cms/press/press-releases/full-text/browse/5/article/the-institute-for-genomic-research-tigr-j-craig-venter-institute-j-craig-venter-science-founda/?tx_ttnews%255BbackPid%255D=67&cHash=d25e6052b9. The Institute for Genomic Research (TiGR) became, in 2006, the J. Craig Venter Institute (JCVI) with the consolidation of a number of his related ventures, as demonstrated at the foregoing website.

[226] Claire M. Fraser and John Wooley, "The Microbe Project: A BIO Advisory Committee Workshop: Report of a Workshop on Microbial Genomics to Develop Recommendations for the BIO Directorate of the National Science Foundation," BIO, August 10–11, 2000; NSF 01–208; 24 ppg., with self covers.

[227] Ibid., pg. 21.

[228] Ibid., ppg. 5–6.

several billion species of microbes on Earth. Most of the 120 were human pathogens, with but few, free-living, non-human pathogenic species targeted. The approximate cost to analyze DNA at $1.00/base, cited earlier above, had dropped to about $0.20 in large batches by 2000. Some 99% of protein- and RNA-coding genes could routinely be found and recognized in any new prokaryote sequenced by that time, too. Determining the function of those proteins and RNAs was the greater challenge, though. A major problem, as well, was the lack of coordination in the bioinformatics community and the need for "standards for genome annotation, curation, and database structure." These did not yet exist and, as alluded to repeatedly in this chapter, such was a significant difficulty for the sequencing community. For these reasons, the recommendations from the committee to BIO included, yet again, emphasis on biosystematics/bioinformatics' tools suitable to genomics research.[229] More will be said about microbial biology in Chapter Nine.

7.11 Closing Views of a COV

A number of Committees of Visitors examined various of the programs and clusters in the Division of Molecular and Cellular Biology during the 1990s and early 2000s.[230] One, however, undertook study of the full division and made its report in 2005 looking back over the 3 years from 2002 through 2004: MCB had "performed at an exemplary level in pursuit of its basic mission."[231] The established "research portfolios…appropriately address[ed] a broad range of NSF and national objectives," the Committee stated. But maintaining the level of excellence that MCB demonstrated would be difficult in coming years, the COV argued, due to declining funds. Biological research (not just for MCB; and excluding environmental areas) was over $512 M in FY02, but only about $437 M for FY03. By FY05, it had rebounded to approximately $578 M, a figure not yet seen by the COV at the time of its report.[232] However, federal funding for all science and engineering R&D had flattened in 2004 (and stayed that way until, at least, 2008).[233]

The COV also worried about maintaining excellence at the division due to the increased staff workload. A larger number of proposals were coming in and there

[229] Ibid., ppg. 7–8.

[230] For instance, a list of all *planned* COVs from FY01-FY04 was published in the NSF BIO GPRA "Performance Report for FY 2001," of December 1, 2001, pg. 19 and covered each of the four divisions per year and groups of clusters to be included. In FY01, for example, the Plant Genome Research Program was assessed by a COV, as was the LTER of DEB, which included three clusters. See Chapter Ten for a discussion of that COV.

[231] "Committee of Visitors Report on the Molecular and Cellular Biosciences Division of the BIO Directorate," 2005; typescript using a standard form approach for BIO COV reports.

[232] Ronald L Meeks, "Federal Funds for Research and Development: Fiscal Years 2002, 2003, and 2004," Volume 52: Detailed Statistical Tables, NSF, February 2005.

[233] See: http://www.nsf.gov/statistics/infbrief/nsf09318/?WT.mc_id=USNSF_178

was an "inefficient workflow in danger of overwhelming the MCB staff," the Committee said. That could also "hinder recruitment of outstanding rotating Program Officers." Finally, "maintaining open lines of communication" was an area the COV saw as a problem that affected organizational lines from one level to the next within MCB, as well as BIO more generally. The Committee came to those conclusions by closely examining some 44 awards and 37 "declinations" (or more properly, declines; proposals not awarded) from a total of 3973 over 3 years. Statistics showed that only 788 were awarded and 3185 were declined—a low success rate.[234] There was also worry about morale in a division under such a heavy workload, yet the POs did not want to change the "labor-intensive review methods." They were particularly opposed to the "NIH-style triage or ranking systems" (earlier chapters).[235] As well, the cluster system "has provided a layer of obfuscation that makes it difficult for PIs to identify program managers and panels," making extra work for the POs.

It was a time of much ramping up in workload at BIO with the rapid expansion of efforts in genomics and even more cross-division and cross-directorate activity. Lines between divisions were, indeed, blurred as the nature of the "New Biology" was such that interdisciplinarity was becoming the driving issue in doing life science. The same situations obtained at IBN and DEB, as shall be seen in the following chapters.

[234] COV, 'MCB Report,' ppg. 1–2.

[235] Ibid., pg. 8.

Chapter 8
Integrating Biology

Make no little plans. They have no magic to stir men's souls.
David Burnham

8.1 After the Reorganization

David Burnham, the architect of Union Station in the nation's capital, may well have had a good point about making "no little plans." Surely, the very considerable plans made to create the new IBN could stir BIO with a new and dynamic mix of disciplines, if not souls. The new division brought *integration* not only into its title but into its structure and management and for the community it served—though only to some degree. The brain is the great integrator for many phyla of animal life, and the nervous system, more broadly, plays a critical role throughout most of that kingdom. Supraorganismal integration occurs in many other ways, too, and at many levels within the organism and both above and below the level of the single creature, whether animal, plant, or microbe. For that reason, IBN dealt not only with animals possessing nervous systems but with plants lacking them and microbes commonly composed of merely a single cell, though working colonially. Integration? I shall return to this notion later in this chapter.

This, the third of the four divisions to issue from the great reorganization was, arguably, the most highly reconstructed of the group because sections of the new Division of Integrative Biology and Neuroscience (IBN) came from widely dispersed ancestry within BBS. And, later, that would provide, in part, some of the roots of a long tension that would finally precipitate IBN's replacement in 2005. Figure 5.1, of earlier, shows assembly of origins and suggests how, especially, the neurosciences were reconfigured; parts of the older BBS structural elements were placed in the new BIO, while others went to SBE, with related sociocultural disciplines, as has been recounted.

Bruce Umminger would be the Director of the New Division for most all of the last decade of the second millennium before leaving to take the leadership of the

D. J. McGraw, *Millennial Biology: The National Science Foundation and American Biology, 1975-2005*, https://doi.org/10.1007/978-3-030-56367-7_8

Office of Science and Technology Centers in 1999 and his full departure from BBS/BIO.[1] As the new decade got under way, Umminger's deputy was Thomas E. Brady[2]; Carter Kimsey became the senior staff associate. By 1995, Brady would become Mary Clutter's Executive Officer, a position Umminger had also once held, and would be replaced by John Fray as DD. Three other staff also worked in the division office.

Programmatically, IBN was endowed with three areas at its creation in 1991: Developmental Biology (changed to Developmental Mechanisms by 1995), under Judith Plesset, a permanent PO; neuroscience, led by Kathie Olsen (another permanent PO, who replaced Brady when he moved into BIO's directorate office; she would eventually become Deputy Director of the Foundation); and, Physiology and Behavior (P and B), under rotator Vincent Gutschick. The last of the three clusters would become the most problematic in an organizational sense. Plesset led the plant and microbial development portion of her program with Sarah Bruce, a rotator, as PD of animal development. Well into the present century, only two permanent persons would occupy the development cluster. Olsen dealt with neuroendocrinology, while five rotators headed up areas from Neural Mechanisms of Behavior to Sensory Systems and three more. (The general Endrocrinology program was in the "P and B" cluster.) The Physiology and Behavior Program saw Gutschick guiding the Functional Physiological Ecology area and long-time NSF employee Fred Stollnitz that for animal behavior. Permanent PO, Machi Dilworth, dealt with Integrative Plant Biology but would become Division Director of DBI from fall of 1997.[3] Two other individuals oversaw yet two more areas in the program. In total, the first incarnation of IBN had 15 scientific staff, 4 to 5 permanent, and about 11 rotators.[4] Finding rotators every year or two was a "very time consuming aspect of my job," Umminger remembered, but he said that the "rotator system, I think, is beautiful." Others have felt similarly, as has been seen in previous chapters. It was not without its downsides, however, as we have seen and shall yet see.

Many of these IBN scientists/administrators have been met in earlier chapters as their respective presence at NSF predated the great reorganization. Yet others were rotators or newer permanent employees. Typical of the period when clusters grew to prominence, permanent persons within the directorate wore more than one hat, that practice continuing into the future. For instance, Dilworth also held a leadership role (by 1996) in the Research Resources Cluster/*Arabidopsis* Genome area, as well as that for the Plant Biology Partnerships Program. Gerald Selzer (of DBI) not only was the Program Director for BIO's part of the Science and Technology Centers

---

[1] He did, however, take a sabbatical from 1993 to 1994 and worked with the renowned biologist Tom Lovejoy at the Smithsonian Institution on biodiversity. Umminger interview, *op. cit.*

[2] Brady was, at various times, dean of life sciences at both the University of Texas, El Paso, and at the University of New Hampshire, Durham, as well as DD for the IOB (successor to the IBN) at NSF.

[3] BIO AC, "Minutes," October 20–21, 1997.

[4] Umminger interview.

Program but PD for the Training Cluster/Research Training Groups and so on. Permanent, senior leadership within BIO was, and continues to be, lean.

By the end of the millennium, James Edwards had assumed the Deputy AD position in BIO, and numerous other programmatic and area assignments had come to exist, though often with individuals drawn from the main four divisions to additionally fill those positions having directorate-wide duties (examples just above). Such would include computation and several plant science specialties, as just two examples of those assignments. While there seemed to be a certain aura of musical chairs in all of those revolving positions, they served the purpose of allowing for the gaining of great breadth in each permanent employee. Indeed, some, such as Judith Verbeke in later years, would serve in a great many roles throughout IBN and its later successor divisions, let alone acting as the first director of Emerging Frontiers. Even as recently as 2004, the "front office" of BIO was still composed of a small staff of about a half dozen persons.

IBN would retain its title into 2005 but would become the Division of Integrative Organismal Biology (IOB), by June: the term neuroscience was no longer seen in the name of the division. But in 1991, Umminger pointed out that he and others "made sure that the neurosciences cluster [was placed] in the integrative biology [division]."[5] Programmatic areas would also be heavily altered again in 2006 and, not much later—after Jim Collins had replaced Mary Clutter—IOB would become the Integrative Organismal Systems (IOS) division as of the time of this writing. Those changes all gave continuing evidence of the ever-increasing levels of integration at BIO and, in many respects, harkened back to both Weiss' *functional* mandate to the NSF of the early 1950s and also to some aspects of an even earlier integrated natural history of the deeper past. However, there was a long-lived general tension about the multifaceted nature of IBN too, and when IOB replaced it in later years, it was believed that a certain unrelatedness of the clusters was eliminated. In fact, one cluster that was seen to be much too polyglot was that of Physiology and Behavior, "P and B," of the early 1990s. This issue is discussed in greater detail below. Suffice to say at this point that one member of a Committee of Visitors said in early 1993, after her review of IBN P and B, "[w]e visitors found the cluster organization a slippery concept."[6] It can be said more positively, however, that if one traces the names of various divisions, programs, and other areas within BBS, but especially BIO, the functional integration nature of the National Science Foundation's biological sciences directorate grew more obvious with time. Though, as seen in the P and B cluster's history, it seemed to have been a bumpy road.

Integrative biology was, about 1990, becoming an intriguing new term in the community, in many respects replacing organismal biology. Though not purely a synonym, the word had considerable advantages and was rapidly being favored by members of the American Society of Zoologists, ASZ, which changed its name to

[5] Ibid.

[6] A university genetics professor to Bruce Umminger, April 6, 1993. Names of living persons are not used in connection with program reviews in this study.

the Society for Integrative and Comparative Biology at that same time. The Society had been Umminger's primary disciplinary society—he studied comparative physiology—and, upon joining NSF, he brought a strong view of organismal biology to his initial position. When the great reorganization occurred, he felt "a strong bent on protecting" the field. As ASZ was in the midst of its own name change, he supported the use of the term *integrative biology* for the NSF: a classic example of the *intricate dance* in which both partners determined to use the same term for the same specialty in biology. He later recalled that "in the eight years I was division director, we made a real case for integrative biology."[7]

At that same time, to Marvalee Wake, the choice of divisional title was essentially clear. She was in the UC, Berkeley, Department of Integrative Biology and was also deeply involved in the NSF reorganization as a member of the BBS AC. Both Berkeley and BBS had been prescient, but quite precocious, in use of the term because it would not be until during the early years of the current century that Wake would publish numerous articles on the nature of integrative biology.[8] In fact, it would not be until 2001 before Wake would say, "integrative biology suddenly has cachet."[9] Just 3 years before, "there [was] no consensus as to what 'integrative biology' really means and how it may be done."[10] Nonetheless, BIO had its new IBN by 1992 and, as such, became a leader on the dance floor for the old/new discipline of integrative biology. Still, with the smaller reorganization of IBN in 2005, the word integrative was not lost, and the term organism even returned.

But IBN was in essence a chimera, which directorate not only began defining a new field but also protected and included an old one, neuroscience. In earlier chapters, we have seen that the history of neuroscience, neurobiology, and its other

---

[7] Ibid. See also Keith R. Benson and Brother C. Edward Quinn, "The American Society of Zoologists, 1889–1989: A Century of Integrating the Biological Sciences" *American Zoologist* (1990):353–396. As the name change to Society for Integrative and Comparative Biology (SICB) came just at the cusp of when Benson and Quinn published, they did not include a discussion of the thinking behind that shift, but it is known from numerous details found during my preparation of this history that the ASZ, and the SICB later, had a close relationship with NSF BBS/BIO. It is possible that mutual discussions were had concerning the virtual simultaneity of the name changes given that Umminger, and others, was a member of ASZ and also affiliated with NSF in varying ways. For example, Marvalee Wake was a member of the ASZ/SICB (see: http://ib.berkeley.edu/labs/mwake/curriculum.html). The *American Zoologist* was the organ of the ASZ and remains that of the SICB. Quinn died suddenly the year this article was published. On a personal note, I had the honor of sharing an office with Benson when both of us were graduate students of the historian of biology Paul Lawrence Farber; I was his first doctoral student.

[8] Marvalee Wake, "Towards an Integrative Biology: A Symposium for the American Association for the Advancement of Science," *Biology International* (2000), Number 38; "Integrative Biology: It's Promise and Perils," *Biology International* (2001):71–74; "What is Integrative Biology?" *Integrative and Comparative Biology* (2003):239–241; see also an earlier work by Camille Ripoll et al., "Defining Integrative Biology," *op. cit.*; and, further, the International Union of Biological Sciences programmatic efforts in integrative biology at: http://www.iubs.org/newiubs/actions/current_prog.php.

[9] Wake, "Promise and Perils," pg. 71.

[10] Ripoll, "Defining," pg. 19.

incarnations had a more than precarious time at the NSF sitting on that fence, as it does, between the biological and the social sciences. "It is a long stretch between the cellular and the thought," and that gap needed bridging in what would become IBN.[11] Umminger was committed to the field, so he saw the new IBN as the sure way to keep the neurosciences highly visible in the nomenclature of structures within the Foundation. Neurosciences would remain so through 2005 and only then be subsumed by full integration.

But in order to determine where all the major and minor research—programmatic—areas, fit within the nascent IBN, a strategic long-range plan was called for in order to set the directions for the new division. At the end of September 1992, such a plan for the future of the division had been drafted.[12] It was thoroughgoing, but it made a number of matters very clear, most especially two. "IBN forms an integrative link between BIO's [MCB] on the one hand and its [DEB] on the other." Further, "IBN serves as BIO's major link to the [SBE]." Those two statements set the place for IBN and the nature of its organizational—and a part of its intellectual—relationships. For the first of those statements, IBN linked areas where "[g]enetic, biochemical, and cellular approaches often drive the developmental, physiological, and neuroscience studies" of the division. Beyond that:

> [t]he study of physiology, environmental adaptation and performance provide a framework for exploring how basic genetic, biochemical, and cellular mechanisms may have evolved, and how they influence evolutionary pathways or the interactions of organisms with their environment.

BIR/DBI was mostly a *service* division of BIO, but DEB and MCB were both primarily *content* areas "linked" at the organism—the intersection of biology of the big and biology of the little: that linkage was the content area of IBN. While that view of IBN as a bridge served in the 1990s, by the early 2000s, the perception of it had changed within both the internal and external community of biologists. That more recent vision was one of IBN being more of a *catchall* than anything else, and the matter would lead to the division being reorganized into IOB.[13] Those roots can be seen in the POs responses to one of the earliest COV reports that came out for the Physiology and Behavior (later Physiology and Ethology) cluster.

The strategic long-range plan of the early 1990s was divided up into a series of major aspects followed by listings of goals and critical achievements. Though too detailed a document to convey properly here, suffice to say that, in part, the plan was as follows. "Intellectual integration: discovery, synthesis, application" was the first of the major points. The goals within it were familiar efforts of the directorate, and the Foundation itself: support base programming and seek the most "creative science regardless of topic." In the process, "identify, encourage, and support new developments" in the areas covered by IBN. The community was, as always,

---

[11] Christopher Platt Interview with the Author, February 19, 2009.

[12] "Division of Integrative Biology and Neuroscience Strategic Long Range Plan," September 30, 1992, draft, nine-page typed document.

[13] Judith Verbeke interview, *op. cit.*, Chapter 8 above.

encouraged to bring forth ideas for new initiatives, but the division was to seek out similar thinking and promote it.[14] Other goals pursued the notions of just what IBN was—*videlicet* the two sentences above concerning linkages—and so there were promulgated a large number of goals and critical achievements. The latter included the more mechanical aspects of both defining and running the division. As but one instance, "[e]ncourage PIs to address possible technology transfer and industrial interest in their research in their annual progress reports."

That first full year of operation of the new Division of Integrative Biology and Neuroscience was productive. Indeed, it gave the largest number of awards of the four divisions in 1992: 1113, as compared to MCB at 1049 and with both DEB and BIR running distant seconds at 878 and 235, respectively. Though the funding levels for the MCB grants, in toto, exceeded that of IBN slightly, $83 M to $77 M, it was clear that the new linking division was itself a significant member of the group within BIO.[15] What did IBN do and become?

8.2 The Neuroscience Cluster

The National Institute of Mental Health (NIMH) of NIH provided funding in 1989 to the NAS's Institute of Medicine to determine whether the state of information technology could offer tools mature enough to study something as complex as the human brain. The NSF joined in that effort and, along with the National Institute of Drug Abuse (NIDA) brought out, in 1991, the report "Mapping the Brain and Its Functions: Integrating Enabling Technologies into Neuroscience Research."[16] The findings led to NIMH creating the Human Brain Project (HBP) in 1993. NSF, however, was already well into the general subject by being a partner in the overall major involvement: the 1990s had already been proclaimed by Congress as "The Decade of the Brain."[17] A Memorandum of Understanding had finally been signed by all of the NIH institutes, by Mary Clutter for NSF, and by others for Office of Naval Research and NASA in 1993.[18] Grant applications were to be sent to the

[14] The question of who led on the dance floor in the partnership between NSF's biology directorate and the community it served has been an overt one in some places in the main text above and in other chapters more implied. While views on the subject varied with different persons, Umminger felt that NSF was more reactive than proactive in the intricate dance. Indeed, he felt that view quite strongly; Umminger interview. Others to whom I have spoken felt just the opposite, and with many averring it was somewhere in the middle ground.

[15] National Science Foundation, "Annual Report, 1992: A Retrospective of Selected Research Funded by NSF," Appendix B, pg. 38.

[16] C.M. Pechura and J.B. Martin, "Mapping the Brain and Its Functions: Integrating Enabling Technologies into Neuroscience Research," (Washington, D.C.: National Academies Press, 1991), *passim*.

[17] BBS, "The National Science Foundation in the Decade of the Brain," NSF-91-58, June 21, 1991, ten pages.

[18] "Memorandum of Understanding for the Human Brain Project," 1993; a brief typescript

Division of Research Grants of the NIMH, which would be the lead agency. Some 50–70 grants were to be supported each year and awards could be 1–5 years in length and could run from $50 k to $1.1 M per year. It was a large project in comparison to many at BIO, but the NIH "empire" was always the wealthier federal brother, by far, and so covered the lion's share of the grant support.

Again, however, NSF was quick at the starting gate as near the same time that the congressional proclamation for the Decade of the Brain was made, BBS had issued a workshop report: "Bio-Control of Neural Networks." The meeting had occurred in May of 1990, and the report came out in September of 1991, making the Foundation an early player among federal agencies involved in the brain decade. In addition, though slightly later in time, the US/European Commission Biotechnology Task Force, co-chaired by Mary Clutter for the United States, led to contacts that established the US/European Commission Committee on Neuroinformatics in 1996. It would be co-chaired by Stephen H. Koslow, who had been chairing the HBP as a figure at NIMH. All of those activities led to a considerable attack on the neurobiology of the brain, as well as neuroinformatics via the World Wide Web, so that by 2004, the HBP was supporting $100 M a year in funded grants, mostly funds from the "empire," but not without NSF being the lead agency for basic research.

A second opportunity arose that cemented together neuroinformatics and brain research when OSTP requested that all federal agencies meet at NIH to promote more pan-global research efforts for the benefit of those in a number of audiences. That was part of the international Organization for Economic Cooperation and Development's Mega Science Forum (later to become the Global Science Forum). The response to the former from the United States then became one in which NIH and NSF joined forces.[19] Jim Edwards of BIO shared biodiversity data with the international biotechnology Task Force, and NIMH's Koslow developed neuroscience data under the new megascience committee he was now chairing. Indeed, the first, or at least a very early, use of the term *neuroinformatics* appeared in that connection, especially for the European community, and soon followed in the United States.[20] In America, the term *biological informatics* (or *bioinformatics*[21]) had been in coinage since the 1970s and had broader content: it originally derived from a Dutch usage. Matters of international management for neuroinformatics spiraled upward from there.

---

document.

[19] "Revised Work Plan following meeting of June 6–7, 1996: Megascience Forum Working Group on Biological Informatics." A six-page, typed document with no clear authorship, but obviously NSF (BIO?).

[20] Wikipedia, "Neuroinformatics." See also Stephen H. Koslow and Michael F. Huerta, eds., *Neuroinformatics: An Overview of the Human Brain Project* (Mahwah, New Jersey: Lawrence Erlbaum Associates, 1997).

[21] See Paulien Hogeweg, "The Roots of Bioinformatics in Theoretical Biology," *PLoS: Public Library of Science*, (2011):Issue 3. Hogeweg and Ben Hesper used it in 1978, but had also used it as early as 1970 in an article in Dutch "not generally accessible."

Returning to the earlier period of the NSF workshop, that effort sought to bring together a perceived drift in engineering research in artificial neural networks and that of physiological control of neural systems. The question of the state of maturation of the biological networks was examined as was the relationship between the biological and the inanimate networks being pursued by the engineering community. Defining what blockages might exist to inhibit advances on either or both fronts was also pursued. Some 32 invited participants were present from NSF, NIH, and the respective engineering and biology research communities. There was a strong orientation toward bioengineering, prosthetics, modeling, and similar matters. Neuromuscular and cardiopulmonary systems were of particular interest. In closing the workshop, it was recommended that more "cross-fertilization" was needed and both postdoctoral and sabbatical support should be made available, not unexpected conclusions.[22] A number of other reciprocal ways in which engineering and biology might aid the opposite field were also suggested. Many of those ideas had both strong theoretical (e.g., modeling) and mechanical (sensors, actuators, and similar) features.

The perpetual mandate for NSF, basic research, implied that much of the work funded would be just that and in model (non-human) systems. However, while NIH clearly had the mandate for human neurobiology, the Foundation had both the social and select areas of the behavioral sciences lodged within the about-to-appear new Directorate for Social, Behavioral, and Economic Sciences, SBE. One wonders if it did not either amuse, or possibly irk, the still then BBS scientists, now moving over to SBE, that while BIO would be going its own way, the Decade of the Brain project implied the necessity for *immediate cross-directorate interaction* to forward new shared programmatic efforts! An irony, at the very least. More conspicuously, the DDs, PDs, POs, and senior leadership of BBS had all come to accept that two new directorates were about to appear on the Foundation landscape. More so than in any organizational chart of the restructuring, one could see the nature of the BIO/SBE split with regard to programs and how they were conveyed in print in NSF's report on its role in the Decade of the Brain.[23]

Bruce Umminger has said that the "last two unsolved problems in biology were how organisms develop…and how…the brain works."[24] (While many in the biology community at that time may have agreed with that assessment, much has happened since that would serve to challenge and even negate that view, certainly with respect to the notion of the "last two."[25] The genomics revolution, so pervasive in *Millennial Biology*, was in its infancy, for instance.) Whatever level of accuracy there was in Umminger's thinking, it was likely that the community overall believed that a Decade of the Brain was a very reasonable thing to proclaim. He was also one of the

---

[22] George A. Bekey and Peter G. Katona, "Bio-Control of Neural Networks: Summary of a Workshop, May 16–18, 1990," NSF-91-17, September 1, 1991, 18 pages. Pg. 8.

[23] "NSF in the DoB," *op. cit.*; *passim.*

[24] Umminger interview.

[25] Personal knowledge.

directorate's permanent staff that was acutely aware of the elephant in the room (NIH) when issues such as research on the central nervous system came up.

Concerning Ummimger's keen desire to see neuroscience in the forefront of the newly established IBN, it can be said that he was of the opinion that Mary Clutter had minimal interest in neuroscience.[26] In fact, Kingsbury had said the same thing of her a few years before (earlier chapters). NIH was spending around $2 *billion* and NSF could only spend around $30 *million* on the subject in the 1990s. Great differences in agency budgets were not necessarily the only explanation for that, either. As evidence of Clutter's presumed views, Umminger noted that over his 8 years as IBN chief his "staff went from six to four program officers[, management] always grinding away at the [number of] people in the neuroscience cluster."[27] He said that "I kept that cluster alive through that whole period,"[28] possibly suggesting an air of desire on Clutter's part to restructure IBN even earlier than when it next occurred division-wide in 2005. The opinion, accurate or not, by Umminger that Clutter was *not* an enthusiast for neuroscience, if true, may have played a role in the elimination of the term in the title of the IOB which replaced the IBN in Clutter's last year of 2005. Holding the cluster together was, in part, to keep grant proposals that intended to investigate cellular aspects of the brain, as but one example, from being directed to MCB and the cell biologists there for adjudication when, as Umminger felt, neuroscientists should be the judges of such proposals and managers of such awards. He was, in that connection, opposed to the level of reductionism that he sensed was a strong feature of BIO at the time. He further assumed that, after he left the DD position in 1999, fragmentation ensued for the neurosciences.[29] His assumption was correct. The move to IOB, in the titular sense, was not the only feature of such changes: the programs were, in fact, restructured, and the neurosciences were dispersed within the new division.

When Umminger became director of the STC Office in 1999, he was followed as IBN division director by Grace Wyngaard, a rotator from James Madison University, Harrisonburg, Virginia. She was an invertebrate geneticist who had come to be Umminger's deputy division director in 1999 and stayed in that position until 2000. Judy Verbeke, a plant biologist and a former PO, followed her for a few months before the arrival of the new IBN division director, Frank Greene, another plant biologist and former PO, who came to NSF as a rotator from the USDA.

But in the early 1990s, as a youthful IBN became a force within the new BIO, the neurosciences were on an upward swing: the Society for Neuroscience already had about 20,000 members at the time and was adding an increasing percentage each year.[30] By the end of 2003, it had climbed to 33,000 members, as Christopher

[26] Umminger interview.

[27] Platt interview. Platt indicated that it was 2001 when the number of POs dropped from six to four.

[28] Umminger interview.

[29] Ibid.

[30] Christopher Platt to Bruce Umminger, "Neuroscience Opportunities for FY95 re Infrastructure," May 18, 1993.

Platt had told Bruce Umminger in the former's capacity as one of the two leading persons in neuroscience at BIO.[31] There were also four rotators in neuroscience for much of its life.[32]

Platt had come to the Foundation with a strong background in neuro*biology* specifically, as opposed to the more purely psychological orientation.[33] He completed his bachelor's degree (1966) at the University of Chicago (after his first 2 years were spent at the University of Michigan) and finished his doctoral work at, what he termed, the "fabulous" Scripps Institution of Oceanography (of UC, San Diego) by 1972. That was accomplished under an NSF fellowship. While there, Platt and his fellow graduate students took a course in a subject then unfamiliar to any of them—it was termed neuroscience. A department for that discipline was being set up at UCSD in the late 1960s by three visiting professors from UC, Los Angeles. Platt then followed his Scripps Ph.D. by 4 years of postdoctoral work at UC, Berkeley, where he learned electron microscopy to expand his knowledge of inner ear balance mechanisms, his specialty. Platt's interests in that area led him to use fish as model systems: "they don't have any neck and they don't have any legs," he said, so they made good test subjects for such studies (both behavioral and physiological). He then became an assistant professor of biological sciences at the University of Southern California, Los Angeles, for several years before moving east. In 1983, he joined Georgetown University in Washington, D.C., and was at a neuroscience meeting in Boston that fall. He dropped by the NSF booth, and in a casual conversation with Nathaniel Pitts (Chapter 2), it was suggested that Platt ought to come to NSF as a 2-year rotator to aid in the Sensory Physiology and Perception Program in BBS. Having expressed interest, the following year Platt got a call from NSF offering him a rotator position. Still new at Georgetown, he sought the University's thinking on the temporary move, and they were in agreement. Platt came to the Foundation in spring 1984 and ended up spending a great portion of his career there. He would have normally returned to Georgetown after his 2 years at BBS, but some eight of that University's faculty and their staffs, with whom Platt had just started to work on shared research interests (sensory neurobiology), decided to leave Georgetown that had left Platt uneasy. He began looking around for another place to go and was offered a third year as a rotator in 1986. Such a situation was not common at the time, and it had to be arranged by Richard Louttit (Chapter 2), then Division Director for Behavioral and Neural Sciences. After that third year, Platt became permanent at NSF in 1987.

James O. Larimer,[34] a rotator, who was an expert in psychophysics of human color vision, was PO for sensory physiology and perception (SPP) in 1982–1984. He had become deeply involved in NSF's new cross directorate initiative on

[31] Huda Akil (Society president) et al. to Rita Colwell, NSF Director, October 7, 2003.

[32] Christopher Platt personal communication with the author June 13, 2012.

[33] Platt interview.

[34] James O. Larimer is not to be confused with James L. Larimer, also a neuroscientist, who spent the greater part of his career at the University of Texas, Austin, and was never a rotator at NSF.

supercomputers and was now swamped with those duties along with the large SPP proposal load (over 100 in some review cycles). Platt's arrival in 1984 added needed help, as well as expertise on ear (hearing and balance) biology to complement Larimar's strength in visual psychology. Larimer and Platt split the total and usually had about 60 proposals to review at any one time. With Pitts, Louttit, and Alan Leshner all present in the BNS of that period, Platt held that it was a "rich [intellectual] environment." When Larimer left (later 1984) and the reorganization later occurred, Platt became PD of the new Sensory Systems Program in the equally new BIO. The later 1980s and early 1990s were a period in which NSF was building its neuroscience program, he recalled. Thus, under Bloch's strong bent toward sabbaticals, still a part of NSF well after the engineer's departure, Platt engaged in more research on the inner ear at the University of Maryland in 1991–1992. In the event, he missed the concluding period of the great reorganization, when Kathie Olsen rose to lead the cluster of six neuroscience POs.

Platt averred that the 1990s saw neuroscience expand very much into cellular, molecular, and developmental areas thanks, in great part, to new technological innovations. "It was a pretty heady time," he reminisced.[35] Such new directions brought a great deal of interdisciplinarity to bear as mathematicians, computer scientists, and chemists, for instance, were crossing the boundary into neuroscience, Platt stated. The Chemistry and Life Processes Program begun under Kingsbury had a strong component in research on neurotransmitters, hormones, and neuromodulators by the early 1990s, with Olsen and Platt as neuroscience representatives to the group. Interest in neurotransmitters would become an extremely important part of biology by the new millennium.

Platt recalled that by the mid-1990s, the neuroscience area had examples of co-funded grants in *every one of the NSF directorates of the time*! IBN was a *functional bridge promoting funding of multidisciplinary projects* and helped to bring scientists from other fields into the neurosciences.[36] But of considerable note was the necessary relationship between neuroscience in BIO and a number of areas within its sister directorate, SBE. It was decided during the reorganization, as has been alluded to earlier, that splitting the former BNS would necessarily *not* please all audiences, either within or without NSF. The split for anthropology, as an example, was done by moving *all* of the discipline's subfields to SBE, including physical anthropology[37] that, almost all agreed, rightly belonged to biology. In the case of

[35] Patt interview.

[36] Platt communication with the author June 13, 2012.

[37] See http://uts.cc.utexas.edu/~bramblet/ant301/one.html, where physical anthropology is defined as follows: "The fundamental questions addressed by physical anthropologists are 'What does it mean to be human?' and 'How did we become human?'…Two concepts provide the fundamental tools to understand the history of our species and our planet. The first concept – uniformitarianism, the idea that the physical processes that shape our landscape today and the biological processes that shape our lives are the same forces that operated in the past- offers a possibility that if we study and understand these processes, we may gain insights about events of the past and the magnitude of time they represent. The second concept is biological continuity, the thread of descent that connects all life. When living beings reproduce, they do not create life – they pass it along." Thus,

another area, and of significance to the present discussion, cognitive neuroscience (CN) moved, as Platt has put it, "lock, stock, and barrel" to SBE. Indeed, CN "took probably 20% of the [grant] proposals we had on perception" with them to SBE. The memory and cognition program, run by Joe Young, also went to SBE, and in that move, the psychological and behavioral aspects of human cognition were largely separated from the biological and neural aspects of cognition, studied invasively in animals, which remained in IBN. Because cognitive neuroscience has behavioral aspects, as well as neural aspects, understanding the relationship of brain to mind makes it difficult to separate the behavior completely from the biology by an arbitrary fence line. The new IBN neuroscience cluster contained behavioral neuroscience and sensory systems programs that included coverage of some cognitive and perceptually related projects, and IBN POs interacted with their counterparts in cognitive and perceptual programs in SBE; the two groups "always found things to talk to each other about," Platt recalled.[38]

Within BIO, the neuroscience cluster was viewed as something of "a clique, a force, with a power beyond what it should have," Platt opined.[39] That was, in part, due to the fact that they began those years with six POs, as Umminger had noted, but had them whittled down to four and then to three by 2001. Indeed, it may have been that the number of those POs was reduced due to the perception of the "clique," as suggested by Platt in his opinion of the views of others. In fact, he has opined that such an observation of the neurosciences was that they were a "drain" on BIO "funneling money out…rather than bringing stuff in." One evidence of their "power" may be found in the fact that the cluster, of its own will, set aside 10% of the budgets of all its relevant activities in order to support particularly interesting and hard to place grant proposals. Platt averred that Umminger, as DD, supported that approach.[40] Funding conferences was one example of the use of the set-aside monies. That 10% budget line was created by Kathie Olsen who headed the cluster in 1993 for a few years until Platt later took it over.

But with time, pressures arose in other divisions over the cluster creating the set-aside, and thus the funds themselves moved through a number of changing locations and definitions. They were tied to the STCs for a while and then to BIO itself.[41]

physical anthropology appears to be strongly biological in character, at least in this definition. In an undergraduate course I took in the subject, the definition centered around the notion that the discipline studied the physical remains (e.g., skeletal) of human (and earlier related groups) and their physical cultural remains (e.g., tools, pottery, structures, etc.).

[38] This paragraph is a mixture of information from Platt derived from both his interview in 2009 and his communication with the author in 2012.

[39] Platt interview.

[40] Ibid.

[41] One of several STCs supported via IBN's neuroscience programs was the Center for Cognitive Science at the University of Pennsylvania. Penn had a formal program in that field since 1978, which had been supported by the Sloan Foundation. In 1991, NSF awarded a grant to Penn to begin the STC. In later years, the Institute for Research in Cognitive Science grew out of the NSF support and became a yet larger entity. See http://www.ircs.upenn.edu/. See also a memo to the members of the NSB and materials from Penn regarding renewal of the original STC for FY94.

Along with the tensions about that fund, there had been the perception that neuroscience POs were seen to be something other than "good team players," and they did not "interact much with the rest of [BIO]," Platt conveyed. However, "I [Platt] said how proud I was that we were able to interact with all the rest of the Foundation." *Interdisciplinarity* is extremely well represented as a historical trend at NSF by the work of the neuroscience cluster in BIO, whatever other views may have also existed. Strangely, as Platt further stated, peers in BIO had said that the neuroscientists were not interdisciplinary enough. The response to that was that neuroscience was already one of the most interdisciplinary areas in biology as it depended upon so many other sciences for its existence.[42] Further, as noted above, the cluster was tied to all other directorates at the Foundation at one period, at least. Much of the foregoing is based upon an interview, so that must be taken into consideration.

But other views were present, too. Platt said of Olsen that "[o]ne of the things I valued tremendously about Kathie Olsen was her political skills."

> As a leader, as somebody who could just go and be right in front and articulate what needed to be done, and fight back. She had skills…I like to think that our pairing as the two permanent people in neuroscience for some time, along with Pitts…and Stollnitz, of course, who was a rock of knowledge, helped out [the building up of] neuroscience here.[43]

Given that Olsen eventually became Deputy Director of NSF, her political skills must have been very authentic and also clearly recognized.

The real or apparent independence of the neuroscientists at NSF had another component—the continuing struggle to make clear to various publics the reason that the Foundation should expend any funds and efforts on the subject when NIH was spending such vast sums on what appeared, to the lay eye, to be the same thing. That NIH was most concerned with human medicine in its ties to the brain, as opposed to the basic biology of neural systems being done by NSF, was very hard to explain to most observers. In fact, the cluster members created a small handout that attempted to clarify the confusion in the minds of those untrained in the area and pointed out that the research and other activities supported by NSF were "things NIH just didn't do," said Platt.[44]

But the "empire" *did* do things that NSF could piggyback upon, such as supporting leading software development for neuroscience databases. In 1994, NSF funded a series of two workshops on "Establishing Identified Neuron Databases."[45] Twenty-four scientists from outside of NSF attended, as well as all those program officers and others from within the neuroscience cluster in BIO. The work in molecular biology and especially in the early period of genomics was leading to the creation of various necessary databases in those fields, and it became apparent that neuroscience was also in need of such interactive data archives.

---

[42] Platt interview.

[43] Ibid.

[44] Ibid.

[45] National Science Foundation, "Establishing Identified Neuron Databases," April 28–30 and June 23–24, 1994. NSF Grant IBN-9411967.

> Access to information about specific neurons, either as individuals or as ensembles, is essential to the coordination and guidance of neuroscience research of all types: molecular and genetic analyses of events occurring within neurons, computational studies of functional circuits, and behavioral studies.[46]

One of the questions asked was whether "it is feasible to design identified neuron databases that would be freely available over the Internet." The Human Brain Project through NIMH, NASA, and others had already funded the creation of NEURON, software for such databases and then a database itself, NeuronDB, managed by Yale University.[47] As well, the GENESIS[48] software existed and the NSF workshop promoted the necessity of making stored data compatible with both the NEURON and GENESIS computer programs.[49] Clearly, The Decade of the Brain was experiencing considerable activity beyond NSF, but the workshops of 1994 at the Foundation sought to encourage work on *non-human* neural cells and ensembles, whether within *model systems* (e.g,, zebrafish or the roundworm *Caenorhabditis*) or at higher levels, such as whole classes and phyla (in this case, insects and mollusks, respectively). That *comparative* neuroscience included "looking at weird beasts like… deep sea shrimp…[that] NIH wouldn't be interested in," Platt held. Indeed, those in computational neuroscience could not "even get in the door there [NIH]," and that led to the fact that "we [BIO] finally developed a computational neuroscience program precisely because of that."[50] So computational neuroscience became one of the six programs in the neuroscience cluster that also included Behavioral Neuroscience, Developmental Neuroscience, Neural and Glial Mechanisms; neuroendocrinology, and sensory systems. Collaborative research that had features of interdisciplinarity and split funding with other directorates was promoted by the neuroscience cluster as "launching platforms" for various investigators, but not as long-term support as was typical of NIH, Platt stated.

Challengingly, by 2001 when the PO numbers had dropped, the total number of programs in the cluster still stood at six. For each of the then three POs, it was necessary to handle not only two programs apiece, but often times such management was required in areas of neuroscience unfamiliar to those POs, virtually all of whom were rotators. Platt had served in 1999–2000 as Acting Deputy Division Director for IBN under Wyngaard, but when he came back to neuroscience, he faced just such a situation by managing, for instance, the program in neuroendocrinology, which was not "even [an area] that I'd been trained in."[51] At the same time, proposal numbers were increasing for the neurosciences, and NIH was in the midst of a massive increase in its own budget with the intent to double it just after the turn of the

[46] Ibid., pg. 1.

[47] See http://senselab.med.yale.edu/neurondb/.

[48] For more on GENESIS, see: http://www.genesis-sim.org/GENESIS/genesis.html.

[49] "Establishing…Databases," pg. 7.

[50] Platt interview.

[51] Ibid.

millennium.[52] In 2003, Platt left NSF for NIH and the leadership of a program involving balance and the inner ear at the National Institute for Deafness and Communication Disorders. He had become "burned out…but never bored" at NSF/BIO where he, like many other POs, often worked numerous extra hours in evenings and on weekends.[53] Though highly dedicated to the Foundation at the time, he left to a better working situation offered at NIH, not just in hours, but for administrative and staff support. Much more of that was available at the "empire" as compared to NSF; it was chronically short at IBN in particular, he held. Also of concern to Platt had been aspects of division of labor and other job-related resources that Platt found inequitable at NSF.

Beginning at the turn of the millennium, a number of COV reports came out during that first half decade assessing the status of IBN and its clusters and finally of the nascent IOB. Since they were division-wide efforts, in most cases, they will be discussed at the end of this chapter highlighting broader aspects of the division's history.

The various historical trends within the neurosciences in BIO in the early 2000s, then, would lead to 2005 and IOB, the division that dropped the term neuroscience from its title. How was that to be explained? By the fall of 2003, the Society for Neuroscience (SfN) had become aware of thinking that had been ongoing at BIO considering changing the name of IBN to IOB and restructuring neuroscience (and other) programs. Three presidents (then present and past) of SfN wrote to Director Colwell asking for a meeting with her and Mary Clutter to "learn more about possible changes" that might affect the neuroscience cluster.[54] The meeting occurred in December 2003. Over the following year, much transpired and the SfN in its *Neuroscience Quarterly* of winter 2004 ran a news article in which Mary Clutter's comments were aired.[55] Concern had been expressed by SfN's membership: Clutter wrote to "allay your fears." NSF would not be terminating funding for neurosciences because "the neurosciences provide a strong model for what is happening in biology and in science overall in the 21st century." While not *terminating* support, the manner of its *distribution* would be significantly altered. As the BIO AC discussed in spring of 2004, "integration of neuroscience across the division" would

[52] "The effort to double the National Institutes of Health (NIH) budget began as a movement among Senate Republicans and has had bipartisan support in both the House and Senate since the first session of the 105th Congress. Based on the substantial increases provided to NIH in fiscal years…1999 through 2002, the effort created a climate supportive of the move to double the NIH budget to a level of $27.221 billion by 2003, and has been responsible for the increases provided in the appropriation measures for NIH. NIH's FY 1999 appropriation was an unprecedented 15-percent increase, or a $2.03 billion increase, over the FY 1998 level, which was claimed by the Chairs of the House and Senate Appropriations Subcommittees on Labor, Health and Human Services…and Education as the first installment in doubling NIH's budget by 2003." See http://olpa.od.nih.gov/legislation/107/pendinglegislation/doubledec.asp.

[53] Platt interview. Much the same comments were made to the author by many other interviewees and during many casual conversations at NSF.

[54] Akil to Colwell letter.

[55] Society for Neuroscience, "Neuroscience Quarterly," Winter, 2004.

occur and the neuroscience cluster would no longer exist.[56] In fact, the "enthusiasm for the intentional 'blurring of boundaries' within and across divisions" was palpable among the members of the AC.

BIO was, very definitely, "considering organizational changes to facilitate" advances in the new century. Clutter welcomed the Society to aid in providing names of potential rotators and panel members, and Diane M. Witt, then PD of the Behavioral Neuroscience Program and also that of the Neuroendocrinology Program (two programs were still under one person, just as it was when Platt had departed) was also quoted in the SfN newsletter. Witt promoted not only BIO but also the SBE and CISE directorates and various education and training programs, as well as select STCs, as having considerable activity in the neurosciences—a wide range of programs in widely scattered areas of the Foundation. She welcomed SfN members to become involved. A question and answer session between Anne Young, the Society's president, and the editor of their newsletter followed Witt's comments. The upshot was that BIO was considering dispersing programs more widely within a new IOB instead of having a single tightly melded cluster as was still then the case in IBN. Possibly, this was more evidence yet for the need, or perceived need, to break up the alleged "clique," though that is merely my surmise. And, further, it may speak to Kingsbury's belief that Clutter was not enamored of the neurosciences; but again, that is an assumption. At any rate, most all of IBN came under the umbrella of reorganization. The SfN membership feared fragmentation of the neurosciences in the reorganization and creation of IOB, but the meeting held the year before led to reassurance for President Young that funding was safe and other changes would not endanger proposals.

By November of 2004, NSF put out a FAQ (frequently asked questions) page on the Internet regarding the reorganization that summarized the planned new structure for the Division of Integrative Organismal Biology, eventually founded in June 2005.[57] Four new clusters were to come into existence: behavioral systems, developmental systems, environmental and structural systems, and functional and regulatory systems. More than ever before, Weiss' functional approach of the early 1950s seemed to have matured to a point where function and not discipline was the leading organizational feature for, at least, a part of biology. In the titular sense, the term *organism* had returned to the center of biology, that pivot upon which all of biology must turn, that midpoint between the molecular and the biospheric realms of the long continuum of life's organization. The "reorganization clearly refocuses IOB on organismal biology," the FAQ stated. The four clusters focused on "processes common to all organisms." What was missing was the word neuroscience.

In 2005, the new IOB division director was Thomas Brady, who had replaced Frank Greene of the former IBN. Greene's deputy, Judith Verbeke, continued on in

---

[56] BIO AC, "Minutes," April 20–21, 2004, pg. 2 of 5 (online version).

[57] The page is no longer active, but a two-page paper version exists. However, see waybackmachine.org.

that position under Brady and would later succeed him. In the fall of that same year, Mary Clutter would retire and be replaced by James P. Collins. It was the end of an era in so many respects.

8.3 The Developmental Mechanisms Cluster

Though beginning as Developmental Biology in 1991, the cluster's title was soon changed to Developmental Mechanisms. Judith Verbeke was destined to play a major role in not only the cluster but at the division level, as well.[58] In respect of rapidity of rise at BIO and of the many jobs she held, hers was a notable career with regard to brisk upward mobility. A native of St. Louis, Missouri, Verbeke completed her bachelor's degree at Rockhurst College in Kansas City (1980). Having gotten a doctorate in plant science under her major professor's NSF grant, she finished that degree in 1985 at UC, Los Angeles. Moving to the University of Chicago as an assistant professor, she quickly won her own Foundation grant. Four years later, having reached her first professorial level, she was interviewing when she was invited to NSF for a symposium for women with similar university status. Shortly thereafter, she accepted a position as associate professor at the University of Arizona, Tucson, and continued to obtain renewals on her grant. She received a call from Judy Plesset asking her to become a rotator within the Visiting Scientist program for plant and microbial development. (Plesset was the program director for both Animal Developmental Mechanisms and Evolution of Developmental Mechanisms, and remained so for many years.) Verbeke accepted and after finishing her 2-year rotator position, she stayed on at BIO but did not immediately relinquish her Tucson faculty post; she commuted for some 6 months. Plesset was temporarily detailed to the NIH later in that period, and so Verbeke moved into an expert position and helped manage the program in Plesset's absence.

Umminger called Verbeke in 1996 offering his deputy division director opening to her. She refused as she and her husband were well established in Tucson, and they also had two children. Also, the deputy position did not appeal to her. She did leave Umminger with the understanding, however, if "something more interesting" arose she would like to hear from him. Consequently, in November of 1997, she was asked to take on the plant genome program which awarded the lump $40 M through the efforts of Senators Bond and Mikulski at just that same time (previous chapter). She became the program's first director but continued her commute back and forth to the University of Arizona. Soon thereafter, Umminger asked her to become a permanent employee. The split positions she was filling in Tucson and Arlington were such that her seventh grade daughter told her to take the NSF offer: the family saw more of her as a commuter than when she was at the University only—and she

[58] Verbeke interview.

was happier with NSF, the child said![59] Verbeke came permanently to the Foundation in 1998, already well-seasoned.

She managed the plant and microbial development area within the cluster and a year later took over the Integrative Plant Biology Program (once led by Machi Dilworth) in the Physiology and Ethology Cluster of the division (below); her expertise extended to plant structure and physiology. Mary Clutter offered Verbeke the opportunity to go for administrative training at the Federal Executive Institute[60] because the AD wanted her to become the Acting Deputy Division Director (DD) for a year in IBN. As it eventuated, upon Verbeke's return from the Institute, the opening became permanent. Verbeke's rise from 2-year rotator in 1993 to an acting division director in 2002 was a rapid one. And upon Greene's departure, she again became acting DD and that time for nearly 18 months.

By mid-2005, IBN had become IOB with four, rather than three, clusters; those four very differently constructed. Plant biology was placed in the behavioral systems cluster, a strange choice, Verbeke averred.[61] Though odd, there had been very great disparities of grant proposal numbers coming into the three IBN clusters for years and so to even out workload the IOB clusters would divide the personnel in new ways to meet, in part, the numbers' challenge. Though beyond the time frame of the current study, it should be said that Verbeke took on the leadership of the new virtual division, Emerging Frontiers, in 2007.[62] Another evidence of Verbeke's talents and flexibility, she having served in so many roles both inside and out of IBN, led her to offer that NSF "is the best job in the universe." She would finally become permanent DD of IOB, in the creation of which she "played a critical role…[it being] a phoenix…kind of a story."[63]

But to return to 1993, Developmental Mechanisms supported investigations into a number of areas. Several examples indicate the wide range of projects: the integration of molecular information, such as the regulatory mechanisms giving rise to the embryonic anterior-posterior axis in animals; evolutionary inferences were examined, too, with the study of gene expression in the fruit fly; and, cell-matrix interactions were investigated with regard to the extracellular matrix holding such things as animal tissues together.[64] But, developmental biology, like other sub-disciplines of biology, has morphed in appearance and content over time, and so findings made in the 1970s and 1980s led to major changes in the nature of the field. What would be supported in the late 1990s and beyond was quite unlike preceding decades. Including such traditional areas of study as animal embryology, NSF went further

[59] Ibid.

[60] For more on the FEI, see https://www.leadership.opm.gov/Locations/FEI/index.aspx.

[61] Verbeke interview.

[62] William Zamer Interview with the Author, February 20, 2009.

[63] Verbeke interview.

[64] Judith Plesset and Sarah A. Bruce to Frank Harris (Acting AD BIO), "Summary of Developmental Mechanisms Programs Presentation of 5/13/93 Meeting of IBN POs and BIO AD."

by including plant embryology (and more) and the "development" in the microbial world in the IBN, thus covering all organisms.

Though from the beginning, but certainly by the turn of the millennium, the perception that IBN was too much of a catchall had become widespread, more of this major issue later. Still, two highly disparate areas in which IBN programs provided considerable support, and which have matured into very significant areas in biology, have been its funding of a number of aspects of research on the microbial world and on the creation and growth of the evolution *of* development—so called "evo-devo." We shall turn to the microbial world further below.

8.3.1 The Evolution of Development: Evo-Devo at BIO

Verbeke has argued that the initiation of evo-devo at NSF was due entirely to Judith Plesset: "It was all Judy."[65] Plesset herself has said that the "study of the evolution of development has experienced a renaissance during the past decade [1990s]."[66] Verbeke believed that during that decade, Plesset was an untiring source of enthusiasm for increasing funding for the field at the Foundation: funding did come to pass.[67] Evo-devo grew in multiple areas of support in programs widely spread across all four divisions of BIO and into the broader area of biocomplexity and other initiatives that penetrated directorates beyond biology.

In 1991, support for the field was only a quarter of a million dollars. By 1999, however, the annual budget for the new Evolution of Development Program in IBN had surpassed $1.6 M annually, and, if all sources for evo-devo Foundation-wide were taken into account, the amount exceeded $3 M.[68] Growth accelerated particularly with the establishment of IBN's evo-devo program after a 1998 workshop that led to its creation. Participants at the workshop reached several conclusions about the field, including the necessity for "robust phylogenetics"[69]: BIO had long been the major supporter of systematic biology in the country. Detailed study of "gene regulation in a developmental/evolutionary context" would also be required for the

---

65 Verbeke interview.

66 Judith Plesset, Samuel Scheiner, and Susan Singer, "Evolution and Development at the National Science Foundation," *Genesis* (2000):45–46. Pg. 45. Scheiner was PD of Population Biology in DEB and Singer was PD of both Plant and Microbial Developmental Mechanisms (the program once directed by Verbeke) and Evolution of Developmental Mechanisms (later headed by Plesset).

67 Verbeke interview.

68 Ibid., ppg. 45 and 46.

69 "Phylogenetic systematics is that field of biology that…deal[s] with identifying and understanding the evolutionary relationships among the many different kinds of life on earth, both living (extant) and dead (extinct). Evolutionary theory states that similarity among individuals or species is attributable to common descent, or inheritance from a common ancestor. Thus, the relationships established by phylogenetic systematics often describe a species' evolutionary history and, hence, its phylogeny, the historical relationships among lineages or organisms or their parts, such as their genes." See http://www.ncbi.nlm.nih.gov/About/primer/phylo.html.

healthy growth of the new discipline: BIO was the lead federal entity in the nation in support of basic research in non-human genetics/genomics. Also, comparative studies were called for, and, too, wide interdisciplinary communication of findings was necessitated given the dispersed nature of the field's practitioners within the greater biology community.[70]

By the fall of 2000, considerable activity was occurring within evo-devo programming at IBN, including a workshop, "Evolution of Development Meets the Tree of Life." The Tree of Life, or Assembling the Tree of Life (AToL), project was based in DEB, but, as with evo-devo, funding came from several sources within NSF, IBN hosting the titular program. The interrelationships among a number of fields had become ever more closely knit once the many genome projects began to produce data that could be shared widely; and such was just then beginning. That led to many new possibilities in systematics, as just one area, and AToL sprang forth from not only genomics and its growing databases but from new thinking in the evo-devo community, also.[71] NSF was there to bring those communities together. As was said in connection with the workshop:

> The [A]ToL community is primarily interested in the *shape* of the Tree of Life, while the Evo-Devo people are primarily concerned with the *mechanisms* that have shaped that tree.[72]

Such specialty orientation—AToL shape versus mechanisms—involved diverse but melding groups of investigators who were dependent upon the same types of data resources, support for which was yet another NSF involvement. Other disciplines could be and were integrated into these growing scientific conglomerates, those in proteomics, as but one example.

The workshop was based upon the fact that:

> [t]hree recent conceptual generalizations in developmental biology have allowed the re-emergence of Evo-Devo and renewed excitement in developmental approaches to phylogeny. First, the genes directing the development of all metazoan organisms are strikingly similar (i.e., [a] common set of genetic tools drives development throughout…extant organisms). Second, this core set of genes (gene families) represents a rather limited set of players. Third, changes in gene expression can cause dramatic changes in animal and plant form.[73]

The workshop in 2000 came just 5 years after the Nobel Prize had been shared by three scientists for their discovery of *gene families*. Two were American biologists, both of whom were NSF-supported.[74] The genetic control of embryonic development studies undertaken by the third, Christiane Nusslein-Volhard, was done in Germany in the 1970s and 1980s. She used the classic model system, *Drosophila*

---

[70] Plesset, "Evolution," pg. 45.

[71] See second indent quotation following this footnote in the main text above for evidence of that thinking.

[72] "Evolution of Development and Tree of Life Workshop Report," September 14, 2000, NSF Grant IBN 0096346, http://www.nsf.gov/pubs/2002/bio012/start.htm?org=NSF, pg. 2. Emphases added.

[73] Ibid., pg. 2.

[74] See http://www.nsf.gov/news/news_summ.jsp?cntn_id=100683.

melanogaster. Little was then known about embryogenesis at the molecular level, but by use of chemical mutagenesis, Nusslein-Volhard was able to create a series of mutants for variations in the body segmentation pattern of the fruit fly. Consequently, others were able to identify specific genes responsible for each segment via the mutants. Now-famous names for those mutants were applied by the German researcher based upon larval appearances. They included, as one important example, "hedgehog" (for a spiny segment in the larval fly), which was later shown to code for a cell-to-cell signaling pathway. The proteins of that pathway provide information to the developing embryo from head to tail (embryonic axis) so that segments along the length develop properly in a stepwise manner. The work of NSF-supported Americans Eric F. Wieschaus[75] (who demonstrated zygotic genes and their role in embryogenesis) and that of Edward B. Lewis (who established the significance of the "bithorax complex"[76]) led to them sharing the Prize with Nusslein-Volhard in 1995.

They all three took our understanding of embryonic development to a new plateau needing only data from the various genomics' projects to push evo-devo into new intellectual territory. The three were not alone, however, as other significant findings included the discovery of the homeobox genes and the HOX complex. Both of those came in the early 1980s and by independent groups. They would play into the expansion of the study of the evolution of development. In that connection, Peter C. M. Molenaar in his review[77] of R. Amundsons's book[78] on the history of evo-devo cogently stated that "one must understand how bodies are built in order to understand how the process of building bodies can be changed, that is, how evolution can occur." The work of the Nobel laureates and the others during that period made understanding "how bodies are built" possible. By 2006, Manfred Laubichler (who has written a leading history of the discipline) in his review of yet two other recent histories of evo-devo could point out that the field not only "continues to rise to scientific prominence" but also that "regulatory evolution" may exist and in "many instances [be] responsible for observable phenotypic changes,"[79] possibly a previously unrecognized mechanism of evolution.

In order better to understand the importance of gene families that began to appear upon the scene in Plesset's "renaissance" of the 1990s, a database of such entities was required. "TreeFam" came to exist to serve that end.[80] While not a project of NSF—its creation and maintenance was funded by several countries and NGOs—the database focuses on animal gene families but includes data on *Arabidopsis* and

[75] See http://nobelprize.org/nobel_prizes/medicine/laureates/1995/wieschaus-autobio.html.

[76] See http://www.sciencemag.org/content/221/4605/23.full.pdf.

[77] See http://www.atypon-link.com/AAP/doi/abs/10.1375/twin.11.1.100.

[78] R. Amundson, *The Changing Role of the Embryo in Evolutionary Thought: Roots of Evo–Devo* (Cambridge, United Kingdom: Cambridge University Press, 2005).

[79] See http://www.mitpressjournals.org/doi/abs/10.1162/biot.2006.1.1.102.

[80] See http://www.treefam.org/, but also particularly see Heng Li, et al., "TreeFam: A Curated Database of Phylogenetic Trees of Animal Gene Families," *Nucleic Acids Research* (2006):D572–D580.

yeast genomes for comparisons. Research, including genomics, was heavily funded for those latter organisms by NSF, as we have seen.

The NSF and BIO were significant players in the establishment and expansion of the field of evolution of development. Given the tightly integrated nature of evo-devo, based primarily in IBN's Developmental Mechanisms Cluster, and that of AToL, located in DEB, the complex of those projects/programs will be considered yet further in the following chapter.

8.3.2 The Microbial World: First Considerations

Fred Sanger sequenced the amino acid order in the insulin protein in 1955. Over two decades later, in 1977, he was able to sequence the nucleic acids in the DNA of the bacteriophage lambda, the first full genome so investigated. Several cellular organellar genomes (e.g., mitochondria) were completed in the early 1990s.[81] But it would not be until 1995 that the first free-living cellular organism had its DNA sequenced. J. Craig Venter and his team announced the feat in July of that year.[82] Venter's first of many organizations, The Institute for Genomic Research (TIGR), was supported by NSF, and others, at that time.[83] Using the bacterium *Haemophilus influenzae*, they found all of the 1,830,137 base pairs of DNA and assembled them in correct order. They had chosen that species as a "pilot project" because:

> its genome size (1.8 Mb) is typical among bacteria, its G+C [guanine and cytosine] base composition (38 percent) is close to that of human[s], and a physical clone map did not exist.[84]

---

[81] For instance, see Kenji Oda et al., "Gene Organization Deduced from the Complete Sequence of Liverwort *Marchantia polymorpha* Mitochondrial DNA: A Primitive Form of Plant Mitochondrial Genome," *Journal of Molecular Biology* (1992):1–7; see also K. Ohyama et al., "Gene Content, Organization and Molecular Evolution of Plant Organellar Genomes and Sex Chromosomes: Insights from the Case of the Liverwort *Marchantia polymorpha*," *Proceeding of the Japan Academy Series B Physical and Biological Sciences* (2009):108–24.

[82] Robert D. Fleishmann et al., "Whole-Genome Random Sequencing and Assembly of *Haemophilus influenzae* Rd.," *Science* (1995):496–512.

[83] "NSF has been a major funder of basic research at TIGR in both plant and microbial genomics. Beginning in 1996, TIGR was the recipient of a multi-year award from the NSF to participate in an international consortium to sequence the first plant genome, *Arabidopsis thaliana*, which serves as a model for 250,000 other plant species. This work was completed in 2000, 4 years ahead of schedule. Because of the continued strong federal investment in plant genome research, TIGR has initiated a number of other NSF-funded, genomics-based research programs on important crop species including rice, potato, tomato, and soybean. In parallel with these studies are related efforts on some of the most important bacterial and fungal plant pathogens that are responsible for millions of dollars in losses each year." See http://energycommerce.house.gov/108/Hearings/05222003hearing927/Venter1471.htm.

[84] Fleishmann, "Whole-Genome," pg. 496.

What was not mentioned was that a major player on the large sequencing team, and next to last author on the publication before Venter's name, was Hamilton O. Smith—long supported by NSF—whose specialty was that very bacterium. Smith had extensive molecular data in the form of a library of restriction enzymes in connection with *H. influenzae* dating back to the 1960s. His work in that area had won him, with two others, the Nobel Prize in 1978.[85] That material and knowledge, brought into the sequencing project by Smith, was crucial to the work of Venter and his team.

Venter's success gave reason for an important Interagency Report on the status of federal funding for the suddenly expanding new field to open with the words: "The *age of microbial genomics* began four years ago with the publication of the complete sequence of *Haemophilus influenzae*."[86] That report of 1999, however, had been preceded by a very rapidly growing effort in sequencing a number of organisms. For that reason, it was incumbent upon the federal establishment to have a grasp of a field growing so astonishingly fast and so heavily dependent on citizen dollars. *H. influenzae* had a good head start due to Smith's contribution and to the fact that its genome was much smaller than many other non-bacterial organisms then under study. As the Venter team pointed out:

> [a]t the present time [1995], there are active genome projects for many organisms, including *Drosophila melanogaster…Escherichia coli…Saccharomyces cerevisiae…Bacillus subtilis…Caenorhabditis elegans…*and *Homo sapiens* [the Human Genome Project].[87]

The reason for the Interagency Report, then, was to ask where the field stood as most of its funding was coming from a cross-section of federal agencies, including NIH, NSF, USDA, DoD, NIST, and DoE. The last of those, as shall be seen, was a major, even founding, player during the 1990s, surely the beginning years of the *genomic revolution*.

BIO and the Foundation had been represented on an extensive visit to Japan, at that point, by IBN PD, Machi Dilworth. It was clear to BIO that this latest revolution in biology was efflorescing at an explosive rate and the directorate needed to keep pace with that phenomenon. Dilworth began a 10-month fellowship at one of

[85] See http://nobelprize.org/nobel_prizes/medicine/laureates/1978/smith-lecture.pdf.

[86] Biotechnology Research Working Group, Subcommittee on Biotechnology [chaired by Mary Clutter], Committee on Science, NSTC, "Interagency Report on the Federal Investment in Microbial Genomics," April 1999. The contributors included 17 persons with the coordinator being Margaret Werner-Washburne of NSF. Emphasis added.

[87] Fleishmann, "Whole-Genome," pg. 496.

the RIKEN[88] laboratories in Japan in November of 1996.[89] Only the month before she left for Asia, the first publication of a eukaryote sequence, that of the common baker's and brewer's yeast *Saccharomyces cerevisiae*, came out in the United States.[90] However, it would not be until some years later that two of the others that the Venter team had cited would be published: the fruit fly in 2000 and the nematode worm, *C. elegans*, in 2003.[91] Both are much more complex animal systems when compared to yeasts and bacteria. Most other sequencing projects would not be published until after the turn of the new millennium. While in Japan, Dilworth was able to attend one particular conference, drawing nearly 500 participants, on microbial genomics.[92] She quoted one scientist as having said that "[m]icroorganisms will give you anything you want if you know how to ask them." It was a notable statement as bacteria, especially, had by then been long viewed as tiny production facilities, and, once their genomes were known, genetic engineering could be undertaken to make specialized living factories for more or less many chemical products or services that may be wanted. The Eli Lilly pharmaceutical company had used bacteria to produce human insulin in the 1980s by employing recombinant DNA—the first production-level use ever coming from "asking" a microorganism's help! From the later 1990s on, such redirecting of the physiology of a given bacterial species could count upon knowledge of full genomes: it would come to be widely termed *genome-enabled science.*

The Department of Energy was deeply involved in microbial genomics from a comparatively early date. DoE established its Microbial Genome Program (MGP) as a companion to the new HGP in 1994.[93] The DoE was particularly interested in

---

88 RIKEN is comparable to NSF and parts of NIH combined. It is interesting what areas are *not* covered in its component sciences, however, as becomes clear in the following: "The mission of RIKEN is to conduct comprehensive research in science and technology (*excluding only the humanities and social sciences*) as provided for under the "RIKEN Law" and to publicly disseminate the results of its scientific research and technological developments. RIKEN carries out high-level experimental and research work in a wide range of fields, including physics, chemistry, medical science, biology, and engineering, covering the entire range from basic research to practical application. RIKEN was first organized in 1917 as a private research foundation and reorganized in 2003 as an independent administrative institution under the Ministry of Education, Culture, Sports, Science and Technology." Emphasis added. See http://www.riken.go.jp/engn/r-world/riken/outline/index.html.

89 Laboratory for Regulation of Plant Functions, Institute of Physical and Chemical Research in Hirosawa, Wako, Saitama, Japan.

90 A. Goffeau et al., "Life with 6000 genes," *Science* (1996):563–567.

91 See all three: http://en.wikipedia.org/wiki/List_of_sequenced_eukaryotic_genomes; http://en.wikipedia.org/wiki/List_of_sequenced_bacterial_genomes; and http://en.wikipedia.org/wiki/List_of_sequenced_archaeal_genomes.

92 Machi Dilworth, "INT 97-11, NSF/Tokyo Report: New Frontiers in Microbiology," March 24, 1997, three typed pages.

93 See http://microbialgenomics.energy.gov/mgp.shtml.

bioremediation and biotechnology, as well as evolution and ecology.[94] By 2000, the MGP was putting nearly $14 M into its efforts and was supporting study of the genomes of 45 organisms, recorded an NSTC report.[95] However, NSF, in that same year, was only putting $3.7 M toward microbial genome sequencing of just four organisms. The total dollars expended that year were over $45 M spread across the four major federal agencies (NIH, DoE, USDA, NSF) then dealing in genomic research; and only large-scale projects were considered in that amount, at that. As lean as the Foundation's share was, it had gone up $2 M in just the one fiscal year from 1999 to 2000. The NIH "empire's" contribution was a very substantial $21.5 M. By the time of that NSTC report, it had become clear to the Council's Subcommittee on Biotechnology's Microbe Project Interagency Working Group that a "coordinated plan" was needed to "address research, infrastructure and human resource gaps" in the many agencies involved.

NSF might have appeared to have been a small player, but the Foundation had awarded "almost 5400 grants…since 1989…in microbial-related research," albeit genomics was but a part of that.[96] The Microbial Genetics Program in MCB (previous chapter) was a significant undertaking, but microbial biology was also strong in the LExEN program, Microbial Observatories, Office of Polar Programs, Oceanography, database and informatics areas (within DBI), the biocomplexity initiative (Foundation-wide), and in two microbe-oriented university STCs supported by BIO. Some of these have already been mentioned in previous chapters; more will be discussed below and in the following chapter.

But NSF sequencing support began not through BIO exclusively rather it had started out in the Directorate for Education and Human Resources (EHR) in the early 1990s. That unit's program employed the common bread mold, *Neurospora crassa*, a notable, long-used *model system* in genetics. The *Neurospora* effort was the first of the "large-scale sequencing efforts" begun by NSF. That had been part of "an undergraduate training program and EST-sequencing project" in the EHR's Research in Minority Institutions Program. By the time of the Interagency Report, the project had moved into MCB's Microbial Genetics Program. At the end of the millennium, the sequence of the archeon,[97] *Halobacterium*, was published.[98] That of

[94] Ibid. By 2011, over 11,000 genome projects were underway worldwide with completed ones approaching 1900; projects of many origins in many countries. See http://www.genomesonline.org/cgi-bin/GOLD/bin/gold.cgi. There is an extraordinary map of locations at this site, which shows where active research is occurring.

[95] Biotechnology Research Working Group, Subcommittee on Biotechnology [co-chaired by Rita Colwell], Committee on Science, NSTC, "The Microbe Project," January 2001, pg. 17.

[96] "Interagency Report," pg. 21–22.

[97] Singular for Archaea, the third of the three great domains (level above kingdom) of life. Though they were once placed with the bacteria, they have been shown to be distant relatives only. See more details at http://www.ucmp.berkeley.edu/archaea/archaea.html.

[98] W. V. Ng, "Genome Sequence of *Halobacterium* species NRC-1," *Proceedings of the National Academy of Sciences* (2000): 12176–12,181.

the cyanobacterium,[99] *Synechococcus*[100] would wait until 2003 for publication. Both of the latter were sequenced via BIO, not EHR, funding.[101]

Crucial to those projects were NSF's database programs, such as the Ribosome Database Project, the Protein Data Bank (NSF along with NIH and DoE), and a microbial biodegradation database.[102] Those involvements came to some $4.5 M of NSF funds. The STCs at the University of Washington and Michigan State University were doing the *Halobacterium* sequencing, and NSF's total investment in those STCs, since 1992, was a striking $38 M, albeit all projects included. NSF's Major Research Instrumentation Program was aiding in funding Oregon State University's, Corvallis, Advanced Microbe Isolation Program, yet another project for microbiological research. The year after the Interagency Report, NSF announced a new post-doctoral program to support study of non-model microorganisms in order to create a new cohort of scientists prepared to enter the truly revolutionary world of microbial DNA sequencing.

The Interagency Report representing a number of federal groups was not the only such study to come out in the last year of the millennium: the American Academy of Microbiology (AAM) also published a review of the new world of microbial genomics. Editors David Relman and Evelyn Strauss commented that "[m]icrobes carry enormous genetic wealth and biological aptitude."[103] They saw the new enterprise as "young and malleable," but with "challenges" that needed molding. Some 29 microbes had been sequenced by 1999 (though many were unpublished).[104] But the dissemination of sequence data, they argued, had been "haphazard," was without standards, and lacked priorities and goals.[105] Simultaneous to the AAM study, the Interagency Report was also being prepared and would appear in print only 1 month after that of the Academy. The combined federal agencies commented that while:

> [a] few of these gaps and opportunities are more salient for some agencies than others, however, the need for infrastructure support, training in genomics, computational biology, and increased integration in policy development was common to all agencies…[and] are common issues for all genomic research [not just microbial].[106]

---

[99] Cyanobacteria were earlier known as blue-green algae, but unlike all true algae, which are eukaryotes, the "blue-greens" are prokaryotes and belong properly with those organisms. For more on these life forms, see http://www.ucmp.berkeley.edu/bacteria/cyanointro.html.

[100] B. Palenik et al., "The Genome of a Motile Marine *Synechococcus*," *Nature* (2003):1037–1042.

[101] "Interagency Report," pg. 21.

[102] Ibid., pg. 22.

[103] David A. Relman and Evelyn Strauss, eds., "Microbial Genomes: Blueprint for Life," A Report from the American Academy of Microbiology (Washington, D.C.: AAM, 1999), 18 ppg. Quotation from pg. 1, Executive Summary.

[104] For a detailed list, see http://en.wikipedia.org/wiki/List_of_sequenced_bacterial_genomes.

[105] Ibid., pg. 3, Introduction.

[106] "Interagency Report," pg. 26.

It was obvious to both the AAM and the relevant federal agencies that coordination and planning were needed to drive, direct, and fund—that is, to coordinate broadly—the many projects then underway and others which would soon be seeking future support. The AAM, in fact, called for a "large-scale, coordinated, multi-agency" effort along with a "central and easily accessible" database. Further, the Academy sought standardization for data release, creation of an ethical framework, and communicating the benefits of the findings to the public. All of that should be integrated into both higher and K–12 education, as well, they said.[107]

The three-dozen microbial genomes sequenced by early 2000 had been placed on TIGR's website (now extinct; but see the Web archive at waybackmachine.org); 11 of those had been discoveries supported by DoE. Energy was one of the first agencies to speak of going beyond the reductionist agenda of the past and to do a "reverse biology" in which a "reconstructionist" approach using sequence data would be the norm: we are "like children with a new set of Lego [building] blocks."[108] In its first Microbial Genome Project report (2000), DoE pointed out that beyond having finished those 11 genomes, 17 more were then in progress—"at least five other U.S. government agencies have more recently initiated microbial sequencing efforts"—and that all of them were establishing firm policies for management of the extensive efforts jointly in process.

The findings of the Interagency Report and of the American Academy of Microbiology's account, and of DoE's first report as the most senior genome program sponsor, would finally arrive in the Office of the President, specifically the NSTC. The premiere report which came out of that Council's Interagency Working Group in January of 2001 would "provid[e] a rationale and interagency plan for the Microbe Project, a coordinated Federal effort in microbial genomics."[109] The many disparate and uncoordinated sequencing investigations would now be unified under the NSTC's aptly and, simply, named Microbe Project. The report examined infrastructure, research, and human resources, all with a view to gaps and opportunities. The summary recommendations had the Microbe Project Interagency Working Group, MPIWG, take on the ownership and coordination of all federally supported microbial genome sequencing research, a familiar approach outlined in previous chapters with other topics.[110] NGOs were players, too. Both Rita Colwell, Co-Chair of the MPIWG, and Mary Clutter, who actually established it in July of 2000, remained major actors over the next several years. It would eventually have its own website, though now discontinued. While DoE was the lead agency, much of the online information is to be found at the NSF site (www.nsf.gov).

Database issues were among the most important, so DoE convened a workshop on computational biology late in 2001 sponsored by TiGR with a second such

---

[107] "Blueprint," pg. 18, Specific Recommendations.

[108] Human Genome Management Information System, Oak Ridge National Laboratory, for the [DoE] Office of Science, Office of Biological and Environmental Research, "Microbial Genome Program," February 2000, DOE/SC-0016, pg. iii.

[109] "Microbe Project," pg. ii, Cover memo.

[110] Ibid., ppg. 13–14.

meeting in later 2003 sponsored by NSF. The third in the series was under the aegis of the AAM in mid-2004.[111] The AAM brought out a new report in that same year which, while it said nothing about organizational infrastructure or its history, stood as a very clear statement on what the *new microbiology* was all about, it having been "enabled," as had become the phrase, by genomics.[112]

The 5-year period from about 1998 to about 2003 saw the birth of the science of microbial genomics and which has continued on into the present era as a prime example of *Millennial Biology*.

8.4 The Physiology and Ethology Cluster: Another Phoenix

When Physiology and Ethology began in IBN, it was first termed Physiology and Behavior and was known in BIO as "P and B." It seems to have been the first cluster to receive a review after 1991 and possibly for good reason. The Committee of Visitors approach had not been fully established when the first of two reviews began in early 1993, so a "panel of experts" was convened and included six scientists from across the country who represented a cross-section of the disciplines covered in P and B. Responses by the individual members of the cluster to that first assessment have not been found. We do know, though, that a COV report, the result of the second review in 1994, was yet a novelty at the time. Responses to that second review do exist, and it became obvious that the Committee's views were not particularly well received by P and B members.[113] Interestingly, both sets of reviewer's comments were very similar and suggested some real difficulties in the cluster that were obvious not only to the panel of experts but to members of the COV, as well. The groups were comprised of two different assemblages of individuals a year apart. More than that, the problems were visible to the biological community.

Machi Dilworth was one of most senior and strongest personalities in BIO and was not happy with what the COV study of P and B had to say.[114] She was one of

[111] "The Microbe Project Interagency Working Group Coordination Plan, March, 2008", "I. Purpose and Accomplishments of the Microbe Project IWG (2001–2006)." Ppg. 8–9. Citation from pg. 9. See http://www.whitehouse.gov/files/documents/ostp/NSTC%20Reports/MPIWG%20plan%2012-17-08FINAL.PDF.

[112] American Academy of Microbiology, "Microbiology in the twenty-first Century: Where Are We and Were Are We Going?" (Washington, D.C.: American Society for Microbiology, 2004), 19 ppg.

[113] See Chapter 7 for a discussion of the COV concept and its relationship to GPRA.

[114] There were two assessments done on P and B. The first was done in the spring of 1993 and eventuated in a series of six separate letter reports to Bruce Umminger. The second was more formally done by a group of nine biologists under a chairman and which produced a single document in mid-August a year later (1994). A transmittal memo covering all of the letters in the 1993 exercise is signed by Tom Brady and indicates that Umminger was on medical leave at the time the letters were arriving, that is, late March and early April, 1993. The note also said that Umminger would likely wish to see them but that they would be shared with cluster members first. Brady and Olsen also appended a note to the single COV document of the following year. Dilworth's response

those "experts" on the first round said about her, "evangelical," for the *contents* of P and B, at least generally, but not its *cluster* structure.[115] Nonetheless, both reviews would prove of great worth, possibly the first less formal one would turn out to be the most valuable given its great detail: each of the six panelists wrote his or her own report. The composite report of the COV was not as detailed but evidently quite disconcerting at P and B. Dilworth was especially upset about the lack of the Visitors addressing the *scientific successes* of the programs in the cluster. However, and very importantly, she averred that the likelihood of that failure was due to there being "no real intellectual links" among the five programs in P and B: the cluster was "unique" in all of BIO for its lack of those links, she opined.[116] One of the earlier "expert" reviewers said that Dilworth "wants out now" from the cluster structure.[117]

But that same reviewer saw Dilworth's Integrative Plant Biology as just that, integrated. It led him to comment that:

> [t]he effort to retain a 'unity of life' theme in the Foundation was noble but at the level of physiology, it has always been a bit forced. I think it is time to recognize this.

Two years later, it was recognized.

The existence of P and B's Functional and Physiological Ecology Program apparently caused the most discord when it was established in 1991, and the reasons became obvious to both the experts and the visitors. Their suggestions would lead to the revamping of that and other programs into the, also suggested, Ecological and Evolutionary Physiology Program of several years later.[118] Further, the Animal Systems Physiology Program was so badly flawed that its director, Barbara Zain, characterized it as the "'what's left over' category." The panel of experts had to agree as they found the program to be "defined by exclusion."[119] The new "element" (program) suggested was Integrated Animal Biology; that suggestion was accepted at BIO, just as was the Ecological and Evolutionary Physiology Program, to become new programs in 1995–1996. Animal Systems Physiology, endrocrinology, and other pieces from elsewhere were integrated into the dynamic new structure of Integrated Animal Biology. The remaining four members of the six in the panel of experts were in substantial agreement and so the new P and *E*—Physiology and Ethology—came to exist once their view's had been essentially seconded by the COV a year later. P and B was laid to rest.

The situation is worthy of examination in even greater detail than in the paragraphs immediately above as it was, first, an early test of the cluster concept and, second, also proved to be a positive test of the still new COV idea. Even more, it

---

is to the second review, but no communications in regard to the first assessment have been found. The main text above brings both of these together in some instances.

[115] A university professor of zoology to Bruce Umminger, April 4, 1993.

[116] Machi Dilworth to Thomas Brady and Kathie Olsen, "P&B Cluster POs Responses to the COV Report," August 23, 1994. Emphasis added.

[117] The university professor of zoology to Bruce Umminger, *op. cit*. Emphasis in original.

[118] Ibid.

[119] Ibid.

further demonstrated how difficult it was to find and retrieve crucial data in the Foundation-wide computer system, a problem not entirely rectified to the time of this writing![120] The Committee of Visitors' assessment in 1994 had led Dilworth to be "disappointed by the tone and emphasis of the CoV report," it is true.[121] But, crucially, "[d]espite the fact that the basic data are already in the system, retrieving them in a format useful to individual needs is very difficult": a statement about how youthful the Foundation's computerization efforts still were in the beginning 1990s.[122]

On the other hand, Fred Stollnitz (Chapter 2 and later below), PD of the Animal Behavior Program, agreed with almost all the comments in the COV report, though he too did note difficulties in retrieving data necessary to judging the effectiveness of each of the programs in the cluster.[123] (The report, however, had been particularly hard on animal behavior, as shall be seen.) As Dilworth said, "[f]rustrations of the CoV about the unavailability of various data [are] evident throughout the report." Further, the "NSF data management system does not offer easy retrieval of…data," Dilworth held and Stollnitz agreed. However, there also appeared to be a tension between those two individuals over animal behavior's "problems" as Dilworth held that too broad a brush, wielded by the COV with regard to the multifaceted P and B, impugned her plant program.

It would take more than a dozen years for IBN to become IOB, but the polyglot nature of the P and B cluster clearly had deep—and lasting—roots. It is possible that the nature of IBN's conception in 1991, even after two robust reviews and with the necessity in the mid-1990s of reorganizing P and B, played a role in the choices for the yet further restructuring of all of IBN to create IOB in 2005. It will be seen that there was a swinging back and forth from functional to disciplinary to functional biology for critical parts of the division.

---

[120] Personal experience.

[121] From a collection of responses to the report by several leaders in the Physiology and Behavior Cluster. Some undated, but those with dates are all in late August, 1994, near the time the COV report had been issued (see earlier footnote). The personnel included Dilworth, Stollnitz, and Barbara Zain (see main text later above), as well as Sharon Emerson and Elvira Doman.

[122] This difficulty persists to the present day. An historian attempting to find many different types of documents either online or in paper format is continually hindered by two issues. First, there has been a long-time behavior pattern at BIO to *not* save documents and/or have those that are surviving input into digital form. While more and more is being done, the emphasis upon the practice seems to me very insufficient even into the present century. Second, the online search engine available to the historian, and others seeking to find data of whatever kind, is stymied by that engine's inadequacies. I argue that the Foundation must give greater thought to its own history and to making information retrieval more user-friendly to historians. The quotation by Dilworth is to be found in her "Comments on COV Report," of late August 1994. The members of the panel of experts had found the same situation to that I have experienced.

[123] James O'Leary, Chair et al., "Report of the Committee of Visitors for the Physiology and Behavior Cluster, Division of Integrative Biology and Neuroscience, National Science Foundation," August 15–17, 1994; 11 pages. Nine persons, including the chair, made up the COV. One of those individuals was Marvalee Wake.

After digesting the COV report in mid-1994 and having the 1993 panel of experts' opinions also available, Machi Dilworth sent a memo to Tom Brady, in his role as deputy assistant director of IBN (Director Umminger was on medical leave), and to Kathie Olsen, along with the responses to the COV report by herself, Stollnitz, Barbara Zain, and remaining PD, Sharon Emerson.[124] There was the general feeling that the science done by the P and B cluster, in the COV's view, was not substantial, Dilworth opined. Also, the cluster wondered what "'interdisciplinary' research" meant specifically for the P and B. Yet further, in regard to "review timetables" (proposal review turnaround), "[Animal Behavior] is the only program element that had problems," Dilworth stated. In regard turnaround, maybe, but there were wider difficulties, as well.

Dilworth found that there were other areas of the report that had disturbed the cluster's membership, Emerson, for instance, concerning use of metrics and numbers of rotators. Dilworth included those comments in her memo to Brady. But with regard to additional rotators, specifically, Dilworth said that "[a]ll permanent POs felt that the CoV report was not complimentary to them…[t]he report generally treated [them] as bureaucrats with no scientific expertise," thus the need for more outside experts: rotators.[125] Unfortunately, any written answer that Brady and Olsen may have provided to Dilworth and the P and B cluster has not been found.

Barbara Zain, of Animal Systems Physiology, also was "disappointed that in the CoV report there is no discussion of the content of the program or the direction which it should take."[126] The program arose from the Physiological Processes Program of the BBS era, that older program having covered both plant and animal physiology in general, and so it was functionally, not disciplinarily or taxonomically, oriented. Once the great reorganization had occurred, any plant, evolution, or ecology proposals had been removed from the program's responsibilities. That left purely animal physiology proposals for the new, but as it eventuated, short-lived entity. It was gone by the middle of the decade along with several others. Outright biophysical and biochemical grant proposals were not considered by Zain's program, either. The lighter load for the new program allowed for just one or two persons to handle it: Zain and a *possible* rotating PO. The COV felt more rotators were needed across P and B. The proposal types no longer coming to what had been the Physiological Processes Program in BBS, and now Animal Systems Physiology in BIO, were being sent to Dilworth or to Stollnitz. Eventually, they would go to Emerson in the new Ecological and Evolutionary Physiology Program that came into being after the mid-decade cluster reorganization.

In slightly later years, Judith Verbeke, who spoke of another phoenix at IBN/IOB, would find NSF/BIO the best place in the universe to work (quoted above).

[124] Dilworth to Umminger. Elvira Doman (earlier footnote) had been away and had not by this date seen the COV report.

[125] Ibid.

[126] "Response to COV Report," Barbara Zain, n.d., but presumed to be in late August 1994.

But Machi Dilworth, early director of the Integrative Plant Biology Program in P and B, was clearly less sure of that at the time of the reviews.

What had been five poorly conceived (retrospectively, at least) programs led to two new and two older, but all fully rethought, programs not long after both the panel of experts and the Visitors made their thoughts known. The initial group of programs included endocrinology, Animal Systems Physiology, Functional and Physiological Ecology, Integrative Plant Biology, and Animal Behavior. As noted above, Dilworth's Integrated Plant Biology was well regarded by the reviewers and would survive the cluster reorganization, as would Stollnitz' animal behavior; but *management* of the latter would come under criticism. Proposals previously sent to one of the other programs, but later viewed as best handled by the surviving two, would now come to one or the other of them. The three programs that did not survive had their contents integrated into Emerson's area and Integrated Animal Biology. Those four, either new or restructured, programs would last until 2005 when the Division of Integrative Organismal Biology was founded and IBN retired.

While the IBN employed the term *integrative* in its division title, it certainly seemed that such thinking did not fully permeate the structure of all parts of it. The P and B cluster was, as Dilworth had said, lacking in intellectual linkage. Recall also that Umminger was set on having neuroscience continue under that name, so the whole of IBN had a patchwork character to it in clusters beyond P and B alone. The two sets of reviewer's comments and thinking had clearly been taken to heart in leading to the cluster restructuring and not just because it was conceptually "slippery," as one expert called it—clusters were working very well widely across BIO as the concept continued into the future—but it had "unique" problems in P and B. One of the reviewers said that, at least in this case, the "organizational idea… sounded better on paper [than] it does in practice."[127]

Finally, the value of the COV concept proved highly valuable in one of its earliest uses and those committees went on to replace the panels of experts when program reviews were called for.

On the other hand, data retrieval at NSF remains, to the time of this writing, difficult.

8.4.1 *The Animal Behavior Program*

Fred Stollnitz had a very long history with NSF having headed up the Psychobiology Program in BBS and had, even earlier, founded the Behavioral and Neural Sciences (BNS) program of BBS at the behest of earlier Director Richard Atkinson.[128] The Psychobiology Program title was changed in 1988 to Animal Behavior (AB) and

[127] The university genetics professor to Bruce Umminger, April 6, 1993; *op. cit.*, earlier footnote.

[128] "Comments by Fred Stollnitz on COV Report," August 22, 1994 document, and Stollnitz interview with the author, *op. cit.* (Chapter 3).

survived under the same name after the great reorganization. It continued as such through the time that P and B became Physiology and Ethology in 1995–1996. Stollnitz was replaced as PD in the fall of 1994, after the COV report appeared. The new AB director became George Uetz, a rotator, and Stollnitz took a year's sabbatical at the National Zoo.[129] A number of other PDs followed to 2005; all were rotators. They were mostly behavioral ecologists, a group that, like the social scientists, felt they were neglected and said so to NSF just prior to the 1995–1996 reorganization or such was Stollnitz' opinion.[130] He had long held additional other duties in IBN and continued as PD of Cross-Directorate Activities, which was part of the division office. Stollnitz, who had been hired in late 1971, did not retire until December of 2006: "35 years, three months and one day," he reminisced.[131]

The "problems" about which Dilworth had spoken in 1994 in response to the COV report, she suggested, had been lodged within the Animal Behavior Program and were not cluster-wide as she thought the COV report seemed to imply. Those difficulties were in getting timely responses to PIs with regard to the acceptance or "declinations" (declines)[132] of a given proposal to the AB program. Stollnitz recognized he was too slow in getting answers out to grant proposers—sometimes running to as much as 6 months—and so made efforts to remedy that problem.[133] When Psychobiology became animal behavior, Stollnitz lost his one PO, yet his proposal load increased. It became even larger, he said, by an additional 30% after 1991.[134] He had to handle all proposals alone, though other PDs were doing the same thing at the time. The reason for his having lost a PO and not having the position replaced, though, is not clear, especially as he stated that Mary Clutter had indicated in a public meeting that the position *would* be replaced; it was not.[135] Funds were tight at the time, the COV noted. He did have two program assistants, however; but they were not qualified to judge proposal offerings.[136] Consultation in August of 1994 with Sharon Emerson after the COV report came out led to suggested new ways for Stollnitz to handle so many proposals. Whatever the technique was, it seemed to be valued by Stollnitz, but he would be gone from his long-held position within a month or so.

---

[129] Stollnitz interview.

[130] Ibid.

[131] Ibid.

[132] The use of the word declination(s) at NSF was of long-standing even though it was not the best choice given that it had a clear, and different, definition with regard to astronomy, one of the large areas served by the Foundation. During the 1990s, the term was dropped in favor of the word decline(s), itself somewhat questionable.

[133] O'Leary, "COV Report."

[134] Stollnitz, "Comments."

[135] Stollnitz interview.

[136] Fred Stollnitz et al., "Annual Report: Animal Behavior, FY 1992," November 30, 1992. Four program assistants were listed on the report with dates of beginning employment at NSF; only two positions existed, however, but all four seemed to have helped in co-authoring the report.

That first COV report for P and B had also been one of the first ever, Clutter having established the concept about 1991 along with that of clusters. It seemed to have been a learning experience all around but was clearly found to be a valuable technique that satisfied GPRA requirements and, with a little maturation, would become a useful device for BIO, in general. As time passed, COVs and their reports became routine, as shall be seen.

The self-produced *annual program reports* were historically informative across BIO, and in the case of those for AB, Stollnitz was very thorough. The program had reached a figure of nearly $3.5 M by FY92 with nearly half going to support 25 continuing grants.[137] In the first of two panels for the year, 49 proposals resulted in 10 awards, and in the second, 62 yielded 16, a 24% success rate.[138] The panel included 13 members coming from colleges and universities from coast to coast, typical of such panels in BIO. The average size for awards was nearly $136 k with a mean duration of 2.5 years. Two areas formerly under a different program had been added to AB: hormones and behavior and ecology and evolution that dealt with behavior. Other similar proposals, but without the behavior content, went to the Ecological and Evolutionary Physiology Program (EEP) after it was created.

Stollnitz' other commitment was to Cross-Directorate Activities and included management of the awards for the BIO versions of Presidential Young Investigators, NSF Young Investigators, and yet others in Career Advancement, REUs (Research Experiences for Undergraduates; long a program at NSF), and RUIs (Research in Undergraduate Institutions; also a significant area of interest to the Foundation). The AB program also supported three meetings, including one international primatological congress in France. FY92 was not atypical for the period, but total awards had increased over those of the previous year.

In 1993, Stollnitz was accorded the first Louttit Award for the Advancement of Science Through Government Service given by the American Psychological Association.[139] Named for Richard T. Louttit (Chapter Two) of BNS and a man with whom Stollnitz had worked for many years, the award must have been a sweet one. FY93 was a good year for AB, too, as it saw its budget increase by about $1.3 M over just the year before, to a new high of $4.8 M; much of that constituted funds transferred from DEB.[140] The proposals submitted continued an upward trend to some 152 in that fiscal year with 35 funded, holding at a 24% success rate. But 1993 was also the year of the P and B panel of experts review, and one of those visitors felt that AB's budget was too lean and suggested a 3-year effort to raise it consonant

---

[137] Ibid

[138] It is interesting that "[s]ince 1980 the rate has never exceeded 40%." Marc Rothenberg, "Making Judgments about Grant Proposals: A Brief History of the Merit Review Criteria at the National Science Foundation," *Technology and Innovation* (2010):189–195; quotation on pg. 190. Rothenberg cites three of his references in regard to the 40% statement.

[139] Fred Stollnitz, "Notes on a Career in Science Administration," August 21, 1993, four-page typed script of a speech delivered at the APA's award reception in Toronto, Canada of that date. The APA no longer gives this award under this name. See: http://www.apa.org/science/awards/index.aspx.

[140] Fred Stollnitz et al., "Annual Report: Animal Behavior, FY 1993," n.d., but likely late in 1993.

with other elements of the cluster.[141] The COV a year later would recommend adding a rotator for AB.

In another sense, not all was well with the subject of animal behavior at NSF in the early 1990s as elements in the greater society, as well as the federal establishment—not unlike those seen in the past—still sought to reduce or eliminate any programming at the Foundation that smacked of social science.[142] Animal behavior, now thoroughly ensconced in BIO, had long been tied to the social sciences at BBS, of course (earlier chapters). A constituent of Senator Barbara Mikulski, which senator had supported work on the plant genome so generously as co-sponsor with Senator Kit Bond (previous chapter), wrote to Mikulski to ask why she was prepared to vote the way she was on Senate Bill 2403 in late 1992.[143] That bill was intended to "rescind certain budget authority" for a great many federal agencies, NSF included. After considerable battering on the floor of Congress, the bill was indefinitely postponed by unanimous consent and so was left to die.[144] Mikulski's constituent, an animal behaviorist, pointed out that funding would have been pulled for AB, with the bill's death that was circumvented. But the incident evidenced the continuing presence, even to the time of this writing, of those in some communities beyond the Foundation opposed to federal support of social science and related areas, even if quite tangential, such as AB.

And that biologist-voter was not alone in her fears, as Benjamin Hart, Chair of the Committee on Issues in Applied Animal Behavior of the Animal Behavior Society, wrote to his Society colleagues in the following year that he was developing a "booklet" documenting the value and importance of the study of animal behavior.

> Immunologists, virologists, molecular biologists, [etc.,] are quick to point out the practical benefits of their research. It is time that animal behaviorists take a comparable pro-active stance.[145]

While it was clear that outside pressures continued to exist, the AB program in BIO continued to flourish just the same. In FY94, the proposal number had climbed to 194 with 59 awards (including 19 renewals) made, a 32% overall success rate (61% for renewals, 25% for new projects).[146] The AB budget held at $4.8 M, but

---

[141] The zoology professor to Umminger, *op. cit.*

[142] Indeed, not only NSF came under such scrutiny, but so also did NIH. In 2003, "some members of the US Congress tried to rescind funding for several National Institutes of Health…funded projects dealing with various aspects of sexual behavior." None of the 250 grants reexamined by NIH were disfunded, but there had been a fear of a "'chilling effect'" in the research community. As late as 2008, some researchers at Rutgers University in New Jersey were "'self-censor[ing]'" their proposals; it was reported. Constance Holden, "Big Chill," *Science* (2008):1169.

[143] Zuleyma Tang-Martinez to Barbara Mikulski, September 5, 1992.

[144] Bill Summary and Status, 102nd Congress (1991–1992), S.2403, All Information. See http://thomas.loc.gov.

[145] Benjamin L. Hart (of U.C., Davis) to Members of the Animal Behavior Society (at their annual meeting), July 24–30, 1993. Noted attachment not present.

[146] Fred Stollnitz et al., "Annual Report: Animal Behavior, FY 1994," n.d., but likely late in 1994.

taking into account other funds at the Foundation, the Program expended over $5.6 M for animal behavior research.

By the turn of the millennium, the COV process had become refined, and the reports were well honed employing a *standard form* (developed by Mary Clutter, or at her behest) making comparisons among programs much easier, as well as getting more substantial assessments. The AB program had been reviewed in 1999 and again in 2002 as part of Visitor groups studying all of P and (by then) E. By the latter date, the COV was able to say that it had been the "unanimous opinion of the [Committee] that the [P and E] Cluster is very effective in carrying out [its] programs" and that AB's work was of "high quality."[147] Indeed, the "vigor" of the research, which AB supported, was excellent based upon the many papers published in *Science* and *Nature* each year by its funded investigators. Such articles in those two premier journals were held to be one of the best indicators of all of programmatic success. At the same time, however, the Committee stated, in bold italics in fact, that the AB program was the "most underfunded program relative to numbers of proposals submitted, percentage funded, and size of awards in IBN"![148] The repercussions for the community were many, the Visitors argued, including constraining the "gateway to new research programs in [the] neurosciences and sensory programs" dependent upon the vitality of animal behavior studies. On the other hand, however, AB demonstrated the "largest proportion of female investigators funded relative to proposals submitted." NSF had worked for years and would continue to do so, with a variety of long-lived and effortful programs to increase the numbers of women and minorities entering upon and continuing in the sciences. AB was one of the success points. Still, AB was shorthanded and the COV recommended the addition of support staff and the conversion of one of the two rotator positions in the program to permanent status. The leadership by rotators having begun when Stollnitz left AB, even with COVs suggesting to the contrary, would not change before IOB was created.

8.5 Maturation of Mechanisms

The COVs, like the BIO Advisory Committee, were extremely helpful in the management of the directorate and would continue to be so to the time of this writing. Coupled with the long-established panel system, as employed by BIO, the strong use of rotators and the cluster concept were all likely elements in stabilizing certain features of the organizational structure of the directorate. With the exception of IBN, overhauled again in 2005, the remaining three divisions held very much the same over the period from the great reorganization until that same year, that is, to

---

[147] "FY 2002 Report, NSF Committee of Visitors (CoVs), February 25–27, 2002, Physiology and Ethology Cluster," 18 ppg., pg. 15.

[148] Ibid.

the conclusion of this study. It did not harm BIO either to have Mary Clutter, a demonstrably strong Assistant Director, on the scene and that for more than half of the time covered in this study.

Among all those mechanisms and steady senior personnel in the directorate office, the COV technique was one of the most useful to BIO's leadership. The 2002 report, for instance, not only found that animal behavior was doing well, but the second of the two programs least changed of the four in the P and B cluster to come through 1992, Integrated Plant Biology (IPB), was thriving over the whole decade: plant genomics was a major part of that success. Though the 1999 COV report found IPB too heavily engaged with *Arabidopsis* as a model, to the exclusion of other such systems, the 2002 Committee found that more balanced IPB had taken to heart suggestions made in 1999.[149] Indeed, IBN's basic goals included the use of as many different model systems as possible: such intention permeated many documents at IBN.[150] IPB was growing over that period as the number of proposals had been increasing from 128, to 136, to 159 in 3 years between those two COVs. Also, IPB supported "over 50% of basic plant research in the country" in those years.[151]

Integrative Animal Biology (IAB) had matured over the more than half decade since it was created from parts of the three more problematic programs present in the initial five of 1991. In fact, like IPB for plants, IAB was a major supporter of animal biology investigations in the country: 70% of all research in "comparative animal physiology and biochemistry (including endocrinology), biomechanics and functional morphology, and comparative immunology" was due to the program.[152] Indeed, it may have been even more lively an area than was plant biology as the number of proposals submitted had jumped by 50% from 1999 to 2002—104 to 155. And support per award had gone from an average of $72 k to $99 k per year funded, a substantial aid to investigators.

The EEP program was getting many proposals in the area of "molecular mechanistic reductionism" and that, the 2002 COV, called a "healthy trend." But in all of the quality found for P and E, the Committee still had to report that finding proposal reviewers was difficult when some 15–20 such scientists had to be polled to find a set of three, say. Also challenging for P and E, the Visitor's felt, was the shortage of personnel, whether permanent or rotator. The workload placed upon the PDs was also "compounded by an inadequate database of email addresses" listing scientists

[149] Ibid., pg. 16.

[150] Each year's annual budget requests, as just one example, gave a short introduction to each of the divisions of BIO and routinely explained a number of their features. For IBN, wide use of model systems was a persistent theme. One sample statement, repeated in many documents, was "the development and use of a wide diversity of organisms as biological models is encouraged to assist both in identifying unifying principles common to all organisms and in documenting the variety of mechanisms that have evolved in specific organisms." "Guide to Programs, Fiscal Year 1998," December 1, 1997, BIO, IBN section.

[151] Ibid.

[152] Ibid., pg. 17.

who might fill the roles of POs, panelists, or even COV members.[153] The Committee suggested a better database and "encourage[d] further automation of soliciting reviews." For all the millions of dollars of support going to databases for all sorts of matters, biological or other scientific, the National Science Foundation seemed to continue to have problems in automating its own internal data infrastructure—as made abundantly clear during the research for the present study.

It was not just the P and E cluster of the three such groupings in IBN that received reviews by COVs but the full division itself, as well. In 2001, the visitor's concluded in an unusual "Preamble" to their report that:

> [t]his meeting of the CoV comes at a particularly important time in the history of the IBN. This division has been extremely effective in identifying the highest quality science *integrating cell function into the larger picture of organisms and their behavior.* Molecular information on gene and cell function is accumulating at a fantastic pace and this information must be integrated into the functioning of organ systems and organisms in order for [it] to be utilized effectively. IBN is the only venue at NSF for which this is the major mission.[154]

We have seen above and in previous chapters how the years circumscribing the turn of the millennium saw genomics and its spin-offs effloresce to become a central aspect of the "new biology." But the COV knew that the "extremely effective" work done by IBN was not without a cost, because:

> IBN…has gone through a particularly difficult period during which its effectiveness may have been compromised by instability and lack of sufficient permanent staff. It is hoped that the appointment of Dr. Frank Greene as the Divisional Director in January [2001] may offer needed stability in a Division characterized by high turnover at the level of senior administration and program officers.

That statement referred, in part, to Umminger's departure from BIO to take over the STC office, next followed by Grace Wyngaard, then Judith Verbeke, and, finally, Frank Greene as permanent DD. But the same could have been said for much of that decade what with the tumult in P and B and its reorganization 5 years before. Indeed, in the year after the division's review report, P and E, and IBN overall, were still understaffed: specifically with regard to programs led by *permanent* PDs. As just seen, IBN would continue to be challenged into the future. In fact:

> [we, the COV,] found that these issues of stability affect every aspect of the ability of the IBN to carry out its mandate and fulfill its role in the BIO Directorate. Since 1997, there have been four changes in senior leadership (Director and Deputy Director). Furthermore, of the current 13 program officers, only *two* are permanent staff members.[155]

Indeed, in 1997, fully half of all of BIO's PDs and POs were rotators, and fear about loss of institutional memory had been discussed seriously by the BIO AC at

---

[153] Ibid., pg. 4.

[154] Philip Meneely, Chair, et al., "NSF Committee of Visitors for IBN," March 20–22, 2001, pg 2 "Preamble to the CoV Report." Emphasis added.

[155] Ibid. Emphasis added.

the time.[156] Of note, too, was the fact that IBN's director, Bruce Umminger, had been the one making the presentation to the AC. While the AC felt the balance between rotator numbers as opposed to permanent personnel was not necessarily unacceptable, the notion of two few permanent PDs in IBN remained fretful to the COV.

The arrangement was not a welcome one, to be sure, but even with multiple COV's suggesting an increase in permanent personnel, the dearth would continue. Not only did it create problems internally for NSF but for its dance partner, the community, too. Building "strong and effective relationships" between the two was compromised, and "preventing dissemination of information on funding opportunities," to say nothing of dissemination of research results, was also a severe problem, the Committee believed.[157] "We strongly urge, as *recommended in several previous CoV reports*" that more personnel be hired, and to do so "immediately," said the Visitors in their Preamble.[158] Other aspects of the report spoke glowingly of the quality of science done through IBN. Nonetheless, suggestions were made in a number of areas. The AC discussed the understaffing of P and E, specifically in its fall 2002 meeting, saying that the small number of personnel and "insufficient resources [cannot] ensure the health and vitality of the science fields it support[s]."[159]

While the committees' concept seemed to be highly valuable in regard to assessing programmatic content and success, their suggestions were seldom taken when it came to certain points: adding more permanent PD personnel was a continuing problem, even after the appearance of IOB.

It should be noted that clusters were instituted in IBN and DEB, as the first two divisions to have them, from the conclusion of the reorganization in 1992 onward. After a couple of years of experimentation, clusters were added as an organizational device to DBI and MCB, as well. The clusters discussed in this chapter under the main headings were the three kept for IBN until 2005. The number of panels was brought into the rationale of the cluster concept by having coordinate panels tied to clusters. Thus the 17 panels present through the reorganization were reduced to the 12 noted in an earlier chapter: three existed for IBN, one for each cluster.[160]

In a Visitor's report of 2003 of IBN's neuroscience cluster, there was a discussion of the *merit review* system for grant proposals.[161] The current system was established in 1997 (Chapter 7) and consisted of only two criteria: (1) scientific merit and (2) broader impact; it had superseded the older four-question approach.[162] The new

---

[156] BIO AC, "Minutes," October 20–22, 1997, pg. 3 of 7 (online version).

[157] Meneely, "CoV Preamble."

[158] Ibid. Emphasis added.

[159] BIO AC, "Minutes," November, 7–8, 2002, pg. 2 of 4 (online version).

[160] Mary Clutter to Constance McLindon, Director of IRM, February 25, 1993.

[161] The term merit review has been the official term at NSF since Erich Bloch was director. See Rothenberg, "Making Judgments," pg. 193.

[162] As stated by Rothenberg, pg. 193, "A number of changes and circumstances led to the next reexamination of the review criteria, beginning in 1996. According to congressional testimony, by Mary E. Clutter, the reexamination was due at least in part by the realization that the four criteria

system was shown to present difficulties for reviewers. Criterion two was, for some reviewers, "ignored…altogether, while most addressed it in a mere sentence or two."[163] At the time of this writing, NSF is studying whether or not to replace the two criteria merit review with some form of revision—nearly a decade after qualms were first expressed by the community over, especially, criterion two of the instrument.

Also in 2003, the COV process came under redesign such that those committees would "address higher order issues," go beyond GPRA expectations, and also include more standardization, streamlining, and changed meeting schedules.[164] Particularly, the COVs would be entirely division level and not specific cluster level.

8.6 The Birth of IOB

When all of IBN came under review in 2005, it had just recently been restructured into the Division of Integrative Organismal Biology as of July 26, 2004.[165] Both critical words, integrative and organism, were to be found in the new title for the first time. The review was very thoroughgoing even though the visitors were charged specifically with studying the "integrity and efficiency of the ***processes*** related to proposal review," as well as to the "quality of ***results*** of NSF's investments in the form of outputs and outcomes" over time.[166] Those two mandates were typical of visitor charges and had been in use for some time but were reemphasized.

The Committee found that IOB was "innovative in its new organizational configuration" and therefore "emerging and creative trends" in the community were supported. The period was one of "transition," they averred, but not unlike most of the previous years, it was also one fraught with "considerable financial constraints." Wars in Iraq and Afghanistan were rapidly escalating at the time. The visitors recommended "[b]etter tools for tracking the Division's ability to support outstanding science and education" and additionally, clarifying the new cluster organization that had come into existence with the restructuring into IOB. They recommended that "[t]erms associated with traditional disciplines should not be avoided"—it was

had been adopted at a time when the NSF had little involvement in education (it coincided with the Regan administration attack on NSF funding of education that resulted in a cut of some 81% of the NSF FY 1982 budget for education, including all of the funds for precollegiate programs) and that the NSF was now very concerned with the integration of research and education and was rethinking the criteria for that reason." There were other reasons, too, which Rothenberg considered.

[163] George L. Liggins, Chair, et al., "Charge to the Committee of Visitors," IBN, Neuroscience Cluster, April 7–8, 2003, pg. 6.

[164] BIO AC, "Minutes," November 13–14, 2003, ppg. 3 of 7 (online version).

[165] BIO AC, "Minutes," November 18–19, 2004, ppg. 2 of 10 (online version). It would take most of a year for the new IOB to develop fully and for the community to understand that change.

[166] Joe Chappell, Chair, et al., "Committee of Visitors (CoV) Review of the Division of Integrative Organismal Biology (IOB)…", June 8–10, 2005. Emphases in original.

possible to push the Weissian functional approach (my phrase) too far in the terminological sense, they implied. The four new clusters were, to reiterate the list from further above: behavioral systems; developmental systems; environmental and structural systems, and functional and regulatory systems. At least the first two seemed tied to "traditional disciplines," but the second two might have been a bit more problematic; at least they appeared to be so for the Committee's members. They wished that all clusters' "intention[s be] more transparent to all persons involved." Nevertheless, it must be said that the four new clusters truly did reflect Weiss's 1950's suggestion probably better than any others to date for a division aimed at integration.

Two other recommendations were made for the new IOB.

> The…division currently operates with a greater percentage of rotating Program Directors [than any other in BIO], and is handling a very significant workload in terms of grant applications received and reviewed.

Apparently, as regarded rotators and funding for personnel, IOB was little improved from its ancestor IBN. The COV "strongly recommend[ed]" more permanent personnel.[167]

8.7 When Did Integration Really Begin for Biology, the Science?

It is a problematic historical question, to say the least. Nevertheless, it may be possible to answer it by arguing that it appeared with the arrival of what I have termed *Millennial Biology* at about the turn of the twenty-first century. It is that time which sees the massive and rapid appearance of genomics, genomics-enabled science, and related areas of study (proteomics, metabolomics, etc.). And it is that time when an understanding of the action of molecules within living things could first truly be tied to activities on the biospheric level at an all-encompassing scale via metagenomics and other avenues. The Minutes of the BIO AC Science Retreat, held in April of 2001, was just one small bit of evidence to say that full integration was very clearly on the way for biology.[168] The retreat topics were Genome-Enabled Science, Biology of Complex Systems, and Beyond the Genome, as mentioned earlier. The word *genome* and the phrase *complex systems* tied together in one meeting of biological minds suggests that genome-enabled investigations were no longer simply on one end of the biological spectrum and complex systems on the other but were interrelated: integration on the grandest scale was at hand.

At the retreat, the topics of the Tree of Life; evo-devo; bacterial artificial chromosomes; establishing cross-disciplinary teams of physical scientists, mathematicians, and theoreticians with biologists; and cells and taxa were discussed, an uncommon

[167] Ibid., pg. 2.

[168] BIO AC, "Minutes," April 26–27, 2001, ppg. 2–4 of 4 (online version).

blend up to that time for any sort of meeting of biologists, to be sure. At the regular semi-annual meeting of the AC after the retreat, the follow-on topics of annotation and standards in microbial genomics, LTER's 20-year review, long-term management of databases and informatics more generally, and major equipment were all discussed. This list is significant as the integrative nature of what was happening in the greater community and, at BIO, was becoming more and more obvious. The early discussions of the National Ecological Observatory Network (NEON) were also to be heard, and even though it would be a full decade before NEON first would be funded, integrative commitment seemed to abound at the AC and beyond.

In the final chapter, big biology and biology of the big will be the central theme. But with the appearance of *genome-enabled science*, the discussion of the Division of Environmental Biology (DEB) will necessarily be different after about 2000 than it might have been should genomics not have arisen as a science. It was no longer possible to speak merely of a well-established discipline of ecology but to speak of big biology *enabled* by discoveries at the smallest scale: Clutter's "New Biology" was "multidisciplinary and multidimensional." Indeed, "[b]iologists are assimilating a new paradigm."[169]

And one program director in BIO was right at the juncture that was the "new biology…the new paradigm": Bill Zamer. Zamer first encountered NSF as a graduate student at William and Mary College, Williamsburg, Virginia, when he was working on a master's degree.[170] He worked as a graduate research assistant under his mentor's Foundation grant. As a doctoral student later at the University of Maine, he once again worked on a grant to his then mentor. He completed his degree in Zoology in 1986, specializing in ecological and evolutionary physiology (EEP) with dissertation research in clonal sea anemones. EEP was a title used both in the community and at NSF reaffirming the *intricate dance*. His postdoctoral work at Iowa State University, Ames, was supported in part by a grant from NSF that he helped to develop with his postdoc mentor.[171] Soon thereafter as a young faculty member, he had two Foundation grants, one for equipment, the other a prestigious Presidential Young Investigator Award (PYI). PYIs were then being given for 5 years at $25 k per year. It was 1990, and he was in his first year as a faculty member at Lake Forest College, Lake Forest, Illinois—a small, private institution. That PYI Award "really boosted my career," he reminisced.[172] Soon after that award began, Zamer was asked to be a panelist for a doctoral dissertation improvement grant (DIG) panel at IBN. He continued as a DIG panelist for about 4 years and during that period inquired of an experienced colleague what it was like to be a rotator. The discussion was not acted upon immediately, though, as during his first

[169] "NSF Annual Budget Request to Congress for Fiscal Year 2002," BIO section, IBN, pg. 104.

[170] Bill Zamer Interview with the Author, February 20, 2009.

[171] Bill Zamer communication with the author in spring 2012. He has clarified certain aspects of the original interview here.

[172] Zamer interview.

sabbatical in 1996–1997, he heard that he was to return to Lake Forest as chair of the biology department. That led to 3 years in the position.

He was, he said, "burned out" after those 3 years and was "looking for a change of venue." In late 1999, he was at a professional, meeting and, though he did not know it at the time, Grace Wyngaard, then DD for IBN, was searching for him at the conference, but to no avail. In early January, after his return from the meeting, he was at his campus office on a Saturday when he decided to email Wyngaard his rotator application, over which decision he had been deliberating. Within 10 min, he received a telephone call from her asking to arrange an interview between the two. By August of 2000, Bill Zamer became the Rotator Program Director for IBN's Integrative Animal Biology of P and E, which cluster was then being led by Judy Verbeke. Verbeke became Zamer's first mentor at NSF/BIO. At the time of Zamer's hiring in 2000, Chris Platt was Acting Deputy Division Director at IBN. It had been he and Wyngaard who had interviewed Zamer for his new position.[173] As it was a rotator post, he was to stay 1 year, but by its end, he was offered a second; later he accepted an offer of staying in the IAB program directorship permanently. But his doctoral degree, and particularly his postdoctoral research, steeped him in evolutionary physiology studies, and, so after a period in IAB, he moved to the more familiar EEP, which he managed for a number of years.

It was in 2003 that Mary Clutter created the virtual division of Emerging Frontiers (EF; earlier chapters). Its first director was Judy Verbeke; she was followed by Maryanna Henkart, formerly the Director of MCB. It would not be until January of 2008, before Zamer took command of EF. A full discussion of EF is left to future historians.

But to go back some years before, since his background in evolutionary physiological studies was so appropriate to what was emerging for EEP with Rita Colwell's imperative of biocomplexity (more formally, Biocomplexity in the Environment, BE), Zamer was approached by Judy Plesset, who was deeply involved with that initiative. Plesset said to Zamer, "Your people [EEP] need to be involved in this [and] I'd like you to be the alternate [for me] on this work."[174] The start-up of BE had led, some years earlier, to a cross-directorate working group, which had created GenEn (also given as GENEN): Genome-Enabled Environmental Science and Engineering. Having already found NSF "a bit like a candy store, intellectually," Zamer was pleased to take Plesset's offer.[175]

GenEn, soon to be co-managed by Zamer, was one of the five programmatic areas crèated in 1999–2000 under the Biocomplexity Initiative, and, by 2003, all five parts of the full project were well underway.

> NSF's Special Competition in Biocomplexity in the Environment [saw] 2003…[as] the fourth year of a multi-year effort. Five subcategories of awards were made: Dynamics of Coupled Natural and Human Systems (CNH), with eight awards totaling $8.5 million;

[173] Ibid.

[174] Ibid.

[175] Ibid.

> Coupled Biogeochemical Cycles (CBC), with ten awards totaling $8.3 million; Genome-Enabled Environmental Science and Engineering (GENEN), with four awards totaling $4.4 million; Instrumentation Development for Environmental Activities (IDEA), with three awards totaling $5.2 million, and Materials Use Science, Engineering, and Society (MUSES), with five awards totaling $ 5.4 million.[176]

The environment, from genes to biosphere, was all fodder for investigation by the four divisions of BIO and, with DEB, the fourth and final division to be discussed, taking on new views by the turn of the new millennium. DEB, like its three confreres, came out of the great reorganization ready to tackle the 1990s, and it is there we first encounter it in its new form.

[176] See http://scienceblog.com/community/older/archives/C/archsf357.html.

Chapter 9
Expanding Big Bio

In conventional Darwinian theory,
the organism proposes and the environment disposes.
Stephen Jay Gould

9.1 After the Reorganization

As DEB entered the more than 2-year reorganization effort that made the BBS directorate into BIO, the division was known as Biotic Systems and Resources (BSR), as it had been during all of the 1980s. In earlier years, though, it had been known as DEB. So, for the unit that dealt with the big end of the biological spectrum, a return to the title of Division of Environmental Biology was a welcome one, the title BSR having confused many in the community. The general *content* in DEB, however, was among the least changed over time as compared to its three sister divisions to come out of 1991–1992. Renaming and restructuring some aspects of the *programs*, however, was notable (see Fig. 5.1). Parts of the Ecology Program of BSR went mostly to the "P and B," Physiology and Behavior (later Physiology and Ethology), cluster in IBN (previous chapter). Much the same was true of several aspects of the BSR's Population Biology and Physiological Ecology (PBPE) Program. The groupings to emanate from the restructuring period recombined previous surviving features to become three clusters: Systematic and Population Biology (SPB); Ecological Studies (ES); and Long-Term Projects in Environmental Biology (LTPEB), under DEB. The last of the three was, of course, a logical addition because the LTER program (among others) would have a more properly named home at the Foundation. As has been seen, it also had been running as an organization with its own management office, semi-independent of NSF, for some years. Biological Research Resources (BRR) of BSR would be placed under the new LTPEB in a newly created format.

The basic programming of the Ecology and Population Biology *section* of the *division* of the older BMS of 1975 could still be seen but in somewhat altered form

D. J. McGraw, *Millennial Biology: The National Science Foundation and American Biology, 1975-2005*, https://doi.org/10.1007/978-3-030-56367-7_9

after the great reorganization 17 years later. DEB was mostly unwavering over three decades, though, especially with regard to its *programmatic contents*.

9.2 The Systematic and Population Biology Cluster

James R. Estes has called the NSF the "broker for systematics" in the United States.[1] That was a correct assessment as BBS had said in its "Adapting to the Future" report in 1991 that 90% of federal funding for the study of biodiversity, phylogeny, and classification came from the Foundation.[2]

Jim Edwards (earlier chapters) began at BBS in 1982 as a rotating PD for the Systematic Biology Program during John Brooks' division directorship. After 2 years, Edwards became permanent, but, by 1985, he was heading the BRR program and would continue to do so until March of 1990.[3] Edwards had replaced Jim Tyler when Tyler left NSF to go to the Smithsonian. (Edwards would himself, in much later years, make a move to the same place.) He then stepped out of BRR into the position of deputy director for BSR until July of 1993, when it had once again become DEB. Finally by that July, he became Mary Clutter's deputy assistant director for the new BIO. The BRR funded natural history collections in museums and botanical gardens, as well as marine laboratories and field stations. It also supported preparation and publication of systematics catalogs, "a really exciting program for me [as a systematist]," he stated. In later years, the cluster added programs such as Assembling the Tree of Life, Microbial Observatories, and others. Edwards' several moves about DEB into areas where systematics penetrated were not so much musical chairs as wise placement of a specialist: he fitted the needs of structure as structure changed to meet the needs of both BIO and the community.

Edwards, like so many other permanent scientific staff at BIO, and BBS, played multiple roles in advancing science both within the Foundation and without. Given his broad experiences and perspectives, Edwards was detailed to the international Organization for Economic Cooperation and Development (OECD) Mega Science group (previous chapter) in the mid-1990s and would spend 5 years, full-time, in such service from 2001 to 2006. "I got Mary Clutter to let me propose to the OECD that biological informatics is a megascience," he told the author. It was not like the big science of telescopic observatories, he stated, but given the large numbers of practitioners of systematics spread worldwide, it was a "distributed megascience."[4] So, in 1996, OECD set up a working group in the discipline, which Edwards chaired

[1] James R. Estes, "The National Science Foundation: Financial, Intellectual, and Knowledge Broker for Systematic Biology," *Annals of the Missouri Botanical Garden* (1993):323–331.

[2] NSF, "Adapting to the Future," *op. cit.*

[3] Edwards interview.

[4] Ibid.

part of the time though employed at the Foundation. This is a good example of NSF's international involvement.

The OECD argued that there was reason to integrate natural history collections' information and bioinformatics on a grand scale. That led to the formation of the Global Biodiversity Information Facility (GBIF), a major force as of this writing, and it was to that institution that Edwards was finally detailed for a half decade. The incubation time was prolonged, but when GBIF finally became operational in 2001, Edwards moved to Europe. The date he was informed of GBIF's start-up was the second "day of infamy" in American history, September 11, 2001, and he was to fly to Paris, France, to become the first Executive Secretary for GBIF. All air traffic in the United States was grounded that fateful day due to the terrorist attacks, and so he did not make the trip. He did, eventually, take on the leadership of the GBIF at its Copenhagen, Denmark, headquarters location, and has argued that NSF was a major element in forwarding *biodiversity informatics* intellectually, if not by direct funding to GBIF.[5] Edwards was paid by GBIF, but had not technically left NSF.

When it was time for Edwards to return to Arlington from Copenhagen, Clutter had already turned over the AD position to Jim Collins (2005), and there was no place for Edwards in the changed organization. While he had continued reporting to NSF annually on his work at GBIF, it was the latter to which he was committed. Finally, at the time of Edwards' return from Denmark and his subsequent retirement from NSF (May 2007), he moved to the Smithsonian National Museum of Natural History (NMNH). He did so having been approached while at GBIF where he was asked to develop a new project, the Encyclopedia of Life. The EoL was, and is as of this writing, sponsored by both the Sloan and the MacArthur Foundations and sited at the NMNH. The EoL had been an idea of the GBIF, but funds had never appeared for its creation by them. It had been Harvard's entomologist and the founder of sociobiology, Edward O. Wilson, who suggested to the MacArthur that they underwrite the establishment of the EoL, Edwards said. Persons at MacArthur approached Jesse Ausubel at the Sloan to join them, and it was he who asked Edwards to lead the Encyclopedia effort.

While the later career activities of Edwards were not at NSF, and they occur beyond the end date of this history, it is notable to mention them because Edwards has said that, "I owe Mary Clutter a real debt, a real thank you." He went on to say that "I learned a hell of a lot from her about managing and…she let me pursue this other career [GBIF]…It was on the international stage where NSF is not a major player." So, with Clutter's aid, Edwards did a great deal to forward bioinformatics as a global science. The NSF, as noted numerous times in this study, has been remarkable as an incubator of science and the scientists who create it, at both home and abroad.

Jim Rodman (earlier) was also a rotator in systematic biology with Edwards in 1983, the two overlapping by a short time after Rodman came to NSF that year.

[5] Ibid. See also A.R. Taylor and R. Braun, "Megascience, the OECD, and the New Global Forum," http://www.questia.com/PM.qst?a=o&d=98558402. The title word is misspelled on the Internet as "Megasceince…" Further, see http://www.gbif.org/.

Their years of service at NSF were nearly the same to the year: Rodman retired in 2006. The systematics program had been run by a single program officer prior to 1976 but became a two-person program from 1976 until FY83. In 1983, a third position was engineered by John Brooks. There had been assistant program managers at some periods during those years. Rodman was the botanist and Edwards the zoologist, and, thanks to Brooks, Lloyd Knutson, detailed from USDA, became the invertebrate zoologist in FY84. The next year he was replaced by Gerald Krantz, and Edwards became a permanent employee and moved laterally to head BRR. Systematics remained fairly small during the later 1980s when molecular aspects of the field were beginning to be seen in the community.[6]

By 1990–1991 and the beginnings of the reorganization, Rodman was on a sabbatical at the University of Wisconsin, Madison, where he was learning the new molecularly based systematics' techniques. He had earlier started his career as a faculty member for a decade at Yale in what had been called biochemical systematics in the 1930s through the 1950s. By his time, the tradition was strong, and he was steeped in it, but the discipline was at that cusp where it moved from electrophoresis and immunological techniques to more modern chromatographic tools and then on to DNA restriction analyses and nucleotide sequencing in the 1970s through the 1990s, respectively. All those areas had been supported by NSF, of course—and genomics would follow. The techniques he learned at Madison he brought back to what was about to become the renamed DEB with its new cluster of Systematic and Population Biology.

Rodman's reference to technologies embraces a profound feature. That is, specialists in widely differing areas of scientific research could use many of the same basic tools and techniques. Chromatography, in its many guises over time and to the more recent period, is used by many types of investigators to solve many disparate questions in biology, chemistry, and many other disciplines. The growth of *interdisciplinarity* can be seen not just on the intellectual level—so many specialists can use genomic information for so many diverse purposes—but also based upon the technological aspects of scientific investigation: a situation of *less is more*. Fewer pieces of equipment, fewer types of technologies, serve many more fields today, particularly as those fields begin to coalesce. This reality underlies much of what can be seen in this study. Less is more.

One of the cluster's efforts, the Biotic Surveys and Inventories Program (BS&I), was begun while Rodman was at Wisconsin. David Schindel was heading up the new program. Its initiation had been announced via a "Dear Colleague" letter and did not get vetted formally via the Office of Management and Budget, the more proper approach.[7] Another, slightly later, program was created through normal channels, though: PEET, Partnerships for Enhancing Expertise in Taxonomy. Both initiatives created the rationale for a fourth position in systematics, usually associated with the BS&I program in the Long-Term Projects cluster along with the

[6] Rodman interview.

[7] Ibid.

collections support program (typically with a rotator PD following Schindel's departure from BIO).[8]

9.2.1 The Biotic Surveys and Inventories Program

In 1990, just before formal efforts to create the new BIO were undertaken, Schindel ran the museum collections support program and also handled marine laboratories and field stations: the latter two were shared with the geoscience and oceanography directorates as "silent partners," he said.[9] All those programs later moved to DBI. Schindel has opined that moving the museum collections support program to DBI was a mistake. The idea for doing so had been discussed for some time prior to that decision with the main question being whether that program fit better under an area that would include Biotic Surveys and Inventories or under the restructured DBI umbrella of research resources. It was moved, but Schindel noted that the budget stagnated after that and it stayed exactly the same for 15 years; it was actually much lower as inflation had not been taken into account. It was his view that the BRR program would have done much better had it stayed in DEB.[10] Schindel also became Head of the BRR unit and stayed in that position until it disappeared in 1992. That was where Rodman found Schindel when the former returned in fall of 1991 from his Wisconsin sabbatical. However, as the permanent program officers, respectively, Rodman headed the systematics programs that year, and Mark Courtney continued to head up the new version of the old BSR's PBPE as the other half of the Systematics and Population Biology cluster. Schindel then left the biology directorate altogether in 1992 with his move to the Directorate of Education and Human Resources. When I interviewed him, he was located at the Smithsonian but was paid from a Sloan Foundation grant to the museum where he functioned (at the time of this writing) as the Executive Secretary of the Consortium for the Barcode of Life, another program in the newer flush of molecularly based systematics' efforts.[11]

In the founding year for BIO, then, there were three PDs in the Systematics area, including Rodman and two rotators, plus a senior program assistant and a program assistant. Similarly, Courtney and two rotators comprised the PD/POs in Population Biology, and they had two assistants.

The Biotic Surveys and Inventories Program (informally termed BS&I), of 1992 and beyond, was created just at the time that the name DEB came back into use and the new structure of both the division and the directorate was congealing. BS&I's *raison d'être* was found in a sentence in one of the NSB's then recent publications.

[8] Personal communication with Jim Rodman, May 7, 2012.

[9] Schindel interview.

[10] Ibid.

[11] For more information on this non-NSF undertaking, see http://www.barcodeoflife.org/.

> We are at a crucial juncture for the conservation and study of biodiversity; such an opportunity will never occur again. Understanding and maintaining that diversity is the key to humanity's continued prosperous and stable existence on Earth.[12]

The statement above was repeated on the opening page of the annual Program Guidelines for BS&I to introduce the reformulated program to the community. The actual purpose of the program was given in a very interdisciplinary commentary that pointed out the wide uses of understanding diversity for not only the biological community, but for others such as the social and behavioral sciences. Such apparently disparate disciplines as conservation, anthropology, paleobiology, and land-use studies were included in a long list of disciplines. In addition, there was employment of the term "global climate change," which, in the early 1990s, had not yet grown to be a major argumentative subject on the national and international political agendas.[13] Funding of proposals that involved

> research to record and document the diversity of life on earth, both past and present, as a prologue to investigations of patterns and processes and the development of plans for conservation of that diversity

would be considered, the call for proposals stated.[14] Biodiversity ran from viruses to the most complex eukaryotes and from fossils to living organisms. Some 15 to 35 awards were to be made annually and for up to 5 years each with levels of $30 k to $150 k per year; most, however, would be approximately $70 k, the Guidelines stated.

Soon (FY96), however, the BS&I had moved into the Long-Term Projects in Environmental Biology cluster of DEB. The LTPEB cluster would also finally include LTER, Long-Term Research in Environmental Biology (LTREB), Land-Margin Ecological Research (LMER), and Research Collections in Systematics and Ecology (RCS&E), the latter being managed by the PD who also handled BS&I after its move. The closing discussion of BS&I, since it started under the Systematics and Population Biology cluster, will conclude here; the others will be considered later in this chapter under the LTPEB cluster. Nonetheless, all of those programs were closely allied and highly interactive no matter the time period or administrative placement.

BS&I had been established before a critical document, *Systematics Agenda 2000*, appeared (below), but the programming of the former was ideal in fitting the needs of the latter: supporting survey and collecting expeditions and electronically publishing inventories of organisms.[15] Especially invited as project proposals were

---

[12] National Science Board, "Loss of Biological Diversity: A Global Crisis Requiring International Solutions," NSB-89-171 (Washington, D.C.: National Science Foundation, 1989).

[13] National Science Foundation, "Biotic Surveys and Inventories: Program Guidelines," NSF 94–66, May 24, 1994. For a history of the subject of global climate change in its early years, see Spencer R. Weart, *The Discovery of Global Warming* (Boston: Harvard University Press, Second Edition, 2008), *passim.*

[14] Ibid.

[15] National Science Foundation, "Biotic Surveys and Inventories (Program Element: 1198): Annual Report FY96," Meredith Lane, PD.

those focusing on the "discovery of species new to science" from "among understudied groups of organisms, and from little-explored regions of the world": surely, the highly romantic end of biology. The program was unique on the planet, its PD, Meredith Lane, held.[16] Indeed, moving into a paperless world, no print publications, save one, would be supported thenceforth. That last one on paper was the "Legumes of the United States: *Lupinus*," a project already long underway. The move to fully electronically published results and interpretations was an aid to many aspects of both science and society. Such online data would contribute to the work of what had been the National Biological Survey, subsumed under the US Geological Survey's Division of Biological Resources; the Convention on Biological Diversity (the CBD mentioned earlier); and the Mega Science Forum, among a host of others.

Included in BS&I support was the massive *Flora of North America* that would eventually cover over 20,000 taxa of higher plants north of Mexico. Also, the ongoing *Flora of China* project was renewed, and other national and international projects such as the *Agaricales of Hawaii* (a fungal taxon) and the *International Kuril Islands Project*, with Russia and Japan as partners, were funded in the mid-1990s. International meetings were also being supported. In FY96, for instance, ten new BS&I project awards were made and six more renewed. Some 63 were declined: a success rate overall, with a few other minor categories included, of 22%. The 30% or greater success rate across the directorate throughout much of the BBS era was less commonly seen in the BIO years of the last decade of the millennium as the total number of proposals, and their quality, went up: competition became very stiff, indeed. The total ongoing awards being administered by the program at that point was 90. Plant diversity studies were awarded the most, as a category, at just under $1 M, with vertebrates a close second and microbes receiving about half-that of plants.[17]

Within a year, the smallest known (up to that time) frog was discovered on a Caribbean island, but, too, more effort was made by NSF to increase the number of proposals from other areas than vertebrate zoology: the microbiological community was one being solicited. And, by year's end, the amount awarded for new grants nearly doubled from that of FY96 to three-quarters of a million dollars for the microbe category. As the total number of proposals (all categories) to the program went up with 13 new ones awarded, the success rate dropped to 19%, an indication of growing competition from the BS&I community. Just under $4 M was expended on new and continuing grants and supplements in FY97.[18]

The following year saw the first Native American panelist serve with BS&I. Another first was that *all* proposals to the program were submitted electronically through the then fully operational FastLane: in fact, BS&I's use of it represented a test case within DEB. And in what *The Washington Post* called "one of the

[16] Ibid.

[17] Ibid.

[18] National Science Foundation, "Biotic Surveys and Inventories (Program Element: 1198): Annual Report FY97," Meredith Lane, PD. Lane was replaced temporarily as PD by James T. Callahan, partway through the year. Lane alone, it appears, authored the Annual Report, however.

most important evolutionary developments of this century," PI Andrew Knoll of Harvard, while working in China, found highly detailed three-dimensional fossil embryos from a Cambrian horizon.[19] Total new awards rose to 15.

9.2.2 *Partnerships for Enhancing Expertise in Taxonomy*

The introductory paragraph of the first PEET call for proposals document apparently had "flowery language," OMB said, and that had to be edited out by Jim Rodman.[20] The OMB objected to, what Rodman characterized as, "grand, E.O. Wilson-style" writing. Such was to the effect that "[t]he stars will be there, but the plants and animals may be gone" if the biodiversity crisis was not addressed by systematists, let alone civilization in general. It was Rodman's opinion that OMB thought that the program announcement draft was possibly "disparaging astronomy," an area that historically received a great deal more funding than systematic biology! Matters were smoothed out and the PEET program began in the mid-1990s. That smoothing began with a call for proposals in 1994 with the first 21 awards in 1995.[21]

The basis for establishing PEET was in response to pressures in the community and in governmental circles to train more systematists in answer to the growing biodiversity crisis, about which E. O. Wilson had begun warning during the 1980s.[22] The systematics communities in the United States, and in Europe, were both concerned about species loss, and a movement arose that included such persons as mammalogist, Terry Yates. Yates ran the division-wide rotator program at DEB and was, specifically, the rotator PD for Systematic Biology for FY91 and FY92. Upon his later return to NSF as division director for DEB, he was instrumental in establishing the AToL initiative. Just after his first 2 years as a rotator, he would demonstrate the reservoir for the mystery disease that was then (the early 1990s) plaguing the Southwestern United States—particularly the "Four Corners" region and deeper into New Mexico, specifically. It was a hantavirus endemic to small rodents. That

[19] National Science Foundation, "Biotic Surveys and Inventories (Program Element: 1198): Annual Report FY98," Joel W. Martin, PD, and Vivian Nolan, science assistant. Douglas Siegel-Causey had come in new and filled a newly created rotatorship. Martin had replaced Lane. See footnote above. The newspaper citation was present in the Annual Report, which also noted that the discovery was reported in *Science* (February, 1998). The discoverer of the smallest frog the year before was cited in *Nature* (in the April 30, 1998, issue) for his creation of a molecular time scale for vertebrate evolution in FY98. Numerous other firsts and significant findings were also listed.

[20] Rodman interview.

[21] James E. Rodman and Jeannine H. Cody, "The Taxonomic Impediment Overcome: NSF's Partnership for Enhancing Expertise in Taxonomy (PEET) as a Model," *Journal of Systematic Biology* (2003):428–435.

[22] Ibid. See also E.O. Wilson, *Biodiversity* (Washington, D.C.: National Academies Press, 1988).

epidemic had led to, at least, 20 human deaths at the time.[23] Yates was then also co-principal investigator at New Mexico's Sevilleta LTER site, which had begun in 1988.

Also, Rita Colwell, a number of years before she became NSF director, also spoke out on the issue that so concerned Wilson and many others.[24] Thus, in 1993 a document, entitled *Systematics Agenda 2000: Charting the Biosphere* alluded to above, appeared from the combined US and European systematics communities. The *Agenda* introduced its three mission items:

> Mission 1: To survey, discover, inventory and describe global species diversity accurately, efficiently and rapidly[;]
> Mission 2: To analyze and synthesize the information derived from this global discovery effort into a predictive classification system that reflects the history of life[; and,]
> Mission 3: To organize the information derived from this global program in an efficiently retrievable form that best meets the needs of science and society.[25]

The intent was to complete the inventory within 25 years, and the authors argued that support for the effort was logical in that it would yield new "usable resources," including new crops; arm conservationists and policy-makers with necessary data for conservation efforts; and provide a structure better to understand global climate and ecosystem change including extinctions, degradation, the proliferation of pests, and the spread of disease. Hanta was an excellent example as it had been tied to unusual weather.[26] British botanist, Sir Ghillean Prance, pointed out in 2001 that:

> [a]n overview of systematics, the naming, classifying and ordering of life, shows that it has never been more relevant than in the present time of environmental crisis and species loss. Only 1.9 million of the many millions of species have been named and classified, making it difficult to use and conserve them. Molecular techniques combined with cladistic methods and high-powered computing ability are producing predictive phylogenies of many groups of organisms and are bringing us closer to working out an accurate tree of life.[27]

---

[23] See http://www.nsf.gov/about/history/nsf0050/environment/solution.htm. The hantavirus group had only been known in Europe and Asia prior to Yates' discovery of its apparent first appearance in the United States in deer mice. It was determined that the El Niño weather pattern was the likely cause for that virus to be seen for the first time in the Four Corners: rodent populations had increased there in 1992–1993 by some 20-fold. The virus had been present in the area all along, but the unusual change in global climate that triggers El Niño events led to the increase in rodent numbers and discovery of the pathogen. More human interaction with the small mammals transferred the virus by human contact with the virus-infested urine of the mice. By a decade later, over 300 cases had been confirmed in the Americas. See http://www.cdc.gov/mmwr/preview/mmwrhtml/rr5109a1.htm.

[24] Rodman interview.

[25] See http://www.bio.net/bionet/mm/bioforum/1993-November/006488.html. See also M. F. Claridge, "Introducing Systematics Agenda 2000[: Charting the Biosphere],"*Biodiversity and Conservation* (1995):451–454.

[26] Ibid.

[27] Ghillean T. Prance, *Systematics: Relevance to the Twenty-first Century* (2001); published online: January 17, 2011; DOI: https://doi.org/10.1002/9780470015902.a0003608. Quotation from the Abstract. The figure of 1.7M species was in use by the Tree of Life research community at that same time. See footnotes for main text later below. The argument about the number of species on

Funding for PEET, and other systematics programs not aimed at training, did not always come in amounts that program officers at NSF may have wanted in order to fund the systematics investigations for which the community hoped. Rodman tells the story that he was "always scrounging for money through joint reviews and joint activities." Indeed, "that's where my energies and mind were through the nineties." Was systematics being short-changed in the budget? His answer was that he believed none of the POs in the group felt that way, but they were all, "I'll use the phrase, 'turf conscious'." He pointed out that areas particularly well funded, *neurosciences*, *developmental biology*, and *genomics*, were making great headway in science and "adding wonderful new technologies."[28] Those were being shared with systematics too, of course, and so he did not feel that the area was being "short-changed." But he did go on to say that he had to "fight hard" for systematics' share of the budget and he wanted more than that—he wanted a "*growing* share." He felt he may have made himself "obnoxious" worrying about support for his discipline.

The intellectual commitment, however, was as great as that of seeking sufficient funding. One of the big three at the time, developmental biology, though at first blush seen through classical biology, did not seem closely related to systematics. It was made to be so because using "comparative study in a phylogenetic context [required that systematics] had to provide a *scaffold* for…the new developmental biology." One "wants to know [about] related organisms" to the one under study, and so building genetic pathways' information plays into both systematics *and* development—an area that came to be called phylogenomic biology.[29] Phylogenomics began to play a role in constructing the Tree of Life in the early 2000s, but the method was not without initial challenges. Phylogenetics, the basis for phylogenomics, was a central construct in the systematic biology program during the 23-year period that Rodman was at NSF. He felt that BIO had "achieved an intellectual coherence around the theme of phylogeny that it had long proclaimed as its ideal."[30] It was a realization implied in the functional organization for biology that Weiss had suggested to the aborning NSF.

In connection with improving the systematic biology programs' financial lot, Rodman argued with POs of other programs that if they were doing "anything comparative," then they would have to be doing some level of systematic biology. He would make the point that he could offer reviewer's names, for instance, for various programs. He found that he got a particularly "sympathetic hearing from the *Arabidopsis* genomics folks," as well as those in plant genomics more broadly.

Earth is an old one with a great many figures having been offered, some as high as 100 M. The latest incarnation of the dispute, offering ~8.7 M (but eukaryotes only!), can be seen at http://www.PLoSBiology.org/article/info%3Adoi%2F10.1371%Fjournal.pbio.1001127. See also Daniel Strain, "8.7 Million: A New Estimate for All the Complex Species on Earth," *Science* (2011):1083; a news article.

[28] Rodman interview.

[29] See http://www.ncbi.nlm.nih.gov/pubmed/15861208. Emphasis is added to the Rodman quotation preceding.

[30] Rodman interview.

Years later, seeking to recall a name, he believed it was with Jane Silverthorne that he did joint work, she having long handled aspects of plant genomics in BIO.[31] He traded his services for funding help for his area: he would do reviews and read proposals, for example. Indeed, he went to Germany to aid its plant biologists in their *Arabidopsis* genomics work, and when they came to America, he would be present to help in this country. He also helped create the joint NSF-DFG (German federal) program, he has stated.

Another spin-off was that graduate students would be employed in these efforts and that the number of future systematists would be increased—the point of PEET. While they may or may not have had PEET funding, the philosophical end was the same: a "win/win situation," he said. Rodman recalled that it was through the encouragement of Machi Dilworth that he was so closely tied into the plant genomics area. But further, he felt, Mary Clutter may have been saying: "This is good…Jim is shaking his tin cup for systematics."[32]

Such team efforts were common at NSF, Rodman and many others have said. And in the case of the *Arabidopsis* genomics efforts, the "significant cross-fertilization" with the systematics program was, for him:

> gratifying and satisfying [because] people doing comparative DNA analyses to create the Tree of Life are using many of the same technologies as the genomics folks for getting comparable data.[33]

As was said many decades ago concerning systematic biology, "knowledge of what a thing is, is the key to all literature on what to do about it."[34] "Literature" may well be expanded to include computerized data—bioinformatics and phylogenetics—in the contemporary era. If these statements are true, the work ongoing for those fields through NSF funding was absolutely critical: creative—"tin cup"—financing was one crucial feature.

But financial support was about to become more available as Rita Colwell became NSF director in 1999, its first woman and first pure biologist director, as we have seen.[35] Not only did her biocomplexity concept help out systematics, but so also did the Computer Science Initiative she began. Rodman noted that the Assembling the Tree of Life project "involved some horrendous computational challenges."[36] The phylogenetic trees with thousands of species, and the nucleic acid trees with millions of data points, required highly sophisticated computational efforts. Indeed, the Biocomplexity Initiative was overshadowed by the computer initiative that was funded through the computer science directorate—and which had

[31] Ibid.

[32] Ibid.

[33] Ibid.

[34] Said by Paul O. Ritcher, former Chair of Entomology at Oregon State University, Corvallis, when this author was a doctoral student there in the early 1970s. For more on Ritcher, see http://www.unl.edu/museum/research/entomology/workers/PRitcher.htm.

[35] Recall that Richard Atkinson was a psychologist.

[36] Rodman interview.

ripple effects in all directorates. (Colwell's computerization effort was Foundation-wide, not just limited to BIO.) Although the systematics PDs found it was somewhat challenging to create joint programs with the computer science group, matters were finally resolved through considerable "jaw-boning and cajoling" by those systematists who had strong computer backgrounds, Rodman recalled. Such individuals, as Schindel and Matthew Kane (more later below), interacted successfully with the other directorate's POs, and so joint programs were established. The first competition for the AToL program, then, was opened in 2002.

9.2.3 Assembling the Tree of Life

Evo-devo (previous chapter) had been developing for several years, but the AToL program was just beginning at the turn of the millennium. Thus evo-devo was to meet the tree of life in a formal way at a mid-September conference in 2000 at NSF. The research advances in genomics that had been so valuable in the early growth of the former were seen to be of application in the latter, and it was that interface that was explored by the attendees. The constraints of one community were found to be the same for the other, and their elimination became the goal. A convergence of research communities was seen to be occurring due to both technical and conceptual advances in each discipline, especially in genomics right at that time. The unifying feature of genomics and associated techniques and technology would provide a synergism which dynamic would help define so much of *Millennial Biology*.[37]

The three discoveries (previous chapter) about genetic similarity in animal development, genes, gene families, and changes in gene expression, were reminiscent of the 1950s' discovery that all living creatures built their DNA with the same set of four nucleic acids. Thus the conservation of fundamental features in living things was seen at higher levels of organization (e.g., gene families), not just the nucleic acid level: a pervasive proof of the interrelatedness of life and of organic evolution—a single tree in evolutionary history seemed likely.[38] Under such circumstances, the convergence and synergism hoped for was fully possible at multiple levels. The conference thus found a shared vision including similar "questions and recommendations for implementing research to address these questions." The three proposals consisted of the following:

- The generation of arrayed BAC [bacterial artificial chromosomes[39]] and cDNA libraries from organisms across the Tree of Life

---

[37] National Science Foundation, "Evolution of Development and the Tree of Life: Workshop Report." See http://www.nsf.gov/pubs/2001/bio012/start.htm?org=NSF. The main text above is a paraphrase of the introductory paragraph in the Report.

[38] Ibid. Report Introduction.

[39] See, for instance, http://www.ornl.gov/sci/techresources/meetings/bacpac/preface2.shtml, for a discussion of the nature of and utility of BACs.

- The extension of molecular techniques to non-model organisms
- The development of a phyloinformatics infrastructure that is easily accessible and flexible enough to answer questions from any area of the biological community[40]

In order to "jump-start" the AToL, the conference attendees determined to choose the major trunk(s) for the tree using an agreed-upon "First 100" non-model organisms. While a number of such candidates had already been sequenced by 2000, and would not be duplicated, many of the others had yet to be chosen. The gathered scientists envisioned 100 grant awards for the 100 organisms resulting in the creation of suitable BACs and cDNA libraries. The repositories for those data would be used by both the evo-devo and the AToL communities. The creation of a robust phyloinformatics infrastructure was already underway in pre-AToL efforts, and the NSF-supported *Drosophila* stock and information center was the model. The phylogenies that AToL was building at the time had begun by including nuclear genome sequences. Finding conserved genetic motifs (e.g., gene families) across wide areas of the tree was of particular interest to the evo-devo community and evidenced just one area of the growing interface between the two. Indeed, "[s]equencing about two dozen major developmental gene families" in the "First 100" was to be a priority for the two communities.[41] A timetable would include competitions to begin in 2001, individual investigator grants for 2002, expansion of informatics and increased evo-devo grants for 2003, and grants for identified evo-devo major questions in 2004, among many other points along the way.

That workshop was hardly the only one on the subject of AToL at millennium's turn, as three others were held at Yale; UC, Davis; and finally at the University of Texas, Austin (December 2000). All three had been aimed at the AToL community and had inspired the meeting at NSF just 2 months before which united it to the evo-devo group.

Over the time of the conferences, at least, the notion of the breadth of applicability of information derivable from a Tree of Life was nothing less than stunning. It was hardly of interest merely to systematists, a group long thought of in some circles of biology, and in more specialties within the physical sciences, as engaging in "stamp collecting." But, as Stephen Jay Gould had sagely said in the defense of the field, "Classifications are theories about the basis of natural order, not dull catalogs compiled only to avoid chaos."[42] Indeed, even the early AToL efforts were leading to such surprising ends as identifying exotic invasive species that could threaten our food supplies, ecosystems, and even human health.[43] Some 26 participants from

---

[40] Report Introduction.

[41] See Web pages in the Report.

[42] Stephen Jay Gould, *Wonderful Life: The Burgess Shale and the Nature of History* (New York: W.W. Norton, 1989), pg. 98.

[43] National Science Foundation, "Tree of Life Workshop III: Developing the Technology and Infrastructure Needed for Assembling the Tree of Life: Report of a Workshop Held at the University of Texas, Austin, December 3 and 4, 2000," citation on pg. 4 of a 31-page typescript report..

across interest groups within and without academia, institutions (botanical gardens, etc.), and other sectors of the greater biological community gathered along with observers from NSF, those including Matt Kane, Jim Rodman, Terry Yates, and two others. They discussed what they termed ToLNets, or groups of investigators that focused on a "particular empirical or theoretical problem." In essence, they would somewhat resemble STCs, but were highly geographically dispersed, and with familiar 5-year funding cycles. They would include ToLNets, or "institutions with a special concentration of, and commitment to, AToL activities" with shared resources, including personnel.[44] Finally, the nets and hubs would be organized by what was termed the PICI, "Phyloinformatics and Coordination Infrastructure." The PICI would be in charge, especially, of shared databases.

The conferees saw the ToLNets constructed along one of two main grounds: taxon-based (a given group of insects, say) or analysis-based (theory or analytical methods for use in assembling the ToL). By the first year of the new century, NSF published a general interest statement about the planned national and international efforts for AToL.

> Resolving the Tree of Life is unquestionably among the most complex scientific problems facing biology and presents challenges much greater than sequencing the human genome.[45]

The slick document containing those words noted that AToL was doing a number of things. Those included

> [u]sing phylogenetic analysis to discover new life forms for biotechnology …improving global agriculture…tracing developmental change…[discovering possible new medicines]…protect[ing] ecosystems from invasive species,

and others.[46] A more refined focus was on emerging diseases (West Nile virus; hantavirus), medical forensics (history of human disease transmission), and reiterating what was slowly becoming obvious once again in the history of biology: *that natural history collections are essential to research* in wider areas than had been comprehended for many recent decades.[47]

It was during the period that AToL was gearing up that Quentin Wheeler came to NSF for an intended 1 year as a rotating program director for systematics.[48] After only some 7 or 8 months in that position, however, Wheeler was offered the chance to become Division Director of DEB, and he stayed on for a total of 3 years. He had replaced Bruce Hayden who was the latest in a line of *rotating division directors*: having such was the philosophy of the division. Joanne Roskoski was the *permanent deputy director* of the division, also the view of how DEB wanted to manage its unit.

---

[44] Ibid., pg. 8–9.

[45] National Science Foundation, "Assembling the Tree of Life: Harnessing Life's History to Benefit Science and Society," n.d., no document number; based on the outcomes of the three workshops; no longer available online; ten-page paper copy.

[46] Ibid., *passim*.

[47] As of the publication date of this study, ToL can be found at http://tolweb.org/tree/phylogeny.html. It is not especially easy to find in a casual Web search.

[48] Quentin Wheeler interview with the author, May 22, 2009.

(However, Samuel Scheiner was acting DD in later 2002.) New ideas would come in with a rotating director, but there would also be a "strong, experienced right hand," for institutional memory and savvy political skills in its deputy, Wheeler recalled. As it happened, the various divisions of BIO had differing views, and hence formulae, for determining the positions that were to be regarded as permanent and those to be reserved for rotators. Those views did change with time, of course, as has been seen in a number of previous chapters. As David Schindel said, in a positive fashion, "[e]ven though they make rules, they then break those rules."[49]

A healthy learning experience for Wheeler came out of his time at NSF and is worth mentioning in regard to the philosophical role that, he opined, was a crucial one for the Foundation to play in and for the community. There were types of data analyses in systematics ongoing in the early years of the present century with which Wheeler did not agree and which he felt were a waste of funds to support. However, he learned an overarching lesson from the *intricate dance* that one gets from a "30,000 foot view" (a term used widely in federal agencies), that is, whether or not a funding officer—program or division director—agrees with something fundamental in a proposal. If that something grew out of years of acceptance within the community through scientific meetings, published papers, and other means of "normal discourse," it had become incorporated into the "standards and practices" of the community. Therefore, a PO or DD then *had* to "respect what the community wants to fund," he said.[50] Should the instincts of the funding officer turn out to have been correct, he argued, then the lesson learned would permeate the community, and so such investments would have served the betterment of all concerned as "necessary intellectual growth." It is not unlike the communal value derived from the publishing of negative data.

Wheeler's background was ideal for a temporary division director of environmental biology as he was an entomologist (comparative anatomy and morphology of insects, in his case) and was concerned with systematic biology more broadly. He had taken all three of his degrees at Ohio State University, Columbus, finishing his doctorate in 1980. He then began his academic career at Cornell University and moved up the faculty ranks to full professor and chair of entomology. He was also the Director of the Liberty Hyde Bailey Hortorium, the de facto US center for systematics of *cultivated* plants. In 2001, then, Wheeler took his leave of absence from Cornell to serve at DEB. Though due to return to the University after his unexpected second, and then third, years at the Foundation, he resigned his professorship to move to London, England, and the position of keeper and head of entomology at Britain's Natural History Museum.[51] Cornell was a "brutal environment," he opined, as he spent his two decades there with his "chair against the wall" to protect himself from the "backstabbing," he felt, that characterized the University. He stayed but

---

[49] Schindel interview.

[50] Wheeler interview.

[51] Prior to the UK Museums and Galleries Act of 1992, the Museum was known as the "British Museum (Natural History)."

2 years in the United Kingdom before his move to Arizona State University, Tempe, where he held such positions as Dean of the College of Liberal Arts and Sciences, university vice president, and, in the later 1990s, Founder of the International Institute for Species Exploration located on that campus.[52] He is Dean of the School of Life Sciences there as of this writing.

Wheeler, not atypically, learned of NSF as a graduate student under his mentor. That scientist was Tod Stuessy at Ohio State, who had a Foundation grant. Also, Stuessy had been Associate Program Director in Systematic Biology there in FY78. Wheeler had his own NSF award once he began his career at Cornell. He also served the Foundation as a proposal reviewer and a COV member, though never served on a panel, something he felt might have been a "unique distinction" for an eventual division director.[53] But with regard to the more philosophical underpinnings of biology, *qua* science, Wheeler recalled his own maturation in the field as a young professor unhappy with the strong movement toward molecular biology over, for example, his more classical comparative morphology. He saw that NSF was not funding much in his field, but rather a great amount of its support was going toward the molecular end of biology. After heading for the library in a very disconcerted mood to look at J. Merton England's *A Patron for Pure Science* (see Chap. 1, footnote 1), he found that the organic act for the Foundation had Congress *mandating* that NSF was to aid in finding *gaps in research support* and fund them. That, he said, made him "very disappointed that it undercut my high horse"![54]

So it was that in 2001, with Wheeler's arrival as a PO and then rapid move into the DD position, he found the draft of a crucial call for proposals ready to be edited and finalized and finally rolled out, all of which fell to him.[55] By the end of the following year, NSF had awarded $17 M to some 25 institutions for research that would allow for beginning to assemble the Tree of Life.[56] Actually preceding the formal AToL project, those awards were part of the Biocomplexity/GenEn competition and functioned as a sort of pre-AToL activity. The first of the grants had been awarded in the fall of 2001 and was still being renewed well after this history's endpoint of 2005. That initial award's total over time came to more than $135 k to study "Comparative Chloroplast Genomics: Integrating Computational Methods, Molecular Evolution, and Phylogeny." It was led by PI Robert Jansen at the University of Texas, Austin. Three others were awarded in that first round. In 2002, the number was seven; five in 2003; eight in 2004; and seven more in 2005. Over the several years from 2001, each of the groups of awardees would be considered "cohorts" for a particular year, and four would speak from the 2001 cohort, seven

52 See http://www.species.asu.edu/index.

53 Wheeler interview.

54 Ibid.

55 Ibid.

56 National Science Foundation, Office of Legislative and Public Affairs, "Discovering the Tree of Life," NSF PR 02–94, November 18, 2002.

from the 2002, and so forth in the annual "All Hands" PI workshops.[57] (A similar term was used by the LTER community: All Scientists.)

But just prior to release of the first call for proposals in 2001, there was at least one issue with regard to how the program was to be structured. The still very new director of DEB, Wheeler, who had to finalize and approve the program content and announcement, took a close look at it and said, "Oh my God, this is all molecular"![58] As an anatomist and morphologist, Wheeler was aghast that there was no requirement for aspects other than the molecular expected in any given AToL grant proposal. In order to avoid even greater "alarm" to himself at what he discovered, he "went upstairs to the geology directorate" and said to them:

> I want some money…so that I can say that Assembling the Tree of Life is jointly supported by the paleontology program and, therefore, we have to include fossils and, therefore, have to include anatomical data or we can't relate to fossils…I said a dollar will do, and as I recall, there was a small amount, I think is was maybe $10,000 [that came from the geology directorate] out of $15 million.[59]

Nevertheless, Wheeler admitted, that was sufficient to "modify the program description and to say that proposals had to include a morphological component and fossils to be relevant." He therefore felt "virtuous" in making the new program "more integrative than it might have been otherwise." When asked what kind of response he got from the more molecularly oriented scientists in BIO, he said the Director of MCB (Maryanna Henkart; earlier chapters) "marched into my office" and a discussion ensued about Wheeler's change to the requirements. Since it eventuated that the proposals were primarily molecular in the end anyway, "I didn't think she'd lost too badly," he recalled.[60]

The NSF made its customary press release upon the establishment of AToL in late fall of 2002 by the time the first $17 M had been awarded.[61] Wheeler was quoted as having said that progress in many areas of biology, such as genomics, evolution, and development, was "encumbered by the lack of a rigorous historical framework to guide research." In a very few words, that summed up the absolute necessity to assemble the Tree of Life and provide the Tree with what is known as the *historical aspect of biology*: biosystematics, historical geology, evolution, etc. But his last words in the press release were of most interest: the work would include "expertise and data sources, from *paleontology to morphology, developmental biology,* and [third in the list!] molecular biology," said Wheeler.[62]

---

[57] National Science Foundation, "Assembling the Tree of Life—PI Meeting [NSF]," November 19–21, 2004, a five-page typed agenda.

[58] Wheeler interview.

[59] Ibid.

[60] Ibid.

[61] Press release NSF PR-02-94 above.

[62] Ibid. Emphasis added.

9.2.4 *Assessing the Systematic Biology Program*

In January of 1991, Marvalee Wake was on a BBS Task Force looking into education planning for future systematic biologists. She submitted a short report to Mary Clutter and Sonya Mallinoff on the subject.[63] She referred to the widely recognized biodiversity crisis, rapid increases in species' extinction rates, and the human role in that growing catastrophe. Many species were believed to be going extinct before science even knew of their existence. For that reason, Wake argued that systematists were needful of training in *specific groups* of organisms. She cited E. O. Wilson's belief that, for insects alone, there was a need for 25,000 specialists, but the entire world could only offer about 1000 then trained in entomological systematics.[64] Many of those were elderly and would be gone before they could aid in training a new generation of such scientists. "We consider it essential that the National Science Foundation sponsor the training of systematists," Wake stated. And it should be international in scope. NSF/BIO did recognize the "immediacy of the problem," and so PEET materialized.

It was not alone among the new and revised programming efforts to come out of the great reorganization. So, too, was the creation of the "Research Collections in Systematics and Ecology Program," the RCSE.[65] PEET would have been insufficient by itself and would require a funding program aimed not at educational grants alone, but at the intellectual heart of systematics where collections are crucial. All sorts of institutions having collections, such as museums, herbaria, botanic gardens, universities, state and local governments, NGOs, and others, could seek grant funding when the RCSE began soon after the reorganization. Similar programs had existed in past years at BBS, of course (earlier chapters), but RCSE was that incarnation under BIO.

So, by 1996, the number of programs had grown in BIO's new systematics' portion of the Systematic and Population Biology cluster. Also, the numbers of persons in that area had grown to seven: three PDs (two of whom were rotators) and four program assistants, the group still under Rodman.[66] In FY97, the seven personnel processed over 400 "actions" of all sorts with 276 proposals reviewed.[67] The success rate was 21% for new awards with renewals running at 29%. Four workshop proposals (out of 6 submitted) were awarded, and categories of individual awards showed an increase in those dealing with fungi (8 of 23 funded), but of those in the

---

[63] Marvalee H. Wake, "Systematics and the Training of Taxon Specialists," one and a half typed pages with handwritten transmittal memo, January 4, 1991.

[64] Ibid. Wake cited Wilson's article in *Science* (1989):754–756.

[65] See, for example, the Program Guidelines for FY93 at http://www.nsf.gov/pubs/stis1993/nsf93116/nsf93116.txt.

[66] National Science Foundation, "Annual Report Fiscal Year 1996, Systematic Biology Program," James E. Rodman, October 1, 1996.

[67] National Science Foundation, "Annual Report Fiscal Year 1997, Systematic Biology Program," James E. Rodman, November 3, 1997.

prokaryote area, only 1 of 10 submitted was funded. Monies used directly in support of research came to over $12 M. It was the period when the "new field of genomics" was growing and the systematics group was pleased that awards to molecular phylogenetics' investigators "now number in the hundreds."[68] The PEET panel reviewed 59 proposals, and other educational programs (e.g., RUI, DDIG, and CAREER) were also receiving and reviewing proposals. There were some 31 PEET grants active by the following year.[69]

The other half of the Systematic and Population Biology cluster was concerned especially with population research on organisms generally known systematically already. The types of biology done in the population area were strongly ecological.

9.2.5 Populations of Organisms

Mark Courtney was the permanent PD of the population biology area of the cluster in which he and Jim Rodman shared management. The great reorganization brought little change to most aspects of DEB, Courtney recalled, but he also said that, "in my program, it had a substantial effect."[70] As we have seen, parts of the BBS incarnation of PBPE were placed with the behavioral and social sciences. Courtney felt that the title Population Biology and Physiological Ecology was unnecessary from the beginning in that population biology "would naturally include physiological ecology." But that latter area was lost when it moved to IBN and their program of ecological and evolutionary physiology. From Courtney's perspective, those changes "removed about a third" of his program.

As far back as the 1975 reorganization, one reason for creating a partially separate area for population biology, however structured, was that it always had high proposal loads, particularly in the physiological areas. Even when, after 1992, the population program budget was reduced over 30%, the proposal numbers dropped for only a short period, and then "there was a real burst" in numbers—especially in the rapidly expanding *molecular* areas and also with regard to *computerization*: population studies are "very data heavy."[71] That combination strongly spurred growth in the area. Figure 9.1 demonstrates some of the fiscal effects of the reorganization, as well as much else over many years.

Courtney had been on sabbatical from 1991 to 1992, and, like Rodman who was on his own such leave the year before, the former missed much of the reorganization efforts; and the two of them were separated for 2 years. Also like Rodman, Courtney studied molecular aspects while away from NSF, particularly the methodologies in

---

[68] Ibid., pg. 9.

[69] National Science Foundation, "Annual Report Fiscal Year 1998, Systematic Biology Program," James E. Rodman, May 16, 2000.

[70] Courtney interview.

[71] Ibid.

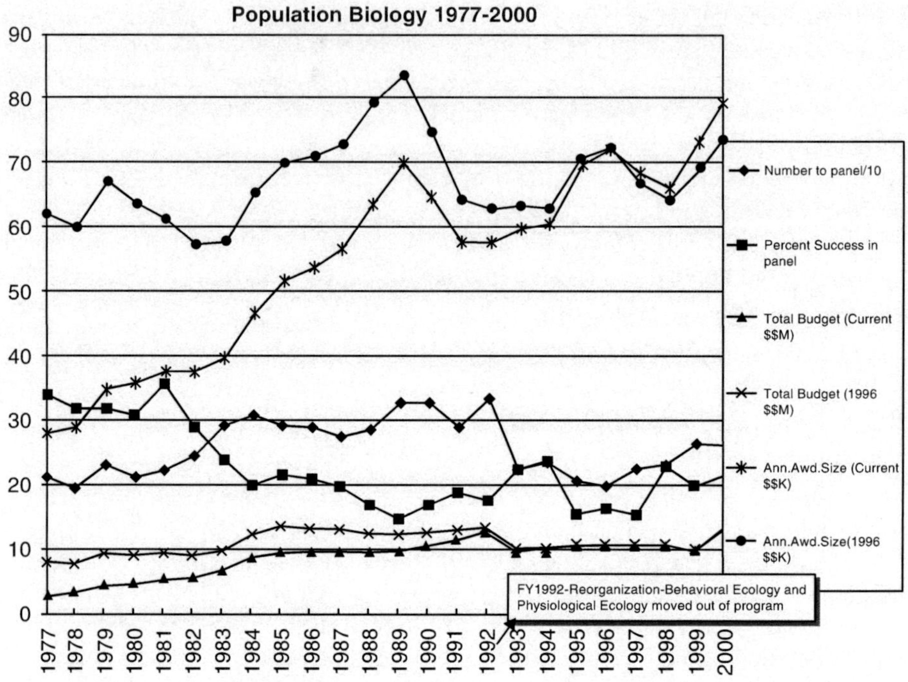

Fig. 9.1 A history of funding for the area of population biology from 1977 to 2000. NSF. Specific provenance unknown

his area. But upon his return to the new BIO, he felt that during the reorganization period, PDs directorate-wide were left out of the process, though the community was well served by the changes, he said. He opined that such feeling was widespread in BIO and that the whole restructuring was a fait accompli driven by the desire of the social and behavioral sciences to separate from biology. Finally, it was his opinion that within the newer, and reduced, population biology portion of the new cluster (clusters were themselves new entities) within DEB, his colleagues had a "siege mentality." With such, they intended to "protect [what was left of]…their small piece of the pie…by all means."[72] Since he was not present at the time, though, he did admit that he was "going out on a projective limb" by telling the author of that particular view. But his opinion would be substantiated by the *Annual Report for FY92*, as shall be seen. The social sciences'/biological sciences' tensions had settled down by the time of his return to NSF, though, he averred. Until the community fully understood the nature of the restructuring of 1992, proposals were sent to old and familiar places at BIO, now no longer extant, and so the PDs had to parcel them out to the new, correct programs. Importantly, the proposals that were felt by the front office of BIO to belong to a given program were the ones upon which their

[72] Ibid.

budgets were built. Should one end up taking a number of proposals from other programs, as their PDs felt they belonged elsewhere and passed them on, it had the effect of reducing a director's funds. Such a situation "leads to a lot of resentment," Courtney felt.

"Now, there's something that's logistically superimposed upon this, and that's the move to this new building [Arlington/Ballston] from the 18th and G Street [location]," the former having been a very open architecture, unlike the latter. Still, there is a "stovepipe…and wall and floor separation" of various directorates and parts of directorates in the Arlington building, he held. *The* "internal geography [led to a] changed ecology" *for the structure of NSF.* While any attempted analysis of the sociology of the Arlington, or previous, locations for NSF, has been explicitly avoided in this study, it remains necessary to point out that numerous scientists at BIO inherent to the move to the new headquarters discussed those challenges with the author. Specifically, Courtney felt that the division's "cohesion" was broken up to a degree, but he also felt that "in the end, [the negative effects were] neutralized because of *very proactive efforts* by division heads to counter that kind of isolation."[73] Further, he remembered that there was the "periodic reclassification of personnel," possibly five times in his career at NSF (1977–2008), and it was disconcerting to him, though he knew it was not uncommon in federal service. He "never saw one [reclassification] that was easy." My personal experience ratifies this.

Regardless of several less-than-ideal features that the PD for population biology had to deal with, Courtney knew that he headed an area with "fuzzy boundaries" (Mary Clutter had often spoken of "leaky borders") and that co-reviews and co-funding were very much a norm for his program. But he also identified over time many topical areas, and subsequent proposals, that he felt to be decidedly within his bailiwick and "which [of those] might be the most productive" scientifically. In the 1990s, when the number of PDs in his area had grown, all of them shared discussions so that no one was isolated in having to make award and funding level decisions. It required much more time in meetings, however, especially as everyone but Courtney was a rotator.

He opined that population biologists represented the "quintessential" example of single investigator researchers in the later 1970s when he arrived at NSF. Over the next several decades, Courtney watched that trend, and finally, in the 2000s, he formulated the idea of a new type of competition, OPUS. The "Opportunities for Promoting Understanding Through Synthesis program" got underway about 2004 and was aimed at senior researchers. The intent was for them to go back over their careers and their mostly single-handed project outcomes—data and interpretations—and prepare a monograph on their particular specialties to promote understanding through such a synthesis. The OPUS program was still in existence at the time of this writing.[74] It will require a close examination by future historians.

---

[73] Ibid. Emphasis added.

[74] See http://www.nsf.gov/pubs/2010/nsf10557/nsf10557.htm.

Returning to FY92, however, the number of PDs in population biology totaled six, a high not to be repeated. Five of them were listed as "Visiting"; that is, they were rotators.[75] Three assistants accounted for the remainder of the group. But it was the *program's annual report itself* for FY92 that was the most interesting aspect of the period. It contained an unusual feature in the title for such documents, which stated that it was "A Year of Turmoil." That was just the beginning, though! Certainly, it was an active year for personnel changes in that three of the rotators were finishing their time at the Foundation and one assistant moved to another position within the agency. Finally, the only permanent PD, Courtney, was on sabbatical. Since no one individual is listed as senior author of the report, it is assumed here that all agreed on the content. The opening paragraph was unlike any I have seen in any other document from NSF/BIO.

> Turmoil in Perspective
>
> Some things in the world are extremely serious: the war in Bosnia, starvation in Somalia and elsewhere, Haitian refugees, a recalcitrant US deficit, AIDS, vanishing biological diversity, drug addiction, too high a rate of illiteracy in the US, toxic environments of many kinds, and much else. Against such imposing troubles, most other things are patently less serious, including the turmoil those in Population Biology experienced during the 1992 reorganization of the Biological Sciences Directorate. But, turmoil it was, to be sure. It is important in the aftermath of this Reorganization, and the decade which preceded it, to examine our current capacity to support basic research in population biology. Fuller comprehension, from various scientific viewpoints, of what the Reorganization has wrought will come in due course, on its own.

Certainly, that was one of the most revealing of any surviving records of the great reorganization with regard to the feelings of some of those most impacted. The population biology *Annual Report* went on to state much else.

> The FY92 year opened calmly in our corner of the Foundation with the Reorganization in a conceptual planning phase. Suddenly, in November 1991 the then Program in Population Biology and Physiological Ecology was directed to serve as such for only one more review cycle. During this interval the Program was to divest itself of past and forthcoming awards in Physiological Ecology and Behavioral Ecology and $3.7 million [$3.1M, as it eventuated] was transferred from the Program's FY92 budget to the newly created Division of Integrative Biology and Neurosciences (IBN), and the Functional and Physiological Ecology and Animal Behavior Programs therein…From January to August 1992 we teetered on the edge of chaos as we tried to sort out an administrative and fiscal-accounting mess. Somehow, we had forgotten to plan the nitty gritty implementation steps of the Reorganization; from January on it was too late; so we flew by the collective seats of our pants until summer.

With no permanent PD on the scene to aid the group of rotators, they struggled through their time of turmoil quite alone. So, to reiterate a reality stated elsewhere in this history, but borrowed here from the all-rotator report authors:

[75] National Science Foundation, "Fiscal Year 1992 Annual Report for Population Biology and the Cluster in Systematics and Population Biology, Division of Environmental Biology: A Year of Turmoil," a 31-page document with 16 pages of text and the remainder being tables and illustrations. This annual report was longer than the average given the large amount of commentary in the text portion.

> Population Biology was probably the most intensely affected of all the programs caught up in the Reorganization, at least within the new Biological Sciences Directorate. And we were simultaneously told to forget about being a program and clusterize at once with Systematics…[the] transmogrification necessarily is not yet complete…[but it] was a time delightfully replete with gallows humor.[76]

The report's introduction, "Turmoil in Perspective," concluded with the remark that "[t]he voices which speak through this report are voices from the trenches of the NSF. We hope someone is listening."

More, but on a different subject, was revealed in later pages of that curious document.

> All this vigor in basic and applied evolutionary science creates a puzzling situation for the Foundation, one which is long-standing. At the root of all this intellectual and practical ferment [i.e., supported research during the year] lies the scientific study of evolution, a subject not to be mentioned in polite political company in Washington, and some parts of the hinterlands. However, code words will only go so far before accusations of chicanery are bound to arise.[77]

Far from the typically dry text seen in virtually all other documents examined for this study, the all-rotator group of authors, for this particular annual program report, were brutally honest. The "great interest generated by studies of molecular evolution" at that time was forcing the issue of use of the word evolution. Members of the systematics and population cluster had been working for several years with Peter Arzberger (earlier chapters) of DBI, where the program for it was lodged. Two workshops on molecular evolution had already occurred, and it would not "be possible to restrict use of references to evolution" in those documents headed for members of Congress.[78]

Therefore:

> [t]he Foundation needs to consider, carefully and formally, how it might better meet the bizarre situation in fostering modern studies of evolution, one more than vaguely reminiscent of battles between clerics and evolutionists more than 100 years ago.

Wonderfully readable, the FY92 annual report for population biology, though not a normally public document (but, by law, open to the public), was a refreshing and wholly atypical work that gave freedom to its authors to "tell it like it is." But in fairness to those authors, it must be noted that they went on in the next sentence to discuss the fact that the evolution situation was truly dire and was indicative of "too little science literacy in the US" and that the Foundation needed to address the issue.[79] The remainder of the Report was a tour de force of well-written prose on the accomplishments of the program for the year.

The following year's annual report, authored by Mark Courtney, would begin with the very first page being a collection of statistics—and not one word of text.

---

[76] Ibid., ppg. 2–3.

[77] Ibid., pg. 5.

[78] Ibid.

[79] Ibid., pg. 6.

The likes of the FY92 document would not be seen again. Indeed, the nine pages of the FY93 Report were mostly lists and data with but two and a half pages of text beginning with a commentary on "Post Re-organization Effects" written in excellent federalese. Among other staccato telegraphic sentences, it noted that the "elements of this remaining entity [after losses of parts of the PBPE to the new SBE] are molecular and quantitative population genetics, life history, population dynamics and demography." It was a skeletal, but complete, capsule statement about the new population biology program in the equally new BIO.[80] However, Courtney has shared with the author that:

> had I been there during that time, I'm pretty sure I would have eagerly joined [my colleagues] in their concern about the changes that occurred during that time.[81]

Some 93 awards were made from the 404 proposals submitted, a 23% success rate. Their average funding level was $60 k with a range of $22 k to $180 k. Six workshops were given. Other awards were made, and women (28% success rate) and minority scientists (24%), as well as disabled investigators (25%), were within the same range as the grand average—a successful year.

As time passed,[82] the population biology program could report other successes, such as supporting the discovery of high mutation levels seen in rodent populations in the vicinity of the Chernobyl, Ukraine, nuclear disaster; punctuated evolution[83] in the bacterium *E. coli*; evolution of phenotypic plasticity[84]; and more.

A workshop in 1998 led to a report on "Frontiers in Population Biology," held at Rutgers University under an NSF grant.[85] The impetus for creating the task force that looked into the subject was a Colwell biodiversity initiative. The initiative's many arms reached widely into areas of interest to the subject, and it included more directorates than just BIO, of course. The three task force (TF) goals included identifying "research areas deserving [focused] investment," mechanisms to incorporate those initiatives, and a summary of the TF's findings more broadly. The TF sought out overarching "research themes" that included aspects of the diversity of genetic architecture in living organisms (e.g., traits open to selective pressures), evolutionary trajectories (influences on rate and direction of change), and other thematic areas, such as evolution of complex genotypes and of symbioses, invasive species, and so on.[86]

---

80 National Science Foundation, "Population Biology Annual Report – FY 1993," nine pages.

81 Mark Courtney communication with the author, June 18, 2012.

82 National Science Foundation, "Annual Report, Population Biology– FY 1996," 11 pages.

83 See an excellent visual interpretation of punctuated evolution at http://evolution.berkeley.edu/evosite/evo101/VIIA1bPunctuated.shtml.

84 T. D. Price, A. Qvarnström, and D.E. Irwin, " *Proceedings in Biological Science* (2003):1433–1440; doi:https://doi.org/10.1098/rspb.2003.2372. PMC 1691402. PMID 12965006.

85 Authored by participants of the workshop, "Frontiers in Population Biology, Report of a Population Biology Task Force," October 1998 (Arlington, Virginia: National Science Foundation, 1998), ten pages.

86 Ibid., ppg. 1–6, and *passim.*

The themes settled out into two categories: "linkages to genomics [and] molecular genetics…[and linkages to] environmental issues." Bruce Hayden, then rotator DD of DEB, suggested to the TF how their findings might be implemented programmatically; the mechanism of a *collaboratory consortium* (earlier chapters) was the device chosen. The consortium could be well accommodated under the extant broad-based Research Coordination Networks program. The TF consisted of 12 leaders in the field and 3 POs from the program: their suggested device and content was accepted by DEB. Within a year, the collaboratory consortium had identified seven areas of research, which were "poised for substantial progress over the next 5 years or so."[87] One of those areas was invasive species where a supplementary grant to an existing one was made to a group of investigators (Ann Sakai, PI) at UC, Irvine, thus beginning one in a trend of successful projects under the "Frontiers in Population Biology" recommendations.

While the overall Population Biology Program had remained steady at one permanent PD, the number of rotators moved slowly from three to four over the decade, but that was by no means a rule. Sometimes rotators shared a portion of a given year, as did the comings and goings of scientific assistants. For example, though four PDs were present during some part of FY99, only three were there at one time. The assistants that year included, by then, four permanent persons, one position of which was the more senior science assistant and the remaining three were program assistants. Finally, one other permanent assistant was on leave. Such groupings allowed for considerable activity in the program at the end of the last millennium: a similar situation held for FY01, but by then a program technology analyst position had also been added.

The science coming out of the program had been centered around three "overlapping conceptual areas," including population ecology; evolutionary genetics; and evolution of phenotypes—all three having been core subjects for some years. Major growth had occurred with *Arabidopsis* over the preceding few years, and awards in that connection were given by the program under one guise or another of the three core areas.[88] The invasive species effort that had started at Irvine had matured into the Collaboratory on the Population Biology of Invasive Species and was, by the opening of the new century, sponsoring its own workshops.

The foregoing are but a very few examples of the direction taken in the Population Biology Program at BIO at and after the great reorganization, and, with but one permanent PD, it must be said that the program was astonishingly active and successful.

[87] National Science Foundation, "Population Biology Program…Annual Report – FY 1999," 10 pages with an additional report termed the "Statistical Summary," of some 19 additional pages (some numbered, others not). See especially pg. 6 in the main text of the Report where the discussion of the collaboratory consortium appears.

[88] National Science Foundation, "Population Biology Program…Annual Report – FY 2001," 13 ppg.; see pg. 4 regarding *Arabidopsis*.

9.3 The Ecological Studies Cluster

In 1992 with the beginning of BIO, the Ecological Studies (ES) area was not nearly so newly restructured as Systematic and Population Biology, though it did represent a combination of the separate Ecology and Ecosystem Studies' programs in the earlier BBS. With the establishment in DEB of the LTPEB cluster, LTER had its own home along with other long-lived programs. Thus, ES would support mostly, but by no means only, individual PIs' smaller and short-term research, not unlike the two closely related programs in the previous decade. And, while NSF would continue to be the primary supporter of environmental biology among federal agencies, the number of advisory panels would be reduced to each of the three clusters in DEB. For the ecology area of BBS, there had been two panels (Ecology and Ecosystem Studies), but, with the newly combined program, only one would be constituted for BIO/DEB's Ecological Studies cluster.[89] The combining of the earlier BBS programs into a fully unified ES, however, would not last. By the mid-1990s, a return to a slight variation of the earlier arrangement was seen with the terms Ecology and Ecosystems being the titles of choice for programs in the cluster, though the overall name of ES remained.

In 1992, there were four PDs and one senior program assistant, a program assistant, and a program clerk. The latter position's occupant later graduated to program assistant in the Research Collections in Systematics and Ecology Program of the LTPEB cluster. Additionally, there were other moves made as the permanent personnel of the new DEB accommodated to the changing incarnations of ES during the first half of the decade. Permanent PD, Scott L. Collins, led three rotators in the ES of the first years, but, by 1996, he had moved over to head the Long-Term Ecological Research Program in the LTPEB cluster that functioned as the NSF office for LTER. (See Chap. 4 for a discussion of the early period of LTER within NSF and also independently of the Foundation as LTER grew. LTER is taken up again later in this chapter.) Two other scientists led the remaining two programs in that cluster. James T. Callahan began as one of the two PDs in the LTPEB but moved over to head the Ecosystems program in ES by the middle of the decade. Callahan had held numerous positions but was best known as the Founder, with several others, of the LTER project itself. By that same time, there were only two PDs, one for each of the programs, Ecology and Ecosystems. Three support personnel were also in ES. But by the later 1990s, the cluster would have grown again.

While much that ES funded was in the smaller awards areas, one notable exception was the establishment and support of the National Center for Ecological Analysis and Synthesis, NCEAS. The Center was a first-of-its-kind STC-like

[89] That panel change was in keeping with the requirements of Executive Order 12838 of February 1993 (see http://www.ncs.gov/library/policy_docs/eo_12838.pdf) in which President Clinton ordered a reduction of federal advisory panels by no less than one-third for each agency based upon the requirements of the Federal Advisory Committee Act; see above-noted website. For BIO, that meant that 12 panels would replace the 17 of BBS. Mary Clutter to Constance McLindon, February 25, 1993.

organization that did not support the creation of new data from field or bench research, but analyzed and synthesized already extant data and so sought new interpretations of those data.

9.3.1 A New Way to Do Ecology: NCEAS

The NCEAS was, indeed, a new approach to do ecology. It represented a breakthrough effort in dealing with the mountains of data being produced in biology overall but especially in the data-rich discipline of ecology (and genomics). The time certainly had arrived in the biological sciences to explore those mountains and analyze and synthesize the data in them in search of what could be learned from data that had been collected though never processed. But also, scientists might discover new information, new hypotheses, maybe even new theories hidden in numbers gathered by so many field and laboratory investigators over the years.

During the preparation of the FY94 budget (which started as early as mid-1992), BIO sought to fund continuing areas of interest such as biotechnology and environmental research broadly. But new initiatives were also being considered for funding that year, so BIO's request included one for $6.3 M to provide for "start-up support for an Ecological Synthesis and Analysis Center."[90] BIO's total request was nearly $312 M for all its activities. The $6.3 M would not be awarded all at once, but was, as it had been for STC start-up years, an earmarking for some 5 years into the future. One million dollars was expected to be spent the first year (FY94) and two million the second. That was a major leap in the budget based on percentage change (100%) as compared to STCs overall (no change) and to LTER sites ($11 M to $15 M, 36% change, from FY94 to FY95). The base figure for the biology-related STCs was to remain at just under $12 M in those years.[91]

A call for proposals was released in calendar 1994, and the winning university that would host NCEAS was UC, Santa Barbara. A sort of birth of ecoinformatics occurred in 1995 as the mountains began to be excavated by PI William W. Murdoch, and Co-PI Michael F. Goodchild, and their colleagues. The NCEAS mission included a search for "general patterns and principles in existing data," as well as the intent to make those data, once analyzed and synthesized, available to resource managers and policy-makers. In the process, they intended that such work would promote a "culture of synthesis, collaboration, and data sharing," let alone "[i]nfluence the way ecological research is conducted."[92] In order to accomplish its mission, the Center "fosters new techniques in mathematical and geospatial modeling, dynamic simulation, and visualization of ecological systems." As of the time of this

---

[90] National Science Foundation, "Justification of Estimates of Appropriations to the Congress, Fiscal Year 1994," pg. 46. See also pg. 45.

[91] National Science Foundation, "Justification of Estimates of Appropriations to the Congress, Fiscal Year 1995," pg. 85.

[92] See http://www.nceas.ucsb.edu/.

writing, NCEAS had hosted over 4000 investigators and supported (by redirection of NSF funds) more than 400 projects. Their outreach program runs from K–12 to graduate education. In 2005, the Center had been included in the top 1% of institutions from around the globe (38,000 of them) working in ecology or environmental science. That standing had come about through total citations as given in the *ISI Essential Science Indicators.*[93] By 2009, some 1800 publications had emanated from NCEAS.

9.3.2 The NSF's Long Interest in Water

The National Science Board has long had an interest in water and its role in nature from many vantage points: hydrology; the marine environment; aquatic ecology; chemistry; engineering; and watersheds, among many others (earlier chapters). When Penelope (Penny) Firth came to NSF in April 1991, she began working in BSR during its final year. The ecology program handled water and watersheds, Firth's primary interest.[94] She came as a rotator, but not without knowledge of the federal environment. She had been a scientist in industry, the Lockheed Corporation, which was working under contract at NASA for some years in South Carolina. She felt she brought an unusual background to her NSF rotator position as "I think you'll find that the vast majority of the people who work at NSF as program officers do not have very much experience in the private sector." Her assumption has been thoroughly borne out in this study: virtually every PO, PD, and higher officer within the biological sciences directorate, among others, has come from academia, as we have seen. As a contractor at NASA, Firth's work was in controlled ecological life support systems for space travel. That was a funding program within NASA to which extramural investigators could apply for support. But she also had strong interests in environmental policy issues and the field of policy itself. At NASA she learned a great deal about both policy- and decision-making, she averred. Additionally, her bent was toward interdisciplinary research.

For those reasons, within a month of her arrival at BBS, she began what she called the Janus Group: an assembly of individuals who shared similar interests. Like Janus, the well-known Greek god of two faces—transitions and beginnings, past and future—the Group wanted to look backward and forward in time and space: to look "outside of the natural phenomena, habitats, ecology [and to, additionally, look at]…political science, sociology and other human behavior," she recalled. They invited non-Foundation specialists, particularly interdisciplinarians, to address the Group. Those individuals included such well-known experts as Robert Costanza,

[93] See online page under heading of "The Impact of NCEAS" at the above-noted website.

[94] Penny Firth interview with the author, Spring, 2009.

an ecological economist.[95] It was Firth's opinion that the conversations the Janus Group held were of a kind that "had not really happened [at] the Foundation before."

She was assigned to a BIO panel her first week on the job, being "thrown into the fire pretty quickly," and found herself working with ecologist O. J. Reichman on policy matters. Reichman would, incidentally, head NCEAS a decade later.[96] But it was in her early work with water and watersheds that she came into contact with the Environmental Protection Agency and, over time, had many dealings with them both formally, as the "architect [for] a partnership between NSF and EPA," and informally. The two agencies shared commitments to water in many connections, including acid rain and mercury pollution, for instance.[97] Her already strong interests in policy- and decision-making were piqued even more so by her close contact with EPA. Though she is an ecologist driven by "curiosity and concern" and not an attorney, as are the "vast majority" of personnel at EPA, all shared concern for the environment, she held. Support in freshwater science was one of the areas where, at least as Firth believed, NSF had been behind up to that time and had not been an equal partner in the *intricate dance* until matters moved suddenly with regard to the subject at the Foundation.

When the so-called water initiative began, Firth said that the community enthused: "this is the best thing you [the Foundation] ever did, why didn't you do it earlier."[98] It was just a short time before that when Fred Bernthal, then Deputy Director of the Foundation, was on Capitol Hill at a hearing and had gotten a question on the status of the freshwater "initiative." Firth recalled that it was not properly an initiative at the NSF, but rather "a completely *ad hoc* activity"; aspects of it had begun about 1992–1993.[99] Bernthal, upon his return from the Hill, apparently gathered persons throughout the Foundation who knew something of aquatic science, which included Firth. That led to her being brought into his office to act as the executive secretary of the committee on Earth and environmental sciences. She had to "sniff the winds" a lot to understand "[w]here the forks in the road seem[ed] to be taking" the community. When she had "capture[d] the winds," she immersed herself in work in watersheds for Bernthal leaving her rotatorship in BIO. She was then hired into a permanent position in the Deputy Director's Office.

The need for interdisciplinarity led to the integration of the social sciences into the water and watershed work along with BIO, engineering, geosciences, and other areas. She wrote the initial Announcement of Opportunity (call for proposals) for the program in 1995, along with a large committee of colleagues from across the

---

[95] See more on Costanza at the Gund Institute for Ecological Economics at the University of Vermont: http://www.uvm.edu/giee/?Page=about/Robert_Costanza.html&SM=about/about_menu.html.

[96] Firth interview.

[97] Ibid.

[98] Ibid.

[99] Penelope Firth, Michelle Kelleher, and Barbara Levinson, "Water and Watersheds Research: Discovery and Broader Impacts," *Journal of Contemporary Water Research and Education* (2007):1–6 (June); pg. 1. See also Firth interview.

Foundation. The first awards were given under the now more formal initiative that same year as the "interagency team worked through an absurdly compressed timeframe."[100] In order for any proposal to be considered under the new efforts in water and watershed science, she said, they had to include aspects of social science.[101] However, NSF "really ran into a lot of heat from the [natural sciences] community," many members of which were resistant to putting into their proposals features of social science. That led to a number of workshops on the fact that watersheds are places humans often live and so humans could not be left out of the equation of water and watershed studies.[102] That activity ran for about 5 years and then became a more formalized program under the name of Dynamics of Coupled Natural and Human Systems (CNH; alluded to earlier above)—a major program to the time of this writing. Firth recalled that after that "everybody loved it," and she also contended that it was the *first actual inter-directorate program with permanent designation* at the Foundation. The matter of *interdisciplinarity* at NSF matured then yet another step with the establishment of CNH. The overall program covered six directorates and ten different programs spread among them. Penny Firth ran the program from its inception in 2000 until 2006, having continued on from its water and watersheds' beginnings.[103]

Initially, many of the directorates and programs under them had interests in water and watershed research, and so it had become a "focus area" for NSF in FY95, after Bernthal's experience.[104] But it also "coincided with the development of a research agenda produced by the academic community: 'The Freshwater Imperative: A Research Agenda.'"[105] The "Freshwater Imperative":

> was endorsed by the relevant professional societies, and was useful in defining specific areas for consideration in the Water and Watersheds competition…[and] in early December, 1994…the Memorandum of Understanding between NSF and EPA was signed.[106]

The staff charged with the proto-CNH program that first year of 1995 estimated that some 400 proposals might be received given the "tremendous outpouring of interest" from the community. The influx quickly began to overwhelm the staff: the final number of proposals received came to 685! A panel of unusually large size was assembled to adjudicate the flood of paper—FastLane was not yet in existence. The panel counted 105 experts from academe, government, and the private sector. NSF had to hire in an administrative support contractor to handle the overload. But the

[100] Ibid., the article.

[101] Firth interview.

[102] It is of fascination to the author how a relationship between the new BIO and the new SBE had often to be reforged on some project or another *after* their split!

[103] Firth interview.

[104] We have seen in previous chapters other evidence of NSF's rapid responses to not only Congressional mandates, but to more casual inquiries from it, as well.

[105] Firth interview. See also Robert Naiman et al., *The Freshwater Imperative: A Research Agenda* (Washington, D.C.: Island Press, 1995), *passim.*

[106] Firth, "Water and Watersheds Research," pg. 1..

contractor's cost overruns led to them being closed down temporarily that summer.[107] In the fall, unrelated political difficulties followed: a full federal government shutdown.[108] But the first of the two struggles led to notable problems as the "contractor had all of the declination jackets at [its] facility...and they could not be touched until the issues were resolved [legally] and the contract restarted."[109] Not having the jackets, both those proposals declined and those to be awarded, kept BIO from disclosing them to the investigators. A "huge volume of calls and emails from PIs" then ensued as the whole proposal cycle had become "intolerabl[y]" prolonged! In the competition 3 years later, USDA joined NSF and EPA, and a rewritten call for proposals led to more targeted ones within the program's new framework: the total received was then manageable.[110] Additionally, more and more of the management responsibilities devolved to EPA, a very much larger organization than NSF.

By 2001, it was clear that lessons learned from the massive, very interdisciplinary and interagency CNH needed to be understood and clarified. Thus, the EPA staged a workshop that led to a final report noting the lessons derived from about a half decade of the undertaking.[111] It was determined that such gargantuan projects as CNH required certain things: both a strong leader and a strong lieutenant; the need to set egos aside; it having to be fully inclusive, especially bringing in young researchers; creation of a sound management plan; and, of particular note, the *substantial* expansion of the time expected to run the project. Only 3 years had been planned for the water and watersheds program.[112] Broader impacts and discoveries were followed up after 2001 by Firth's colleague, Michelle Kelleher, who was a science assistant at DEB for a number of years.

---

[107] Ibid., ppg. 1–2.

[108] See http://en.wikipedia.org/wiki/United_States_federal_government_shutdown_of_1995_and_1996. "The **United States federal government shutdown of 1995 and 1996** [emphasis in original] was the result of a conflict between Democratic President Clinton and the Republican-controlled Congress over funding for Medicare, education, the environment, and public health. It took place after Clinton vetoed the spending bill that Congress sent him. Thereupon, the federal government of the United States put non-essential government workers on furlough and suspended non-essential services from November 14 through November 19, 1995 and from December 16, 1995 to January 6, 1996. The major players were President Bill Clinton and the Speaker of the U.S. House of Representatives Newt Gingrich."

[109] Firth, "Water and Watersheds Research," pg. 2.

[110] Ibid.

[111] Environmental Protection Agency, National Center for Environmental Research, "Managing Interdisciplinary Research: *Lessons Learned*," a workshop summary and final report, March 19, 2001, 17 typed pages plus 3 pages of addenda.

[112] Ibid. and Firth, "Water and Watersheds Research," ppg. 2–3.

9.3.3 *Earth's Changing Climate … and More*

Penny Firth's interests went beyond water and watersheds to other large environmental matters. She had a strong interest in the issue of *climate change*. That matter had become undesirably known by the ill-chosen epithet of *global warming*, thus causing untold difficulties for the scientific—and public—communities for years to come. While the latter phrase is undoubtedly true and both phrases are, in substance, correct, weather expression at more regional and local scales has proven to be highly varied, some cooling from earlier regimes: it is global only in the *accumulation* of events. Historian of science, Spencer R. Weart, has considered this subject in detail.[113] Firth arrived early in the presidency of George Bush, *pater*, and found that he wanted to know more on the subject—but his administration did not wish to learn about it too fast, it seemed.

> Governments in Western Europe and Japan did grasp that they should consider serious policy changes, although they were satisfied with studying how to go about it. Some U.S. agencies also wanted to get to work restraining greenhouse gases. But President George H.W. Bush's [that is, *pater*] administration, its beliefs shaped by industry lobbyists, denied any need to address climate change. A White House memorandum, inadvertently released in 1990, endorsed the conservative's view that the best way to address concern about global warming was 'to raise the many uncertainties' [Weart wrote].[114]

Firth opined that when Bill Clinton replaced the elder Bush as President, Clinton wanted information about climate change straight away. NSF, if Firth is correct, "had to sort of immediately spring into action." Evidence, as shown below, certainly tends to support Firth's contention. Walter Massey had been the Foundation director when Firth arrived in the spring of 1991 but left shortly thereafter leaving Fred Bernthal in the acting position. She said that Bernthal was a steadying influence for NSF and did not allow the Foundation to "fly back and forth with Bush and Clinton."[115] Neal Lane relieved Bernthal of the latter's second acting directorship (earlier chapters) in October of 1993. And it was Firth's opinion that Lane "took the bounce out of Fred's approach, which was let many, many, many flowers bloom." Penny Firth called Neal Lane "the best director we've ever had." And, further, "he brought in some of the very best ADs," as well as Joe Bordogna (earlier chapters) later as his deputy, she recounted. Firth worked for Anne Petersen, the first deputy director Lane hired. To add to all of this, Lane's more than 5-year tenure was one of growing budgets.

---

[113] Spencer R. Weart, *The Discovery of Global Warming* (Cambridge: Harvard University Press, Revised and Expanded Edition, 2008). Originally published in 2003. See pg. 161 (2008) for the quotation.

[114] Ibid. The internal quotation from Weart's text comes from Philip Shabecoff, "Bush Denies Putting Off Action on Averting Global Climate Shift," *New York Times* April 19, 1990, pg. B4. Shabecoff's words became a national catchphrase and were widely used. A Google online search using the phrase in quotations will demonstrate the many places it has been employed.

[115] Firth interview.

Lane moved into the areas of the environment and *global climate change* with apparent gusto. Within a year of his arrival, he "had been reviewing the science, engineering, education, and mathematics activities of the Foundation" and so introduced a presentation to the NSB in October of 1994 emphasizing global change.[116] Mary Clutter initiated the discussion and introduced Thomas J. Baerwald, a PD in SBE, to chair the session. (Baerwald would later head the SBE's portion of the CNH program.) He pointed out to the Board four areas where NSF staff had been involved with the NSTC's Committee on Environment and Natural Resources (CENR): Global Change; Biodiversity and Ecosystems Dynamics; Social and Economic Sciences; and Environmental Technology. He argued that their work "illustrate[d] five major messages." The first of those concerned fundamental processes working in the environment and used an illustration from the GEO directorate concerning tropical ocean-atmospheric interactions: El Niño/Southern Oscillation (widely known as ENSO, or simply, El Niño), soon to become famous in its own right.[117]

The second message was illustrated via James Gosz's work in DEB on LTER. The network was then composed of 18 (later 25) sites spread from Earth's north to south poles, and, while it was "a very sparse network," Gosz noted that NSF and LTER's "leadership role" would expand the sites into other countries; that did come to pass with ILTER (later below).[118] The third message concerned SBE with its Human Dimensions of Global Change Program regarding population trends, integrated systems assessments, adaptation and mitigation, and patterns of trade and global economic activity.

The fourth message came from EHR and their efforts in education of future scientists and engineers (especially in developing countries) to prepare them to address issues of global environmental change under the program Global Learning and Observations to Benefit the Earth (GLOBE). The final message was delivered from an acting division director in engineering concerning efforts to alleviate chemical pollution problems. That had become a model program for other agencies and was well regarded by Congress.[119] The AD for GEO, Bob Corell, closed the unusually long and full agenda in that Board session.

DEB's budget request for FY94 had been for more than $83 M, with the ES area having the highest figure of the division's three clusters. It was set at just under $30 M, $2.5 M of that for global change and high-performance computing (so necessary with the level of data flooding in from environmental science studies). Systematics and Population Biology's budget was to be something over $25 M, and

---

[116] NSB, "Minutes," October 13–14, 1994, NSB-94-182, 10–94:6.

[117] See http://press.princeton.edu/titles/7707.html. "Until 1997, few people had heard of the seasonal current that Peruvians nicknamed El Niño [as it arrives around Christmas time]. But when meteorologists linked it to devastating floods in California, severe droughts in Indonesia, and strange weather everywhere, its name became entrenched in the common parlance faster than a typhoon making landfall."

[118] NSB, "Minutes," October 13–14, 1994, NSB-94-182, pg. 10–94:7.

[119] Ibid., pg. 10–94:8.

LTPEB was nearly equal to that of ES. Indeed, yet another $2 M was requested for LTER specifically for the start-up for NCEAS.

Foundation deputy director, Anne Petersen, asked that same fall that Mary Clutter and GEO's Bob Corell, as well as Cora Marrett (Chap. 2), by then AD of SBE, prepare a white paper on issues, options, and recommendations on *combining* the two then current environmental and global change strategic areas.[120] Environmental research and global change had been marked as two of eight strategic areas of focus for the Foundation in the FY95 budget plan. An examination of overall environmental research done at the Foundation had been undertaken just a year before.[121] Apparently, it was considered reasonable to combine that broad area of NSF's interests with the growing realities of the need for yet more global change research, *per se*, because Clutter and her two colleagues made recommendations along those lines. The two large research agendas had evolved in different ways and at different times and were managed as separate programs. Global climate change (GCC) research had roots dating back to the 1980s at the Foundation, and the multi-agency budget was expanding rapidly by the next decade: in FY91 it was $954 M, but by FY95, it had nearly doubled to $1.8 billion. That budget represented the combined contributions of the initial nine federal entities that were part of the US Global Change Research Program (GCRP).[122] NSF's portion of the full budget for the Program moved upward from $87 M to $170 M during those same years. It was an intense period across the federal scene for the subject, and would remain so, for the most part.

Environmental Research (ER), under that epithet, had been established as a formal strategic effort at the Foundation in 1991 due to two interests of the NSB. One was their 1989 publication of "Loss of Biological Diversity: A Global Crisis Requiring International Solutions," and the other was their formation of a Task Force on the Environment in 1990.[123] One may wonder if the "inadvertent" leak of

---

[120] Mary Clutter, Bob Corell, and Cora Marrett, "Environment and Global Change: Management of a Combined Strategic Area: Issues, Options, and Recommendations," September 12, 1994; an 11-typed page document with 1 page transmittal letter. Attached to the white paper was an appendix: James Hay, Chair; Penelope Firth, Executive Secretary; et al., "Report of the Director's Task Force on Organizing Environmental Research," August, 1993; also 11 pages, including its own appendices.

[121] Ibid., the Hay report.

[122] Thirteen federal agencies are part of the Program as of this writing. For more information, see http://www.globalchange.gov/. "The U.S. Global Change Research Program (USGCRP) coordinates and integrates federal research on changes in the global environment and their implications for society. The USGCRP began as a presidential initiative in 1989 and was mandated by Congress in the Global Change Research Act of 1990 (P.L. 101-606), which called for '*a comprehensive and integrated United States research program which will assist the Nation and the world to understand, assess, predict, and respond to human-induced and natural processes of global change*'." Emphasis in original.

[123] Clutter et al., September 12, 1994 report, *op. cit.*, pg. 2. The NSB "Loss" document cited in the main text above was a Report of the NSB's Committee on International Science's Task Force on Global Biodiversity (NSB-89-122), draft of August 3, 1989, 44 pages. The Committee consisted of some 11 members, including Peter Raven, E.O. Wilson, and Rita Colwell, among others.

the White House memo in 1990, noted above, may not have been a third contributing factor! ER's budget grew from $104 M in its first year of FY92 to $160 M by FY95. Clearly, the Board, the Foundation, and BIO—along with, at least, GEO and SBE—were deeply into a multipronged effort that included aspects of environmental change from biodiversity loss to climatic disruptions, and much else under that broad umbrella. The two big areas of ER and GCC had been somewhat muddled together in practice in those first couple of years, so Petersen's request likely represented a welcome airing of whether or not to combine them in the formal sense. Indeed, as Clutter and her colleagues said, "there is no significant intellectual basis to justify continued separation of the strategic areas."[124] GCC tended to use geologically longer timelines in its research projects, but little else made the two areas "intellectual[ly]" distinct. Merging the two would have many benefits, the three ADs averred, including avoiding possible duplication of effort, and also reduction of confusion on the part of Congress, academe, and the interested public. Further, the combination would begin to resemble the NSTC's CENR, making investigator support via NSF and the other agencies involved a sensible complement to that standing body of the administration. A great many more management suggestions were made by the three ADs, though ending with a plan for a phased merger over the period from fall of 1994 to the same time a year later.[125]

A half decade beyond that, the fusion of the two strategic areas proved to have been beneficial as Neal Lane, who had by then moved on from Director of NSF to Director of President Clinton's OSTP, was able to present to Congress the latest edition of the annual document *Our Changing Planet: The FY 2001 U.S. Global Change Research Program*, GCRP.[126] Since 1990, the yearly report had been required under the Global Change Research Act[127] and is still compiled to the time of this writing. In 2000, Lane promised Congress the upcoming first national assessment of the potential consequences of climate variability and change, a publication also required under the Act.[128] Over the years, Kathi Olsen, Joanne Roskoski, and others of NSF have served on the committees and subcommittees of the GCRP along with their colleagues from the 13 agencies finally involved.

---

[124] Ibid.

[125] Ibid., ppg. 2–10, and *passim.*

[126] Although Weart's book, *op. cit.*, is aimed more at the early discovery of GCC, he does not, even in the revised 2008 edition, cite any of the *Our Changing Planet* series. For a complete listing of those annual reports, see http://www.globalchange.gov/publications/our-changing-planet-ocp.

[127] P.L. 101–606.

[128] National Assessment Synthesis Team, *Climate Change Impacts on the United States:The Potential Consequences of Climate Variability and Change* (Washington, D.C.: Global Change Research Program, 2000).

9.3.4 A Companion Issue: The Biodiversity Crisis

For BIO, among the other NSF directorates and the rest of the interagency group making up the US Global Change Research Program, the issue of biodiversity loss was a significant one. Its close intellectual relatives of conservation and restoration biology, biodiversity as a basic biological research area (beyond just loss studies), and other relevant disciplines, such as systematics and evolution (and then, soon to come, bioinformatics), were all of concern to the directorate. The GCRP was only a portion of that greater picture of investigational support. So the NSB report on biodiversity loss of 1989 played across a number of programmatic venues that would have distinct effects on the research support structure for BIO as it matured, affecting DEB most heavily.

The report had set forth a number of recommendations.[129] First, NSF should take the leadership in inventorying the world's biodiversity. We already have seen some of the programmatic areas that developed in the young BIO/DEB that would start to undertake such a mandate. The NSB committee further recommended a decided increase in funding for such efforts for BSR (DEB) of \$5 M and later to climb to \$20 M. Also, a "significant expansion of support of microbial systematics and ecology" was to be made from an initial level of \$8 M to approximately \$20 M annually. This was the basis of much of the microbial ecology activity that would later include Microbial Observatories and early parts of what became The Microbe Project. Finally, as part of its first recommendation, the committee argued for those institutions receiving funds from BRR for biodiversity research that they be provided \$5 M annually to handle the increase in specimens generated by the program, building and maintaining data banks, for GIS, and for data dissemination—a wise plan in an era of massive data influx.

The second recommendation noted that for the creation of inventories and for associated training of biologists, especially in developing countries, the value would be significant. It would aid a number of the "emerging fields" noted above and others impacting the environmental arena.[130] The committee was especially keen on US educational efforts in the K–12, informal science education disciplines, and pre- and postdoctoral training in the fields they had listed. As they said, "NSF has virtually the full responsibility for the health of these fields…in the U.S." Their fourth recommendation was to support, at about \$1 M per year, the study of social and economic causes and consequences and also recommendations via theoretical and empirical studies of the biodiversity crisis. The funds would be added to SBE's budget. In its fifth recommendation, the committee urged finding new avenues to aid those types of efforts in other, especially developing, countries through collaborations with US scientists and institutions. Those efforts were to focus on educational and research infrastructure.

[129] NSB, "Loss" document, ppg. 31–40.

[130] Ibid., pg. 35.

The effects of the report were finally to be seen widely within the Foundation, but especially BIO, for years to come. For instance, the budget request for NSF for FY93, the drafting of which began in 1991, reflected the NSB report. It noted that biodiversity research was one of four major new and continuing "research emphases," along with basic research on the environment, manufacturing, and cognitive science: two, almost three, of the four were biological sciences or nearly so. The funding sought for the "emphasis" was over $4.5 M, a nearly 18% increase in the budget for DEB. (That was also the year that the Foundation's budget request would break the $3B mark. Incidentally, the NIH total budget figure for FY93 was $10.3B!) The amount of $135 M would be NSF's share of the interagency GCRP effort for the year, too.

In 1993, not only did the budget reflect massive new efforts that would impact DEB and its ability to support even more research in the areas of its charge but the Board unanimously accepted the Report of its own recent Task Force on the Environment presented by NSB members Warren J. Baker and Charles E. Hess.[131] Franklin Harris, of BIO, followed up on the Report the next month by telling Board members about the many activities to be seen across federal agencies in environmental research efforts then new or just coming on line.[132] Clearly, much came to pass in the whole governmental apparatus since the "inadvertent" release of the White House memo in 1990. The same was true with regard to scientific societies' support of more work on environmental science, too, Harris stated. Peter Raven, then a Board member and a player in many areas of BBS and BIO over a considerable number of years:

> presented a strong case for the NSF to include a major initiative in environmental research in the FY 95 budget. He noted that since 1989 [and the biodiversity loss report] funds have been unavailable to adopt recommendations from [that] report, even though it was a two-year study unanimously endorsed by the Board.[133]

Hess seconded Raven's comments. The Board discussed the matter with some concern that beginning another new major initiative in the field and seeking to place NSF at the head of all agencies working in the discipline may not please Congress. That was also a hiatus period in top leadership at the Foundation with Fred Bernthal only functioning as acting director. But just before the Board's mid-October meeting that year, Vice President Al Gore would swear in Neal Lane as the new director. Lane then attended the NSB's meeting.[134] And, while Raven's suggestion for another new initiative in environmental research was still being considered by the Board, the creation of the NCEAS was unanimously approved by the members at that same meeting. Lane was present to see the Board's strong interest in environmental science, and, as noted above, he moved most rapidly into those crucial areas under the global change issue.

---

[131] NSB, "Minutes," May 12–14, 1993, NSB-93-131, 5–93:9.

[132] Ibid., June 17–18, 1993, NSB-93-159, 6–93:4.

[133] Ibid., pg. 6–93:5.

[134] Ibid., October 14–15, 1993, NSB-93-243, 10–93:2.

The whole period of the early 1990s was one of rapid expansion in the global climate change (GCC) and environmental research (ER) strategic areas. That is difficult to separate from biodiversity loss studies and *many* other areas of interest to DEB, and also from so many other governmental and non-governmental organizations: threads of many colors and characters made the fabric one of considerable complexity. As John Muir once said, "When we try to pick out anything by itself, we find it hitched to everything else in the Universe."[135] The statement is not only true with regard to the actual environment, of which Muir so eloquently spoke, but also to America's efforts to understand and save that environment through research supported by NSF and others beginning with such intensity after about 1989. Indeed, in mid-1994, a large collection of interim reports on strategic initiatives, of which GCC and ER were two, was created at NSF indicating the plans and milestones for the following 5 years for those two programs.[136] The first of those reports, which was heavily involved with GEO, was the Processes and Consequences of Earth System Change. Among the programs managed by BIO were the Land Margin Ecosystems Research (LMER), a part of the LTPEB cluster (below); Global Ocean Ecosystem Dynamics (GLOBEC); Ecological Rates of Change (EROC); and Antarctic Ecosystems.[137] Environmental Research was also covered through four integrating themes (earlier above), one of which was Water and Watersheds and another was Biodiversity. The interim reports were extensive and predated the combination of GCC and ER but included many other areas, such as biotechnology.

9.3.5 Conservation and Restoration

Inextricably tied to the biodiversity crisis was the need for conservation and restoration of natural ecosystems: DEB, through the Ecological Studies cluster, had answers for the research community. The NSB's 1989 report cast ripples across BIO, and so DEB created a new competition in conservation biology and restoration

[135] John Muir, *My First Summer in the Sierra* (Boston: Houghton Mifflin, 1911), pg. 110 of the Sierra Club Books' 1988 edition. This quotation is one of the most misquoted of all of Muir's well-known phrases: variants of it populate all sorts of literary genres. See http://www.sierraclub.org/john_muir_exhibit/writings/misquotes.aspx, for a full discussion.

[136] A memo to many of the ADs and other interested individuals at NSF, from the Assistant to the Director for Science Policy and Planning, "Interim Reports on Strategic Initiatives," June 24, 1994; six of eight such documents were then attached; two were still in preparation.

[137] While little has been said in this study about NSF's involvement in Antarctica, a short history can be found at http://www.nsf.gov/pubs/1997/antpanel/antpan05.pdf. A full history of *biological* research there, especially that done under NSF aegis, has yet to be written. However, William James Miller has written the *Exploring Polar Frontiers: A Historical Encyclopedia*; see http://www.abc-clio.com/product.aspx?id=61993, for a good outline of the explorations. But for the science, the starting point is Gordon E. Fogg and Margaret Thatcher, *A History of Antarctic Science* (Cambridge: University Press, 1992).

ecology (CRB). By mid-1990, the initial group of awards numbered 16 projects.[138] The 2- to 3-year-long studies totaled just over $3 M and sought to do:

> assessments of the diversity contained within the gene pools of different species, the fate over generations of individual gene variants within species, the effects of species diversity within ecosystems, and hormonal and behavioral factors affecting animal reproduction.

Specific examples of investigation would include breeding and reproduction in Nicaraguan fishes, in Great Plains bison, and in monitor lizards in Africa. It also supported investigations of the immune systems in cranes and condors. Plants were not forgotten: some 24 species of ferns in Hawaii would be examined to determine the role of invasive species upon the native ferns' habitats. Tropical plants' use of soil and space was to be studied from an ecosystem point of view: the new program in conservation biology and restoration ecology was off to a broad-based beginning.

By the program's third year, another 16 projects were established, but funds had not kept pace: in 1992, only $2.8 M was expended on CRB. But that was a recovery from 1991, when barely over $2 M had been awarded.[139] Under the heading of restoration ecology, so-called "derelict" landfills in metropolitan areas were under study as normal plant succession did not seem to occur in those locales: the reasons were unknown. Researchers Steven Handel and Joan Ehrenfeld of Rutgers University, New Brunswick, New Jersey, investigated the Fresh Kills, New York, landfill as normal succession by woody plants was desirable for many reasons. The intent was to design recommendations for restoration on such lands once the causes of the disrupted succession cycle were understood. Other projects under CRB included work on amphibian decline (a continuing planet-wide disaster), endangered sea turtles, and surveying the habitat of red-cockaded woodpeckers. The awards for the year would run 1–4 years in duration and had monetary levels of $60 k to nearly $300 k per project. They included not only universities, but such NGOs as the Zoological Society of San Diego, California, and Wildlife Conservation International.[140]

That same year, the National Academies Press came out with its *Conserving Biodiversity: A Research Agenda for Development Agencies*, in which the National Research Council's Panel on Biodiversity Research Priorities discussed the fact that the "potential impact on the human condition, on the fabric of the Earth's living systems, and on the process of evolution is immense."[141] The Panel made that abundantly clear when it said that the "conservation of biodiversity is a global responsibility," but NSF would bear the primary burden for that research. While much had

---

[138] National Science Foundation, Office of Legal and Public Affairs, "From Mouse DNA to Whale Migrations: Researchers Seek Basic Knowledge to Support Conservation and Restoration on All Scales," NSF PR 90–50, August 13, 1990.

[139] Ibid., "Researchers Seek Basic Knowledge to Support Conservation and Restoration on All Scales," NSF PR 92–82, October 27, 1992.

[140] Ibid., attached List of Awards.

[141] NRC, *Conserving Biodiversity: A Research Agenda for Development Agencies* (Washington, D. C.: National Academies Press, 1992). Quotation from pg. 1, Executive Summary.

yet to be considered both nationally and globally, the Panel had some specific recommendations. "[S]trengthening institutional capacity in developing countries" was foremost among them. Second, "[d]evelopment agencies should support long-term ecological research sites (comparable to La Selva in Costa Rica or Barro Colorado Island in Panama)." Indigenous knowledge, so called gray literature, should be emphasized and systematized; the US Agency for International Development (AID, or USAID) and the World Bank were to be called upon to assist in that area.[142] "[M]ore and broader-based research" was necessary, as was multidisciplinary research. Finally, investigation into the "operation of economic systems as they affect biological diversity" was also a recommendation of the NRC Panel.[143]

By 1993 and the fourth year of CRB, the Program Guidelines that BIO released was for a "Selected List of Research Programs in Biodiversity."[144] They included ones in which NSF had taken the NRC's recommendation by partnering with AID "to fund meritorious research on biodiversity in developing countries." Both agencies issued the guidelines: the document noted that research in biodiversity was funded not only through a number of programs in BIO (within and without DEB) but also via SBE, GEO, and others, such as oceanography and polar programs. Within DEB, BS&I was one source of support, as was CRB, of course, along with a number of initiatives through the Division of International Programs (part of the Director's Office). In a separate announcement on biodiversity efforts as independent of CRB, BIO/DEB was promoting more international research under AID. Also, it included the US National Committee for Man and the Biosphere Program, another joint effort of NSF.[145]

By 1994, the CRB program funds were down again to about $2 M, but several of the first year's multi-year grants were reaching maturity. For example, the work on Colombia's cotton-top tamarin monkey had produced data that was allowing for a long-term management program to assist the species both in captivity and in the wild. Conservation was happening for the cichlid fishes in Nicaraguan lakes. There, African *Tilapia* spp. had been introduced as a protein source for the human population, but they were displacing the native fishes and thus threatening an ecological collapse. Funding was continuing for some projects as yet more new ones came on line. The CRB program, however, came to an end with the 1997 competition, at least as it had been embodied since 1990. Support for conservation and restoration

---

[142] See a small group of documents concerning the AID relationship examined for this study that includes some miscellaneous single sheets and others but most notably a memo from Jim Rieger and Sy Sohmer of the Bureau for Research and Development, Office of Environment and Natural Resources, USAID, to the AID/NSF Steering Committee, October 27, 1992, which included Minutes of their September 14 meeting and some addenda. The memo provides some sense of the nature of the AID/NSF relationship.

[143] Ibid. The recommendations can be found in the Summary on ppg. 9–11.

[144] National Science Foundation, BIO, "A Selected List of Research Programs in Biodiversity: Program Guidelines," NSF 93–160, November 23, 1993.

[145] National Science Foundation, BIO, "International Research Opportunities in Biodiversity: Program Guidelines," NSF 93–158, November 18, 1993.

biology, however, would hardly end at DEB. A newly structured program would be created for the 1998 competition, and the fields would continue to be funded under a somewhat rewritten set of guidelines.[146]

The NRC continued its long interest in biodiversity issues and published again on the subject in 1999.[147] The Council's study had its roots in a request to its authoring committee by DoD in 1995, and it had been in the works since that time. Not the product of the Panel, which authored the 1992 report, the latter effort was undertaken by a different group at the NRC: the Committee on Noneconomic and Economic Value of Biodiversity in the Board on Biology of the NRC's Commission on Life Sciences. Management methods issuing from the views generated from both noneconomic and economic standpoints would, of course, dominate that study. The new findings and recommendations would be different from the Panel's report 7 years earlier. The Committee found an "urgent need for more *information* about biodiversity": the work by DEB's several programs, while apparently of value, was not enough.[148] "Traditional and emerging benefit-cost approaches to valuing biodiversity" were needed but differed from place to place globally; it was difficult to make comparisons. Without the latter, however, broader management schemes and models were harder to create. Local mechanisms then would dominate, and more and better data on local biodiversity thus was still needed in order to preserve it and to ally it logically with the various human populations.

As mentioned in previous chapters, reports from the National Research Council had clear influence at NSF. They would have a decided effect on the thinking about the steps that would be taken by the federal partner across the dance floor of research. Those reports, such as that of 1992, affected how the guidelines were written for any given program(s), certainly not least of which were those in conservation and restoration biology. The changes brought about by the restructuring of the CRB program in 1997–1998 would reflect some of the features of the NRC report in that conservation and restoration biology would thenceforth be altered to:

> invest the responsibility for CRB research with all Advisory Panels, reviewers, and Program Directors, thereby broadening the support base for basic research [in the fields, and hence generating more of the needed information].[149]

Under the revised program, numerous specialists throughout the NSF structure might have an opportunity to see or review a given proposal, as they had to be "prefaced with ~CRB~" on their covers to alert the Foundation as to how to route them. So, after 1997, new proposals could therefore be adjudicated by more persons in more disciplines within the NSF.

During the 1990s, a great many agencies, committees, and offices were created and studies released concerning biodiversity and its values, environmental "services"

---

[146] Thomas Brady, "Dear Colleague" letter of December 24, 1996 (NSF 97–35).

[147] NRC, *Perspectives on Biodiversity: Valuing its Role in an Everchanging World* (Washington, D. C.: National Academies Press, 1999).

[148] Ibid. Emphasis added.

[149] Brady letter.

(detoxification, decomposition, erosion control, and so forth), and the various crises facing the Earth's environments.[150] One of those, the NSTC's CENR (above), was chaired by Mary Clutter. That Committee spoke in terms of "ecological forecasting," as had PCAST (earlier chapters) before them.[151] Such forecasting could be done the better after the creation of a National Biological Information Infrastructure. The NBII grew out of the NRC report of 1994, "A Biological Survey for the Nation."[152] While Clutter did not play a part in that, Peter Raven chaired the effort, which bore fruit in the eventual establishment of the NBII. It would be managed by the US Geological Survey, but the National Biological Survey recommended by the NRC never came to pass, as we have seen. And the NBII itself, however, was not funded until 2001, even though two federal reports, that of 1994 and another of 1998, urged its formal establishment.[153] The NBII exists as of this writing and is, in essence, a database of databases concerning biological information on a vast scale. As noted earlier, the Global Biodiversity Information Facility (GBIF) was formed at about the same time that NBII became available.

Also in the later 1990s, NSB and the Foundation had created the Knowledge and Distributed Intelligence (KDI) initiative (Chap. 6). KDI was a perfect program under which the various directorates could make awards for database development, such as NBII. The same year it came online, a grant was made under DBI to the Center for Research at the University of Kansas, Lawrence, to pursue "KDI: Knowledge Networking of Biodiversity Information."[154] The supplement to Kansas' already ongoing project was nearly $82 k, and the total for the project was just under $2.3 M.[155] Biodiversity bioinformatics was given a dramatic boost in the event.

The US-EC held a workshop by the end of the decade, co-chaired by Clutter, which was a policy-level forum under the aegis of the international Task Force (TF) on Biotechnology Research (Chap. 5). It concluded that the TF:

---

150 See http://www.gcrio.org/USGCRP/CENR/sgcr.html.

151 Committee on Environment and Natural Resources [CENR], Subcommittee on Ecological Systems [Mary Clutter, Chair], NSTC, "Ecological Forecasting: Agenda for the Future," n.d., but mid-1990s, an eight-page slick publication.

152 See http://www.nap.edu/catalog/2243.html for the PDF file. See the discussion and associated footnotes in Chap. 5 concerning Wagner's question of "Whatever Happened to the National Biological Survey?" The NBII had a troubled infancy, a brief history of which can be seen at http://en.wikipedia.org/wiki/National_Biological_Information_Infrastructure. See also information on the National Spatial Data Infrastructure Executive Order 12906 (President Clinton) of 1994 at http://www.nan.usace.army.mil/business/buslinks/geo-gis/execord.html. Also see the report, "Teaming With Life: Investing in Science to Understand and Use America's Living Capital," at a page on the NBII website http://www.nbii.gov/images/uploaded/8496_1139316568259_twl.pdf.

153 Ibid., the Wikipedia entry.

154 See http://www.nhm.ku.edu/desktopgarp/Printer.html.

155 Paulette L. Green, Grants Officer, to Barbara Armbrister, Director of Finance, August 17, 2001. Both KU employees.

recognizes that bioinformatics provides a critical element of the infrastructure required for the effective storage, analysis, visualization, and public dissemination of information derived from the study of plant, microbial, and animal genomes.[156]

The broad reach of Ecological Studies programs seemed a success going into the new millennium as a product of the 1992 reorganization. For the greatest part, they were.

9.3.6 A COV's View of the Ecological Studies (ES) Cluster

In May of 2002, a Committee of Visitors convened to assess how well the Ecological Studies cluster, with its wide variety of programs, was doing on the dance floor of research. It made a series of recommendations, the first being that ES and/or DEB should increase its support for postdoctoral research.[157] Such suggestions were common in many COV reports across BIO and over time. Minorities and women—also regular recommendation subjects—should be given special attention, the COV members held. But mid-career researchers were also mentioned: fewer programs had been created for that class of investigators at NSF over the years. It was hoped that the latter group would "forge innovative links between research and education" and that it would allow a "'retool[ing]'" for that group. A number of years had passed since the NRC asked for more information from federal research agencies—read bioinformatics from ES. Still the COV noted that the project jackets they reviewed were short in the area of evidence for data management and "explicit plans and funds" for that issue were needed.[158] As massive an effort as bioinformatics had long been at BIO, that one ES COV, at least, did not feel that it was sufficient. They recommended that the cluster take notice of the extensive data management aspects of the LTER program as a model and increase funding and efforts for ES programs and relevant proposals.

The Committee also found that the cluster's success rate under the Foundation's CAREER program was "low." CAREER was and is (at the time of this writing) a program that awards support to junior faculty members to promote "teacher-scholars."[159] The COV felt that making mentors available would be one way to

---

[156] National Science Foundation, "Workshop on Plant and Animal Bioinformatics: U.S.-E.C. Task Force on Biotechnology Research, Final Report," September 2000. The workshop was held on September 13–14, 1999.

[157] Taber D. Allison, Chair, and Members of the Committee of Visitors, "Report to the National Science Foundation: Review of the Ecological Studies Cluster for FY 1999, 2000, 2001; May 20–22, 2002," pg. 23ff.

[158] Ibid., pg. 24.

[159] "*CAREER*: The Faculty Early Career Development (CAREER) Program is a Foundation-wide activity that offers the National Science Foundation's most prestigious awards in support of junior faculty who exemplify the role of teacher-scholars through outstanding research, excellent education and the integration of education and research within the context of the mission of their organizations. Such activities should build a firm foundation for a lifetime of leadership in integrating education and research." See http://www.nsf.gov/funding/pgm_summ.jsp?pims_id=503214.

improve the success rate for the program. At the same time, the total number of proposals had been rising, and the staff and panels were being stressed, though morale remained high and FastLane was proving its worth. Nevertheless, the Committee recommended a "management review" of the entire ES area.[160] Finally, the long struggle with the two-criteria review process (earlier chapters), particularly the most thorny of those, "broader impact," arose again: PIs and reviewers alike were not meeting that criterion. More work was called for, stated the COV.

Those Committee findings were considered by them to be general, but they detailed others specific to ES at the level of more pure science infrastructure. They found that the cluster was sound and did not need reorganization, that its programs were distinct, but that reviews were not always as outstanding as would be desired. They also found that the Population Ecology program was ending up with some of what should be ES awards and asked if that program should not be transferred to Population Biology; it would not be. Finally, they found that review panels had gotten too big, at about 20 members, and that they should be reduced to 15–18 members.[161] The COV's study was a thorough-going one.

9.4 Long-Term Investments: The LTPEB Cluster

Long-Term Projects in Environmental Biology (LTPEB) contained one of the true gems of the biology directorate at the National Science Foundation: the Long-Term Ecological Research Project, LTER. But a number of other programs were also included in that third and final cluster of the division. As LTPEB began its life after the great reorganization, James Callahan and Leonard Krishtalka were the two PDs; they were aided by just one senior program assistant and a single program clerk. Small beginnings, but a future expansion in programs, though barely in personnel, would occur within the next few years. By 1996, the cluster would become exactly that: a *grouping of related programs* that included LTER but also several others. Scott Collins had become the PD for LTER and had one assistant; Callahan and Krishtalka, as discussed earlier, had moved on to other assignments in BIO. The Research Collections in Systematics and Ecology (RCS&E) was headed by Meredith Lane, with two assistants. Biotic Surveys and Inventories (BS&I) would move into LTPEB from its original home in Systematics and Population Biology (SPB). Since BS&I began under SPB, its history was discussed earlier in this chapter. Meredith Lane headed it in 1996, with one assistant. She also headed RCS&E. Those two related programs would be managed in that fashion into the future by Lane's successors. The last program as part of LTPEB in its 1996 incarnation was Penny Firth's Special Projects. Much of the structure and personnel would remain the same through the change of the millennium.

---

160 COV report on ES, pg. 24.

161 Ibid., ppg. 24–26.

LMER was an early program that came into the new DEB and would last for only several more years.

9.4.1 A Cross-Directorate Program: LMER

The Land-Margin Ecological Research (LMER) Program had begun in 1988 and was cooperatively managed by the Biological Oceanography Program in the Division of Ocean Sciences of the GEO directorate and BSR; after 1992, it fell under DEB's LTPEB. The program was to increase understanding of a major Earth feature in light of two goals: the "structure and function of land-margin ecosystems," in essence, oceanic and other shorelines, and, "the linkages between these systems at the land-sea interface" with special reference to global change, particularly into the early 1990s.[162] The Great Lakes were included as were estuaries, coastal wetlands, tidal portions of rivers, and coral reefs.

Though not a large program in total numbers of awards per year, it was one that offered large individual grants. Five or six awards might be made each year, but grantees also could request as much as 6 years of support, and an award might reach as high as $500 k *per annum*. Not surprisingly, LMER, not unlike LTER, came to have a program office outside of NSF. Located at the Marine Biological Laboratory at Woods Hole, Massachusetts, it was directed for some years by John Hobbie, a scientist at MBL.[163]

When James Callahan was still a PD in LTPEB, he was the point of contact for the program at BIO, along with Phillip R. Taylor in Biological Oceanography. While research venues (field sites) were always the choice of the PI, there was encouragement from NSF for researchers to consider certain established locations. Those included a score or more over the years of field laboratories administered by NOAA on US coastal areas. The Great Lakes were among them as they were part of the National Estuarine Research Reserve (established in 1972), an ideal location for LMER-supported studies.[164] NOAA's dozen or more National Marine Sanctuaries system also offered superlative sites for study as many of them offered data sets of long-term duration. The guidelines promoted use of all of those areas and provided contact information at NOAA. This type of cross-agency involvement, as we have seen, was common in program guidelines at NSF over the years and did not represent something unique within the LMER program. Still, it is a good example of such inter-directorate and interagency arrangements.

As of the FY95 budget, LMER was to be funded at just under $3 M with Michael Allen having replaced Callahan as the DEB contact some time before. But that

---

[162] Those two goals were stated annually in the Program Guidelines. See, for instance, the 1993 issue: "LMER Land Margin Ecosystems Research: Program Guideline[s]," NSF 93–35, September 20, 1993, pg. 1.

[163] Ibid.; see also other years' issues of the Guidelines, as well.

[164] See http://www.nerrs.noaa.gov/BGDefault.aspx?ID=61.

would change by 1997 when a joint decision was made between BIO and GEO to discontinue LMER and replace it with an LTER program termed Land/Ocean Margin Ecosystems (LOME).[165] The LOME program was not terribly unlike LMER, but had a much stronger component of sociological features as part of it. That was likely due to the increasing global climate change interest that was so permeating all of NSF by the later 1990s. LOME's initial competition was administered by Biological Oceanography in GEO.

9.4.2 Where Biology of Big and Little Met

None of the incarnations of the biological sciences at the Foundation, that is, BMS/BBS/BIO, ever had a specific organizational category of programmatic activity devoted especially to microbiology, strangely, though many programs with such a focus did exist over the years. But by the mid-1990s, that oversight began to change and would do so at an almost dizzying pace once microbiologist Rita Colwell became Director of NSF. Program development just prior to her expected arrival led to successes among grantees such that the American Society for Microbiology (ASM), which was a member of the Coalition for National Science Funding,[166] strongly supported NSF's request for a 10% increase in its budget for FY99, along with other Coalition member organizations.[167] The NSF FY98 budget had been slightly over $3.4B. The increase of $344 M would allow for a variety of things, including increases in programs aimed at the microbial world. While it only received a 7.7% increase (to just under $3.7B), new programs were indeed begun in the microbiological realm.

Few programmatic efforts at NSF, prior to Colwell's arrival and her concept of biocomplexity, *per se*, were quite as powerful at mixing the big and little ends of the spectrum of life as were the Life in Extreme Environments (LExEn) initiative (already several years old then) and the aborning Microbial Observatories (MO) Program. MO would eventually be placed under MCB's Cellular Systems cluster but is discussed here with DEB.

LExEn was the first large initiative that began in support of big biology investigations in the field of microbiology. It premiered in the mid-1990s when some 20 research awards were made to about 40 investigators in the 1996–1997 period.[168] It was a cross-directorate initiative and was under the overall aegis of Michael Purdy, Director of the Division of Ocean Sciences in the GEO directorate. Phil Harriman of BIO (earlier chapters), its PO there, called Purdy a "kind of motivating force" for

---

[165] See http://www.nsf.gov/pubs/1997/nsf97126/prognew.htm.

[166] See http://www.cnsfweb.org/.

[167] See http://www.asm.org/index.php/policy/nsffy-1999.html.

[168] National Science Foundation, "Limits of Life on Earth: Are They the Key to Life on Other Planets?", NSF PR 97–61, October 10, 1997, pg. 1 of 5.

LExEn.[169] Included were GEO, MPS, ENG, BIO, and the Office of Polar Programs. Joint activities were being discussed with NOAA, DoE, and NASA at the time, too, because the NSF group was developing future plans for the FY98 program offerings that would include more agencies.[170] Among the initial grantees, there were those looking at life in:

> hydrothermal systems [the already much-studied Yellowstone hot springs, for example], sea ice and ice sheets [Antarctica], anoxic habitats [certain deep marine areas], hypersaline lakes [e.g., Great Salt Lake, Utah], high altitude (Atacama Desert in Chile) or polar deserts, and human engineered environments such as those created for industrial processes.[171]

Creatures living under such conditions are known as extremophiles.[172] The program foci that first year were upon methods for microbial isolation and culture; creation of field methods, including sensors and probes; protocols for the avoidance of contamination; methods for studying ancient microbial life; paleoenvironmental conditions on Earth; and ways to study life on other planets (exobiology, or astrobiology).[173] Harriman remembered the LExEn projects as "some of the most exciting" in his career at NSF.[174]

Work was undertaken in one of Earth's driest environments, the Atacama Desert,[175] by grantees at Louisiana State University. Also, some of the earliest work on deep Earth crustal environments (at 9200 feet, in this case) in the gold mines of South Africa and their microbial autochthons was accomplished by Tullis Onstott of Princeton University. He found intact, respiring bacteria in the mineral-bearing facies at that extreme depth. The discovery was widely showcased in the popular press around the world, and he was later added to *Time* magazine's 100 most influential persons.[176] Glaciologists at Ohio State University, Columbus, were "resuscitating" microbes which were "Lazarus-like, viable but entombed in ice sheets," and in similar situations. Douglass Bartlett at Scripps Institution of Oceanography was studying hyper-barophiles—here, bacteria living in extreme pressures in the benthos of the deep-sea Japan Trench—to learn more of their physiology. Others of the initial 20 awards took on similar studies, all areas in which little research had been done theretofore.[177]

---

[169] Harriman interview.

[170] See http://www.nsf.gov/pubs/1997/nsf97126/life.htm.

[171] "Limits of Life," pg. 1.

[172] See the excellent general introduction to these organisms, which includes some historical commentary on them, in the semi-popular article by Michael T. Madigan and Barry L. Marrs, "Extremophiles," *Scientific American* (1997):82–87.

[173] NSF 97–126, *op. cit.*

[174] Harriman interview.

[175] Incidentally, see the author's book review of "From Atacama to Makalu: A Journey to Extreme Environments on Earth and Beyond," R.T. Arrieta (Panama City, Florida: Coqui Press, 1997), *Science Books & Films* 33(9):263–264.

[176] See http://www.time.com/time/specials/2007/time100/article/0,28804,1595326_1595329_1615985,00.html.

[177] "Limits of Life," pg. 3.

By the final year of the past millennium, the LExEn program guidelines had placed more focus on methods' design and had also established a connection with NASA.[178] That arrangement sought to have investigators propose projects that would be decadal-length research. The "LExEn Sites for Long-term Interdisciplinary Studies" component of LExEn was established to relate the NSF grants to NASA's interest in astrobiology.[179] Other expectations in the later guidelines were more explicit and more prescriptive than in the first several years of the program. Such general to specific guidance over several years of many programs was typical: find the primary hotspots in the first or second year, and then direct funds more specifically to the major trend areas in coming years. Such seemed to be a winning approach at NSF/BIO, and possibly other directorates.

Closely allied in many respects, but still an independent initiative, the Microbial Observatories Program was begun in 1999, tied to Colwell's arrival. It was one that the ASM had strongly supported as the MOs would play a signal role in biodiversity studies. Indeed, PCAST had recommended an additional $130 M for 3 years in the Discovery of New Species work at BIO.[180] The MO program was, as its second director Matt Kane said, "foreshadowed" by a series of events. Harriman, along with his duty for LExEN, had also been MO's first director (though the program had been planned by Charles Liarakos, then Deputy Director of MCB) until his retirement in 2001.[181] Kane recounted the early history of MO by noting that, in 1994, a discussion was held at the Smithsonian jointly sponsored by ASM, the American Institute of Biological Sciences, and the American Phytopathological Society.[182] The purpose of the meeting was to "emphasize the importance of focusing U.S. federal agency research agendas on microbial diversity science."[183] The American Academy of Microbiology followed that discussion a short time later with a Critical Issues Colloquium, "Microbial World: Foundation of the Biosphere," authored by a group that included not only Kane but also Colwell.[184]

---

[178] National Science Foundation and the National Aeronautics and Space Administration, "Life in Extreme Environments (LExEn): Announcement of Opportunity and Special Competition for FY 1999," NSF 99–43; see especially ppg. 4–5.

[179] See more on this subject at http://astrobiology.nasa.gov/.

[180] http://www.asm.org/index.php/policy/nsffy-1999.html, *op. cit.*

[181] Harriman interview and Kane article acknowledgments (see following footnote for citation).

[182] M. D. Kane, "Microbial Observatories: Exploring and Discovering Microbial Diversity in the 21st Century," *Microbial Ecology* (2004):447–448. This brief article is an introduction to a special issue of the journal in which are reported about a dozen MO projects.

[183] Ibid.

[184] Kane lists the Colloquium report, but the website he provided has been superseded. See the newest Web page at http://academy.asm.org/index.php/colloquium-program/browse-all-reports/230-the-microbial-world-foundation-of-the-biosphere-1997-b. The Executive Summary is worthy of repeating in full here, especially for the non-biologist reader. "A revolution has occurred in microbiology, the ramifications of which are not yet widely known or appreciated, but which provides new opportunities for agriculture, medicine, and industry. Using new techniques and approaches—in particular sequencing and analysis of ribosomal RNA genes—microbiologists have discovered that there are at least 20 major evolutionary groups of the microbial life forms on Earth—bacteria,

The publication (of the same name) resulting from that Colloquium provided a road map to what the MO program should be and "helped foreshadow the birth" of the initiative 2 years later.[185] Some characteristics of Colwell's biocomplexity concept can be seen clearly in that publication. Harriman had but 2 years left at NSF before his retirement, and he remembered them as extremely busy but "fun." Nevertheless, he felt he had very little time to put any "thought" into the programs as his plate was full with the mechanics of managing them. That was especially so, what with rising expectations concerning biocomplexity along with the MO and other duties, such as his work in the interagency metabolic engineering working group. He retired in October of 2011, "just after 9/11" and the terrorist attacks. Indeed, he recalled that he could see from his office window the smoke rising from the Pentagon. Consequently, he was "happy to get out of Washington," at least as

fungi, algae, and protozoa—that are even more diverse than the better known plant and animal kingdoms. Furthermore, recent evidence indicates that millions of microorganisms still remain to be discovered. These new microorganisms provide a vast untapped reservoir of genetic and metabolic diversity, the harvesting and study of which will have far-reaching, positive effects for society in areas such as enhanced food production, medicine (e.g., antibiotic discovery), bioremediation of waste materials, and agriculture. Part of the revolution in microbiology stems from development of methods that now make it possible to inventory the diversity of microbial life on Earth and to do so cost-effectively. The power of the new technologies on which these methods are based will reveal the extraordinary range of metabolic, physiological, and evolutionary diversity of microbial life on Earth. Programs to harvest this bountiful, yet invisible, microbial diversity have already begun in Japan and Europe. The United States scientific community, largely responsible for creating the revolution, is uniquely positioned to mount a comprehensive effort that takes full advantage of newly evolving technologies. A microbial exploration and harvesting program to assess the variety and commercial potential of microbial life forms on Earth will reap great benefits for society. Furthermore, a microbial exploration and harvesting program employing existing and developing technology would launch a voyage of discovery and exploration that would quantify the diversity of microbial life forms on Earth and provide, at least in part, an answer to the question of how much microbial diversity has yet to be discovered. Current estimates, based on partial and random sampling, are certainly gross underestimates of the actual diversity of microbial life extant on Earth. An inventory would identify environments where the greatest diversity exists, and those areas can be productively targeted for more intensive exploration. Harvesting and study of novel microbial types will doubtless lead to the discovery of new biomaterials, development of new commercial processes and enterprises, and acquisition of knowledge of practical importance to society. To grasp the full extent of microbial diversity, iterative steps need to be taken. For example, sequencing the genome of at least one microbial species from each of the major phyla of microbial life can provide information about another type of diversity: the biochemical potential of microorganisms comprising each of the major groups. DNA sequence analyses of microbial genomes obtained to date have revealed an unanticipated richness of novel genes; this approach can be expected to identify completely novel processes and biochemical compounds that will find timely application in biotechnology and health care. In order to delineate effectively the extent of microbial diversity and identify potential applications, several recommendations are made, addressing infrastructure necessary to ensure attainment of the goals, and include initiatives in education and training, instrumentation needs, establishment of regional and national research centers of excellence, development of databases, and enhancement of genetic resources in culture collections."

[185] Kane, "Microbial Observatories," pg. 447.

regards the matter of the attacks.[186] And while pleased to return to Northern California, he did say that he missed much else about the capital area.[187]

The MO goal sought to "develop a number of sites in different habitats to study and understand microbial diversity over time and across environmental gradients," Kane averred. It was expected that researchers would discover unknown microbes, singly and as part of biological consortia (metagenomics); their habitats; and novel physiological and biochemical features of life. The first MOs were developing to study terrestrial, marine, and freshwater environments: many were sited at LTER project locations. About one-fourth were located in particularly extreme environments such as the types noted above under the LExEn program. Sites included such exotic locales as, among others, Galapagos, Kamchatka, and Antarctica—mostly new areas for ***field microbiology***. Some nine awards were made in the program's first year with the largest being nearly $1.2 M over 5 years. That project was led by Eric Triplett, then of the University of Wisconsin, Madison, along with four co-PIs.[188] It was to create a MO at the North Temperate Lakes LTER site.

LExEn and MO had been in operation for several years when, in 2002, the first workshop was held for the two programs. The authors of the report noted that the:

> investigators funded by these programs have extended the frontiers of microbial diversity and microbial biogeochemistry research, discovering novel microbial lineages, describing the complexity of natural microbial communities, and linking microbial taxa to critical ecosystems functions.[189]

Further, by using modern molecular techniques, such as 16S rRNA,[190] some 13,000 new prokaryotes were discovered by projects in the two programs. Metagenomics was very new at the time, and the two programs were just then seeing the first projects to make use of the techniques, they having been first applied to eukaryotes.[191] Other newer techniques were employed by the grantees, including environmental microarrays and proteomics.

The workshop report made a number of recommendations. Attendees desired that environmental genomics be further developed with emphasis on "capturing

---

[186] Harriman interview.

[187] Personal communication with the author, May 2012.

[188] See http://www.nsf.gov/bio/pubs/awards/mo.htm#1999.

[189] Mary Ann Moran and Sherry L. Cady, "Microbial Research: Progress and Potential, Principal Investigator's Workshop September 22–24, 2002," NSF Grant (MCB-0234272) to the University of Georgia.

[190] A portion (16S) of ribosomes (rRNA) in bacteria turns out to be an excellent tool for identifying and preparing phylogenies for a given bacterium. The moiety weighs 16 Svedberg units (16S) when centrifuged, hence the name, and differs from one species to the next. For more information, see http://www.ncbi.nlm.nih.gov/pubmed/15489351.

[191] Two exemplar publications are K.L. Vergin et al.'s "Screening of a Fosmid Library of Marine Environmental Genomic DNA Fragments Reveals Four Clones Related to Members of the Order Planctomycetales," *Applied and Environmental Microbiology* (1998):3075–3078, and, M.R. Rondon et al.'s "Cloning the Soil Metagenome: A Strategy for Accessing the Genetic and Functional Diversity of Uncultured Microorganisms," *Applied and Environmental Microbiology* (2000):2541–2547.

ecological heterogeneity"; development of array technologies for greater sensitivity and better understanding of complexity; more work in bioinformatics and modeling from mixed populations with undefined species; and better sampling, culturing, collections work, and dissemination of data.[192] Participants at the workshop did judge the MO program to be "overwhelmingly successful."

The first decade of the new millennium would be, in every sense, one dominated in the life sciences by the use of and discoveries made by genomics and, in many cases, metagenomics. "Next generation DNA sequencing has transformed microbial ecology studies [by deriving data from a] soil sample, the gut, even a computer keyboard."[193] Specifically:

> [t]he microbial makeup of our gut is a case in point. In the past decade, scientists have come to realize that animal intestines naturally harbor diverse microbial communities that help provide nutrients and sustain good health. A *landmark* 2005 study by Stanford University's David Relman and colleagues...concluded that the bacterial communities in the human gut vary tremendously from one individual to another.[194]

The authors of the National Academies Press book on metagenomics[195] would have certainly agreed with the extraordinary importance of the field, as Relman showed. And, for what is termed here *Millennial Biology*, it is easy to agree that "[m]icrobes run the world." Further, that is so because their study using metagenomics "has already produced a wealth of knowledge about the uncultured microbial world."[196]

During the same general period, Phase II of the Biocomplexity in the Environment programs came out under four headings from several areas within the Foundation. Within a single program solicitation, then, broad-based programs carried *interdisciplinarity* to greater levels of integration at NSF—arguably the greatest seen in its history to date.[197] The solicitation of 2001 included programs titled Dynamics of Coupled Natural and Human Systems (discussed above), Coupled Biogeochemical Cycles, Genome-Enabled Environmental Science and Engineering, and Instrumentation Development for Environmental Activities (IDEA). The next year, a fifth program was added into the mix—Materials Use: Science, Engineering, and Society.[198] Across the country, 47 awards were made totaling just under $38 M for those biocomplexity-related programs that first year. Director Colwell was quoted as having said that:

---

[192] Moran, "Microbial Research," pg. 4.

[193] News Focus, "Digging Deep into the Microbiome," *Science* (2011):1008–1009.

[194] Quoted words, with emphasis added here, are ibid., but see the Relman paper at Paul Eckburg's hotlink: http://nature.berkeley.edu/brunslab/pmb220/papers/eckburg2005.pdf.

[195] *Op. cit.*, Chap. 6.

[196] Ibid., quotations from ppg. 1 and 3, respectively.

[197] National Science Foundation, "Biocomplexity in the Environment (BE): Integrated Research and Education in Environmental Systems," NSF 01-34, 2001.

[198] Ibid., NSF-02-010, 2002.

> [r]esearch on biocomplexity in the environment…provides science with a more complete understanding of natural processes, of human behaviors and decisions in the natural world, and of ways to use technology effectively to observe the environment and to sustain the diversity of life on earth.[199]

By the following year of 2003, a second, closely related agenda was added to MO, the Microbial Interaction and Processes (MIP) Program. MIP projects did not require site basing and/or could be of shorter duration than MO projects.[200] MIP was a joint program with USDA and rose to some 250 proposals per year within a few years of its establishment. During its short existence in the first decade of the new millennium, it had a strikingly low funding success rate of just 10–15%, so intense was the competition.[201] In 1975, 50% success rates were common in BMS. By the 1980s, 30% was the more typical level in BBS. By the early years of BIO, rates were dropping and 20% was not unusual. After 2000, rates at 10–15% were increasingly more common. But:

> [w]hen the success rate drops to some critical value (perhaps below 10% or even 20%), future solicitations lose significant community interest and participation[, they placing] a burden on the entire system of proposal writers, proposal reviewers, and program staff.[202]

Though that was said about biocomplexity programming specifically, the same statement may well have fit many large initiatives at BIO at and near the turn of the century.

By late 2008, both MO and MIP programs were discontinued by BIO. The directorate had spread responsibility for program content widely within BIO and so informed the community.[203] Work along the lines initially begun under MO, and MIP continues at the time of this writing under such other programs as Genes and Genome Systems (DEB); Symbiosis, Defense and Self-recognition (MCB); Advances in Biological Systematics (IOS); and, under DBI, AToL and a newer program termed Life in Transition. There were others, also. All of these changes came about or were expanded post-2005 but deserve mention due to the signal standing of the Microbial Observatories Program, in particular. That same year, the Center for Microbial Ecology at Michigan State University, one of the original 11 STCs supported by NSF, published the latest edition of a continuing online resource that represented an excellent example of the legacy of MOs and the rise of metagenomics: "The Taxonomic Outline of Bacteria and Archaea."[204] Microbiology had moved into

---

[199] NSF, OLPA, "NSF Awards $37.9 Million in Grants to Study Biocomplexity in the Environment," NSF PR 02-83, October 15, 2002. Eamon Kelly was, at the time, the NSB chair.

[200] Kane, "Microbial Observatories," pg. 447.

[201] Kane interview.

[202] Elizabeth J. Kelley and Mary Jane Perry, Co-chairs, National Science Foundation, "2004 Committee of Visitors Report for the Biocomplexity in the Environment Competition (2001–2003)," February 25–27, 2004, pg. 2. Margaret Leinen, NSF Coordinator for ERE and AD for GEO, added a memo of transmittal that charged the COV with its task.

[203] See http://www.nsf.gov/pubs/2008/nsf08077.jsp.

[204] See http://www.taxonomicoutline.org/index.php/toba.

new realms not only with regard to systematics but also with regard to ecology on a very large scale: big biology of the little.

It remains difficult to separate many of the activities and programs at BIO, and it to other directorates, once very aggressive interdisciplinarity was undertaken through the Biocomplexity in the Environment initiative. That, though, was as it should have been for an organization striving toward integrative biology—and integrative science more broadly. *Biocomplexity in the Environment was, in so many senses, a tremendously expansive concept, as well as program crucible and driver.* Complexity of programmatic offerings at NSF became more typical, and integration of them, one to the next, and within a web of interconnected disciplinary areas, became commonplace at the Foundation. It was just as much so of biology in the community itself as it became unified in many areas due to the adoption of metagenomics, bioinformatics, and other areas that came to characterize the state of *Millennial Biology*.

So, how did Biocomplexity fair in the first half decade of the twenty-first century in the eyes of the cogent Committee of Visitors? As it happened, the initiative had become somewhat subsumed under another major activity then newly ongoing at the Foundation: Environmental Research and Education. That matter requires some consideration because the COV report brought these programmatic areas together: the two cannot easily be separated, at any rate.

9.5 ERE: Environmental Research and Education

In mid-1998, the NSB's new Task Force on the Environment was established to consider the matters that would later come into existence as the ERE program. ERE emanated out of the TF report, "Environmental Science and Engineering for the 21st Century: The Role of the National Science Foundation."[205] Jane Lubchenco was the chair of the TF, Mary Clutter one of the five members, and its executive secretary was Penny Firth. GEO was also represented on the committee as so much of ERE was tied to that directorate. Twelve extensive recommendations were made, including a 5-year plan to expand the budget, for matters that could logically fall under ERE, intending to bring the initiative's worth to some $1.6 *b*illion Foundation-wide by 2007.[206] A working group (WG) had been established for BE within the TF and later was renamed the WG for the new ERE. An advisory committee was also birthed at that time.[207] Arguably, the NSB's report was one of its most significant in

[205] National Science Board, National Science Foundation, "Environmental Science and Engineering for the 21st Century: The Role of the National Science Foundation," NSB 00-22, February 2, 2000. A 106-page report.

[206] Ibid., pg. xiii of the Executive Summary. See also the NSF Press Release NSF PR 00-4 of February 4, 2000.

[207] See http://www.nsf.gov/geo/ere/ereweb/about.cfm.

its existence given the sheer grandeur of the integration of science—and education—that it sought to craft.

The primary purpose of ERE was, in essence, to continue and more fully institutionalize already existing realities at NSF: the Foundation had been deeply involved, via disciplinary programs, in environmental research and education for much of its life. Now, however, biocomplexity had appeared on the scene, and it would be worked into what became ERE. Since a great many POs were seeing that ERE-like issues had come to "cut across extant disciplines," the federal partner on the floor of the *intricate dance* was "respon[ding]" to the community's moves. Thus, NSF "sought to integrate holistic multidisciplinary investments with disciplinary-intensive opportunities": an elegant statement—and concept—indeed![208]

The first AC-ERE meeting was given its charge by Director Colwell in October of 2000. By 2003, that Foundation-wide AC had released two of its own reports. They considered such matters as "Complex Environmental Systems: Synthesis for Earth, Life, and Society in the 21st Century," and "Environmental Cyberinfrastructure: Tools for the Study of Complex Environmental Systems"[209]—interdisciplinarity incarnate.

So broad was ERE that it listed many of the already familiar programs discussed to this point under its encompassing umbrella. LTER was there, as was systematic biology and biodiversity inventories, PEET, AToL, and others. The first award year was FY99 with seven grants made for biocomplexity. Thus, from 1999 to 2007–2008, the ERE program listed them as being included in its "portfolio."[210] Just 1 year later, the number of grants made was 65 and those under 2 headings: Integrated Research and Education Awards and Awards for Exploratory and Developmental Activities. The numbers climbed over following years.

In early 2004, the Biocomplexity in the Environment (BE) COV reported on the status of the initiative over the period of 2001–2003.[211] That report necessarily tied ERE and BE together, both of those initiatives covering seven directorates and two offices in the competition's five broad programmatic areas. The BE priority areas were intended to end in 2005, the COV noted, but went on after that, as it eventuated, until 2008. Indeed, so big was the undertaking that that COV was the first ever at NSF to evaluate such a massive programmatic effort.[212] In fact, the Committee found that, through BE, the Foundation had "fostered an important new area of multidisciplinary research" taking into account not only the natural sciences but the human factor also. Reminiscent of the old BBS? That feature was not unalloyed, however, as the report found that "in many instances social sciences were interpreted to be 'societal impacts' in proposals," thus leading to improperly constructed such documents. Beyond that, the Committee felt that the competition was too narrowly

[208] Ibid.

[209] Ibid.

[210] See http://www.nsf.gov/geo/ere/ereweb/fund-biocomplex.cfm.

[211] Kelley and Perry, "BE COV".

[212] Ibid., pg. 1.

phrased in that the social sciences were viewed mostly through quantitative lenses: a broader interpretation should be used in program guidelines, the COV held. And while there was initial confusion as to what biocomplexity actually meant, the COV stated that it had "evolved into truly a cross-cutting competition" under its five program subheadings due, in large part, to the leadership and dedication of the staff. Intriguingly, the Committee found that there had been demonstrated the:

> development of new multidisciplinary communities, the realignment of departments and resources in universities, the establishment of new centers, the creation of a new journal dedicated to biocomplexity, the establishment [of] a Gordon Conference series in biocomplexity, and the inclusion of workshops and special sessions on BE at professional society meetings.[213]

If there is truth in the argument, mentioned elsewhere in this history, that "scientists are opportunists," the foregoing comments would suggest that one way that NSF might have profound effects on the communities it served was to offer large, complex, and non-disciplinary programming to those communities. It would appear, according to the COV, that just such a thing happened. Historians need to pursue this issue in greater detail in years to come. By then, more perspective will allow a better understanding of how organizational structure and program offerings made by funding institutions may affect intellectual areas within the new millennium. The COV was well aware of some of this when it stated that needs were met in the community for just this type of competition, at least with regard to the integration of education into research proposals.[214] Thus, just how biology was *intellectually and organizationally* affected as a broad discipline needs to be considered in greater detail in future studies.

The COV wondered how NSF would "sustain the new communities once they germinate." Happily, they did believe that educational institutions "are moving in the direction of BE," partly due to NSF's offerings and also due to the "challenges of the future [that] demands this structure."[215] Such movement remains to be demonstrated.

The inevitable crossing of disciplinary lines wrought by ERE/BE that would include an LTER/ILTER (International LTER) may, in many ways, allow us to use that global program to represent so much of what happened to NSF and BIO from LTER's own beginnings in the 1980s to the development of BIO as the third millennium opened. Thus, this history will conclude with a discussion of LTER and ILTER from the early 1990s to 2005 and shed some light on a new idea: NEON.

---

[213] Ibid.

[214] Ibid., pg. 2.

[215] Ibid.

9.6 The Long-Term Ecological Research Program: LTER After the Reorganization and Views of a COV

About the same time as the great reorganization that led to BIO, the LTER 10-year review was being written and became available in mid-1993.[216] Commissioned by BBS, though when it came out BIO's imprimatur would be upon it, the report was prepared by a committee made up of ten leading US and foreign scientists. The group was charged with examining the status of LTER with regard to both individual projects and the network taken as a whole. Accountability had grown to be a major theme in government (GPRA), and the report was offered by Mary Clutter to a broad readership as evidence of such.[217]

While LTER began with 5 sites and a total budget of $1.2 M in 1980, it had risen to some 18 sites and $11 M in support by the time of the review. The mainland US sites had been joined by the first Antarctic location, too. About 200 scientific personnel aided by 300 technicians and students made up the cadre of the LTER when BIO came into existence. The Committee averred that the program had produced a "distinguished and extensive literature" over that first decade, a portion of that body of knowledge leading the US Forest Service, for instance, to establish its "New Perspectives in Forestry" series. That set of literature came out of science produced at three of the sites: the Andrews Forest, Coweeta, and Hubbard Brook (all discussed in Chap. 4).[218] Though the Committee found that LTER exhibited "unquestionable success," the "intentional flexibility" for grant proposers written into the competition announcements had "led to some ambiguity" for investigators. At the sixth year of an 11-year cycle, the long-term sites had to seek renewal of their grants, and that implied not only funding certain continuing projects but also adding new ones. Since the then-current figure of $500 k per site per year was the maximum allowable award, supporting both continuing *and* new programs was not always possible financially. The "intentional flexibility" was, in fact, a constraint: "funding has increased, but not at the rate of the increasing demands on the science in the LTER program."[219]

Further, "[s]ome intersite comparisons have been conducted, but the power of the *network* of coordinated research sites has not yet been fully realized."[220] A problem, the program had not included all the *types* of sites necessary. So Great Basin; southern swamps and wetlands[221]; large rivers; certain types of marine systems; and—intriguingly—urban sites were missing. Also, some saw the 11 sites of the period as

[216] Paul Risser and Jane Lubchenko, Co-chairs, National Science Foundation, Biological Sciences Directorate, "Ten-Year Review of the National Science Foundation Long-Term Ecological Research (LTER) Program," NSF 94–26, July 1993, 44 pages with 3 appendices.

[217] Mary Clutter, transmittal memo to the "Ten-Year Review".

[218] See http://andrewsforest.oregonstate.edu/pubs/pdf/pub1388.pdf.

[219] Ibid., pg. 2, Executive Summary.

[220] Emphasis added.

[221] The Okefenokee LTER had withdrawn from the program some years earlier. See Chap. 5.

being biased toward *ecosystem-level* research at the expense of other aspects/levels of ecological work. And even though the reviews by peers for site proposals ensured that the quality of science done at a given locale was the premier criterion, the Committee felt that individual reviewers were more "accustomed to evaluating hypothesis-driven, single-investigator proposals"—those which were more inclined to short rather than long-term outcomes. Moreover, reviewers often focused on achievements at single sites and failed to see cross-site, or network, contributions.[222] That erroneus focus assured, in many respects, that broad system-wide aspects were not given consideration and the LTER *network* features were being lost to sight. In fact, it would be some years before that value of the entire program was more fully recognized and exhibited to full advantage, as shall be seen.

The Committee, therefore, made certain suggestions for modification of the then-current program. While the safeguards and high expectations for success must remain in place, they argued, several "steps" would insure improvements. The goals needed to be reframed, which the Committee did. The five core research topics shouldn't be used as organizing principles, they stated, but rather objectives, which ensured that "breadth, continuity of measurements, and comparability" should be foremost. Those original core areas included:

- Pattern and control of primary production
- Spatial and temporal distribution of populations selected to represent trophic structures
- Pattern and control of organic matter accumulation in surface layers and sediments
- Pattern of inorganic inputs and movements of nutrients through soils, groundwater, and surface waters
- Pattern and frequency of disturbance to the research site

The influence of the original Hubbard Brook watershed experiment can clearly be seen in several of those core areas. Each LTER project must have a "defined core of site-appropriate observations and measurements," the review stated.[223] Beside the strong emphasis on ecosystem-level research, the Committee wanted to see work done on the levels of populations, communities, landscapes, and global views, also. Further, the review Committee wished to see accomplished: a newer, expanded role for the Network Office; more cross-site patterns and processes study; and emphasis on anthropogenic influences.[224]

As early as 1981, the LTER hosted a series of All-Scientists' Meetings to discuss the progress of the program. Though they were held around the country at different places once every 3 years, they most commonly met at a large, national YMCA site in Estes Park, Colorado, the gateway to Rocky Mountain National Park. Coleman

[222] Ibid., ppg. 2–3. Much of this paragraph is a paraphrase of the Review's contents at this page location.

[223] The five core research topics list is on pg. 1 of the Review and the quotation is on pg. 3.

[224] Coleman, *op. cit.*, pg. 132.

noted that the 1993 meeting was a memorable one for quality of presentations and that some scientific facts arose during one talk that demonstrated that certain LTER research on climate showed its impact at the global scale, something which would become of great importance later when global climate change issues became a dominant, and very public, feature of environmental science.[225] At that same meeting, Jim Callahan (Chap. 4 especially) reviewed the history of LTER dipping back as far as 1965, where its roots could be discovered. Jerry Franklin (also Chap. 4) discussed a next-decade vision termed LTER 2000. Two years after that meeting, the NSF would formally request the Coordinating Committee of LTER to write up a proposal for the future.[226] That plan was to expand the network to a fully global-wide program with sites around the planet. The 10-year review document was also available at that meeting of 1993.

The LTER program announcement calling for proposals the following year specifically sought plans for regionalization, comprehensive site histories, and increased disciplinary scope.[227] And, during the next several years, continuing expansion in number and types of sites was sought. By FY95, cross-site competition to broaden research and increase investigator numbers occurred, and the plan for FY97 was to bring in the first two urban sites, a direct result of recommendations issuing from the 10-year review. More coastal sites, one at the Great Lakes, were to be listed for LTER in FY98.[228] By the first year of the new millennium, BIO had provided a slick publication on the entire network of the period and was ready to begin LTER's 20-year review process.[229]

Mary Clutter initiated contact with potential review committee members in early 2001. She had also established with Franklin Harris, former NSF biologist at Oak Ridge National Laboratory and a member of the BIO AC, to be a co-chair for the upcoming work.[230] Eventually, the committee was comprised of 17 members with Harris and former BIO staffer Leonard Krishtalka, then at the University of Kansas, as the two co-chairs. A much bigger network—24 sites by then—required a larger review committee than at its first decadal anniversary. The group met in August of 2001 and began its work on an LTER that had not only gone global with sites in Canada, Taiwan, China, and the United Kingdom—20 nations, in all—but which had inspired other countries to set up their own LTER-like projects. China, for instance, took a $25 M loan from the World Bank as early as 1993 to begin building its own 29 long-term ecological study sites to investigate erosion and other matters.[231]

---

[225] Ibid., ppg. 127–128.

[226] 1993 LTER All-Scientists' Meeting Agenda, September 18–22, 1993, Estes Park, Colorado, and, Coleman, pg. 132.

[227] National Science Foundation, "LTER Program Guidelines," NSF 94–60, April 12, 1994.

[228] One-page ephemeron on highlights of the LTER program; provenance unknown.

[229] See the overall publication at http://www.nsf.gov/about/history/nsf0050/environment/environment.htm.

[230] Mary Clutter to Barbara Pecharsky, February 27, 2001

[231] Jocelyn Kaiser, "An Experiment for All Seasons," *Science* (2001):624–627.

In her charge to the Committee, Clutter made one particularly salient point. Not only had a great deal of science been learned by the work under LTER over two decades, but an "entire generation of ecologists has been trained with access to LTER sites"[232]—an excellent example of what comes of the *intricate dance.* The Committee met a total of three times and submitted its draft report to BIO in April of 2002. The $11 M budget at the 10-year mark had moved up to just under $18 M by the 20th anniversary. Some 1100 investigators and students comprised the cadre of scientists then part of the international LTER/ILTER network.[233] While the first decade was primarily a period of data collection, the second decade began to look at larger-scale patterns and processes and anthropogenic influences, Krishtalka's report to BIO stated. The projected third decade would be aimed at "[s]ynthesis science for ecological forecasting." That was to be via a "bold decade of synthesis…to serve science and society."[234] As Coleman interpreted the 20-year review, the LTER program:

> should become a research collaborator, conducting research along a seamless, integrated continuum from site-specific to cross-site to network- and systems-level ecological studies [and, also,] partnering with social scientists would pay big dividends in increasing the understanding of the interrelationships and reciprocal impacts of natural and human ecosystems in order to inform environmental policy.[235]

It would seem that the social sciences, once allied in the same directorate with biology, were needed in close association with it once again to explain certain aspects of the nature of nature! Those who opposed the split at the reorganization of NSF in 1991–1992 may have had a good point in arguing against it. Mary Clutter (who was one of those opposed) would be quoted by the Foundation's public relations office as saying that the next 10 years for LTER would be the "'Decade of Synthesis.'"[236] Should interested parties wish to contact the right individual to learn more about the LTER program, the news release offered up its newest person at NSF, a very experienced hand at LTER field work: Henry Gholz.

Gholz had been a doctoral student at Oregon State University (OSU), Corvallis, during the period 1973 to 1975 and had done research at the H.J. Andrews research forest site before it became one of the first six in the network. His PhD research involved an ecological transect study across the State of Oregon. When the first competition was announced in 1979 that led to founding the network the following

---

[232] Mary Clutter, "Charge to Twenty-Year Review of the LTER Program," OAD/BIO:3-2-01; one page with transmittal memo and early draft list of committee members then to date. A handwritten note on the bottom of the cover memo indicated that "Dr. Wake," Marvalee Wake of UC, Berkeley, had "declined to serve as chair," she having apparently been asked to do so by Clutter. The note was signed by Peggy Weber of the BIO AD front office.

[233] Data taken from paper copies of overhead slides presented at a meeting at BIO on April 26, 2002 by Krishtalka.

[234] Ibid.

[235] Coleman, pg. 133.

[236] Cheryl Dybas, "Twenty-Years of Long-Term Ecological Research: National Science Foundation Releases Review Report," NSF PR 02–65, August 2, 2002.

year, Gholz was ready to leave Corvallis for a faculty position at the University of Florida, Gainesville. His major professor at OSU had been Richard Waring, and Gholz was introduced, as Waring's student, to the reviewing of proposals headed for NSF. The young doctoral candidate was not, however, involved with the Andrews' proposal for LTER site selection, he recalled.[237] He did, however, meet Frank Harris when Harris was a PO at the Foundation and did a site visit at the Oregon research forest.

Gholz took up his professorship at Gainesville in the Department of Forestry and found himself to be a "lone wolf" when it came to seeking his first NSF grant as a member of the department: none of the faculty were used to seeking funding from the Foundation, but rather from USDA and other sources more familiar to foresters. He held that it was second nature to him and he thought "every person in science" made proposals to NSF for support, not so his forestry colleagues, however, as they had extension service appointments and Gholz did not; his was purely teaching and research.[238] He was, though, quick to receive NSF funding (beginning in 1982) for ecosystem research and had Frank Harris as his PO contact at BIO. He even wrote two proposals for setting up his own LTER sites, but they were not successful. Nevertheless, he continued his research with a number of Foundation grants over the years in non-LTER site ecosystem studies. Though the Everglades was an LTER site, it was a long way from northern Florida where he was located, and so Gholz did not do research there. He did, however, spend 23 years on the faculty of the University of Florida and did research from 1985 to 1995 at the famed La Selva Biological Station in Costa Rica, let alone other work in both the New and Old World tropics: Brazil, Bolivia, and Madagascar. Some of that was done with the USAID, and some as an AAAS fellow.

Gholz' AID work was done at NSF during a sabbatical year in 1994–1995. It was his first time at the Foundation, though long a grantee of it. About that same time, one of his multilingual graduate students at Gainesville did his research on the Japanese agricultural colonies in Brazil. Gholz found that work extremely fascinating as the student tied together biological aspects of his ecological studies with sociological features. Gholz held that the student's work "sort of fed into my susceptibility" to appreciate the social aspects of ecology. That would later have definite effects on how he would deal with his future job of program director for LTER. While working with NSF/AID on his sabbatical, he was aware of the growing knowledge concerning Earth's carbon budget and the role of carbon dioxide as a climate change agent. Indeed:

> we joked that the motto for AID changed overnight from how many mouths did you feed today, to how many kilograms [of] carbon did you sequester today [laughs].[239]

He went back to Florida with, as he said, "my horizons broadened." But, having found his wife-to-be in Washington while at NSF and given the fact that she had a

[237] Henry Gholz interview with the author, May 5, 2009.

[238] Ibid.

[239] Ibid.

good job in the capital, they dated long-distance for several years until an offer of a rotatorship at BIO arose in 2000. They then married and he relocated north, never returning permanently to Gainesville.

He came into the Foundation not long after Rita Colwell brought in her many ideas, including biocomplexity, and a program then trying to get started: NEON, the National Ecological Observatory Network, of which more below. The then PD of LTER, Scott Collins, was trying to get NEON going, but the program, opined Gholz, was "sputtering," and so Collins needed some time away. He took a sabbatical and that thrust Gholz into heading up LTER; he had already been aiding Collins with that work as PO and so was familiar with it. Besides that, he had a career full of knowledge of LTER or similar sites as a researcher. Collins decided to leave NSF during his sabbatical and that left leadership for LTER in question. At first, an attempt was made to have Gholz, still a rotator, run the program under that sort of appointment, but it really required a permanent person. When the job did open, he applied for and got the position. He was still PD for LTER when I interviewed him in 2009. Although he was hired to run the Ecosystems Studies Program as his primary duty, he was quickly given LTER, and, by 2004, he also added NCEAS to his responsibilities.[240]

In that same half decade beginning the new century, two COV reports were done, one on the LTREB cluster and the other on DEB, as a whole.

> From our assessment, the COV has every confidence that the highest quality science is being solicited and funded by the LTER/LTREB programs...The Program Officers and staff for long-term projects have demonstrated a very high level of dedication to the success of these programs,

said the members of the first of the two Committees of Visitors.[241] The LTREB programs included not only LTER but also LTREB itself and a number of what were termed cross-site investigations: many used LTER sites for studies, whether funded under the actual LTER program or one of the others. Particular successes included data from Hubbard Brook Forest being used in guiding Clean Air Act legislation.[242] And at the Cary Institute of Ecosystem Studies in Millbrook, New York, which had been founded by Gene Likens of Hubbard Brook fame (Chap. 4) in 1983, significant investigation was done on Lyme disease.[243] At the Sevilleta LTER site in New Mexico, work on hantavirus had "important applications for public health policy." Much new synthesis work was done, in part, by and through NCEAS, which was both a "model for and a facilitator of [such] efforts." And the "development of an increasingly integrated culture of teamwork" had brought sociological aspects into

[240] Henry Gholz communication with the author, May 17, 2012.

[241] "FY 2001 Report for NSF Committees of Visitors (COVs): Long-Term Projects in Environmental Biology," April 2–4, 2001.

[242] Ibid., pg. 2.

[243] See http://www.ecostudies.org/about.html.

the entire program, not just in connection with the two urban sites (Baltimore and Phoenix).[244]

Even with all the positive comments, a COV's duty was to be objective; hence the entire membership of those committees was chosen from the community and not from within NSF. The LTREB COV found that greater funding and increased emphasis were needed on "environmental informatics," avoidance of "underreporting" of work accomplished through grants at system sites, more equity of funding among sites, ever more synthesis, and greater outreach efforts.

The second COV, which was charged with the review of the entirety of the Division of Environmental Biology, accomplished its work in June of 2003. The typical basic charge for COVs across the Foundation was, and had been for many years, two-pronged: first, assessment of the quality and integrity of the programs and, second, discovering the "degree to which outputs and outcomes" contributed to NSF's mission and goals.[245] Mary Clutter, however, asked the DEB COV to also examine three additional, and more specific, questions: "[h]as DEB appropriately balanced their education portfolio"; has it "sufficiently supported synthesis activities"; and where was DEB headed?

While the Committee found that international standards of assessment methods were met, "the low number of *ad hoc* reviews that are received for each proposal concerns the COV." In some cases such reviews were non-existent. The COV guessed that the situation "may be due to increased workloads of successful scientists, as well as the diminishing culture of professional service."[246] Requiring evidence that a PI had provided such service to funding agencies her- or himself might be an answer to the problem, the members suggested. Another problem that the COV discovered was that the SGER program, Small Grants for Exploratory Research, not restricted to DEB nor even to BIO, had very few proposals, only ten in 2002 for the entire division. The SGER program was created Foundation-wide specifically for "high-risk" projects, and, given the low numbers and the fact that it was "well-known that PI's often divert funds from existing grants (potentially weakening those projects) to gather preliminary data," DEB should award "substantially" more SGERs.[247] That was one of the more worrisome aspects to the COV.

The report was a typically large and complex tool as it had to cover an entire division's work over 3 years. Thus, it was filled with many suggestions, including

---

[244] COV report of April 2001, pg. 3.

[245] David Young, Chair, "Report to the National Science Foundation: Review of the Division of Environmental Biology, 11–13 June 2003," pg. 2.

[246] Ibid., pg. 5.

[247] Ibid., pg. 15. See also pg. 32 where the issue was restated and emphasized. The SGER program began in 1990 and did not require that a proposal undergo external review. "Characteristics of activities that can be supported by an SGER award include: preliminary work on untested and novel ideas; application of new approaches to 'old' topics; ventures into emerging research areas; and narrow windows of opportunity for data collection, such as natural disasters and infrequent phenomena [the Mt. St. Helens' eruption and Hurricane Katrina were classic examples]." See http://www.nsf.gov/nsb/documents/2001/nsb0136/nsb0136_5.pdf.

answers to Clutter's three additional questions. The COV found that the "post-docs [and] young and mid-career faculty" training of both graduate and undergraduate students "could be better supported by DEB." With regard to the synthesis question, the Committee noted that the division was doing well in that area, but they "encourag[ed] DEB to continue to foster" that effort. Finally, the COV lauded the fact that "multidisciplinarity" had long been at the core of the division's science and that the "trend will certainly intensify in coming years." Nevertheless, to assure that such an eventuality came to pass:

> rapid growth of multidisciplinarity necessitates the growth of new partnerships and mechanisms for information sharing, synthesis and access among scientists, administrators, and the public…The COV feels that as DEB programs move forward in the next few years, some of their greatest challenges will be to build and strengthen effective institutional partnerships and infrastructure and to maintain community input to generate and evaluate emerging areas of study.[248]

LTER itself had become a much more integrated network with many cross-site activities in the early 2000s, and it also brought the social sciences deeply into the mix. One group of authors even viewed LTER becoming what they termed LTSER: Long-Term *Socioecological* Research.[249] While LTER is still LTER, it does remain one of the most significant biological research programs in the history of the NSF. Dave Coleman's book, *Big Ecology: The Emergence of Ecosystem Science*, cited both in this discussion and that in Chap. 4, is only the first work to examine LTER. The story of this global adventure lies open to investigation by both present and future historians of science and will be shown to be one of the *major milestones of Millennial Biology*.

9.7 First Light: The Founding of NEON

"I could see it [NEON] was a political and complicated issue. Rita [Colwell] had this idea and tried to implement it top down, from Congress all the way down," said Henry Gholz.[250] Thinking back, Gholz was not entirely sure the idea was totally Colwell's, but may have been the product of a number of persons then in the LTER program, especially that of Bruce Hayden, an ecologist at the University of Virginia.[251] At any rate, Hayden would (and still did at the time of this writing) have a leading role in the effort to establish the National Ecological Observatory Network.

[248] Ibid., pg. 34 of the Young Report.

[249] Helmut Haberl et al., "From LTER to LTSER: Conceptualizing the Socioeconomic Dimension of Long-term Socioecological Research," *Ecology and Society* (2006); online at http://www.ecologyandsociety.org/vol11/iss2/art13.

[250] Gholz interview.

[251] Ibid. See also http://www.uvacres.org/faculty_page.php?id=5. "In addition, to work at the VCR [Virginia Coast Reserve] LTER program, he also leads a team for the nascent NEON program that will develop a nation-wide network of ecological observatories."

Gholz opined that Colwell "ran into a huge brick wall on the Hill, and she ran into massive complications with the ecological community out there." That was not just the opinion of Gholz, however, as the community was much more than just a little leary about NEON and its implications—particularly from the LTER cadre. The "foundering" that Scott Collins saw was because the whole enterprise "definitely did not have a groundswell of…public support, you might say, at either end," community or Congress.[252]

The second workshop on establishing NEON had occurred in March of 2000 at the UC, San Diego Supercomputer Center, where its 25 participants had argued that the Network would respond to the NSB's Task Force on the Environment's twenty-first century report.[253] It had followed rapidly after the first workshop, which had taken place at the Archbold Biological Station in Lake Placid, Florida, just 2 months before.[254] Some 26 researchers from the community were present at the initial meeting, including Terry Yates, Scott Collins, and two other observers from NSF. In their report, they had stated that

> NEON represents an unprecedented opportunity for infrastructure to address environmental questions ranging over multiple levels of biological organization over broad ecoregions,

and the members of the second workshop felt the need to reiterate that statement, and more, as it opened its own commentary following the March assembly.[255] The crucial word in that statement used by members of both workshops was "infrastructure." There was, according to Gholz, the reality that "there's always been confusion right from the start…between LTER and NEON." LTER, however, even though "we always talked about [it as a group of] sites," is, in fact, "a research project." But it was always, he felt, "undercapitalized" such that proposed projects under the LTER program, *per se*, had to be done at established sites, facilities, field stations, etc.[256] LTER never included funding for building new field facilities, nor even much in the way for scientific equipment; it was "always based on science questions," averred Gholz. He summed it up very well by stating that "NEON is for capital…and LTER [is] for research projects." NEON sites are to be data collection points and typically will be fully automated—no researchers present—some will even be mobile. The use of the future tense here is well considered as it would be more than 6 years after

[252] Gholz interview.

[253] "Report to the National Science Foundation from the Second Workshop on the Development of a National Ecological Observatory Network (NEON) held at the San Diego Supercomputer Center, La Jolla," March 9–13, 2000. See pg. 3 re the NSB report. A third workshop occurred in Santa Fe, New Mexico, on May 3–4, 2000, that included 13 players, some of whom had been at NSF at different times over the years and have been mentioned in this history, such as Peter Arzberger, Jim Gosz, Leonard Krishtalka, and others. See http://ibrcs.aibs.org/reports/pdf/NEON3_May2000.pdf. The fourth workshop was held, appropriately, at the Millennium Hotel in Boulder, Colorado, in June of 2002 when very effortful planning was seen with regard to needs of the Network. See http://ibrcs.aibs.org/reports/pdf/NEON4_June2002.pdf.

[254] For the first report, see http://ibrcs.aibs.org/reports/pdf/NEON1_Jan.2000.pdf.

[255] Second NEON workshop report, pg. 3.

[256] Gholz interview.

the end date of this study (2005) before Congress was *set* to authorize the first funds for the NEON project; but see below.

The first workshop members were given a charge from the Foundation and stated them thus:

> [a] framework of initial assumptions for these scenarios was provided by the NSF officials for the workshop. These assumptions include:
>
> (1) The overall goal of NEON is to develop distributed infrastructure for environmental biology, ranging from molecular genetics to landscape-level study.
> (2) This distributed infrastructure should advance the study of continental- and regional-scale issues.
> (3) A figure of $10 million for each of 10 observatories, each with suites of sites, was proposed as a target infrastructure budget with $1million/site/year for operating expenses. Additional funds for research would also be sought.
> (4) Consortia of multiple institutions, including biological field stations, need to be developed for each site. Developing strong links with monitoring networks, field stations and other facilities of State and Federal agencies and non- governmental organizations is also encouraged.
> (5) NEON sites should provide the infrastructure that would offer facilities for researchers not directly associated with the network.
> (6) NEON sites should be designed to last for at least 30 years.[257]

Highly specialized facilities—infrastructure—was, as Macbeth had it, the be all and end all for environmental/ecological field research in the NEON project. LTER investigators could use the NEON sites, as well, as Gholz stated: "about a third of LTER could use NEON at any one…or all…the time." But, like so much else about NEON, even approaching a decade later, Gholz also said (in 2009): "Do I think that NEON [and LTER] relationships have been sufficiently well defined and articulated? Absolutely not."[258] In fact, from the LTER community, "there still is this huge, in some cases, indifference, in other cases, animosity" toward NEON, opined Gholz.

In the pages of *Science, BioScience*, *Nature*, and in other leading journals over the first full decade of the new millennium, numerous articles had been penned about the status of NEON. Also, a number of other reports and white papers came out about it, including a book from the National Academies Press in 2003. By the winter of 2008, the first number of the *NEON Quarterly* had come to light, but Congress had *still not funded* the project! In 2011, the community was warned that "[r]eady or not, the era of big data is coming to ecology."[259]

> After years of discussion and debate, the United States is moving forward with an environmental monitoring network that promises to help transform a traditionally small-scale, local science into a continental-scale group enterprise.[260]

---

[257] First NEON workshop report, pg. 6.

[258] Gholz interview.

[259] Jeff Tollefson, "US Launches Eco-Network," *Nature* (2011):135.

[260] Ibid.

However, on September 11, 2011, an otherwise extremely notable anniversary in American history, there also occurred the following:

> A Senate proposal that would reduce next year's budget for the National Science Foundation (NSF) by $162 million includes unprecedented flexibility for the agency to decide how best to spend a portion of that shrinking pot of money. It's a small but important silver lining in a year when even NSF's best friends in Congress have turned their backs on the Obama Administration's request for a 13% increase for the basic research agency as part of the political consensus on reducing the federal deficit.
>
> Yesterday the Senate Appropriations Committee voted to give NSF $6.7 billion in the 2012 fiscal year that begins next month. The money is contained in a $52.7 billion spending bill (S. 1572) to fund dozens of federal agencies, most of which would also see their budgets shrink. The NSF figure, 2.4% below its current level, would reduce funding for NSF's six research directorates by $120 million, to $5.44 billion. It would lower spending on education by $32 million, to $829 million. And while the committee would hold steady the $117 million now available for large new facilities, flat funding would actually mean a significant reduction in plans to scale up work on half a dozen projects in that account, for which NSF had requested $224 million.
>
> That squeeze prompted legislators to take the unusual step of giving NSF the option of moving $100 million from its research account to erase most of the shortfall in the facilities account. The shift, if adopted, would allow NSF to stay on schedule in building out the Ocean Observatories Initiative (OOI) and begin major construction on the National Ecological Observatory Network (NEON). The committee had clearly done its arithmetic: It specified that the $117 million should be used to fully fund the relatively small amounts requested for three projects close to completion, plus $83 million of the $103 million requested for OOI. *But that left no money for NEON, for which NSF had requested $88 million.*[261]

But like so many years of the half century plus in the life of the National Science Foundation and its directorate for biological sciences, Congress reversed its thinking at the last minute, and, in November 2011, a fiscal reprieve occurred for several science agencies, and, at BIO, it would appear that NEON's lights would soon be switched on.

[261] Jeffrey Mervis, "Senate Plan Gives NSF a Choice on Facilities vs. Research," *Science Insider*, September 16, 2011; http://news.sciencemag.org/scienceinsider/2011/09/senate-plan-gives-nsf-a-choice.html. Emphasis added at last sentence.

Epilogue

E.1 An Era Ends

The last year of Mary Clutter's leadership of the Directorate for Biological Sciences, 2005, ended in a crescendo of change as she retired and James P. "Jim" Collins became AD. Clutter told the BIO AC in April that the FY06 budget request for BIO would be almost $600 M (the eventual figure was $580 M).[1] But her Emerging Frontiers, EF, was still with no permanent director or staff, though it would inherit the Microbial Biology Program, which was a conglomerate of several other programs and funds. The Frontiers in Integrative Biological Research, FIBR, had already been located in EF for over 2 years and had awarded its first six 5-year projects totaling some $30 M.[2] Also in the "virtual division" of EF, a commitment in the Ecology of Infectious Diseases actively contributed to NSF's portion of America's still new programming under the Homeland Security portfolio, a consequence of the terror attacks on America in 2001.

Programs which had begun in the later 1990s in information science, biocomplexity, and other areas, were reported to the AC by Joann Roskoski, long the Executive Officer of BIO, as doing well. And she emphasized to the Committee that they did not take funding from individual PIs' smaller grants—long a worry in the community. Some of the cross-directorate, high priority initiatives had recently ended, but ones such as biocomplexity were not scheduled to move into core programs until 2007, Roskoski stated. NEON, in development then for almost a decade, was finally allocated $6 M by Congress for the FY05 budget for planning and design.[3] Under a grant to the American Institute of Biological Sciences, the society

[1] BIO AC "Minutes," April 7–8, 2005.

[2] Ibid., and Sean Kearns, NSF, OLPA, "NSF's FIBR to Mix Disciplines, Use Breakthroughs on 5-Year Explorations into Biology's Mysteries," NSF PR 03-106.

[3] BIO AC, 'Minutes,' pg. 2.

D. J. McGraw, *Millennial Biology: The National Science Foundation and American Biology, 1975-2005*, https://doi.org/10.1007/978-3-030-56367-7

would undertake such preparations. Six more years were yet to pass until Congress would finally provide the first financing for NEON, however.

Arden L. Bement, then the Foundation director, was present at the April AC meeting and addressed education efforts, especially for minority students and most particularly in the STEM areas of science, technology, engineering, and mathematics. STEM education would continue to grow over the following years and would experience a "crisis," by the second decade of the new millennium.[4] International assessments of student learning continued to place the United States well below the top echelon of nations demonstrating successes among their own students in the STEM disciplines.

A number of other matters were considered by the AC, reports were heard, and items discussed, with the escalating growth in cyberinfrastructure, continuing its already long upward spiral, being foremost among the topics. An extensive appendix was added to the minutes with regard to the matter and a report was given by Michael R. Willig, Division Director of DEB. Willig told the AC that a CIBS-WG, that is, Cyberinfrastructure in the Biological Sciences Working Group, had been formed that included familiar names, such as Bill Zamer of EF and Matt Kane, then of MCB. Other figures not mentioned in this history were newer PDs from each of the four BIO divisions and who would be members of the WG. One of Willig's overhead slides contained a graphic that seconds what has been discussed often in previous chapters (see Fig. E.1). Cyberinfrastructure played at all levels along the two arrows in the graphic uniting biology from both the big and little ends of the spectrum. The CIBS' vision stated that, "cyberinfrastructure serves as a universal integrator."[5] The WG was tasked with reporting on its efforts at the November meeting. But October 15 saw the change of command from Clutter to Collins (The organization chart for BIO at the end of Clutter's career as AD is seen in Fig. E.2.) The latter would inherit a directorate that could hardly be recognized from the one that Mary Clutter took over in the late 1980s and restructured to meet the needs of and play a major role in creating her "new biology."

The new BIO AD, Jim Collins, had earned his bachelor's degree in 1969 from Manhattan College, Riverdale, New York and his doctorate from the University of Michigan, Ann Arbor, in 1975.[6] He became an Assistant Professor at Arizona State University (ASU), Tempe, in zoology, and was an NSF grantee. He moved up the ranks to become the Ullman Professor of Natural History and the Environment in the School of Life Sciences. Chairing several areas in biology during those years, he was a visiting professor at Duke University and then Director of the Population Biology and Physiological Ecology Program at BBS in 1985–1986. He maintained his laboratory at ASU while rotating at NSF and returned to a distinguished career at the university, including yet further NSF service on the BIO AC, until he became

---

[4] So severe had matters become that one university, California Polytechnic State University, San Luis Obispo, placed its School of Education *under* its College of Science and Mathematics!

[5] Taken from a paper copy of a slide used in the Willig address to the AC in April, 2005.

[6] For more on Collins, see: http://sols.asu.edu/people/faculty/jcollins.php and http://www.nsf.gov/news/news_summ.jsp?cntn_id=104341.

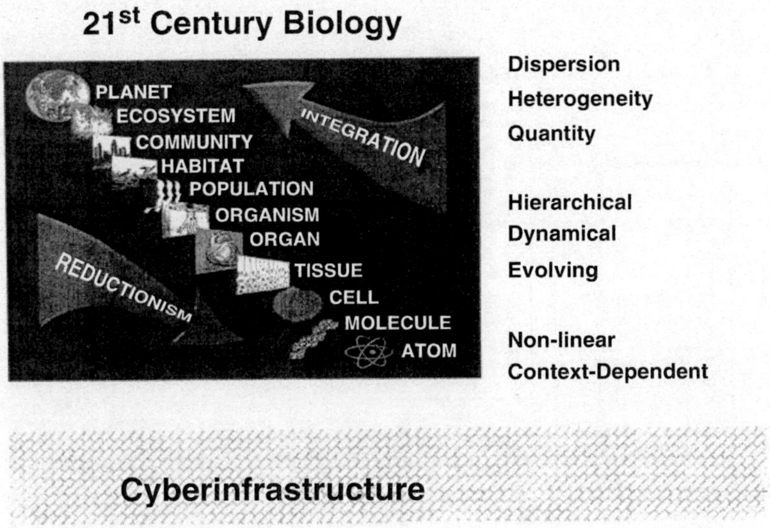

Fig. E.1 The role of cyberinfrastructure in the biological sciences as practiced at NSF and reflective of biology in the community. NSF. Taken from a visual used at the BIO Advisory Committee meeting of April 2005

AD. He returned again to ASU after leaving the Foundation in 2009 and remains there as of this writing. Collins was succeeded by Joann Roskoski, who was acting director until 2011 when the next permanent AD, John C. Wingfield assumed the leadership.

When Collins came aboard, however, an editor at *Science* magazine stated that with Collins' assuming the BIO leadership position, within about 8 months none of the seven ADs at NSF would have had more than 1 year in their positions![7] Institutional memory was under threat. But that was not surprising at the Foundation given its long history of dependency upon the rotator system. Among the new ADs at the time was that for SBE, David Lightfoot, who had a long external association with NSF.

Collins brought a strong attitude of interdisciplinarity and education orientation with him to his new position. His interests in scientific ethics grew while in the office. In an interview for the journal *BioScience*, Collins stated that

> [w]e need to have a very clear sense of, not so much the integration between a research agenda and an education agenda, but what an integrated agenda should look like…[they should] become one of a piece. [Students will] becom[e] researchers in the context of becoming educated.[8]

[7] Yudhijit Bhattacharjee, ed., "Random Samples: People," *Science* (2005):1275.

[8] Editor, "James P. Collins: Seeking Biology's Growing Points," *BioScience* (2006):106. (February).

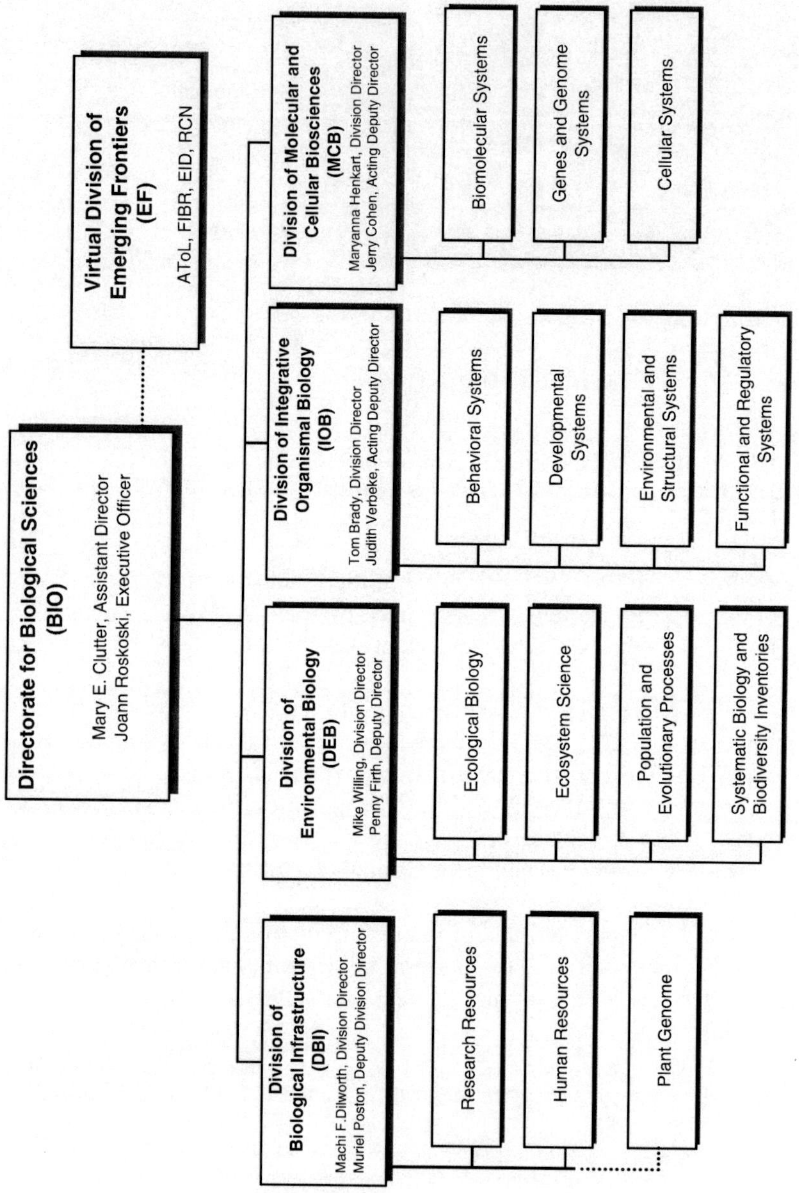

Fig. E.2 Organizational chart for BIO at the end of the period covered in this history. NSF. Specific provenance unknown

Collins opened the November 2005 meeting of the BIO AC just a month after his arrival. He gave a thorough-going illustrated presentation concerning how to move biology into the twenty-first century. Using evolutionary metaphors, Collins asked about the "fitness landscapes" of the directorate's research programs. A series of scheduled reports on the plant genome, NEON, the Foundation-wide cyberinfrastructure strategic plan (the initiative had its own Office), the CIBS-WG plans, and others all followed. Director Bement presented on the second day, along with COV statements on two recently assessed divisions. It was an action-packed meeting that did much to set the direction for BIO into the immediate future.

It remains for future historians of biology to take the next step in telling the story of BIO.

E.2 What Has Been Discovered?

A number of behavioral, technological, and organizational matters, as well as aspects of basic science, have arisen in this study of the National Science Foundation and its two most recent directorates for biological sciences. These points have become obvious during this study and may ***possibly be definitional for a new era of biology***. Summing up these features in the history of biology as viewed from within the NSF and as created or modified by the *intricate dance* between the Foundation and the biological community, takes us back to the problems this study sought to solve—and considerably more. What has been discovered?

During the middle to later 1980s, Erich Bloch was director of the Foundation, but was interviewed by the Historian of the NSF, Marc Rothenberg, only many years later (2008). Rothenberg asked Bloch how the agency should measure its success. Bloch had little apparent problem in his retrospective view that, by necessity, covered both his own directorship and that of all who followed him to more recent times. First, he held, is education. "How many people does NSF touch and contribute to their education in a meaningful way, starting with kindergarten through the undergraduate [and] all the way to post-Ph.D.?" Second, Bloch put forward the question of "how many fields has NSF really, if not started, at least was instrumental in getting going?" He noted that, in his tenure, it was obviously computer science, as but one example. His third query was "how well have we funded the universities to allow them to move forward into new research arenas"; a question he saw as difficult of answering. Finally, he asked, "what role does the NSF play in the government? Are we leading the government or are we strictly the followers?" What most intrigued Rothenberg, as the then Foundation's historian, was the fact that engineer Bloch never brought up technological aspects as a form of measure.[9]

That may be as it should be because so much that was and is the National Science Foundation, in biology or otherwise, seeks to fund the more theoretical, the *basic* or

[9] Bloch interview, *op. cit.*

"pure" aspects of science. The *applied* aspects, though long a part of NSF—and an important one, as we have seen—have always been viewed as late-comers and secondary to basic scientific research. And it is basic research, primarily, that has done so much to advance biology over the three decades from 1975 to the early years of the third millennium. Yet, crucially, recent biology *unaided by a number of critical technologies* would not have matured to the point at which we experience it today.

It is my contention, and the evidence has been put forward here to support it, that there were two crucial periods in the years covered in this study. The first of them was when both Erich Bloch and David Kingsbury were in the leadership of the Foundation and BBS, respectively: that is, 1984–1988. John Wooley is not to be left out as a significant scientist here, however. There was a sort of dynamic synergy between Bloch and Kingsbury. Very closely associated with them was Mary Clutter and the many roles she played as a member of what may be viewed as a triumvirate of the time. The three did much to create what I have called *Millennial Biology*. And they did it during what, I posit, was the post-Watson and Crick *third great era in biology*. The first was the long history from antiquity (the Pre-Socratics, Aristotle and beyond) to Darwin (and *The Origin* of 1859), the second from Darwin to Watson and Crick (1953), and the third from that latter year through the present.

The second crucial period for the history of biology as documented in this study and as seen through the eyes of NSF, BBS, and BIO, was that beginning about 1998 and the directorship of Rita Colwell and her widely encompassing concept of biocomplexity, let alone her commitment to computational biology and computerization more broadly. By the time of Colwell's arrival, Mary Clutter had been for some years AD for BBS, then led its change to BIO, and finally remained as the BIO AD through the Colwell years and beyond. There was, once again, a synergy between those two leaders. They were a pair of thinkers that were at the nexus of the rise of the oncoming century of biology; Clutter's "new biology." The triumvirate and the pair, then, were central personalities. But they worked along with many brilliant colleagues in science, both in the community and at the Foundation, as well as many of those same individuals who additionally functioned as accomplished and astute *administrators* of BBS and BIO. We have met only a few of that number through interview transcripts especially, but many of that cadre' were significant figures in the greater history of biology.

Specifically, what was discovered? A number of definitional terms, concepts and related milestones in the history of biology arose from about 1975 onward, though a number of them had very deep roots. They may be summed up in a brief manner, though not necessarily in any particular order of importance, time of appearance, or maturation.

- The refinements of and continuingly closer partnership within the *intricate dance*;
- The increase of *interdisciplinarity*;
- Rapidly increasing *quantization*;
- The rise of *computerization* and the appearance of *computational biology;*
- The appearance of the many forms of *bioinformatics*;

- The application of increasingly *complex technologies*, including laboratory devices (e.g., the various types of chromatography, and the inventions of gene and nucleic acid sequencers) and field technologies (e.g., the appearance of GPS, GIS, and satellite imagery);
- The use of new *model systems* (e.g., *Arabidopsis, Caenorhabditis*, zebra fish, and others) that complemented more classic examples;
- The rise of *databases* (Protein Data Base, GenBank, etc.) and the consequences;
- A renaissance of *systematic biology,* especially molecular;
- A renewed recognition of the significance of *collections* of myriad types;
- The increasing significance and swift growth of *ecology* and *environmental biology* (e.g., LTER);
- The important, and changing, *relationship between reductionism and holism* in biology;
- The appearance of (or return of nineteenth century?) *integrative biology*;
- The rise of *big biology*, as defined in several ways;
- The rise of a *new neo-Darwinism* (clearer recognition of evolutionary mechanisms beyond natural selection alone; genetics integrating with evolution);
- The new partnership between *academic biology and industry*;
- The appearance and spiraling growth of *biotechnology;*
- The provision and use of *genomic and metagenomic* data for studies of all sorts;
- The melding of microbiology and ecology; and,
- The clearer realization, and its proof, of the effects of microbes on all aspects of the planet's environment and its inhabitants.

In association with the above list, certain internal *management features* at NSF played strong roles in how biology developed, not only at the Foundation but also in the academy/community via the *intricate dance*. They included the:

- Rotator system;
- Panel system as employed by BIO;
- Cluster concept as created in BIO;
- Mandated rise of applied research;
- Use of the Committee of Visitors as an assessment tool; and,
- The establishment of the Science and Technology Centers, and related such mechanisms.

The foregoing bulleted points in both of the two groupings above require some further elaboration.

While the lists synopsize much of what is recorded in the main text, the fact that BBS included not only the biological, but also the social sciences, remains important to the history of the former for present purposes. For instance, the STCs began within the social science arena and later, when adapted to the biological sciences at NSF, made numerous leaps in discovery possible for those disciplines (e.g., new types of light microscopes). The borderline behavioral sciences were long at the center of difficulty when organizational structuring/restructuring was undertaken.

That the behavioral sciences *do* belong to both the social and biological sciences continued to show itself in various ways after the great reorganization. That many projects and programs had to be shared over that boundary—and, indeed, also that of the lesser related purely *social* sciences (linguistics, economics, etc.)—*after* 1992 and into the present millennium, raises the question of the wisdom of the split that Mary Clutter and a minority of others opposed. The rise of *interdisciplinarity* to the very high levels now seen across the Foundation—BIO and SBE, for one pairing—evidences the fact that it is, indeed, challenging to separate the life sciences broadly interpreted and as constituted in BIO and SBE. And *interdisciplinarity* grew to intense levels in so many other ways, both internal to the biology directorates and externally into other organizational hierons of the Foundation (cross-directorate activities). The same was occurring, though very slowly, in the community, too. A completed *integrative biology* is still in the future, the findings of this study suggest.

Recently, SBE held a meeting concerning "Rebuilding the Mosaic" to further foster research in the directorate and to examine its "[f]uture, [which] will be interdisciplinary, data-intensive, and collaborative."[10] Thinking in both directorates travels along similar paths.

The triumvirate members were the central players in the computerization of both the Foundation and the biological sciences: the "data deluge" in the "-omics" arena began its inexorable rise, as has recently been considered.[11] That torrent of data has been seen in not only genomics and its related disciplines, but in ecology, as well. The rise of *bioinformatics,* then, is a child of the efflorescence of both the "-omics" *and* the environmental sciences. (Bioinformatics *did* begin with the former, however, as Robert Strausberg's work was one of the first to examine expression patterns of the *whole mitochondrial genome* in yeast as early as the late 1970s.[12] Influenced by James Watson as a scientist on the Human Genome Project, Strausberg spent his career in genomics. Not alone, Eric Schadt of the industrial arm of that science, held recently that "'I don't know if it's appreciated yet, but a major revolution is happening.'"[13]).

It was *complex technology* that, in so many ways, allowed for and drove the rise of genomics and other areas that profited by such "sophistication inflation" in relevant laboratory and field instrumentation. The increasingly more powerful types of chromatographies, microscopes, other devices and similar *technologies*, and the *techniques* such as PCR, probes or tags and their visualization, and related molecular feats, not only powered little biology, but made its use in big biology/ecology a possibility not theretofore fully envisioned. Ecology and environmental science with their own increased sophistication (GPS, satellite-enabled data transfer, etc.), as nicely exhibited in LTER, began to truly *integrate* areas of biology in striking

---

[10] Myron P. Gutman, AD, SBE, "Rebuilding the Mosaic: Fostering Research in the [SBE] at [NSF] in the Next Decade," NSF 11-086; online version.

[11] Megan Scudellari, "Omics—Data Deluge," *The Scientist*, (2011):44–45.

[12] Ibid., pg. 44.

[13] Ibid., Schadt quoted by Scudellari.

new ways. Not only did practical biology at the bench and in the field lead to new factual discovery by the coming together of the two ends of the biological spectrum, but so did the rise of theory profit as evidenced by better understanding of the role of the little in the realm of the big: Microbial Observatories and environmental metagenomics, being just two outstanding examples.

The rebirth of *systematics* and the new uses to which *collections* have been put interweaves with genomics, LTER-like science, AToL, and related projects and programs to recall an earlier biology when systematics was the central part of natural history. It was older science that was renewed for both older purposes—which still needed and need to be answered (new vertebrate, including mammalian, species are still being discovered, for instance)—and new science for which new understanding still requires proofs. *Databases* and *model systems* have arisen that aid in the work of systematics at all hieronic levels of biological study, little and big. *Computerization, quantification, bioinformatics*—all tie into systematics and collections, too. Data management leads to new understanding and new theory production.

Much of the "new biology" as Clutter had it, or *Millennial Biology* as here, is seen in the establishment of *biotechnology* and biology's relationship to industry. The mandated journey into *applied science* in the NSF of 1968 changed the Foundation—and biology—forever.

Even as all parts of the lists given above stand out in characterizing *Millennial Biology*, several aspects rise to the top, I would argue. The role of *reductionism* and its ties to *holism* did a great deal to *integrate* biology and to, in some senses, return it to the highly integrated period of natural history in the nineteenth century before the era of specialization (beginning at about that century's last two decades). But it was not until genomics blossomed in the first decade of the third millennium that a high degree of *holism* could once again return. And microbiology and ecology could integrate at levels that nineteenth century holistic natural historians might never have conceived.

Reductionism, holism, and integration—as well as quantitization, computerization and bioinformatics—have become central concepts in the "new biology." But can a fully integrated *view* and understanding of *Millennial Biology* become complete by examining *only one partner in the intricate dance*? Clearly, the answer is in the negative as this history is one-sided in its analysis. It remains for the historian to tackle the other side, to assess the roles of the *community*, the other partner in the dance to make whole an understanding of *Millennial Biology*.

Author Index

D. J. McGraw, *Millennial Biology: The National Science Foundation and American Biology, 1975-2005*, https://doi.org/10.1007/978-3-030-56367-7

C

Subject Index

D. J. McGraw, *Millennial Biology: The National Science Foundation and American Biology, 1975-2005*, https://doi.org/10.1007/978-3-030-56367-7

D

E